T0094016

Undergraduate Texts in Mathematics

Emanuel Fischer

Intermediate Real Analysis

With 100 Illustrations

Springer-Verlag
New York Heidelberg Berlin

Emanuel Fischer
Department of Mathematics
Adelphi University
Garden City, NY 11530
U.S.A.

AMS Subject Classifications (1980): 26-01

Library of Congress Cataloging in Publication Data

Fischer, Emanuel.
Intermediate real analysis.

Bibliography: p.
Includes index.
1. Mathematical analysis. 2. Functions of real variables. I. Title.
QA300.F635 1982 515 82-19433

Typeset by Computype, Inc., St. Paul, MN.

9 8 7 6 5 4 3 2 1

ISBN-13:978-1-4613-9483-9 e-ISBN-13:978-1-4613-9481-5
DOI: 10.1007/978-1-4613-9481-5

To
Wilhelm Magnus

Contents

Chapter XV
Improper Integrals. Elliptic Integrals and Functions 710

Preface

There are a great deal of books on introductory analysis in print today, many written by mathematicians of the first rank. The publication of another such book therefore warrants a defense. I have taught analysis for many years and have used a variety of texts during this time. These books were of excellent quality mathematically but did not satisfy the needs of the students I was teaching. They were written for mathematicians but not for those who were first aspiring to attain that status. The desire to fill this gap gave rise to the writing of this book.

This book is intended to serve as a text for an introductory course in analysis. Its readers will most likely be mathematics, science, or engineering majors undertaking the last quarter of their undergraduate education. The aim of a first course in analysis is to provide the student with a sound foundation for analysis, to familiarize him with the kind of careful thinking used in advanced mathematics, and to provide him with tools for further work in it. The typical student we are dealing with has completed a three-semester calculus course and possibly an introductory course in differential equations. He may even have been exposed to a semester or two of modern algebra. All this time his training has most likely been intuitive with heuristics taking the place of proof. This may have been appropriate for that stage of his development. However, once he enters the analysis course he is subject to an abrupt change in the point of view and finds that much more is demanded of him in the way of rigorous and sound deductive thinking. In writing the book we have this student in mind. It is intended to ease him into his next, more mature stage of mathematical development.

Throughout the text we adhere to the spirit of careful reasoning and rigor that the course demands. We deal with the problem of student adjustment to the stricter standards of rigor demanded by slowing down the pace at which topics are covered and by providing much more detail in the proofs than is customary in most texts. Secondly, although the book contains its share of abstract and general results, it concentrates on the specific and concrete by applying these theorems to gain information about some of the important functions of analysis. Students are often presented and even have proved for them theorems of great theoretical significance without being given the opportunity of seeing them "in action" and applied in a non-trivial way. In our opinion, good pedagogy in mathematics should give substance to abstract and general results by demonstrating their power.

This book is concerned with real-valued functions of one real variable. There is a chapter on complex numbers, but these play a secondary role in the development of the material, since they are used mainly as computational aids to obtain results about trigonometric sums.

For pedagogical reasons we avoid "slick" proofs and sacrifice brevity for straightforwardness.

The material is developed deductively from axioms for the real numbers. The book is self-contained except for some theorems in finite sets (stated without proof in Chapter II) and the last theorem in Chapter XIV. In the main, any geometry that is included is there for purposes of visualization and illustration and is not part of the development. Very little is required from the reader in the way of background. However, we hope that he has the desire and ability to follow a deductive argument and is not afraid of elementary algebraic manipulation. In short, we would like the reader to possess some "mathematical maturity." The book's aim is to obtain all its results as logical consequences of the fifteen axioms for the real numbers listed in Chapter I.

The material is presented sequentially in "theorem–proof–theorem" fashion and is interspersed with definitions, examples, remarks, and problems. Even if the reader does not solve all the problems, we expect him to read each one and to understand the result contained in it. In many cases the results cited in the problems are used as proofs of later theorems and constitute part of the development. When the reader is asked, in a problem, to prove a result which is used later, this usually involves paralleling work already done in the text.

Chapters are denoted by Roman numerals and are separated into sections. Results are referred to by labeling them with the chapter, section, and the order in which they appear in the section. For example, Theorem X.6.2 refers to the second theorem of section 2 in Chapter X. When referring to a result in the same chapter, the Roman numeral indicating the chapter is omitted. Thus, in Chapter X, Theorem X.6.2 is referred to as Theorem 6.2.

We also mention a notational matter. The open interval with left endpoint a and right endpoint b is written in the book as $(a;b)$ using a semi-colon between a and b, rather than as (a, b). The latter symbol is reserved for the ordered pair consisting of a and b and we wish to avoid confusion.

I owe a special debt of gratitude to my friend and former colleague Professor Abe Shenitzer of York University in Ontario, Canada, for patiently reading through the manuscript and editing it for readability.

My son Joseph also deserves special thanks for reading most of the material, pointing out errors where he saw them, and making some valuable suggestions.

Thanks are due to Professors Eugene Levine and Ida Sussman, colleagues of mine at Adelphi University, and Professor Gerson Sparer of Pratt Institute, for reading different versions of the manuscript.

Ms. Maie Croner typed almost all of the manuscript. Her skill and accuracy made the task of readying it for publication almost easy.

I am grateful to the staff at Springer-Verlag for their conscientious and careful production of the book.

To my wife Sylvia I give thanks for her patience through all the years the book was in preparation. תושלב״ע

Adelphi University E. F.
Garden City, L. I., N. Y.
November, 1982

CHAPTER I
Preliminaries

1. Sets

We think of a *set* as a collection of objects viewed as a single entity. This description should not be regarded as a definition of a set since in it "set" is given in terms of "collection" and the latter is, in turn, in need of definition. Let us rather consider the opening sentence merely as a guide for our intuition about sets.

The objects a set consists of are called its *members* or its *elements*. When S is a set and x is one of its members we write $x \in S$, and read this as "x belongs to S" or as "x is a member of S" or as "x is an element of S." When $x \in S$ is false, we write $x \notin S$.

To define a set whose members can all be exhibited we list the members and then put braces around the list. For example,

$$M = \{\text{Peano, Dedekind, Cantor, Weierstrass}\}$$

is a set of mathematicians. We have Cantor $\in M$, but Dickens $\notin M$.

When a set theory is applied to a particular discipline in mathematics, the elements of a set come from some fixed set called the *domain of discourse*, say U. In plane geometry, for example, the domain of discourse is the set of points in some plane. In analysis, the domain of discourse may be $\mathbb{R}$, the set of real numbers, or $\mathbb{C}$, the set of complex numbers.

As an intuitive crutch, it may help to picture the domain of discourse U as a rectangle in the plane of the page and a set S in this domain as a set of points bounded by some simple closed curve in this rectangle (Fig. 1.1). The figure suggests that $x \in S$, but $y \notin S$.

A *singleton* is a set consisting of exactly one member as in $A = \{b\}$. We have

$$x \in \{b\} \quad \text{if and only if} \quad x = b. \tag{1.1}$$

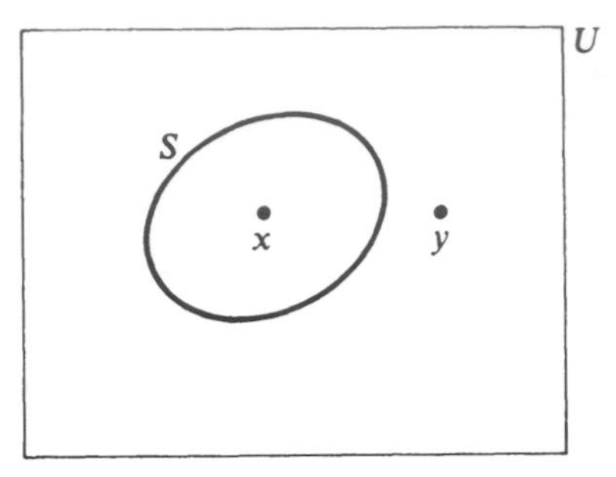

Figure 1.1

We distinguish between the set $\{b\}$ and its member b. Thus, we write

$$b \neq \{b\} \qquad \text{for each } b. \tag{1.2}$$

For example, 2 is a number, but $\{2\}$ is a certain set of numbers.

$S = \{a, b\}$ is a set whose members are a and b. We refer to it as the *unordered pair* consisting of a and b.

Unfortunately, the sets usually dealt with in mathematics are such that their members cannot all be exhibited. Therefore, we describe sets by means of a property common to all their members. Let $P(x)$ read "x has property P." The set B of elements having property P is written

$$B = \{x \mid P(x)\}. \tag{1.3}$$

This is read as "B is the set of all x such that x has property P." For example, the set $\mathbb{R}$ of real numbers will be written

$$\mathbb{R} = \{x \mid x \text{ is a real number}\}. \tag{1.4}$$

Here, $P(x)$ is the sentence "x is a real number." If U is the domain of discourse, the set of members of U having property P is often written

$$B = \{x \in U \mid P(x)\}. \tag{1.5}$$

This is read as "the set of all x belonging to U such that x has property P."

A set A is called a *subset* of a set B and we write $A \subseteq B$, if and only if each element of A is an element of B, i.e., if and only if $x \in A$ implies $x \in B$. We visualize this in Fig. 1.2. Each part of this figure suggests that

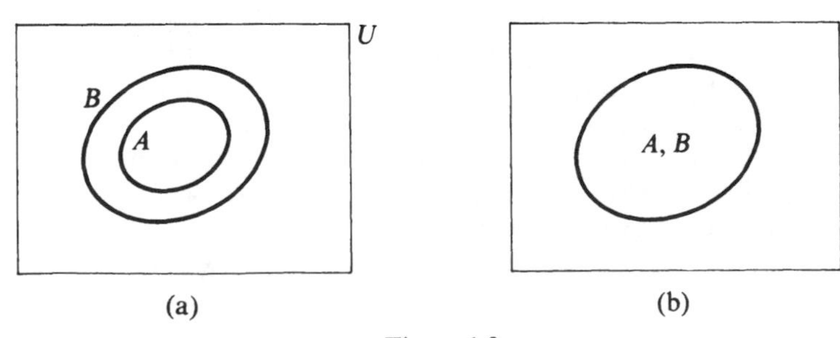

Figure 1.2

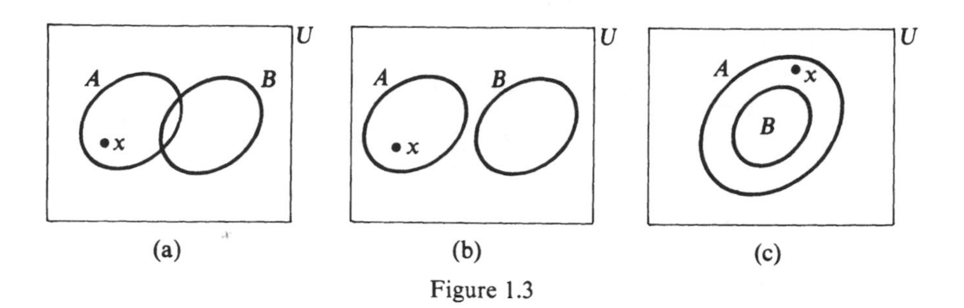

(a) (b) (c)

Figure 1.3

each element of A is an element of B; in (a) there are supposed to be elements of B not in A, where as in (b) every element of B is also an element of A. Thus, $A \subseteq B$ holds also when A and B have the same members. If $A \subseteq B$ is false, we write $A \nsubseteq B$.

$$A \nsubseteq B \text{ is equivalent to "there exists } x \in A \text{ such that } x \notin B." \tag{1.6}$$

Each of the diagrams in Fig. 1.3 portrays the situation $A \nsubseteq B$.

Sets A and B are called *equal* and we write $A = B$, if and only if both

$$A \subseteq B \quad \text{and} \quad B \subseteq A$$

hold. Thus, $A = B$, if and only if A and B have the same members.

When A and B are sets such that $A \subseteq B$ but $A \neq B$, we call A a *proper* subset of B and write

$$A \subset B.$$

This means that every element of A is an element of B, but there exists an $x \in B$ such that $x \notin A$.

One should distinguish carefully between the notions "$\in$" and "$\subseteq$." Thus, $x \in A$ means that x is an element of A, while $A \subseteq B$ means that $x \in A$ implies $x \in B$. The distinction is perhaps more noticeable when we deny these relations. For example, $x \notin A$ means x is not a member of A, whereas $A \nsubseteq B$ means that there exists $x \in A$ such that $x \notin B$. The distinction is important. The two relations have different properties. Thus, if $A \subseteq B$ and $B \subseteq C$, then $A \subseteq C$ (cf. Prob. 1.3). Because of this $\subseteq$ is called a *transitive relation*. On the other hand, the relation "$\in$" is not transitive. For consider

$$X = 1, \quad A = \{1\}, \quad \text{and} \quad B = \{\{1\}\}.$$

Here B is a singleton set whose member is $A = \{1\}$ (nothing prevents us from having sets whose members are sets). We have $X \in A$ and $A \in B$, but $X \in B$ is false since this would imply $X = \{1\}$ or $1 = \{1\}$ and this is false (cf. (1.2)).

When $x \in A$ and $A \subseteq B$, we write $x \in A \subseteq B$. This clearly implies $x \in B$. Similarly, when $A \subseteq B$ and $B \subseteq C$, we write $A \subseteq B \subseteq C$.

A set having no members is said to be *empty*. Such a set is also called a *null* set. Sometimes, in the course of a mathematical discussion, a set is defined by some property. When no elements exist which have this property we call the set *empty*. An empty set is written $\emptyset$. We prove that for any set A we have

$$\emptyset \subseteq A. \tag{1.7}$$

Were this false, i.e., $\emptyset \nsubseteq A$, there would exist $x \in \emptyset$ such that $x \notin A$. This is impossible since no $x \in \emptyset$ exists.

PROB. 1.1. Prove: $\{a, b\} = \{a, b, a\}$.

PROB. 1.2. Prove: If A is a set, then $A \subseteq A$ and $A = A$.

PROB. 1.3. Prove: If $A \subseteq B \subseteq C$, then $A \subseteq C$.

PROB. 1.4. Prove: If A and B are sets, then $A = B$ implies $B = A$.

PROB. 1.5. When A, B, and C are sets such that $A = B$ and $B = C$, we write $A = B = C$. Prove: $A = B = C$ implies $A = C$.

PROB. 1.6. Prove: If $A \subset B \subseteq C$ or $A \subseteq B \subset C$, then $A \subset C$.

PROB. 1.7. Prove: If $A \subseteq \emptyset$ for some set A, then $A = \emptyset$ (cf. (1.7)).

Remark 1.1. Examine the sets $A = \{a, b\}$ and $B = \{b, c\}$, where a, b, and c are distinct. Clearly $A \nsubseteq B$ and $B \nsubseteq A$ hold. Thus, not all sets are related by the subset relation.

2. The Set $\mathbb{R}$ of Real Numbers

We shall treat the real numbers axiomatically and list 15 axioms for them. In this section we cite only 14 of the 15 axioms. The fifteenth axiom will be called the *completeness* axiom and will be stated in Section 8.

The set $\mathbb{R}$ of real numbers is postulated to have the properties:

(I) **(Axiom 0_1).** There are at least two real numbers.

(II) **(Axiom 0_2).** There is a relation called *less than*, written as $<$, between real numbers such that if x and y are real numbers, then exactly one of the following alternatives holds: Either (1) $x = y$ or (2) $x < y$ or (3) $y < x$;

PROB. 2.1. Prove: If x is a real number, then $x < x$ is false.

[We need not postulate the existence of a *greater than* relation between real numbers since this relation can be defined in terms of the "less than" relation.]

Def. 2.1. We define $x > y$ to mean $y < x$, reading this as "x is greater than y." We can now reformulate Axiom 0_2 as

(II′) **(Axiom $0_2'$).** If x and y are real numbers, then exactly one of the following alternatives holds: Either (1) $x = y$ or (2) $x < y$ or (3) $x > y$.

Def. 2.2. When $x < y$ or $x = y$, we write $x \leqslant y$.

Prob. 2.2. Prove: If x and y are real numbers such that $x \leqslant y$ and $x \geqslant y$, then $x = y$.

(III) **(Axiom 0_3).** If x, y and z are real numbers such that $x < y$ and $y < z$, then $x < z$.

Def. 2.3. When $x < y$ and $y < z$ both hold, we write $x < y < z$. Thus, by Axiom 0_3, $x < y < z$ implies $x < z$.

Prob. 2.3. Prove: (a) Either of $x < y \leqslant z$ or $x \leqslant y < z$ imply $x < z$; (b) $x \leqslant y \leqslant z$ implies $x \leqslant z$.

[We now introduce postulates for addition and multiplication. The lowercase Latin letters x, y, z appearing in the axioms below will represent real numbers.]

(IV) **(Axiom A_1)** (Closure for Addition). If x and y are real numbers, there is a unique real number $x + y$ called the *sum* of x and y.

(V) **(Axiom A_2)** (Associativity for Addition)

$$(x + y) + z = x + (y + z); \tag{2.1}$$

(VI) **(Axiom A_3)** (Commutativity for Addition)

$$x + y = y + x; \tag{2.2}$$

[The next axiom relates addition to the "less than" relation in R.]

(VII) **(Axiom 0_4).** $x < y$ implies $x + z < y + z$.

Prob. 2.4. Prove: $x < y$ and $u < v$ imply $x + u < y + v$.

(VIII) **(Axiom S).** If x and y are real numbers, there is a real c such that $y + c = x$;

(IX) **(Axiom M_1)** (Closure for Multiplication). If x and y are real numbers, there is a real number xy (also written as $x \cdot y$) called the *product* of x and y;

(X) **(Axiom M_2)** (Associativity for Multiplication)

$$(xy)z = x(yz). \tag{2.3}$$

(XI) **(Axiom M_3)** (Commutativity for Multiplication)

$$xy = yx. \tag{2.4}$$

(XII) **(Axiom D)** (Distributive Law)

$$x(y+z) = xy + xz. \tag{2.5}$$

[The next axiom relates multiplication to the "less than" relation in $\mathbb{R}$.]

(XIII) **(Axiom 0_5).** $x < y$ and $u < v$ imply $xu + yv > xv + yu$.
(XIV) **(Axiom Q).** If x and y are real numbers, where $z + y \neq z$ holds for some real z, then there exists a real number q such that $yq = x$.

Thus far, 14 axioms were cited. As mentioned earlier, the fifteenth and last one will be stated in Section 8.

The axioms indicate that addition and multiplication are *binary operations*, that is, we add and multiply two numbers at a time. We define $x + y + z$ and xyz by means of

$$x + y + z = (x + y) + z \quad \text{and} \quad xyz = (xy)z. \tag{2.6}$$

Axioms A_2 and M_2 respectively imply that

$$x + y + z = x + (y + z) \quad \text{and} \quad xyz = x(yz). \tag{2.7}$$

Having defined $x + y + z$ and xyz, we define $x + y + z + u$ and $xyzu$ as

$$\begin{aligned} x + y + z + u &= (x + y + z) + u, \\ xyzu &= (xyz)u. \end{aligned} \tag{2.8}$$

Prob. 2.5. Prove: If x, y, z and u are real numbers, then
(a) $x + y + z + u = (x + y) + (z + u) = x + (y + z + u)$ and
(b) $xyzu = (xy)(zu) = x(yzu)$.

Prob. 2.6. Prove: (a) $x + z < y + z$ implies $x < y$
(b) $x + z = y + z$ implies $x = y$. The result in part (b) is called the *cancellation law* for addition.

Prob. 2.7. Prove: The c such that $y + c = x$, of Axiom S, is unique.

Theorem 2.1. *There exists a real number z such that $x + z = x$ holds for all $x \in \mathbb{R}$. This z is the only real number with this property.*

Proof. Let b be some real number. By Axiom S, there exists a real z such that $b + z = b$. We prove that $x + z = x$ for *all* $x \in \mathbb{R}$. From $b + z = b$, we obtain for $x \in \mathbb{R}$,

$$(b + z) + x = b + x \quad \text{and hence,} \quad b + (z + x) = b + x. \tag{2.9}$$

In the second equality, we "cancel" the b on both sides to obtain $z + x = x$. This proves the existence of z. Next we prove its uniqueness.

Assume that there also exists a $z' \in \mathbb{R}$ such that $x + z' = x$ for all $x \in \mathbb{R}$. It follows that $z + z' = z$. Similarly, in view of the property of z, $z' + z = z'$. By Axiom A_3 we have $z + z' = z' + z$ and we conclude that $z' = z$. This completes the proof.

Def. 2.4. The z in $\mathbb{R}$ such that $x + z = x$ for all $x \in \mathbb{R}$ is called *zero* and is written as 0. Thus

$$x + 0 = x = 0 + x \qquad \text{for all} \quad x \in R. \tag{2.10}$$

Theorem 2.2. *If* $x \in \mathbb{R}$, *then* $x0 = 0$.

PROOF. For any y in $\mathbb{R}$

$$xy + x0 = x(y + 0) = xy = xy + 0,$$

so that $xy + x0 = xy + 0$. "Cancelling" xy we obtain $x0 = 0$ as claimed.

Given $x \in \mathbb{R}$, there exists (Axiom S) a real y such that $x + y = 0$. y is the only real number with this property (why?).

Def. 2.5. For each $x \in \mathbb{R}$, the unique y such that $x + y = 0$ is called the *negative* (or *additive inverse*) of x and is written as $-x$. Accordingly, we have

$$x + (-x) = 0 \qquad \text{for each} \quad x \in R. \tag{2.11}$$

PROB. 2.8. Prove: $-0 = 0$.

PROB. 2.9. Prove: $-(-x) = x$ for each $x \in R$.

Def. 2.6. Define $x - y$ as the c such that $y + c = x$ and call it x *minus* y.

PROB. 2.10. Prove: (a) $y + (x - y) = x$ and (b) $x - y = x + (-y)$.

PROB. 2.11. Prove: $-(x - y) = y - x$.

PROB. 2.12. Prove: $z + y \neq z$ if and only if $y \neq 0$.

Def. 2.7. A real p such that $p > 0$ is called *positive*. A real n such that $n < 0$ is called *negative*.

PROB. 2.13. Prove: If $x > 0$ and $y \geqslant 0$, then $x + y > 0$.

PROB. 2.14. Prove: If $x > y$, then (a) $z > 0$ implies $xz > yz$ and (b) $z < 0$ implies $xz < yz$. (Hint: use Axiom 0_5)

PROB. 2.15. Prove: $z \neq 0$ and $xz = yz$ imply $x = y$. This is called the *cancellation law for multiplication* of real numbers.

Using Prob. 2.12 we reformulate Axiom Q as:

Axiom Q′. If x and y are real numbers and $y \neq 0$, then there is a real q such that $yq = x$.

PROB. 2.16. Prove: If $y \neq 0$, then the q of Axiom Q′ is unique.

Theorem 2.3. *There exists a real number $e \neq 0$ such that $xe = x$ for all $x \in \mathbb{R}$, and e is the only such number.*

PROOF. Because of Axiom 0_1, there exists a real number a such that $a \neq 0$. By Axiom Q′ there exists a real number e such that $ae = a$. We prove that $xe = x$ for all $x \in \mathbb{R}$. If $x \in \mathbb{R}$, then $(ae)x = ax$. This implies that $a(ex) = ax$. Since $a \neq 0$, we can "cancel" the a to obtain $ex = x$, i.e., $xe = x$.

We prove that $e \neq 0$. Were $e = 0$, we would have for each $x \in \mathbb{R}$, $x = xe = x0 = 0$. Thus, 0 would be the only real number, contradicting Axiom 0_1. Hence $e \neq 0$. We leave the proof of the last statement of the theorem to the reader (Prob. 2.17).

PROB. 2.17. Complete the proof of Theorem 2.3 by proving that e is the only real number such that $xe = x$ holds for all $x \in \mathbb{R}$.

Def. 2.8. The $e \in \mathbb{R}$ such that $xe = x$ for all $x \in \mathbb{R}$ is called *one* and written as 1. Thus we have

$$x1 = x \qquad \text{for all} \quad x \in \mathbb{R}. \tag{2.12}$$

PROB. 2.18. Prove: (a) $(-1)x = -x$ for all $x \in \mathbb{R}$ and (b) $(-1)(-1) = 1$.

PROB. 2.19. Prove: (a) $(-x)y = -(xy) = x(-y)$ and (b) $(-x)(-y) = xy$.

PROB. 2.20. Prove: $z(x - y) = zx - zy$.

PROB. 2.21. Prove: $-x - y = -(x + y)$.

PROB. 2.22. Prove: $xy = 0$ if and only if $x = 0$ *or* $y = 0$.

Remark 2.1. The "or" in Prob. 2.22, as in all of mathematics, is used in the sense of and/or.

PROB. 2.23. Prove: $xy \neq 0$ if and only if $x \neq 0$ *and* $y \neq 0$.

Def. 2.9. If x and y are real numbers, where $y \neq 0$, then the real q such that $yq = x$ (Axiom Q′ and Prob. 2.16) is called x *over* y or x *divided* by y and is written as x/y.* If $x \neq 0$, then $1/x$ is called the *reciprocal* or the *multiplicative inverse* of x and is also written as x^{-1}, so that $1/x = x^{-1}$ when $x \neq 0$.

PROB. 2.24. Prove: If $y \neq 0$, then (a) $y(x/y) = x$, (b) $yy^{-1} = 1 = y(1/y)$, and (c) $(y^{-1})^{-1} = y$.

PROB. 2.25. Prove: If $y \neq 0$, then $xy^{-1} = x(1/y) = x/y$.

PROB. 2.26. Prove: (a) $x/1 = x$, (b) $0/x = 0$ if $x \neq 0$.

PROB. 2.27. Prove: If $x \neq 0$, then $x/x = 1$.

PROB. 2.28. Prove: If $b \neq 0$ and $c \neq 0$, then

$$\frac{ac}{bc} = \frac{a}{b}.$$

PROB. 2.29. Prove: If $b \neq 0$ and $d \neq 0$, then

$$\frac{a}{b} \cdot \frac{c}{d} = \frac{ac}{bd}.$$

PROB. 2.30. Prove: If $c \neq 0$, then

$$\frac{a+b}{c} = \frac{a}{c} + \frac{b}{c}.$$

PROB. 2.31. Prove: If $b \neq 0$ and $d \neq 0$, then

$$\frac{a}{b} + \frac{c}{d} = \frac{ad+bc}{bd}.$$

PROB. 2.32. Prove: If $b \neq 0$ and $d \neq 0$, then

$$\frac{a}{b} = \frac{c}{d}.$$

if and only if $ad = bc$.

PROB. 2.33. Prove:

$$\frac{-a}{b} = -\frac{a}{b} = \frac{a}{-b} \qquad \text{if} \quad b \neq 0.$$

Remark 2.2 (Do not divide by 0). We defined x/y for $y \neq 0$ as the q such that $yq = x$. If $y = 0$ and $x \neq 0$, then no such q exists since its existence would imply $0 = 0q = x$ which contradicts $x \neq 0$. If $y = 0 = x$, then the q such that $yq = x$, i.e., such that $0q = 0$, is not unique (explain). In the last case, q exists but is not unique. In any case, we do not divide by 0.

* This is written as a built-up fraction; however, because of its position in the text, the solidus notation is used.

3. Some Inequalities

Prob. 3.1. Prove: (a) $x > 0$, $y > 0$ imply $xy > 0$, (b) $x > 0$ and $y < 0$ imply $xy < 0$, and (c) $x < 0$ and $y < 0$ imply $xy > 0$.

Prob. 3.2. Prove: $1 > 0$.

Prob. 3.3. Prove: $x < y$ implies $-y < -x$.

Prob. 3.4. Prove: (a) If $p > 0$, then $x + p > x$ and if $n < 0$, then $x + n < x$.

Prob. 3.5. Prove: (a) $x > 0$ if and only if $-x < 0$ and (b) $x < 0$ if and only if $-x > 0$.

Prob. 3.6. Define $2 = 1 + 1$, $3 = 2 + 1$, $4 = 3 + 1$. Prove: $-2 < -1 < 0 < 1 < 2$ and that $2 + 2 = 4$.

Prob. 3.7. Define $x^2 = x \cdot x$, $x^3 = x \cdot x \cdot x$. Prove: If $x \in \mathbb{R}$, then $x^2 \geqslant 0$ and if $x \neq 0$, then $x^2 > 0$.

Prob. 3.8. Prove: (a) If $0 \leqslant x$ and $0 \leqslant y$, then $x^2 < y^2$ if and only if $x < y$; (b) $x^3 < y^3$ if and only if $x < y$.

Prob. 3.9. Prove: (a) If $x > 0$, then $x^{-1} > 0$ and (b) if $x < 0$, then $x^{-1} < 0$.

Prob. 3.10. Prove: If $x < y$, and $xy > 0$, then $1/y < 1/x$.

Prob. 3.11. Prove: $x < y$ implies $x < (x + y)/2 < y$. Thus, there exists a real number between any two real numbers.

Prob. 3.12. Prove: If $x < y + \epsilon$ for all $\epsilon > 0$, then $x \leqslant y$. In particular, prove: If $x < \epsilon$ for all $\epsilon > 0$, then $x \leqslant 0$.

Prob. 3.13. Let $a > 0$ be given. Prove: If $x < y + \epsilon$ for all ϵ such that $0 < \epsilon < a$, then $x \leqslant y$.

Prob. 3.14. Prove: If $x \leqslant y + \epsilon$ for all $\epsilon > 0$, then $x \leqslant y$.

Prob. 3.15. Prove: (a) If $x > 1$, then $x^2 > x$ and (b) if $0 < x < 1$, then $x^2 < x$.

Miscellaneous Problems

Prob. 3.16. Note that $(ax + b)(cx + d) = acx^2 + (ad + bc)x + bd$ and $4a^2x^2 + 4abx + 4ac = (2ax + b)^2 + 4ac - b^2$. Prove: If $a > 0$, then $ax^2 + bx + c \geqslant 0$ for all $x \in \mathbb{R}$ if and only if $4ac - b^2 \geqslant 0$.

Prob. 3.17. Prove: (a) $x > 0$ implies $x + 1/x \geqslant 2$ and $x < 0$ implies $x + 1/x \leqslant -2$.

Prob. 3.18. Prove: If x and y are real numbers, then

$$xy \leqslant \frac{x^2 + y^2}{2}.$$

Prob. 3.19. Prove: If $x \in \mathbb{R}$, then $x - x^2 \leqslant 1/4$.

Prob. 3.20. Prove $x > 0$, $y > 0$ and $x + y = 1$ imply $(1 + 1/x)(1 + 1/y) \geqslant 9$.

Prob. 3.21. Prove $xy + yz + zx \leqslant x^2 + y^2 + z^2$ for x, y, and z in $\mathbb{R}$.

Prob. 3.22. Verify $(x + y + z)(x^2 + y^2 + z^2 - xy - yz - zx) = x^3 + y^3 + z^3 - 3xyz$ and prove: If $x \geqslant 0, y \geqslant 0, z \geqslant 0$, then $xyz \leqslant (x^3 + y^3 + z^3)/3$.

Prob. 3.23. Prove: $x > 0$, $y > 0$, $z > 0$, and $x + y + z = 1$ imply $1/x + 1/y + 1/z \geqslant 9$.

4. Interval Sets, Unions, Intersections, and Differences of Sets

[The set of real numbers may be visualized by "spreading them out" on a line l in a manner familiar to the reader from his earlier mathematical education (see Fig. 4.1).

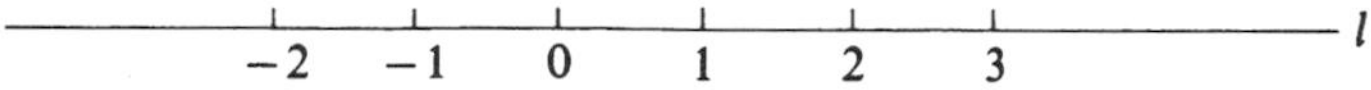

−2 −1 0 1 2 3 l]

Figure 4.1

Certain subsets of $\mathbb{R}$, called *intervals*, will play an important role in our later work. Let a and b be real numbers such that $a < b$, then by the *open interval* with *left endpoint* a and *right endpoint* b, written here as $(a; b)$,* we mean the set

$$(a; b) = \{x \in \mathbb{R} \mid a < x < b\}. \tag{4.1}$$

a b

Figure 4.2

* Many texts use (a, b) to denote the open interval just introduced. This, however, could be confused with the ordered pair (a, b).

Along with $(a;b)$ there are three other types of intervals with the same endpoints. These are written respectively as $[a,b]$, $(a,b]$, and $[a,b)$ and are defined as

$$[a,b] = \{x \in \mathbb{R} \mid a \leqslant x \leqslant b\}, \tag{4.2}$$

$$(a,b] = \{x \in \mathbb{R} \mid a < x \leqslant b\}, \tag{4.3}$$

$$[a,b) = \{x \in \mathbb{R} \mid a \leqslant x < b\}. \tag{4.4}$$

Figure 4.3 Figure 4.4 Figure 4.5

The interval $[a,b]$ in (4.2) is called a *closed* interval. The last two types of intervals may be called *half open*, or *half-closed*. The interval sets defined thus far are called *finite intervals*. There are also five types of *infinite* intervals, written respectively as $(a; +\infty)$, $[a, +\infty)$, $(-\infty; a)$, $(-\infty, a]$, and $(-\infty; +\infty)$, where a is a real number, and defined as

$$(a; +\infty) = \{x \in \mathbb{R} \mid x > a\}, \tag{4.5}$$

$$[a, +\infty) = \{x \in \mathbb{R} \mid x \geqslant a\}, \tag{4.6}$$

$$(-\infty; a) = \{x \in \mathbb{R} \mid x < a\}, \tag{4.7}$$

$$(-\infty, a] = \{x \in \mathbb{R} \mid x \leqslant a\}, \tag{4.8}$$

$$(-\infty; +\infty) = \mathbb{R} \qquad \text{(see Fig. 4.1).} \tag{4.9}$$

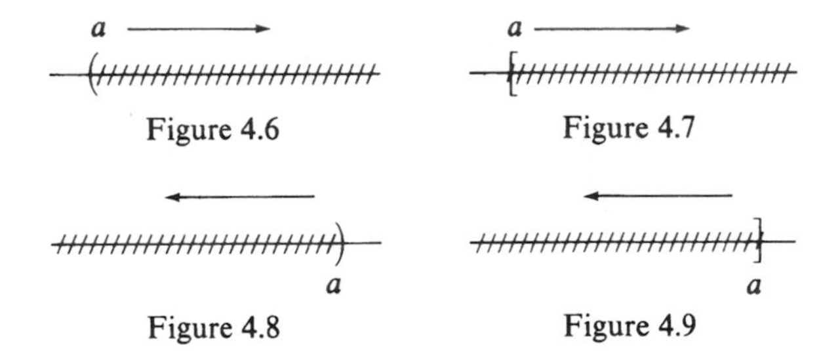

Figure 4.6 Figure 4.7

Figure 4.8 Figure 4.9

Unions of Sets

Sets may be combined in certain ways to yield other sets. If A and B are sets, then by the *union* of A and B we mean the set $A \cup B$, where

$$A \cup B = \{x \mid x \in A \text{ or } x \in B\}. \tag{4.10}$$

Thus, $A \cup B$ is the set of elements x such that x is in at least one of A or B (see Fig. 4.10). In the figure, A is shaded with horizontal lines and B with vertical ones. $A \cup B$ is the set of points having one or the other shading.

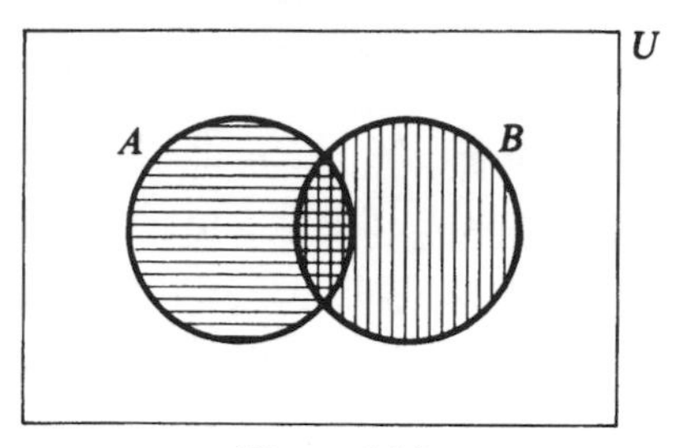

Figure 4.10

For example, if $a \in \mathbb{R}$, then

$$(-\infty, a] = (-\infty; a) \cup \{a\}.$$

If $a \in \mathbb{R}$ and $b \in \mathbb{R}$, where $a < b$, then

$$[a, b] = (a; b) \cup \{a, b\}.$$

Prob. 4.1. Prove: If a and b are real numbers such that $a < b$, then $(-\infty; b) \cup (a; +\infty) = \mathbb{R}$.

Prob. 4.2. Prove: If A and B are sets, then $A \subseteq A \cup B$ and $B \subseteq A \cup B$.

Prob. 4.3. Prove: If A and B are sets, then $A \cup B = B$ if and only if $A \subseteq B$. Also prove $A \cup \emptyset = A$ and $A \cup A = A$.

If $\mathbb{S}$ is a class of sets, then its union is defined as the set of all x such that $x \in S$ holds for some $S \in \mathbb{S}$. We write the union of $\mathbb{S}$ as

$$\bigcup \mathbb{S} \quad \text{or as} \quad \bigcup_{S \in \mathbb{S}} S. \tag{4.11}$$

In symbols,

$$\bigcup \mathbb{S} = \{x \mid x \in S \text{ for some } S \in \mathbb{S}\}. \tag{4.12}$$

For the special case where $\mathbb{S} = \{A, B\}$, where A and B are sets, we have

$$\begin{aligned}\bigcup \mathbb{S} = \bigcup \{A, B\} &= \{x \mid x \in S \text{ for some } S \in \{A, B\}\} \\ &= \{x \mid x \in A \text{ or } x \in B\} \\ &= A \cup B.\end{aligned}$$

If $\mathbb{S} = \{A, B, C\}$, where A, B, and C are sets, we write $\bigcup \mathbb{S} = \bigcup \{A, B, C\}$ as $A \cup B \cup C$. For example, we write the set $\mathbb{R}$ of reals as

$$\mathbb{R} = \mathbb{R}_+ \cup \{0\} \cup \mathbb{R}_-, \tag{4.13}$$

where $\mathbb{R}_+$ and $\mathbb{R}_-$ are respectively the sets of positive and negative real numbers.

Prob. 4.4. Prove: (a) If $A \subseteq S$ and $B \subseteq S$, then $A \cup B \subseteq S$.
(b) More generally prove: If $S \subseteq T$ for all $S \in \mathbb{S}$, then $\bigcup \mathbb{S} \subseteq T$.

Intersections of Sets

By the *intersection* $A \cap B$ of sets A and B we mean the set

$$A \cap B = \{x \mid x \in A \text{ and } x \in B\}. \tag{4.14}$$

Thus, $A \cap B$ is the set of all elements x such that x is in both A and B (see Fig. 4.11). In the figure, $A \cap B$ is the set of elements having both horizontal and vertical shading. For example, if $a < b$, then

$$(a;b) = (-\infty;b) \cap (a;+\infty) = (a;+\infty) \cap (-\infty;b)$$

and if $a \geqslant b$, then

$$(a;+\infty) \cap (-\infty;b) = \emptyset.$$

When $A \cap B = \emptyset$, we say that A and B are *disjoint* sets.

PROB. 4.5. Prove: If A and B are sets, then $A \cap B \subseteq A$ and $A \cap B \subseteq B$.

PROB. 4.6. Prove: $A \subseteq B$ if and only if $A \cap B = A$. Also prove $A \cap \emptyset = \emptyset$ and $A \cap A = A$.

If $\mathbb{S}$ is a class of sets, then its *intersection* is defined as the set of all x such that $x \in S$ for each $S \in \mathbb{S}$. We write the intersection of $\mathbb{S}$ as

$$\bigcap \mathbb{S} \quad \text{or as} \quad \bigcap_{S \in \mathbb{S}} S. \tag{4.15}$$

In symbols,

$$\bigcap \mathbb{S} = \{x \mid S \in \mathbb{S} \text{ implies } x \in S\}. \tag{4.16}$$

PROB. 4.7. Let $\mathbb{S}$ be a class of sets: Prove: If $S \in \mathbb{S}$, then $\bigcap \mathbb{S} \subseteq S$.

For the special class $\mathbb{S} = \{A, B\}$, where A and B are sets, we have

$$\begin{aligned}\bigcap \mathbb{S} = \bigcap \{A, B\} &= \{x \mid S \in \{A, B\} \text{ implies } x \in S\}\\ &= \{x \mid x \in A \text{ and } x \in B\}\\ &= A \cap B.\end{aligned}$$

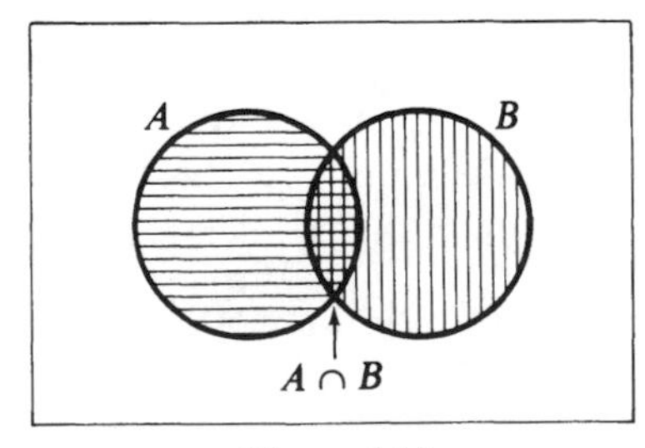

Figure 4.11

If $\mathbb{S} = \{A, B, C\}$, where A, B, and C are sets, we write $\bigcap \mathbb{S} = \bigcap\{A, B, C\}$ as $A \cap B \cap C$.

Prob. 4.8. Prove: (a) If $S \subseteq A$ and $S \subseteq B$, then $S \subseteq A \cap B$. (b) More generally, prove: If $T \subseteq S$, for all $S \in \mathbb{S}$, where $\mathbb{S}$ is a class of sets, then $T \subseteq \bigcap \mathbb{S}$.

When $\mathbb{S}$ is a class of sets, we say that $\mathbb{S}$ is a *pairwise disjoint* class when $S \in \mathbb{S}$, $T \in \mathbb{S}$, and $S \neq T$ imply $S \cap T = \emptyset$.

It is clear that we have for sets A, B, and C

$$A \cup B \cup C = (A \cup B) \cup C = A \cup (B \cup C) \tag{4.17}$$

and

$$A \cap B \cap C = (A \cap B) \cap C = A \cap (B \cap C). \tag{4.18}$$

Prob. 4.9. Prove: If A, B, and C are sets, then $(A \cup B) \cup C = (A \cup C) \cup (B \cup C)$ and $(A \cap B) \cap C = (A \cap C) \cap (B \cap C)$.

Prob. 4.10. Prove: If A, B, and C are sets, then $A \cap (B \cup C) = (A \cap B) \cup (A \cap C)$ and $A \cup (B \cap C) = (A \cup B) \cap (A \cup C)$.

Difference and Complements

If A and B are sets, then by $A - B$ we mean the set of all elements of A which are not in B (see Fig. 4.12).

$$A - B = \{x \mid x \in A \text{ and } x \notin B\}. \tag{4.19}$$

Note, $A - B \subseteq A$ and $(A - B) \cap B = \emptyset$.

When $S \subseteq U$, then $U - S$ is called the *complement of S with respect to U* and is written as $C_U(S)$. When all the sets under discussion are subsets of some domain of discourse U, then we write $C_U(S)$ simply as $C(S)$ and call it the *complement* of S.

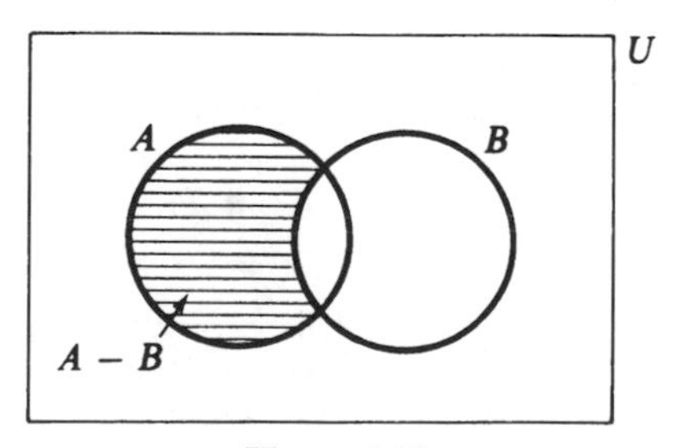

Figure 4.12

Prob. 4.11. Prove: If $A \subseteq U$ and $B \subseteq U$, where U is the domain of discourse, then $A - B = A - AB = A \cap C(B)$.

Prob. 4.12. Assume $A \subseteq U$ and $B \subseteq U$. Prove:

(a) $C(A) \cup A = U, \quad C(B) \cup B = U.$
(b) $C(A) \cap A = \emptyset, \quad C(B) \cap B = \emptyset.$
(c) $C(A \cup B) = C(A) \cap C(B).$
(d) $C(A \cap B) = C(A) \cup C(B).$
(e) $C(C(A)) = A.$

Prob. 4.13. Prove: If A, B, and C are sets such that $A \cup B = C$ and $A \cap B = \emptyset$, then $A = C - B$ and $B = C - A$.

5. The Non-negative Integers

We defined $2 = 1 + 1$, $3 = 2 + 1$; $4 = 3 + 1$. We wish to single out the set of real numbers arrived at by continuing this procedure. To this end we introduce the notion of an *inductive* set of reals

Def. 5.1. A set $I \subseteq \mathbb{R}$ is called an *inductive* set of reals if and only if (i) $0 \in I$ and (ii) $x \in I$ implies $x + 1 \in I$ for each $x \in \mathbb{R}$.

It is clear that the set $\mathbb{R}$ itself is inductive.

Prob. 5.1. Prove that the following sets of real numbers are all inductive: (a) $[0, +\infty)$, (b) $[-1, +\infty)$, (c) $\{0\} \cup [1, +\infty)$, (d) $\{0, 1\} \cup [2, +\infty)$.

By Prob. 5.1 we see that there are many inductive sets of real numbers. Let $\mathcal{I}$ be the class of all inductive sets of reals. We define $\mathbb{Z}_0$ as the set of all real numbers which belong to *all* the inductive sets of real numbers. Thus

$$\mathbb{Z}_0 = \bigcap \mathcal{I} = \{x \in \mathbb{R} \mid I \in \mathcal{I} \text{ implies } x \in I\}. \tag{5.1}$$

Theorem 5.1. *The set* $\mathbb{Z}_0$ *defined in* (5.1) *is an inductive set of reals.*

Proof. Assume $I \in \mathcal{I}$, so that I is an inductive set of reals. Then $0 \in I$. Thus, $I \in \mathcal{I}$ implies that $0 \in I$ and hence (i) $0 \in \mathbb{Z}_0$. We prove that (ii) $x \in \mathbb{Z}_0$ implies $x + 1 \in \mathbb{Z}_0$. Let $x \in \mathbb{Z}_0$. Assume that $I \in \mathcal{I}$ so that I is an inductive set of reals. Since $\mathbb{Z}_0 = \bigcap \mathcal{I} \subseteq I$ (Prob. 4.7), it follows that $x \in I$ and hence that $x + 1 \in I$. Thus, $I \in \mathcal{I}$ implies $x + 1 \in I$ and we have $x + 1 \in \bigcap \mathcal{I} = \mathbb{Z}_0$. This proves: $x \in \mathbb{Z}_0$ implies $x + 1 \in \mathbb{Z}_0$. By Def. 5.1, $\mathbb{Z}_0$ is an inductive set of real numbers.

Corollary 1. *$\mathbb{Z}_0$ is the "smallest" inductive set of reals in the sense that $\mathbb{Z}_0 \subseteq I$ holds for all inductive sets I.*

PROOF. Exercise.

Corollary 2. *If S is an inductive set of reals such that $S \subseteq \mathbb{Z}_0$, then $S = \mathbb{Z}_0$.*

PROOF. Since S is an inductive set of reals, we have by Corollary 1, $\mathbb{Z}_0 \subseteq S$. But $S \subseteq \mathbb{Z}_0$ holds by hypothesis. We conclude that $S = \mathbb{Z}_0$. This completes the proof.

Note that $n \in \mathbb{Z}_0$ implies $n + 1 \in \mathbb{Z}_0$ since $\mathbb{Z}_0$ is an inductive set of reals. Accordingly, since $0 \in \mathbb{Z}_0$, we have $1 = 0 + 1 \in \mathbb{Z}_0$, $2 = 1 + 1 \in \mathbb{Z}_0$, and $3 = 2 + 1 \in \mathbb{Z}_0$. Corollary 2 of Theorem 5.1 states that the only inductive subset of $\mathbb{Z}_0$ is $\mathbb{Z}_0$ itself. It is easy to see that Corollary 2 of Theorem 5.1 may be reformulated as

Corollary 3 (of Theorem 5.1). *If $S \subseteq \mathbb{Z}_0$, where* (i) $0 \in S$ *and* (ii) $n \in S$ *implies $n + 1 \in S$, then $S = \mathbb{Z}_0$.*

Theorem 5.2. *If $n \in \mathbb{Z}_0$, then $n \geqslant 0$.*

PROOF. The set $[0, +\infty)$ is an inductive set of reals (Prob. 5.1). Hence $\mathbb{Z}_0 \subseteq [0, +\infty)$ (Corollary 1 of Theorem 5.1). Consequently, if $n \in \mathbb{Z}_0$, then $n \in [0, +\infty)$ and hence $n \geqslant 0$.

Remark 5.1. According to the last theorem, a nonzero element of $\mathbb{Z}_0$ is positive.

Def. 5.2. We call 0 an *integer*. Moreover, each nonzero element of $\mathbb{Z}_0$ will be called a *positive integer*. Elements of $\mathbb{Z}_0$ will be called *nonnegative integers*. The set $\mathbb{Z}_+ = \mathbb{Z}_0 - \{0\}$ is the *set* of positive integers. Since $1 \in \mathbb{Z}_0$ and $1 > 0$, we see that 1 is a positive integer.

Theorem 5.3. *If n is a positive integer, then $n \geqslant 1$.*

PROOF. The set $A = \{0\} \cup [1, +\infty)$ is an inductive set of real numbers (Prob. 5.1(c)). Hence, $\mathbb{Z}_0 \subseteq A$. If n is a positive integer, then $n \in \mathbb{Z}_0 \subseteq A = \{0\} \cup [1, +\infty)$. Hence $n \in \{0\} \cup [1, +\infty)$. Since $n > 0$, $n \in [1, +\infty)$. But then $n \geqslant 1$. This completes the proof.

We now formulate the *principle of mathematical induction*. First, we define the notion of a *statement about nonnegative integers*. This is a sentence $P(n)$ containing n which becomes a true or false statement when n is replaced by some specific nonnegative integer. For example, $n = 0$ is a statement about nonnegative integers since it is true when n is replaced by 0 and false if n is replaced by a positive integer.

Theorem 5.4 (Principle of Mathematical Induction). *If $P(n)$ is a statement about non-negative integers such that*: (i) $P(0)$ *is true and* (ii) $P(k)$ *implies* $P(k+1)$ *for each* $k \in \mathbb{Z}_0$, *then* $P(n)$ *is true for all* $n \in \mathbb{Z}_0$.

Proof. Let

$$S = \{n \in \mathbb{Z}_0 \mid P(n) \text{ is true}\}.$$

Using the hypothesis on $P(n)$, it is easy to prove that S is an inductive set of reals. Also $S \subseteq \mathbb{Z}_0$. By Corollary 2 of Theorem 5.1, $S = \mathbb{Z}_0$. Therefore, $n \in \mathbb{Z}_0$ implies that $n \in S$ and hence that $P(n)$ is true.

Another principle of mathematical induction is based on

Theorem 5.5. *If* $S \subseteq \mathbb{Z}_+$, *where* (i) $1 \in S$ *and* (ii) $k \in S$ *implies* $k+1 \in S$ *for each* $k \in \mathbb{Z}_+$, *then* $S = \mathbb{Z}_+$.

Proof. Let $T = \{0\} \cup S$. We have $T \subseteq \mathbb{Z}_0$. (i) It is clear that $0 \in T$ holds. Next assume that $n \in T$, so that $n = 0$ or $n \in S$. Either of these implies that $n + 1 \in T$. For, if $n = 0$, then $n + 1 = 0 + 1 = 1 \in S$ and if $n \in S$, then $n + 1 \in S \subseteq T$ and hence $n + 1 \in T$. This proves: (ii) $n \in T$ implies $n + 1 \in T$. We conclude that $T = \mathbb{Z}_0$. Thus, $\{0\} \cup S = \{0\} \cup \mathbb{Z}_+$. Since $0 \notin S$ and $0 \notin \mathbb{Z}_+$, it follows that $S = \mathbb{Z}_+$ (explain).

Corresponding to this theorem we state a principle of induction:

Theorem 5.6. *If $P(n)$ is a statement about positive integers* such that* (i) $P(1)$ *is true and* (ii) $P(k)$ *implies* $P(k+1)$ *for each positive integer k, then $P(n)$ is true for all positive integers n.*

Proof. Exercise.

We illustrate the use of the last theorem by proving

Theorem 5.7. *If m and n are positive integers, then so is $m + n$.*

Proof. $m + 1$ is a positive integer for all positive integers m (why?). Assume that for some positive integer n, $m + n$ is a positive integer for all positive integers m. Hence $(m + n) + 1$ is a positive integer for all positive integers m and since $m + (n + 1) = (m + n) + 1$, so is $m + (n + 1)$. By the principle of mathematical induction stated in Theorem 5.6, for each positive integer

* The reader should explain what is meant by "a statement $P(n)$ about positive integers" using as a guide our definition of a "statement about negative integers" in the paragraph following Theorem 5.3 and its proof.

n, $m + n$ is a positive integer for all positive integers m. Hence, if m and n are positive integers, so is $m + n$.

PROB. 5.2. Prove: If m and n are positive integers, so is their product mn.

Theorem 5.8. *If m is a positive integer such that $m > 1$, then $m - 1$ is a positive integer.*

PROOF. Suppose, for the sake of obtaining a contradiction, that there exists some positive integer $m > 1$, such that $m - 1$ is not a positive integer. Let S be the set defined as

$$S = \{n \in \mathbb{Z}_+ \mid n \neq m\}.$$

Then $S \subseteq \mathbb{Z}_+$. By the hypothesis on m, $m > 1$ holds, so $1 \in S$. Second, assume $n \in S$, so that $n \in \mathbb{Z}_+$ and $n \neq m$. Since $m - 1 \notin \mathbb{Z}_+$ by our present assumption on m, it follows that $n \neq m - 1$ for our n. This implies that $n + 1 \neq m$. Since $n + 1 \in \mathbb{Z}_+$ we obtain $n + 1 \in S$. S is thus a set of positive integers satisfying the hypothesis of Theorem 5.5, and we conclude that $S = \mathbb{Z}_+$. But this is impossible since $m \notin S$ and $m \in \mathbb{Z}_+$. We must therefore conclude that if $m \in \mathbb{Z}_+$, where $m > 1$, then $m - 1 \in \mathbb{Z}_+$.

PROB. 5.3. Prove: If m and n are positive integers such that $m > n$, then $m - n$ is a positive integer (Hint: use Theorem 5.8 and induction on n).

Theorem 5.9. *If n is a non-negative integer, then no non-negative integer m exists such that $n < m < n + 1$.*

PROOF. If m were a non-negative integer such that $n < m < n + 1$, we would conclude that $0 < m - n < 1$. If n is a positive integer, then by Prob. 5.3, under the conditions on m and n, so is $m - n$. This implies $m - n \geqslant 1$ (see Theorem 5.3)—a contradiction. If $n = 0$, then $0 < m < 0 + 1 = 1$, where m is a positive integer, which is impossible, again by Theorem 5.3. In either case, $n < m < n + 1$ cannot hold when m and n are non-negative integers.

Corollary. *If m and n are non-negative integers such that $m > n$, then $m \geqslant n + 1$.*

PROOF. If $m < n + 1$, it would follow from the hypothesis that $n < m < n + 1$. This would contradict Theorem 5.9. Hence $m \geqslant n + 1$ as claimed.

We state another important theorem about elements of $\mathbb{Z}_0$. It is referred to as the *well-ordered property* of the non-negative integers.

Theorem 5.10. *Every nonempty set of nonnegative integers has a least member.*

Proof. The proof we give is indirect. Suppose a set $S \subseteq \mathbb{Z}_0$ exists which has *no* least number. Let

$$T = \{n \in \mathbb{Z}_0 \mid n < k \text{ holds for all } k \in S\}.$$

Since $T \subseteq \mathbb{Z}_0$, we know that $k \geqslant 0$ holds for all $k \in \mathbb{Z}_0$. If $0 \in S$ holds, then 0 would be the least member of S. This would contradict our assumption on S. Hence, $0 \notin S$, so that $k > 0$ holds for all $k \in S$. This implies that $0 \in T$. Now assume that $n \in T$ so that $n < k$ holds for all $k \in S$. Since n and k are non-negative integers, $n + 1 \leqslant k$ for all $k \in S$ (corollary of Theorem 5.9). If $n + 1 \in S$ holds, it would be the least member of S, and this would contradict the assumption on S. We conclude that $n + 1 \notin S$, so that $n + 1 < k$ holds for all $k \in S$. This implies $n + 1 \in T$. Thus, $n \in T$ implies $n + 1 \in T$. All this implies that $T = \mathbb{Z}_0$. In turn, this implies that $S = \emptyset$. For, if there exists a $k_0 \in S$, it would follow that $k_0 \in \mathbb{Z}_0$ which implies $k_0 \in T$. But then $k_0 < k_0$, which is impossible. Hence, $S = \emptyset$. We have proved: If a subset S of $\mathbb{Z}_0$ has no least number, then it is empty. This implies that a nonempty subset of $\mathbb{Z}_0$ must have a least member.

Prob. 5.4. Prove: Every nonempty set of positive integers has a least number. (Thus, there is a well-orderedness principle for the positive integers also.)

Certain subsets of $\mathbb{Z}_0$ are important for applications.

Def. 5.3. If $n \in \mathbb{Z}_0$, we define ω_n to be the set

$$\omega_n = \{k \in \mathbb{Z}_0 \mid k < n\}.$$

Each ω_n is called an *initial* segment of $\mathbb{Z}_0$.

For example, $\omega_0 = \emptyset$, $\omega_1 = \{0\}$, $\omega_2 = \{0, 1\}$.

Prob. 5.5. Prove: If n is a nonnegative integer, then $\omega_{n+1} = \omega_n \cup \{n\}$.

When n is a positive integer, we often write ω_n as

$$\omega_n = \{0, 1, \ldots, n-1\} \quad \text{or} \quad \omega_n = \{0, \ldots, n-1\}. \tag{5.2}$$

Def. 5.4. An *initial segment* of $\mathbb{Z}_+$ is defined as follows: Let n be some positive integer, then (n) is the set

$$(n) = \{k \in \mathbb{Z}_+ \mid k \leqslant n\}, \tag{5.3}$$

(n) is called an *initial segment* of positive integers.

Clearly,

$$(1) = \{1\}, \qquad (2) = \{1,2\}, \qquad (3) = \{1,2,3\}.$$

Prob. 5.6. Prove: If n is a positive integer, then $(n + 1) = (n) \cup \{n + 1\}$.

If n is some positive integer, we often write (n) as

$$(n) = \{1, 2, \ldots, n\} \quad \text{or} \quad (n) = \{1, \ldots, n\}. \tag{5.4}$$

It is clear that if $n = 0$, then $(n) = (0) = \emptyset$.

It is now possible to formulate another principle of induction which we refer to as *complete induction* (Theorem 5.12).

Theorem 5.11. *If $S \subseteq \mathbb{Z}_0$, where $\omega_n \subseteq S$ implies $n \in S$, then $S = \mathbb{Z}_0$.*

PROOF. We have $\emptyset \subseteq S$. This implies $\omega_0 \subseteq S$. By hypothesis, $0 \in S$. Consider the complement $C(S) = \mathbb{Z}_0 - S$ relative to $\mathbb{Z}_0$. Suppose that $C(S) \neq \emptyset$. Since $C(S) \subseteq \mathbb{Z}_0$, we know then that $C(S)$ has a least member, n_0 say. By our first observation, this implies that $n_0 \neq 0$. (Recall that $0 \in S$. Since $n_0 \notin S$, $n_0 \neq 0$). This implies that $n_0 > 0$. Thus, $\omega_{n_0} \not\subseteq S$. (If $\omega_{n_0} \subseteq S$, the hypothesis implies that $n_0 \in S$, contradicting $n_0 \in C(S)$.) Therefore, there exists a $k \in \omega_{n_0}$ such that $k \notin S$. This shows that there exists $k < n_0$ such that $k \in C(S)$. In turn, this contradicts the definition of n_0 as the least member of $C(S)$. We conclude that $C(S) = \emptyset$. But then $S = \mathbb{Z}_0$, as claimed.

Theorem 5.12. *If $P(n)$ is a statement about nonnegative integers such that $P(0)$ is true, and for each $n \in \mathbb{Z}_0$, the truth of $P(k)$ for $k \in \mathbb{Z}_0$ and $k < n$ implies the truth of $P(n)$, then $P(n)$ is true for all $n \in \mathbb{Z}_0$.*

PROOF. Exercise.

6. The Integers

Def. 6.1. A real number m is called a *negative* integer if and only if its negative $-m$ is a positive integer. Note, since $m = -(-m)$, m is a negative integer if and only if it is the negative of a positive integer. The set of negative integers will be written as $\mathbb{Z}_-$. The set $\mathbb{Z}$ where $\mathbb{Z} = \mathbb{Z}_0 \cup \mathbb{Z}_-$ will be called the *set* of integers and each of its members an *integer*.

Thus, -1, -2, and -3 are negative integers since they are negatives of positive integers. We have

$$\mathbb{Z} = \mathbb{Z}_0 \cup \mathbb{Z}_- = \mathbb{Z}_+ \cup \{0\} \cup \mathbb{Z}_-. \tag{6.1}$$

PROB. 6.1. Prove: If n is an integer, then so is $-n$.

PROB. 6.2. If n is an integer, then so are $n + 1$ and $n - 1$.

We now state an induction principle for integers (Theorem 6.2).

Theorem 6.1. *If* $S \subseteq \mathbb{Z}$, *where* (i) $0 \in S$ *and* (ii) $n \in S$ *implies* $n+1 \in S$ *and* $-n \in S$, *then* $S = \mathbb{Z}$.

PROOF. We first prove $\mathbb{Z}_0 \subseteq S$. Let S_0 be the set

$$S_0 = \{n \in S \mid n \geqslant 0\}.$$

Clearly, $S_0 \subseteq S$ and $S_0 \subseteq \mathbb{Z}_0$. From the properties of S and from the fact that $S_0 \subseteq S$, it is easy to prove that $0 \in S_0$ and that $n \in S_0$ implies $n+1 \in S_0$ (do this). Using Corollary 3 of Theorem 5.1, we conclude that $S_0 = \mathbb{Z}_0$. It now follows that $\mathbb{Z}_0 = S_0 \subseteq S$, and hence that $\mathbb{Z}_0 \subseteq S$.

Now we prove $\mathbb{Z}_- \subseteq S$. Assume $n \in \mathbb{Z}_-$. This implies that $-n \in \mathbb{Z}_+ \subset \mathbb{Z}_0 \subseteq S$ and hence that $-n \in S$. By the hypothesis on S this yields $n = -(-n) \in S$. Thus $n \in \mathbb{Z}_-$ implies $n \in S$ so that $\mathbb{Z}_- \subseteq S$. This, the result obtained in the first paragraph, and properties of sets imply that

$$Z = \mathbb{Z}_0 \cup \mathbb{Z}_- \subseteq S.$$

Thus, $\mathbb{Z} \subseteq S$. But $S \subseteq \mathbb{Z}$ by hypothesis. Hence, $S = \mathbb{Z}$.

Theorem 6.2. *If* $P(n)$ *is a statement about integers such that* (i) $P(0)$ *is true and* (ii) $P(n)$ *implies* $P(-n)$ *and* $P(n+1)$, *then* $P(n)$ *is true for all integers* n.

PROOF. Exercise.

We use Theorem 6.2 to prove:

Theorem 6.3. *If* m *and* n *are integers, so is* $m + n$.

PROOF. We use the induction principle of the last theorem and perform induction on n. The theorem holds for $n = 0$ and for all integers m. Assume the theorem holds for some integer n — so that $m + n$ is an integer for all integers m. Therefore $(m + n) + 1$ is an integer for all integers m and since $m + (n + 1) = (m + n) + 1$ that $m + (n + 1)$ is an integer for all integers m. Using the n of the last sentence and the induction hypothesis, we see that $n - m = n + (-m)$ is an integer for all integers m (explain). We can now claim that $m + (-n) = -(n - m)$ is an integer for all integers m. We have proved that if $m + n$ is an integer for all integers m, then $m + (n + 1)$ and $m + (-n)$ are integers for all integers m. By the principle of induction for integers, we conclude that for each n, $m + n$ is an integer for all integers m. This proves the theorem.

PROB. 6.3. Prove: If m and n are integers, then so is $m - n$.

PROB. 6.4. Prove: If m and n are integers, so is their product mn.

PROB. 6.5. Prove: If n is an integer, then no integer m exists such that $n < m < n + 1$.

Prob. 6.6. Prove: If m and n are integers such that $m > n$, then $m \geqslant n + 1$.

Prob. 6.7. Prove: If $S \subseteq \mathbb{Z}_-$ where (i) $-1 \in S$ and (ii) $n \in S$ implies $n - 1 \in S$, then $S = \mathbb{Z}_-$.

Using the result cited in this problem we obtain still another principle of induction for the integers (Theorem 6.5).

Theorem 6.4. *If $S \subseteq \mathbb{Z}$, where* (i) $0 \in S$ *and* (ii) $n \in S$ *implies $n + 1 \in S$ and $n - 1 \in S$, then $S = Z$.*

Proof. Exercise.

Theorem 6.5. *If $P(n)$ is a statement about integers such that* (i) *$P(0)$ is true and* (ii) *$P(n)$ implies $P(n + 1)$ and $P(n - 1)$ for each $n \in Z$, then $P(n)$ is true for all $n \in \mathbb{Z}$.*

Proof. Exercise.

7. The Rational Numbers

Def. 7.1. A *rational* number r is one that can be written

$$r = \frac{p}{q}, \tag{7.1}$$

where p and q are integers such that $q \neq 0$.

The set of rational numbers will be written as $\mathbb{Q}$. We write $\mathbb{Q}_0$, $\mathbb{Q}_+$, and $\mathbb{Q}_-$ respectively for the sets of nonnegative, positive, and negative rationals.

Remark 7.1. Since each integer n can be written as $n/1$ and is therefore of the form (7.1), we see that each integer is a rational number. The converse of the last statement is false. For example, $\frac{1}{2}$ is not an integer (why?) but is a rational number and we see that not every rational number is an integer. Thus, $Z \subset \mathbb{Q}$, the inclusion being proper.

Prob. 7.1. Prove: If r and s are rational numbers, then so are $r + s$, $r - s$, and rs. Moreover, if $s \neq 0$, then prove r/s is rational.

Prob. 7.2. Prove: Between any two rational numbers there exists another rational number (see Prob. 3.11).

Remark 7.2. If in our first 14 axioms we replace the reals by the rationals and the set $\mathbb{R}$ by the set $\mathbb{Q}$, the axioms will hold (the reader can check this).

As a matter of fact, any set $\mathfrak{F}$ in which a "less than" relation is defined and for which there are two operations called "addition" and "multiplication" such that all the 14 axioms hold with the real numbers replaced by elements of $\mathfrak{F}$ and the set $\mathbb{R}$ replaced by $\mathfrak{F}$ is called an *ordered field*. Accordingly, $\mathbb{Q}$ and $\mathbb{R}$ are both examples of ordered fields. All the work done so far was based on the first 14 axioms characterizing an ordered field. Consequently, these results hold in any ordered field. In the next section we state the *Axiom of Completeness*. This fifteenth axiom for the real numbers will be seen not to hold for the ordered field of rational numbers.

8. Boundedness: The Axiom of Completeness

Def. 8.1. A set $S \subseteq \mathbb{R}$ is called *bounded from above* if some real number u exists such that

$$x \in S \quad \text{implies} \quad x \leqslant u. \tag{8.1}$$

The number u is called an *upper bound* for S. Similarly, $S \subseteq \mathbb{R}$ is said to be *bounded from below* if some real number l exists such that

$$x \in S \quad \text{implies} \quad x \geqslant l. \tag{8.2}$$

The number l being called a *lower bound* for S. Finally, a set $S \subseteq \mathbb{R}$ is called *bounded* if and only if there exist real l and u such that

$$x \in S \quad \text{implies} \quad l \leqslant x \leqslant u. \tag{8.3}$$

For example, if $a \in \mathbb{R}$, then the intervals $(-\infty; a)$ and $(-\infty, a]$ are bounded from above, but not from below. In both cases a is an upper bound and any real number greater than a is also an upper bound for the set. Similarly, the intervals $(a; +\infty)$ and $[a, +\infty)$ are each bounded from below but not from above. In both cases, a is a lower bound and any real number less than a is also a lower bound for the set. The finite intervals $(a; b)$, $[a, b]$, and $[a, b)$, where a and b are real numbers, are all examples of bounded sets of reals. We say of a finite interval that it is *bounded* and of an infinite interval that it is *unbounded*.

Def. 8.2. If $S \subseteq \mathbb{R}$ and S has a greatest member M, then M is called the *maximum* of S. On the other hand, if S has a least member m, then m is called the *minimum* of S. When M is the maximum of S, we write

$$M = \max S \quad \text{or} \quad M = \max_{x \in S} x. \tag{8.4}$$

Similarly, if m is the minimum of S, we write

$$m = \min S \quad \text{or} \quad m = \min_{x \in S} x. \tag{8.5}$$

For example, for $a \in \mathbb{R}$, we have a $\max(-\infty; a]$. Note that the set $(\infty; a)$ has *no* maximum (prove this). As another example, consider the set $\{a, b\}$, where a and b are real numbers and $a \leqslant b$. We have, $\max\{a, b\} = b$ and $\min\{a, b\} = a$.

The notions of least upper bound and greatest lower bound are very important and we introduce them now.

Def. 8.3. If $S \subseteq \mathbb{R}$, then a real number μ is called the *least upper bound* or *supremum* of S if and only if (a) μ is an upper bound for S and (b) no upper bound for S is less than μ. Similarly, a real number λ is called the *greatest lower bound* or the *infimum* of $S \subseteq \mathbb{R}$ if and only if (c) λ is a lower bound for S and (d) no lower bound for S is greater than λ.

Remark 8.1. Note that an upper bound μ of a set $S \subseteq \mathbb{R}$ is a least upper bound or supremum of S, when $\mu \leqslant u$ for each upper bound u of S. Similarly, a lower bound λ of a set $S \subseteq \mathbb{R}$ is a greatest lower bound or infimum of S when $\lambda \geqslant l$ for each lower bound l of S.

Theorem 8.1. *A set of real numbers has at most one supremum and at most one infimum.*

Proof. Assume that $S \subseteq \mathbb{R}$. Suppose μ and μ' are each suprema of S. Thus, μ' is an upper bound for S and μ is the supremum of S. It follows from this that $\mu \leqslant \mu'$. Using the same reasoning we arrive at $\mu' \leqslant \mu$. We therefore conclude that $\mu = \mu'$. The proof for the infimum is similar.

Notation. If μ is the supremum of S, we write

$$\mu = \sup S \quad \text{or} \quad \mu = \sup_{x \in S} x, \tag{8.6}$$

and if λ is the infimum of S, we write

$$\lambda = \inf S \quad \text{or} \quad \lambda = \inf_{x \in S} x. \tag{8.7}$$

Remark 8.2. As an example we consider the infinite interval $I = (-\infty; a)$. We prove $a = \sup(-\infty; a)$. Clearly, a is an upper bound for I. Let u be some upper bound for I. Suppose that $u < a$. Since a real number x_0 exists such that $u < x_0 < a$, an $x_0 \in I$ exists with $x_0 > u$. This contradicts the assumption that u is an upper bound for I. Hence, $a \leqslant u$ holds for each upper bound of I. This completes the proof. Note that $a \notin I = (-\infty; a)$, so a cannot be the maximum of I. This shows that the supremum of a set, when it exists, is not necessarily its maximum.

Prob. 8.1. Prove: If $a \in \mathbb{R}$, then $a = \inf(a; +\infty)$.

PROB. 8.2. Prove: If $M = \max S$ for $S \subseteq \mathbb{R}$, then $M = \sup S$. Similarly, if $m = \min S$, then $m = \inf S$. (Note that the converses of these statements do not hold in view of Remark 8.2.)

PROB. 8.3. (a) Let $\mu = \sup S$, where $S \subseteq \mathbb{R}$ and $\epsilon > 0$. Prove: There exists $x_0 \in S$ such that $\mu - \epsilon < x_0 \leqslant \mu$. (b) Let $\lambda = \inf S$ and $\epsilon > 0$. Prove there exists $y_0 \in S$ such that $\lambda \leqslant y_0 < \lambda + \epsilon$.

PROB. 8.4. Assume that $S \subseteq \mathbb{R}$. Prove: (a) S is not bounded from above if and only if for each real B there exists an $x_0 \in S$ such that $x_0 > B$. (b) S is not bounded from below if and only if for each real B there exists a $y_0 \in S$ such that $y_0 < B$.

PROB. 8.5. Let $a \in \mathbb{R}$. Prove: Neither of $[a, +\infty)$ or $(a; +\infty)$ is bounded from above and that neither of $(-\infty; a)$ or $(-\infty, a]$ is bounded from below.

We now state the fifteenth axiom for the real number system.

(XV) **(Axiom C)** (The Completeness Axiom). Every nonempty set of real numbers which is bounded from above has a real supremum.

Theorem 8.2. *Every nonempty set of real numbers which is bounded from below has a real infimum.*

PROOF. Let S be a nonempty set of real numbers which is bounded from below. Therefore, a real number exists which is a lower bound for S. Let B be the set of all lower bounds of S. We have that $B \neq \emptyset$. Since $S \neq \emptyset$, there exists $x_0 \in S$, and it follows that $b \in B$ implies $b \leqslant x_0$. Thus, x_0 is an upper bound for the set B and B is bounded from above. By Axiom C there exists a real number λ such that $\lambda = \sup B$. We prove next that $\lambda = \inf S$.

Assume that $x^* \in S$ exists such that $x^* < \lambda$. This implies that x^* is not an upper bound for B and that there exists a $b^* \in B$ such that $b^* > x^*$. This is impossible because $x^* \in S$ and $b^* \in B$ implies $b^* \leqslant x^*$. Thus, $x \in S$ implies that $x \geqslant \lambda$ and we see that λ is a lower bound for S. Assume l is some lower bound for S. This implies that $l \in B$ and, hence, that $l \leqslant \lambda$ since $\lambda = \sup B$. This completes the proof that $\lambda = \inf S$.

9. Archimedean Property

Theorem 9.1 (Archimedean Property for $\mathbb{R}$). *If a and b are real numbers such that $a > 0$ and $b > 0$, then there exists a positive integer such that $na > b$.*

PROOF. Let $\mathfrak{M}$ be the set

$$\mathfrak{M} = \{ na \mid n \text{ is a positive integer} \} = \{ a, 2a, 3a, \dots \}.$$

Clearly, $\mathfrak{M} \neq \emptyset$. If $na \leqslant b$ holds for all positive integers n, then $\mathfrak{M}$ is bounded from above by b, and, since it is not empty it has a supremum, μ say (Axiom C). Since μ is necessarily an upper bound for $\mathfrak{M}$, $na \leqslant \mu$ for all positive integers n. But $n+1$ is a positive integer whenever n is, so that $(n+1)a \leqslant \mu$ holds for all positive integers n. This implies

$$na \leqslant \mu - a \qquad \text{for all positive integers } n$$

so that $\mu - a$ is an upper bound for $\mathfrak{M}$. Accordingly, $\mu - a \geqslant \mu$. But this is impossible, because $a > 0$ implies that $\mu - a < \mu$. Thus, b is not an upper bound for $\mathfrak{M}$ and there exists a positive integer n such that $na > b$, as claimed.

Corollary 1. *If $\epsilon > 0$, there exists a positive integer n such that $1/n < \epsilon$.*

PROOF. Since $\epsilon > 0$ and $1 > 0$, there exists by the last theorem, a positive integer n such that $n\epsilon > 1$. The conclusion follows from this.

Corollary 2. *If x is a real number, there exists a positive integer n such that $n > x$.*

PROOF. If $x \leqslant 0$, there is nothing to prove, since then $x \leqslant 0 < 1$. Suppose that $x > 0$. Since $1 > 0$, the last theorem implies the existence of a positive integer n such that $n = n \cdot 1 > x$.

Theorem 9.2. *If x is a real number, there exists a unique integer n such that $n \leqslant x < n+1$.*

PROOF. By Corollary 2 of Theorem 9.1 there exists an integer m such that $x < m$. It is also easy to see that there exists an integer p such that $p < x$. Indeed, there exists an integer k such that $-x < k$. Hence $-k < x$. But then $p = -k$ is a required integer. Thus, $p < x < m$, where p and m are integers. Since $m - p$ is a positive integer, it follows that $p < x < p + (m - p)$. Let S be the set such that

$$S = \{ l \in \mathbb{Z}_+ \mid x < p + l \}.$$

Clearly $S \neq \emptyset$ since $m - p \in S$. Thus, S is a nonempty set of positive integers. As such, S has a least member, n_0 say. If $n_0 = 1$, we have $p < x < p + 1$, so p is the integer n in the conclusion of the theorem. If $n_0 > 1$, then $n_0 - 1$ is a positive integer. Since $n_0 - 1 < n_0$, the fact that n_0 is the least member of S implies that $p + n_0 - 1 \leqslant x < p + n_0$. Putting $n = p + n_0 - 1$, we obtain $n + 1 = p + n_0$ and $n \leqslant x < n + 1$. Here, too, n is an integer. This proves the existence of n.

We prove that the integer n such that $n \leqslant x < n+1$ is unique. Assume that $n_1 \leqslant x < n_1 + 1$ for some integer n_1. If $n_1 \neq n$, we have $n_1 < n$ or $n < n_1$. In the first case, $n_1 < n \leqslant x < n_1 + 1$. But then $n_1 < n < n_1 + 1$, where n and n_1 are integers. This is impossible (Prob. 6.5). Thus, $n_1 < n$ is false. A similar argument shows that $n < n_1$ is false. Thus, $n_1 \neq n$ is false and we have $n_1 = n$. This completes the proof.

Def. 9.1. If $x \in \mathbb{R}$, then the unique integer n such that $n \leqslant x < n+1$ is called the *greatest integer* $\leqslant x$. It is written as $[x]$. Thus,

$$[x] \leqslant x < [x] + 1 \qquad \text{for each} \quad x \in \mathbb{R} \tag{9.1}$$

and

$$0 \leqslant x - [x] < 1 \qquad \text{for each} \quad x \in \mathbb{R}. \tag{9.2}$$

For example, $[n] = n$ if and only if n is an integer. We have $[5/3] = 1$, $[-7/2] = -4$, $[1/3] = 0$.

Prob. 9.1. Prove: If x is a real number, then $[x+1] = [x] + 1$.

Prob. 9.2. Prove: If x is a real number, there exists a unique integer n such that $n - 1 < x \leqslant n$.

10. Euclid's Theorem and Some of Its Consequences

Theorem 10.1 (Euclid's Theorem). *If a and b are integers and $b > 0$, there exist unique integers q and r such that*

$$a = bq + r, \qquad \textit{where} \quad 0 \leqslant r < b. \tag{10.1}$$

Proof. Let $q = [a/b]$, the greatest integer $\leqslant a/b$. Then

$$q = \left[\frac{a}{b}\right] \leqslant \frac{a}{b} < \left[\frac{a}{b}\right] + 1 = q + 1.$$

Hence,

$$bq \leqslant a < bq + b.$$

This implies that $0 \leqslant a - bq < b$. Put $r = a - qb$. It is clear that q and r are integers such that $a = bq + r$, where $0 \leqslant r < b$.

We prove the uniqueness of the q and r in (10.1). Assume that $a = bq_1 + r_1$, where q_1 and r_1 are integers and $0 \leqslant r_1 < b$. The equality

$$bq_1 + r_1 = bq + r$$

holds. This implies

$$b(q_1 - q) = r - r_1. \tag{10.2}$$

Since $0 \leqslant r < b$ and $-b < -r_1 \leqslant 0$, we have $-b < r - r_1 < b$. This and

(10.2) imply

$$-b < b(q - q_1) < b.$$

In turn, this implies $-1 < q - q_1 < 1$. Since $q - q_1$ is integer, it follows that $q - q_1 = 0$. Thus, $q = q_1$. This and (10.2) yield $r = r_1$, which completes the proof.

PROB. 10.1. Prove: If a and b are integers and $b \neq 0$ (here b could be negative), there exist integers q and r such that $a = bq + r$, where $0 \leqslant r < b$, if $b > 0$, and $0 \leqslant r < -b$, if $b < 0$.*

Def. 10.1. If a and b are integers, $b \neq 0$, then the integers q and r of Prob. 10.1 are called respectively the *quotient* and *remainder upon dividing* a by b.

Def. 10.2. If a and b are integers and an integer q exists such that $a = bq$, then we say that a is a *multiple* of b, or that a is *divisible* by b, or that b *divides* a, or that b is a *factor* of a and we write $b \mid a$. When $b \mid a$ is false, we write $b \nmid a$. We will never write $b \mid a$ or $b \nmid a$ when b and a are not integers. Note that if $b \neq 0$ and b divides a, then the remainder upon dividing a by b is zero.

We have, for example, $18 = (-5)(-3) + 3$. The quotient and remainder upon dividing 18 by -5 are -3 and 3 respectively. Also, since $6 = 2 \cdot 3$, we have $2 \mid 6$ and also $3 \mid 6$.

Remark 10.1. When $x \neq 0$, $x/0$ does not exist. Since $0/0$ is not unique, we do not divide 0 by 0. Nevertheless, $0 \mid 0$ in the sense of Def. 10.2; indeed, $0 = 0 \cdot m$ for any integer m.

PROB. 10.2. Prove: If n is an integer, then $n \mid 0$ and $0 \nmid n$ when $n \neq 0$.

PROB. 10.3. Prove: (a) $(-1) \mid n$, (b) $1 \mid n$, (c) $n \mid n$ if n is an integer.

PROB. 10.4. Prove: $a \mid b$ and $b \mid c$ imply $a \mid c$.

PROB. 10.5. Prove: If $b \neq 0$, then $b \nmid a$ if and only if the remainder upon dividing a by b is positive.

Def. 10.3. An integer which is divisible by 2 is called *even*. Otherwise, we say it is *odd*. Thus, n is even if and only if there exists an integer m such that $n = 2m$.

* In terms of the notion of *absolute value*, which will be introduced in the last section, the condition $0 \leqslant r < b$, if $b > 0$, and $0 \leqslant r < -b$, if $b < 0$, can be formulated at once as: $0 \leqslant r < |b|$. Here $|b|$ is read as the *absolute value* of b. It is defined as

$$|b| = \begin{cases} b, & \text{if } b \geqslant 0 \\ -b, & \text{if } b < 0. \end{cases}$$

PROB. 10.6. Prove: An integer n is odd if and only if an integer m exists such that $n = 2m + 1$.

PROB. 10.7. Prove: If n is an integer, then

$$\left[\frac{n}{2}\right] = \begin{cases} \dfrac{n}{2} & \text{if } n \text{ is even} \\ \dfrac{n-1}{2} & \text{if } n \text{ is odd.} \end{cases}$$

PROB. 10.8. Prove: If $b \mid a$, where $b > 0$ and $a > 0$, then $0 < b \leqslant a$.

PROB. 10.9. Prove: If a and b are nonnegative integers and both $a \mid b$ and $b \mid a$ hold, then $a = b$.

PROB. 10.10. Prove: If $d \mid 1$, where d is a positive integer, then $d = 1$.

The GCD of Integers a and b

Def. 10.4. If a, b, and m are integers and $m \mid a$ and $m \mid b$, then m is called a *common divisor* of a and b. By the *greatest common divisor* or GCD of a and b we mean the nonnegative integer d such that

$$d \mid a \quad \text{and} \quad d \mid b \tag{10.3a}$$

and

$$m \mid a \quad \text{and} \quad m \mid b \quad \text{imply} \quad m \mid d. \tag{10.3b}$$

In other words, the GCD of integers a and b is the nonnegative divisor of a and b divisible by all the common divisors of a and b. The GCD of integers a and b is written as (a,b). We call integers a and b *relatively prime*, and say that each is *prime* to the other if and only if $(a,b) = 1$.

PROB. 10.11. Prove: $(a,b) = 1$ if and only if the only common divisors of a and b are 1 and -1.

PROB. 10.12. Let a be a nonnegative integer. Prove: $(a,0) = a$.

PROB. 10.13. Prove: If $b \mid a$, where $b \geqslant 0$, then $(a,b) = b$.

Theorem 10.2. *If a and b are integers, then there exist integers x_0 and y_0 such that*

$$ax_0 + by_0 = (a,b). \tag{10.4}$$

PROOF. If $a = 0 = b$, then $(a,b) = 0$ and there is nothing to prove for then any integers x and y will satisfy (10.4). Consider the case where one of a or

b is not 0. Suppose for definiteness that $a \neq 0$. Let S be the set

$$S = \{ax + by > 0 \mid x \in \mathbb{Z} \text{ and } y \in \mathbb{Z}\}. \tag{10.5}$$

If $x = 1$ and $y = 0$, we have $ax + by = a$. If $x = -1$ and $y = 0$, we have $ax + by = -a$. Since $a \neq 0$, one of a or $-a$ is positive and is in S. Hence $S \neq \emptyset$. Therefore, S is a nonempty set of positive integers. As such S has a least member, d say. $d > 0$ and integers x_0 and y_0 exist such that

$$ax_0 + by_0 = d. \tag{10.6}$$

Now, integers q and r exist such that

$$a = dq + r, \tag{10.7}$$

where $0 \leqslant r < d$. Substitute from (10.6) into (10.7) for d to obtain

$$a = (ax_0 + by_0)q + r,$$

so that

$$a(1 - x_0q) + b(-y_0q) = r. \tag{10.8}$$

Suppose that $r > 0$. Then $0 < r < d$. In view of (10.8) we could claim that there is in S a positive integer less than its least member d. This is impossible. Hence, we must conclude $r = 0$. Using (10.7), we see that $a = dq$, and, hence, that $d \mid a$. Similar reasoning establishes $d \mid b$. Thus, d is a common divisor of a and b.

Next assume that d' is a common divisor of a and b so that integers m and n exist such that $a = d'm$ and $b = d'n$. Substituting these expressions into (10.6) we see that

$$d'(mx_0 + ny_0) = d.$$

This implies that $d' \mid d$. Thus, d is a positive common divisor of a and b which is divisible by every common divisor of a and b. By Def. 10.4, we see that $d = (a,b)$. This and (10.6) establish (10.4).

Corollary 1. *If a and b are integers, not both 0, then (a,b) exists and is the minimum of the set S defined in* (10.5).

Proof. Obvious from the proof of the theorem above.

Corollary 2. *If a and b are integers, then $(a,b) = 1$ if and only if integers x_0 and y_0 exist such that*

$$ax_0 + by_0 = 1.$$

Proof. Exercise.

Theorem 10.3. *If a, b and m are integers, then $(a,m) = 1 = (b,m)$ if and only if $(ab,m) = 1$.*

PROOF. If $(ab, m) = 1$, then, by the last corollary, integers x_1 and y_1 exist such that

$$(ab)x_1 + my_1 = 1. \tag{10.9}$$

Writing $x_2 = bx_1$ we have from (10.9)

$$ax_2 + my_1 = 1,$$

where x_2 and y_1 are integers. By the last corollary, we have $(a, m) = 1$. A similar argument shows that $(b, m) = 1$.

Conversely, assume $(a, m) = 1 = (b, m)$. There then exist integers x and y such that

$$ax + my = 1.$$

Multiply both sides by b and obtain

$$abx + bmy = b. \tag{10.10}$$

There exist integers k and l such that

$$ab = (ab, m)k \quad \text{and} \quad m = (ab, m)l. \tag{10.11}$$

Substituting these expressions in (10.10) we obtain

$$(ab, m)(kx + bly) = b.$$

This implies that $(ab, m) \mid b$. Since $(ab, m) \mid m$ also holds, we find that (ab, m) is a nonnegative common divisor of b and m. Accordingly, $(ab, m) \mid (b, m)$. Since $(b, m) = 1$, this implies that $(ab, m) = 1$ and this completes the proof.

Theorem 10.4. *If a, b, and m are integers such that $(a, m) = 1$ and $m \mid ab$, then $m \mid b$.*

PROOF. There exist integers x and y such that

$$ax + my = 1.$$

Multiply both sides here by b to obtain

$$abx + bmy = b. \tag{10.12}$$

Since $m \mid ab$, there exists an integer k such that $ab = km$. Substituting in (10.12) we obtain

$$m(kx + by) = b.$$

This implies that $m \mid b$.

PROB. 10.14. If a and b are nonzero integers such that $a = (a, b)k$ and $b = (a, b)l$, then $(k, l) = 1$.

PROB. 10.15. Prove: If a, b, c, and d are integers such that $(a, b) = 1 = (c, d)$ and $ad = bc$, where $b > 0$ and $d > 0$, then $a = c$ and $b = d$.

Def. 10.5. An integer $p > 1$ is called a *prime* if and only if its only factors are 1, -1, p, and $-p$. An integer $m > 1$ which is not a prime is called *composite*.

For example, 2, 3, 5, 7, 11, 13, 17, 19, 23, and 29 are primes. Note: Since $6 = 3 \cdot 2$ and $91 = 7 \cdot 13$, 6 and 91 are composite.

Remark 10.2. An integer $m > 1$ is composite if and only if integers a and b exist such that $a > 1$ and $b > 1$ and $m = ab$.

PROB. 10.16. Prove: If a prime p does not divide an integer a, then $(a, p) = 1$.

PROB. 10.17. Prove: If a and b are integers and p is a prime such that $p \mid ab$, then either $p \mid a$ or $p \mid b$.

Def. 10.6. We say that the nonzero rational number

$$r = \frac{p}{q},$$

where p and q are integers and $q \neq 0$ is written in *lowest terms* if and only if $q > 0$ and $(p, q) = 1$.

Theorem 10.5. *Each nonzero rational number can be written in lowest terms in a unique way.*

PROOF. Let r be a nonzero rational number, then integers p_1 and q_1 exist such that

$$r = \frac{p_1}{q_1},$$

where $q \neq 0$. Multiplying p_1 and q_1 by -1 if necessary, we can write

$$r = \frac{p}{q},$$

where p and q are integers and $q > 0$. We have $p = (p, q)a$ and $q = (p, q)b$, where a and b are integers and $b > 0$. By Prob. 10.14 we know that $(a, b) = 1$. Hence,

$$r = \frac{p}{q} = \frac{(p, q)a}{(p, q)b} = \frac{a}{b}, \tag{10.13}$$

where $(a, b) = 1$. Thus, r can be written in lowest terms.

We prove the uniqueness of the representation, in lowest terms, of r. Assume

$$r = \frac{c}{d},$$

where c/d is in lowest terms. It follows from this and (10.13) that

$$\frac{a}{b} = \frac{c}{d} \quad \text{and hence} \quad ad = bc. \tag{10.14}$$

Here $(c,d) = 1 = (a,b)$ and $b > 0$, $d > 0$. Since $ad = bc$, it follows that (Prob. 10.15) $a = c$ and $b = d$. This completes the proof.

11. Irrational Numbers

Lemma 11.1. *Let k be a positive integer. If there is a rational number r such that*

$$r^2 = k, \tag{11.1}$$

then r is an integer.

PROOF. Let r be a rational number satisfying (11.1). Clearly, $r \neq 0$. Write

$$r = \frac{p}{q}, \tag{11.2}$$

where p/q is in lowest terms. Thus, p and q are integers, $q > 0$, and $(p,q) = 1$. There exist, therefore, integers x_0 and y_0 such that

$$px_0 + qy_0 = 1. \tag{11.3}$$

Since

$$\left(\frac{p}{q}\right)^2 = r^2 = k,$$

we have

$$p^2 = kq^2. \tag{11.4}$$

Since $(px_0 + qy)^2 = 1$, (11.3) implies that

$$p^2x_0^2 + 2pqx_0y_0 + q^2y_0^2 = 1.$$

Using (11.4), we obtain

$$q^2kx_0^2 + 2pqx_0y_0 + q^2y_0^2 = 1$$

so that

$$q\left(qkx_0^2 + 2px_0y_0 + qy_0^2\right) = 1.$$

This implies that $q \mid 1$. Since $q > 0$, it follows that $q = 1$. But then

$$r = \frac{p}{q} = \frac{p}{1} = p, \qquad p \text{ an integer.}$$

This completes the proof.

Theorem 11.1. *If k is a positive integer which is not the square of an integer, then no rational number r exists such that $r^2 = k$.*

PROOF. If a rational number r exists such that $r^2 = k$, then, by the lemma, r is an integer. This implies k is the square of an integer and contradicts the hypothesis. We conclude then that no rational number r can exist such that $r^2 = k$.

Corollary (Pythagoras). *No rational number r exists such that $r^2 = 2$.*

PROOF. We prove that 2 is not the square of an integer. Suppose p is an integer such that $p^2 = 2$. There is no loss of generality if we assume that $p > 0$. It is clear that $p > 1$. This implies that $p \geqslant 2$, so that $p^2 \geqslant 4 > 2$. Thus, $p^2 = 2$ is false and we have a contradiction. Hence, 2 is not the square of an integer. By the theorem, no rational number r can exist such that $r^2 = 2$.

Theorem 11.2. *If $y \in \mathbb{R}$ where $y > 0$, there exists exactly one real, positive number x such that $x^2 = y$.*

PROOF. Let S be the set

$$S = \{x \in \mathbb{R} \mid x > 0 \text{ and } x^2 < y\}.$$

Let

$$x_0 = \frac{y}{y+1}.$$

We have

$$0 < x_0 < 1 \quad \text{and} \quad 0 < x_0 < y \qquad \text{(why?)}.$$

These imply $x_0^2 < x_0 < y$. This proves $x_0 \in S$, so that $S \neq \emptyset$.

We prove that S is bounded from above by $y + 1$. Indeed, let $x \geqslant y + 1$, so that $x^2 \geqslant (y+1)^2 = y^2 + 2y + 1 > y$. Then, $x > 0$ and $x^2 > y$ and $x \notin S$. We conclude that $x \in S$ implies $x \leqslant y + 1$. Therefore S is a nonempty set of real numbers bounded from above. By Axiom C, S has a real supremum, μ say. We have $\mu = \sup S$.

Since $x_0 \in S$ and $x_0 > 0$ and μ is an upper bound for S, we know that $0 < x_0 \leqslant \mu$, so that $\mu > 0$. Let $\epsilon \in \mathbb{R}$, where $0 < \epsilon < 1$. We have $\mu < \mu + \epsilon$. Therefore $\mu + \epsilon$ is a positive real number not in S (otherwise $\mu + \epsilon \leqslant \mu$ and $\epsilon \leqslant 0$). But then $(\mu + \epsilon)^2 \geqslant y$ and, therefore, $\mu^2 + 2\mu\epsilon + \epsilon^2 \geqslant y$. This implies that

$$\epsilon(2\mu + 1) > \epsilon(2\mu + \epsilon) \geqslant y - \mu^2$$

and that

$$\frac{y - \mu^2}{2\mu + 1} < \epsilon \qquad \text{for all} \quad 0 < \epsilon < 1.$$

By Prob. 3.13 this implies

$$\frac{y-\mu^2}{2\mu+1} \leqslant 0$$

and hence that $y \leqslant \mu^2$.

Next take $0 < \epsilon < \mu$, so that $0 < \mu - \epsilon < \mu$. Since $\mu = \sup S$, this implies that there exists $x_0 \in S$ such that $0 < \mu - \epsilon < x_0$. It follows that $(\mu - \epsilon)^2 < x_0^2 < y$. We obtain $\mu^2 - 2\epsilon\mu + \epsilon^2 < y$, and hence that

$$\mu^2 - y < \epsilon(2\mu - \epsilon) = \epsilon(\mu + \mu - \epsilon) < 2\epsilon\mu.$$

Thus,

$$\frac{\mu^2 - y}{2\mu} < \epsilon \qquad \text{for each } \epsilon \text{ such that} \quad 0 < \epsilon < \mu.$$

Applying Prob. 3.13 once more we obtain from this that

$$\frac{\mu^2 - y}{2\mu} \leqslant 0$$

so that $\mu^2 \leqslant y$. Thus $\mu^2 \leqslant y$ and $y \leqslant \mu^2$ and, therefore, $\mu^2 = y$. We now know that there exists a positive real number μ such that $\mu^2 = y$.

We prove the uniqueness of the positive μ such that $\mu^2 = y$. If z is a positive number such that $z^2 = y$, then we prove $\mu = z$. If $\mu \neq z$, we would have $0 < \mu < z$ or $0 < z < \mu$. In the first case, we have $y = \mu^2 < z^2$ and in the second, $z^2 < \mu^2 = y$. In either case we obtain the contradiction $z^2 \neq y$. Hence $\mu = z$.

PROB. 11.1. Prove: If $y > 0$, there exists exactly one *negative* real number x such that $x^2 = y$.

Remark 11.1. By Theorem 11.2 and the result cited in the last problem we find that for each positive real number y there exist exactly two real numbers whose square is y, one being positive and the other negative. As a matter of fact, if $x > 0$ where $x^2 = y$, then $-x < 0$ and $(-x)^2 = x^2 = y$.

Def. 11.1. If $y > 0$, then any real x such that $x^2 = y$ is called a *square root* of y. The $x > 0$ such that $x^2 = y$ is called the *positive square root* and is written as $x = \sqrt{y}$. The *negative square* root of y is $-\sqrt{y}$. We also define $\sqrt{0} = 0$.

PROB. 11.2. Prove: If $y_1 \geqslant 0$ and $y_2 \geqslant 0$, then $\sqrt{y_1 y_2} = \sqrt{y_1}\sqrt{y_2}$.

Def. 11.2. A real number which is not rational is called *irrational*.

Remark 11.2. By the corollary of Theorem 11.1, we know that $\sqrt{2}$ is not rational. By Theorem 11.2 we see that it is real. Thus, irrational numbers exist. $\sqrt{2}$ is an example of one.

PROB. 11.3. Prove: If r is rational and c is irrational, then (a) $r+c$ and $r-c$ are irrational; (b) if in addition $r \neq 0$, then rc, c/r, and r/c are irrational.

PROB. 11.4. Prove: $\sqrt{3}$ and $3+\sqrt{5}$ are irrational.

Theorem 11.3. *Between any two distinct real numbers, there exists a rational number.*

PROOF. Assume $x<y$ for real numbers x and y. Then $y-x>0$. By Corollary 1 of Theorem 9.1, there exists a positive integer q such that

$$0<\frac{1}{q}<y-x, \tag{11.5}$$

so that $qx<qx+1<qy$. There exists an integer p such that $p \leqslant qx+1 < p+1$. The second inequality implies that $qx<p$. This, $p \leqslant qx+1$, and the inequalities following (11.5) imply that

$$qx<p \leqslant qx+1<qy.$$

It follows that $qx<p<qy$, and, hence, that

$$x<\frac{p}{q}<y.$$

We see that p/q is a rational number between x and y.

Theorem 11.4. *Between any two real numbers there exists an irrational number.*

PROOF. Assume $x<y$ for some real numbers x and y. We have $\sqrt{2}\,x<\sqrt{2}\,y$. There exists a rational number r such that $\sqrt{2}\,x<r<\sqrt{2}\,y$, which implies that

$$x<\frac{r}{\sqrt{2}}<y.$$

If $r \neq 0$, then $r/\sqrt{2}$ is an irrational number between x and y. If $r=0$, we have $x<0<y$ and $\sqrt{2}\,x<0<\sqrt{2}\,y$. There exists a rational number s such that $0<s<\sqrt{2}\,y$. We now have $\sqrt{2}\,x<s<\sqrt{2}\,y$, where s is rational and $s>0$. But then

$$x<\frac{s}{\sqrt{2}}<y, \qquad s \text{ rational and } \quad s>0.$$

Hence $s/\sqrt{2}$ is an irrational number between x and y.

12. The Noncompleteness of the Rational Number System

We first ask the reader to solve:

PROB. 12.1. Prove: If $S \subseteq T \subseteq \mathbb{R}$, where $S \neq \varnothing$, then (a) if T is bounded from above, then $\sup S \leqslant \sup T$; (b) If T is bounded from below, then $\inf T \leqslant \inf S$; (c) if T is bounded, then $\inf T \leqslant \inf S \leqslant \sup S \leqslant \sup T$.

Theorem 12.1. *There exist nonempty sets of rationals which are bounded from above but have no* rational *supremum.*

PROOF. Let

$$T = \{x \in \mathbb{Q} \mid x > 0 \text{ and } x^2 < 2\}$$

(Recall, $\mathbb{Q}$ is the set of rational numbers.) We have $T \subseteq \mathbb{Q}$ and $T \neq \varnothing$ since $1 \in T$. T is bounded from above by 2. In fact, if x is a rational number such that $x > 2$, then $x^2 > 2$, and hence $x \notin T$. We conclude: If $x \in T$, then $x \leqslant 2$ so that T is bounded from above. By Axiom C, T has a real supremum. Let $\mu = \sup T$. Let

$$S = \{x \in \mathbb{R} \mid x > 0, x^2 < 2\}. \tag{12.1}$$

We proved earlier (see proof of Theorem 11.2) that $\sup S = \sqrt{2}$. Since $T \subseteq S \subseteq \mathbb{R}$, we obtain (Prob. 12.1)

$$\sup T \leqslant \sup S = \sqrt{2} \quad \text{and hence} \quad \sup T \leqslant \sqrt{2}\,.$$

Now $\sup T > 0$ since $1 \in T$ implies $0 < 1 \leqslant \sup T$. Suppose that $\sup T < \sqrt{2}$. Then there exists a rational number r such that $\sup T < r < \sqrt{2}$. But $\sup T < r$ implies $r \notin T$ and since $r > 0$, $r^2 \geqslant 2$. On the other hand, $0 < r < \sqrt{2}$. This implies $r^2 < 2$ and we obtain a contradiction. Hence, $\sup T = \sqrt{2}$. The set T therefore has an irrational supremum and not a rational one. This completes the proof.

Remark 12.1. Theorem 12.1 demonstrates by example that the system $\mathbb{Q}$ of rationals does not enjoy the completeness property. Since the real numbers have the completeness property, they form a *complete ordered field* (cf. Remark 7.2). The rational numbers constitute an ordered field which is not complete.

Remark 12.2. In later chapters we will encounter the notion of *Cauchy completeness*. "Completeness" in the sense of Axiom C will be referred to as *order-completeness*. Thus, we say that the real numbers constitute an ordered field which is *order-complete*.

13. Absolute Value

Def. 13.1. We define the *absolute value* of $x \in R$, written as $|x|$, as follows:

$$|x| = \begin{cases} x & \text{if} \quad x \geqslant 0 \\ -x & \text{if} \quad x < 0. \end{cases} \tag{13.1}$$

For example, $|3| = 3$, $|0| = 0$, and $|-3| = -(-3) = 3$.

Prob. 13.1. Prove: (a) $|x| \geqslant 0$; (b) $|x| = 0$ if and only if $x = 0$. Note: $x = |x|$ if and only if $x \geqslant 0$ and $x = -|x|$ holds if and only if $x \leqslant 0$.

Prob. 13.2. Prove: (a) $-|x| \leqslant x \leqslant |x|$ and (b) $-|x| \leqslant -x \leqslant |x|$.

Prob. 13.3. Prove: $|-x| = |x|$.

Prob. 13.4. Prove: If $\epsilon \geqslant 0$, then $|x - a| = \epsilon$ if and only if $x = a + \epsilon$ or $x = a - \epsilon$.

Theorem 13.1. *If $\epsilon > 0$, then $|x| < \epsilon$ if and only if $-\epsilon < x < \epsilon$.*

Proof. Assume that $-\epsilon < x < \epsilon$. If $0 \leqslant x$, we have $0 \leqslant x < \epsilon$, so $|x| < \epsilon$. If $x < 0$, then from the hypothesis we have $-\epsilon < x < 0$, and therefore $0 < -x < \epsilon$ so that again $|x| < \epsilon$. In either case, $-\epsilon < x < \epsilon$ implies $|x| < \epsilon$.

Conversely, assume that $|x| < \epsilon$. Consider the cases (1) $x \geqslant 0$ and (2) $x < 0$. In case (1), we have $0 \leqslant x < \epsilon$ so that $-\epsilon < 0 \leqslant x < \epsilon$ and therefore $-\epsilon < x < \epsilon$. In case (2), $x = -|x|$ so we have from $|x| < \epsilon$ that $-\epsilon < -|x| = x < 0 < \epsilon$ and again $-\epsilon < x < \epsilon$. Thus, $|x| < \epsilon$ implies $-\epsilon < x < \epsilon$. This completes the proof.

Remark 13.1. It is clear from this theorem and Prob. 13.4 that if $\epsilon \geqslant 0$, then $|x| \leqslant \epsilon$ if and only if $-\epsilon \leqslant x \leqslant \epsilon$.

Theorem 13.2. *If x and y are real numbers, then*

$$|x + y| \leqslant |x| + |y|. \tag{13.2}$$

Proof. We have (Prob. 13.2)

$$-|x| \leqslant x \leqslant |x| \quad \text{and} \quad -|y| \leqslant y \leqslant |y|.$$

Adding, we obtain

$$-(|x| + |y|) = -|x| - |y| \leqslant x + y \leqslant |x| + |y|.$$

Hence,

$$-(|x| + |y|) \leqslant x + y \leqslant |x| + |y|. \tag{13.3}$$

Let $\epsilon = |x| + |y|$ so that $\epsilon \geqslant 0$. In view of (13.3)

$$|x+y| \leqslant \epsilon = |x| + |y|.$$

This proves (13.2).

Prob. 13.5. Prove: $|x-y| \leqslant |x| + |y|$.

Prob. 13.6. Prove: $\big||x| - |y|\big| \leqslant |x-y|$ (Hint: note that $x = y + (x-y)$, so that $|x| \leqslant |y| + |x-y|$, etc.)

Prob. 13.7. Prove: $|xy| = |x||y|$.

Prob. 13.8. Prove: $|x|^2 = x^2$ and $\sqrt{x^2} = |x|$.

Prob. 13.9. Prove: If $y \neq 0$, then

$$\left|\frac{x}{y}\right| = \frac{|x|}{|y|}.$$

Prob. 13.10. Prove that $|x+y| = |x| + |y|$ if and only if $xy \geqslant 0$, and that $|x+y| < |x| + |y|$ if and only if $xy < 0$.

Prob. 13.11. Prove:

$$\frac{|x+y|}{1+|x+y|} \leqslant \frac{|x|}{1+|x|} + \frac{|y|}{1+|y|}.$$

Prob. 13.12. Prove

$$\sqrt{x^2+y^2} \leqslant |x| + |y|.$$

Prob. 13.13. Prove: $\sqrt{|x+y|} \leqslant \sqrt{|x|} + \sqrt{|y|}$.

Remark 13.2. When dealing with the real numbers, it is often helpful to adopt a geometric point of view. This is why we sometimes refer to the elements of $\mathbb{R}$ as *points*.

Def. 13.2. By the *Euclidean distance* between real numbers x and y we mean $d(x, y)$, where

$$d(x, y) = |x-y|. \tag{13.4}$$

For example, $d(3, -5) = |3-(-5)| = 8$. The Euclidean distance between a point $x \in \mathbb{R}$ and 0 is

$$d(x, 0) = |x-0| = |x|. \tag{13.5}$$

Prob. 13.14. Prove: If x, y, and z are real numbers, then

(1) $d(x, y) \geqslant 0$,
(2) $d(x, y) = 0$, if and only if $x = y$,
(3) $d(x, y) = d(y, x)$,
(4) (triangle inequality) $d(x, z) \leqslant d(x, y) + d(y, z)$.

Prob. 13.15. Prove: If x, y, and z are real numbers, then

$$|d(x, y) - d(x, z)| \leqslant d(y, z).$$

Def. 13.3. By the *length* of a finite interval $(a; b)$, we mean $b - a$. By the *midpoint* of this interval we mean the point $m \in (a; b)$ such that

$$d(m, a) = d(m, b). \tag{13.6}$$

Sometimes we refer to the midpoint of $(a; b)$ as its *center*.

Prob. 13.16. Prove: If m is the midpoint of the interval $(a; b)$, then $m = (a + b)/2$.

It is interesting to observe that $\max\{a, b\}$ and $\min\{a, b\}$, where a and b are real numbers can be expressed by using $|a - b|$ (see Prob. 13.17 below).

Prob. 13.17. Prove: If a and b are real numbers, then

$$\max\{a, b\} = \frac{a + b + |a - b|}{2} \quad \text{and} \quad \min\{a, b\} = \frac{a + b - |a - b|}{2}.$$

Def. 13.4. By the *positive part* of the real number x, we mean x^+, where $x^+ = \max\{0, x\}$. The *negative part* x^- of x is defined as $x^- = \min\{0, x\}$.

Prob. 13.18. Prove:

$$x^+ + x^- = x \quad \text{and} \quad x^+ - x^- = |x|.$$

Prob. 13.19. Prove:

$$\max\{-a, -b\} = -\min\{a, b\} \quad \text{and} \quad \min\{-a, -b\} = -\max\{a, b\}.$$

Prob. 13.20. Prove: (a) If $a < b$, then

$$-\max\{|a|, |b|\} \leqslant a < b \leqslant \max\{|a|, |b|\}.$$

(b) A set $S \subseteq \mathbb{R}$ is bounded if and only if there exists an $M > 0$ such that $|x| \leqslant M$ holds for all $x \in S$.

CHAPTER II
Functions

1. Cartesian Product

If $\{a,b\}$ and $\{c,d\}$ are given sets, then $\{a,b\}=\{c,d\}$ implies $a=c$ or $a=d$ and $b=c$ or $b=d$. This was why we referred to $\{a,b\}$ (cf. Section I.1) as the unordered pair consisting of a and b. By the *ordered pair** (a,b) of elements a and b, we mean the set $\{a,b\}$ together with the ordering of its members in which a is first and b second. We call a the *first component* or *coordinate* of (a,b) and b its *second*.

For ordered pairs, we have

$$(a,b)=(c,d) \quad \text{if and only if} \quad a=c \quad \text{and} \quad b=d. \tag{1.1}$$

If A and B are sets, then the *Cartesian Product* $A\times B$ of A and B (in that order) is the set.

$$A\times B=\{(x,y)\,|\,x\in A \text{ and } y\in B\}. \tag{1.2}$$

For example, let $A=\{0,1\}$ and $B=\{2,3,4\}$, then

$$A\times B=\{(0,2),(0,3),(0,4),(1,2),(1,3),(1,4)\}$$

and

$$B\times A=\{(2,0),(2,1),(3,0),(3,1),(4,0),(4,1)\}.$$

In $A\times B$, we may have $A=B$. The Cartesian product $A\times A$ is the set

$$A\times A=\{(x,y)\,|\,x\in A \text{ and } y\in A\}. \tag{1.3}$$

For example, let $A=\{0,1\}$, we have

$$A\times A=\{(0,0),(0,1),(1,0),(1,1)\}.$$

* The ordered pair (a,b) can be defined in set theoretic terms by means of $(a,b)=\{\{a\},\{a,b\}\}$.

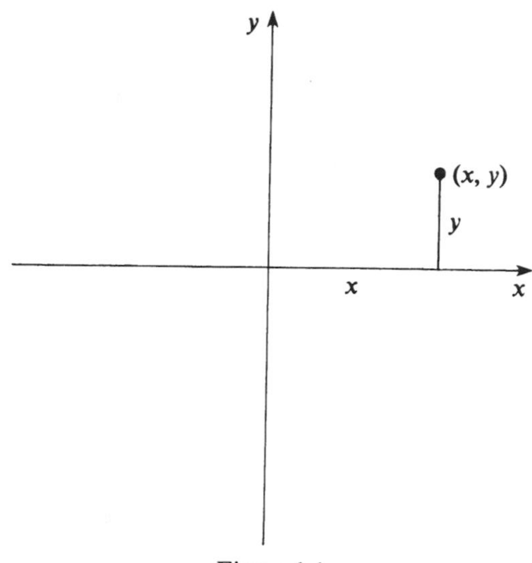

Figure 1.1

The Cartesian product $\mathbb{R} \times \mathbb{R}$ is the set

$$\mathbb{R} \times \mathbb{R} = \{(x, y) \mid x \times \mathbb{R} \text{ and } y \times \mathbb{R}\}. \tag{1.4}$$

We visualize $\mathbb{R} \times \mathbb{R}$ as the set of points in a plane (Fig. 1.1) provided with a rectangular coordinate system, familiar to the reader from his earlier mathematics education.

PROB. 1.1. Prove: If A and B are sets, then $A \times B \neq \emptyset$ if and only if $A \neq \emptyset$ and $B \neq \emptyset$.

PROB. 1.2. Prove: If A, B, C, and D are sets such that $A \subseteq C$ and $B \subseteq D$, then $A \times B \subseteq C \times D$.

PROB. 1.3. Prove: If $A \times B \neq \emptyset$, then $A \times B \subseteq C \times D$ implies $A \subseteq C$ and $B \subseteq D$.

2. Functions

One of the most important ideas in mathematics is that of a function. Intuitively, a *function* or a *mapping* from a set X to a set Y is a correspondence which assigns to each $x \in X$ exactly one $y \in Y$. The set X is called the *domain* of f and Y its *codomain*. This is pictured in Fig. 2.1. The figure is meant to suggest that it is possible for a function to assign distinct x's to the same y. This definition of a function is "intuitive" because it defines "function" in terms of "correspondence"—a term which is itself in need of

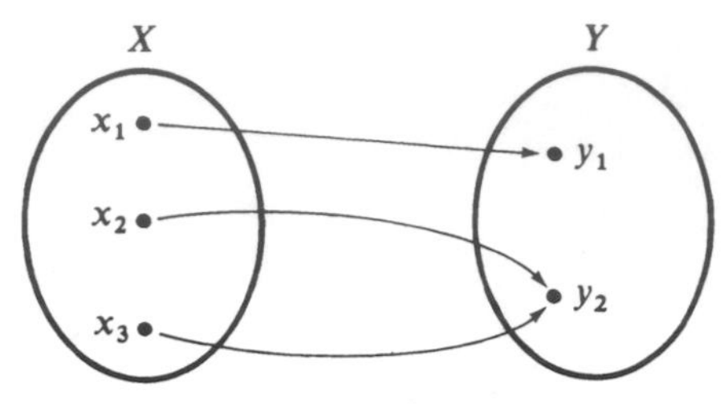

Figure 2.1

definition. It is customary today to define functions in terms of sets. We begin by defining "correspondence."

Def. 2.1. A *correspondence* γ between a set X and a set Y is a subset of $X \times Y$. If $(x, y) \in \gamma$, we say that y *corresponds* to x under γ or that γ *assigns* y to x. A correspondence is also called a *relation*.

Having defined "correspondence," "assigns to," and "corresponds to," we proceed to define *function*.

Def. 2.2. A *function* or *mapping* with *domain* X and *range* in Y is a correspondence between sets X and Y which assigns to each $x \in X$ exactly one $y \in Y$. The unique y assigned to x is called the *image* of x under f and is written as

$$y = f(x). \tag{2.1}$$

We write $\mathfrak{D}(f)$ for the domain of f, so that $\mathfrak{D}(f) = X$. The set of all the images of the x's in X is called the *range* of f and we write it as $\mathfrak{R}(f)$. When f is a function with domain X and range in Y we write

$$f: X \to Y \tag{2.2}$$

and also say that f *maps* X into Y. Y is called the *codomain* of f. To indicate that x is mapped into $y = f(x)$ we also write

$$x \mapsto f(x). \tag{2.3}$$

Thus, if $f: X \to Y$, we have $f \subseteq X \times Y$, f being a certain kind of correspondence between X and Y and, therefore, a subset of $X \times Y$. Since f assigns to each $x \in X$ a $y \in Y$, we have $(x, y) \in f$, or $(x, f(x)) \in f$. What makes a correspondence a function is the property:

$$(x, y) \in f \quad \text{and} \quad (x, y') \in f \quad \text{imply} \quad y = y' \qquad \text{for each} \quad x \in X = \mathfrak{D}(f). \tag{2.4}$$

This states that f assigns exactly one y to each $x \in \mathfrak{D}(f)$.

Let $(x, y) \in f$ and $(x', y') \in f$, where f is a function. Then

$$y = f(x) \quad \text{and} \quad y' = f(x'). \tag{2.5}$$

If $x = x'$, x' can be replaced by x to yield $(x, y') \in f$, so that $y = y'$. Therefore,

$$f(x') = y' = y = f(x).$$

Thus, if the correspondence f is a function, then

$$x = x' \quad \text{implies} \quad f(x) = f(x') \qquad \text{for } x \text{ in } \mathfrak{D}(f). \tag{2.6}$$

A function will also be referred to as a *single-valued* correspondence.

By the definition of the range $\mathfrak{R}(f)$ of a function, we have: $y \in \mathfrak{R}(f)$ if and only if some $x \in X$ exists such that $y = f(x)$. Therefore,

$$\begin{aligned} \mathfrak{R}(f) &= \{ y \in Y \mid y = f(x) \text{ for some } x \in \mathfrak{D}(f)\} \\ &= \{ f(x) \mid x \in \mathfrak{D}(f)\}. \end{aligned} \tag{2.7}$$

Clearly, we always have

$$\mathfrak{R}(f) \subseteq Y. \tag{2.8}$$

When X is a set, then a subset of $X \times X$ is called *a correspondence between X and itself* or *a correspondence on X*. If f is a function $f: X \to X$ whose domain and codomain are equal, then we say that f is a *function* or *mapping* from X *into* itself. We also say that f is a *function* on X, or a *mapping* on X. A function on X is also called a *transformation* of X.

A function having $\mathbb{R}$ as its codomain is called a *real-valued* function. When the domain of f is in $\mathbb{R}$ we say that f is a *function of a real variable*. When both the domain and the range of f are in $\mathbb{R}$, we call f a *real-valued* function of a *real* variable (see Fig. 2.2). In this volume we deal with real-valued functions of a real variable.

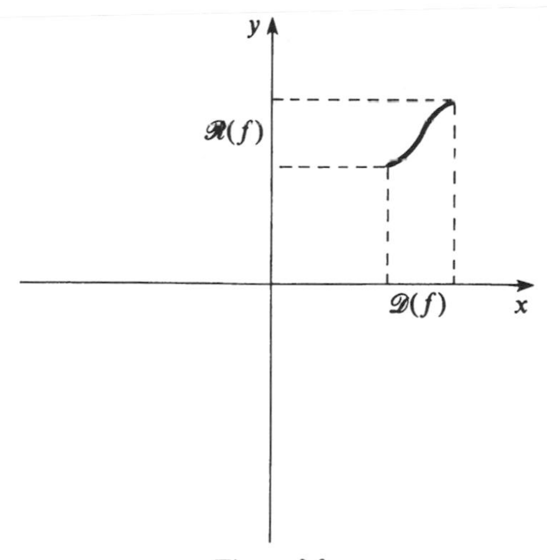

Figure 2.2

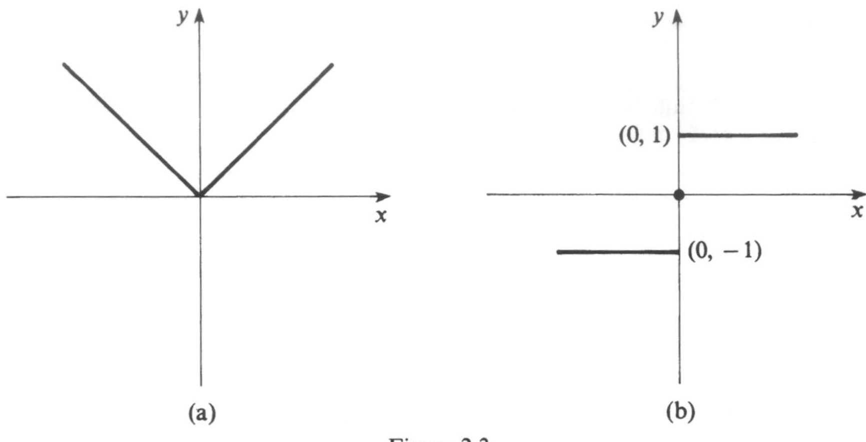

Figure 2.3

EXAMPLE 2.1 (The Absolute Value Function). This is the function $f: \mathbb{R} \to \mathbb{R}$, defined as $f(x) = |x|$ for each $x \in \mathbb{R}$ (see Fig. 2.3(a)). We write it as Abs. Thus, $\operatorname{Abs} x = |x|$ for each $x \in \mathbb{R}$.

EXAMPLE 2.2 (The *Signum* Function). The function $g: \mathbb{R} \to \mathbb{R}$ defined as:

$$g(x) = \begin{cases} 1 & \text{if} \quad x > 0 \\ 0 & \text{if} \quad x = 0 \\ -1 & \text{if} \quad x < 0 \end{cases}$$

is called the *signum* function. We write it as sig. Thus, $\operatorname{sig} x = 1$ if $x > 0$, $\operatorname{sig} 0 = 0$, $\operatorname{sig} x = -1$ if $x < 0$. Note, $|x| = x \operatorname{sig} x$ for each $x \in \mathbb{R}$.

PROB. 2.1. Let $f: X \to Y$ and $g: X \to Y$ be functions with the same domain and codomain. Prove: $f = g$ if and only if $f(x) = g(x)$ for each $x \in X$.

EXAMPLE 2.3 (The Empty Correspondence). We prove that the only function $f: \emptyset \to Y$, where Y is some set, is $f = \emptyset$. Since $\emptyset \subseteq \emptyset \times Y$, we know that $\emptyset$ is a correspondence between $\emptyset$ and Y. Since $\emptyset \times Y = \emptyset$ (Prob. 1.1) if f is any correspondence at all between $\emptyset$ and Y, we have $f = \emptyset$. Is this correspondence a function? Suppose it is not. There would then exist an $x \in \emptyset$ such that no unique y with $(x, y) \in \emptyset$ exists. This is impossible, since no $x \in \emptyset$ exists. Hence $\emptyset: \emptyset \to Y$ is a function.

PROB. 2.2. Prove: If $X \neq \emptyset$, then no function $f: X \to \emptyset$ exists. Thus, if f is a function defined on the nonempty set, its range is never empty.

Remark 2.1. We shall not deal with functions having empty domains unless we explicitly say so.

EXAMPLE 2.4 (A Constant Function). $f: X \to Y$ defined as $f(x) = b$ for all $x \in X$ and a fixed $b \in Y$ is called a *constant* function with value b.

Example 2.5 (The Identity Function). Let $f: X \to X$ be defined as follows: $f(x) = x$ for all $x \in X$. We call f the *identity* function on X, writing it as I_X. Thus, $I_X(x) = x$ for all $x \in X$.

Image of a Set Under Mapping. Restrictions and Extensions

Given a function $f: X \to Y$ and a set $A \subseteq X$, we define the image $f(A)$ of A as the set of images of the x's in A. Thus,

$$f(A) = \{ f(x) \mid x \in A \}. \tag{2.9}$$

For example,

$$\mathfrak{R}(f) = f(\mathfrak{D}(f)). \tag{2.10}$$

Again let $f: X \to Y$ and $A \subseteq X$. Consider the correspondence

$$g = \{ (x, y) \in f \mid x \in A \}. \tag{2.11}$$

This correspondence is a function $g : A \to Y$ (explain) such that

$$g(x) = f(x) \qquad \text{for all} \quad x \in A. \tag{2.12}$$

If $A \subseteq X$, then we call g the *restriction* of f to A and write $g = f|_A$. f is called an *extension* of g.

Prob. 2.3. Let $g = f|_A$ be the restriction of f to $A \subseteq X$. It is clear that g is uniquely determined by f. Show that the extension of $g = f|_A$ to X is not unique.

Prob. 2.4. Prove: If $f: X \to Y$ is a function and $A \subseteq B \subseteq X$, then $f(A) \subseteq f(B)$.

3. Sequences of Elements of a Set

By an infinite sequence of elements of a set $S \neq \varnothing$ we mean a function $f: \mathbb{Z}_+ \to S$ from the positive integers into S. If $f(n) = a$, we write $f(n) = a_n$, calling a_n the nth *term* or nth *coordinate* of the sequence. f itself is written as $f = \langle a_n \rangle_{n \in \mathbb{Z}_+}$ or as $f = \langle a_n \rangle_{n \geqslant 1}$. The set $\mathbb{Z}_+$ is called the *index set* and the n in a_n is called the *index* or *subscript* of a_n. Usually, when $\mathbb{Z}_+$ is understood, we write the sequence simply as $\langle a_n \rangle$. Sometimes we write out the terms as in

$$\langle a_n \rangle = \langle a_1, a_2, a_3, \ldots \rangle. \tag{3.1}$$

By the *range* of $\langle a_n \rangle$ we mean the range $f(\mathbb{Z}_+)$. We have

$$f(\mathbb{Z}_+) = \{ a_n \mid n \in \mathbb{Z}_+ \}. \tag{3.2}$$

For example, let $\langle a_n \rangle$ be a *constant* sequence in that $a_n = c$ for all $n \in \mathbb{Z}_+$. We then have $f(\mathbb{Z}_+) = \{c\}$ and

$$\langle a_n \rangle = \langle c, c, c, \ldots \rangle. \tag{3.3}$$

By a *real* sequence we mean a sequence $\langle a_n \rangle$ such that $a_n \in \mathbb{R}$ for each $n \in \mathbb{Z}_+$.

Sometimes $\mathbb{Z}_0$ is used as an index set. In this case a function $f: \mathbb{Z}_0 \to S$ is also called an *infinite sequence* of elements of S and we write $f = \langle a_n \rangle_{n \in \mathbb{Z}_0}$, or

$$\langle a_n \rangle_{n \geq 0} = \langle a_0, a_1, a_2, \ldots \rangle. \tag{3.4}$$

This does not overly abuse terminology since $\langle a_n \rangle$ can be reindexed by thinking of it as the sequence $\langle b_n \rangle_{n \geq 1}$, where $b_n = a_{n-1}$ for $n \in \mathbb{Z}_+$. In fact, let n_0 be some fixed integer and H_{n_0} be the set

$$H_{n_0} = \{ n \in \mathbb{Z} \mid n \geq n_0 \}. \tag{3.5}$$

A function $f: H_{n_0} \to S$ is also called an *infinite sequence* of elements of S and we write $f = \langle a_n \rangle_{n \geq n_0}$ and

$$f = \langle a_n \rangle_{n \geq n_0} = \langle a_{n_0}, a_{n_0+1}, a_{n_0+2}, \ldots \rangle \tag{3.6}$$

for the sequence. Here, too, we can reindex and view $\langle a_n \rangle_{n \geq n_0}$ as $\langle b_n \rangle_{n \geq 1}$, where $b_n = a_{n+n_0-1}$, so that $b_1 = a_{n_0}$, $b_2 = a_{n_0+1}$, $b_3 = a_{n_0+2}, \ldots$.

We sometimes define sequences by induction as in Examples 3.1 and 3.2 below.

Example 3.1 (Exponents Which Are Nonnegative Integers). Let x be some real number. We define the sequence $\langle x^n \rangle_{n \geq 0}$ as follows: (i) $x^0 = 1$ and (ii) if x^n is defined for some nonnegative integer n, define $x^{n+1} = x^n x$. For example, $x^1 = x^{0+1} = x^0 x = 1x = x$; $x^2 = x^{1+1} = x^1 x = xx$; $x^3 = x^{2+1} = x^2 x = (xx)x = xxx$.

Prob. 3.1. Prove: If $x \in \mathbb{R}$, $y \in \mathbb{R}$, and m, n are nonnegative integers, then

(a) $x^m x^n = x^{m+n}$,
(b) $(x^m)^n = x^{mn}$,
(c) $1^n = 1$,
(d) $x^n y^n = (xy)^n$,
(e) if $y \neq 0$, then $x^n / y^n = (x/y)^n$.

Prob. 3.2. Prove: If $a \geq 1.5$, then for each positive integer n, $a^n > n$.

Prob. 3.3. Prove: (a) If $x > 1$, then $x^{n+1} > x^n \geq x$ for each positive integer n; (b) if $0 < x < 1$, then $x^{n+1} < x^n \leq x$ for each positive integer n.

Example 3.2. Here we extend the definition of x^n to the case where n is a negative integer and $x \neq 0$. If $x \in \mathbb{R}$, $x \neq 0$ and n is a negative integer, we define x^n as

$$x^n = (x^{-1})^{(-n)}. \tag{3.7}$$

This equality holds even if n is a nonnegative integer. Indeed, it holds trivially if $n = 0$. Now suppose n is a positive integer, so that $-n$ is a negative integer. If $x \neq 0$, then

$$(x^{-1})^{(-n)} = \left((x^{-1})^{-1}\right)^{(-(-n))} = \left((x^{-1})^{-1}\right)^n = x^n.$$

In Prob. 3.6 below we ask the reader to establish the laws of exponents for integer exponents. In Prob. 3.1 the reader was asked to establish these laws for exponents which were nonnegative integers. The lemmas and problems which follow will serve to facilitate the task.

Example 3.2 implies

Lemma 3.1. *If $x \neq 0$, then* (3.7) *holds for any integer n.*

PROB. 3.4. Prove: If $x \in \mathbb{R}$ and $x \neq 0$, then, for n an integer,

$$x^{-n} = (x^{-1})^n. \tag{3.8}$$

Lemma 3.2. *If $x \in \mathbb{R}$, $x \neq 0$, then*

$$x^{n-1} = x^n x^{-1} \qquad \textit{for any integer } n. \tag{3.9}$$

PROOF. Using induction on n, (3.9) can easily be proven for $n \in \mathbb{Z}_0$. We therefore assume that n is a negative integer and obtain:

$$x^n x^{-1} = (x^{-1})^{(-n)} x^{-1}. \tag{3.10}$$

Here $-n$ is a positive integer. By the laws of exponents for positive integers (Prob. 3.1) it follows from (3.10) that

$$x^n x^{-1} = (x^{-1})^{(-n)} x^{-1} = (x^{-1})^{((-n)+1)} = (x^{-1})^{(-(n-1))}. \tag{3.11a}$$

By Lemma 3.1, we have

$$(x^{-1})^{(-(n-1))} = x^{n-1}.$$

This, (3.11a), and (3.10) establish (3.9) for n a negative integer. Thus, the conclusion holds also for the case where n is a negative integer and the proof is complete.

PROB. 3.5. Prove: If $x \in \mathbb{R}$, $x \neq 0$, and $n \in \mathbb{Z}$, then $x^{n+1} = x^n x$.

Theorem 3.1. *If $x \neq 0$ and $y \neq 0$, then $x^n y^n = (xy)^n$ for each integer n.*

PROOF. We use induction on n and the induction principle for integers (Theorem I.5.5). First note: $(xy)^0 = 1 \cdot 1 = x^0y^0$. Next assume x^ny^n holds for some integer n. We have

$$(xy)^{n+1} = (xy)^n(xy) = (x^ny^n)(xy) = (x^nx)(y^ny) = x^{n+1}y^{n+1}$$

and, similarly, that

$$(xy)^{n-1} = x^{n-1}y^{n-1}.$$

By the second principle of induction for integers (Theorem 6.5), we conclude that $x^ny^n = (xy)^n$ holds for any integer n.

Corollary 1. *If n is an integer and $x \in \mathbb{R}$, $x \neq 0$, then*

$$(x^n)^{-1} = (x^{-1})^n = x^{-n}. \tag{3.11b}$$

PROOF. The second equality here follows from Prob. 3.4. We prove the first equality in (3.11b). We have

$$1 = 1^n = (xx^{-1})^n = x^n(x^{-1})^n.$$

This implies that $(x^{-1})^n$ is the reciprocal of x^n, and hence $(x^n)^{-1} = (x^{-1})^n$.

PROB. 3.6. Prove: If $x \in \mathbb{R}$, $x \neq 0$ and m, n are integers, then

(a) $x^mx^n = x^{m+n}$,
(b) $(x^m)^n = x^{mn}$,
(c) $x^m/x^n = x^{m-n}$.

Finite Sequences of Elements of a Set

Here the sets $(n) = \{1, 2, \ldots, n\}$ of Def. I.5.4 play a role. Recall that these are the *initial segments* of the set $\mathbb{Z}_+$ of positive integers.

By a *finite sequence* of n *terms* of a set S, where n is a positive integer, we mean a function $f: (n) \to S$. Here, too, just as in the case of an infinite sequence, if $f(k) = x$, then we write $x = x_k$. We write f as

$$f = \langle x_k \rangle_{1 \leqslant k \leqslant n} = \langle x_1, x_2, \ldots, x_n \rangle. \tag{3.12}$$

We call x_k the kth *term*, x_1 the *first term*, and x_n the *last term* of the finite sequence $\langle x_k \rangle_{1 \leqslant k \leqslant n}$.

A finite sequence of n terms is also called an *ordered n-tuple*. By properties of functions (Prob. 2.1), we have for the ordered n-tuples $x_1, \ldots, x_n$ and $y_1, \ldots, y_n$,

$$\langle x_1, \ldots, x_n \rangle = \langle y_1, \ldots, y_n \rangle \text{ if and only if } x_k = y_k, \quad \text{for all } k \in (n) \tag{3.13}$$

Sum and Product of Functions

If f and g are real-valued functions with a domain $\mathfrak{D}$, their sum $f+g$ is defined by means of

$$(f+g)(x)=f(x)+g(x) \qquad \text{for} \quad x\in\mathfrak{D}. \tag{3.14}$$

Similarly, their *product* fg is defined by means of

$$(fg)=f(x)\,g(x) \qquad \text{for} \quad x\in\mathfrak{D}. \tag{3.15}$$

If c is some constant, the function cf, called a *real multiple* of f is defined by means of

$$(cf)(x)=cf(x) \qquad \text{for} \quad x\in\mathfrak{D}. \tag{3.16}$$

The function $(-1)f$ is usually written as $-f$, so that we have

$$((-1)f)(x)=(-1)f(x)=-f(x)=(-f)(x) \qquad \text{for} \quad x\in\mathfrak{D}.$$

That is, we have $(-1)f=-f$. The function $f-g$ is defined as $f-g=(f+(-g))$. Clearly, if $x\in\mathfrak{D}$, then

$$(f-g)(x)=(f+(-g))(x)=f(x)+(-g(x))=f(x)-g(x). \tag{3.17}$$

4. General Sums and Products

If $a_1, a_2, \ldots$ are real numbers, we define the general sum $\sum_{k=1}^{n} a_k$ inductively as follows:

$$\begin{aligned} &\sum_{k=1}^{0} a_k = 0, \\ &\sum_{k=1}^{n+1} a_k = \left(\sum_{k=1}^{n} a_k\right) + a_{n+1} \qquad \text{for each} \quad n\in\mathbb{Z}_0. \end{aligned} \tag{4.1}$$

$\sum_{k=1}^{n} a_k$ is called a *sum of n terms*.

We have

$$\sum_{k=1}^{1} a_k = \left(\sum_{k=1}^{0} a_k\right) + a_1 = 0 + a_1 = a_1,$$

$$\sum_{k=1}^{2} a_k = \left(\sum_{k=1}^{1} a_k\right) + a_2 = a_1 + a_2,$$

$$\sum_{k=1}^{3} a_k = \sum_{k=1}^{2} a_k + a_3 = (a_1 + a_n) + a_3 = a_1 + a_2 + a_3.$$

If n is a positive integer, then we write $\sum_{k=1}^{n} a_k = a_1 + a_2 + \cdots + a_n$ or

$$\sum_{k=1}^{n} a_k = a_1 + \cdots + a_n. \tag{4.2}$$

If $a_k = x$ for all $k \in (n) = \{1, \ldots, n\}$, we write

$$\sum_{k=1}^{n} a_k = \sum_{k=1}^{n} x \qquad \text{for each nonnegative integer } n. \tag{4.3}$$

Prob. 4.1. Prove: If $x \in \mathbb{R}$, then for all $n \in \mathbb{Z}$

$$\sum_{k=1}^{n} x = nx.$$

Prob. 4.2. Prove: If a and $b_1, \ldots, b_n$ are real numbers, then

$$\sum_{k=1}^{n} (ab_k) = a \sum_{k=1}^{n} b_k .$$

Prob. 4.3. Prove: If $s, t, x_1, \ldots, x_n; y_1, \ldots, y_n$ are real numbers, then

$$\sum_{k=1}^{n} (sx_k + ty_k) = s \sum_{k=1}^{n} x_k + t \sum_{k=1}^{n} y_k .$$

Prob. 4.4. Note, if n is a positive integer and a and b are real numbers, then

$$a^{n+1} - b^{n+1} = a^n(a-b) + b(a^n - b^n).$$

Prove: If n is a positive integer and a and b are real numbers, then

(a)
$$\begin{aligned} a^n - b^n &= (a-b) \sum_{i=0}^{n-1} a^{n-1-i} b^i \\ &= (a-b)(a^{n-1} + a^{n-2}b + \cdots + ab^{n-2} + b^{n-1}), \end{aligned}$$

(b)
$$a^n - 1 = (a-1) \sum_{i=0}^{n-1} a^{n-1-i} = (a-1)(a^{n-1} + a^{n-2} + \cdots + a + 1).$$

Prob. 4.5. Prove: (a) If $0 \leqslant a < b$, then for $n \in \mathbb{Z}_+$, $a^n < b^n$; (b) if $a < b$, then $a^{2n+1} < b^{2n+1}$ for $n \in \mathbb{Z}_0$.

Prob. 4.6. Prove: If n is a positive integer, then

(a)
$$1 + 2 + \cdots + n = \sum_{k=1}^{n} k = \frac{n(n+1)}{2},$$

(b)
$$1^2 + 2^2 + \cdots + n^2 = \sum_{k=1}^{n} k^2 = \frac{n(n+1)(2n+1)}{6},$$

(c)
$$1^3 + 2^3 + \cdots + n^3 = \sum_{k=1}^{n} k^3 = \frac{n^2(n+1)^2}{4},$$

(d)
$$1^4 + 2^4 + \cdots + n^4 = \sum_{k=1}^{n} k^4 = \frac{n(n+1)(2n+1)(3n^2+3n-1)}{30}.$$

Prob. 4.7. Prove: If n is a positive integer, then

(a) $$1 \cdot 2 + 2 \cdot 3 + \cdots + n(n+1) = \frac{n(n+1)(n+2)}{3},$$

(b) $$1 \cdot 2 \cdot 3 + 2 \cdot 3 \cdot 4 + \cdots + n(n+1)(n+2) = \frac{n(n+1)(n+2)(n+3)}{4}.$$

Prob. 4.8. Prove: If x is *not* a negative integer and n is a positive integer, then

$$\sum_{k=1}^{n} \frac{1}{(x+k)(x+k+1)} = \frac{n}{(x+1)(x+n+1)}.$$

Prob. 4.9. By an *arithmetic progression* of real numbers we mean a finite sequence $\langle a, a+d, \ldots, a+(n-1)d \rangle$ of n terms, where n is a positive integer. a is called the *first term* and $a+(n-1)d = l$ the *last term* of the progression. By the *sum* S_n of the progression, we mean

$$S_n = a + (a+d) + \cdots + a + (n-1)d = \sum_{x=1}^{n} (a + (k-1)d).$$

Prove:

$$S_n = \frac{n(a+l)}{2}.$$

Prob. 4.10. By a *geometric progression* of real numbers, we mean a finite sequence $\langle a, ar, ar^2, \ldots, ar^{n-1} \rangle$, where a and r are real numbers and n is a positive integer. By the *sum* σ_n of the geometric progression, we mean

$$\sigma_n = \sum_{k=1}^{n} ar^{k-1} = a + ar + \cdots + ar^{n-1}.$$

Note, if $r = 1$, then

$$\sum_{k=1}^{n} ar^{k-1} = \sum_{k=1}^{n} a(1)^{k-1} = \sum_{k=1}^{n} a = na = \underbrace{a + a + \cdots + a}_{(n \text{ terms})}$$

Prove: If $r \neq 1$, then

$$\sigma_n = \sum_{k=1}^{n} ar^{k-1} = a\frac{r^n - 1}{r - 1}.$$

In analogy with the general sum $\sum_{k=1}^{n} a_k$, we have the *general product* $\prod_{k=1}^{n} a_k$, where $a_1, a_2, \ldots$ are real numbers. We define it inductively:

$$\prod_{k=1}^{0} a_k = 1,$$
$$\prod_{k=1}^{n+1} a_k = \left(\prod_{k=1}^{n} a_k \right) a_{n+1} \qquad \text{for each} \quad n \in \mathbb{Z}_0. \tag{4.4}$$

We call $\prod_{k=1}^{n} a_k$ the *product* of n *terms*.

For example,

$$\prod_{k=1}^{1} a_k = \left(\prod_{k=1}^{0} a_k\right)a_1 = 1a_1 = a_1,$$

$$\prod_{k=1}^{2} a_k = \left(\prod_{k=1}^{1} a_k\right)a_2 = a_1a_2,$$

$$\prod_{k=1}^{3} a_k = \left(\prod_{k=1}^{2} a_k\right)a_3 = (a_1a_2)a_3 = a_1a_2a_3.$$

When n is a positive integer, then we write

$$\prod_{k=1}^{n} a_k = a_1a_2 \ldots a_n \quad \text{or} \quad \prod_{k=1}^{n} a_k = a_1 \ldots a_n. \tag{4.5}$$

If $a_k = x$ for all $k \in (n)$, then we write

$$\prod_{k=1}^{n} a_k = \prod_{k=1}^{n} x. \tag{4.6}$$

Prob. 4.11. Prove: If x is a real number, then

$$\prod_{k=1}^{n} x = x^n.$$

If n is a nonnegative integer, then we define

$$\prod_{k=1}^{n} k = n!. \tag{4.7}$$

Thus,

$$0! = \prod_{k=1}^{0} k = 1,$$

$$1! = \prod_{k=1}^{1} k = 1,$$

$$2! = \prod_{k=1}^{2} k = 1 \cdot 2,$$

$$3! = \prod_{k=1}^{3} k = 1 \cdot 2 \cdot 3,$$

and so on. Note that

$$(n+1)! = \prod_{k=1}^{n+1} k = \left(\prod_{k=1}^{n} k\right)(n+1) = n!\,(n+1) \tag{4.8}$$

for each $n \in \mathbb{Z}_0$.

PROB. 4.12. By the *product* P_n of the geometric progression $\langle a, ar, \ldots, ar^{n-1}\rangle$ (cf. Prob. 4.10), we mean:

$$P_n = \prod_{k=1}^{n} ar^{k-1} = a(ar)\ldots ar^{n-1} \qquad \text{for} \quad n \in \mathbb{Z}_+ .$$

Prove:

$$P_n^2 = (al)^n \qquad \text{for} \quad n \in \mathbb{Z}_+ ,$$

where a and l are respectively the first and last terms of the geometric progression.

PROB. 4.13. Prove: If $a_1, \ldots, a_n$ are nonzero real numbers, then

$$\frac{a_1 - 1}{a_1} + \frac{a_2 - 1}{a_1 a_2} + \cdots + \frac{a_n - 1}{a_1 a_2 \ldots a_n} = 1 - \frac{1}{a_1 a_2 \ldots a_n} .$$

PROB. 4.14. Prove: For all positive integers n, if one of $a_1, \ldots, a_n$ is 0, then

$$\prod_{k=1}^{n} a_k = 0.$$

5. Bernoulli's and Related Inequalities

The inequalities below will play a very important role in our work.

Theorem 5.1. *If $h \in \mathbb{R}$, $h > -1$, $h \neq 0$ and n is a positive integer, $n \geqslant 2$, then*

$$1 + nh(1 + h)^{n-1} > (1 + h)^n > 1 + nh. \tag{5.1}$$

PROOF. We first prove that, $(1 + h)^n > 1 + nh$ for $n \geqslant 2$. We use induction on n for $n \geqslant 2$. Since $h^2 > 0$, then for $n = 2$

$$(1 + h)^2 = 1 + 2h + h^2 > 1 + 2h.$$

This proves that $(1 + h)^2 > 1 + 2h$. Assume that for some integer $n \geqslant 2$, $(1 + h)^n > 1 + nh$. Multiply both sides by $1 + h$. Since $1 + h > 0$ and $nh^2 \geqslant 2h^2 > 0$, it follows that

$$(1 + h)^{n+1} > (1 + nh)(1 + h) = 1 + (n + 1)h + nh^2 > 1 + (n + 1)h.$$

Invoking the principle of induction, we see that $(1 + h)^n > 1 + nh$ for all integers $n \geqslant 2$.

We now prove $1 + nh(1 + h)^{n-1} > (1 + h)^n$ for all integers $n \geqslant 2$. We have

$$(1 + h)^2 = 1 + 2h + h^2 < 1 + 2h + 2h^2 = 1 + 2h(1 + h)$$

so that

$$1 + 2h(1 + h) > (1 + h)^2. \tag{5.2}$$

Now assume that $1 + nh(1 + h)^{n-1} > (1 + h)^n$ holds for some integer $n \geqslant 2$ under the stated conditions on h. Multiply both sides by $1 + h > 0$. Then

$$1 + h + nh(1 + h)^n > (1 + h)^{n+1}. \tag{5.3}$$

If $h > 0$, then $(1 + h)^n > 1$, and therefore $h(1 + h)^n > h$. If $-1 < h < 0$, then $(1 + h)^n < 1$, and, since $h < 0$, we again have $h(1 + h)^n > h$. In either case, if $h > -1$ and $h \neq 0$, then $h(1 + h)^n > h$. This implies $1 + h(1 + h)^n > 1 + h$. Adding $nh(1 + h)^n$ to both sides yields

$$1 + (n + 1)h(1 + h)^n > 1 + h + nh(1 + h)^n.$$

This and (5.3) imply

$$1 + (n + 1)h(1 + h)^n > (1 + h)^{n+1}. \tag{5.4}$$

Using induction on n for $n \geqslant 2$, we conclude that $1 + nh(1 + h)^{n-1} > (1 + h)^n$ for all integers $n \geqslant 2$, and the proof is complete.

Remark 5.1. The inequality

$$(1 + h)^n > 1 + nh, \tag{5.5}$$

where $h > -1$, $h \neq 0$, and n is an integer $n \geqslant 2$ is known as the *strict Bernoulli* inequality. When we omit the condition $h \neq 0$ and merely require $h > -1$, we obtain

$$(1 + h)^n \geqslant 1 + nh, \tag{5.6}$$

where $h > -1$ and n is a positive integer. This is known as *Bernoulli*'s *inequality*.

Note that we also have

$$1 + nh(1 + h)^{n-1} \geqslant (1 + h)^n, \tag{5.7}$$

where $h > -1$ and n is a positive integer.

Remark 5.2. When in (5.1) we write $x = 1 + h$, so that $h = x - 1$, we have an alternate form of that inequality,

$$n(x - 1)x^{n-1} > x^n - 1 > n(x - 1), \tag{5.8}$$

where $x > 0$, $x \neq 1$ and n is an integer $n \geqslant 2$. The weaker form of (5.8) is

$$n(x - 1)x^{n-1} \geqslant x^n - 1 \geqslant n(x - 1), \tag{5.9}$$

where $x > 0$ and n is a positive integer.

Prob. 5.1. Prove: If n is a positive integer, then

$$\left(1 + \frac{1}{n}\right)^n \geqslant 2.$$

Prob. 5.2. Prove: If n is a positive integer, then

$$\left(1-\frac{1}{n}\right)^n<\frac{1}{2}.$$

Prob. 5.3. Prove: If n is a positive integer, then

$$\text{(a)}\left(1-\frac{1}{(n+1)^2}\right)^{n+1}>1-\frac{1}{n+1}\text{ and (b)}\left(1+\frac{1}{n+1}\right)^{n+1}>\left(1+\frac{1}{n}\right)^n.$$

Prob. 5.4. Prove: If n is a positive integer, then

$$\frac{1}{n!}\leqslant\frac{n!}{n^n}\leqslant\frac{1}{2^{n-1}}.$$

6. Factorials

Def. 6.1. If $x\in\mathbb{R}$ and r is a nonnegative integer, then by a *factorial* of *degree r*, we mean:

$$(x)_r=\prod_{k=1}^{r}(x-k+1). \tag{6.1}$$

It follows that

$$\begin{aligned}(x)_0&=\prod_{k=1}^{0}(x-k+1)=1,\\(x)_r&=\prod_{k=1}^{r}(x-k+1)=x(x-1)\dots(x-r+1)\qquad\text{if}\quad r\geqslant 1.\end{aligned}\tag{6.2}$$

For example, $(x)_1=x$, $(x)_2=x(x-1)$, and $(x)_3=x(x-1)(x-2)$. Note that if $x=n$, where n is a nonnegative integer, then there are two cases: (1) $0\leqslant n<r$ or (2) $0\leqslant r\leqslant n$. In case (1) we have $r\geqslant 1$ and $n\in\omega_r=\{0,1,\dots,r-1\}$. Thus, a $j\in\{0,1,\dots,r-1\}$ exists such that $n=j$, and we know that one of the $n, n-1,\dots,n-r+1$ is zero. Consequently

$$(n)_r=\prod_{k=1}^{r}(n-k+1)=0\qquad\text{if}\quad 0\leqslant n<r. \tag{6.3}$$

In case (2)

$$(n)_r=(n)_0=1\qquad\text{if}\quad r=0. \tag{6.4}$$

Continuing with case (2) we consider $1\leqslant r\leqslant n$ and obtain

$$(n)_r=\prod_{k=1}^{r}(n-k+1)=n(n-1)\dots(n-r+1)\qquad\text{if}\quad 1\leqslant r\leqslant n. \tag{6.5}$$

Note:

$$(n)_n = \prod_{k=1}^{n} (n-k+1) = n(n-1)\ldots 2\cdot 1 = n!. \tag{6.6}$$

We multiply both sides in (6.5) by $(n-r)!$. Then

$$(n)_r(n-r)! = n(n-1)\ldots(n-r+1)(n-r)! = n!.$$

Hence,

$$(n)_r = \frac{n!}{(n-r)!} \qquad \text{if } n \text{ is an integer and } 0 \leqslant r \leqslant n. \tag{6.7}$$

In summary: If $n \in \mathbb{Z}_0$ and $r \in \mathbb{Z}_0$, then

$$(n)_r = \begin{cases} 0, & \text{if } \ 0 \leqslant n < r \\ \dfrac{n!}{(n-r)!} & \text{if } \ 0 \leqslant r \leqslant n. \end{cases} \tag{6.8}$$

PROB. 6.1. Prove: If $x \in \mathbb{R}$ and $r \in \mathbb{Z}_+$, then

$$(x+1)_r = (x)_r + r(x)_{r-1}.$$

We now define:

Def. 6.2. If $x \in \mathbb{R}$ and $r \in \mathbb{Z}_0$, then

$$\binom{x}{r} = \frac{(x)_r}{r!}. \tag{6.9}$$

This definition yields

$$\binom{x}{0} = \frac{(x)_0}{0!} = \frac{1}{1} = 1 \tag{6.10}$$

and

$$\binom{x}{r} = \frac{x(x-1)\ldots(x-r+1)}{r!} \qquad \text{if } \ r \geqslant 1. \tag{6.11}$$

For example,

$$\binom{x}{1} = \frac{(x)_1}{1!} = x, \qquad \binom{x}{2} = \frac{(x)_2}{2!} = \frac{x(x-1)}{2!},$$

$$\binom{x}{3} = \frac{(x)_3}{3!} = \frac{x(x-1)(x-2)}{3!}.$$

Other examples are:

$$\binom{-1}{5} = \frac{(-1)_5}{5!} = \frac{(-1)(-1-1)(-1-2)(-1-3)(-1-4)}{5!} = -\frac{5!}{5!} = -1$$

and

$$\binom{\frac{1}{2}}{4} = \frac{\frac{1}{2}(\frac{1}{2}-1)(\frac{1}{2}-2)(\frac{1}{2}-3)}{4!} = -\frac{1 \cdot 3 \cdot 5}{2^4 \cdot 4!}.$$

Prob. 6.2. Prove: If $x \in \mathbb{R}$ and $r \in \mathbb{Z}_+$, then

$$\binom{x+1}{r} = \binom{x}{r} + \binom{x}{r-1}.$$

(See Prob. 6.1.)

Prob. 6.3. Prove: If $n \in \mathbb{Z}_0$, then

(a) $\binom{n}{r} = 0$, if $0 \leqslant n < r$
(b) $\binom{n}{r} = (n!/r!(n-r)!)$, if $0 \leqslant r \leqslant n$.

Prob. 6.4. Prove: If $n \in \mathbb{Z}_0$, $r \in \mathbb{Z}_0$, then $\binom{n}{r}$ is a nonnegative integer.

Prob. 6.5. Prove: If $n \in \mathbb{Z}_0$, then

$$\binom{n}{r} = \binom{n}{n-r} \qquad \text{for} \quad r \in \mathbb{Z}_0.$$

Prob. 6.6 (Binomial Theorem). Prove: If x and y are real numbers and $n \in \mathbb{Z}_0$, then

$$(x+y)^n = \sum_{k=0}^{n} \binom{n}{k} x^{n-k} y^k$$

(Hint: use induction on n and Prob. 6.2 with $x = n$).

Prob. 6.7. Prove: If $n \in \mathbb{Z}_0$, then

$$\text{(a)} \quad 2^n = \sum_{k=0}^{n} \binom{n}{k}, \qquad \text{(b)} \quad 0 = \sum_{k=0}^{n} (-1)^k \binom{n}{k}.$$

Prob. 6.8. Prove: If $x_1, \ldots, x_{n+1}$ are real numbers, then

(a) $$\sum_{k=1}^{n} (x_{k+1} - x_k) = x_{n+1} - x_1,$$

(b) $$(k+1)! - k! = k \cdot k! \qquad \text{for} \quad k \in \mathbb{Z}_0,$$

(c) $$(n+1)! - 1 = 1 \cdot 1! + 2 \cdot 2! + \cdots + n \cdot n! \qquad \text{for} \quad n \in \mathbb{Z}_+.$$

Prob. 6.9. Prove: If $n \in \mathbb{Z}_+$, then

(a)
$$\sum_{k=1}^{n} k\binom{n}{k} = n2^{n-1},$$

(b)
$$\sum_{k=1}^{n} k^2\binom{n}{k} = n(n+1)2^{n-2} \qquad \text{if} \quad n \geqslant 2.$$

Prob. 6.10. Prove: If $n \in \mathbb{Z}_0$, $n \geqslant 2$, then

(a)
$$\sum_{k=1}^{n} (-1)^k k\binom{n}{k} = 0.$$

More generally, prove:
(b) If $1 \leqslant j < n$, where j and n are integers, then

$$\sum_{k=1}^{n} (-1)^k k^j\binom{n}{k} = 0$$

and

(c)
$$\sum_{k=1}^{n} (-1)^k k^n\binom{n}{k} = (-1)^n n!$$

Prob. 6.11. Prove: If n and k are nonnegative integers, then

$$\sum_{i=1}^{n} \binom{i}{k} = \binom{n+1}{k+1}.$$

Prob. 6.12. Prove: If $k \in \mathbb{Z}_0$, then

(a)
$$\binom{-1}{k} = (-1)^k,$$

(b)
$$\binom{\frac{1}{2}}{k} = (-1)^{k-1}\frac{\prod_{i=2}^{k}(2i-3)}{2^k k!} = (-1)^{k-1}\frac{1\cdot 3\cdot 5\ldots(2k-3)}{2^k k!}.$$

Prob. 6.13. Prove: If $n \in \mathbb{Z}_0$ and $\alpha \in \mathbb{R}$, then

$$1+\alpha+\frac{\alpha(\alpha+1)}{2!}+\cdots+\frac{\alpha(\alpha+1)\ldots(\alpha+n-1)}{n!}$$
$$=\frac{(\alpha+1)(\alpha+2)\ldots(\alpha+n)}{n!},$$

i.e.,

$$\sum_{k=0}^{n}\binom{\alpha+n-1}{k} = \binom{\alpha+n}{n}.$$

PROB. 6.14. Prove: If $\alpha \in \mathbb{R}$ and $k \in \mathbb{Z}_0$, then

$$\binom{-\alpha}{k} = (-1)^k \binom{\alpha + k - 1}{k}.$$

7. Onto Functions. nth Root of a Positive Real Number

If $f: X \to Y$ is a function that maps the nonempty set X into the set Y, then

$$\mathcal{R}(f) \subseteq Y. \tag{7.1}$$

Def. 7.1. If the range of the function $f: X \to Y$ above is Y, i.e., if

$$\mathcal{R}(f) = Y,$$

then we say that f maps X *onto* Y and call f an *onto* function. Onto functions are also called *surjective* functions. When $\mathcal{R}(f) \subset Y$, we say that f maps X into *but not onto* Y.

For example, the function $f: \mathbb{R} \to \mathbb{R}$ defined as $f(x) = x^2$ for all $x \in \mathbb{R}$ does *not* map $\mathbb{R}$ onto $\mathbb{R}$ since $f(x) \geqslant 0$ for each $x \in \mathbb{R}$ here. The range of this f is a subset of $[0, +\infty)$. When we change the codomain (Def. 2.1) to $[0, +\infty)$, the resulting function is an onto function. This follows from the fact that if $y \in [0, +\infty)$, then there exists an $x \in \mathbb{R}$ such that $x^2 = y$ (Theorem I.11.2) so that $[0, +\infty) \subseteq \mathcal{R}(f)$. Since $\mathcal{R}(f) \subseteq [0, +\infty)$ it follows that $\mathcal{R}(f) = [0, +\infty)$. The function defined here will usually be written as $f = (\ )^2$.

Investigating the ontoness of a function $f: X \to Y$ amounts to asking whether for each $y \in Y$ there exists some $x \in X$ such that $f(x) = y$. This is an *existence* question. The answer can be provided in two ways. One way is to produce the x in question by "solving" for it. Another way is to prove that it must exist without necessarily exhibiting it.

PROB. 7.1. Prove: If a and b are real numbers and $a \neq 0$, then the function $f: \mathbb{R} \to \mathbb{R}$ given by $f(x) = ax + b$ for each $x \in \mathbb{R}$ is an onto function.

We now ask whether the function $f: \mathbb{R} \to \mathbb{R}$, given by $f(x) = x^n$, $x \in \mathbb{R}$, and n a positive integer, is an onto function. We shall usually write this function as $(\ )^n$. It is clear that for $n = 1$, $(\ )^n$ is onto since it is the identity function $I_{\mathbb{R}}$ on $\mathbb{R}$ (Example 2.5), i.e., $(x)^1 = x$ for each $x \in \mathbb{R}$. We consider $(\ )^n$ for integral values of n, $n \geqslant 2$. If n is even, then $x^n \geqslant 0$ for all x and the range of $(\ )^n$ is a subset of $[0, \infty)$. Therefore, it does not map $\mathbb{R}$ onto $\mathbb{R}$. Below, we prove that $(\ )^n$ for $n \geqslant 2$ maps $(0; \infty)$ onto itself.

Theorem 7.1. *If y is a positive real number and n an integer such that $n \geqslant 2$, there exists exactly one positive real number x such that $x^n = y$.*

PROOF. In many respects the proof is similar to the proof of Theorem I.11.2. Construct the set

$$S = \{x \in \mathbb{R} \mid x > 0 \text{ and } x^n < y\}. \tag{7.2}$$

Let

$$x_0 = \frac{y}{1+y}. \tag{7.3}$$

As in the proof of Theorem I.11.2, it is easy to see that $0 < x_0 < 1$ and $0 < x_0 < y$. Since $0 < x_0 < 1$, we have $x_0^n \leqslant x_0 < y$ from which we conclude that $x_0 \in S$, so that $S \neq \varnothing$.

We prove that S is bounded from above. Take $x > 1 + y$. We have $x > 1 + y > 0$. In view of Bernoulli's inequality,

$$x^n > (1+y)^n \geqslant 1 + ny > ny \geqslant 2y > y.$$

This implies that $x \notin S$. Hence, $x \in S$ implies $x \leqslant 1 + y$. Thus, S is bounded from above by $1 + y$.

Since S is a nonempty set of real numbers which is bounded from above, it has a real supremum. Let $\mu = \sup S$. Since the x_0 defined in (7.3) is in S, $0 < x_0 \leqslant \mu$.

We prove that $y \leqslant \mu^n$. Take $\epsilon \in \mathbb{R}$ such that $0 < \epsilon < 1$. Then $0 < \mu < \mu + \epsilon$. This implies that $\mu + \epsilon \notin S$, and, since $\mu + \epsilon > 0$, it follows that $(\mu + \epsilon)^n \geqslant y$, and that

$$(\mu + \epsilon)^n - \mu^n \geqslant y - \mu^n. \tag{7.4}$$

Since $\epsilon/\mu > 0$, Theorem 5.1 implies that

$$n\left(\frac{\epsilon}{\mu}\right)\left(1 + \frac{\epsilon}{\mu}\right)^{n-1} > \left(1 + \frac{\epsilon}{\mu}\right)^n - 1.$$

Multiply both sides here by μ^n. Then

$$n\epsilon(\mu + \epsilon)^{n-1} > (\mu + \epsilon)^n - \mu^n.$$

This and (7.4) imply that

$$n\epsilon(\mu + 1)^{n-1} > n\epsilon(\mu + \epsilon)^{n-1} > y - \mu^n.$$

It follows that

$$\epsilon > \frac{y - \mu^n}{n(\mu + 1)^{n-1}} \quad \text{for all} \quad 0 < \epsilon < 1.$$

By Prob. I.3.13, we conclude from this that

$$\frac{y - \mu^n}{n(\mu + 1)^{n-1}} \leqslant 0,$$

so that $y - \mu^n \leqslant 0$, and $y \leqslant \mu^n$.

We now prove that $\mu^n \leqslant y$. Consider a real ϵ such that $0 < \epsilon < \mu$. We have $0 < \mu - \epsilon < \mu$. Therefore, there exists a $z \in S$ such that $\mu - \epsilon < z \leqslant \mu$. This yields

$$(\mu - \epsilon)^n < z^n < y,$$

so that $(\mu - \epsilon)^n < y$. Therefore,

$$(\mu - \epsilon)^n - \mu^n < y - \mu^n. \tag{7.5}$$

Now, $0 < \epsilon/\mu < 1$. Hence, by Theorem 5.1,

$$-n\frac{\epsilon}{\mu} < \left(1 - \frac{\epsilon}{\mu}\right)^n - 1.$$

Multiply both sides here by μ^n. In view of (7.5),

$$-n\epsilon\mu^{n-1} < (\mu - \epsilon)^n - \mu^n < y - \mu^n.$$

This implies that

$$\frac{\mu^n - y}{n\mu^{n-1}} < \epsilon \qquad \text{for all } \epsilon \text{ such that} \quad 0 < \epsilon < \mu.$$

By Prob. I.3.13, we conclude that

$$\frac{\mu^n - y}{n\mu^{n-1}} \leqslant 0,$$

so that $\mu^n - y \leqslant 0$ or $\mu^n \leqslant y$. This and the already established inequality $y \leqslant \mu^n$ yield $\mu^n = y$. Thus, μ is an $x > 0$ such that $x^n = y$.

Finally, we prove that at most one $x > 0$ exists such that $x^n = y$. Suppose, that x_1 also has the properties: $x_1 > 0$ and $x_1^n = y$. This implies that $x^n = x_1^n$. If $x \neq x_1$, we would have either $0 < x_1 < x$ or $0 < x < x_1$. The first of these inequalities implies that $x_1^n < x^n$ and the second implies that $x^n < x_1^n$. In either case, $x \neq x$ implies $x^n \neq x_1^n$. This would contradict $x^n = x_1^n$. We must conclude that $x_1 = x$. This completes the proof.

Prob. 7.2. Prove: If n is an odd positive integer and y is a real number, then there exists exactly one real number x such that $x^n = y$. Accordingly, the function $(\ )^n : \mathbb{R} \to \mathbb{R}$ where n is an odd positive integer maps $\mathbb{R}$ onto $\mathbb{R}$.

Def. 7.2. If $y \in \mathbb{R}$, $y > 0$ and n is a positive integer, then a real x such that $x^n = y$ is called an nth *root* of y. The unique positive u such that $u^n = y$ is written as $u = \sqrt[n]{y}$. Since $x^n = 0$ if and only if $x = 0$, we define $0 = \sqrt[n]{0}$. When $n = 1$, we write $y = \sqrt[1]{y}$ for $y \in \mathbb{R}$. We always write $\sqrt{y}$ instead of $\sqrt[2]{y}$. When n is an odd positive integer and $y \in \mathbb{R}$ (note that y can be negative), then the unique x such that $x^n = y$ is written as $x = \sqrt[n]{y}$.

Prob. 7.3. Let n be a positive integer. Prove: (a) If $y \geqslant 0$, then $(\sqrt[n]{y})^n = y$; (b) if $x \geqslant 0$, then $\sqrt[n]{x^n} = x$.

PROB. 7.4. Prove: If n is an *even* positive integer and $y > 0$, then $\sqrt[n]{y}$ is the only negative real number x such that $x^n = y$. In this case, there are exactly two real nth roots of y each being the negative of the other.

PROB. 7.5. Let n be a positive integer and $y_1, y_2, \ldots, y_m$ be m nonnegative numbers. Prove: (a) $\sqrt[n]{y_1}\ \sqrt[n]{y_2} \cdots \sqrt[n]{y_m} = \sqrt[n]{y_1 y_2 \cdots y_m}$. (b) $(\sqrt[n]{y_1}\,)^m = \sqrt[n]{y_1^m}$.

8. Polynomials. Certain Irrational Numbers

Def. 8.1. A *polynomial* on $\mathbb{R}$ is a function $f: \mathbb{R} \to \mathbb{R}$ defined by means of

$$f(x) = a_0 x^n + a_1 x^{n-1} + \cdots + a_{n-1} x + a_n \qquad \text{for each} \quad x \in \mathbb{R}, \tag{8.1}$$

where n is a nonnegative integer and $a_0, a_1, \ldots, a_n$ are real numbers. The latter are called the *coefficients* of the polynomial and is a_0 called its leading coefficient. If $a_0 \neq 0$, then f is said to be of *degree n*. Clearly, a polynomial of degree 0 is a nonzero constant a_0 on $\mathbb{R}$. The polynomial g such that $g(x) = 0$ for all $x \in \mathbb{R}$ is called the *zero* polynomial. It is assigned no degree. The equation

$$a_0 x^n + a_1 x^{n-1} + \cdots + a_{n-1} x + a_n = 0, \tag{8.2}$$

is called a *polynomial equation* and an r satisfying it is called a *root* of the equation or a *zero* of the polynomial in (8.1). By a *rational root* of the equation we mean a root which is rational. Similarly, a *real root* is defined as a root which is real.

Before proving the next theorem, we cite a lemma.

Lemma 8.1. *If a and m are integers such that $(a, m) = 1$, then $(a^n, m) = 1$ for each positive integer n.*

PROOF. This is a corollary of Theorem I.10.3 and we leave its proof to the reader as an exercise.

Theorem 8.1. *The polynomial equation (8.2) is of degree $n \geqslant 1$ and has integer coefficients. If $r = p/q$, where p and q are relatively prime integers, is a rational root of the equation, then $p \mid a_n$ and $q \mid a_0$.*

PROOF. Substitute $x = p/q$ in the equation and multiply both sides of the equation by q^n. Then

$$a_0 p^n + a_1 p^{n-1} q + \cdots + a_{n-1} p q^{n-1} + a_n q^n = 0 \tag{8.3}$$

or

$$p\left(a_0 p^{n-1} + a_1 p^{n-2} q + \cdots + a_{n-1} q^{n-1}\right) = -a_n q^n. \tag{8.4}$$

The second factor on the left is an integer, so this implies that $p \mid (-a_n q^n)$. Since $(p, q) = 1$, it follows (by Lemma 8.1) that (p, q^n). By Theorem I.10.4, $p \mid a_n$.

It also follows from (8.3) that

$$-a_0 p^n = q\left(a_1 p^{n-1} + \cdots + a_{n-1} q^{n-2} + a_n q^{n-1}\right). \tag{8.5}$$

Reasoning as we did immediately above, we see that $q \mid a_0$. The proof is now complete.

Corollary. *If, in the equation of the theorem, $a_0 = 1$, then a rational root of the equation is necessarily an integer.*

PROOF. Exercise.

PROB. 8.1. Prove: (a) If k and n are positive integers and a rational number r exists such that $r^n = k$, then r is an integer; (b) if k is not the nth power of an integer, then $\sqrt[n]{k}$ is irrational.

Theorem 8.2. *If n is an integer such that $n \geqslant 2$, then $\sqrt[n]{n}$ is irrational.*

PROOF. Suppose $\sqrt[n]{n}$ is rational. Let $r = \sqrt[n]{n}$, so that $r^n = n$. By Prob. 8.1(a), $r = p$ where p is an integer. Now $p = r = \sqrt[n]{n}$ is a *positive* integer, so that $p \geqslant 1$ and $p^n = n$. Clearly, $p \neq 1$, so $p \geqslant 2$. This implies (Prob. 3.2) that $p^n > n$, contradicting $p^n = n$. Hence $\sqrt[n]{n}$ is rational.

PROB. 8.2. Prove $\sqrt{2} + \sqrt{3}$ and $\sqrt{2} + \sqrt[3]{2}$ are irrational.

9. One-to-One Functions. Monotonic Functions

Def. 9.1. A function $f: X \to Y$ is called *one-to-one* or *injective*, when it maps distinct elements of X into distinct elements of Y. In more detail: f is one-to-one if and only if $x_1 \in X$, $x_2 \in X$, and $x_1 \neq x_2$ imply $f(x_1) \neq f(x_2)$.

Remark 9.1. Often one proves that f is one-to-one by proving that $f(x_1) = f(x_2)$ implies $x_1 = x_2$. Questions concerning the one-to-oneness of a function are actually uniqueness questions, whereas questions concerning the ontoness of functions are existence questions (see the paragraph preceding Prob. 7.1).

Example 9.1. The function $(\ )^2 : \mathbb{R} \to [0, +\infty)$ is not one-to-one since $(-1)^2 = 1^2$. However, the restriction of $(\ )^2$ to $[0, +\infty)$ is one-to-one (Theorem 7.1).

Example 9.2. The identity function $I_X : X \to X$ on a nonempty set X (Example 2.5) is obviously one-to-one (explain).

Remark 9.2. We saw (Prob. 2.4) that $f: X \to Y$ and $A \subseteq B \subseteq X$ imply $f(A) \subseteq f(B)$. It is possible to have $A \subset B \subseteq X$ and $f(A) = f(B)$. This occurs for the function $(\ )^2 : \mathbb{R} \to \mathbb{R}$. It maps both $A = [0, +\infty)$ and $B = \mathbb{R}$ onto $[0, +\infty)$, in spite of the fact that $A \subset B$. However, if $f: X \to Y$ is one-to-one, then $A \subset B$ implies $f(A) \subset f(B)$. We prove this. Assume $A \subset B \subseteq X$, so that there exists $b \in B$ such that $b \notin A$. We know that $f(A) \subseteq f(B)$. Assume that $f(A) = f(B)$. Since $b \in B$, $f(b) \in f(B) = f(A)$ so that $f(b) \in f(A)$. This implies that $a \in A$ exists such that $f(b) = f(a)$. Since f is one-to-one, we obtain $b = a \in A$. This contradicts $b \notin A$. We conclude $f(A) \neq f(B)$. Hence, $f(A) \subset f(B)$, as claimed.

An important class of functions is the class of *monotonic* functions.

Def. 9.2. Let $f: X \to \mathbb{R}$ be a real-valued function of a real variable, so that not only is its range $f(X)$ a subset of $\mathbb{R}$, but also $X \subseteq \mathbb{R}$. We call such an f *monotonically increasing* if and only if

$$x_1 \in X, \quad x_2 \in X, \quad \text{and} \quad x_1 < x_2 \quad \text{imply} \quad f(x_1) \leqslant f(x_2). \tag{9.1}$$

Similarly f is called *monotonically decreasing* if and only if

$$x_1 \in X, \quad x_2 \in X, \quad \text{and} \quad x_1 < x_2 \quad \text{imply} \quad f(x_1) \geqslant f(x_2). \tag{9.2}$$

f is called *strictly monotonically increasing* if

$$x_1 \in X, \quad x_2 \in X, \quad \text{and} \quad x_1 < x_2 \tag{9.3}$$

implies

$$f(x_1) < f(x_2) \tag{9.4a}$$

and *strictly monotonically decreasing* if (9.3) implies

$$f(x_1) > f(x_2). \tag{9.4b}$$

A function which is monotonically increasing or monotonically decreasing is called *monotonic*. We write $f\uparrow$ when f is increasing monotonically and $f\downarrow$ when f is monotonically decreasing. We say that f has any of the above properties on a set A when its restriction $f|_A$ to A has that property.

Example 9.3. If n is an even positive integer, the function $(\ )^n : [0, +\infty] \to [0, +\infty)$ is strictly monotonically increasing (Prob. 4.5(a)), but $(\ )^n : \mathbb{R} \to [0, +\infty)$ is neither monotonically increasing nor monotonically decreasing (explain). (See Fig. 9.1(a).)

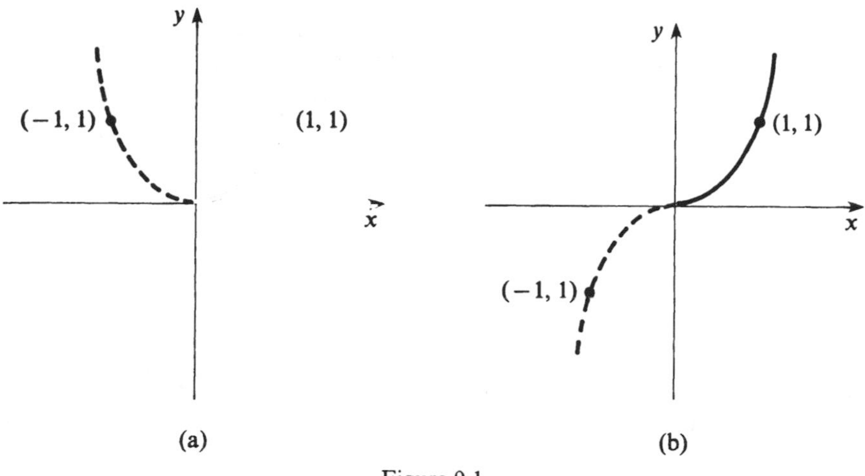

Figure 9.1

EXAMPLE 9.4. If n is an *odd* positive integer, then $(\ )^n : \mathbb{R} \to \mathbb{R}$ is strictly monotonically increasing (Prob. 4.5(b)). Note, this function is also an onto function (Prob. 7.2) (see Fig. 9.1(b)).

PROB. 9.1. Prove: $x_1 < x_2$ in $\mathbb{R}$ implies $[x_1] \leqslant [x_2]$. Thus, prove that the greatest integer function [ ] is monotonically increasing (Fig. 9.2). Note, [ ] is not strictly monotonically increasing (explain).

The result in the next problem gives a relation between strict monotonicity and one-to-oneness for real-valued functions of a real variable.

PROB. 9.2. Prove: A real-valued function of a real variable which is strictly monotonic is necessarily a one-to-one function.

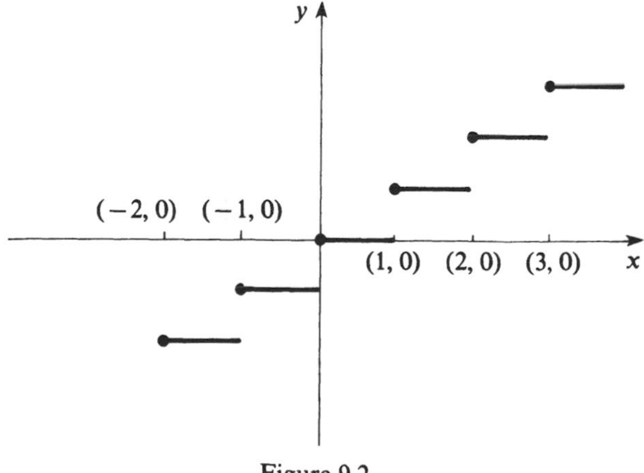

Figure 9.2

EXAMPLE 9.5. The converse of the result in the last problem does not hold. For, let $D = \{x \in \mathbb{R} \mid x \neq 0\} = (-\infty; 0) \cup (0; +\infty)$. The function $f: D \to \mathbb{R}$ defined as $f(x) = x^{-1}$ for $x \neq 0$ is one-to-one but not monotonic (prove this).

Monotonic Real Sequences

An infinite real sequence $\langle a_n \rangle$ is a real-value function of a real variable. This justifies the use of the terminology of Def. 9.2 for real sequences.

PROB. 9.3. Let $\langle a_n \rangle$ be a real sequence. Prove that $\langle a_n \rangle \uparrow$ if and only if $a_n \leqslant a_{n+1}$ for each $n \in \mathbb{Z}_+$ and that $\langle a_n \rangle \downarrow$ if and only if $a_n \geqslant a_{n+1}$ for each $n \in \mathbb{Z}_+$.

PROB. 9.4. Let $\langle n_k \rangle_{k \geqslant 1}$ be a sequence of *positive integers* which is strictly monotonically increasing. Prove: $n_k \geqslant k$ holds for all $k \in \mathbb{Z}_+$.

10. Composites of Functions. One-to-One Correspondences. Inverses of Functions.

We first define the composite of correspondences γ and δ.

Def. 10.1. Let γ be a correspondence between sets B and C. The *composite* of δ and γ (in that order), written as $\delta \circ \gamma$ is defined as the correspondence η between A and C such that $(a, c) \in \delta \circ \gamma = \eta$ if and only if some $b \in B$ exists such that $(a, b) \in \gamma$ and $(b, c) \in \delta$. In symbols

$$\delta \circ \gamma = \{(a, c) \in A \times C \mid (a, b) \in \gamma \text{ and } (b, c) \in \delta \text{ for some } b \in B\}. \tag{10.1}$$

We now define the composite $f \circ g$ of functions f and g.

Def. 10.2. Let $g: A \to B$ and $f: B \to C$ be functions, then their *composite* $h = f \circ g$ is defined as the composite of f and g in the sense of the last definition.

Theorem 10.1(a). *If $g: A \to B$ and $f: B \to C$, then $f \circ g$ is a function with domain A and codomain C and*

$$(f \circ g)(x) = f(g(x)) \qquad \text{for each} \quad x \in A. \tag{10.2}$$

PROOF. Let $x \in A$. There exists a unique $y \in B$ such that $(x, y) \in g$. There exists a unique $z \in C$ such that $(y, z) \in f$. Thus, for each $x \in A$, there exists a unique $z \in C$ such that $(x,z) \in f \circ g$. Therefore, $f \circ g$ is a function with domain A and codomain C and $(f \circ g)(x) = z$. But $y = g(x)$ and $z = f(y) = f(g(x))$. Hence

$$(f \circ g)(x) = f(g(x)) \qquad \text{for each} \quad x \in A. \tag{10.3}$$

PROB. 10.1(a). Let $g: A \to B$ and $f: B \to C$ be functions. Prove: (a) If f and g is one-to-one, so is $f \circ g$; (b) if f and g are onto functions, so is $f \circ g$.

Def. 10.3. If a function $f: X \to Y$ is both one-to-one and onto, we call it a *one-to-one correspondence* between X and Y. A one-to-one correspondence is also called a *bijection*.

EXAMPLE 10.1. The identity function I_X on a set X is a one-to-one correspondence between two copies of X.

EXAMPLE 10.2. The function $(\ )^n : [0, +\infty) \to [0, +\infty)$, where n is a positive integer is a one-to-one correspondence between two copies of $[0, +\infty)$. When n is an odd positive integer, then the function $(\ )^n : \mathbb{R} \to \mathbb{R}$ is a one-to-one correspondence between two copies of $\mathbb{R}$.

EXAMPLE 10.3. Let a and b be real numbers such that $a < b$. Define $f: [0,1] \to \mathbb{R}$ by means of

$$f(t) = a + t(b-a) \qquad \text{for each} \quad t \in [0,1]. \tag{10.4}$$

Since $a \leqslant a + t(b-a) \leqslant b$ for each $t \in [0,1]$, $a \leqslant f(t) \leqslant b$ for $0 \leqslant t \leqslant 1$. Thus, $f([0,1]) \subseteq [a,b]$, and we see that f maps $[0,1]$ into $[a,b]$. Now take $z \in [a,b]$, so that $a \leqslant z \leqslant b$. This implies that $0 \leqslant z - a \leqslant b - a$ and hence that

$$0 \leqslant \frac{z-a}{b-a} \leqslant 1.$$

Thus, $(z-a)/(b-a)$ is an element of $[0,1]$. We have

$$f\left(\frac{z-a}{b-a}\right) = a + \frac{z-a}{b-a}(b-a) = z.$$

Each z in $[a,b]$ is the image of the element $(z-a)/(b-a)$ in $[0,1]$ and is therefore in the range of f. Hence, $f([0,1]) = [a,b]$. f is also a one-to-one function. Consequently, f is a one-to-one correspondence between the interval $[0,1]$ and the interval $[a,b]$ where $a < b$.

Inverse of a Function

We define the inverse of a correspondence and then the inverse of a function.

Def. 10.4(a). If γ is a correspondence between X and Y, then its *inverse* γ^{-1} is defined as that correspondence between Y and X which assigns $x \in X$ to

$y \in Y$ if and only if γ assigns y to x. In symbols,

$$\gamma^{-1} = \{(y,x) \mid (x,y) \in \gamma\}. \tag{10.5}$$

When we apply this definition to the special correspondences we call functions we find that a function f always has an inverse correspondence, but this correspondence is not always a function. Take, for example, a function f which is *not* one-to-one. There exist *distinct* x_1 and x_2 in X such that $y = f(x_1) = f(x_2)$. Thus, $(x_1, y) \in f$ and $(x_2, y) \in f$. By the definition of the correspondence f^{-1},

$$(y, x_1) \in f^{-1} \quad \text{and} \quad (y, x_2) \in f^{-1},$$

where $x_1 \neq x_2$. Thus, different x's are assigned to the same y, and, in this case, the correspondence f^{-1} is not a function.

Another difficulty arises when a function $g: X \to Y$ is not onto Y. In this case, there exists some $y_1 \in Y$ such that $(x, y_1) \notin f$ for each $x \in X$. But then g^{-1} assigns no $x \in X$ to y_1. Here, g^{-1} is a function with domain differing from Y. In the theorem below we consider the inverse f^{-1} of a one-to-one correspondence.

Theorem 10.1(b). *If f is a one-to-one correspondence between X and Y, then the inverse correspondence f^{-1} is a function with domain Y and range X.*

Proof. Let $y \in Y$. Since f is one-to-one and maps X *onto* Y, there exists exactly one $x \in X$ such that $(x, y) \in f$. By the definition of f^{-1}, $(y, x) \in f^{-1}$. Thus, f^{-1} assigns to each $y \in Y$ exactly one $x \in X$, and is a function $f^{-1}: Y \to X$ with domain Y and codomain X. But each $x \in X$ has a y corresponding to it under f, so that $(x, y) \in f$ and $(y, x) \in f^{-1}$. Accordingly, x is the image of this y under f^{-1}. Thus, each $x \in X$ is the image, under f^{-1}, of a $y \in Y$. This proves that f^{-1} maps Y onto X. Thus, the range of f^{-1} is X. This completes the proof.

Def. 10.4(b). Let $f: X \to Y$ be a one-to-one correspondence between X and Y. By the previous theorem, the correspondence f^{-1} is a function. We call it the *inverse* of f, or the function *inverse to* f.

Remark 10.1. When $f: X \to Y$ is a one-to-one correspondence between X and Y, then

$$f^{-1}(y) = x \quad \text{if and only if} \quad y = f(x), \qquad \text{where} \quad x \in X. \tag{10.6}$$

It follows that

$$f(f^{-1}(y)) = f(x) = y \quad \text{and hence} \quad f(f^{-1}(y)) = y \qquad \text{for each} \quad y \in Y. \tag{10.7}$$

Also,

$$f^{-1}(f(x)) = f^{-1}(y) = x, \tag{10.8}$$

so that $f^{-1}(f(x)) = x$ for each $x \in X$. Using the notation of composition of functions, we can express (10.7) and (10.8) as

$$f \circ f^{-1} = I_Y \tag{10.9a}$$

$$f^{-1} \circ f = I_X . \tag{10.9b}$$

PROB. 10.1(b). Prove: If $f: X \to Y$ and $g: Y \to X$ are functions such that

$$f \circ g = I_Y \quad \text{and} \quad g \circ f = I_X ,$$

then f is a one-to-one correspondence between X and Y and $f^{-1} = g$. (Note that g is also a one-to-one correspondence and $g^{-1} = f$. Thus, f and g are each other's inverses.) Here, f is the inverse of its inverse. To avoid confusion, we call f the *direct* function.

Apropos the remark in the last problem, we now consider f^{-1} as the direct function. We have

$$f^{-1} = \{(y, f^{-1}(y)) \mid y \in \mathcal{R}(f) = \mathcal{D}(f^{-1})\}. \tag{10.10}$$

Here, y is a "dummy" variable, and, therefore,

$$\begin{aligned} f^{-1} &= \{(x, f^{-1}(x)) \mid x \in \mathcal{R}(f)\} \\ &= \{(x, y) \mid y = f^{-1}(x), \text{ where } x \in \mathcal{R}(f)\} \\ &= \{(x, y) \mid f(y) = x, \text{ where } x \in \mathcal{R}(f)\}. \end{aligned}$$

Thus, we see that to graph f^{-1} we interchange x and y in $y = f(x)$. The result is $x = f(y)$. This amounts to reflecting the graph of $y = f(x)$ in the line $y = x$ (see Fig. 10.1).

Remark 10.2. Having defined the inverse of a one-to-one correspondence, we proceed to define the inverse of a one-to-one function. Here, we drop

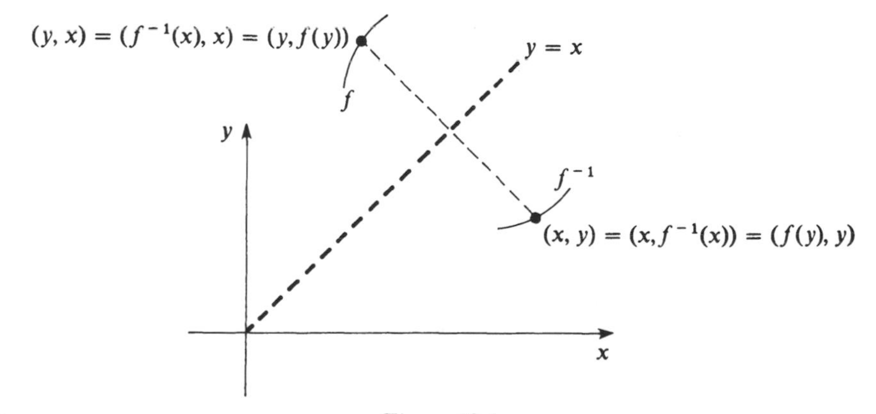

Figure 10.1

the assumption that the function is onto. We note that any function maps its domain onto its range and is, therefore, always a correspondence between its domain and its range. Accordingly, if f is one-to-one, the function $f: X \to \mathfrak{R}(f)$ is a one-to-one correspondence between X and $\mathfrak{R}(f)$. The latter function has an inverse $f^{-1}: \mathfrak{R}(f) \to X$. It is this function that we call the inverse of the one-to-one function $f: X \to Y$. For each $y \in \mathfrak{R}(f)$, $f^{-1}(y)$ is the unique $x \in X$ such that $y = f(x)$. We have $\mathfrak{D}(f^{-1}) = \mathfrak{R}(f)$ and $\mathfrak{R}(f^{-1}) = \mathfrak{D}(f)$. Also

$$f^{-1}(f(x)) = x \qquad \text{for each} \quad x \in \mathfrak{D}(f) \tag{10.11}$$

and

$$f(f^{-1}(y)) = y \qquad \text{for each} \quad y \in \mathfrak{R}(f). \tag{10.12}$$

Example 10.4. If n is a positive integer, then the function $(\ )^n: [0, +\infty) \to [0, +\infty)$ is a one-to-one correspondence two copies of $[0, +\infty)$. For each $y \in [0, +\infty)$, $\sqrt[n]{y}$ is the unique x such that $y = x^n$. The inverse of our function is the nth root function, which we write as $\sqrt[n]{y}$. We have

$$\left(\sqrt[n]{y}\right)^n = y \qquad \text{for each} \quad y \in \left[0, +\infty\right) \tag{10.13}$$

and

$$\sqrt[n]{x^n} = x \qquad \text{for each} \quad x \in \left[0, +\infty\right). \tag{10.14}$$

Thus, $(\ )^n$ and $\sqrt[n]{\ }$ are each other's inverses. When we wish to treat $\sqrt[n]{\ }$ as the direct function we interchange x and y and write $y = \sqrt[n]{x}$ for each $x \in [0, +\infty)$.

Equipotent Sets

Def. 10.5. We call sets X and Y *equipotent* if and only if there exists a one-to-one correspondence $f: X \to Y$ between X and Y. When the sets X and Y are equipotent, we write

$$X \simeq Y. \tag{10.15}$$

The diagram in Fig. 10.2 suggests that the circles C and C' are equipotent, and so are the segments $\overline{\mathrm{AB}}$ and $\overline{\mathrm{A'B'}}$.

In Example 10.3 we saw that there exists a one-to-one correspondence between the closed intervals $[0, 1]$ and $[a, b]$ (where $a < b$). Thus, we have: if $a < b$ for real numbers a and b, then $[0, 1] \simeq [a, b]$.

Prob. 10.2. Prove: $\emptyset \simeq \emptyset$.

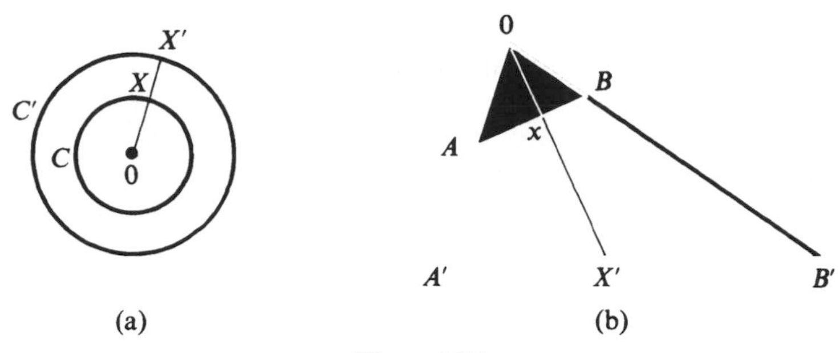

Figure 10.2

Prob. 10.3. Prove: If X, Y, and Z are sets, then

(a) $X \simeq X$,
(b) $X \simeq Y$ implies $Y \simeq X$,
(c) $X \simeq Y$ and $Y \simeq Z$ imply $X \simeq Z$.

Prob. 10.4. Prove: If n is a nonnegative integer, then $\omega_n \simeq (n)$ (cf. Defs. I.5.3 and I.5.4).

Prob. 10.5. Prove: $\mathbb{Z}_+ \simeq \mathbb{Z}_0$. (This means that $\mathbb{Z}_0$ is equipotent with one of its proper subsets!)

Remark 10.3(a). We do our counting by tacitly using the notion of equipotent sets. The set $\mathbb{Z}_0$ and its initial segments ω_n (Def. I.5.3) are taken as standard sets.

Def. 10.6. We call a set S *finite* if and only if $S \simeq \omega_n$ for some $n \in \mathbb{Z}_0$. Otherwise we call it *infinite*.

When we "count" the elements of a finite set S, we are really defining a one-to-one correspondence between S and the set $(n) = \{1, 2, \ldots, n\}$ for some positive integer n. Since $(n) \simeq \omega_n$ for each n, we have $S \simeq (n)$ if and only if $S \simeq \omega_n$.

Remark 10.3(b). We state some properties of finite sets:

(1) Each subset of a finite set is finite.
(2) No finite set is equipotent with any of its proper subsets.
(3) If m and n are nonnegative integers such that $m < n$, then ω_m and ω_n are not equipotent and neither are (m) and (n).
(4) If S is a finite set then the $n \in \mathbb{Z}_0$ such that $S \simeq \omega_n$ is unique. This n is also called the *number of elements* of S and we write it as $N(S)$.
(5) If A and B are finite sets, so are $A \cap B$ and $A - B$.
(6) If $A \subset B$ where B is finite, then $N(A) < N(B)$.

(7) (a) Each finite set of real numbers is bounded. (b) A nonempty finite set of real numbers has a maximum and a minimum.
(8) A nonempty set of nonnegative integers which is bounded has a maximum. (This is a consequence of Theorem I.5.10.)
(9) (The pigeon-hole principle). If A and B are finite sets such that $N(A) < N(B)$, then a function with domain B and range in A is not one-to-one.

Def. 10.7. A set S, such that $S \simeq \mathbb{Z}_0$ is called *denumerable*. Such a set is said to have *cardinal* $\aleph_0$. (Here, $\aleph$, read as *Aleph*, is the first letter of the Hebrew alphabet. $\aleph_0$ is read as *Aleph-null*.)

Remark 10.4. A denumerable set is necessarily infinite. If S is a denumerable set, so that $S \simeq \mathbb{Z}_0$, there exists a one-to-one correspondence $f: S \to \mathbb{Z}_0$. Let $f(n) = a_n$ for each $n \in \mathbb{Z}_0$. The set $S - \{a_0\} = B$ is equipotent with $\mathbb{Z}_+$. Since $\mathbb{Z}_+ \simeq \mathbb{Z}_0$, we have $B \simeq \mathbb{Z}_0$ and $\mathbb{Z}_0 \simeq S$, so that $B \simeq S$. Thus, S is equipotent with one of its proper subsets, B. As such, it cannot be finite (property (2) of finite sets). Hence, S is infinite.

Def. 10.8. A set which is either finite or denumerable is called a *countable* set. A denumerable set is also called a *countably infinite* set.

Theorem 10.2. *Every infinite set of nonnegative integers is denumerable.*

Proof. Let S be an infinite set of nonnegative integers. Clearly, S has a least member. Let $x_0 = \min S$. We have $\{x_0\} \subseteq S$. Since S is infinite, $S \neq \{x_0\}$, so $S - \{x_0\} \neq \emptyset$. Again, since the last set is a nonempty set of nonnegative integers, it has a last member. Let $x_1 = \min(S - \{x_0\})$. Clearly, $x_0 < x_1$ (explain). We have $\{x_0, x_1\} \subseteq S$, and, since S is infinite, $\{x_0, x_1\} \subset S$ and $S - \{x_0, x_1\} \neq \emptyset$. We proceed inductively. Suppose that for some positive integer n, we have $\{x_0, x_1, \ldots, x_n\} \subseteq S$, where $x_n = \min(S - \{x_0, \ldots, x_{n-1}\})$ and $x_0 < x_1 < \cdots < x_n$. Since S is infinite, we know that $S - \{x_0, x_1 \ldots, x_n\} \neq \emptyset$. The last is a nonempty set of nonnegative integers and has a least member. Let $x_{n+1} = \min(S - \{x_0, x_1, \ldots, x_n\})$. It is clear that $x_{n+1} > x_n$. This proves the existence of a sequence $\langle x_n \rangle$ of distinct nonnegative integers which is strictly monotonically increasing. By Prob. 9.4 we know that $x_n \geqslant n$ holds for each positive integer n. Since x_0 is also a nonnegative integer, it follows that $x_0 \geqslant 0$. Thus, $x_n \geqslant n$ for all nonnegative integers n. Let $g: \mathbb{Z}_0 \to S$ be the function defined as $g(n) = x_n$ for each $n \in \mathbb{Z}_0$. g is a one-to-one function (show this). We prove that g maps $\mathbb{Z}_0$ onto S.

$\{x_0, x_1, \ldots, x_n\} \subset S$ for each positive integer n. Assume that $x \in S$. Suppose $x \in S - \{x_0, x_1, \ldots, x_n\}$ for each positive integer n. Since x_{n+1} is the least member of $S - \{x_0, x_1, \ldots, x_n\}$ it would follow that $x \geqslant x_{n+1} \geqslant n + 1 > n$ for each nonnegative integer n. This is impossible since x itself is necessarily a nonnegative integer. Therefore, $x \in \{x_0, x_1, \ldots, x_n\}$ for some nonnegative integer n. Thus, a nonnegative integer k exists such that $k \leqslant n$ and $x = x_k = g(k)$. This implies that x is in the range of g. Thus, each $x \in S$ is in the range of g and g is an onto mapping. Since g is also one-to-one, it is a one-to-one correspondence between $\mathbb{Z}_0$ and S, so that $\mathbb{Z}_0 \simeq S$. This proves that S is denumerable, as claimed.

PROB. 10.6. Prove: Every *infinite* subset of a denumerable set is denumerable.

PROB. 10.7. Prove: Every subset of a denumerable set is countable.

PROB. 10.8. Prove: The union of two sets, one of which is countable and the other denumerable, is denumerable.

PROB. 10.9. Prove: The union of two countable sets is countable.

Theorem 10.3. *Every infinite set contains a denumerable subset.*

PROOF.* Let S be an infinite set. Since $S \neq \emptyset$, it contains some element. Write this element as a_0. We have $\{a_0\} \subsetneq S$. Since S is infinite, we know that $S \neq \{a_0\}$ and hence that $S - \{a_0\} \neq \emptyset$. The last set contains some element. Write this element as a_1. We have $a_1 \in S - \{a_0\}$, so that $a_0 \neq a_1$ and $\{a_0, a_1\} \subset S$. We continue this procedure inductively. If, for some positive integer n, $\{a_0, a_1, \ldots, a_n\} \subsetneq S$, where $a_i \neq a_j$ for $i \neq j$, we know, since S is infinite, that so is $S - \{a_0, a_1, \ldots, a_n\}$. Thus, $S - \{a_0, a_1, \ldots, a_n\} \neq \emptyset$ and contains some element. Write this element as a_{n+1}. We have $a_{n+1} \in S - \{a_0, a_1, \ldots, a_n\}$ so that $a_{n+1} \neq a_k$ for all integers k such that $0 \leqslant k \leqslant n$. This shows the existence of a sequence $\langle a_n \rangle$ of distinct elements of S. Let $g : \mathbb{Z}_0 \rightarrow S$ be defined as $g(n) = a_n$ for all $n \in Z_0$. g is one-to-one (why?) and maps $\mathbb{Z}_0$ onto its range $g(Z_0) = \{a_0, a_1, \ldots\}$ which is the range of the sequence of $\langle a_n \rangle$. Therefore, $\mathbb{Z}_0$ and $g(\mathbb{Z}_0)$ are equipotent. Accordingly, $g(\mathbb{Z}_0)$ is a denumerable subset of S.

11. Rational Exponents

The following theorem will be useful in the sequel.

Theorem 11.1. *A real-valued function f of a real variable which is strictly monotonic has a strictly monotonic inverse f^{-1}. If f is strictly monotonically increasing, so is f^{-1}. If f is strictly monotonically decreasing, so is f^{-1}.*

* This proof uses the *Axiom of Choice*. It would carry us too far afield to discuss this axiom here. The interested reader could consult P. Halmos, *Naive Set Theory*.

Proof. We prove the theorem for the monotonically increasing case leaving the proof of the decreasing case to the reader (Prob. 11.1). Let $f: X \to \mathbb{R}$, where $X \subseteq \mathbb{R}$, be strictly monotonically increasing. This implies that f is one-to-one (Prob. 9.2). Using Remark 10.2, we define the inverse f^{-1} of f as follows: If $y \in \mathfrak{R}(f)$, then $f^{-1}(y)$ is the x such that $y = f(x)$. Take $y_1 < y_2$ in $\mathfrak{R}(f)$. There exist x_1 and x_2 in X such that $y_1 = f(x_1)$ and $y_2 = f(x_2)$, so that $x_1 = f^{-1}(y_1)$ and $x_2 = f^{-1}(y_2)$. We prove $x_1 < x_2$. Suppose that there exist x_1 and x_2 in X such that $x_1 \geqslant x_2$. Since $f\uparrow$, it follows that $y_1 = f(x_1) \geqslant f(x_2) = y_2$. This contradicts $y_1 < y_2$. We see that $x_1 < x_2$ or $f^{-1}(y_1) < f^{-1}(y_2)$ for any y_1 and y_2 in $\mathfrak{R}(f)$, with $y_1 < y_2$. Therefore, f^{-1} is strictly monotonically increasing.

Prob. 11.1. Complete the proof of the last theorem by proving that a function which is strictly monotonically decreasing has a strictly monotonically decreasing inverse.

Prob. 11.2. Prove: For each positive integer n the function $\sqrt[n]{\ }: [0, +\infty) \to [0, +\infty)$ is strictly monotonically increasing.

Prob. 11.3. Let n be an odd positive integer. Prove: the function $\sqrt[n]{\ }: R \to R$ is strictly monotonically increasing.

Def. 11.1. Let $r = m/n$, where m and n are integers and $n > 0$. Define:

(a) $x^r = x^{m/n} = (\sqrt[n]{x}\,)^m$ for $x > 0$.
(b) $0^r = 0$ for $r > 0$.
(c) $x^r = x^{m/n}$ for $x < 0$ and odd n as in (a).

Remark 11.1. If, in the above definition, $n = 1$, then $x^{m/1} = (\sqrt[1]{x}\,)^m = x^m$. If m/n is an integer k so that $m = nk$, we have for $x > 0$

$$x^{m/n} = x^{nk/n} = \left(\sqrt[n]{x}\right)^{nk} = \left(\left(\sqrt[n]{x}\right)^n\right)^k = x^k. \tag{11.1}$$

Also, if n is a positive integer and $x > 0$, then

$$x^{1/n} = \left(\sqrt[n]{x}\right)^1 = \sqrt[n]{x}\,. \tag{11.2}$$

Remark 11.2. From Def. 11.1 parts (a) and (c), we obtain

$$x^0 = x^{0/1} = \left(\sqrt[1]{x}\right)^0 = 1 \qquad \text{if} \quad x \neq 0.$$

This leaves 0^0 undefined. We do not know how this should be defined. If we wish to preserve $x^0 = 1$, then we must define $0^0 = 1$. However, if we wish to preserve $0^r = 0$, for $r > 0$, then we must define $0^0 = 0$. We continue the usage $0^0 = 1$ adopted in Example 3.1.

Note that if $x < 0$, then we do not define $x^{m/n}$, where m and n are integers and n is even.

Prob. 11.4. Prove: If m and n are integers, $n > 0$, then $\sqrt[n]{p^m} = (\sqrt[n]{p})^m$ whenever both sides are defined.

Prob. 11.5. Prove: (a) If $x > 0$ and r is rational, then $x^r > 0$; (b) if m and n are integers, where n is odd and positive, and $x < 0$, then $x^{m/n} > 0$ if m is even, and $x^{m/n} < 0$ if m is odd.

Prob. 11.6. Prove: If $x > 0$, and m, n, p, and q are integers such that $n > 0$, $q > 0$ and $m/n = p/q$, then

$$x^{m/n} = x^{p/q}.$$

Prob. 11.7. Prove: If $x > 0$ and r and s are rational numbers, then

$$\text{(a) } x^r x^s = x^{r+s} \quad \text{and} \quad \text{(b) } (x^r)^s = x^{rs}.$$

Prob. 11.8. Prove: If $x > 0$ and r is rational, then

$$x^{-r} = \frac{1}{x^r}.$$

Prob. 11.9. Assume that $x > 1$ and t is rational. Prove that $x^t > 1$ if $t > 0$, and $x^t < 1$ if $t < 0$.

Theorem 11.2. *If $a > 0$, then the function $f_a : \mathbb{Q} \to \mathbb{R}$, defined as*

$$f_a(r) = a^r \qquad \text{for each} \quad r \in \mathbb{Q}. \tag{11.3}$$

(recall, $\mathbb{Q}$ is the set of rational numbers) is strictly monotonically increasing if $a > 1$ and strictly monotonically decreasing if $0 < a < 1$.

Proof. Consider first the case $a > 1$. Assume that $r_1 < r_2$ for some rational numbers r_1 and r_2. We have $r_2 - r_1 > 0$. By Prob. 11.9,

$$a^{r_2 - r_1} > 1.$$

Multiplying both sides by a^{r_1} gives

$$a^{r_2} = a^{r_1 + (r_2 - r_1)} > a^{r_1}.$$

Thus, $r_1 < r_2$ implies that $a^{r_1} < a^{r_2}$.

If $0 < a < 1$, then $a^{-1} > 1$. Hence, if r_1 and r_2 are rationals such that $r_1 < r_2$, then

$$(a^{-1})^{r_2} > (a^{-1})^{r_1}.$$

This implies that

$$\frac{1}{a^{r_2}} = a^{-r_2} > a^{-r_1} = \frac{1}{a^{r_1}},$$

that is, $a^{r_1} > a^{r_2}$. Thus, $r_1 < r_2$ implies $a^{r_2} < a^{r_1}$ for r_1 and r_2 in $\mathbb{Q}$. This completes the proof.

PROB. 11.10. Prove: If r is rational, then

(a) $r > 0$ and $0 < x_1 < x_2$ imply $x_1^r < x_2^r$,
(b) $r < 0$ and $0 < x_1 < x_2$ imply $x_1^r > x_2^r$.

PROB. 11.11. Prove: If $a > 0$, $b > 0$, and n is a positive integer, then

$$a - b = (a^{1/n} - b^{1/n}) \sum_{k=0}^{n-1} a^{k/n} b^{1-(k+1)/n}.$$

Special cases of this are (a) $a - b = (\sqrt{a} - \sqrt{b})(\sqrt{a} + \sqrt{b})$ and (b) $a - b = (\sqrt[3]{a} - \sqrt[3]{b})(\sqrt[3]{a^2} + \sqrt[3]{a}\sqrt[3]{b} + \sqrt[3]{b^2})$.

12. Some Inequalities

PROB. 12.1. Prove: If $x \geqslant 0$ and $y \geqslant 0$, then

$$\sqrt{xy} \leqslant \frac{x+y}{2}.$$

In this problem $(x + y)/2$ is called the *arithmetic mean* of x and y and $\sqrt{xy}$ their *geometric mean*. The result in Prob. 12.1 is that the geometric mean of nonnegative real numbers does not exceed their arithmetic mean. The *harmonic mean* of two positive real numbers is defined as the reciprocal of the arithmetic mean of their reciprocals. If $H(x, y)$ is the harmonic mean of the positive numbers x and y, then

$$H(x, y) = \frac{1}{\frac{1}{2}(1/x + 1/y)} = \frac{2xy}{x+y}. \tag{12.1}$$

PROB. 12.2. Prove: If $x > 0$ and $y > 0$, then

$$\frac{2xy}{x+y} \leqslant \sqrt{xy}.$$

PROB. 12.3. Prove: If x, y, and z are nonnegative real numbers, then

$$\sqrt[3]{xyz} \leqslant \frac{x+y+z}{3}.$$

Generalizing the terminology following Prob. 12.1, we see that the geometric mean of three nonnegative real numbers does not exceed their arithmetic mean.

Prob. 12.4. Prove: If $x > 0$, $y > 0$, and $z > 0$, then

$$\frac{3xyz}{xy + yz + zx} \leqslant \sqrt[3]{xyz}\,.$$

Prob. 12.5. Prove: If $x \geqslant 0$ and $y \geqslant 0$, then

$$\sqrt{xy} \leqslant \left(\frac{x^{1/2} + y^{1/2}}{2}\right)^2 \leqslant \frac{x + y}{2}\,.$$

The inequalities given in Theorem 12.1 below will be useful in our later work. Before proving this theorem we cite a lemma.

Lemma 12.1.* *If p and q are integers such that $p > q \geqslant 1$, and $x > 0$, $x \neq 1$, then*

$$\frac{x^p - 1}{p} > \frac{x^q - 1}{q}\,.$$

Proof. *Case* 1. $p > 1 = q$. By Remark 5.2, if p is an integer $p \geqslant 2 > 1 = q$, then

$$x^p - 1 > p(x - 1) \quad \text{so that} \quad \frac{x^p - 1}{p} > x - 1 = \frac{x^1 - 1}{1} = \frac{x^q - 1}{q}\,.$$

Case 2. $p > q > 1$. Note that

$$q(x^p - 1) - p(x^q - 1) = qx^q(x^{p-q} - 1) - (p - q)(x^q - 1). \quad (12.2)$$

Since $p - q > 0$ and $q - 1 > 0$ it follows that $q \geqslant 2$. Hence, by Remarks 5.1 and 5.2, we obtain

$$x^{p-q} - 1 \geqslant (p - q)(x - 1) \quad \text{and} \quad x^q - 1 < qx^{q-1}(x - 1).$$

Using these inequalities on the right-hand side of (12.2) gives us for that side

$$\begin{aligned} qx^q(x^{p-q} - 1) - (p - q)(x^q - 1) &> qx^q(p - q)(x - 1) \\ &\quad - (p - q)qx^{q-1}(x - 1) \\ &= q(p - q)(x - 1)(x^q - x^{q-1}) \\ &= q(p - q)(x - 1)^2 x^{q-1} > 0. \end{aligned}$$

This and (12.2) imply that

$$q(x^p - 1) - p(x^p - 1) > 0$$

and the desired conclusion follows in this case also.

* Chrystal, *Algebra*, Vol. 2, pp. 42–43, Dover, New York.

Theorem 12.1.* If r is rational,† $x > 0$, and $x \neq 1$, then

(a) $r > 1$ or $r < 0$ imply

$$r(x-1) < x^r - 1 < rx^{r-1}(x-1). \tag{12.3}$$

(b) $0 < r < 1$ implies

$$r(x-1) > x^r - 1 > rx^{r-1}(x-1). \tag{12.4}$$

PROOF. We consider three cases: (1) $r > 1$, (2) $0 < r < 1$, or (3) $r < 0$.

Case 1. $r > 1$, where $r = p/q$, and p and q are integers such that $p > q \geqslant 1$. It follows that

$$x^{1/q} > 0 \quad \text{and} \quad x^{1/q} \neq 1. \tag{12.5}$$

Apply Lemma 12.1 and obtain:

$$\frac{(x^{1/q})^p - 1}{p} > \frac{(x^{1/q})^q - 1}{q} = \frac{x-1}{q}.$$

Putting $r = p/q$, we can write this inequality as

$$x^r - 1 > r(x-1). \tag{12.6}$$

This proves the leftmost inequality in (a) for $r > 1$.

Now note that $x^{-1} > 0$ and $x^{-1} \neq 1$. Replace x in (12.6) by x^{-1}. Then

$$(x^{-1})^r - 1 > r(x^{-1} - 1)$$

which can be written as $x^{-r} - 1 > rx^{-1}(1-x)$. Multiply both sides of the latter inequality by x^r. Since $x^r > 0$, this yields

$$1 - x^r > rx^{r-1}(1-x). \tag{12.7}$$

Hence,

$$x^r - 1 < rx^{r-1}(x-1).$$

This (12.7) and (12.6) complete the proof of part (a) for the case (1).

Case 2. $0 < r < 1$. Then $r^{-1} > 1$, and we can use the result in case (1) with r^{-1} instead of r. Replace r with r^{-1} in (12.6). Then

$$(x^r)^{1/r} - 1 > \frac{1}{r}(x^r - 1).$$

This yields

$$x^r - 1 < r(x-1) \quad \text{if} \quad 0 < r < 1. \tag{12.8}$$

Now use this inequality with x^{-1} replacing x to obtain

$$(x^{-1})^r - 1 < r(x^{-1} - 1) = rx^{-1}(1-x).$$

* Chrystal, *Algebra*, Vol. 2, pp. 43–44, Dover, New York.

† The theorem holds also if r is real (see Theorem V.7.1).

Multiply both sides here by x^r and obtain

$$1 - x^r < rx^{r-1}(1 - x).$$

But then $x^r - 1 > rx^{r-1}(x - 1)$ for $0 < r < 1$. This and (12.8) complete the proof in case (2).

Case 3. $r < 0$. This implies that $1 - r > 1$. We use the right-hand inequality in (12.3) with $1 - r$ in place of r and obtain

$$x^{1-r} - 1 < (1 - r)x^{-r}(x - 1) = (1 - r)(x^{1-r} - x^{-r}).$$

Now add $x^{-r} - x^{1-r}$ to both sides. Then

$$x^{-r} - 1 < -r(x^{1-r} - x^{-r}).$$

Multiply both sides by x^r. Then

$$1 - x^r < -r(x - 1).$$

But then

$$x^r - 1 > r(x - 1), \tag{12.9}$$

where $r < 0$. Finally replace x by x^{-1} to find, as in case (1), that the rightmost inequality in (a) holds also for $r < 0$. The proof is now complete.

Prob. 12.6. Prove: If a and b are *distinct* positive real numbers and* r is a rational number, then

(a) $r > 1$ or $r < 0$ each imply

$$rb^{r-1}(a - b) < a^r - b^r < ra^{r-1}(a - b)$$

and

(b) $0 < r < 1$ implies

$$rb^{r-1}(a - b) > a^r - b^r > ra^{r-1}(a - b).$$

Prob. 12.7. Prove: If a and b are distinct negative real numbers and $r = m/n$, where m and n are nonzero integers and n is odd, then one of the inequalities in Prob. 12.6 holds: (1) if m is even then (a) holds if $r < 0$ or $r > 1$ and (b) holds if $0 < r < 1$; (2) if m is odd, then (a) holds if $0 < r < 1$, while (b) holds if $r < 0$ or $r > 1$.

Theorem 12.2 (Young's Inequality). *If a and b are positive real numbers*† *and r and s are rational numbers such that $r + s = 1$, then*

(a) $0 < r < 1$ *implies*

$$a^r b^s \leqslant ra + sb \tag{12.10}$$

* The statement also holds if r is real (see Prob. V.7.1).

† The theorem also holds if r and s are real.

and

(b) $r < 0$ *or* $r > 1$ *imply*

$$a^r b^s \geqslant ra + sb. \tag{12.11}$$

The equality in each of (a) *and* (b) *holds if and only if* $a = b$. *If* $0 < r < 1$, *then the hypothesis can be extended to* $a \geqslant 0$ *and* $b \geqslant 0$.

PROOF. Use Prob. 12.6 and the leftmost inequalities in parts (a) and (b) there and add b^r to both sides to obtain

$$rb^{r-1}(a-b) + b^r < a^r \qquad \text{if} \quad \text{(a)} \quad r > 1 \quad \text{or} \quad r < 0 \tag{12.12}$$

and

$$rb^{r-1}(a-b) + b^r > a^r \qquad \text{if} \quad \text{(b)} \quad 0 < r < 1. \tag{12.13}$$

Now multiply each of these relations by b^s and note that $s + r - 1 = 0$. Then

$$r(a-b) + b < a^r b^s \qquad \text{if} \quad r > 1 \quad \text{or} \quad r < 0 \tag{12.14}$$

and

$$r(a-b) + b > a^r b^s \qquad \text{if} \quad \text{(a)} \quad 0 < r < 1. \tag{12.15}$$

Since $r + s = 1$, we see that

$$ra + sb < a^r b^s \qquad \text{if} \quad \text{(a)} \quad r > 0 \quad \text{or} \quad r < 0 \tag{12.16}$$

and

$$ra + sb > a^r b^s \qquad \text{if} \quad \text{(b)} \quad 0 < r < 1. \tag{12.17}$$

Note that $a \neq b$, $a > 0$, and $b > 0$ imply the strict inequalities in (12.10) and (12.11). Consequently, the equality in (12.10) and (12.11) implies that $a = b$. Conversely, if $a = b$, it is easily seen that equality holds in each of (12.10) and (12.11). The last statement in this theorem is obvious.

Theorem 12.3* (Generalization of Part (a) of Theorem 12.2). *If* $x_1, \ldots, x_n$ *are nonnegative real numbers and* $\alpha_1, \alpha_2, \ldots, \alpha_n$ *are positive rationals*† *such that*

$$\sum_{k=1}^{n} \alpha_k = 1, \tag{12.18}$$

then

$$x_1^{\alpha_1} x_2^{\alpha_2} \ldots x_n^{\alpha_n} \leqslant \alpha_1 x_1 + \alpha_2 x_2 + \cdots + \alpha_n x_n. \tag{12.19}$$

The equality in (12.19) *holds if and only if* $x_1 = x_2 = \cdots = x_n$.

* Beckenbach and Bellman, *Inequalities*. "Ergebnisse der Mathematik und Ihrer Grenz gebiete Neue Folge," Band 30, Springer-Verlag, New York, 1965.

† The theorem holds if $\alpha_1, \ldots, \alpha_n$ are all real.

PROOF. We use induction on n for $n \geqslant 2$. By Part (a) of Theorem 12.2, the theorem holds for $n = 2$. Assume it holds for some integer $n \geqslant 2$. Take distinct positive rational numbers $\alpha_1, \ldots, \alpha_n, \alpha_{n+1}$ such that

$$\alpha_1 + \alpha_2 + \cdots + \alpha_n + \alpha_{n+1} = 1. \tag{12.20}$$

Then

$$\frac{\alpha_1}{1-\alpha_{n+1}} + \frac{\alpha_2}{1-\alpha_{n+1}} + \cdots + \frac{\alpha_n}{1-\alpha_{n+1}} = 1. \tag{12.21}$$

The numbers

$$\frac{\alpha_1}{1-\alpha_{\alpha+1}}, \frac{\alpha_2}{1-\alpha_{n+1}}, \ldots, \frac{\alpha_n}{1-\alpha_{n+1}} \tag{12.22}$$

are n positive rational numbers satisfying (12.21). By the induction hypothesis, we, therefore, have for nonnegative real $x_1, \ldots, x_n$

$$x_1^{\alpha_1/(1-\alpha_{n+1})} x_2^{\alpha_2/(1-\alpha_{n+1})} \ldots x_n^{\alpha_n/(1-\alpha_{n+1})} \leqslant \frac{\alpha_1}{1-\alpha_{n+1}} x_1 + \frac{\alpha_2}{1-\alpha_{n+1}} x_2 + \cdots + \frac{\alpha_n}{1-\alpha_{n+1}} x_n, \tag{12.23}$$

where the equality holds if and only if $x_1 = \cdots = x_n$. Next let

$$a = x_1^{\alpha_1/(1-\alpha_{n+1})} x_2^{\alpha_2/(1-\alpha_{n+1})} \ldots x_n^{\alpha_n/(1-\alpha_{n+1})} \quad \text{and} \quad b = x_{n+1}, \tag{12.24}$$

a and b are nonnegative real numbers, $\alpha_{n+1} + (1 - \alpha_{n+1}) = 1$, and $0 < \alpha_{n+1} < 1$. By Young's inequality we obtain

$$a^{1-\alpha_{n+1}} b^{\alpha_{n+1}} \leqslant (1-\alpha_{n+1})a + \alpha_{n+1}b. \tag{12.25}$$

This and (12.24) imply that

$$x_1^{\alpha_1} x_2^{\alpha_2} \ldots x_n^{\alpha_n} x_{n+1}^{\alpha_{n+1}} \leqslant (1-\alpha_{n+1})a + \alpha_{n+1}b. \tag{12.26}$$

But the left-hand side of (12.23) is equal to a (see (12.24)), so (12.23) may be written

$$a \leqslant \frac{\alpha_1 x_1 + \cdots + \alpha_n x_n}{1-\alpha_{n+1}}. \tag{12.27}$$

This and the fact that $b = x_{n+1}$ imply that

$$(1-\alpha_{n+1})a + \alpha_{n+1}b \leqslant \alpha_1 x_1 + \cdots + \alpha_n x_n + \alpha_{n+1} x_{n+1}.$$

This inequality and (12.26) yield

$$x_1^{\alpha_1} x_2^{\alpha_2} \ldots x_{n+1}^{\alpha_{n+1}} \leqslant \alpha_1 x_1 + \alpha_2 x_2 + \cdots + \alpha_{n+1} x_{n+1}. \tag{12.28}$$

It is clear that if $x_1 = x_2 = \cdots = x_{n+1}$, then the equality holds here. Conversely, suppose that two of $x_1, \ldots, x_n, x_{n+1}$ differ. If the two that differ are among $x_1, \ldots, x_n$, then, by the induction hypothesis, (12.23) is a strict inequality. Hence (12.28) is a strict inequality. On the other hand, if $x_1 = \cdots = x_n$, but $x_n \neq x_{n+1}$, then (12.24) implies

$$a = x_n \quad \text{and} \quad b \neq a.$$

This time (12.25) is a strict inequality. This implies that (12.28) is strict. Therefore, if $x_1 = \cdots = x_n = x_{n+1}$ is false, then the strict inequality in (12.28) holds. Hence, if equality holds in (12.28), then $x_1 = \cdots = x_n = x_{n+1}$. Invoking induction, we find that the theorem holds for all n.

Corollary (The Geometric-Arithmetic Inequality). *Let n be some positive integer. If $x_1, x_2, \ldots, x_n$ are nonnegative real numbers then*

$$(x_1 x_2 \ldots x_n)^{1/n} \leqslant \frac{x_1 + \cdots + x_n}{n}. \tag{12.29}$$

PROOF. In the theorem, let $\alpha_1 = \alpha_2 = \cdots = \alpha_n = 1/n$, so that

$$x_1^{\alpha_1} x_2^{\alpha_2} \ldots x_n^{\alpha_n} = x_1^{1/n} x_2^{1/n} \ldots x_n^{1/n} = (x_1 x_2 \ldots x_n)^{1/n}$$

and

$$\alpha_1 x_1 + \alpha_2 x_2 + \cdots + \alpha_n x_n = \frac{1}{n} x_1 + \frac{1}{n} x_2 + \cdots + \frac{1}{n} x_n = \frac{x_1 + \cdots + x_n}{n}.$$

The conclusion is an immediate consequence of the theorem.

Theorem 12.4. *Suppose r is rational, $r^2 \neq r$, and $s = r/(r-1)$, so that*

$$\frac{1}{r} + \frac{1}{s} = 1. \tag{12.30}$$

Let A and B be positive real numbers. Then

(a) $r > 1$ *implies*

$$AB \leqslant \frac{A^r}{r} + \frac{B^s}{s} \tag{12.31}$$

and

(b) $r < 1$ *implies*

$$AB \geqslant \frac{A^r}{r} + \frac{B^s}{s}. \tag{12.32}$$

Equality holds in (12.31) *and* (12.32) *if and only if*

$$B = A^{r-1} \tag{12.33}$$

which is equivalent to $A = B^{s-1}$. If $r > 0$, the hypothesis may be extended to $A \geqslant 0$, $B \geqslant 0$.

PROOF. Let $a = A^r$ and $b = B^s$, so that $A = a^{1/r}$ and $B = b^{1/s}$. If $r > 1$, then $0 < 1/r < 1$. By Theorem 12.2 and (12.30) we obtain

$$AB = a^{1/r} b^{1/s} \leqslant \frac{1}{r} a + \frac{1}{s} b = \frac{A^r}{r} + \frac{B^s}{s}. \tag{12.34}$$

If $r < 1$, since $r^2 \neq r$, we have $r \neq 0$. Therefore, $0 < r < 1$ or $r < 0$. This implies $1/r > 1$ or $1/r < 0$. This time Theorem 12.2 yields:

$$AB = a^{1/r} b^{1/s} \geqslant \frac{1}{r} a + \frac{1}{s} b = \frac{A^r}{r} + \frac{B^s}{s}. \tag{12.35}$$

Equality holds in (12.34) and (12.35) if and only if $a = b$; that is, if and only if $A^r = B^s$, or $A^{1/s} = B^{1/r}$. Since $(1/r) + (1/s) = 1$, this implies that

$$A = A^{1/r}A^{1/s} = A^{1/r}B^{1/r} = (AB)^{1/r} \quad \text{or} \quad A^r = AB$$

and

$$(AB)^{1/s} = A^{1/s}B^{1/s} = B^{1/s}B^{1/r} = B \quad \text{or} \quad AB = B^s.$$

These yield $B = A^{r-1}$ and $A = B^{s-1}$ as necessary and sufficient conditions for the equality to hold in (12.31) and (12.32).

Theorem 12.5 (Hölder's Inequality). *If r is rational* and $a_1, \ldots, a_n$, $b_1, \ldots, b_n$ are real and nonnegative and $r^2 \neq r$ and $s = r/(r-1)$, so that*

$$\frac{1}{r} + \frac{1}{s} = 1, \tag{12.36}$$

then (a) $r > 1$ *implies*

$$\sum_{i=1}^{n} a_i b_i \leqslant \left(\sum_{i=1}^{n} a_i^r\right)^{1/r} \left(\sum_{i=1}^{n} b_i^s\right)^{1/s} \tag{12.37}$$

and (b) $r < 1$ *implies*

$$\sum_{i=1}^{n} a_i b_i \geqslant \left(\sum_{i=1}^{n} a_i^r\right)^{1/r} \left(\sum_{i=1}^{n} b_i^s\right)^{1/s}. \tag{12.38}$$

Equality will hold in (12.37) *and* (12.38) *if and only if*

$$\text{there exists a real } \lambda \geqslant 0 \text{ such that } a_i^r = \lambda b_i^s \text{ for each } i \tag{12.39}$$

or

$$\text{there exists a real } \eta \geqslant 0 \text{ such that } b_i^s = \eta a_i^r. \tag{12.40}$$

Proof. We prove (a) using Theorem 12.4, part (a). The proof of (b) uses part (b) of Theorem 12.4 and we leave it to the reader (Prob. 12.8). We begin by assuming that $r > 1$. If

$$\sum_{i=1}^{n} a_i^r = 0 \quad \text{or} \quad \sum_{i=1}^{M} b_i^s = 0, \tag{12.41}$$

then either $a_i = 0$ for all i or $b_i = 0$ for all i and we have equality in (12.37). We, therefore, assume that

$$\sum_{i=1}^{n} a_i^r > 0 \quad \text{and} \quad \sum_{i=1}^{n} b_i^s > 0. \tag{12.42}$$

Let $u = (\sum_{i=1}^{n} a_i^r)^{1/r}$ and $v = (\sum_{i=1}^{n} b_i^s)^{1/s}$. Using (12.31) we obtain

$$\frac{a_i}{u} \cdot \frac{b_i}{v} \leqslant \frac{1}{r} \frac{a_i^r}{u^r} + \frac{1}{s} \frac{b_i^s}{v^s} \qquad \text{for each } i \text{ with } \quad 1 \leqslant i \leqslant n. \tag{12.43}$$

* The theorem also holds if r is real.

Sum over i. Then

$$\frac{1}{uv}\sum_{i=1}^{n} a_i b_i \leqslant \frac{1}{r}\,\frac{\sum_{i=1}^{n} a_i^r}{u^r} + \frac{1}{s}\,\frac{\sum_{i=1}^{n} b_i^s}{v^s} = \frac{1}{r}\cdot 1 + \frac{1}{s}\cdot 1 = 1.$$

This implies

$$\sum_{i=1}^{n} a_i b_i \leqslant uv = \left(\sum_{i=1}^{n} a_i^r\right)^{1/s}\left(\sum_{i=1}^{n} b_i^s\right)^{1/r}$$

Thus, if $r > 1$, then (12.37) holds.

We examine the conditions under which the equality in (12.37) holds. Suppose that a $\lambda \geqslant 0$ exists such that $a_i^r = \lambda b_i^s$ for all i. This implies that $a_i = \lambda^{1/r} b^{s/r}$ for each i. Hence,

$$\sum_{i=1}^{n} a_i b_i = \sum_{i=1}^{n} \lambda^{1/r} b_i^{s/r} b_i = \lambda^{1/r} \sum_{i-1}^{n} b_i^{(s+r)/r}. \tag{12.44}$$

But (12.36) implies $r + s = rs$. It follows from this and (12.44) that

$$\sum_{i=1}^{n} a_i b_i = \lambda^{1/r} \sum_{i=1}^{n} b_i^s. \tag{12.45}$$

Also,

$$\left(\sum_{i=1}^{n} a_i^r\right)^{1/r}\left(\sum_{i=1}^{n} b_i^s\right)^{1/s} = \left(\sum_{i=1}^{n} \lambda b_i^s\right)^{1/r}\left(\sum_{i=1}^{n} b_0^s\right)^{1/s} = \lambda^{1/r}\left(\sum_{i=1}^{n} b_i^s\right)^{1/r+1/s}$$

$$= \left(\lambda^{1/r} \sum_{i=1}^{n} b_i^s\right). \tag{12.46}$$

Comparing the right-hand side with (12.41), we arrive at

$$\sum_{i=1}^{n} a_i b_i = \left(\sum_{i=1}^{n} a_i^r\right)^{1/r}\left(\sum_{i=1}^{n} b_i^s\right)^{1/s}, \tag{12.47}$$

which is the equality in (12.37). Conversely, assume that we have equality in (12.37), so that (12.47) holds.

If $u^r = \sum_{i=1}^{n} a_i^r = 0$ or $v^s = \sum_{i=1}^{n} b_i^s = 0$, then either $a_i = 0$ for all i or $b_i = 0$ for all i and

$$\text{either } a_i^r = 0 b_i^s \text{ for all } i \text{ or } b_i^s = 0 a_i^r \text{ for all } i. \tag{12.48}$$

If $u > 0$ and $v > 0$, we divide both sides of (12.47) by uv and obtain

$$\sum_{i=1}^{n}\left(\frac{a_i}{u}\right)\left(\frac{b_i}{v}\right) = 1. \tag{12.49}$$

Suppose a k exists such that

$$\frac{a_k^r}{u^r} \neq \frac{b_k^s}{v^s}. \tag{12.50}$$

Write

$$A_i = \frac{a_i}{u} \quad \text{and} \quad B_i = \frac{b_i}{v} \qquad \text{for all } i. \tag{12.51}$$

Note that for the k for which (12.50) holds, we have

$$A_k^r \neq B_k^s \quad \text{and hence} \quad A_k \neq B_k^{s-1} \quad \text{or, equivalently,} \quad B_k \neq A_k^{r-1},$$

so that by Theorem 12.4, part (a),

$$A_k B_k < \frac{A_k^r}{r} + \frac{B_k^s}{s}.$$

But then

$$\sum_{i=1}^{n} A_i B_i < \frac{\sum_{i=1}^n A_k^r}{r} + \frac{\sum_{i=1}^n B_i^s}{s}. \tag{12.52}$$

Since

$$\sum_{i=1}^{n} A_i^r = \frac{\sum_{i=1}^n a_i^r}{u^r} = 1 \quad \text{and} \quad \sum_{i=1}^{n} B_i^s = \frac{\sum_{i=1}^n b_i^s}{v^r} = 1, \tag{12.53}$$

(12.52) implies that

$$\sum_{i=1}^{n} A_i B_i < \frac{1}{r} + \frac{1}{s} = 1.$$

This contradicts (12.49). (In view of (12.51), (12.49) states $\sum_{i=1}^n A_i B_i = 1$.) Thus, if (12.47) holds, then we must have (see (12.50))

$$\frac{a_i^r}{u^r} = \frac{b_i^s}{v^s} \quad \text{and therefore} \quad a_i^r = \lambda b_i^s \qquad \text{for all } i,$$

where $\lambda = u^r/v^s$. Note that necessarily $\lambda > 0$. This proves (12.39).

Prob. 12.8. Complete the proof of the last theorem by proving part (b).

Theorem 12.6. (Cauchy–Schwarz Inequality). *If $a_1, \ldots, a_n; b_1, \ldots, b_n$ are all real numbers (not necessarily nonnegative), then*

$$\left|\sum_{i=1}^{n} a_i b_i\right| \leqslant \left(\sum_{i=1}^{n} a_i^2\right)^{1/2} \left(\sum_{i=1}^{n} b_i^2\right)^{1/2}. \tag{12.54}$$

Here the equality holds if and only if either there exists a real t such that $b_i = ta_i$ for all i or there exists a real s such that $a_i = sb_i$ for all i.

Proof. Use Hölder's inequality with $r = s = 2$ and obtain

$$\left|\sum_{i=1}^{n} a_i b_i\right| \leqslant \sum_{i=1}^{n} |a_i||b_i| \leqslant \left(\sum_{i=1}^{n} |a_i|^2\right)^{1/2} \left(\sum_{i=n}^{n} |b_i|^2\right)^{1/2}$$

$$= \left(\sum_{i=1}^{n} a_i^2\right)^{1/2} \left(\sum_{i=1}^{n} b_i^2\right)^{1/2}.$$

Thus, (12.50) holds.

Now examine the conditions for the equality in (12.54) to hold. If either a real t exists such that $b_i = ta_i$ for all i or there exists a real s such that

$a_i = sb_i$ for all i, then easy calculations prove that the equality holds in (12.54).

Conversely, suppose the equality in (12.54) holds, so that

$$\left|\sum_{i=1}^{n} a_i b_i\right| = \left(\sum_{i=1}^{n} a_i^2\right)^{1/2}\left(\sum_{i=1}^{n} b_i^2\right)^{1/2}. \tag{12.55}$$

If $\sum_{i=1}^{n} a_i^2 = 0$ or $\sum_{i=1}^{n} b_i^2 = 0$, then, as shown earlier, we have $b_i = 0a_i$ for all i or $a_i = 0b_i$ for all i. If $\sum_{i=1}^{n} a_i^2 > 0$ and $\sum_{i=1}^{n} b_i^2 > 0$, put

$$t = \frac{\sum_{i=1}^{n} a_i b_i}{\sum_{i=1}^{n} a_i^2}.$$

We have:

$$\begin{aligned}
\sum_{i=1}^{n} (b_i - ta_i)^2 &= \sum_{i-1}^{n} (b_i^2 - 2ta_i b_i + t^2 a_i^2) \\
&= \sum_{i=1}^{n} b_i^2 - 2t\sum_{i=1}^{n} a_i b_i + t^2 \sum_{i=1}^{n} a_i^2 \\
&= \sum_{i=1}^{n} b_i^2 - 2\frac{(\sum_{i=1}^{n} a_i b_i)^2}{\sum_{i=1}^{n} a_i^2} + \frac{(\sum_{i=1}^{n} a_i b_i)^2}{(\sum_{i=1}^{n} a_i^2)^2}\sum_{i=1}^{n} a_i^2 \\
&= \sum_{i=1}^{n} b_1^2 - \frac{(\sum_{i=1}^{n} a_i b_i)^2}{\sum_{i=1}^{n} a_1^2}.
\end{aligned}$$

In view of (12.55), the expression on the right-hand side is equal to 0. This implies that $\sum_{i=1}^{n}(b_1 - ta_i)^2 = 0$. In turn, this implies $b_i - ta_i = 0$ for all i and, hence, $b_i = ta_i$ for all i.

Theorem 12.7 (Minkowski's Inequality). *If $a_1, \ldots, a_n; b_1, \ldots, b_n$ are non-negative real numbers and r is a rational number,* then*

(a) $r > 1$ *implies*

$$\left(\sum_{i=1}^{n} (a_i + b_i)^r\right)^{1/r} \leqslant \left(\sum_{i=1}^{n} a_1^r\right)^{1/r} + \left(\sum_{i=1}^{n} b_i^r\right)^{1/r} \tag{12.56}$$

and

(b) $r < 1$ *implies*

$$\left(\sum_{n=1}^{n} (a_i + b_i)^r\right)^{1/r} \geqslant \left(\sum_{i=1}^{n} a_i^r\right)^{1/r} + \left(\sum_{i=1}^{n} b_i^r\right)^{1/r}. \tag{12.57}$$

(If $r < 0$, we assume in the hypothesis that $a_i > 0$ and $b_i > 0$ for all i.) The equality holds in each of (12.56) *and* (12.57) *if and only if* (1) *there exists a*

* The theorem also holds if r is real.

$u \geqslant 0$ *such that* $b_i = ua_i$ *for all* i *or* (2) *there exists a* $v \geqslant 0$ *such that* $a_i = vb_i$ *for all* i.

PROOF. We prove (a) and leave the proof of (b) to the reader (Prob. 12.9). Assume $r > 1$ and define

$$s = \frac{r}{r-1} \tag{12.58}$$

which implies $1/s = 1 - 1/r$ or $1/r + 1/s = 1$. Note that

$$r - 1 = \frac{r}{s}. \tag{12.59}$$

For each i, we have

$$\begin{aligned}(a_i + b_i)^r &= (a_i + b_i)(a_i + b_i)^{r-1} = (a_i + b_i)(a_i + b_i)^{r/s} \\ &= a_i(a_i + b_i)^{r/s} + b_i(a_i + b_i)^{r/s}.\end{aligned} \tag{12.60}$$

Using Hölder's Inequality, we have

$$\begin{aligned}\sum_{i=1}^n a_i(a_i + b_i)^{r/s} &\leqslant \left(\sum_{i=1}^n a_i^r\right)^{1/r}\left(\sum_{i=1}^n \left((a_i + b_i)^{r/s}\right)^s\right)^{1/s} \\ &= \left(\sum_{i=1}^n a_i^r\right)^{1/r}\left(\sum_{i=1}^n (a_i + b_i)^r\right)^{1/s}.\end{aligned} \tag{12.61}$$

Similarly,

$$\sum_{i=1}^n b_i(a_i + b_i)^{r/s} \leqslant \left(\sum_{i=1}^n b_i^r\right)^{1/r}\left(\sum_{i=1}^n (a_i + b_i)^r\right)^{1/s}. \tag{12.62}$$

Summing in (12.60) and using (12.61) and (12.62), we obtain

$$\sum_{i=1}^n (a_i + b_i)^r \leqslant \left(\left(\sum_{i=1}^n a_i^r\right)^{1/r} + \left(\sum_{i=1}^n b_i^r\right)^{1/r}\right)\left(\sum_{i=1}^n (a_i + b_i)^r\right)^{1/s}. \tag{12.63}$$

If $(\sum_{i=1}^n (a_i + b_i)^r)^{1/s} = 0$, then $a_1 + b_i = 0$ for all i so that $a_1 = 0 = b_i$ for all i. In this case the equality in (12.56) holds trivially. If $(\sum_{i=1}^n (a_i + b_i)^r)^{1/s} > 0$, we divide both sides in (12.63) by $(\sum_{i=1}^n (a_i + b_i)^r)^{1/s}$ and obtain

$$\left(\sum_{i=1}^n (a_i + b_i)^r\right)^{1-1/s} \leqslant \left(\sum_{i=1}^n a_i^r\right)^{1/r} + \left(\sum_{i=1}^n b_i^r\right)^{1/r}. \tag{12.64}$$

Since $1 - 1/s = 1/r$, (12.56) follows.

We now investigate the conditions under which we have equality in (12.56). If a $u \geqslant 0$ exists such that $b_i = ua$ for all i or a $v \geqslant 0$ exists such that $a_i = vb_i$ for all i, then easy calculations show that we have equality in (12.56). Conversely, assume

$$\left(\sum_{i=1}^n (a_i + b_i)^r\right)^{1/r} = \left(\sum_{i=1}^n a_i^r\right)^{1/r} + \left(\sum_{i=1}^n b_i^r\right)^{1/r}. \tag{12.65}$$

Examine inequalities (12.61) and (12.62). They are consequences of Hölder's Inequality for $r > 1$ and necessarily hold. If the strict inequality in one of (12.61) or (12.62) holds, then the strict inequality in (12.63) would hold. Thus, (12.65) would be false. Consequently, (12.61) and (12.62) are equalities. That is,

$$\sum_{i=1}^{n} a_i(a_i + b_i)^{r/s} = \left(\sum_{i=1}^{n} a_i^r\right)^{1/r}\left(\sum_{i=1}^{n}(a_i + b_i)^r\right)^{1/s} \tag{12.66}$$

and

$$\sum_{i=1}^{n} b_i(a_i + b_i)^{r/s} = \left(\sum_{i=1}^{n} b_i^r\right)^{1/r}\left(\sum_{i=1}^{n}(a_i + b_i)^r\right)^{1/s}. \tag{12.67}$$

Apply the condition for the equality in the Hölder inequality to (12.66). It follows that: either a $\lambda \geqslant 0$ exists such that

$$a_i^r = \lambda\left((a_i + b_i)^{r/s}\right)^s = \lambda(a_i + b_i)^r \qquad \text{for all } i, \tag{12.68}$$

or a $\mu \geqslant 0$ exists such that

$$(a_i + b_i)^r = \mu a_i^r \qquad \text{for all } i. \tag{12.69}$$

If (12.68) holds and $\lambda = 0$, then $a_i = 0$ for all i and hence $a_i = 0b_i$ for all i. If $\lambda > 0$, then (12.68) yields

$$a_i = \lambda^{1/r}(a_i + b_i) \quad \text{and hence} \quad (\lambda^{-1/r} - 1)a_i = b_i .$$

Thus $b_i = \nu a_i$ for all i and some ν. A similar conclusion is arrived at if (12.69) holds. This completes the proof of part (a).

Prob. 12.9. Complete the proof of Theorem 12.7 by proving part (b).

Prob. 12.10. Prove: If $a_1, \ldots, a_n; b_1, \ldots, b_n$ are real numbers (not necessarily nonnegative), then

$$\left(\sum_{i=1}^{n}(a_i + b_i)^2\right)^{1/2} \leqslant \left(\sum_{i=1}^{n} a_i^2\right)^{1/2} + \left(\sum_{i=1}^{n} b_i^2\right)^{1/2},$$

where the equality holds if and only if either a real u exists such that $b_i = ua_i$ for all i or a real v exists such that $a_i = vb_i$ for all i.

Prob. 12.11. Prove: If $a_1, \ldots, a_n$ are positive real numbers, then

$$\left(\sum_{k=1}^{n} a_k\right)\left(\sum_{k=1}^{n} \frac{1}{a_k}\right) \geqslant n^2.$$

Prob. 12.12. Prove: If $a_1, \ldots, a_n$ are nonnegative real numbers and m_n, M_n are defined as

$$m_n = \min\{a_1, \ldots, a_n\}, \qquad M_n = \max\{a_1, \ldots, a_n\},$$

then

$$m_n \leqslant (a_1 a_2 \ldots a_n)^{1/n} \leqslant \frac{a_1 + a_2 + \cdots + a_n}{n} \leqslant M_n .$$

PROB. 12.13. Prove: Let $a_1, \ldots, a_n; b_1, \ldots, b_n$ be positive real numbers. Put

$$m_n = \min\left\{ \frac{a_1}{b_1}, \ldots, \frac{a_n}{b_n} \right\}, \qquad M_n = \max\left\{ \frac{a_1}{b_1}, \ldots, \frac{a_n}{b_n} \right\}.$$

Then

$$\text{(1)} \quad m_n \leqslant \frac{a_1 + a_2 + \cdots + a_n}{b_1 + b_2 + \cdots + b_n} \leqslant M_n \quad \text{and} \quad \text{(2)} \quad m_n^n \leqslant \frac{a_1 a_2 \ldots a_n}{b_1 b_2 \ldots b_n} \leqslant M_n^n .$$

(In (1), $a_1, \ldots, a_n$ need not be positive.)

The results cited in the next two problems are useful in the theory of infinite products.

PROB. 12.14.* If $a_1, a_2, \ldots, a_n$ are nonnegative, then

$$\prod_{k=1}^{n} (1 + a_k) \geqslant 1 + a_1 + a_2 + \cdots + a_n .$$

PROB. 12.15.† Prove: If $0 < a_i < 1$ for $1 \leqslant i \leqslant n$, then

$$\text{(a)} \quad \prod_{k=1}^{n} (1 - a_k) \geqslant 1 - \sum_{k=1}^{n} a_k ,$$

$$\text{(b)} \quad \prod_{k=1}^{n} (1 - a_k) < \frac{1}{\prod_{k=1}^{n}(1 + a_k)} ,$$

$$\text{(c)} \quad \prod_{k=1}^{n} (1 - a_k) < \frac{1}{1 + \sum_{k=1}^{n} a_k} ,$$

$$\text{(d)} \quad \prod_{k=1}^{n} (1 + a_k) < \frac{1}{1 - \sum_{k=1}^{n} a_k} \quad \text{if} \quad \sum_{k=1}^{n} a_k < 1.$$

PROB. 12.16. Prove:

$$\frac{1}{2} \cdot \frac{3}{4} \cdot \frac{5}{6} \cdots \frac{99}{100} < \frac{1}{10} .$$

* T. J. I. Bromwich, *Infinite Series*, Macmillan, New York, 1942.

† *Loc. cit.*

CHAPTER III

Real Sequences and Their Limits

1. Partially and Linearly Ordered Sets

The subset relation $\subseteq$ between sets has the following properties:

(a) If A is a set, then $A \subseteq A$.
(b) If A and B are sets such that $A \subseteq B$ and $B \subseteq A$, then $A = B$.
(c) If A, B, and C are sets such that $A \subseteq B$ and $B \subseteq C$, then $A \subseteq C$.

Note that sets A and B may exist which are not related by $\subseteq$. Thus, it may happen that $A \not\subseteq B$ and $B \not\subseteq A$; Fig. 1.1 makes this clear.

Def. 1.1. If S is a nonempty set of elements and $\leqslant$ a relation between the elements of S such that if a, b, and c are in S, then

(a) $a \leqslant a$,
(b) $a \leqslant b$ and $b \leqslant a$ imply $a = b$,
(c) $a \leqslant b$ and $b \leqslant c$ imply $a \leqslant c$.

We call the system consisting of S together with the relation $\leqslant$ a *partially ordered system*. Such a system will often be written $(S, \leqslant)$. Usually, we call the system a *partially ordered set* and abbreviate by referring to it as a POS. The relation $\leqslant$ is called an *ordering* of S. When the relation $\leqslant$ on S has the additional property

(d) If $a \in S$ and $b \in S$, then $a \leqslant b$ or $b \leqslant a$,

we call the POS a *linearly ordered set* and the ordering $\leqslant$, a *linear ordering* of S. In a linearly ordered set any two elements are related by the ordering.

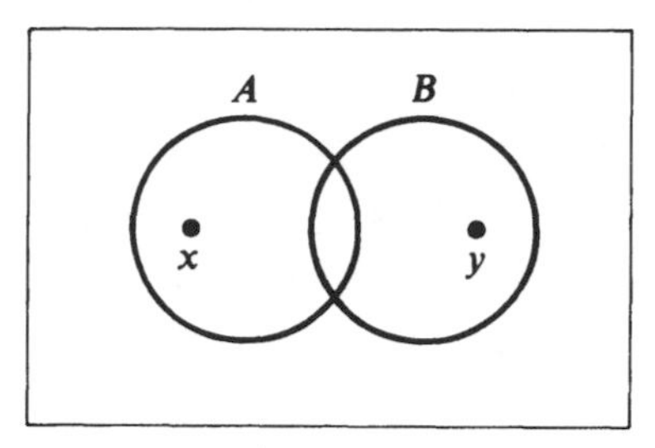

Figure 1.1

Remark 1.1. If $(S, \leqslant)$ is a POS, we define $a \geqslant b$ to mean $b \leqslant a$. It is easy to see that the relation $\geqslant$ is also an ordering of S. The resulting POS $(S, \geqslant)$ is called the *dual* of the original system.

Remark 1.2. The real number system with the relation $\leqslant$ as defined in Def. I.2.2, which we write as $(\mathbb{R}, \leqslant)$, is a linearly ordered system. The relation $\geqslant$ on $\mathbb{R}$ is the dual of $\leqslant$. We call the ordering $\leqslant$ the *natural ordering* of $\mathbb{R}$. The system $(\mathbb{Q}, \leqslant)$ consisting of the set $\mathbb{Q}$ of rationals together with the natural ordering $\leqslant$ of $\mathbb{Q}$ is also a linearly ordered system.

Def. 1.2. In a partially ordered system $(S, \leqslant)$ define $a < b$ to mean $a \leqslant b$ but $a \neq b$.

It is easy to see that $a < a$ is false, and that $a < b \leqslant c$ implies $a < c$.

2. The Extended Real Number System $\mathbb{R}^*$

We extend the real number system as follows: If S is a set of real numbers which is *not* bounded from above, we write

$$\sup S = +\infty \quad \text{or} \quad \sup_{x \in s} x = +\infty. \tag{2.1}$$

Similarly, if S is *not* bounded from below, we write

$$\inf S = -\infty \quad \text{or} \quad \inf_{x \in S} x = -\infty. \tag{2.2}$$

This introduces two symbols, $-\infty$ and $+\infty$, called respectively *minus infinity* and *plus infinity*. The set

$$\mathbb{R}^* = \{-\infty\} \cup \mathbb{R} \cup \{+\infty\} \tag{2.3}$$

is called the set of *extended* real numbers and each of its members is called an *extended real number*. We order $\mathbb{R}^*$ as follows: (a) $-\infty < +\infty$, (b) if $x \in \mathbb{R}$, then $-\infty < x < +\infty$, (c) if x and y are real numbers, they have the same order in $\mathbb{R}^*$ as in $\mathbb{R}$.

The advantage gained by introducing $\mathbb{R}^*$ is that each of the subsets of the latter is bounded from above and from below and thus is bounded. Moreover, if $S \subseteq \mathbb{R}^*$, $S \neq \emptyset$, then

$$z \in S \quad \text{implies} \quad \inf S \leqslant z \leqslant \sup S.$$

Basically, we are interested in sets of real numbers and $\mathbb{R}^*$ is introduced as an aid in the study of these.

Terminology. Although each set in $\mathbb{R}^*$ is bounded, a set of real numbers which is not bounded from above in $\mathbb{R}$ will still be called *not* bounded from above. Similarly, a set of real numbers not bounded from below in $\mathbb{R}$ will still be called not bounded from below (even though it is bounded from below, in $\mathbb{R}^*$). In the same spirit we say that a set of real numbers which is bounded from above or from below or simply bounded, when it is so as a subset of $\mathbb{R}$.

Functions Whose Ranges Are in Linearly Ordered Sets

Let S be a nonempty linearly ordered set. If $f: X \to S$, where $X \neq \emptyset$, then we say that f is *bounded from above* or that f is *bounded from below*, or simply *bounded*, if and only if its range is so. An *upper bound*, or for that matter a *lower bound* of f, is defined as an upper, or respectively a lower bound of the range $\mathfrak{R}(f)$ of f. Similarly, the *supremum* of f is defined as the supremum of $\mathfrak{R}(f)$ and the *infimum* of f as the infimum of $\mathfrak{R}(f)$. If $\mu = \sup f$, then we write

$$\mu = \sup_X f = \sup_{x \in X} f(x), \tag{2.4}$$

and if $\lambda = \inf f$, then we write

$$\lambda = \inf_X f = \inf_{x \in X} f(x). \tag{2.5}$$

If S is also *order-complete* (every nonempty subset of S which is bounded from above has a supremum in S), then, since $X \neq \emptyset$, $\mathfrak{R}(f) \neq \emptyset$, we have: If f is bounded from above, then $\sup f$ exists and is in S and if f is bounded from below, then $\inf f$ exists and is in S. If $f: X \to S$ has $u \in S$ as an upper bound, then

$$x \in X \quad \text{implies} \quad f(x) \leqslant u, \tag{2.6}$$

and, dually, if $f: X \to S$ has l as a lower bound, then

$$x \in X \quad \text{implies} \quad f(x) \geqslant l. \tag{2.7}$$

Remark 2.1. Since sequences (Section II.3) of elements of a set are functions, the terminology just adopted for functions whose ranges are subsets of a linearly ordered set S is applicable to sequences of elements of S. We can, therefore, speak of sequences of elements of S which are bounded from above or from below and of the suprema or infima of such sequences when they exist. Hence, if $\langle a_n \rangle$ is a sequence of elements of a linearly ordered set S, we write

$$\sup_{n \geqslant 1} a_n \quad \text{or} \quad \inf_{n \geqslant 1} a_n$$

for the supremum or the infimum of $\langle a_n \rangle$ when these exist. These are of course respectively the supremum or infimum of the range of $\langle a_n \rangle$.

It should also be clear that just as we spoke of monotonically increasing or monotonically decreasing sequences in $\mathbb{R}$ (see end of Section II.9), so too we can speak of monotonically increasing or monotonically decreasing sequences of elements of a linearly ordered set S. The result cited in Prob. II.9.3 carries over to sequences of elements of a linearly ordered set S.

Prob. 2.1. Let $f: X \to S$ be a function whose codomain S is linearly ordered. Prove: (a) $\mu = \sup f$ if and only if (1) $f(x) \leqslant \mu$ for all $x \in X$ and (2) if $u < \mu$ for some $u \in S$, then there exists $x_0 \in X$ such that $u < f(x_0) \leqslant \mu$. (b) $\lambda = \inf f$ if and only if (1) $f(x) \geqslant \lambda$ for all $x \in X$ and (2) if $l > \lambda$ for some $l \in S$, then there exists an $x_1 \in X$ such that $l > f(x_1) \geqslant \lambda$.

Remark 2.2. When applied to a sequence $\langle x_n \rangle$ of elements of a linearly ordered set S the result in the last problem becomes: (a) $\mu = \sup x_n$ if and only if (1) $x_n \leqslant \mu$ for all $n \in \mathbb{Z}_+$ and (2) if $u < \mu$ for some $u \in S$, there exists an n_0 such that $u < x_{n_0} \leqslant \mu$; (b) $\lambda = \inf x_n$ if and only if (1) $x_n \geqslant \lambda$ for all $n \in \mathbb{Z}_+$ and (2) if $l > \lambda$ for some $l \in S$, then there exists an n_1 such that $l > x_{n_1} \geqslant \lambda$.

Prob. 2.2. Let $\langle a_n \rangle$ be a sequence of elements of a linearly ordered set S. Prove: (a) If $\langle a_n \rangle\uparrow$, then $\langle a_n \rangle$ is bounded if and only if it is bounded from above; (b) $\langle a_n \rangle\downarrow$ if and only if it is bounded from below.

Prob. 2.3. Let $f: X \to \mathbb{R}$ be a real-valued function: Prove: f is bounded if and only if $|f(x)| \leqslant M$ for some real $M > 0$ (Prob. I.13.20).

Def. 2.1 (*Subsequence* of a Sequence). Let $\langle x_n \rangle$ be a sequence of elements of a set Y and $\langle n_k \rangle_{k \geqslant 1}$ be a strictly monotonic increasing sequence of positive integers. Then the sequence $\langle b_k \rangle_{k \geqslant 1}$, where $b_k = x_{n_k}$ for all $k \in \mathbb{Z}_+$, is called a *subsequence* of $\langle x_n \rangle$. This subsequence is written as $\langle x_{n_k} \rangle$.

Thus if $\langle x_n \rangle$ is a sequence, then the sequence

$$\langle x_{n_k} \rangle = \langle x_{n_1}, x_{n_2}, x_{n_3}, \ldots \rangle,$$

where $n_1 < n_2 < n_3 < \ldots$, is a subsequence of $\langle x_n \rangle$. As an example, consider $\langle x_n \rangle = \langle 1, 2, 1, 2, \ldots \rangle$, where $x_n = 1$ if n is odd, and $x_n = 2$ if n is even. Letting $n_k = 2k$ for each $k \in \mathbb{Z}_+$, we obtain $\langle n_k \rangle = \langle 2, 4, 6, \ldots \rangle$. Here, $\langle n_k \rangle$ is a strictly monotonic increasing sequence of positive integers. The sequence $\langle x_{2k} \rangle = \langle 2, 2, 2, \ldots \rangle$ is a subsequence of $\langle 1, 2, 1, 2, \ldots \rangle$.

Remark 2.3. If $\langle x_n \rangle$ is a sequence of elements of a linearly ordered set S and $\mu = \sup x_n$, where $\mu \in S$, then it may happen that there exists a subsequence $\langle x_{n_k} \rangle$ of $\langle x_n \rangle$ which does not have μ as its supremum. For example, let $\langle x_n \rangle = \langle 1, 2, 1, 2, \ldots \rangle$. Then $\langle x_{2k-1} \rangle = \langle 1, 1, 1, \ldots \rangle$, $2 = \sup x_n$, and $1 = \sup x_{2k-1}$. For certain classes of sequences of elements of a linearly ordered set it is true that if $\mu = \sup x_n$, then $\mu = \sup_{k \geqslant 1} x_{n_k}$ holds for all subsequences $\langle x_{n_k} \rangle$ of $\langle x_n \rangle$. Below we prove that this is the case for monotonic sequences.

Theorem 2.1. *If $\langle x_n \rangle$ is a monotonic increasing sequence of elements of a linearly ordered set S and $\mu = \sup_{n \geqslant 1} x_n$, where $\mu \in S$, then for each subsequence $\langle x_{n_k} \rangle$ of $\langle x_n \rangle$ we have $\mu = \sup_{k \geqslant 1} x_{n_k}$.*

Proof. Since each term of $\langle x_{n_k} \rangle$ is a term of $\langle x_n \rangle$, (1) $x_{n_k} \leqslant \mu$ for all $k \in \mathbb{Z}_+$. (2) Assume $u < \mu$ for some $u \in S$. This implies that u is not an upper bound of $\langle x_n \rangle$ and hence that there exists a k_0 such that $u < x_{k_0} \leqslant \mu$. Since $\langle n_k \rangle$ is a strictly monotonic increasing sequence of positive integers, we know (Prob. II.9.4) that $n_{k_0} \geqslant k_0$. Since $\langle x_n \rangle \uparrow$, by hypothesis, it follows

$$u < x_{k_0} \leqslant x_{n_{k_0}} \leqslant \mu \quad \text{so that} \quad u < x_{n_{k_0}} \leqslant \mu.$$

By Remark 2.2, we have $\mu = \sup_{k \geqslant 1} x_{n_k}$.

Prob. 2.4. Prove: If $\langle x_n \rangle$ is a monotonic decreasing sequence of elements of a linearly ordered set S and $\lambda = \inf_{n \geqslant 1} x_n$, where $\lambda \in S$, then for each subsequence $\langle x_{n_k} \rangle$ of $\langle x_n \rangle$ we have $\lambda = \inf_{k \geqslant 1} x_{n_k}$.

Prob. 2.5. Let $\langle x_n \rangle$ be a monotonic increasing sequence of a linearly ordered set S and $\mu = \sup_{n \geqslant 1} x_n$. Then we know $x_n \leqslant \mu$ for all n. Prove: If $\langle x_n \rangle$ is *strictly* monotonically increasing, then $x_n < \mu$ for all n. Dually, let $\langle x_n \rangle$ be a *strictly* monotonically decreasing sequence of elements of a linearly ordered set S and $\lambda = \inf x_n$. Prove that $x_n > \lambda$ for all n.

Remark 2.4. If S is a linearly ordered set where $S \neq \emptyset$, then the range of a sequence of elements of S is not empty. If, in addition, S is order-complete (every nonempty subset of S which is bounded from above has a supremum in S), then, if $\langle x_n \rangle$ is bounded from above, it has a supremum in S. The dual of this is easily formulated and holds also.

Remark 2.5. The sets $\mathbb{R}$ and $\mathbb{R}^*$ are linearly ordered, so all the theorems proved about linearly ordered sets hold in $\mathbb{R}$ and in $\mathbb{R}^*$.

3. Limit Superior and Limit Inferior of Real Sequences

Since every sequence $\langle x_n \rangle$ of elements of $\mathbb{R}^*$ is bounded in the sense that

$$-\infty \leqslant x_n \leqslant +\infty \tag{3.1}$$

holds for all n, we see that each real sequence $\langle x_n \rangle$ ($x_n \in \mathbb{R}$ for each n) has an infimum and supremum in $\mathbb{R}^*$. We have for the kth term x_k of $\langle x_n \rangle$

$$-\infty \leqslant \inf_{n \geqslant 1} x_n \leqslant x_k \leqslant \sup_{n \geqslant 1} x_n \leqslant +\infty. \tag{3.2}$$

If our sequence is bounded from above, then it has a real supremum and we have

$$-\infty < x_k \leqslant \sup_{n \geqslant 1} x_n < +\infty. \tag{3.3}$$

If it is bounded from below, then

$$-\infty < \inf_{n \geqslant 1} x_n \leqslant x_k < +\infty. \tag{3.4}$$

If the sequence is either not bounded from above or not bounded from below, then in the respective cases

$$\sup_{n \geqslant 1} x_n = +\infty \quad \text{and} \quad \inf_{n \geqslant 1} x_n = -\infty. \tag{3.5}$$

As an example, consider the sequence $\langle x_n \rangle$, where $x_n = n$ for all n, so that $\langle x_n \rangle = \langle 1, 2, 3, \ldots \rangle$. Clearly, here

$$\inf_{n \geqslant 1} x_n = 1 \quad \text{and} \quad \sup_{n \geqslant 1} x_n = +\infty.$$

Notation. We adopt the following notation: Given a sequence $\langle x_n \rangle$ of real numbers, we write for each $j \in \mathbb{Z}_+$

$$\underline{A}_j = \inf_{n \geqslant j} x_n \quad \text{and} \quad \overline{A}_j = \sup_{n \geqslant j} x_n . \tag{3.6}$$

For the ranges of $\langle x_n \rangle_{n \geqslant j}$ and $\langle x_n \rangle_{n \geqslant j+1}$, we have

$$\{x_n\}_{n \geqslant j+1} \subseteq \{x_n\}_{n \geqslant j} \qquad \text{for each} \quad j \in \mathbb{Z}_+ . \tag{3.7}$$

These imply (cf. Prob. I.12.1; the result cited in that problem is extendable to $\mathbb{R}^*$) that

$$\inf_{n \geqslant j} x_n \leqslant \inf_{n \geqslant j+1} x_n \leqslant \sup_{n \geqslant j+1} x_n \leqslant \sup_{n \geqslant j} x_n \qquad \text{for each} \quad j \in \mathbb{Z}_+ ,$$

or, in terms of the notation just adopted, that

$$\underline{A}_j \leqslant \underline{A}_{j+1} \leqslant \overline{A}_{j+1} \leqslant \overline{A}_j \qquad \text{for each} \quad j \in \mathbb{Z}_+ . \tag{3.8}$$

These inequalities show that the sequences $\langle \underline{A}_j \rangle_{j \geqslant 1}$ and $\langle \overline{A}_j \rangle_{j \geqslant 1}$ are respectively monotonically increasing and monotonically decreasing sequences of $\mathbb{R}^*$.

We next observe that

$$\underline{A}_j \leqslant \overline{A}_k \qquad \text{for any positive integers } j \text{ and } k. \tag{3.9}$$

To see this, first take $j < k$. Since $\langle \underline{A}_k \rangle \uparrow$, we have

$$\underline{A}_j \leqslant \underline{A}_k \leqslant \overline{A}_k. \tag{3.10}$$

On the other hand, if $j \geqslant k$, then since $\langle \overline{A}_k \rangle \downarrow$, we have

$$\underline{A}_j \leqslant \overline{A}_j \leqslant \overline{A}_k.$$

In either case, (3.9) holds.

In terms of the notation just adopted

$$\underline{A}_1 \leqslant \underline{A}_j \leqslant \overline{A}_k \leqslant \overline{A}_1.$$

Here $\underline{A}_1 = \inf_{n \geqslant 1} x_n$ and $\overline{A}_1 = \sup_{n \geqslant 1} x_n$. We shall drop $n \geqslant 1$ in both of these equalities when there is no danger of confusion and write $\inf x_n$ instead of $\inf_{n \geqslant 1} x_n$ and $\sup x_n$ instead of $\sup_{n \geqslant 1} x_n$. Hence

$$\inf x_n \leqslant \underline{A}_j \leqslant \overline{A}_k \leqslant \sup x_n \quad \text{if } j \text{ and } k \text{ are positive integers.} \tag{3.11}$$

Def. 3.1. If $\langle x_n \rangle$ is a sequence of real numbers and for each k, $\underline{A}_k$ and $\overline{A}_k$ are defined as in (3.6), then we call $\inf \overline{A}_k$ the *limit superior* of $\langle x_n \rangle$ and $\sup \underline{A}_k$ the *limit inferior* of $\langle x_n \rangle$ and write

$$\lim_{n \to +\infty} \inf x_n = \sup \underline{A}_k = \sup_{k \geqslant 1} \left(\inf_{n \geqslant k} x_n \right),$$
$$\lim_{n \to +\infty} \sup x_n = \inf \overline{A}_k = \inf_{k \geqslant 1} \left(\sup_{n \geqslant k} x_n \right) \tag{3.12}$$

for the limit inferior and limit superior of $\langle x_n \rangle$. We also use the notation

$$\underline{\lim}_{n \to +\infty} x_n = \lim_{n \to +\infty} \inf x_n \quad \text{and} \quad \overline{\lim}_{n \to +\infty} x_n = \lim_{n \to +\infty} \sup x_n. \tag{3.13}$$

Sometimes, when there is no danger of confusion, we omit $n \to +\infty$.

Theorem 3.1. *If $\langle x_n \rangle$ is a real sequence, then limits superior and inferior of the sequence are unique extended real numbers and we have the result: If j and k are positive integers, then*

$$\underline{A}_j \leqslant \underline{\lim}_{n \to +\infty} x_n \leqslant \overline{\lim}_{n \to +\infty} x_n \leqslant \overline{A}_k. \tag{3.14}$$

PROOF. For each k, $\inf_{n \geqslant k} x_n$ is a unique extended real number. (That it is an extended real number follows from the properties of $\mathbb{R}^* = \{-\infty\} \cup \mathbb{R} \cup \{+\infty\}$.) Since each sequence of elements of a linearly ordered set has at most one infimum, we conclude that there exists exactly one extended real number $\underline{A}_k$ that is the infimum of $\{x_n\}_{n \geqslant k}$.) Consider next the sequence $\langle \underline{A}_k \rangle$. This is a sequence of extended real numbers and as such its supremum (which is $\underline{\lim}\, x_n$) is a uniquely determined extended real number. Similar reasoning shows that $\overline{\lim}\, x_n$ is a unique extended real number.

Next we prove that the inequalities (3.14) hold. The leftmost inequality is a consequence of $\underline{\lim}\, x_n = \sup \underline{A}_k$, while the rightmost one is a consequence of $\overline{\lim}\, x_n = \inf \overline{A}_k$. It remains to prove that the middle inequality in (3.14) holds. In view of (3.9), each $\underline{A}_j$ is a lower bound (in $\mathbb{R}^*$) of the sequence $\langle \overline{A}_k \rangle$. Hence

$$\underline{A}_j \leqslant \inf \overline{A}_k = \overline{\lim}\, x_n \qquad \text{for each } j.$$

Here $\overline{\lim}\, x_n$ is an upper bound (in $\mathbb{R}^*$) of the sequence $\langle \underline{A}_j \rangle$. Hence,

$$\underline{\lim}\, x_n = \sup \underline{A}_j \leqslant \overline{\lim}\, x_n$$

so that $\underline{\lim}\, x_n \leqslant \overline{\lim}\, x_n$. This completes the proof.

Theorem 3.2. *If $\langle x_n \rangle$ is a real sequence, then*

$$\overline{\lim}\, x_n = +\infty \quad \text{if } \langle x_n \rangle \text{ is not bounded from above} \tag{3.15}$$

and

$$\underline{\lim}\, x_n = -\infty \quad \text{if } \langle x_n \rangle \text{ is not bounded from below.} \tag{3.16}$$

We prove the first part and leave the proof of the second part to the reader (Prob. 3.1).

Proof. Assume that there exists a positive integer k such that $\overline{A}_k < +\infty$. Since $x_m \in \mathbb{R}$ for each m, it follows that

$$-\infty < x_m \leqslant \sup_{n \geqslant k} x_n = \overline{A}_k < +\infty \qquad \text{if} \quad m \geqslant k.$$

Therefore, $\overline{A}_k \in \mathbb{R}$. If $k = 1$, then we have $x_n \leqslant \overline{A}_1$ for all n and, hence, $\langle x_n \rangle$ is bounded from above. Let $k > 1$. Put

$$B = \max\{x_1, \ldots, x_k, \overline{A}_k\}.$$

$\{x_1, \ldots, x_k, \overline{A}_k\}$ is a finite set of real numbers and, hence, has a maximum B (Remark II.10.3(b)(7)). We have $x_n \leqslant B$ for all n (why?). Thus $\langle x_n \rangle$ is bounded from above also in the case $k > 1$. It follows that if $\langle x_n \rangle$ is *not* bounded from above, then $\overline{A}_k = +\infty$ for all k and so $\overline{\lim}\, x_n = \inf \overline{A}_k = +\infty$.

Prob. 3.1. Complete the proof of Theorem 3.2 by proving that: If the real sequence $\langle x_n \rangle$ is not bounded from below, then $\underline{\lim}\, x_n = -\infty$.

Prob. 3.2. Let $\langle x_n \rangle$ be a sequence of real numbers. Prove: (a) If for some positive integer j, $\overline{A}_j = +\infty$, then $\overline{A}_k = +\infty$ for all k and $\langle x_n \rangle$ is not bounded from above; (b) if for some positive integer j, $\underline{A}_j = -\infty$, then $\underline{A}_k = -\infty$ for all k and $\langle x_n \rangle$ is not bounded from below.

Prob. 3.3. Let $\langle x_n \rangle$ be a real sequence. Prove: (a) $\overline{\lim}\, x_n = +\infty$ if and only if $\langle x_n \rangle$ is not bounded from above, and (b) $\underline{\lim}\, x_n = -\infty$ if and only if $\langle x_n \rangle$ is not bounded from below. (See Theorem 3.2 for the "if" part.)

EXAMPLE 3.1. Let $\langle x_n \rangle$ be defined by means of: $x_n = -n$ for each positive integer n. Thus, $\langle x_n \rangle = \langle -1, -2, -3, \ldots \rangle$. If $n \geqslant k$, then we have $-n \leqslant -k$, so that

$$\overline{A}_k = \sup_{n \geqslant k} x_n = \max_{n \geqslant k} x_n = -k.$$

If $B \in \mathbb{R}$, we know that a positive integer k' exists such that $-B < k'$. This implies that

$$\overline{\lim}\, x_n = \inf \overline{A}_k \leqslant \overline{A}_{k'} = -k' < B.$$

Thus $\overline{\lim}\, x_n < B$ for all real B. The only element of $\mathbb{R}^*$ having this property is $-\infty$. Hence, $\overline{\lim}\, x_n = -\infty$.

EXAMPLE 3.2. Let $\langle x_n \rangle$ be given by $x_n = (-1)^{n+1} n$ for each positive integer n. We have

$$\langle x_n \rangle = \langle (-1)^{n+1} n \rangle_{n \geqslant 1} = \langle 1, -2, 3, -4, \ldots \rangle$$

and

$$\begin{aligned} +\infty &= \overline{A}_1 = \sup x_n, & -\infty &= \underline{A}_1 = \inf x_n \\ +\infty &= \overline{A}_2 = \sup_{n \geqslant 2} x_n, & -\infty &= \underline{A}_2 = \inf_{n \geqslant 2} x_n \\ &\vdots & &\vdots \end{aligned}$$

Thus $\overline{A}_k = +\infty$ for all k and $\underline{A}_k = -\infty$ for each k. It follows that

$$\overline{\lim}\, x_n = \inf \overline{A}_k = +\infty, \qquad \underline{\lim}\, x_n = \sup \underline{A}_k = -\infty.$$

4. Limits of Real Sequences

Def. 4.1. We shall say that a real sequence has a *limit in the extended sense* if and only if

$$\underline{\lim}_{n \to +\infty} x_n = \overline{\lim}_{n \to +\infty} x_n. \tag{4.1}$$

When this equality holds and there is no danger of confusion, then we often, simply say that $\langle x_n \rangle$ *has a limit*. The L in $\mathbb{R}^*$ such that $L = \overline{\lim}\, x_n = \underline{\lim}\, x_n$ is called the *limit* of x_n, and we write

$$L = \lim_{n \to +\infty} x_n. \tag{4.2}$$

Other notations used when $\langle x_n \rangle$ has a limit are

$$x_n \to L \quad \text{as} \quad n \to +\infty \qquad \text{and} \qquad x_n \underset{n}{\to} L. \tag{4.3}$$

When the limit of $\langle x_n \rangle$ is a real number, we say that it *converges* or that it is *convergent*. In this case we also say that $\langle x_n \rangle$ has a *finite limit*. A sequence which does not converge is said to *diverge* or to be *divergent*. When $\langle x_n \rangle$ has a limit in the extended sense but diverges, then we say that it has an *infinite limit* and also that it *diverges to that limit*.

Remark 4.1. If a real sequence does not *converge*, then there are three possibilities: (a) $\underline{\lim}\, x_n < \overline{\lim}\, x_n$, (b) $\lim x_n = -\infty$, or (c) $\lim x_n = +\infty$. Thus, a divergent sequence may not have a limit at all or have an infinite limit.

EXAMPLE 4.1. In Example 3.1, we saw that $\overline{\lim}_{n\to+\infty}(-n) = -\infty$. Since (Theorem 3.1)

$$\underline{\lim}_{n\to+\infty}(-n) \leqslant \overline{\lim}_{n\to+\infty}(-n),$$

it follows that $\underline{\lim}(-n) \leqslant -\infty$. Hence, $\underline{\lim}(-n) = -\infty$ and $\underline{\lim}(-n) = \overline{\lim}(-n)$. Thus, $\lim_{n\to+\infty}(-n)$ exists and $\lim_{n\to+\infty}(-n) = \underline{\lim}_{n\to+\infty} = -\infty$.

EXAMPLE 4.2. Let $x_n = (-1)^{n+1}$ for each $n \in \mathbb{Z}_+$, so that $\langle x_n \rangle = \langle 1, -1, 1, \ldots \rangle$. Clearly, for each k,

$$\underline{A}_k = \inf_{n\geqslant k}(-1)^{n+1} = -1 \quad \text{and} \quad \overline{A}_k = \sup_{n\geqslant k}(-1)^{n+1} = 1.$$

Hence,

$$\underline{\lim}_{n\to+\infty}(-1)^{n+1} = -1 \quad \text{and} \quad \overline{\lim}_{n\to+\infty}(-1)^{n+1} = 1.$$

This implies that our sequence diverges. Since $-1 \leqslant (-1)^{n+1} \leqslant 1$ for each n, we know that the sequence is bounded. This furnishes us with an example of a bounded sequence which diverges.

PROB. 4.1. What are $\underline{\lim}_{n\to+\infty}(1+(-1)^{n+1})$ and $\overline{\lim}_{n\to+\infty}(1+(-1)^{n+1})$?

Theorem 4.1. *If a real sequence has a limit, then this limit is a unique extended real number.*

PROOF. The theorem follows directly from the definition of limit and Theorem 3.1.

Theorem 4.2. *If* $\langle x_n \rangle$ *is a real sequence, then*

(a) $-\infty \leqslant \overline{\lim}\, x_n < +\infty$ *if and only if* $\langle x_n \rangle$ *is bounded from above.*
(b) $-\infty < \underline{\lim}\, x_n \leqslant +\infty$ *if and only if* $\langle x_n \rangle$ *is bounded from below.*

PROOF. The leftmost inequality in (a) and the rightmost inequality in (b) always hold for a real sequence. By Prob. 3.3, the rightmost inequality in (a) holds if and only if $\langle x_n \rangle$ is bounded from above and the leftmost inequality in (b) holds if and only if $\langle x_n \rangle$ is bounded from below.

Theorem 4.3. *If a real sequence converges, then it is bounded.*

PROOF. This is an immediate consequence of Theorem 4.2 and the definition of a convergent sequence. In fact, the limit and hence the limits inferior and superior of a convergent sequence, are in $\mathbb{R}$ and are therefore neither $+\infty$ nor $-\infty$.

Remark 4.2. In view of Example 4.2, the converse of Theorem 4.3 does not hold. The sequence $\langle x_n \rangle = \langle (-1)^{n+1} \rangle$ is bounded but does not converge.

PROB. 4.2. Prove: A sequence $\langle x_n \rangle$ of real numbers is bounded if and only if there exists an $M > 0$ such that $|x_n| \leqslant M$ holds for all $n \in \mathbb{Z}_+$ (Hint: see Prob. I.13.20). Using this criterion for the boundedness of a real sequence we see that $\langle x_n \rangle = \langle (-1)^{n+1} \rangle$ is bounded. In fact, $|x_n| = |(-1)^{n+1}| = 1 \leqslant 1$ holds for all $n \in \mathbb{Z}_+$.

Theorem 4.4. *(a) A monotonic sequence of real numbers has a limit (possibly an infinite one). (b) A bounded monotonic sequence converges. (c) If $\langle x_n \rangle$ is monotonically increasing, then*

$$\lim_{n \to +\infty} x_n = \sup x_n ,$$

while if $\langle x_n \rangle$ is monotonically decreasing, then

$$\lim_{n \to \infty} x_n = \inf x_n .$$

PROOF. We prove the theorem for the monotonically increasing case, leaving the decreasing case to the reader (Prob. 4.3). Let $\langle x_n \rangle$ be a monotonically increasing sequence of real numbers. Let $\mu = \sup x_n$. For each positive integer k, $\langle x_n \rangle_{n \geqslant k}$ is a subsequence of $\langle x_n \rangle$ (explain). By Theorem 2.1,

$$\mu = \sup_{n \geqslant k} x_n = \overline{A}_k \qquad \text{for each } k.$$

This implies that

$$\overline{\lim_{n \to +\infty}} x_n = \inf \overline{A}_k = \mu. \tag{4.4}$$

(The sequence $\langle \overline{A}_k \rangle_{k \geqslant 1}$ is a constant sequence, i.e., $\mu = \overline{A}_k$ for all k. Hence, its range is the singleton set $\{\mu\}$. But then, $\inf \overline{A}_k = \inf\{\mu\} = \mu$.) Now $n \geqslant k$ implies that $x_n \geqslant x_k$ and x_k is an element of the range of $\langle x_n \rangle_{n \geqslant k}$. This implies

$$x_k = \min\{x_n\}_{n \geqslant k} = \inf_{n \geqslant k} x_n = \underline{A}_k \qquad \text{for each} \quad k \in \mathbb{Z}_+ . \tag{4.5}$$

It follows that

$$\underline{\lim}_{n \to +\infty} x_n = \sup \underline{A}_k = \sup_{k \geqslant 1} x_k = \sup x_n = \mu. \tag{4.6}$$

This result and (4.4) imply that $\lim_{n\to+\infty}x_n$ exists and is an element of $\mathbb{R}^*$, and

$$\lim x_n = \mu = \sup x_n . \tag{4.7}$$

If, in addition, $\langle x_n\rangle$ is bounded from above, then it has a real supremum, i.e., the μ here is real. By (4.7), $\langle x_n\rangle$ converges.

PROB. 4.3. Complete the proof of the last theorem by proving the results stated there for the case of a monotonically decreasing sequence.

Corollary (of Theorem 4.4). *If $\langle x_n\rangle$ is a monotonic sequence of real numbers which is not bounded, then*

$$\lim x_n = +\infty \quad or \quad \lim x_n = -\infty$$

according as to whether $\langle x_n\rangle$ is increasing or decreasing.

PROOF. In the increasing case, if $\langle x_n\rangle$ is not bounded, then it is not bounded from above (Prob. 2.2). But then (Theorem 3.2) $\overline{\lim}\, x_n = +\infty$. By Theorem 4.4,

$$\lim x_n = \overline{\lim}\, x_n = +\infty.$$

The proof for the case where $\langle x_n\rangle$ is decreasing is similar (with the appropriate modifications of course).

PROB. 4.4. Prove: (a) If $\langle x_n\rangle\uparrow$ and there exists $M \in \mathbb{R}^*$ such that $x_n \leqslant M$ for all n, then $\lim x_n \leqslant M$; (b) if $\langle x_n\rangle\downarrow$, and there exists an $m \in \mathbb{R}^*$ such that $x_n \geqslant m$ for all n, then $\lim x_n \geqslant m$.

PROB. 4.5. Prove: (a) If $\langle x_n\rangle\uparrow$, then for each x_k, $x_k \leqslant \lim x_n$. If, moreover, $\langle x_n\rangle\uparrow$ strictly, then $x_k < \lim x_n$ for each k. (b) If $\langle x_n\rangle\downarrow$, then we have for each term x_k, $x_k \geqslant \lim x_n$. If moreover $\langle x_n\rangle\downarrow$ strictly, then $x_k > \lim x_n$.

PROB. 4.6(a). Prove: If $\langle x_n\rangle$ is a constant sequence, i.e., if $x_n = c$ for all n for some $c \in \mathbb{R}$, then $\langle x_n\rangle$ converges and $c = x_n \underset{n}{\to} c$.

Remark 4.3. The notions of limit superior can be extended to any linearly ordered set, in particular in $\mathbb{R}^*$. Consequently, if $\langle x_n\rangle$ is a real sequence, then its associated sequence $\langle \underline{A}_k\rangle$ is a monotonically increasing sequence of extended real numbers, while the sequence $\langle \overline{A}_k\rangle$ is a monotonically decreasing set of extended real numbers. Therefore,

$$\lim_{k\to+\infty} \underline{A}_k = \sup \underline{A}_k \quad \text{and} \quad \lim \overline{A}_k = \inf \overline{A}_k .$$

Hence,

$$\underline{\lim}_{n\to+\infty}\, x_n = \sup \underline{A}_k = \lim_{k\to+\infty} \underline{A}_k = \lim_{k\to+\infty}\left(\inf_{n\geqslant k} x_n\right) \tag{4.8}$$

and

$$\overline{\lim_{n\to+\infty}}\, x_n = \inf \bar{A}_k = \lim_{k\to+\infty} \bar{A}_k = \lim_{k\to+\infty}\left(\sup_{n\geqslant k} x_k\right).$$

EXAMPLE 4.3. We examine $\langle x_n\rangle$, where $x_n = n^{-k}$ for each $n \in \mathbb{Z}_+$ and k is a fixed positive integer. Since

$$0 < \frac{1}{(n+1)^k} < \frac{1}{n^k} \qquad \text{for each} \quad n \in \mathbb{Z}_+$$

we see that $\langle x_n\rangle$ is strictly monotonically decreasing and bounded from below by 0. Therefore, it converges (Theorem 4.4) and

$$\lim_{n\to+\infty} x_n = \lim_{n\to+\infty} \frac{1}{n^k} = \inf_{n\geqslant 1} \frac{1}{n^k} \geqslant 0.$$

Given $\epsilon > 0$, there exists a positive integer n_0 such that $n_0^{-1} < \epsilon$, so

$$n_0^{-k} \leqslant n_0^{-1} < \epsilon.$$

Thus, no $\epsilon > 0$ is a lower bound for $\langle x_n\rangle$. Accordingly $\inf n^{-k} > 0$ is impossible and we conclude that $\inf n^{-k} = 0$. Therefore,

$$\lim_{n\to+\infty} n^{-k} = \inf_{n\geqslant 1} n^{-k} = 0. \tag{4.9}$$

PROB. 4.6(b). Prove:

$$\lim_{n\to+\infty}\left(1 - \frac{1}{n}\right) = 1 = \lim_{n\to+\infty}\left(1 + \frac{1}{n}\right).$$

Theorem 4.5. *If $\langle x_{n_k}\rangle$ is a subsequence of the real sequence $\langle x_n\rangle$ (here $\langle x_n\rangle$ is not necessarily monotonic), then*

$$\underline{\lim}_{n\to+\infty}\, x_n \leqslant \underline{\lim}_{x\to+\infty}\, x_{n_k} \leqslant \overline{\lim_{k\to+\infty}}\, x_{n_k} \leqslant \overline{\lim_{n\to+\infty}}\, x_n\,. \tag{4.10}$$

It follows that if $\lim x_n$ *exists, then*

$$\lim_{n\to+\infty} x_n = \lim_{k\to+\infty} x_{n_k}\,. \tag{4.11}$$

This last result states that if $\langle x_n\rangle$ has a limit, then all its subsequences have the same limit.

PROOF. Since $\langle x_{n_k}\rangle$ is a subsequence of $\langle x_n\rangle$, we have $n_k \geqslant k$. For the respective ranges

$$\{x_{n_j}\}_{j\geqslant k} \subseteq \{x_n\}_{n\geqslant k} \qquad \text{for each } k. \tag{4.12}$$

(Here we reindexed $\langle x_{n_k}\rangle$ to avoid confusion.) Put

$$\bar{B}_k = \sup_{j\geqslant k} x_{n_j}\,, \qquad \underline{B}_k = \inf_{j\geqslant k} x_{n_j} \tag{4.13}$$

and

$$\overline{A}_k = \sup_{n \geqslant k} x_n, \qquad \underline{A}_k = \inf_{n \geqslant k} x_n. \tag{4.14}$$

By properties of infima and suprema of sets, we obtain from (4.12), (4.13), and (4.14),

$$\underline{A}_k \leqslant \underline{B}_k \leqslant \sup \underline{B}_k = \underline{\lim}_{k \to +\infty} x_{n_k} \tag{4.15}$$

and

$$\overline{A}_k \geqslant \overline{B}_k \geqslant \inf \overline{B}_k = \overline{\lim}_{k \to +\infty} x_{n_k}. \tag{4.16}$$

By (4.15) we obtain

$$\underline{\lim}_{n \to +\infty} x_n = \sup \underline{A}_k \leqslant \underline{\lim}_{k \to +\infty} x_{n_k} \tag{4.17}$$

and by (4.16) we obtain

$$\overline{\lim}_{n \to +\infty} x_n = \inf \overline{A}_k \geqslant \overline{\lim}_{k \to +\infty} x_{n_k}. \tag{4.18}$$

Since we also have

$$\underline{\lim}_{k \to +\infty} x_{n_k} \leqslant \overline{\lim}_{k \to +\infty} x_{n_k}, \tag{4.19}$$

we see that inequalities (4.10) follow from ((4.19)), (4.17), and (4.18).

Finally, if $\langle x_n \rangle$ has a limit, then $\underline{\lim}\, x_n = \overline{\lim}\, x_n = \lim x_n$. This and (4.10) imply that equality holds in (4.19) and that (4.11) holds. The proof is now complete.

Theorem 4.6. *If $\langle x_n \rangle$ and $\langle y_n \rangle$ are real sequences such that $x_n \leqslant y_n$ for all $n \in \mathbb{Z}_+$, then*

$$\underline{\lim}\, x_n \leqslant \underline{\lim}\, y_n \tag{4.20a}$$

and

$$\overline{\lim}\, x_n \leqslant \overline{\lim}\, y_n. \tag{4.20b}$$

Proof. We prove (b) and leave the proof of (a) to the reader (Prob. 4.7). Put

$$\overline{A}_k = \sup_{j \geqslant k} x_j \quad \text{and} \quad \overline{B}_k = \sup_{j \geqslant k} y_j \qquad \text{for each positive integer } k.$$

We have

$$x_n \leqslant y_n \leqslant \sup_{j \geqslant k} y_j = \overline{B}_k \qquad \text{for} \quad n \geqslant k.$$

Thus, for each k, $\overline{B}_k$ is an upper bound for the sequence $\langle x_n \rangle_{n \geqslant k}$ and,

therefore,

$$\bar{A}_k = \sup_{n \geqslant k} x_n \leqslant \bar{B}_k \qquad \text{for each } k.$$

This implies that

$$\overline{\lim}\ x_n \leqslant \inf \bar{A}_k \leqslant \bar{A}_k \leqslant \bar{B}_k \qquad \text{for each } k.$$

Thus, $\overline{\lim}\, x_n$ is a lower bound for the sequence $\langle \bar{B}_k \rangle$. This implies that

$$\overline{\lim}\ x_n \leqslant \inf \bar{B}_k = \overline{\lim}\ y_n\,;$$

this proves (4.20b).

PROB. 4.7. Complete the proof of Theorem 4.6 by proving (4.20a).

Corollary 1 (of Theorem 4.6). *If $\langle x_n \rangle$ and $\langle y_n \rangle$ are real sequences such that $x_n \leqslant y_n$ for all n and if* $\lim x_n$ *and* $\lim y_n$ *exist, then* $\lim x_n \leqslant \lim y_n$.

PROOF. This is an immediate consequence of the definition of limit and the theorem.

Corollary 2 (of Theorem 4.6). *If $b \in \mathbb{R}^*$ and $\langle x_n \rangle$ is a real sequence such that $x_n \leqslant b$ holds for all n, then $\overline{\lim}\, x_n \leqslant b$. If $x_n \geqslant b$ for all n, then $\underline{\lim}\, x_n \geqslant b$.*

PROOF. If $b = \pm\infty$, then $b = +\infty$ in the first case and $b = -\infty$ in the second and the conclusions are obvious. Assume $b \in \mathbb{R}$ and let $\langle y_n \rangle$ be the constant sequence where $y_n = b$ for all n. Since $\lim_{n\to+\infty} y_n = \lim_{n\to+\infty} b = b$, the theorem yields, in the first case,

$$\overline{\lim}\ x_n \leqslant \overline{\lim}\ y_n = \lim y_n = b$$

and in the second,

$$\underline{\lim}\, x_n \geqslant \underline{\lim}\ y_n = \lim y_n = b.$$

Remark 4.4. In Theorem 4.6, the strict inequality $x_n < y_n$ for all n does not warrant the strict $\overline{\lim}\, x_n < \overline{\lim}\, y_n$ or $\underline{\lim}\, x_n < \underline{\lim}\, y_n$ in the conclusion. For example, let $x_n = 1/(n+1)$ and $y_n = 1/n$ for all n. Since $\langle x_n \rangle$ is a subsequence of $\langle y_n \rangle$ here, we have

$$\lim \frac{1}{n+1} = \lim x_n = \lim \frac{1}{n} = \lim \frac{1}{n} = 0.$$

This observation is also relevant to Corollary 2 above. We have $0 < 1/n$ for all $n \in \mathbb{Z}_+$, and yet $\underline{\lim}(1/n) = \overline{\lim}(1/n) = \lim(1/n) = 0$.

Corollary 3 (of Theorem 4.6). *Let $\langle x_n \rangle$, $\langle y_n \rangle$, and $\langle z_n \rangle$ be real sequences such that $x_n \leqslant z_n \leqslant y_n$. Then*

(a) $\bar{L} = \overline{\lim}\, x_n = \overline{\lim}\ y_n$ *implies* $\bar{L} = \overline{\lim}\, z_n$ *and*
(b) $\underline{L} = \underline{\lim}\, x_n = \underline{\lim}\ y_n$ *implies* $\underline{L} = \underline{\lim}\, z_n$.

PROOF. Exercise.

Corollary 4 (of Theorem 4.6). *If $\langle x_n \rangle$, $\langle y_n \rangle$, and $\langle z_n \rangle$ are real sequences such that $x_n \leqslant z_n \leqslant y_n$ and $L = \lim x_n = \lim y_n$, then $L = \lim z_n$.*

PROOF. Exercise.

Remark 4.5. This last corollary is known as the *Sandwich Theorem.*

EXAMPLE 4.4. We prove that

$$\lim_{n\to+\infty} \frac{1}{\sqrt{1+1/n}} = 1. \tag{4.21}$$

This follows from the fact that

$$1 < \sqrt{1+\frac{1}{n}} < 1+\frac{1}{n}, \qquad \text{for each} \quad n \in \mathbb{Z}_+$$

so that

$$1-\frac{1}{n} < 1-\frac{1}{n+1} = \frac{n}{n+1} < \frac{1}{\sqrt{1+1/n}} < 1$$

and hence

$$1-\frac{1}{n} < \frac{1}{\sqrt{1+1/n}} < 1 \qquad \text{for each} \quad n \in \mathbb{Z}_+ . \tag{4.22}$$

Since (Prob. 4.6) $\lim_{n\to+\infty}(1-1/n) = 1$, we obtain from (4.22) and the sandwich theorem that (4.21) holds.

PROB. 4.8. Prove: (a) $\lim \sqrt{1+1/n} = 1$, (b) $\lim_{n\to+\infty} \sqrt[n]{1+1/n} = 1$. (c) Prove: $\lim_{n\to+\infty}(1/\sqrt{1+(1/n^2)}\,) = 1$.

PROB. 4.9. Prove

(a) $\displaystyle\lim_{n\to+\infty} \sum_{k=1}^{n} \frac{1}{n^2+k} = 0;$

(b) $\displaystyle\lim_{n\to+\infty} \sum_{k=1}^{n} \frac{1}{\sqrt{n^2+k}} = 1.$

5. The Real Number e

We shall use Theorem 4.4 to prove that the sequence $\langle x_n \rangle$,

$$x_n = \left(1+\frac{1}{n}\right)^n \qquad \text{for each} \quad n \in \mathbb{Z}_+ \tag{5.1}$$

is convergent.

Prob. II.5.3(b) implies that the sequence (5.1) is strictly monotonic increasing. Next we prove that it is bounded. By the Binomial Theorem

$$\begin{aligned} x_n = \left(1+\frac{1}{n}\right)^n &= \sum_{k=0}^{n}\binom{n}{k}\left(\frac{1}{n}\right)^k (1)^{n-k} \\ &= 1 + \sum_{k=1}^{n}\binom{n}{k}\frac{1}{n^k} \\ &= 1 + \sum_{k=1}^{n}\frac{n(n-1)\cdots(n-k+1)}{k!\,n^k} \\ &= 1 + \sum_{k=1}^{n}\frac{1}{k!}\left(1-\frac{1}{n}\right)\left(1-\frac{2}{n}\right)\cdots\left(1-\frac{k-1}{n}\right). \end{aligned} \tag{5.2}$$

Since for $1 \leqslant k \leqslant n$, $k \in \mathbb{Z}_+$, we have

$$\left(1-\frac{1}{n}\right)\left(1-\frac{2}{n}\right)\cdots\left(1-\frac{k-1}{n}\right) \leqslant 1,$$

it follows that

$$\frac{1}{k!}\left(1-\frac{1}{n}\right)\left(1-\frac{2}{n}\right)\cdots\left(1-\frac{k-1}{n}\right) \leqslant \frac{1}{k!}\,. \tag{5.3}$$

This and (5.2) imply

$$x_n = \left(1+\frac{1}{n}\right)^n \leqslant 1 + \sum_{k=1}^{n}\frac{1}{k!}\,. \tag{5.4}$$

The reader could prove that

$$\frac{1}{k!} \leqslant \frac{1}{2^{k-1}} \qquad \text{if} \quad k \in \mathbb{Z}_+\,.$$

This and (5.4) imply

$$x_n \leqslant 1 + \sum_{k=1}^{n}\frac{1}{k!} \leqslant 1 + \sum_{k=1}^{n}\frac{1}{2^{k-1}}\,. \tag{5.5}$$

But

$$\sum_{k=1}^{n}\frac{1}{2^{k-1}} = \sum_{k=1}^{n}\left(\frac{1}{2}\right)^{k-1} = \frac{1-\left(\frac{1}{2}\right)^n}{1-\frac{1}{2}} = 2\left(1-\left(\tfrac{1}{2}\right)^n\right) < 2.$$

From this and (5.5) it follows

$$x_n \leqslant 1 + \sum_{k=1}^{n}\frac{1}{2^{k-1}} < 1 + 2 = 3 \qquad \text{for each} \quad n \in \mathbb{Z}_+\,. \tag{5.6}$$

Thus, $\langle x_n \rangle$ is bounded from above by 3. This together with the fact that $\langle x_n \rangle$ is monotonic increasing implies, by Theorem 4.4, that $\langle x_n \rangle$ converges. We define

$$e = \lim_{n\to+\infty}\left(1+\frac{1}{n}\right)^n. \tag{5.7}$$

This definition and Theorem 4.4 imply that

$$e = \sup_{n \geqslant 1} \left(1 + \frac{1}{n}\right)^n. \tag{5.8}$$

Since 3 is an upper bound of $\langle x_n \rangle$, it follows that

$$\left(1 + \frac{1}{n}\right)^n \leqslant e \leqslant 3 \qquad \text{for each} \quad n \in \mathbb{Z}_+ .$$

Since $\langle x_n \rangle$ is strictly increasing, this result and the result cited in Prob. 2.5 imply that

$$\left(1 + \frac{1}{n}\right)^n < e \leqslant 3 \qquad \text{for each} \quad n \in \mathbb{Z}_+ . \tag{5.9}$$

Hence

$$2 < e \leqslant 3. \tag{5.10}$$

PROB. 5.1. Prove: $n \in \mathbb{Z}_+$ and $n \geqslant 3$, then

$$n^{n+1} > (n+1)^n.$$

PROB. 5.2. (a) Prove: If n is a positive integer, then

$$\left(1 + \frac{1}{n(n+2)}\right)^{n+2} > 1 + \frac{1}{n} .$$

(b) Let $\langle y_n \rangle$ be the sequence defined by $y_n = (1 + 1/n)^{n+1}$ for each $n \in \mathbb{Z}_+$. Prove: $\langle y_n \rangle$ is strictly monotonically *decreasing*. (c) Prove $\langle y_n \rangle$ converges.

PROB. 5.3. Prove: The sequence $\langle z_n \rangle$, where $z_n = (1 - 1/n)^n$ for $n \in \mathbb{Z}_+$, is strictly monotonically increasing.

PROB. 5.4. Prove: (a)

$$e \leqslant \lim_{n \to +\infty} \left(1 + \frac{1}{n}\right)^{n+1}.$$

(Actually, this inequality can be replaced by an equality, but we cannot prove this yet.) (See Prob. 8.9.) (b) $e < (1 + 1/n)^{n+1}$ for each $n \in \mathbb{Z}_+$.

Remark 5.1. From Prob. 5.4 and formula (5.9), we have

$$\left(1 + \frac{1}{k}\right)^k < e < \left(1 + \frac{1}{k}\right)^{k+1} \qquad \text{for each} \quad k \in \mathbb{Z}_+ . \tag{5.11}$$

This implies first of all, using $k = 5$ on the right, that $e < (1 + \frac{1}{5})^6 = 2.985984 < 3$. Thus, (5.10) may be strengthened to

$$2 < e < 3. \tag{5.12}$$

We now use $k \in \{1,2,\ldots,n-1\}$, $n > 1$, and obtain

$$2 = \left(1+\frac{1}{1}\right)^1 < e < \left(1+\frac{1}{1}\right)^2 = 2^2,$$

$$\left(1+\frac{1}{2}\right)^2 < e < \left(1+\frac{1}{2}\right)^3,$$

$$\vdots$$

$$\left(1+\frac{1}{n-1}\right)^{n-1} < e < \left(1+\frac{1}{n-1}\right)^n.$$

Multiplying these together gives us

$$2 \cdot \frac{3^2}{2^2} \cdots \frac{n^{n-1}}{(n-1)^{n-1}} < e^{n-1} < 2^2 \cdot \frac{3^3}{2^3} \cdots \frac{n^n}{(n-1)^n}.$$

This implies, after cancellation that

$$\frac{1}{2} \cdot \frac{1}{3} \cdots \frac{n^{n-1}}{n-1} < e^{n-1} < \frac{1}{2} \cdot \frac{1}{3} \cdots \frac{n^n}{n-1} \qquad \text{if} \quad n > 1$$

and therefore that

$$\frac{n^{n-1}}{(n-1)!} < e^{n-1} < \frac{n^n}{(n-1)!} \quad \text{or that} \quad \frac{n^n}{n!} < e^{n-1} < \frac{n^{n+1}}{n!}.$$

We conclude from the second set of inequalities above that

$$e(n^n e^{-n}) < n! < (en)(n^n e^{-n}) \qquad \text{for} \quad n > 1. \tag{5.13}$$

6. Criteria for Numbers To Be Limits Superior or Inferior of Real Sequences

Theorem 6.1. *Let $\bar{L} = \overline{\lim}\, x_n$. The following results hold:*

(a) *If $\bar{L} < +\infty$, then for each B such that $\bar{L} < B$ there exists a positive integer N such that $n \geqslant N$ implies $x_n < B$;*
(b) *If $L \in \mathbb{R}^*$, $L < +\infty$ and if for each real B such that $L < B$, there exists a positive integer N such that $n \geqslant N$ implies $x_n \leqslant B$, then $\bar{L} \leqslant L$;*
(c) *If $-\infty < \bar{L}$ and L is an extended real number such that $-\infty < L$, then $L \leqslant \bar{L}$ holds if and only if for each real B with $B < L$, we have that $x_n > B$ holds for infinitely many n's.*

Proof. Proof of (a). Assume $\bar{L} < +\infty$ and $\bar{L} < B$. Then

$$\inf \bar{A}_k = \bar{L} < B.$$

It follows that B is not a lower bound for the sequence $\langle \bar{A}_k \rangle$. Hence, a positive integer N exists such that $\bar{A}_N < B$ and we have: If $n \geqslant N$, then

$$x_n \leqslant \sup_{j \geqslant N} x_j = \bar{A}_N < B.$$

This proves (a).

Proof of (b). Assume $L \in \mathbb{R}^*$, $L < +\infty$ and that for each real B with $L < B$ there exists a positive integer N such that $n \geqslant N$ implies $x_n \leqslant B$. Such a B is then an upper bound for the sequence $\langle x_n \rangle_{n \geqslant N}$. Hence

$$\bar{L} = \inf \bar{A}_k \leqslant \bar{A}_N = \sup_{n \geqslant N} x_n \leqslant B,$$

so that $\bar{L} \leqslant B$. Thus, $L < B$ implies $\bar{L} \leqslant B$. We conclude from this that $\bar{L} \leqslant L$ (otherwise $L < \bar{L}$, and there exists a real B such that $L < B < \bar{L}$, contradicting what was just proved).

Proof of (c). This is in "if and only if" form. First assume $-\infty < \bar{L}$, $-\infty < L$. If $L \leqslant \bar{L}$, let $B < L$ for a real B. Then

$$B < L \leqslant \bar{L} \leqslant \inf \bar{A}_k \leqslant \bar{A}_k \qquad \text{for each} \quad k \in \mathbb{Z}_+ ,$$

so that

$$B < \bar{A}_k = \sup_{n \geqslant k} x_n \qquad \text{for each} \quad k \in \mathbb{Z}_+ .$$

Therefore, for each $k \in \mathbb{Z}_+$ there exists an $n \geqslant k$ such that $B < x_n$. This implies that the set

$$I_B = \{ n \in \mathbb{Z}_+ \mid x_n > B \} \tag{6.1}$$

is an unbounded set of positive integers and as such is infinite. Thus, $x_n > B$ holds for infinitely many n's for each $B < L$. Conversely, assume that for each real $B < L$, the set I_B is an infinite set. It follows that I_B is an unbounded set of positive integers. Thus for each real $B < L$ and given $k \in \mathbb{Z}_+$ there exists a positive integer $n > k$ such that $x_n > B$. For such an n

$$B < x_n \leqslant \sup_{n \geqslant k} x_n = \bar{A}_k \qquad \text{for each } k.$$

Since $B < \bar{A}_k$ for each k,

$$B \leqslant \inf \bar{A}_k = \overline{\lim}\, x_n = \bar{L}.$$

Thus, $B < L$ implies $B \leqslant \bar{L}$. This implies that $L \leqslant \bar{L}$. (Otherwise $\bar{L} < L$ and a B exists such that $\bar{L} < B < L$. This contradicts: $B < L$ implies $B \leqslant \bar{L}$.) This completes the proof.

Corresponding to Theorem 6.1, there is a dual theorem for $\underline{\lim}\, x_n$.

Theorem 6.2. *Let* $\underline{L} = \underline{\lim}\, x_n$. *The following results hold*:

(a) *If* $-\infty < \underline{L}$, *then for each real* B *such that* $B < \underline{L}$, *there is a positive integer* N *such that* $n \geqslant N$ *implies* $x_n > B$;

(b) *If $L \in \mathbb{R}^*$, $-\infty < L$ and if for each real B such that $B < L$, there exists a positive integer N such that $n \geqslant N$ implies $x_n \geqslant B$, then $\underline{L} \geqslant L$;*
(c) *If $\underline{L} < +\infty$ and L is an extended real number such that $L < +\infty$, then $L \geqslant \underline{L}$ holds if and only if for each real B with $B > L$, $x_n < B$ holds for infinitely many n's.*

Prob. 6.1. Prove Theorem 6.2.

Theorem 6.3(a). *If x_n is a real sequence, then*

(a) $\lim x_n = +\infty$ *if and only if* $\underline{\lim}\, x_n = +\infty$,
(b) $\lim x_n = +\infty$ *if and only if for each real B there exists a positive integer N such that $n \geqslant N$ implies $x_n > B$.*

Proof. Let P, Q, and R be the statements:

P: $\lim x_n = +\infty$
Q: $\underline{\lim}\, x_n = +\infty$
R: For each real B there exists a positive integer N such that $n \geqslant N$ implies $x_n > B$.

It is easy to see that P implies Q. We prove that Q implies R. Let Q hold for a sequence $\langle x_n \rangle$ of real numbers. Let B be some real number. Then $B < +\infty = \underline{\lim}\, x_n$. By Theorem 6.2(a) there exists a positive integer such that $n \geqslant N$ implies $x_n > B$. Thus, Q implies R. Next we prove that R implies P. Suppose R holds for a real sequence $\langle x_n \rangle$. Let L be an extended real number with $-\infty < L$, and B be a real number such that $B < L$. Since R holds for $\langle x_n \rangle$, there exists a positive integer N such that $n \geqslant N$ implies $x_n > B$. By Theorem 6.2(b), $\underline{L} \geqslant L$. This holds for each $L \in \mathbb{R}^*$ with $-\infty < L$, and so it holds for $L = +\infty$, that is, $\underline{L} \geqslant +\infty$. But then $\underline{L} = +\infty$. Since $\overline{L} \geqslant \underline{L}$, we have $\overline{L} = +\infty = \underline{L}$. Hence, $\underline{\lim}\, x_n = \overline{\lim}\, x_n$ for $\langle x_n \rangle$, $\lim x_n$ exists, and $\lim x_n = \underline{\lim}\, x_n = +\infty$. This proves that R implies P. Thus, P implies Q, Q implies R, and R implies P. Since P implies Q and Q implies P, P holds if and only if Q holds. Similarly P holds, if and only if R holds. This completes the proof.

Theorem 6.3(b). *If $\langle x_n \rangle$ is a real sequence, then*

(a) $\lim x_n = -\infty$ *if and only if* $\overline{\lim}\, x_n = -\infty$,
(b) $\lim x_n = -\infty$ *if and only if for each real B there exists a positive integer N such that $n \geqslant N$ implies $x_n < B$.*

Prob. 6.2. Prove Theorem 6.3.

Prob. 6.3. Prove: If $\langle x_n \rangle$ is a real sequence, then

(a) $\lim x_n = +\infty$ if and only if there exists a real N (not necessarily a positive integer) such that $n > N$ implies $x_n < B$;

(b) $\lim x_n = -\infty$ if and only if for each real B, there exists a real N (not necessarily an integer) such that $n > N$ implies $x_n < B$.

The result cited in this last problem constitutes a practical way of showing that $x_n \underset{n}{\to} +\infty$ or $x_n \underset{n}{\to} -\infty$.

EXAMPLE 6.1. We prove: If $a \in \mathbb{R}$, $a > 1$, then $\lim_{n\to+\infty} a^n = +\infty$. Now $a > 1$. Put $h = a - 1$. Then $a = 1 + h$, where $h > 0$. By Bernoulli's inequality, we obtain

$$a^n = (1+h)^n \geqslant 1 + nh > nh \qquad \text{for each} \quad n \in \mathbb{Z}_+ . \tag{6.2}$$

Given a real B, take N such that $N \geqslant B/h$ and $n > N$, $n \in \mathbb{Z}_+$. We have

$$a^n > nh > Nh \geqslant \frac{B}{h} h = B \qquad \text{for} \quad n > N, \quad n \in \mathbb{Z}_+ .$$

By Prob. 6.3(a), $\lim_{n\to+\infty} a^n = +\infty$.

EXAMPLE 6.2. We prove $\lim_{n\to+\infty}(n^2 - n + 1) = +\infty$. We first analyze the problem. We must prove: For each real B, there exists an $N \in \mathbb{R}$ such that $n > N$, $n \in \mathbb{Z}_+$, implies that $n^2 - n + 1 > B$.

PROOF. Given $B \in \mathbb{R}$, take $N \geqslant \max\{1, B - 1\}$ and $n \in \mathbb{Z}_+$, $n > N$. We then have $n > 1$ and $n > B - 1$. Since $n > 1$ and n is a positive integer, we have $n \geqslant 2$. This implies that

$$n^2 \geqslant 2n = n + n > B - 1 + n.$$

Hence,

$$n^2 - n + 1 > B \qquad \text{for} \quad n \in \mathbb{Z}_+, \quad n > N \geqslant \max\{1, B - 1\}.$$

PROB. 6.4. Prove: $n^2/(n+1) \underset{n}{\to} +\infty$.

PROB. 6.5. Prove: $\sqrt{n} \to +\infty$ as $n \to +\infty$.

PROB. 6.6. Prove: (a) If n is a positive integer and $h > 0$, then $(1+h)^n \geqslant 1 + nh + n((n-1)/2!)h^2$; (b) use the result in part (a) to prove: If $a > 1$, then

$$\lim_{n\to+\infty} \frac{a^n}{n} = +\infty.$$

EXAMPLE 6.3. We prove $\lim_{n\to+\infty} \sqrt[n]{n!} = +\infty$. By Prob. II.5.4, we have

$$\frac{1}{n!} \leqslant \frac{n!}{n^n},$$

so that $n^n \leqslant (n!)^2$ and $\sqrt{n} \leqslant \sqrt[n]{n!} < +\infty$ for $n \in \mathbb{Z}_+$. Since $\sqrt{n} \to +\infty$ as $n \to +\infty$ (Prob. 6.5), we have

$$+\infty = \lim_{n\to+\infty} \sqrt{n} = \underline{\lim}_{n\to+\infty} \sqrt{n} \leqslant \underline{\lim}_{n\to+\infty} \sqrt[n]{n!} .$$

This implies that $\underline{\lim} \sqrt[n]{n!} = +\infty$, and hence, by Theorem 6.3(a), that $\lim_{n\to+\infty} \sqrt[n]{n!} = +\infty$.

Theorem 6.4. *If $\langle x_n \rangle$ is a real sequence and $\bar{L} = \overline{\lim}\, x_n$, where $\bar{L} \in \mathbb{R}$, then*

(a) *for each real $\epsilon > 0$ there exists a positive integer N such that $n \geqslant N$ implies $x_n < \bar{L} + \epsilon$;*
(b) *if L is a real number such that for each real $\epsilon > 0$, there exists a positive integer N such that $n \geqslant N$ implies $x_n < L + \epsilon$, then $\bar{L} \leqslant L$;*
(c) *if L is a real number, then $L \leqslant \bar{L}$ holds if and only if for each $\epsilon > 0$, $L - \epsilon < x_n$ holds for infinitely many n's.*

Proof. This theorem is merely a reformulation of Theorem 6.1 for the case where $\bar{L} \in \mathbb{R}$.

We prove (a). Given $\epsilon > 0$, $\bar{L} < \bar{L} + \epsilon$. Put $B = \bar{L} + \epsilon$, and hence $\bar{L} < B$. By Theorem 6.1, (a) there exists a positive integer N such that $n \geqslant N$ implies $x_n < B = \bar{L} + \epsilon$. This proves (a).

We now prove (b). Let L be a real number such that for each $\epsilon \in \mathbb{R}$, $\epsilon > 0$, there exists a positive integer N such that $n \geqslant N$ implies $x_n < L + \epsilon$. Take any real B such that $L < B$. Put $\epsilon = B - L$, so that $B = L + \epsilon$, where $\epsilon > 0$. By the present assumption, there exists a positive integer N such that $n \geqslant N$ implies $x_n < L + \epsilon = B$. Thus, for each real B with $L < B$ there exists a positive integer such that $n \geqslant N$ implies $x_n < B$. By Theorem 6.1, part (b), this implies that $\bar{L} \leqslant L$. This proves (b).

We prove (c) in a similar manner. Each real B such that $B < L$ may be written as $B = L - \epsilon$, where $\epsilon = L - B > 0$. Hence, for each real $\epsilon > 0$, $L - \epsilon < x_n$ holds for infinitely many n's, if and only if $x_n > B$ holds for infinitely many n's for each real B such that $B < L$. By Theorem 6.1, part (c), we have: $L \leqslant \bar{L}$ holds if and only if, for each real $\epsilon > 0$, $L - \epsilon < x_n$ holds for infinitely many n's. This completes the proof.

A dual statement holds for $\underline{L} = \underline{\lim}\, x_n$.

Theorem 6.5. *If $\langle x_n \rangle$ is a real sequence and $\underline{L} = \underline{\lim}\, x_n$, $\underline{L} \in \mathbb{R}$, then the following results hold:*

(a) *for each real $\epsilon > 0$, there exists a positive integer N such that $n \geqslant N$ implies $\underline{L} - \epsilon < x_n$;*
(b) *if L is a real number, and if for each real $\epsilon > 0$ there exists a positive integer N such that $n \geqslant N$ implies $L - \epsilon < x_n$, then $L \leqslant \underline{L}$;*
(c) *if L is a real number, then $L \geqslant \underline{L}$ if and only if for each real $\epsilon > 0$, $x_n < L + \epsilon$ for infinitely many n's.*

Prob. 6.7. Prove Theorem 6.5.

Prob. 6.8. Prove:

(1) The sentence: there exists a positive integer N such that $n \geqslant N$ implies $x_n < \bar{L} + \epsilon$, in parts (a) and (b) of Theorem 6.4, can be replaced by the sentence: there exists a real number N such that $n > N$ implies $x_n < \bar{L} + \epsilon$.
(2) The sentence: there exists a positive integer N such that $n \geqslant N$ implies $\underline{L} - \epsilon < x_n$, in parts (a) and (b) of Theorem 6.5, can be replaced by the sentence: there exists a real n such that $n > N$ implies $\underline{L} - \epsilon < x_n$.

The following theorem is important in the evaluation of limits of sequences.

Theorem 6.6. *If $\langle x_n \rangle$ is a real sequence, then $\langle x_n \rangle$ is convergent and has the real number L as its limit if and only if for each real $\epsilon > 0$, there exists a positive integer N such that*

$$n \geqslant N \quad \textit{implies} \quad |x_n - L| < \epsilon. \tag{6.3}$$

PROOF. Let L be a real number such that for each real $\epsilon > 0$ there exists a positive integer N such that (6.3) holds. Consequently,

$$n \geqslant N \quad \text{implies} \quad L - \epsilon < x_n < L + \epsilon. \tag{6.4}$$

We prove that this implies that $\langle x_n \rangle$ is bounded. Take $\epsilon = 1$, then there exists a positive integer N_1 such that

$$n \geqslant N_1 \quad \text{implies} \quad L - 1 < x_n \leqslant L + 1. \tag{6.5}$$

If $N_1 = 1$, then (6.5) holds for all n. It is clear that in this case $\langle x_n \rangle$ is bounded. If $N_1 > 1$, let

$$M = \max\{x_1, \dots, x_{N_1 - 1}, L + 1\} \quad \text{and} \quad m = \min\{x_1, \dots, x_{N_1 - 1}, L - 1\},$$

so that

$$m \leqslant x_n \leqslant M \qquad \text{for all} \quad n \in \mathbb{Z}_+ .$$

Thus, $\langle x_n \rangle$ is bounded in this case also. Since $\langle x_n \rangle$ is bounded, we have (Theorem 4.2)

$$-\infty < \underline{\lim}\, x_n \leqslant \overline{\lim}\, x_n < +\infty. \tag{6.6}$$

Put

$$\underline{L} = \underline{\lim}\, x_n \quad \text{and} \quad \bar{L} = \overline{\lim}\, x_n . \tag{6.7}$$

Then (6.6) becomes

$$-\infty < \underline{L} \leqslant \bar{L} < +\infty. \tag{6.8}$$

From (6.4) we conclude that for each $\epsilon > 0$ there exists a positive integer N such that

$$n \geqslant N \quad \text{implies} \quad x_n < L + \epsilon. \tag{6.9}$$

By Theorem 6.4, part (b), $\overline{L} \leqslant L$. Similar reasoning tells us that for each $\epsilon > 0$ there exists a positive integer N_1 such that

$$n \geqslant N_1 \quad \text{implies} \quad L - \epsilon < x_n. \tag{6.10}$$

Using part (b) of Theorem 6.5 we conclude that $L \leqslant \underline{L}$. Therefore, $\overline{L} \leqslant L \leqslant \underline{L}$, and hence $\overline{L} \leqslant \underline{L}$. Since $\underline{L} \leqslant \overline{L}$ is always true, it follows that $\underline{L} = \overline{L} = L$. We have: $\langle x_n \rangle$ converges with limit L.

Conversely, let $\langle x_n \rangle$ converge and let $L = \lim x_n$. This implies that

$$L = \underline{\lim}\, x_n = \overline{\lim}\, x_n, \qquad L \in \mathbb{R}. \tag{6.11}$$

By part (a) of Theorem 6.5 there exists a positive integer N_1 such that

$$n \geqslant N_1 \quad \text{implies} \quad L - \epsilon = \underline{L} - \epsilon < x_n. \tag{6.12}$$

By part (a) of Theorem 6.4 there exists a positive integer N_2 such that

$$n \geqslant N_2 \quad \text{implies} \quad x_n < \overline{L} + \epsilon = L + \epsilon. \tag{6.13}$$

Let $N = \max\{N_1, N_2\}$ and take $n \in \mathbb{Z}_+$, $n > N$, so that $n > N_1$ and $n > N_2$. By (6.12) and (6.13) we have

$$L - \epsilon < x_n < L + \epsilon \quad \text{or} \quad |x_n - L| < \epsilon$$

if $n \geqslant N$. This completes the proof.

Prob. 6.9. Prove: If $\langle x_n \rangle$ is a real sequence, then it converges to $L = \lim x_n$ if and only if for each $\epsilon > 0$ there exists a real N such that $n > N$ implies $|x_n - L| < \epsilon$.

Remark 6.1. The result cited in Prob. 6.9 is the one most convenient to use in proving that a real number L is the limit of a real sequence. As a matter of fact, the usual treatment of limits begins with the result in Prob. 6.9 as a definition of the limit L of a real sequence $\langle x_n \rangle$.

Example 6.4. We prove: If $|a| < 1$, then $\lim_{n \to +\infty} a^n = 0$. This is trivially true when $a = 0$ (why?). Assume $a \neq 0$, so that $0 < |a| < 1$, and hence

$$\frac{1}{|a|} > 1 \quad \text{or} \quad \frac{1}{|a|} - 1 > 0.$$

Analysis. We wish to prove that for each $\epsilon > 0$, there exists an N such that

$$n > N \quad \text{implies} \quad |a^n - 0| = |a^n| < \epsilon. \tag{6.14}$$

Write $h = 1/|a| - 1$, so that $1/|a| = 1 + h$, where $h > 0$. By Bernoulli's inequality,

$$\frac{1}{|a^n|} = \frac{1}{|a|^n} = \left(\frac{1}{|a|} \right)^n = (1 + h)^n \geqslant 1 + nh > nh \tag{6.14$'$}$$

for each positive integer n. This implies that

$$|a^n| < \frac{1}{nh} \quad \text{for each positive integer } n. \tag{6.15}$$

PROOF. Given $\epsilon > 0$, take

$$N \geqslant \frac{1}{\epsilon h} \quad \text{and} \quad n > N, \quad n \in \mathbb{Z}_+ .$$

For such n, $n > 1/\epsilon h$, so that

$$|a^n - 0| = |a^n| < \frac{1}{nh} = \frac{1}{n}\frac{1}{h} < (\epsilon h)\frac{1}{h} = \epsilon.$$

Using Prob. 6.9 we obtain $\lim_{n\to+\infty} a^n = 0$.

PROB. 6.10. Prove: If $|a| < 1$, then (see Prob. 6.6) $\lim_{n\to+\infty} na^n = 0$.

PROB. 6.11. Prove: $1/\sqrt{n} \to 0$ as $n \to +\infty$.

EXAMPLE 6.5. To evaluate

$$\lim_{n\to+\infty} \frac{n^2+n+1}{2n^2-n+1},$$

we note that

$$\frac{n^2+n+1}{2n^2-n+1} = \frac{1+1/n+1/n^2}{2-1/n+1/n^2}.$$

Later theorems about sums, differences, products, and quotients of limits will enable us to prove that the right-hand side has a limit equal to $\frac{1}{2}$ as $n \to +\infty$. Meanwhile, our intuition tells us the same thing. Indeed, if "n is large," then $1/n$ and $1/n^2$ "become small" so the quotient on the right gets close to $\frac{1}{2}$ as "n gets large." We guess that

$$\lim_{n\to+\infty} \frac{n^2+n+1}{2n^2-n+1} = \frac{1}{2}.$$

To prove that our guess is right we prove that given $\epsilon > 0$ there exists an N such that if $n > N$, then

$$\left| \frac{n^2+n+1}{2n^2-n+1} - \frac{1}{2} \right| < \epsilon, \tag{6.16a}$$

that is,

$$\left| \frac{3n+1}{2(2n^2-n+1)} \right| < \epsilon. \tag{6.16b}$$

We first take $n > 2$, so that $n^2 > 2n$. Adding $n^2 - n + 1$ to both sides gives us

$$2n^2 - n + 1 > n^2 + n + 1 > n^2.$$

Hence,

$$0 < \frac{1}{2n^2-n+1} < \frac{1}{n^2} \qquad \text{if} \quad n > 2.$$

This implies that

$$0<\frac{3n+1}{2n^2-n+1}<\frac{3n+1}{n^2}=\frac{3}{n}+\frac{1}{n^2}. \tag{6.17}$$

But since n is a positive integer, $n \geqslant 1$ and $n^2 \geqslant n$ or $1/n^2 \leqslant 1/n$. This and (6.30) imply that

$$0<\frac{3n+1}{2n^2-n+1}<\frac{3}{n}+\frac{1}{n^2}\leqslant\frac{3}{n}+\frac{1}{n}=\frac{4}{n}.$$

This proves: If $n>2$, then

$$\left|\frac{n^2+n+1}{2n^2-n+1}-\frac{1}{2}\right|=\frac{3n+1}{2(2n^2-n+1)}<\frac{2}{n}. \tag{6.18}$$

This "analysis" indicates that the second inequality in (6.16) holds if $2/n<\epsilon$. We, therefore, take for given $\epsilon>0$, $n>N\geqslant\max(2,2/\epsilon)$, so that $n>2$ and $n>2/\epsilon$. This and (6.18) yield

$$\left|\frac{n^2+n+1}{2n^2-n+1}-\frac{1}{2}\right|=\frac{3n+1}{2(2n^2-n+1)}<\frac{2}{n}=\frac{1}{n}2<\frac{\epsilon}{2}2=\epsilon \qquad \text{if} \quad n>N.$$

We conclude

$$\lim_{n\to+\infty}\frac{n^2+n+1}{2n^2-n+1}=\frac{1}{2}.$$

Prob. 6.12. Prove:

(a) $\lim(\sqrt{n+1}-\sqrt{n})=0$,
(b) $\lim((n+1)^2-n^2)=+\infty$,
(c) $\lim_{n\to+\infty}\sqrt{n}(\sqrt{n+1}-\sqrt{n})=\frac{1}{2}$.

Prob. 6.13. Prove:

$$\lim_{n\to+\infty}\frac{n!}{n^n}=0$$

(Hint: see Prob. II.5.4).

Example 6.6. We prove

$$\lim_{n\to+\infty}\sqrt[n]{n}=1. \tag{6.19}$$

Note that if $n>1$, then $\sqrt[n]{n}-1>0$. Put $h_n=\sqrt[n]{n}-1$. Then $h_n>0$ if $n>1$. Since $\sqrt[n]{n}=1+h_n$, where $h_n>0$ for $n>1$, we find, using Prob. 6.6, part (a) that

$$n=(1+h_n)^n\geqslant 1+nh_n+\frac{n(n-1)}{2}h_n^2>\frac{n(n-1)}{2}h_n^2$$

when $n>1$. This implies that

$$0<\sqrt[n]{n}-1=h_n<\sqrt{\frac{2}{n-1}} \qquad \text{if} \quad n>1.$$

Given $\epsilon > 0$, take

$$n > 1 + \frac{2}{\epsilon^2}.$$

For such n

$$|\sqrt[n]{n} - 1| = \sqrt[n]{n} - 1 < \sqrt{\frac{2}{n-1}} < \epsilon.$$

This proves (6.19).

7. Algebra of Limits: Sums and Differences of Sequences

To give our theorems generality, it is useful to define algebraic operations in $\mathbb{R}^*$. We do this first for addition and subtraction. (Multiplication and division in $\mathbb{R}^*$ are defined in Defs. 7.3 and 7.4.)

Def. 7.1. Let x and y be in $\mathbb{R}^*$. If (1) $x \in \mathbb{R}$, $y \in \mathbb{R}$, then $x \pm y$ retain the value they have in $\mathbb{R}$. If one of x or y is in $\mathbb{R}^*$ but not in $\mathbb{R}$, then we define:

$$x + (+\infty) = (+\infty) + x = +\infty \quad \text{for} \quad x \in \mathbb{R}^* \quad \text{and} \quad -\infty < x. \tag{7.1}$$

$$x + (-\infty) = (-\infty) + x = -\infty \quad \text{for} \quad x \in \mathbb{R}^* \quad \text{and} \quad x < +\infty. \tag{7.2a}$$

$$-(+\infty) = -\infty \quad \text{and} \quad -(-\infty) = +\infty. \tag{7.2b}$$

If $x \in \mathbb{R}^*$, and $y \in \mathbb{R}^*$ and $x + (-y)$ is defined as in (7.1) or in (7.2), then

$$x - y = x + (-y). \tag{7.3}$$

According to this definition,

$$\begin{array}{l} (+\infty) + (+\infty) = +\infty, \\ (-\infty) + (-\infty) = -\infty, \\ (+\infty) - (-\infty) = +\infty, \\ (-\infty) - (+\infty) = -\infty, \end{array} \quad \text{and} \quad \begin{array}{l} 0 \pm \infty = \pm\infty \\ 0 - (\pm\infty) = \mp\infty. \end{array} \tag{7.4}$$

Note, however, that $(+\infty) + (-\infty)$, $(+\infty) - (+\infty)$, $(-\infty) - (-\infty)$ are not defined.

Remark 7.1. There is no closure in $\mathbb{R}^*$ for addition and subtraction. As remarked above, there exist x and y in $\mathbb{R}^*$ for which $x \pm y$ are not defined. We also note that not every element x of $\mathbb{R}^*$ has an additive inverse. Since $(+\infty) + y = +\infty$ for all $y \in \mathbb{R}^*$ such that $y \neq -\infty$ and $(+\infty) + (-\infty)$ is not defined, no $y \in \mathbb{R}^*$ exists such that $(+\infty) + y = 0$, and $+\infty$ has no additive inverse. Similarly $-\infty$ has no additive inverse.

The systems $\mathbb{R}$ and $\mathbb{R}^*$ differ from each other with respect to their order properties. For, assume $x < y$ for some x and y in $\mathbb{R}^*$. Normally, if we add z to both sides of $x < y$ then we obtain $x + z < y + z$. But if $z = \pm\infty$, then one of $x + z$ or $y + z$ may not be defined. Hence, we qualify and state:

(a) If $-\infty < x < y$ and $z \in \mathbb{R}$, then $x + z < y + z$ and if $z = +\infty$, then $x + z = y + z = +\infty$;
(b) if $x < y < +\infty$ and $z \in \mathbb{R}$, then $x + z < y + z$ and if $z = -\infty$, then $x + z = y + z = -\infty$.

Thus, in case (a), if $z \in \mathbb{R}^*$, $z > -\infty$, then the most that can be concluded is $x + z \leqslant y + z$ and in case (b), if $z \in \mathbb{R}^*$, $z < +\infty$, then $x + z \leqslant y + z$. However, from

$$-\infty < x < u \quad \text{and} \quad -\infty < y < v,$$

we can still conclude the strict

$$x + y < u + v.$$

To see this, note that $-\infty < y < v$ implies $y \in \mathbb{R}$. Since $-\infty < x < u$, it follows that $-\infty < x + y < u + y$. Since u may be $+\infty$, we obtain from $-\infty < y < v$ that $u + y \leqslant u + v$. This implies that $x + y < u + v$. Similarly, we note that

$$x < u < +\infty \quad \text{and} \quad y < v < +\infty$$

imply

$$x + y < u + v.$$

Def. 7.2(a). The *absolute value* $|x|$, of $x \in \mathbb{R}^*$, we define as

$$|x| = \begin{cases} x & \text{if} \quad 0 \leqslant x \leqslant +\infty \\ -x & \text{if} \quad -\infty \leqslant x < 0. \end{cases}$$

PROB. 7.1. Prove: (a) $x \in \mathbb{R}^*$ implies $0 \leqslant |x| \leqslant +\infty$, (b) $|-x| = |x|$, (c) $|-x| \leqslant x \leqslant |x|$ and $-|x| \leqslant -x \leqslant |x|$.

PROB. 7.2. Assume that $x \in \mathbb{R}^*$ and $0 < A \in \mathbb{R}^*$. Prove: (a) $-A < x < A$ if and only if $|x| < A$; (b) $|x| = A$, if and only if $x = A$ or $x = -A$; (c) if $0 \leqslant A \in \mathbb{R}^*$, then $-A \leqslant x \leqslant A$ if and only if $|x| \leqslant A$.

PROB. 7.3. Prove: if $x \in \mathbb{R}^*$ and $y \in \mathbb{R}^*$ and $x + y$ is defined, then $|x + y| \leqslant |x| + |y|$. (Note, $|x| + |y|$ is always defined for x and y in $\mathbb{R}^*$.)

Def. 7.2(b). If $\langle x_n \rangle$ and $\langle y_n \rangle$ are real sequences, (a) define their *sum* $\langle x_n \rangle + \langle y_n \rangle$ as the sequence $\langle c_n \rangle$, where $c_n = x_n + y_n$ for all n. Thus, $\langle x_n \rangle + \langle y_n \rangle = \langle x_n + y_n \rangle$. (b) If $c \in \mathbb{R}$, define $c\langle y_n \rangle$ as the sequence $\langle z_n \rangle$, where $z_n = cy_n$ for all n, so that $c\langle y_n \rangle = \langle cy_n \rangle$. In particular, we have: $(-1)\langle y_n \rangle = \langle -y_n \rangle$. We write $(-1)\langle y_n \rangle$ as $-\langle y_n \rangle$. (c) The *difference*

$\langle x_n\rangle - \langle y_n\rangle$ of the above sequence is defined as $\langle x_n\rangle - \langle y_n\rangle = \langle x_n\rangle + (-\langle y_n\rangle)$, so that $\langle x_n\rangle - \langle y_n\rangle = \langle x_n - y_n\rangle$.

Theorem 7.1. *If $\langle x_n\rangle$ and $\langle y_n\rangle$ are sequences of real numbers, then*

(a)
$$\overline{\lim_{n\to+\infty}} (x_n + y_n) \leqslant \overline{\lim_{n\to+\infty}} x_n + \overline{\lim_{n\to+\infty}} y_n$$

and

(b)
$$\underline{\lim_{n\to+\infty}} (x_n + y_n) \geqslant \underline{\lim_{n\to+\infty}} x_n + \underline{\lim_{n\to+\infty}} y_n ,$$

provided that the sums on the right in (a) and (b) are defined.

Proof. We prove (a) and leave the proof of (b) to the reader (Prob. 7.4). First consider the case where one of $\overline{L} = \overline{\lim}\, x_n$, $\overline{M} = \overline{\lim}\, y_n$ is $+\infty$ and the other is *not* $-\infty$. Then $\overline{L} + \overline{M} = +\infty$ and we have

$$\overline{\lim}\,(x_n + y_n) \leqslant +\infty = \overline{L} + \overline{M} = \overline{\lim}\, x_n + \overline{\lim}\, y_n . \tag{7.5a}$$

Now suppose one of $\overline{L}$ or $\overline{M}$ is $-\infty$ and the other is not $+\infty$. For definiteness, let $\overline{L} = -\infty$ and $\overline{M} < +\infty$. Let B be some real number. There exists some B_1 such that $\overline{M} < B_1 < +\infty$. By Theorem 6.1, part (a), there exists a positive integer N_1 such that

$$n \geqslant N_1 \quad \text{implies} \quad y_n < B_1 . \tag{7.5b}$$

Since $\overline{\lim}\, x_n = \overline{L} = -\infty$, $\lim x_n = -\infty$, this implies (Theorem (6.3(a)) that a positive integer N_2 exists such that

$$n \geqslant N_2 \quad \text{implies} \quad x_n < B - B_1 . \tag{7.6}$$

Let $n \geqslant \max\{N_1, N_2\}$. Then $n \geqslant N_1$ and $n \geqslant N_2$, and (7.5b) and (7.6) imply that

$$x_n + y_n < B \qquad \text{for} \quad n \geqslant N. \tag{7.7a}$$

Thus, for each real B, there exists a positive integer N such that $n \geqslant N$ implies that $x_n + y_n < B$. But then (Theorem 6.3(b)) $\lim(x_n + y_n) = -\infty$. Since here $\overline{L} + \overline{M} = -\infty$, it follows that in this case,

$$\overline{\lim}\,(x_n + y_n) = \lim(x_n + y_n) = -\infty = \overline{L} + \overline{M}$$

and (a) holds with equality.

The remaining case where $\overline{L} + \overline{M}$ is defined is: $-\infty < \overline{L} < +\infty$ and $-\infty < \overline{M} < +\infty$. We consider this case now: Given $\epsilon > 0$, we have $\overline{L} < \overline{L} + \epsilon/2$ and $\overline{M} < \overline{M} + \epsilon/2$. By Theorem 6.4(a), there exist positive integers N_1 and N_2 such that

$$x_n < \overline{L} + \frac{\epsilon}{2} \qquad \text{if} \quad n \geqslant N \quad \text{and} \quad y_n < \overline{M} + \frac{\epsilon}{2} \qquad \text{if} \quad n \geqslant N_2 . \tag{7.7b}$$

Put $N = \max\{N_1, N_2\}$ and take $n \geqslant N$. Then $n \geqslant N_1$ and $n \geqslant N_2$, and (7.7b) yields

$$x_n + y_n < \overline{L} + \overline{M} + \epsilon \qquad \text{if} \quad n \geqslant N. \tag{7.8}$$

Thus, for each $\epsilon > 0$ there exists a positive integer N such that (7.8) holds. By Theorem 6.4(b), we conclude that

$$\overline{\lim}\,(x_n + y_n) \leqslant \bar{L} + \bar{M} = \overline{\lim}\; x_n + \overline{\lim}\; y_n .$$

This completes the proof of part (a).

Prob. 7.4. Complete the proof of the last theorem by proving part (b) there.

Remark 7.2. We use notation adopted in the proof of Theorem 7.1. Let $\bar{L} = -\infty$ and $\bar{M} < +\infty$. Theorem 6.3(b) implies that $\lim x_n = \overline{\lim}\, x_n = \bar{L} = -\infty$ and that

$$\bar{L} + \bar{M} = (-\infty) + \bar{M} = -\infty.$$

Thus (Theorem 7.1),

$$\overline{\lim}\,(x_n + y_n) \leqslant -\infty.$$

This implies that $\lim(x_n + y_n) = -\infty$. In this case the inequality in Theorem 7.1, part (a) becomes an equality. We may therefore state: If $\overline{\lim}\, x_n = -\infty$ and $\langle y_n \rangle$ is bounded from above, then $\lim x_n$ and $\lim(x_n + y_n)$ each exist and

$$\lim(x_n + y_n) = -\infty = \lim x_n + \overline{\lim}\; y_n . \tag{7.9}$$

Similarly, with respect to part (b) of Theorem 7.1, we state: If $\underline{\lim}\, x_n = +\infty$ and $\langle y_n \rangle$ is bounded from below, then $\lim x_n$ and $\lim(x_n + y_n)$ each exist and we have the equality

$$\lim(x_n + y_n) = \lim x_n + \underline{\lim}\; y_n = +\infty. \tag{7.10}$$

Corollary (*of Theorem* 7.1). *If* $\lim x_n$ *and* $\lim y_n$ *each exist and their sum is defined in* $\mathbb{R}^*$, *then* $\lim(x_n + y_n)$ *exists and*

$$\lim(x_n + y_n) = \lim x_n + \lim y_n . \tag{7.11}$$

Proof. Exercise.

Def. 7.3 (Multiplication in $\mathbb{R}^*$). Let x and y be in $\mathbb{R}^*$. If $x \in \mathbb{R}$, $y \in \mathbb{R}$, then xy retains the value it has in $\mathbb{R}$. If $x \in \mathbb{R}^*$, $x > 0$, define

$$\begin{aligned} x(+\infty) &= (+\infty) = +\infty, \\ x(-\infty) &= (-\infty) = -\infty. \end{aligned} \tag{7.12}$$

If $x \in \mathbb{R}^*$, $x < 0$, define

$$\begin{aligned} x(+\infty) &= (+\infty)x = -\infty, \\ x(-\infty) &= (-\infty)x = +\infty. \end{aligned} \tag{7.13}$$

We do not define $0(\pm\infty)$.

Def. 7.4 (Division in $\mathbb{R}^*$). Let x and y be in $\mathbb{R}^*$. If $x \in \mathbb{R}$ and $y \in \mathbb{R}$, where $y \neq 0$, then x/y retains the value it has in $\mathbb{R}$. If $x \in \mathbb{R}$, then we define

$$\frac{x}{\pm\infty} = 0. \tag{7.14}$$

If $x \in \mathbb{R}$, $x \neq 0$, define

$$\frac{\pm\infty}{x} = \frac{1}{x}(\pm\infty). \tag{7.15}$$

We do not define x/y if $y = 0$ or if $|x| = |y| = +\infty$.

Remark 7.3. As was the case for addition in $\mathbb{R}^*$, there is no closure for multiplication in $\mathbb{R}^*$ since $0(\pm\infty)$ are not defined. Also, $+\infty$ and $-\infty$ do not have multiplicative inverses (explain) even though they are nonzero elements of $\mathbb{R}^*$.

We add some remarks about the preservation of order under multiplication in $\mathbb{R}^*$. Assume $z > 0$ and $x < y$. If one of x or y is equal to 0 and $z = +\infty$, then one of xz or yz is not defined, so they are not comparable. However, if $x < y$, neither x nor y is equal to 0, and $z > 0$, then we can conclude that $xz \leqslant yz$, but cannot infer that the inequality here is necessarily strict. In fact $x < y < 0$ implies that $x(+\infty) = -\infty = y(+\infty)$ and $0 < x < y$ implies that $x(+\infty) = +\infty = y(+\infty)$. On the other hand, $x < 0 < y$ implies that $x(+\infty) = -\infty < +\infty = y(+\infty)$.

Note also that in $\mathbb{R}^*$ we still have: (a) $x > 0$, $y > 0$ imply $xy > 0$, (b) $x > 0$, $y < 0$ imply $xy < 0$, and (c) $x < 0$ and $y < 0$ imply $xy > 0$.

We also have: If $x \in \mathbb{R}^*$, $y \in \mathbb{R}^*$, and xy is defined, then $|xy| = |x||y|$. If x/y is defined, we have $|x/y| = |x|/|y|$.

Theorem 7.2. *If $\langle x_n \rangle$ is a real sequence, then (a) $0 < c < +\infty$ implies $\overline{\lim}(cx_n) = c\,\overline{\lim}\,x_n$ and $\underline{\lim}(cx_n) = c\,\underline{\lim}\,x_n$ and (b) $-\infty < c < 0$ implies $\overline{\lim}(cx_n) = c\,\underline{\lim}\,x_n$ and $\underline{\lim}(cx_n) = c\,\overline{\lim}\,x_n$. For the case $c = 0$, we have $\lim(cx_n) = c\,\overline{\lim}\,x_n$ if $\overline{\lim}\,x_n$ is a real number, and $\lim(cx_n) = c\,\underline{\lim}\,x_n$ if $\underline{\lim}\,x_n$ is a real number.*

Proof. Suppose $0 < c < +\infty$. If $\overline{L} = \overline{\lim}\,x_n = +\infty$, then $\langle x_n \rangle$ is not bounded from above. It follows that $\langle cx_n \rangle$ is not bounded from above (why?). Therefore,

$$\overline{\lim}\; cx_n = +\infty = c(+\infty) = c\overline{L} = c\;\overline{\lim}\; x_n,$$

in this case. If $\overline{L} < +\infty$, then $\langle x_n \rangle$ is bounded from above. Hence, $\langle cx_n \rangle$ is (why?) and we have $\overline{\lim}(cx_n) < +\infty$. Take a real B such that $\overline{\lim}(cx_n) < B$. There exists a positive integer N_1 such that

$$n \geqslant N_1 \quad \text{implies} \quad cx_n < B, \quad \text{so that} \quad x_n < \frac{B}{c}.$$

This implies that

$$\bar{L} = \overline{\lim}\, x_n = \inf A_k \leqslant A_{N_1} \leqslant \frac{B}{c}, \qquad \text{so that} \quad c\bar{L} \leqslant B.$$

Thus,

$$c\bar{L} \leqslant B \qquad \text{for all real } B \text{ such that } \overline{\lim}\,(cx_n) < B. \tag{7.16}$$

This implies that

$$c\,\overline{\lim}\, x_n = c\bar{L} \leqslant \overline{\lim}\,(cx_n). \tag{7.17}$$

Now put

$$\bar{D}_k = \sup_{n \geqslant k}\,(cx_n) \qquad \text{for each} \quad k \in \mathbb{Z}_+ .$$

For each $k \in \mathbb{Z}_+$,

$$x_n \leqslant \bar{A}_k \qquad \text{for} \quad n \geqslant k \quad \text{and hence} \quad cx_n \leqslant c\bar{A}_k \qquad \text{for} \quad n \geqslant k.$$

It follows that

$$\overline{\lim}\,(cx_n) \leqslant \bar{D}_k \leqslant c\bar{A}_k \qquad \text{for each } k.$$

Accordingly,

$$\frac{1}{c}\,\overline{\lim}\,(cx_n) \leqslant \bar{A}_k \qquad \text{for each} \quad k \in \mathbb{Z}_+ .$$

This implies that

$$\frac{1}{c}\,\overline{\lim}\,(cx_n) \leqslant \overline{\lim}\, x_n = \inf \bar{A}_k$$

and finally that

$$\overline{\lim}\,(cx_n) \leqslant c\,\overline{\lim}\, x_n . \tag{7.18}$$

This and (7.17) prove that $\overline{\lim}(cx_n) = c\,\overline{\lim}\, x_n$ for the case when $\overline{\lim}\, x_n < +\infty$ also.

We now prove: If $-\infty < c < 0$, then $\underline{\lim}(cx_n) = c\,\overline{\lim}\, x_n$. First assume $\overline{\lim}\, x_n = +\infty$, so that $c\,\overline{\lim}\, x_n = c(+\infty) = -\infty$. Clearly, $\langle x_n \rangle$ is not bounded from above, so $\langle cx_n \rangle$ is not bounded from below and we have $\underline{\lim}(cx_n) = -\infty = c\,\overline{\lim}\, x_n$. Now suppose $\overline{\lim}\, x_n < +\infty$, so that $\langle x_n \rangle$ is bounded from above and $\langle cx_n \rangle$ is bounded from below. Let

$$\bar{L} = \overline{\lim}\, x_n \quad \text{and} \quad \underline{L}^* = \underline{\lim}(cx_n).$$

We have $\bar{L} < +\infty$, $\underline{L}^* > -\infty$. Let B be a real number such that $\underline{L}^* > \text{B}$. By Theorem 6.2, part (a) there exists a positive integer N such that

$$n \geqslant N \quad \text{implies} \quad cx_n > B.$$

This implies that

$$\text{if} \quad n \geqslant N, \quad \text{then} \quad x_n < \frac{B}{c}.$$

Thus, $\bar{A}_N \leqslant B/c$ and, therefore,

$$\overline{\lim}\ x_n = \inf \bar{A}_k \leqslant \bar{A}_N \leqslant \frac{B}{c}.$$

This implies that

$$c\bar{L} = c\ \overline{\lim}\ x_n \geqslant B.$$

We have proved: for each real B such that $B < \underline{L}^*$, $B \leqslant c\bar{L}$. This proves (explain)

$$\underline{L}^* \leqslant c\bar{L}. \tag{7.19}$$

Now, for each $k \in \mathbb{Z}_+$ put

$$\underline{D}_k = \inf_{n \geqslant k} (cx_n).$$

We have

$$x_n \leqslant \bar{A}_k \quad \text{and hence} \quad cx_n \geqslant c\bar{A}_k \quad \text{for} \quad n \geqslant k.$$

The last inequalities imply that

$$c\bar{A}_k \leqslant \underline{D}_k \leqslant \underline{\lim}(cx_n) \qquad \text{for each } k$$

and, therefore, that

$$\bar{A}_k \geqslant \frac{1}{c}\ \underline{\lim}(cx_n) \qquad \text{for each } k.$$

This implies that

$$\bar{L} = \overline{\lim}\ x_n = \inf \bar{A}_k \geqslant \frac{1}{c}\ \underline{\lim}(cx_n) = \frac{1}{c}\ \underline{L}^*,$$

from which it follows that

$$c\bar{L} \leqslant \underline{L}^*.$$

This and (7.19) yield

$$\underline{\lim}(cx_n) = L^* = c\bar{L} = c\ \overline{\lim}\ x_n,$$

also when $\overline{\lim}\, x_n < +\infty$.

To complete the proof, we note that if $-\infty < c < 0$, then applying what we just proved

$$\underline{\lim}\, x_n = \underline{\lim}\left(\frac{1}{c}(cx_n)\right) = \frac{1}{c}\ \overline{\lim}\,(cx_n),$$

so that

$$\overline{\lim}\,(cx_n) = c\ \underline{\lim}\, x_n \qquad \text{if} \quad -\infty < c < 0.$$

This completes the proof of part (b).

We now complete the proof of part (a). Let $0 < c < +\infty$. By what was just proved, we know that

$$\overline{\lim}\,(-cx_n) = -c\ \underline{\lim}\, x_n \tag{7.20}$$

and that

$$\overline{\lim}\,(-cx_n) = -\underline{\lim}(cx_n). \tag{7.21}$$

The second conclusion in (a) now follows by equating the two right-hand sides in (7.20) and (7.21). The last part of the theorem is obvious. The proof is now complete.

PROB. 7.5(a). Prove: If $c \in \mathbb{R}$ and $\lim x_n$ exists, then $\lim_{n\to+\infty}(cx_n) = c \lim x_n$.

Remark 7.4. Note that special cases of Theorem 7.2 are $\overline{\lim}(-x_n) = -\underline{\lim}\, x_n$ and $\underline{\lim}(-x_n) = -\overline{\lim}\, x_n$.

PROB. 7.5(b). Prove: (a) $\overline{\lim}(x_n - y_n) \leqslant \overline{\lim}\, x_n - \underline{\lim}\, y_n$ and (b) $\underline{\lim}(x_n - y_n) \geqslant \underline{\lim}\, x_n - \overline{\lim}\, y_n$ when the differences on the right are defined.

PROB. 7.6. Prove: If $\lim x_n$ and $\lim y_n$ exist, then $\lim(x_n - y_n) = \lim x_n - \lim y_n$, provided the difference on the right is defined.

EXAMPLE 7.1. We point out that if $\lim x_n = +\infty$, $\lim y_n = -\infty$, then $\lim(x_n + y_n)$ can be any value in $\mathbb{R}^*$. Thus,

(a) If $a \in \mathbb{R}$ and we write $x_n = n + a$, $y_n = -n$ for each $n \in \mathbb{R}$, then $\lim(x_n + y_n) = \lim a = a$. (Here $\lim x_n = +\infty$ and $\lim y_n = -\infty$.)
(b) Let $x_n = \sqrt{n+1}$, $y_n = -\sqrt{n}$ for each $n \in \mathbb{Z}_+$. Then $\lim x_n = +\infty$, $\lim y_n = -\infty$ and $\lim(x_n + y_n) = \lim_{n\to+\infty}(\sqrt{n+1} - \sqrt{n}) = 0$ (Prob. 6.12, part (a)).
(c) Let $x_n = n^2$ and $y_n = -n$. Then $\lim x_n = \lim n^2 = +\infty$, $\lim y_n = \lim(-n) = -\infty$, and $\lim(x_n + y_n) = \lim(n^2 - n) = +\infty$.
(d) Let $x_n = n$ and $y_n = -n^2$. Then $\lim x_n = \lim(n) = +\infty$, $\lim y_n = \lim(-n)^2 = -\infty$, and $\lim(x_n + y_n) = \lim(n - n^2) = -\infty$.

Theorem 7.3. *If* $\lim x_n$ *exists, then* $\lim|x_n| = |\lim x_n|$.

PROOF. The proof for the case $\lim x_n = L \in \mathbb{R}$ follows readily from

$$||x_n| - |L|| \leqslant |x_n - L|$$

as the reader can check.

Let $\lim x_n = +\infty$. Since $x_n \leqslant |x_n|$ for all n, it is clear that $\lim|x_n| = +\infty$.

Let $\lim x_n = -\infty$. Since $-x_n \leqslant |x_n|$ for all n and $\lim(-x_n) = +\infty$, then it follows that $\lim|x_n| = +\infty = |-\infty| = |\lim x_n|$.

Remark 7.5. There are some important observations to be made concerning the relation between $\lim x_n$ and $\lim|x_n|$. First, we note that $\lim|x_n|$ may exist but $\lim x_n$ may not. This is evident from the sequence $\langle x_n \rangle$, where $x_n =$

$(-1)^{n+1}$ for each n. Clearly, $|x_n| = 1$ for all n so $\lim|x_n| = 1$ in this case. However, note that $\langle x_n \rangle$ itself diverges here.

We also observe that even if $\langle x_n \rangle$ has a limit, it may happen that $\lim|x_n| = |A|$ but that $\lim x_n \neq A$. In fact,

$$\lim\left|\frac{1}{n} - 1\right| = 1 = |1| \quad \text{and} \quad \lim\left(\frac{1}{n} - 1\right) = -1.$$

An important special case is cited in Prob. 7.7 below.

PROB. 7.7. Prove: $\lim x_n = 0$, if and only if $\lim|x_n| = 0$.

PROB. 7.8. Prove: If $\langle x_n \rangle$ is a real sequence, then $-\overline{\lim}|x_n| \leqslant \underline{\lim}\, x_n \leqslant \overline{\lim}\, x_n \leqslant \overline{\lim}|x_n|$.

8. Algebra of Limits: Products and Quotients of Sequences

Def. 8.1. By the product $\langle x_n \rangle\langle y_n \rangle$ of real sequences $\langle x_n \rangle$ and $\langle y_n \rangle$, we mean the sequence $\langle p_n \rangle$, where $p_n = x_n y_n$ for all n. Thus, $\langle x_n \rangle\langle y_n \rangle = \langle x_n y_n \rangle$.

Theorem 8.1. *If $\langle x_n \rangle$ and $\langle y_n \rangle$ are real sequences, where $\lim x_n = 0$, and $\langle y_n \rangle$ is bounded, then $\lim(x_n y_n) = 0$.*

PROOF. Since $\langle y_n \rangle$ is bounded, there exists a real $M > 0$ such that $|y_n| \leqslant M$ holds for all n. Let $\epsilon > 0$. Since $\lim x_n = 0$, there exists a positive integer N such that

$$n \geqslant N \quad \text{implies} \quad |x_n| = |x_n - 0| < \frac{\epsilon}{M}.$$

Therefore,

$$|x_n y_n - 0| = |x_n y_n| = |x_n||y_n| < \frac{\epsilon}{M} M = \epsilon \qquad \text{for} \quad n \geqslant N.$$

It follows that $\lim x_n y_n = 0$.

PROB. 8.1. Let $\langle x_n \rangle$ be a real sequence such that $x_n \geqslant 0$ for all n and $\overline{A}_k = 0$ for some positive integer k. (Here, as usual, $\overline{A}_k = \sup_{n \geqslant k} x_n$.) Prove: $\lim x_n = 0$ and for any real sequence $\langle y_n \rangle$ we have $\lim x_n y_n = 0$.

Theorem 8.2. *If $\langle x_n \rangle$ is a real sequence such that $\underline{\lim}\, x_n > 0$, then there exists a positive integer N such that $n \geqslant N$ implies $x_n > 0$.*

Proof. There exists a B such that $\underline{\lim}\, x_n > B > 0$. By Theorem 6.2, part (c), there exists a positive integer N such that $n \geqslant N$ implies $x_n > B > 0$.

Theorem 8.3. *If $\langle x_n \rangle$ and $\langle y_n \rangle$ are real sequences such that* $\lim x_n = +\infty$ *and* $\underline{\lim}\, y_n > 0$, *then* $\lim(x_n y_n) = +\infty$.

Proof. There exists an L such that $\underline{\lim}\, y_n > L > 0$. By the hypothesis and Theorem 6.2, there exists a positive integer N_1 such that

$$n \geqslant N_1 \quad \text{implies} \quad y_n > L > 0. \tag{8.1}$$

Let B be some real number. Since $\lim x_n = +\infty$, there exists a positive integer N_2 such that

$$x_n > \frac{1 + |B|}{L} \qquad \text{if} \quad n \geqslant N_2. \tag{8.2}$$

Let $N = \max\{N_1, N_2\}$ and $n \geqslant N$. Then $n \geqslant N_1$ and $n \geqslant N_2$. Since (8.1) and (8.2) hold for such n, it follows that

$$x_n y_n > L \frac{1 + |B|}{L} = 1 + |B| > |B| \geqslant B \qquad \text{if} \quad n \geqslant N.$$

Thus, for each real B there exists a positive integer N such that $n \geqslant N$ implies $x_n y_n > B$. This yields $\lim x_n y_n = +\infty$.

Prob. 8.2. Prove: (a) $\lim x_n = -\infty$ and $\underline{\lim}\, y_n > 0$ imply $\lim x_n y_n = -\infty$; (b) $\lim x_n = +\infty$, $\overline{\lim}\, y_n < 0$ imply $\lim x_n y_n = -\infty$; (c) $\lim x_n = -\infty$ and $\overline{\lim}\, y_n < 0$ imply $\lim x_n y_n = +\infty$.

Theorem 8.4. *If $\langle x_n \rangle$ and $\langle y_n \rangle$ are real sequences such that $x_n \geqslant 0$, $y_n \geqslant 0$ for all n, then*

(a) $\overline{\lim}(x_n y_n) \leqslant \overline{\lim}\, x_n \overline{\lim}\, y_n$, *and*
(b) $\underline{\lim}(x_n y_n) \geqslant \underline{\lim}\, x_n \underline{\lim}\, y_n$,

provided that the products on the right in (a) and (b) are defined.

Proof. We first prove (a). The hypothesis implies that $\overline{\lim}\, x_n \geqslant 0$, $\overline{\lim}\, y_n \geqslant 0$, and $\overline{\lim}(x_n y_n) \geqslant 0$. The product $\overline{\lim}\, x_n \overline{\lim}\, y_n$ is defined if one of $\overline{\lim}\, x_n$ or $\overline{\lim}\, y_n$ is not 0 and the other is not $+\infty$. Consider first the case where one of $\overline{\lim}\, x_n$ or $\overline{\lim}\, y_n$ is $+\infty$ and the other is > 0, so that $\overline{\lim}\, x_n \overline{\lim}\, y_n = +\infty$. We then have

$$\overline{\lim}\,(x_n y_n) \leqslant +\infty = \overline{\lim}\, x_n\, \overline{\lim}\, y_n,$$

so the theorem holds in this case.

Next consider the case $0 \leqslant \overline{\lim}\, x_n < +\infty$, $0 \leqslant \overline{\lim}\, y_n < +\infty$. Put

$$\bar{A}_k = \sup_{n \geqslant k} x_n, \qquad \bar{B}_k = \sup_{n \geqslant k} y_n, \qquad \bar{C}_k = \sup_{n \geqslant k} (x_n y_n)$$

for each $k \in \mathbb{Z}_+$. If j and k are positive integers, then

$$0 \leqslant x_n \leqslant \bar{A}_k \quad \text{for} \quad n \geqslant k \quad \text{and} \quad 0 \leqslant y_n \leqslant \bar{B}_j \quad \text{for} \quad n \geqslant j. \tag{8.3}$$

If a positive integer k_1 exists such that $\bar{A}_{k_1} = 0$, then by Prob. 8.1, we have $\lim x_n = 0 = \lim(x_n y_n)$, so that we obtain $\overline{\lim}\, x_n \,\overline{\lim}\, y_n = \lim x_n \,\overline{\lim}\, y_n = 0 = \underline{\lim}(x_n y_n) = \overline{\lim}(x_n y_n)$. Thus, (a) holds with equality. Similarly (a) holds if $\bar{B}_{j_1} = 0$ for some positive integer j_1. We therefore assume that $\bar{A}_k > 0$ and $\bar{B}_j > 0$ for any positive integers j and k. Suppose $j \geqslant k$. Take $n \geqslant j$ so that $n \geqslant k$. It follows from (8.3) that

$$0 \leqslant x_n y_n \leqslant \bar{A}_k \bar{B}_j \quad \text{if} \quad n \geqslant j.$$

But then

$$\overline{\lim}\,(x_n y_n) = \inf \bar{C}_k \leqslant \bar{C}_j = \sup_{n \geqslant j} x_n y_n \leqslant \bar{A}_k \bar{B}_j .$$

Hence

$$\overline{\lim}\,(x_n y_n) \leqslant \bar{A}_k \bar{B}_j \quad \text{for} \quad j \geqslant k.$$

It follows that

$$\frac{1}{\bar{A}_k}\,\overline{\lim}\,(x_n y_n) \leqslant \bar{B}_j \quad \text{for} \quad j \geqslant k. \tag{8.4}$$

Now $\langle \bar{B}_k \rangle$ is a monotonic decreasing sequence of elements of $\mathbb{R}^*$. Hence, we have (Prob. 2.4),

$$\overline{\lim}\; y_n = \inf \bar{B}_j = \inf_{j \geqslant k} \bar{B}_j \quad \text{for each } k.$$

This and (8.4) imply

$$\frac{1}{\bar{A}_k}\,\overline{\lim}\,(x_n y_n) \leqslant \overline{\lim}\; y_n \quad \text{for each } k. \tag{8.5}$$

Assume that $\overline{\lim}\, y_n = 0$. Since $y_n \geqslant 0$ for all n, it follows that $\lim y_n = 0$. Also $\overline{\lim}\, x_n < +\infty$ here. It follows that $\langle x_n \rangle$ is a bounded sequence and therefore that $\lim(x_n y_n) = 0$ (Theorem 8.1). This implies that (a) holds. If $\overline{\lim}\, y_n > 0$, then (8.5) yields

$$\frac{\overline{\lim}\,(x_n y_n)}{\overline{\lim}\; y_n} \leqslant \bar{A}_k \quad \text{for each } k.$$

But then

$$\frac{\overline{\lim}\,(x_n y_n)}{\overline{\lim}\,(y_n)} \leqslant \overline{\lim}\; x_n = \inf \bar{A}_k .$$

Thus, the conclusion in (a) holds in this case also. This completes the proof of (a).

We prove (b). As before $\underline{\lim}\, x_n \geqslant 0$, $\underline{\lim}\, y_n \geqslant 0$, and $\underline{\lim}\, x_n y_n \geqslant 0$. We consider several cases. If $\underline{\lim}\, x_n = +\infty$ and $\underline{\lim}\, y_n > 0$, then (b) holds with equality (Theorem 8.3). The same is true if $\underline{\lim}\, x_n > 0$ and $\underline{\lim}\, y_n = +\infty$. It is left to consider the case $0 \leqslant \underline{\lim}\, x_n < +\infty$, $0 \leqslant \underline{\lim}\, y_n < +\infty$. If $\underline{\lim}\, x_n = 0$ or $\underline{\lim}\, y_n = 0$, we have $\underline{\lim}\, x_n \; \underline{\lim}\, y_n = 0 \leqslant \underline{\lim}\, x_n y_n$ and (a) holds in this case.

It remains to consider the case $0 < \underline{\lim}\, x_n < +\infty$ and $0 < \underline{\lim}\, y_n < +\infty$. Put

$$\underline{A}_k = \inf_{n \geqslant k} x_n, \qquad \underline{B}_k = \inf_{n \geqslant k} y_n, \qquad \underline{C}_k = \inf x_n y_n \qquad \text{for each } k.$$

We have: $0 \leqslant \underline{A}_k$, $0 \leqslant \underline{B}_k$, $0 \leqslant \underline{C}_k$ for each k. If $\underline{A}_k = 0$ for all k, we would have $\underline{\lim}\, x_n = \sup \underline{A}_k = 0$, contradicting $\underline{\lim}\, x_n > 0$. Hence, a positive integer k_1 exists such that $\underline{A}_{k_1} > 0$. Similarly, there exists a positive integer j_1 such that $\underline{B}_{j_1} > 0$. We write $l = \max\{k_1, j_1\}$ and take $j \geqslant k \geqslant l$. Since $\langle \underline{A}_k \rangle \uparrow$ and $\langle \underline{B}_k \rangle \uparrow$,

$$\underline{A}_j \geqslant \underline{A}_k \geqslant \underline{A}_{k_1} > 0 \quad \text{and} \quad \underline{B}_j \geqslant \underline{B}_k \geqslant \underline{B}_{j_1} > 0.$$

Also,

$$\underline{A}_j \leqslant x_n \quad \text{and} \quad \underline{B}_k \leqslant y_n \qquad \text{for} \quad n \geqslant j.$$

Hence,

$$0 < \underline{A}_j \underline{B}_k \leqslant x_n y_n \qquad \text{for} \quad n \geqslant j.$$

This implies that

$$\underline{A}_j \underline{B}_k \leqslant \underline{C}_j \leqslant \underline{\lim}(x_n y_n).$$

Therefore,

$$\underline{A}_j \leqslant \frac{\underline{\lim}(x_n y_n)}{\underline{B}_k} \qquad \text{for} \quad j \geqslant k \geqslant l. \tag{8.6}$$

Since $\langle \underline{A}_j \rangle \uparrow$ we have (Theorem 2.1)

$$\underline{\lim}\, x_n = \sup \underline{A}_j = \sup_{j \geqslant k} \underline{A}_j \qquad \text{for} \quad k \geqslant l.$$

Because of (8.6), this implies

$$\underline{\lim}\, x_n \leqslant \frac{\underline{\lim}(x_n y_n)}{\underline{B}_k} \qquad \text{for} \quad k \geqslant l$$

and hence that

$$\underline{B}_k \leqslant \frac{\underline{\lim}(x_n y_n)}{\underline{\lim}\, x_n} \qquad \text{for} \quad k \geqslant l.$$

By reasoning as before, we conclude that

$$\underline{\lim}\, y_n = \sup \underline{B}_k = \sup_{k \geqslant l} \underline{B}_k < \frac{\underline{\lim}(x_n y_n)}{\underline{\lim}\, x_n}.$$

Thus, (b) follows in this last case also. This completes the proof.

Products of Sequences Which Have Limits

Lemma 8.1. *If $\langle x_n \rangle$ is a real sequence and a positive integer N exists such that* $\lim_{k\to+\infty} x_{k+N-1} = L \in \mathbb{R}^*$, *then* $\lim x_n = L$.

Proof. We prove the lemma for the case $L \in \mathbb{R}$ and ask the reader to prove it for $L = \pm\infty$ (Prob. 8.3). Given $\epsilon > 0$, there exists a positive integer N_1 such that

$$k \geqslant N_1 \quad \text{implies} \quad |x_{k+N-1} - L| < \epsilon. \tag{8.7}$$

We take $n > N + N_1 - 1$, so that $n = k + N - 1$, where $k > N_1$. By (8.7),

$$|x_n - L| < \epsilon \qquad \text{for} \quad n > N + N_1 - 1.$$

Thus, $\lim x_n = L$ when $L \in \mathbb{R}$.

Prob. 8.3. Prove Lemma 8.1 for the cases $L = \pm\infty$.

Theorem 8.5. *If* $\lim x_n$ *and* $\lim y_n$ *each exist, then* $\lim(x_n y_n)$ *exists and*

$$\lim(x_n y_n) = \lim x_n \lim y_n \tag{8.8}$$

provided that the product on the right is defined in $\mathbb{R}^*$.

Proof. If $\lim x_n = 0$ and $\lim y_n \in \mathbb{R}$, the conclusion follows from Theorem 8.1. This is also the case if $\lim x_n \in \mathbb{R}$ and $\lim y_n = 0$. Suppose $\lim x_n = X > 0$ and $\lim y_n = Y > 0$. By Theorem 8.2, there exist positive integers N_1 and N_2 such that

$$n \geqslant N_1 \text{ implies } x_n > 0 \text{ and } n \geqslant N_2 \text{ implies } y_n > 0. \tag{8.9}$$

Put $N = \max\{N_1, N_2\}$. Then

$$x_n > 0 \quad \text{and} \quad y_n > 0 \qquad \text{if} \quad n \geqslant N. \tag{8.10}$$

Consider the respective subsequences $\langle u_k \rangle$ and $\langle v_k \rangle$ of $\langle x_n \rangle$ and $\langle y_n \rangle$, where

$$u_k = x_{k+N-1} \quad \text{and} \quad v_k = y_{k+N-1} \quad \text{for each positive integer } k.$$

We have $u_k > 0$ and $v_k > 0$ for each k, and also

$$\lim_{k\to+\infty}(x_{k+N-1}) = X \quad \text{and} \quad \lim_{k\to+\infty}(y_{k+N-1}) = Y. \tag{8.11}$$

By Theorem 8.4, we know that

$$\overline{\lim_{k\to+\infty}}(u_k v_k) \leqslant \overline{\lim_{k\to+\infty}}\, u_k \,\overline{\lim_{k+}}\, v_k = XY \tag{8.12}$$

and

$$\underline{\lim_{k\to+\infty}}(u_k v_k) \geqslant \underline{\lim}\, u_k \,\underline{\lim}\, v_k = XY. \tag{8.13}$$

It follows that

$$\overline{\lim}\,(u_k v_k) \leqslant XY \leqslant \underline{\lim}(u_k v_k)$$

and hence that (explain)

$$XY = \overline{\lim}\,(u_k v_k) = \underline{\lim}(u_k v_k) = \lim(u_k v_k). \tag{8.14}$$

By Lemma 8.1, we conclude from this that

$$\lim_{n\to+\infty}(x_n y_n) = \lim_{k\to+\infty}(u_k v_k) = XY = \lim x_n \lim y_n$$

in the case under consideration.

If $X < 0$ and $Y < 0$, then $-X > 0$ and $-Y > 0$, so

$$\lim(x_n y_n) = \lim(-x_n)(-y_n) = \lim(-x_n)\lim(-y_n) = (-X)(-Y) = XY.$$

If $X > 0$ and $Y < 0$, then

$$\lim(x_n(-y_n)) = \lim x_n \lim(-y_n) = X(-Y) = -(XY)$$

so

$$\lim(x_n y_n) = -\lim(x_n(-y_n)) = XY.$$

The proof is now complete.

Quotients of Real Sequences

Lemma 8.2. *If $\langle x_n \rangle$ is a real sequence such that $x_n \neq 0$ for all n and $\lim x_n \neq 0$, then*

$$\lim \frac{1}{x_n} = \frac{1}{\lim x_n}. \tag{8.15}$$

PROOF. Let $\lim x_n = +\infty$ and let $\epsilon > 0$ be given. There exists a positive integer N such that

$$n \geqslant N \quad \text{implies} \quad x_n > \frac{1}{\epsilon} > 0.$$

This implies that

$$0 < \frac{1}{x_n} < \epsilon \qquad \text{if} \quad n \geqslant N.$$

Hence

$$\left| \frac{1}{x_n} - 0 \right| = \left| \frac{1}{x_n} \right| = \frac{1}{x_n} < \epsilon \qquad \text{if} \quad n \geqslant N.$$

But then

$$\lim \frac{1}{x_n} = 0 = \frac{1}{+\infty} = \frac{1}{\lim x_n}. \tag{8.16}$$

Similarly, if $\lim x_n = -\infty$, then for a given $\epsilon > 0$ there exists a positive integer N such that

$$x_n < -\frac{1}{\epsilon} < 0 \qquad \text{if} \quad n \geqslant N.$$

This implies that

$$|x_n| = -x_n > \frac{1}{\epsilon} \qquad \text{if} \quad n \geqslant N.$$

Hence

$$\left|\frac{1}{x_n} - 0\right| = \left|\frac{1}{x_n}\right| = \frac{1}{|x_n|} < \epsilon \qquad \text{if} \quad n \geqslant N.$$

Again we conclude that

$$\lim \frac{1}{x_n} = 0 = \frac{1}{-\infty} = \frac{1}{\lim x_n}.$$

We now consider the case where $\lim x_n$ exists, is finite, and $\neq 0$. Put $L = \lim x_n$. We have $L \neq 0$ and $L \in \mathbb{R}$. There exists a positive integer N_1 such that

$$|x_n - L| < \frac{|L|}{2} \qquad \text{if} \quad n \geqslant N_1,$$

so

$$|L| - |x_n| < \frac{|L|}{2} \qquad \text{if} \quad n \geqslant N_1$$

and hence

$$|x_n| > \frac{|L|}{2}, \quad \text{i.e.,} \quad \frac{1}{|x_n|} < \frac{2}{|L|} \qquad \text{if} \quad n \geqslant N_1. \tag{8.17}$$

Let $\epsilon > 0$ be given. There exists a positive integer N_2 such that

$$|x_n - L| < \frac{|L|^2}{2}\epsilon \qquad \text{if} \quad n \geqslant N_2. \tag{8.18}$$

Put $N = \max\{N_1, N_2\}$ and take $n \geqslant N$, so that $n \geqslant N_1$ and $n \geqslant N_2$. This implies that

$$\left|\frac{1}{x_n} - \frac{1}{L}\right| = \frac{|x_n - L|}{|x_n||L|} < \frac{2}{L^2}|x_n - L| < \frac{2}{L^2}\left(\frac{L^2}{2}\epsilon\right) = \epsilon \qquad \text{if} \quad n \geqslant N.$$

We conclude that

$$\lim \frac{1}{x_n} = \frac{1}{L}.$$

This completes the proof.

Theorem 8.6. *If* $\lim x_n = X$ *and* $\lim y_n = Y \neq 0$, *where* $y_n^{-1} \neq 0$ *for all* n, *then*

$$\lim_{n \to +\infty} \frac{x_n}{y_n} = \frac{\lim x_n}{\lim y_n} = \frac{X}{Y}, \tag{8.19}$$

provided X *and* Y *are not infinite together.*

PROOF. Since $Y \neq 0$, $1/Y \neq \pm\infty$ (explain). By properties of real numbers and Def. 7.4,

$$\frac{X}{Y} = X\left(\frac{1}{Y}\right).$$

The hypothesis can be phrased as follows: Either X is not infinite or $1/Y \neq 0$, so that $X(1/Y)$ is defined. Applying Theorem 8.5 and Lemma 8.2, we obtain

$$\lim \frac{x_n}{y_n} = \lim\left(x_n \frac{1}{y_n}\right) = \lim x_n \lim \frac{1}{y_n} = X\left(\frac{1}{Y}\right) = \frac{X}{Y} = \frac{\lim x_n}{\lim y_n}.$$

Theorem 8.7. *If $\langle x_n \rangle$ is a real sequence such that $x_n > 0$ for all n, then*

(a) $$\overline{\lim} \frac{1}{x_n} = \frac{1}{\underline{\lim}\, x_n} \quad \text{if} \quad \underline{\lim}\, x_n > 0$$

and

(b) $$\underline{\lim} \frac{1}{x_n} = \frac{1}{\overline{\lim}\, x_n} \quad \text{if} \quad \overline{\lim}\, x_n > 0.$$

PROOF. We begin by proving that

$$\overline{\lim} \frac{1}{x_n} \leqslant \frac{1}{\underline{\lim}\, x_n} \quad \text{if} \quad \underline{\lim}\, x_n > 0. \tag{8.20}$$

The hypothesis implies that $0 \leqslant \underline{\lim}\, x_n \leqslant \overline{\lim}\, x_n$. If $\underline{\lim}\, x_n = +\infty$, $\langle x_n \rangle$ has a limit and $\lim x_n = +\infty$. By Lemma 8.2, $\langle 1/x_n \rangle$ has a limit and

$$\overline{\lim} \frac{1}{x_n} = \lim \frac{1}{x_n} = \frac{1}{\lim x_n} = \frac{1}{\underline{\lim}\, x_n},$$

in this case. Thus, the equality in (8.20) holds here. Next assume $0 < \underline{\lim}\, x_n < +\infty$. Put

$$\underline{A}_k = \inf_{n \geqslant k} x_n \quad \text{and} \quad \overline{B}_k = \sup_{n \geqslant k} \frac{1}{x_n} \qquad \text{for each } k.$$

Since $0 < \underline{\lim}\, x_n < +\infty$ and $x_n > 0$ for all n, it follows that $0 \leqslant \underline{A}_k < +\infty$ for each k. As a matter of fact, $0 < \underline{\lim}\, x_n$ implies that a positive integer k_1 exists such that $A_{k_1} > 0$. Therefore

$$0 < \underline{A}_{k_1} \leqslant \underline{A}_k \leqslant x_n \qquad \text{if} \quad n \geqslant k \geqslant k_1,$$

so that

$$\frac{1}{x_n} \leqslant \frac{1}{\underline{A}_k} \leqslant \frac{1}{\underline{A}_{k_1}} \qquad \text{if} \quad n \geqslant k \geqslant k_1.$$

This implies that

$$\overline{B}_k = \sup_{n \geqslant k} \frac{1}{x_n} \leqslant \frac{1}{\underline{A}_k} \qquad \text{if} \quad k \geqslant k_1. \tag{8.21}$$

The sequences $\langle \overline{B}_k \rangle$ and $\langle 1/\underline{A}_k \rangle$ are monotonic decreasing and, therefore, have limits. Since

$$\lim_{k\to+\infty} \overline{B}_k = \lim_{k\to+\infty}\left(\sup_{n\geq k}\frac{1}{x_n}\right) = \varlimsup_{n\to+\infty}\frac{1}{x_n}$$

and

$$\lim \overline{A}_k = \lim_{k\to+\infty}\left(\inf_{n\geq k} x_n\right) = \varliminf_{n\to+\infty} x_n,$$

we obtain from (8.21) and Lemma 8.2

$$\varlimsup \frac{1}{x_n} = \lim \overline{B}_k \leqslant \lim_{k\to+\infty}\frac{1}{\underline{A}_k} = \frac{1}{\lim_{k\to+\infty}\underline{A}_k} = \frac{1}{\varliminf x_n}.$$

Thus, (8.20) holds in this case also.

We now prove that under the hypothesis,

$$\varliminf \frac{1}{x_n} \geqslant \frac{1}{\varlimsup x_n} \qquad \text{if} \quad \varlimsup x_n > 0. \tag{8.22}$$

Put

$$\overline{A}_k = \sup_{n\geq k} x_n \quad \text{and} \quad \underline{B}_k = \inf_{n\geq k}\frac{1}{x_n} \qquad \text{for each } k.$$

We have

$$0 < x_n \leqslant \overline{A}_k \qquad \text{for} \quad n \geqslant k$$

and hence

$$0 < \frac{1}{\overline{A}_k} \leqslant \frac{1}{x_n} \qquad \text{if} \quad n \geqslant k.$$

This implies that

$$0 < \frac{1}{\overline{A}_k} \leqslant \inf_{n\geq k}\left(\frac{1}{x_n}\right) = \underline{B}_k \qquad \text{for each } k. \tag{8.23}$$

The sequences $\langle 1/\overline{A}_k \rangle$ and $\langle \underline{B}_k \rangle$ are monotonic increasing and therefore have limits. Also

$$\lim_{k\to+\infty} \overline{A}_k = \lim_{k\to+\infty}\left(\sup_{n\geq k} x_n\right) = \varlimsup x_n > 0$$

and

$$\lim_{k\to+\infty} \underline{B}_k = \lim\left(\inf_{n\geq k}\frac{1}{x_n}\right) = \varliminf_{n\to+\infty}\frac{1}{x_n}.$$

It therefore follows from this, (8.23), and Lemma 8.2 that

$$\frac{1}{\lim \overline{A}_k} = \lim \frac{1}{\overline{A}_k} \leqslant \lim \underline{B}_k = \varliminf \frac{1}{x_n},$$

i.e., that

$$\frac{1}{\overline{\lim}\, x_n} \leqslant \underline{\lim}\, \frac{1}{x_n}.$$

This proves (8.22).

We prove (b). Assume $\overline{\lim}\, x_n > 0$ so that (8.22) holds. If $\underline{\lim}(1/x_n) = 0$, we obtain from $\overline{\lim}\, x_n \geqslant 0$ and (8.22) that $1/\overline{\lim}\, x_n = 0$ so that (b) holds in this case. If $\underline{\lim}(1/x_n) > 0$, we apply (8.20) to the sequence $\langle 1/x_n \rangle$ to obtain

$$0 < \overline{\lim}\, x_n \leqslant \frac{1}{\underline{\lim}(1/x_n)}.$$

This implies that $\underline{\lim}(1/x_n) < +\infty$. Thus, $0 < \underline{\lim}(1/x_n) < +\infty$. Therefore, $0 < \overline{\lim}\, x_n < +\infty$ and

$$\underline{\lim}\, \frac{1}{x_n} \leqslant \frac{1}{\overline{\lim}\, x_n}.$$

This and (8.22) imply that the equality in (8.22) holds and prove (b) in this case ($\underline{\lim}(1/x_n) > 0$) also.

We now prove (a). Begin with $\underline{\lim}\, x_n > 0$. If $\underline{\lim}\, x_n = +\infty$, we obtain from (8.20) that $\overline{\lim}(1/x_n) = 0$. Thus, both sides of (a) are equal to 0 in this case and (a) holds. If $\underline{\lim}\, x_n < +\infty$, we have $0 < \underline{\lim}(1/x_n) < +\infty$. We can now apply (b) to the sequence $\langle 1/x_n \rangle$ to obtain

$$\underline{\lim}\, x_n = \frac{1}{\overline{\lim}\,(1/x_n)}, \qquad \text{so that} \quad \overline{\lim}\, \frac{1}{x_n} = \frac{1}{\underline{\lim}\, x_n}.$$

Hence, (a) holds in this case also. This completes the proof.

Theorem 8.8. *If* $0 \leqslant x_n$ *and* $0 < y_n$, *then*: (*a*) *If* $0 < \underline{\lim}\, y_n$ *and* $\overline{\lim}\, x_n$, $\underline{\lim}\, y_n$ *are not both equal to* $+\infty$ *together, then*

$$\overline{\lim}\, \frac{x_n}{y_n} \leqslant \frac{\overline{\lim}\, x_n}{\underline{\lim}\, y_n}. \tag{8.24}$$

(*b*) *If* $0 < \overline{\lim}\, y_n$ *and* $\underline{\lim}\, x_n$, $\overline{\lim}\, y_n$ *are both not equal to* $+\infty$ *together, then*

$$\underline{\lim}\, \frac{x_n}{y_n} \geqslant \frac{\underline{\lim}\, x_n}{\overline{\lim}\, y_n}.$$

PROOF. Exercise.

Remark 8.1. The theorems on limits superior and inferior and limits of sums and products of real sequences can be extended to the cases of sums and products of finitely many sequences by induction on the number n of sequences. Thus, if $\langle x_{n1} \rangle, \langle x_{n2} \rangle, \ldots, \langle x_{nm} \rangle$ are finitely many sequences where m is some positive integer, then

$$\overline{\lim_{n\to+\infty}}(x_{n1}+x_{n2}+\cdots+x_{nm}) \leqslant \overline{\lim_{n\to+\infty}} x_{n1} + \overline{\lim_{n\to+\infty}} x_{n2} + \cdots + \overline{\lim_{n+}} x_{nm} \tag{8.25}$$

and

$$\underline{\lim_{n\to+\infty}}(x_{n1}+x_{n2}+\cdots+x_{mm}) \geqslant \underline{\lim_{n\to+\infty}} x_{n1} + \underline{\lim_{n\to+\infty}} x_{n2} + \cdots + \underline{\lim} x_{nm} \tag{8.26}$$

provided that the sums on the right exist in $\mathbb{R}^*$. Therefore, if $\langle x_{n1}\rangle, \langle x_{n2}\rangle, \ldots, \langle x_{nm}\rangle$ are m sequences which have limits, then

$$\lim_{n\to+\infty}(x_{n1}+x_{n2}+\cdots+x_{nm}) = \lim_{n\to+\infty} x_{n1} + \lim_{n\to+\infty} x_{n2} + \cdots + \lim x_{nm} \tag{8.27}$$

if the sum on the right is defined in $\mathbb{R}^*$, and

$$\lim_{n\to+\infty}(x_{n1}x_{n2}\ldots x_{nm}) = \lim_{n\to+\infty} x_{n1} \lim_{n\to+\infty} x_{n2} \ldots \lim x_{nm} \tag{8.28}$$

if the product on the right exists in $\mathbb{R}^*$. We also have: If $x_{n1} \geqslant 0$, $x_{n2} \geqslant 0, \ldots, x_{nm} \geqslant 0$ for all n, then

$$\overline{\lim_{n\to+\infty}}(x_{n1}x_{n2}\ldots x_{nm}) \leqslant \overline{\lim_{n\to+\infty}} x_{n1} \overline{\lim_{n\to+\infty}} x_{n2} \ldots \overline{\lim_{n\to+\infty}} x_{nm} \tag{8.29}$$

and

$$\underline{\lim_{n\to+\infty}}(x_{n1}x_{n2}\ldots x_{nm}) \geqslant \underline{\lim_{n\to+\infty}} x_{n1} \underline{\lim_{n\to+\infty}} x_{n2} \ldots \underline{\lim_{n\to+\infty}} x_{nm} \tag{8.30a}$$

provided that the products on the right exist in $\mathbb{R}^*$.

We use these results to evaluate $\lim x_n$, where

$$x_n = \frac{2n^2-3n+1}{3n^2+4}.$$

We have

$$\lim x_n = \lim \frac{2-3/n+2/n^2}{3+4/n^2} = \frac{\lim(2-3/n+1/n^2)}{\lim(3+4/n^2)}$$

$$= \frac{2-3(\lim(1/n))^2}{3+4(\lim(1/n))^2} = \frac{2-3(0)+0^2}{3+4(0)^2} = \frac{2}{3}.$$

Remark 8.2. We observe that the equalities

$$\lim(x_n y_n) = \lim x_n \lim y_n \quad \text{and} \quad \lim \frac{x_n}{y_n} = \frac{\lim x_n}{\lim y_n} \tag{8.30b}$$

break down when the operations on the right are not defined in $\mathbb{R}^*$. Here the limit on the left can be any value. For example, let $a \in \mathbb{R}$ and

$x_n = an^{-2}$, $y_n = n^2$ for all n. We have $x_n \to 0$ and $y_n \to +\infty$ as $n \to +\infty$, but $x_n y_n = a \to a$ as $n \to +\infty$. On the other hand, if $x_n = n^{-1}$ and $y_n = n^2$ for all n, we still have $x_n \to 0$ and $y_n \to +\infty$ as $n \to +\infty$ and $x_n y_n = n \to +\infty$ as $n \to +\infty$. Similar examples illustrate the same point with respect to the second equality in (8.30b). Thus, if $x_n = an$ for each n and $y_n = n$ for each n and $a > 0$, we have $x_n \to +\infty$ and $y_n \to +\infty$ as $n \to +\infty$ but $x_n/y_n \to a$ as $n \to +\infty$. Similarly, if $x_n = n^2$ and $y_n = n^3$ for each n, we have $x_n \to +\infty$, $y_n \to +\infty$ and $x_n/y_n \to 0$ as $n \to +\infty$. On the other hand, writing $x_n = n^3$, $y_n = n^2$ for each n, we have $x_n \to +\infty$, $y_n \to +\infty$ and $x_n/y_n \to +\infty$.

EXAMPLE 8.1. We prove: If $p > 0$, then $\lim_{n\to+\infty} p^{1/n} = 1$. This is obvious if $p = 1$. Hence, assume first that $p > 1$. We have $1 < p^{1/n}$. This and Theorem II.12.1 yield

$$0 < p^{1/n} - 1 < \frac{1}{n}(p - 1)$$

where $p - 1 > 0$. It is clear from this that $\lim_{n\to+\infty} p^{1/n} = 1$ (explain). If $0 < p < 1$, then we have $1/p > 1$. By the result just proved, we have

$$\lim_{n\to+\infty} \frac{1}{p^{1/n}} = \lim_{n\to+\infty} \left(\frac{1}{p}\right)^{1/n} = 1.$$

Thus,

$$\lim_{n\to+\infty} p^{1/n} = \frac{1}{\lim_{n\to+\infty} (1/p^{1/n})} = \frac{1}{\lim_{n\to+\infty} (1/p)^{1/n}} = \frac{1}{1} = 1$$

if $0 < p < 1$. Hence, for all $p > 0$, $\lim_{n\to+\infty} p^{1/n} = 1$.

PROB. 8.4. Evaluate

$$\text{(a)} \quad \lim_{n\to+\infty} \frac{3n^3 - 4n + 1}{4n^4 - 3n + 2}, \quad \text{(b)} \quad \lim_{n\to+\infty} \frac{2n^4 - n}{3n^4 + n^2}, \quad \text{(c)} \quad \lim_{n\to+\infty} \frac{3 - n^2}{n + 1}.$$

PROB. 8.5. Prove: If $p > 0$, then $\lim_{n\to+\infty} p^{1/10^n} = 1$.

PROB. 8.6. Let $\langle x_n \rangle$ be defined as follows: $x_1 = \sqrt{2}$ and $x_{n+1} = \sqrt{2 + x_n}$ for each positive integer n. For example, $x_2 = \sqrt{2 + x_1} = \sqrt{2 + \sqrt{2}}$, $x_3 = \sqrt{2 + x_2} = \sqrt{2 + \sqrt{2 + \sqrt{2}}}$. Prove: $\langle x_n \rangle$ converges and $\lim x_n = 2$.

PROB. 8.7. Let $\langle x_n \rangle$ be defined as follows: $x_1 = \sqrt{2}$ and $x_{n+1} = \sqrt{2x_n}$ for each positive integer n. For example, $x_2 = \sqrt{2x_1} = \sqrt{2\sqrt{2}}$, $x_3 = \sqrt{2x_2} = \sqrt{2\sqrt{2\sqrt{2}}}$. Prove: $\langle x_n \rangle$ converges and $\lim x_n = 2$.

PROB. 8.8. Let $a>0$ be given. Define $\langle x_n \rangle$ as follows: $x_1>0$, $x_{n+1} = \frac{1}{2}(x_n + a/x_n)$. Prove: $\lim_{n\to+\infty} x_n = \sqrt{a}$.

PROB. 8.9 (See Probs. 5.2(b), 5.3, and 5.4). Prove: $\lim_{n\to+\infty}(1+1/n)^{n+1} = e$.

PROB. 8.10. Prove: $\lim_{n\to+\infty}(1-1/n)^n = 1/e$.

PROB. 8.11. Define sequences $\langle x_n \rangle$ and $\langle y_n \rangle$ as follows: (1) $0 < x_1 \leqslant y_1$ and (2) $x_{n+1} = \sqrt{x_n y_n}$ and $y_{n+1} = (x_n + y_n)/2$ for each n. Prove: $\langle x_n \rangle$ and $\langle y_n \rangle$ converge to the same limit L. (Gauss called this L the *arithmetic-geometric mean* of x_1 and y_1.)

PROB. 8.12.* Define the sequences $\langle x_n \rangle$ and $\langle y_n \rangle$ as follows: (1) $0 < x_1 \leqslant y_1$ and (2) $x_{n+1} = 2x_n y_n/(x_n + y_n)$, $y_{n+1} = (x_n + y_n)/2$ for each n (Note: x_{n+1} is the harmonic mean and y_{n+1} is the arithmetic mean of x_n and y_n (Prob. II.12.1). Prove: $\langle x_n \rangle$ and $\langle y_n \rangle$ each converge to $\sqrt{x_1 y_1}$ (Hint: note $x_{n+1} y_{n+1} = x_n y_n$ for each n).

PROB. 8.13. Given $\alpha \in \mathbb{R}$, $\alpha > 1$, and some positive integer k. Prove: (1) $\lim_{n\to+\infty}(\alpha^{n/k}/n) = +\infty$ and (2) $\lim_{n\to+\infty}(\alpha^n/n^k) = +\infty$.

PROB. 8.14. Prove: If $0<p<1$ and k is some positive integer, then (1) $\lim_{n\to+\infty}(np^{n/k}) = 0$ and (2) $\lim_{n\to+\infty}(n^k p^n) = 0$.

PROB. 8.15. Give some examples showing that $\lim(x_n + y_n)$ may exist, but $\lim x_n$ and $\lim y_n$ need not exist.

PROB. 8.16. Give some examples showing that $\lim x_n y_n$ may exist, but $\lim x_n$ and $\lim y_n$ need not exist.

PROB. 8.17. Evaluate $\lim_{n\to+\infty}(n+1)^{1/n}$.

9. L'Hôpital's Theorem for Real Sequences

Theorem 9.1. *If $\langle a_n \rangle$ is a real sequence and $\langle b_n \rangle$ is a real sequence such that $b_{n+1} > b_n > 0$ for all n and $\lim b_n = +\infty$, then*

$$\lim_{n\to+\infty} \frac{a_{n+1} - a_n}{b_{n+1} - b_n} = L \quad \textit{implies} \quad \lim \frac{a_n}{b_n} = L. \tag{9.1}$$

* T. J. I. Bromwich: *Introduction to the Theory of Infinite Series*, Macmillan, 1942, p. 23, Prob. 9.

PROOF. First suppose that $L = +\infty$. Because of (9.1), if a real B is given, then there exists a positive integer N such that

$$\frac{a_{n+1} - a_n}{b_{n+1} - b_n} > 2|B| + 1 \qquad \text{if} \quad n \geqslant N. \tag{9.2}$$

For each positive integer k define

$$m_{N,k} = \min\left\{ \frac{a_{N+1} - a_N}{b_{N+1} - b_N}, \frac{a_{N+2} - a_{N+1}}{b_{N+2} - b_{N+1}}, \dots, \frac{a_{N+k} - a_{N+k-1}}{b_{n+k} - a_{N+k-1}} \right\}. \tag{9.3}$$

This and (9.2) yield

$$1 + 2|B| < m_{N,k} \qquad \text{for each } k. \tag{9.4}$$

By Prob. II.12.13, part (1),

$$\begin{aligned} m_{n,k} &\leqslant \frac{(a_{N+1} - a_N) + (a_{N+2} - a_{N+1}) + \cdots + (a_{N+k} - a_{N+k-1})}{(b_{N+1} - b_N) + (b_{n+2} - b_N) + \cdots + (b_{N+k} - b_{N+k-1})} \\ &= \frac{a_{N+k} - a_N}{b_{N+k} - b_N} \end{aligned} \tag{9.5}$$

(Note: this is where the condition $b_{n+1} > b_n$ is used). Thus,

$$1 + 2|B| < \frac{a_{N+k} - a_N}{b_{N+k} - b_N} \qquad \text{for each } k. \tag{9.6}$$

Now consider the sequence $\langle u_k \rangle$, where

$$u_k = \frac{a_{N+k} - a_N}{b_{N+k} - b_N} = \frac{1}{1 - b_N / b_{N+k}} \left(\frac{a_{N+k} - a_N}{b_{N+k}} \right) \qquad \text{for each } k. \tag{9.7}$$

Since $\lim b_n = +\infty$, $\lim_{k \to +\infty}(b_n / b_{N+k}) = 0$. Therefore, a positive integer N_2 exists such that

$$0 < \frac{b_N}{b_{N+k}} < \frac{1}{2}, \quad \text{i.e.,} \quad 1 - \frac{b_N}{b_{N+k}} > \frac{1}{2} \qquad \text{for} \quad k \geqslant N_2. \tag{9.8}$$

This, (9.7), and (9.6) yield

$$1 + 2|B| < 2\frac{a_{N+k} - a_N}{b_{N+k}}.$$

Hence

$$\frac{2a_N}{b_{N+k}} + 1 + 2|B| < \frac{2a_{N+k}}{b_{N+k}} \qquad \text{for} \quad k \geqslant N_2.$$

Taking $\underline{\lim}$ of both sides and using the fact that on the left-hand side the limit exists and is equal to $1 + 2|B|$, we find that

$$1 + 2|B| = \lim_{k \to +\infty} \left(\frac{2a_N}{b_{N+k}} + 1 + 2|B| \right) \leqslant 2 \varliminf_{k \to +\infty} \frac{a_{N+k}}{b_{N+k}}.$$

From this it follows that

$$2|B| < 1 + 2|B| \leqslant 2 \varliminf \frac{a_{N+k}}{b_{N+k}},$$

so that

$$B \leqslant |B| \leqslant \lim_{k\to+\infty} \frac{a_{N+k}}{b_{N+k}} \qquad \text{for each real } B.$$

This implies that

$$+\infty = \lim_{k\to+\infty} \frac{a_{N+k}}{b_{N+k}} \quad \text{and, hence, that} \quad \lim_{k\to+\infty} \frac{a_{N+k}}{b_{N+k}} = +\infty.$$

By Lemma 8.1 we obtain

$$\lim_{n\to+\infty} \frac{a_n}{b_n} = +\infty.$$

This proves the theorem for the case $L = +\infty$. We leave the proof for the case $L = -\infty$ to the reader (Prob. 9.1).

We now consider the case $L \in \mathbb{R}$. By the hypothesis in (9.1), for each $\epsilon > 0$ there exists a positive integer N such that

$$L - \frac{\epsilon}{2} < \frac{a_{n+1} - a_n}{b_{n+1} - b_n} < L + \frac{\epsilon}{2} \qquad \text{for} \quad n \geqslant N. \tag{9.9}$$

We now define $m_{N,k}$ as in (9.3) and $M_{N,k}$ as

$$M_{N,k} = \max\left\{ \frac{a_{N+1} - a_N}{b_{N+1} - b_N}, \frac{a_{N+2} - a_{N+1}}{b_{N+2} - b_{N+1}}, \ldots, \frac{a_{N+k} - a_{N+k-1}}{b_{N+k} - b_{N+k-1}} \right\}$$

for each k and obtain by (7.9) and Prob. 12.13

$$L - \frac{\epsilon}{2} < m_{N,k} \leqslant \frac{(a_{N+1} - a_N) + (a_{N+2} - a_{N+1}) + \cdots + (a_{N+k} - a_{N+k-1})}{(b_{N+1} - b_N) + (b_{N+2} - b_{N+1}) + \cdots + (b_{N+k} - b_{N+k-1})}$$

$$\leqslant M_{N,k} < L + \frac{\epsilon}{2},$$

so that

$$L - \frac{\epsilon}{2} < \frac{a_{N+k} - a_N}{b_{N+k} - b_N} < L + \frac{\epsilon}{2} \qquad \text{for each positive integer } k. \tag{9.10}$$

This is equivalent to

$$\left(1 - \frac{b_N}{b_{N+k}}\right)\left(L - \frac{\epsilon}{2}\right) < \frac{a_{N+k} - a_N}{b_{N+k}} < \left(1 - \frac{b_N}{b_{N+k}}\right)\left(L + \frac{\epsilon}{2}\right)$$

for each k. This can be written as

$$-\left(1 - \frac{b_N}{b_{N+k}}\right)\frac{\epsilon}{2} < \frac{a_{N+k}}{b_{N+k}} - L - \left(\frac{a_N - Lb_n}{b_{N+k}}\right) < \left(1 - \frac{b_N}{b_{N+k}}\right)\frac{\epsilon}{2}.$$

Using properties of absolute value, we obtain from this

$$\left|\frac{a_{N+k}}{b_{N+k}} - L\right| - \left|\frac{a_N - Lb_N}{b_{N+k}}\right| \leqslant \left|\frac{a_{N+k}}{b_{N+k}} - L - \left(\frac{a_N - Lb_N}{b_{N+k}}\right)\right|$$

$$< \left(1 - \frac{b_N}{b_{N+k}}\right)\frac{\epsilon}{2} < \frac{\epsilon}{2}$$

for each k. This yields

$$\left|\frac{a_{N+k}}{b_{N+k}} - L\right| < \frac{\epsilon}{2} + \left|\frac{a_N - Lb_N}{b_{N+k}}\right| \qquad \text{for each } k. \tag{9.11}$$

Taking $\overline{\lim}$ on both sides and noting that the limit of the right-hand side exists and is equal to $\epsilon/2$, we arrive at

$$\overline{\lim_{k\to+\infty}} \left|\frac{a_{N+k}}{b_{N+k}} - L\right| \leqslant \frac{\epsilon}{2} < \epsilon.$$

Thus,

$$0 \leqslant \underline{\lim_{k\to+\infty}} \left|\frac{a_{N+k}}{b_{N+k}} - L\right| \leqslant \overline{\lim_{k\to+\infty}} \left|\frac{a_{N+k}}{b_{N+k}} - L\right| < \epsilon \qquad \text{for all} \quad \epsilon > 0.$$

This implies (explain) that

$$\lim_{k\to+\infty} \left|\frac{a_{n+k}}{b_{n+k}} - L\right| = 0, \quad \text{and, hence, that} \quad \lim_{k\to+\infty} \frac{a_{n+k}}{b_{n+k}} = L.$$

Applying Lemma 8.1 we obtain the desired conclusion for the case $L \in \mathbb{R}$.

PROB. 9.1. Prove Theorem 9.1 for the case $L = -\infty$.

EXAMPLE 9.1. We prove that if $\lim a_n = L \in \mathbb{R}^*$, then

$$\lim_{n\to+\infty} \frac{a_1 + \cdots + a_n}{n} = L. \tag{9.12}$$

(This result is due to Cauchy.) Let $S_n = \sum_{k=1}^{n} a_k = a_1 + \cdots + a_n$ and $b_n = n$ for each positive integer n. We have $S_{n+1} - S_n = a_n$ for each n and hence

$$\lim_{n\to+\infty} \frac{S_{n+1} - S_n}{b_{n+1} - b_n} = \lim_{n\to+\infty} a_{n+1} = L.$$

By Theorem 9.1,

$$\lim_{n\to+\infty} \frac{a_1 + \cdots + a_n}{n} = \lim_{n\to+\infty} \frac{S_n}{n} = L.$$

PROB. 9.2. Prove:

$$\lim_{n\to+\infty} \frac{1 + 1/2 + \cdots + 1/n}{n} = 0.$$

10. Criteria for the Convergence of Real Sequences

In this section we present two criteria for the convergence of sequences of real numbers. It is important to know whether or not a sequence converges even if we do not know its limit. We encountered such a criterion for bounded monotonic sequences and proved that such sequences converge. We now state such criteria for general real sequences.

Theorem 10.1. *If $\langle x_n \rangle$ is a real bounded sequence and we put, as usual*

$$\overline{A}_k = \sup_{n \geqslant k} x_n, \qquad \underline{A}_k = \inf_{n \geqslant k} x_n \qquad \textit{for each } k,$$

then $\langle x_n \rangle$ converges if and only if for each $\epsilon > 0$, there exists a positive integer N such that

$$\overline{A}_N - \underline{A}_N < \epsilon. \tag{10.1}$$

PROOF. Let $\underline{L} = \underline{\lim}\, x_n$ and $\overline{L} = \overline{\lim}\, x_n$. Since $\langle x_n \rangle$ is bounded, we know that

$$-\infty < \inf x_n = \underline{A}_1 \leqslant \overline{A}_1 = \sup x_n < +\infty$$

and, hence, that

$$-\infty < \underline{A}_k \leqslant \underline{L} \leqslant \overline{L} \leqslant \overline{A}_k < +\infty \qquad \text{for each } k. \tag{10.2}$$

This implies that

$$0 \leqslant \overline{L} - \underline{L} \leqslant \overline{A}_k - \underline{A}_k \qquad \text{for each } k. \tag{10.3}$$

Now assume that for each $\epsilon > 0$ there exists a positive integer N such that (10.1) holds. By (10.4) this implies

$$0 \leqslant \overline{L} - \underline{L} < \epsilon \qquad \text{for each} \quad \epsilon > 0. \tag{10.4}$$

It follows from this that $\overline{L} - \underline{L} = 0$ (explain) and, hence, that $\underline{L} = \overline{L}$. But then $\langle x_n \rangle$ has the limit $L = \underline{L} = \overline{L}$ and also that $-\infty < \underline{L} = \overline{L} < +\infty$. Hence, $\langle x_n \rangle$ converges.

Conversely, assume that $\langle x_n \rangle$ converges so that $-\infty < \underline{L} = \overline{L} < +\infty$. Let $L = \underline{L} = \overline{L}$. Then $\langle x_n \rangle$ is necessarily bounded now. Given $\epsilon > 0$, we have

$$L - \frac{\epsilon}{2} < L = \underline{L} \quad \text{and} \quad \overline{L} = L < L + \frac{\epsilon}{2}.$$

There exists positive integers N_1 and N_2 such that

$$L - \frac{\epsilon}{2} < \underline{A}_{N_1} \quad \text{and} \quad \overline{A}_{N_2} < L + \frac{\epsilon}{2}.$$

Put $N = \max\{N_1, N_2\}$. This implies that $N \geqslant N_1$ and $N \geqslant N_2$ so that $\underline{A}_{N_1} \leqslant \underline{A}_N$ and $\overline{A}_N \leqslant \overline{A}_{N_2}$ and, therefore,

$$L - \frac{\epsilon}{2} < \underline{A}_{N_1} \leqslant \underline{A}_N \leqslant \overline{A}_N \leqslant \overline{A}_{N_2} < L + \frac{\epsilon}{2}.$$

Hence,

$$\overline{A}_N - \underline{A}_N < \epsilon.$$

This completes the proof.

Def. 10.1. A real sequence $\langle x_n \rangle$ is called a *Cauchy sequence* if and only if for each $\epsilon > 0$ there exists an $N \in \mathbb{R}$ such that

$$\text{if} \quad m > N \quad \text{and} \quad n > N, \quad \text{then} \quad |x_m - x_n| < \epsilon. \tag{10.5}$$

For example, the sequence $\langle x_n \rangle$ where $x_n = n^{-1}$ for each positive integer n, is a Cauchy sequence since

$$\left| \frac{1}{m} - \frac{1}{n} \right| \leqslant \frac{1}{m} + \frac{1}{n}.$$

Let $\epsilon > 0$ be given. Take $N = 2/\epsilon$. If $m > N$ and $n > N$, we have

$$\left| \frac{1}{m} - \frac{1}{n} \right| \leqslant \frac{1}{m} + \frac{1}{n} < \frac{\epsilon}{2} + \frac{\epsilon}{2} = \epsilon.$$

The main theorem on Cauchy sequences will be Theorem 10.2 below. Before stating and proving this theorem we ask the reader to do Probs. 10.1–10.4 below. These are problems concerning equivalent formulations of the notion of a Cauchy Sequence.

PROB. 10.1. Prove: $\langle x_n \rangle$ is a Cauchy sequence if and only if for each $\epsilon > 0$ there exists an $N \in \mathbb{R}$ such that if $m > n > N$, then $|x_m - x_n| < \epsilon$.

PROB. 10.2. Prove: $\langle x_n \rangle$ is a Cauchy sequence if and only if for each $\epsilon > 0$ there exists a positive integer N such that if $m \geqslant N$ and $n \geqslant N$, then $|x_m - x_n| < \epsilon$.

PROB. 10.3. Prove: $\langle x_n \rangle$ is a Cauchy sequence if and only if for each $\epsilon > 0$ there exists a positive integer N such that if $m > n \geqslant N$, then $|x_m - x_n| < \epsilon$.

PROB. 10.4. Prove: $\langle x_n \rangle$ is a Cauchy sequence if and only if for each $\epsilon > 0$ there exists a positive integer N such that if n is an integer such that $n \geqslant N$ and p is any positive integer, then $|x_{n+p} - x_n| < \epsilon$.

Theorem 10.2. *A sequence of real numbers converges if and only if it is a Cauchy sequence.*

PROOF. Let $\langle x_n \rangle$ be a real sequence which converges. There exists a real number L such that $\lim x_n = L$. Given $\epsilon > 0$, there exists an N such that if $n > N$, then $|x_n - L| < \epsilon/2$. If m is a positive integer such that $m > N$, then $|x_m - L| < \epsilon/2$. Hence, if $m > N$ and $n > N$, then

$$|x_m - x_n| = |x_m - L - (x_n - L)| \leqslant |x_m - L| + |x_n - L| < \frac{\epsilon}{2} + \frac{\epsilon}{2} = \epsilon.$$

This proves: If $\langle x_n \rangle$ converges, then it is a Cauchy sequence.

Conversely, assume $\langle x_n \rangle$ is a Cauchy sequence. Given $\epsilon > 0$, there exists a positive N (Prob. 10.2) such that

$$-\frac{\epsilon}{3} < x_m - x_n < \frac{\epsilon}{3} \qquad \text{if} \quad m \geqslant N \quad \text{and} \quad n \geqslant N.$$

It follows that

$$x_N - \frac{\epsilon}{3} < x_n < x_N + \frac{\epsilon}{3} \qquad \text{if} \quad n \geqslant N.$$

This implies that

$$x_N - \frac{\epsilon}{3} \leqslant \underline{A}_N \leqslant \overline{A}_N \leqslant x_N + \frac{\epsilon}{3} \tag{10.6}$$

so that $-\infty < \underline{A}_N \leqslant \overline{A}_N < +\infty$. This, in turn, implies that $\langle x_n \rangle$ is bounded (why?). Inequality (10.6) also yields

$$\overline{A}_N - \underline{A}_N \leqslant \frac{2\epsilon}{3} < \epsilon.$$

Hence, by Theorem 10.1, $\langle x_n \rangle$ converges. This completes the proof.

Def. 10.2. An ordered field (Remark I.7.2) $\mathfrak{F}$ in which every Cauchy sequence converges to an element of $\mathfrak{F}$ is said to be *Cauchy-complete*.

Remark 10.1. According to the above terminology, Theorem 10.2 states that the real numbers constitute a Cauchy-complete ordered field. This result may be formulated as follows: Every order-complete (cf. Remarks I.12.2) ordered field is Cauchy-complete.

The rational numbers form an ordered field which is not order-complete (Remark I.12.1 and Theorem I.12.1). It is not difficult to prove that the rational numbers are not Cauchy-complete. Consider the sequence $\langle x_1, x_2, \ldots \rangle$ in which

$$x_1 = 1, \qquad x_{n+1} = \frac{1}{2}\left(x_n + \frac{2}{x_n}\right)$$

for each positive integer n. This sequence is a sequence of rational numbers (explain) and moreover it converges to the limit $\sqrt{2}$ (Prob. 8.8). It, therefore, is an example of a Cauchy sequence (because it converges) of rational elements of $\mathbb{Q}$ which does not converge to an element of $\mathbb{Q}$.

Def. 10.3. An ordered field $\mathfrak{F}$ is called an *Archimedean-ordered* field if it has the Archimedean property (Theorem I.9.1), that is, if a and b are elements of $\mathfrak{F}$ such that $a > 0$ and $b > 0$, then there exists a positive integer n such that $na > b$.

The system $\mathbb{R}$ of real numbers is Archimedean-ordered (Theorem I.9.1). In the next problem we ask the reader to prove that the system $\mathbb{Q}$ of rational numbers is Archimedean-ordered.

PROB. 10.5. First prove: If i and j are positive *integers*, then there exists a positive integer n such that $ni > j$, then prove: If r and s are positive *rational* numbers, there exists a positive integer n such that $nr > s$.

In Theorem 10.3 below we prove that an Archimedean-ordered field which is Cauchy-complete is also order-complete. We state some lemmas first.

Lemma 10.1. *If $\mathfrak{F}$ is an Archimedean-ordered field and $\mathfrak{A}$ is a nonempty subset of $\mathfrak{F}$ which is bounded from above, there exists a unique integer n in $\mathfrak{F}$ such that $n+1$ is an upper bound for $\mathfrak{A}$ but n is not.*

PROOF. By hypothesis $\mathfrak{A} \subseteq \mathfrak{F}$ is bounded from above so that $\mathfrak{F}$ contains an upper bound b of $\mathfrak{A}$. Since $\mathfrak{F}$ is Archimedean-ordered, there exists an integer N in $\mathfrak{F}$ such that $b < N$ (see the proof of Corollary 2 of Theorem I.9.1.) Thus, N is an *integral* upper bound of $\mathfrak{A}$. Let S be the set of all integral upper bounds of $\mathfrak{A}$. Since $N \in S$, it follows that $S \neq \emptyset$. Also, there exists $x_0 \in \mathfrak{A}$. Clearly, x_0 is a lower bound for S. Since $[x_0] \leqslant x_0$, the integer $[x_0]$ is a lower bound for S. S is now seen to be a nonempty set of integers which is bounded from below. As such S has a least number, m_0 say. Thus, m_0 is an integer and is an upper bound for $\mathfrak{A}$ but $m_0 - 1$ is not an upper bound for $\mathfrak{A}$. Writing $n = m_0 - 1$, we have: the integer $n + 1 = m_0$ is an upper bound for $\mathfrak{A}$, but $n = m_0 - 1$ is not. We leave the proof of the uniqueness of n to the reader.

The next lemma is stated in terms of the digits 0, 1, 2, 3, 4, 5, 6, 7, 8, 9.

Lemma 10.2. *If $\mathfrak{F}$ is an Archimedean-ordered field and $\mathfrak{A}$ is a nonempty set of elements of $\mathfrak{F}$ which is bounded from above, there exists an integer N, an infinite sequence $\langle d_n \rangle$ of digits such that $d_n \neq 0$ for infinitely many n's and such that the sequences $\langle r_n \rangle$ and $\langle q_n \rangle$, defined by means of*

$$r_n = N + \sum_{k=1}^{n} d_k 10^{-k} \quad \text{and} \quad q_n = r_n + 10^{-n} \tag{10.7}$$

for each nonnegative integer n, have the property: for each n, q_n is an upper bound for $\mathfrak{A}$ but r_n is not.

PROOF. By Lemma 10.1, there exists an integer N in $\mathfrak{F}$ such that $N + 1$ is an upper bound for $\mathfrak{A}$ but N is not. Write $r_0 = N$ and $q_0 = r_0 + 1 = N + 1$. We prove there exists a digit d_1 such that

$$r_1 = N + d_1 10^{-1} \text{ is not an upper bound for } \mathfrak{A}, \text{ but } q_1 = r_1 + 10^{-1} \text{ is.} \tag{10.8}$$

Take all digits d such that

$$N + (d+1)10^{-1} \qquad \text{is an upper bound for } \mathfrak{A}. \tag{10.9}$$

$d = 9$ is such a digit. Let d_1 be the least digit d for which (10.9) holds. We have $0 \leqslant d_1 \leqslant 9$ and $1 \leqslant d_1 + 1 \leqslant 10$. Here $d_1 + 1$ is the least integer $d + 1$

with $1 \leqslant d+1 \leqslant 10$ for which (10.9) holds. Since $d_1 < d_1 + 1$, it follows that $r_1 = N + d_1 10^{-1}$ is not an upper bound for $\mathfrak{A}$ and that $q_1 = r_1 + 10^{-1} = N + (d_1 + 1)10^{-1}$ is. Thus, d_1 is a digit for which (10.8) holds. Using $n = 1$ in (10.7), we see $r_n = r_1$ is not an upper bound for $\mathfrak{A}$ but $q_n = q_1$ is.

We proceed inductively and assume that for some positive integer n there exist digits $d_1, d_2, \ldots, d_n$ such that for r_n and q_n as defined in (10.7); r_n is not an upper bound for $\mathfrak{A}$ and q_n is. Now take all the digits d such that

$$r_n + (d+1)10^{-n-1} \qquad \text{is an upper bound for } \mathfrak{A}. \tag{10.10}$$

Note that $d = 9$ is such a digit. Let d_{n+1} be the least digit d for which (10.10) holds. Then we have $0 \leqslant d_{n+1} \leqslant 9$, and $1 \leqslant d_{n+1} + 1 \leqslant 10$. Here $d_{n+1} + 1$ is the least integer $d + 1$ with $1 \leqslant d + 1 \leqslant 10$ for which (10.10) holds. Define r_{n+1} and q_{n+1} as

$$r_{n+1} = r_n + d_{n+1}10^{-n-1}$$

and

$$q_{n+1} = r_n + (d_{n+1} + 1)10^{-n-1}.$$

Since $d_{n+1} < d_{n+1} + 1$, r_{n+1} is not an upper bound for $\mathfrak{A}$ and q_{n+1} is. By the principle of mathematical induction, there exists a sequence of digits $\langle d_n \rangle$ and sequences $\langle r_n \rangle$ and $\langle q_n \rangle$ defined as in (10.7), such that for each n, r_n is not an upper bound for $\mathfrak{A}$ and q_n is, as claimed.

We prove next that $d_n \neq 0$ for infinitely many n's. In order to obtain a contradiction, suppose that there exists positive integer j such that if $n > j$, then $d_n = 0$. We have: $n \geqslant j$ implies $r_n = r_j$ and $q_n = r_j + 10^{-n}$. Here, if $n \geqslant j$,

$$r_n = r_j \text{ is not an upper bound for } \mathfrak{A} \text{ but } q_n = r_j + 10^{-n} \text{ is.} \tag{10.11}$$

Take $\epsilon > 0$. There exists a positive integer n' such that $1/n' < \epsilon$ so that

$$10^{-n'-1} = \frac{1}{10^{n'}} < \frac{1}{n'} < \epsilon.$$

Let $M = \max\{n', j\}$ so that M is an integer such that $M \geqslant n'$ and $M \geqslant j$ and, therefore,

$$10^{-M} \leqslant 10^{-n'} < \epsilon$$

and

$$r_j + 10^{-M} < r_j + \epsilon. \tag{10.12}$$

Here, because of (10.11), the left-hand side is an upper bound for $\mathfrak{A}$, so the right-hand side is. Assume $x \in \mathfrak{A}$, so that $x \leqslant r_j + \epsilon$. Since this is true for all $\epsilon > 0$, $x \leqslant r_j$. Thus, r_j is an upper bound for $\mathfrak{A}$. But this contradicts the fact that r_j is not an upper bound for $\mathfrak{A}$ (cf. (10.11)). We, therefore, conclude that for each positive integer j there exists an integer $n > j$ such that $d_n > 0$. Accordingly, infinitely many d_n's are positive and, hence, $\neq 0$. This completes the proof.

Remark 10.2. We observe that if j is a nonnegative integer and n is an integer such that $n > j$ and $d_{j+1}, d_{j+2}, \dots, d_n$ are digits where $d_{j+1} > 0$, then

$$\frac{1}{10^{j+1}} \leqslant \frac{d_{j+1}}{10^{j+1}} + \frac{d_{j+2}}{10^{j+2}} + \cdots + \frac{d_n}{10^n} < \frac{1}{10^j}. \tag{10.13}$$

The inequality on the left is obvious. To establish the right-hand inequality note that $0 \leqslant d_k \leqslant 9$ for $j+1 \leqslant k \leqslant n$ so

$$\begin{aligned}
\frac{d_{j+1}}{10^{j+1}} + \cdots + \frac{d_n}{10^n} &\leqslant \frac{9}{10^{j+1}} + \cdots + \frac{9}{10^n} \\
&= \frac{9}{10^{j+1}}\left(1 + \frac{1}{10} + \cdots + \frac{1}{10^{n-j-1}}\right) \\
&= \frac{9}{10^{j+1}}\left(\frac{1 - 1/10^{n-j}}{1 - \frac{1}{10}}\right) \\
&= \frac{1}{10^j}\left(1 - \frac{1}{10^{n-j}}\right) \\
&< \frac{1}{10^j}.
\end{aligned}$$

PROB. 10.6. Prove that the sequence $\langle r_n \rangle$ of Lemma 10.2 is monotonic increasing, and that the sequence $\langle q_n \rangle$ there is monotonic decreasing.

Theorem 10.3. *An Archimedean-ordered field $\mathfrak{F}$ that is Cauchy-complete is also order-complete.*

PROOF. Assume that $\mathfrak{A}$ is a nonempty subset of $\mathfrak{F}$ which is bounded from above. By Lemma 10.2, there exists an integer N and a sequence $\langle d_n \rangle$ of digits such that $d_n \neq 0$ for infinitely many n's, and such that the sequences $\langle r_n \rangle$ and $\langle q_n \rangle$ defined for each n by the equations

$$r_n = N + \sum_{k=1}^{n} d_k 10^{-k}$$

and

$$q_n = r_n + 10^{-n}$$

have the property

$$r_n \text{ is not an upper bound for } \mathfrak{A} \text{ but } q_n \text{ is.}$$

Take $\epsilon > 0$. There exists a positive integer M such that $10^{-M} < \epsilon$. For integers m and n such that $m > n > M$, we have

$$|r_m - r_n| = \frac{d_{n+1}}{10^{n+1}} + \cdots + \frac{d_m}{10^m} < \frac{1}{10^n} < \frac{1}{10^M} < \epsilon. \tag{10.14}$$

This implies that $\langle r_n \rangle$ is a Cauchy sequence of elements of $\mathfrak{F}$. Since $\mathfrak{F}$ is Cauchy-complete by hypothesis, there exists $L \in \mathfrak{F}$ such that $\lim r_n = L$. Since $\langle r_n \rangle$ is monotonic increasing, we know $r_n \leqslant L$ for all n (why?), so that L is an upper bound for $\langle r_n \rangle$. If M is a real number such that $M < L$, then $M = L - (L - M)$, where $L - M > 0$, so an integer n_0 exists s such that $|r_n - L| < L - M$ for $n \geqslant n_0$ and, hence, $M - L < r_n - L < L - M$ for $n \geqslant n_0$. This implies that $M < r_n$ for $n \geqslant n_0$ and, hence, that M is not an upper bound for $\langle r_n \rangle$. Thus, any upper bound M for $\langle r_n \rangle$ is $\geqslant L$. Accordingly,

$$\sup r_n = L = \lim r_n .$$

We prove that $\sup \mathfrak{U} = L$. Assume $x \in \mathfrak{U}$. The properties of the sequence $\langle q_n \rangle$ imply that

$$x \leqslant q_n = r_n + 10^{-n} \qquad \text{for each } n. \tag{10.15}$$

Since $\lim q_n = \lim(r_n + 10^{-n}) = \lim r_n + \lim 10^{-n} = L + 0 = L$, it follows from (10.15) that $x \leqslant L$. This proves that L is an upper bound for the set $\mathfrak{U}$. Now take $B \in \mathfrak{F}$ such that $B < L$ and put $\epsilon = L - B$, so that $\epsilon > 0$ and $B = L - \epsilon$. Since $L = \sup r_n$, there exists an integer n_1 such that $B = L - \epsilon < r_{n_1}$. But r_{n_1} is not an upper bound for $\mathfrak{U}$. Since $B < r_{n_1}$, it follows that B is not an upper bound for $\mathfrak{U}$. Thus, no $B \in \mathfrak{F}$ such that $B < L$ is an upper bound for $\mathfrak{U}$. Since L is an upper bound for $\mathfrak{U}$, we conclude that $L = \sup \mathfrak{U}$. Thus, each nonempty subset $\mathfrak{U}$ of $\mathfrak{F}$ which is bounded from above has a supremum in $\mathfrak{F}$. $\mathfrak{F}$ is therefore order-complete, as claimed.

Theorem 10.4. *Corresponding to each real number x, there exists a unique integer N and a unique sequence $\langle d_n \rangle$ of digits such that $d_n \neq 0$ for infinitely many n's and such that for the sequence $\langle r_n \rangle$, where*

$$r_n = N + \sum_{k=1}^{n} d_k 10^{-k} \qquad \textit{for each } n, \tag{10.16}$$

we have

$$r_n < x \leqslant r_n + 10^{-n} \qquad \textit{for each } n. \tag{10.17}$$

PROOF. Apply Lemma 10.2 to the set $\mathfrak{U} = (-\infty, x]$. This is a nonempty set of real numbers which is bounded from above by x. There exists an integer N and a sequence $\langle d_n \rangle$ of digits such that the sequence $\langle r_n \rangle$ defined in (10.16) and the sequence $\langle q_n \rangle$, where $q_n = r_n + 10^{-n}$ for each n have the property that for each n, r_n is not an upper bound for $\mathfrak{U}$ but q_n is. Since $x = \sup \mathfrak{U}$, this implies that (10.17) holds.

We prove the uniqueness of N and the $\langle d_n \rangle$. Let N' be an integer and $\langle d_n' \rangle$ be a sequence of digits such that for the sequence $\langle r_n' \rangle$, where $r_n' = N' + \sum_{k=1}^{n} d_k' 10^{-k}$, we have

$$r_n' < x \leqslant r_n' + 10^{-n} \qquad \text{for each } n.$$

If $N > N'$, then $N \geqslant N' + 1$. Since $r_n > N \geqslant N' + 1 \geqslant x$, we obtain a contradiction to $r_n < x$. $N' > N$ is also impossible. Hence, $N = N'$. Suppose $d_n \neq d'_n$ holds for some n. Let j be the least positive integer such that $d_j \neq d'_j$. Assume $d_j < d'_j$ so that $d_j + 1 \leqslant d'_j$. For $k < j$, we have $d_k = d'_k$. Hence,

$$\begin{aligned} r'_j &= N + \sum_{k=1}^{j} d'_k 10^{-k} = N + \sum_{k=1}^{j-1} d_k 10^{-k} + \frac{d'_j}{10^j} \\ &\geqslant N + \sum_{k=1}^{j-1} d_k 10^{-k} + \frac{(d_j + 1)}{10^j} \\ &= N + \sum_{k=1}^{j} d_k 10^{-k} + \frac{1}{10_j} \\ &= r_j + \frac{1}{10^j} \\ &= q_j . \end{aligned}$$

Thus, $r'_j \geqslant q_j$ which is impossible (explain). Therefore, we have $N = N'$ and $d_n = d'_n$ for all n so that $r_n = r'_n$ for all n.

Remark 10.3. Let $x \in \mathbb{R}$. The sequence $\langle r_n \rangle$ of the last theorem, where $r_n = N + \sum_{k=1}^{n} d_k 10^{-k}$ for each n such that $d_n \neq 0$ for infinitely many n's and

$$r_n < x \leqslant r_n + 10^{-n},$$

has the property

$$|x - r_n| < \frac{1}{10^n} \qquad \text{for each } n.$$

Hence, $\lim r_n = x$.

CHAPTER IV

Infinite Series of Real Numbers

1. Infinite Series of Real Numbers. Convergence and Divergence

The sums

$$\sum_{k=1}^{n} a_k = a_1 + \cdots + a_n, \qquad \sum_{k=0}^{n} a_k = a_0 + \cdots + a_n,$$

where n is some positive integer, were defined in Chapter II. They are examples of finite sums. Now we define the "sum" of the infinite series

$$\sum_{n=1}^{\infty} a_n = a_1 + a_2 + \cdots \quad \text{or} \quad \sum_{k=0}^{\infty} a_k = a_0 + a_1 + \cdots .$$

Def. 1.1. If $\langle a_n \rangle$ is a real sequence, then the sequence $\langle S_n \rangle$, where for each n,

$$S_n = a_1 + \cdots + a_n,$$

is called the *infinite series* of terms of the sequence $\langle a_n \rangle$. The nth term of the sequence $\langle S_n \rangle$ is called the nth *partial sum* of the series. The series $\langle S_n \rangle$ is written $\sum_{n=1}^{\infty} a_n$ and a_n is called the nth *term* of this series.

According to this definition, $\sum_{n=1}^{\infty} a_n$ is merely a notation for the sequence $\langle S_n \rangle$ of partial sums S_n, where

$$S_1 = \sum_{k=1}^{1} a_k = a_1, \quad S_2 = \sum_{k=1}^{2} a_k = a_1 + a_2,$$

$$S_3 = \sum_{k=1}^{3} a_k = a_1 + a_2 + a_3, \quad \text{etc.}$$

The partial sums are defined inductively from the sequence $\langle a_n \rangle$ of terms of $\sum_{n=1} a_n$ by means of

$$\left.\begin{aligned} S_1 &= a_1 \\ S_{n+1} &= S_n + a_{n+1} \end{aligned}\right. \qquad \text{for each } n.$$

If N is some positive integer, then $\sum_{k=N}^{\infty} a_k$ is the infinite series of terms of $\langle a_n \rangle_{n \geqslant N} = \langle a_N, a_{N+1}, \ldots \rangle$.

Def. 1.2. The series $\sum_{k=1}^{\infty} a_k$ is said to *converge* if and only if the sequence $\langle S_n \rangle$ of its partial sums converges. Otherwise it is said to *diverge*. If

$$S = \lim_{n \to +\infty} S_n = \lim_{n \to +\infty} \sum_{k=1}^{n} a_k ,$$

then S is called the *sum* of the series. We also write

$$S = \sum_{n=1}^{\infty} a_n .$$

Thus, when $\sum_{n=1}^{\infty} a_n$ converges, the symbol $\sum_{n=1}^{\infty} a_n$ has two meanings: In Def. 1.1 it denotes the sequence $\langle S_n \rangle$ of partial sums, and in another sense, Def. 1.2, it denotes the sum S of the series. We hope that the sense in which $\sum_{n=1}^{\infty} a_n$ will be used will be clear from the context.

According to Def. 1.2, the series has sum $S \in \mathbb{R}$ if and only if for each $\epsilon > 0$ there exists an N such that $n > N$ implies

$$|S - S_n| < \epsilon. \tag{1.1}$$

Using the Cauchy criterion for the convergence of $\langle S_n \rangle$, we see that $\sum_{n=1}^{\infty} a_n$ converges if and only if for each $\epsilon > 0$ there exists an N such that if $n > m > N$, then

$$\left| \sum_{k=m+1}^{m} a_k \right| = |a_{n+1} + \cdots + a_n| = |S_m - S_n| < \epsilon. \tag{1.2}$$

This is equivalent to saying that $\sum a_n$ converges if and only if for each $\epsilon > 0$ there exists an N such that if $n > N$, then

$$|a_{n+1} + \cdots + a_{n+p}| = |S_{n+p} - S_n| < \epsilon \qquad \text{for all positive integers } p. \tag{1.3}$$

We refer to (1.2), or its equivalent form (1.3), as the *Cauchy criterion for the convergence of series*.

If n is a nonnegative integer, then $\sum_{k=n+1}^{\infty} a_k$ is the series whose terms are those of the sequence $\langle a_k \rangle_{k \geqslant n+1}$. Writing $S_{n,m}$ for the nth partial sum of this series, we have

$$S_{n,m} = \sum_{k=n+1}^{m} a_k \qquad \text{where} \quad m > n. \tag{1.4}$$

If the above series converges, then we write its sum as R_n, so that

$$R_n = \lim_{m\to+\infty} S_{n,m} = \lim_{m\to+\infty} \sum_{k=n+1}^{m} a_k. \tag{1.5}$$

We call R_n the *remainder after n terms* of $\sum_{k=1}^{\infty} a_k$.

Prob. 1.1. Prove: (a) If $\sum_{k=1}^{\infty} a_k$ converges, then for each n, $\sum_{k=n+1}^{\infty} a_k$ converges; (b) if for some n, $\sum_{n+1}^{\infty} a_k$ converges, then so does $\sum_{k=1}^{\infty} a_k$. Also prove: $\sum_{k=1}^{\infty} a_k = S = S_n + R_n$ for each n and $R_n \to 0$ as $n \to +\infty$ in this case.

Example 1.1 (The Telescoping Series). The series

$$\sum_{n=1}^{\infty} \frac{1}{n(n+1)} = \frac{1}{1\cdot 2} + \frac{1}{2\cdot 3} + \cdots$$

converges and its sum is 1. We prove this. First note that if k is a positive integer, then

$$\frac{1}{k(k+1)} = \frac{1}{k} - \frac{1}{k+1}.$$

The nth partial sum for this series is S_n, where

$$S_n = \sum_{k=1}^{n} \frac{1}{k(k+1)} = \sum_{k=1}^{n} \frac{1}{k} - \frac{1}{k+1} = 1 - \frac{1}{n+1} = \frac{n}{n+1}.$$

Hence,

$$S = \lim_{n\to+\infty} S_n = \lim_{n\to+\infty} \frac{n}{n+1} = 1.$$

Prob. 1.2. Prove: If x is not a nonnegative integer, then

$$\sum_{n=1}^{\infty} \frac{1}{(x+n)(x+n+1)} = \frac{1}{1+x}.$$

Prob. 1.3. Prove:

(a)
$$\sum_{n=1}^{\infty} \frac{1}{n(n+1)(n+2)} = \frac{1}{4},$$

(b)
$$\sum_{n=1}^{\infty} \frac{1}{n(n+1)(n+2)(n+3)} = \frac{1}{18}.$$

We now present a necessary, but not sufficient, condition for the convergence of an infinite series.

Theorem 1.1. *If $\sum a_n$ converges, then $a_n \to 0$.*

PROOF. If n is a positive integer, then

$$S_n = S_{n-1} + a_n \quad \text{and} \quad S_n - S_{n-1} = a_n .$$

If S is the sum of the series, then $S = \lim S_n$, where for each n, S_n is the nth partial sum of the series. It follows (explain) that $S = \lim S_{n-1}$, so

$$0 = S - S = \lim_{n\to+\infty} S_n - \lim_{n\to+\infty} S_{n-1} = \lim_{n\to+\infty} (S_n - S_{n-1}) = \lim_{n\to+\infty} a_n .$$

This proves the theorem.

We shall see later (Example 1.3) that the converse of the above theorem is false. We apply this theorem in the next example to prove the divergence of certain series.

EXAMPLE 1.2 (The Geometric Series). The series

$$\sum_{n=0}^{\infty} x^n = 1 + x + x^2 + \cdots \tag{1.6}$$

is called the *geometric series*. We prove that it converges for $|x| < 1$ and diverges for $|x| \geqslant 1$. First note that if $x \neq 1$, then

$$S_n = 1 + x + x^2 + \cdots + x^n = \frac{1 - x^{n+1}}{1 - x} = \frac{1}{1-x} - x^n \frac{x}{1-x} . \tag{1.7}$$

If $|x| < 1$, then $\lim_{n\to+\infty} x^n = 0$. Hence

$$\lim_{n\to+\infty} S_n = \lim_{n\to+\infty} \left(\frac{1}{1-x} - x^n \frac{x}{1-x} \right) = \frac{1}{1-x} .$$

It follows that

$$1 + x + x^2 + \cdots = \frac{1}{1-x} \qquad \text{if} \quad |x| < 1. \tag{1.8}$$

However, if $|x| \geqslant 1$, then $\lim_{n\to+\infty} x^n \neq 0$. Therefore, by Theorem 1.1 $\sum x^n$ diverges for $|x| \geqslant 1$.

EXAMPLE 1.3 (The Converse of Theorem 1.1 Is False). Here, we give an example of a series for which $a_n \to 0$ as $n \to +\infty$ holds but such that $\sum a_n$ diverges. The series

$$\sum_{n=1}^{\infty} \frac{1}{n} = 1 + \frac{1}{2} + \cdots + \frac{1}{n} + \cdots \tag{1.9}$$

is called the *harmonic* series. We have $a_n = 1/n \to 0$ as $n \to +\infty$. We prove, however, that this series diverges. Let $\langle H_n \rangle$ be the sequence of partial sums of the harmonic series (1.9). Then

$$H_n = 1 + \frac{1}{2} + \cdots + \frac{1}{n} = \sum_{k=1}^{n} \frac{1}{k} . \tag{1.10}$$

Note that

$$H_1 = 1, \qquad H_2 = 1 + \tfrac{1}{2},$$

$$H_{2^2} = H_4 = 1 + \tfrac{1}{2} + \tfrac{1}{3} + \tfrac{1}{4} > 1 + \tfrac{1}{2} + \tfrac{1}{4} + \tfrac{1}{4} = 1 + 2(\tfrac{1}{2}).$$

We claim that, quite generally,

$$H_{2^n} \geqslant 1 + \frac{n}{2} \qquad \text{for each nonnegative integer } n. \tag{1.11}$$

We prove this by induction on n. Inequality (1.11) holds for $n = 0$, 1, and 2. Let (1.11) hold for some nonnegative integer n. Now

$$H_{2^{n+1}} = H_{2^n} + \sum_{k=2^n+1}^{2^{n+1}} \frac{1}{k}. \tag{1.12}$$

In the second sum on the right

$$\frac{1}{k} \geqslant \frac{1}{2^{n+1}} \qquad \text{for all integers } k \text{ such that } \quad 2^n + 1 \leqslant k \leqslant 2^{n+1} = 2^n + 2^n.$$

This implies that

$$\sum_{k=2^n+1}^{2^{n+1}} \frac{1}{k} \geqslant \sum_{k=2^n+1}^{2^n+2^n} \frac{1}{2^{n+1}} = \frac{1}{2^{n+1}} \sum_{k=2^n+1}^{2^n+2^n} 1 = \frac{1}{2^{n+1}}(2^n) = \frac{1}{2}.$$

This, (1.12), and the induction hypothesis yield

$$H_{2^{n+1}} = H_{2^n} + \sum_{k=2^n+1}^{2^{n+1}} \geqslant 1 + \frac{n}{2} + \frac{1}{2} = 1 + \frac{n+1}{2}.$$

By the principle of mathematical induction it follows that (1.11) holds. It is clear from (1.12) that $\lim_{n\to+\infty} H_{2^n} = +\infty$. Thus, the subsequence $\langle H_{2^n}\rangle$ diverges. This implies that $\langle H_n\rangle$ diverges. But then the harmonic series diverges, even though $\lim a_n = \lim(1/n) = 0$.

Remark 1.1. The *sum* and *difference* of $\sum a_n$ and $\sum b_n$ are defined respectively as

$$\sum(a_n + b_n) \quad \text{and} \quad \sum(a_n - b_n).$$

We also define $\sum(ca_n)$ as the series whose nth term for each n is ca_n, where c is some constant.

PROB. 1.4. Prove: If $\sum a_n$ and $\sum b_n$ converge, then so do $\sum(a_n + b_n)$, $\sum(a_n - b_n)$ and $\sum(ca_n)$ and we have (a) $\sum(a_n + b_n) = \sum a_n + \sum b_n$, (b) $\sum(a_n - b_n) = \sum a_n - \sum b_n$, (c) $\sum(ca_n) = c\sum a_n$.

2. Alternating Series

Def. 2.1. An *alternating* series is a series of the form

$$\sum_{n=1}^{\infty} (-1)^{n+1} a_n = a_1 - a_2 + a_3 - a_4 + \cdots, \tag{2.1}$$

where (1) $a_n > 0$ for all n, (2) $\langle a_n\rangle$ is strictly decreasing, and (3) $a_n \to 0$ as $n \to +\infty$.

For example, the series

$$\sum_{n=1}^{\infty}(-1)^{n+1}\frac{1}{n}=1-\frac{1}{2}+\frac{1}{3}-\frac{1}{4}+\cdots \tag{2.2}$$

is an alternating series.

Theorem 2.1. *Each alternating series $\sum_{n=1}^{\infty}(-1)^{n+1}a_n$ converges. Moreover, if S is its sum, we have for each partial sum S_n,*

$$|S-S_n|<a_{n+1}. \tag{2.3}$$

Proof. We first prove that the subsequence $\langle S_{2k}\rangle$ of even-indexed partial sums of the sequence $\langle S_n\rangle$ is strictly monotonic increasing. Note that $S_2=a_1-a_2>0$ and $S_4=S_2+(a_3-a_4)>S_2$. More generally, since $a_{2k+1} > a_{2k+2}$, we have

$$S_{2k+2}=S_{2k}+(a_{2k+1}-a_{2k+2})>S_{2k} \qquad \text{for each } k. \tag{2.4}$$

Since $a_{2k}>a_{2k+1}$, the subsequence $\langle S_{2k-1}\rangle$ of odd-indexed partial sums is strictly monotonic decreasing, as we see from

$$S_{2k+1}=S_{2k-1}-(a_{2k}-a_{2k+1})<S_{2k-1} \qquad \text{for each } k. \tag{2.5}$$

It also follows from

$$S_{2k}=S_{2k-1}-a_{2k}<S_{2k-1} \qquad \text{for each } k \tag{2.6}$$

that $S_{2k}<S_{2k-1}$ for each k.

We prove that each even-indexed partial sum is less than *every* odd-indexed partial sum. If $m<n$, since $\langle S_{2k}\rangle$ is increasing and (2.6) holds, we obtain

$$S_{2m}<S_{2n}<S_{2n-1}.$$

If $m\geqslant n$, we obtain from (2.6) and the fact that $\langle S_{2k-1}\rangle$ is decreasing that

$$S_{2m}<S_{2m-1}\leqslant S_{2n-1}.$$

Thus, we have

$$S_{2m}<S_{2n-1} \qquad \text{for any positive integers } m \text{ and } n. \tag{2.7}$$

It follows that each S_{2n-1} is an upper bound for $\langle S_{2k}\rangle$. Hence, putting $S=\sup S_{2k}$, we have

$$S_{2m}\leqslant S\leqslant S_{2n-1} \qquad \text{for } m \text{ and } n.$$

Since $\langle S_{2k}\rangle$ is strictly increasing and $\langle S_{2k-1}\rangle$ is strictly decreasing, this inequality can be strengthened to read

$$S_{2m}<S<S_{2n-1} \qquad \text{for } m \text{ and } n, \tag{2.8}$$

It follows from (2.8) that for each m

$$S_{2m}<S<S_{2m+1} \tag{2.9a}$$

$$S_{2m}<S<S_{2m-1}. \tag{2.9b}$$

The first of these inequalities yields

$$0 < S - S_{2m} < S_{2m+1} - S_{2m} = a_{2m+1} \tag{2.10}$$

and the second yields

$$-a_{2m} = S_{2m} - S_{2m-1} < S - S_{2m-1} < 0. \tag{2.11}$$

We use absolute values in (2.10) and (2.11) and obtain for each m

$$|S - S_{2m}| < a_{2m+1} \quad \text{and} \quad |S - S_{2m-1}| < a_{2m}.$$

Together these imply

$$|S - S_n| < a_{n+1} \qquad \text{for } any \text{ positive integer } n,$$

which proves (2.3). Since $\lim a_{n+1} = 0$, (2.3) yields $\lim S_n = S$ and that the alternating series we began with converges.

According to this theorem, the series (2.2), which is alternating, converges. On the other hand, the series whose terms are the absolute values of the terms of the series (2.2) is the harmonic series and so diverges. Contrast this behavior with that of the series

$$\sum_{n=1}^{\infty} (-1)^{n+1} \frac{1}{2^{n-1}} = 1 - \frac{1}{2} + \frac{1}{2^2} - \frac{1}{2^3} + \cdots. \tag{2.12}$$

This series converges (why?). The series

$$\sum_{n=1}^{\infty} \frac{1}{2^{n-1}} = 1 + \frac{1}{2} + \frac{1}{2^2} + \frac{1}{2^3} + \cdots, \tag{2.13}$$

whose terms are the absolute values of the previous series, also converges (why?). Series that behave like the series (2.12) are called *absolutely converging*.

Def. 2.2. Given $\sum a_n$. If $\sum |a_n|$ converges, we call $\sum a_n$ an *absolutely converging* series. If $\sum a_n$ converges and $\sum |a_n|$ diverges, then we say that $\sum a_n$ is *conditionally convergent*.

According to this definition, the series (2.12) converges absolutely and the series (2.2) is conditionally convergent.

Theorem 2.2. *An absolutely converging series converges.*

PROOF. Assume that $\sum a_n$ converges absolutely so that $\sum |a_n|$ converges. Let $\langle S_n \rangle$ be the partial sum sequence of $\sum a_n$ and $\langle T_n \rangle$ the sequence of partial sums of $\sum |a_n|$. We know that $\langle T_n \rangle$ is a Cauchy sequence. We shall prove that $\langle S_n \rangle$ is a Cauchy sequence. Let $\epsilon > 0$ be given. There exists an N such that if $m > n > N$, then

$$\sum_{k=n+1}^{m} |a_k| = |T_m - T_n| < \epsilon. \tag{2.14}$$

By the properties of absolute value, we have

$$|S_m - S_n| = \left|\sum_{k=n+1}^{m} a_k\right| \leqslant \sum_{k=n+1}^{m} |a_k| = |T_m - T_n| < \epsilon, \qquad \text{for} \quad m > n > N.$$

Thus, $\langle S_n \rangle$ is a Cauchy sequence and is, therefore, convergent. This implies that $\sum a_n$ converges.

PROB. 2.1. Prove: If $\sum_{n=1}^{\infty}(-1)^{n+1}a_n$ is alternating, then for each n, $0 < a_n - a_{n+1} + a_{n+2} - a_{n+3} + \cdots$.

3. Series Whose Terms Are Nonnegative

Theorem 3.1. *If $\sum_{n=1}^{\infty} a_n$ is a series such that $a_n \geqslant 0$ for all n, then it converges if and only if the sequence $\langle S_n \rangle$ of its partial sums is bounded from above.*

PROOF. Assume that $\sum a_n$ converges, so $\langle S_n \rangle$ converges and, therefore, is bounded.

Conversely, assume that $\langle S_n \rangle$ is bounded. Since $a_n \geqslant 0$ for all n, it follows that

$$S_{n+1} = S_n + a_{n+1} \geqslant S_n$$

so that $S_{n+1} \geqslant S_n$ for each n. Thus, $\langle S_n \rangle$ is monotonic increasing. Since it is also bounded, it converges.

Remark 3.1. Suppose that $\sum a_n$ has nonnegative terms and its sequence of partial sums is not bounded. Since the partial sum sequence is monotonic increasing,

$$\sum_{n=1}^{\infty} a_n = \lim_{n \to +\infty} S_n = +\infty. \tag{3.1}$$

If the above $\sum a_n$ converges, then

$$0 \leqslant \sum a_n < +\infty. \tag{3.2}$$

PROB. 3.1. Prove: If $\sum a_n$ has nonnegative terms and some $M \in \mathbb{R}$ exists such that $S_n \leqslant M$ for all the partial sums S_n, then $S \leqslant M$ for its sum S. Note that $S_n \leqslant S$ for each n. Also prove: If $a_n > 0$ holds for infinitely many n's, then $S_n < S$ holds for each n.

EXAMPLE 3.1. If $\langle d_n \rangle$ is an infinite sequence of digits and N is an integer, then the series

$$N + \sum_{n=1}^{\infty} d_n 10^{-n} = N + \frac{d_1}{10} + \frac{d_2}{10^2} + \cdots \tag{3.3}$$

has the sequence $\langle r_n \rangle$, where

$$r_n = N + \sum_{k=1}^{n} d_k 10^{-k} = N + \frac{d_1}{10} + \cdots + \frac{d_n}{10^n} \quad \text{for each } n \tag{3.4}$$

as its sequence of partial sums. As is customary we write

$$.d_1 d_2 \ldots d_n = \sum_{k=1}^{n} d_k 10^{-k}, \tag{3.5}$$

so that $r_n = N + .d_1 \ldots d_n$ for each n.) Then

$$\begin{aligned} N \leqslant r_n \leqslant N + \underbrace{.99\ldots 9}_{n \text{ nines}} &= N + \frac{9}{10} + \cdots + \frac{9}{10^n} \\ &= N + \frac{9}{10}\left(1 + \frac{1}{10} + \cdots + \frac{1}{10^{n-1}}\right) \\ &= N + \frac{9}{10}\left(\frac{1 - 1/10^n}{1 - 1/10}\right) \\ &= N + 1 - \frac{1}{10^n} < N + 1. \end{aligned}$$

Hence

$$N \leqslant r_n < N + 1 \qquad \text{for each } n. \tag{3.6}$$

This tells us that the sequence of partial sums of the infinite series (3.3) is bounded. Since its terms are nonnegative, Theorem (3.1) implies that it converges. Writing S for the sums of the series, we have

$$S = N + \sum_{n=1}^{\infty} d_n 10^{-n}. \tag{3.7}$$

By Prob. 3.1 and by (3.6) we have $N \leqslant r_n \leqslant S \leqslant N + 1$. Hence $N \leqslant S \leqslant N + 1$. When $d_j > 0$ for some $j \geqslant 0$, then $N < N + r_j$. Hence, $N < r_j \leqslant S \leqslant N + 1$. But then

$$N < S \leqslant N + 1$$

holds when not all the terms of $\langle d_n \rangle$ are equal to 0. We write

$$S = N + .d_1 d_2 \ldots \tag{3.8}$$

for the sum of the series (3.7), and call $N + .d_1 d_2 \ldots$, an *infinite decimal*. When infinitely many of the digits d_n are not equal to 0, we call the infinite decimal *nonterminating*. Numbers of the form (3.5) are called *terminating decimals*. When the decimal in (3.8) is nonterminating, we call it a *nonterminating decimal representation* of S. We refer to $r_n = N + .d_1 \ldots d_n$, the nth *truncation* of the infinite decimal and $\langle r_n \rangle$ as the *sequence* of its truncations.

By Theorem III.10.4, to each real number x there corresponds a unique integer N and a unique sequence $\langle d_n \rangle$ of digits d_n such that $d_n \neq 0$ for infinitely many n's and such that the sequence $\langle r_n \rangle$, where $r_n = N + .d_1 \ldots d_n$ for each n, has the property

$$r_n < x \leqslant r_n + 10^{-n} \qquad \text{for each } n. \tag{3.9}$$

As observed in Remark III.10.2, this implies $x = \lim r_n$ and we have

$$x = N + .d_1 d_2 \ldots . \tag{3.10}$$

Using the terminology just introduced, we state this result as follows: Each real number x has a unique non-terminating decimal representation.

PROB. 3.2. Prove: (a) $N + .99\ldots = N + 1$. In the decimal on the left the decimal point is followed by 9's only, so that $d_n = 9$ for all n. More generally, (b)

$$N + .\underbrace{0\ldots0}_{j \text{ zeros}}\overbrace{99\ldots}^{\text{all nines}} = N + .\underbrace{0\ldots0}_{j-1 \text{ zeros}}1 = N + 10^{-j}. \tag{3.11}$$

PROB. 3.3. Prove: The terminating decimal $x = N + .d_1 \ldots d_j$, where $d_j > 0$, has the nonterminating decimal representation

$$x = N + .d_1 \ldots d_j = N + .\underbrace{d_1 \ldots (d_j - 1)}_{j \text{ places}}\overbrace{99\ldots}^{\text{all nines}} . \tag{3.12}$$

Thus, we see that although each real number has a unique nonterminating decimal representation, some also have a terminating one.

Remark 3.2. Let $x = N + .d_1 d_2 \ldots$, where the decimal on the right is nonterminating. Let $\langle r_n \rangle$ be the sequence of its truncations and $\langle q_n \rangle$ the sequence defined as $q_n = r_n + 10^{-n}$ for each n. For r_{n+1} and q_{n+1} we have $r_{n+1} = r_n + d_{n+1}10^{-n-1}$ and $q_{n+1} = r_{n+1} + 10^{-n-1}$. The sequence $\langle r_n \rangle$ is monotonic increasing. Since it converges,

$$x = \lim r_n = \sup r_n . \tag{3.13}$$

Notice that $\langle q_n \rangle$ is monotonic decreasing. In fact, for each n,

$$\begin{aligned} q_{n+1} &= r_{n+1} + 10^{-n-1} = r_n + (d_{n+1} + 1)10^{-n-1} \\ &\leqslant r_n + (10)10^{-n-1} = q_n . \end{aligned}$$

Consequently $q_{n+1} \leqslant q_n$ for each n. Clearly,

$$x = \lim r_n = \lim(r_n + 10^{-n}) = \lim q_n = \inf q_n . \tag{3.14}$$

EXAMPLE 3.2. Consider the series

$$\sum_{n=1}^{\infty} \frac{1}{n^p} = 1 + \frac{1}{2^p} + \frac{1}{3^p} + \cdots , \tag{3.15}$$

where p is rational and $p > 1$. We limit ourselves to rational $p > 1$ because, thus far, powers with real exponents have not been defined. Once these are defined (see Section 10) it will be seen that the results proved here also hold for real $p > 1$. We consider the subsequence $\langle S_{2^n - 1} \rangle$ of the sequence $\langle S_n \rangle$

of partial sums of our series

$$S_{2^1-1} = S_1 = 1,$$
$$S_{2^2-1} = S_3 = 1 + \frac{1}{2^p} + \frac{1}{3^p} < 1 + \frac{2}{2^p} = 1 + 2^{-(p-1)}. \tag{3.16}$$

More generally, if $n \geqslant 2$ and $p > 0$, we prove using induction on n that

$$S_{2^n-1} < 1 + \sum_{k=1}^{n-1} 2^{-(p-1)k}$$
$$= 1 + 2^{-(p-1)} + \cdots + 2^{-(p-1)(n-1)}. \tag{3.17}$$

We saw in (3.16) that (3.17) holds for $n = 2$. Assume that (3.17) holds for some integer $n \geqslant 2$. We have

$$S_{2^{n+1}-1} = S_{2^n-1} + \sum_{k=2^n}^{2^{n+1}-1} \frac{1}{k^p}. \tag{3.18}$$

In the sum on the right, $k \geqslant 2^n$. Since $p > 0$, this implies that $k^p \geqslant (2^n)^p = 2^{pn}$. Therefore $k^{-p} \leqslant 2^{-pn}$. Now

$$\sum_{k=2^n}^{2^{n+1}-1} k^{-p} \leqslant \sum_{k=2^n}^{2^{n+1}-1} 2^{-pn} = 2^{-pn} \sum_{k=2^n}^{2^{n+1}-1} 1 = 2^{-pn}\big(2^{n+1} - 1 - (2^n - 1)\big)$$
$$= 2^{-pn}2^n = 2^{-(p-1)n}.$$

This and (3.18) yield

$$S_{2^{n+1}-1} \leqslant S_{2^n-1} + 2^{-(p-1)n}. \tag{3.19}$$

This and the induction hypothesis (3.17) imply that

$$S_{2^{n+1}-1} < 1 + \sum_{k=1}^{n-1} 2^{-(p-1)k} + 2^{-(p-1)n}$$
$$= 1 + \sum_{k=1}^{n} 2^{-(p-1)k} \tag{3.20}$$

for $p > 0$ and $n \geqslant 2$. In view of the induction assumption, (3.17) holds for each $n \geqslant 2$ and $p > 0$. Now note that, for $p > 0$,

$$1 + \sum_{k=1}^{n-1} 2^{-(p-1)k} = 1 + 2^{-(p-1)} + \cdots + \left(2^{-(p-1)}\right)^{(n-1)}$$
$$= \frac{1 - 2^{-(p-1)n}}{1 - 2^{-(p-1)}}. \tag{3.21}$$

Using the assumption $p > 1$, we have $2^{-(p-1)} < 1$. Hence (3.21) yields

$$1 + \sum_{k=1}^{n-1} 2^{-(p-1)k} < \frac{1}{1 - 2^{-(p-1)}} = \frac{2^{p-1}}{2^{p-1} - 1} \qquad \text{if} \quad p > 1, \quad n \geqslant 2.$$

This and (3.17) imply that

$$S_{2^n-1} < \frac{2^{p-1}}{2^{p-1} - 1} \qquad \text{if} \quad p > 1, \quad n \geqslant 2. \tag{3.22}$$

Using (3.16) and noting that the number on the right is greater than 1, we conclude that (3.22) also holds for $p > 1$ and n a positive integer. Now $2^n > n$ holds for each positive integer n. Therefore, we have $2^n \geqslant n + 1$ and $2^n - 1 \geqslant n$ for n a positive integer. Accordingly, we have

$$S_n \leqslant S_{2^n-1} < \frac{2^{p-1}}{2^{p-1}-1} \quad \text{if} \quad p > 1 \quad \text{and } n \text{ a positive integer.} \tag{3.23}$$

This proves that the partial sums of our series are bounded. By Theorem 3.1, we see that the series (3.15) converges for $p > 1$. Note that if $p = 1$, the series diverges with sum $S = +\infty$.

We write

$$\zeta(p) = \sum_{n=1}^{\infty} \frac{1}{n^p} \quad \text{if} \quad p > 1. \tag{3.24}$$

The function ζ just defined is called the *Riemann zeta* function. It plays an important role in the study of prime numbers.

4. Comparison Tests for Series Having Nonnegative Terms

Theorem 4.1. *If $\sum a_n$ and $\sum b_n$ are series for which there exists a positive integer N such that $0 \leqslant a_n \leqslant b_n$ holds for $n \geqslant N$, then the convergence of $\sum b_n$ implies the convergence of $\sum a_n$, and the divergence of $\sum a_n$ implies the divergence of $\sum b_n$.*

PROOF. Assume $\sum b_n$ converges. This implies that $\sum_{n=N}^{\infty} b_n$ converges. Let T_N be the sum of the last series and $T_{N,n}$ be its nth partial sum. Let $S_{N,n}$ be the nth partial sum of $\sum_{n=N}^{\infty} a_n$. Since $0 \leqslant a_n \leqslant b_n$ for $n \geqslant N$, it is clear that

$$S_{N,n} \leqslant T_{N,n} \leqslant T_N \quad \text{for} \quad n \geqslant N.$$

Thus, the sequence $\langle S_{N,n} \rangle_{n \geqslant N}$ of $\sum_{n=N}^{\infty} a_n$ is bounded from above by T_N. Hence, $\sum_{n=N}^{\infty} a_n$ converges. This implies that $\sum_{n=1}^{\infty} a_n$ does.

Next assume $\sum a_n$ diverges, so that $\sum_{n=N}^{\infty} a_n$ does. Since the terms of the latter are nonnegative, we know that the sequence $\langle S_{N,n} \rangle$ of its partial sums is not bounded from above. Writing $T_{N,n}$ again for the nth partial sum of $\sum_{n=N}^{\infty} b_n$, we see that

$$S_{N,n} \leqslant T_{N,n} \quad \text{for} \quad n \geqslant N.$$

This implies that $\langle T_{N,n} \rangle$ is not bounded. Accordingly, $\sum_{n=N}^{\infty} b_n$ diverges. It follows that $\sum^{\infty} b_n$ also does.

PROB. 4.1. Prove: If $p \leqslant 1$, then $\sum_{n=1}^{\infty} 1/n^p$ diverges.

PROB. 4.2. Prove: If $\sum a_n$ and $\sum b_n$ are infinite series of real numbers and there exist M and N where $M > 0$ and N is a positive integer such that

$0 \leqslant a_n \leqslant Mb_n$ for $n \geqslant N$, then the convergence of $\sum b_n$ implies that of $\sum a_n$ and the divergence of $\sum a_n$ implies that of $\sum b_n$.

Def. 4.1. If $\langle a_n \rangle$ and $\langle b_n \rangle$ are real sequences for which there exists an $M > 0$ and a positive integer N such that

$$|a_n| \leqslant Mb_n \qquad \text{for} \quad n \geqslant N, \tag{4.1}$$

then we write

$$a_n = O(b_n), \qquad n \to +\infty \tag{4.2}$$

and read this as: a_n is *big O* of b_n as $n \to +\infty$.

EXAMPLE 4.1. The reader can show that

$$\frac{1}{n^2 - 1} \leqslant \frac{2}{n^2} \qquad \text{if} \quad n \geqslant 2. \tag{4.3}$$

Therefore we can write

$$\frac{1}{n^2 - 1} = O\left(\frac{1}{n^2}\right) \qquad n \to +\infty.$$

EXAMPLE 4.2. If a sequence $\langle a_n \rangle$ is bounded, then

$$a_n = O(1), \qquad n \to +\infty.$$

Indeed, there exists an $M > 0$ such that $|a_n| \leqslant M = M \cdot 1$ for $n \geqslant 1$. Let $\langle b_n \rangle$ be the constant sequence $b_n = 1$ for all n. Then $|a_n| \leqslant Mb_n = M \cdot 1$ for $n \geqslant 1$.

PROB. 4.3. Let $\langle b_n \rangle$ be a sequence of positive numbers. Prove: If $\lim_{n \to +\infty} (a_n / b_n) = 0$, then $a_n = O(b_n)$, $n \to +\infty$.

Def. 4.2. If $\langle a_n \rangle$ and $\langle b_n \rangle$ are real sequences such that $b_n > 0$ for all n and $\lim_{n \to +\infty}(a_n / b_n) = 0$, then we write

$$a_n = o(b_n) \qquad \text{as} \quad n \to +\infty \tag{4.4}$$

reading this as: a_n is *little o* of b_n as $n \to +\infty$.

Remark 4.1. By Prob. 4.3 we have: If $a_n = o(b_n)$ as $n \to +\infty$, then $a_n = O(b_n)$ as $n \to +\infty$. The converse does not hold (give an example).

PROB. 4.4. Prove: If $a_n = O(b_n)$ as $n \to +\infty$, and $b_n = O(c_n)$ as $n \to +\infty$, then $a_n = O(c_n)$ as $n \to +\infty$.

Def. 4.3. If $a_n = O(b_n)$ as $n \to +\infty$ and $b_n = O(a_n)$ as $n \to +\infty$, we say that a_n and b_n are of the *same order of magnitude* as $n \to +\infty$ and write $a_n \asymp b_n$ as $n \to +\infty$. This occurs if and only if there exist $M_1 > 0$, $M_2 > 0$, and a positive integer N such that

$$0 < M_1 \leqslant \frac{a_n}{b_n} \leqslant M_2 \qquad \text{for} \quad m \geqslant N. \tag{4.5}$$

Def. 4.4 (Asymptotic Equivalence). If $\lim_{n\to+\infty}(a_n/b_n) = 1$, we say that a_n and b_n are *asymptotically equivalent* as $n \to +\infty$ and write

$$a_n \sim b_n \qquad \text{as} \quad n \to +\infty. \tag{4.6}$$

EXAMPLE 4.3. Since $\lim_{n\to+\infty}((n+1)/n) = 1$, we write $n+1 \sim n$ as $n \to +\infty$.

Using the "big O" notation we can extend the result in Prob. 4.2 somewhat.

Theorem 4.2. *If $\langle b_n \rangle$ is a real sequence for which there exists a positive integer N such that $b_n \geqslant 0$ for all $n \geqslant N$ and $a_n = O(b_n)$ as $n \to +\infty$, then the convergence of $\sum b_n$ implies that of $\sum a_n$ and the divergence of $\sum |a_n|$ implies that of $\sum b_n$.*

PROOF. By the hypothesis on $\langle b_n \rangle$, there exists a real $M > 0$ and a positive integer N_1 such that

$$|a_n| \leqslant M b_n \qquad \text{for} \quad n \geqslant N_1. \tag{4.7}$$

Assume that $\sum b_n$ converges. By the Cauchy criterion for the convergence of series (Section 1), we have: for each $\epsilon > 0$, there exists an N_2 such that

$$|b_{n+1} + b_{n+2} + \cdots + b_m| < \frac{\epsilon}{M} \qquad \text{if} \quad m > n > N_2. \tag{4.8}$$

Put $N_3 = \max\{N, N_1, N_2\}$. From the hypothesis on $\langle b_n \rangle$, (4.7) and (4.8), it follows that if $m > n > N_3$, then

$$|a_{n+1}| + |a_{n+2}| + \cdots + |a_m| \leqslant M(b_{n+1} + b_{n+2} + \cdots + b_m) < \epsilon.$$

By the Cauchy criterion for the convergence of series, $\sum |a_n|$ converges. Therefore $\sum a_n$ converges.

Next assume that $\sum |a_n|$ diverges. Using (4.7) we see that $\sum b_n$ diverges (Prob. 4.2).

Theorem 4.3. *If $\sum a_n$ and $\sum b_n$ are series with a nonnegative term, where $b_n > 0$ for all n and*

$$0 < \underline{\lim}\, \frac{a_n}{b_n} \leqslant \overline{\lim}\, \frac{a_n}{b_n} < +\infty,$$

then $\sum a_n$ and $\sum b_n$ converge together or diverge together.

PROOF. Put

$$\underline{L} = \underline{\lim}\, \frac{a_n}{b_n} \quad \text{and} \quad \overline{L} = \overline{\lim}\, \frac{a_n}{b_n},$$

so that

$$0<\underline{L}\leqslant\overline{L}<+\infty \quad \text{and, hence,} \quad 0<\frac{\underline{L}}{2}<\underline{L}\leqslant\overline{L}<\tfrac{3}{2}\overline{L}.$$

There exist positive integers N_1 and N_2 such that

$$0<\frac{\underline{L}}{2}<\frac{a_n}{b_n} \qquad \text{for} \quad n\geqslant N_1 \tag{4.9}$$

and

$$\frac{a_n}{b_n}<\frac{3}{2}\overline{L} \qquad \text{for} \quad n\geqslant N_2. \tag{4.10}$$

We put $N=\max\{N_1,N_2\}$ and obtain from (4.9) and (4.10)

$$0<\tfrac{1}{2}\underline{L}b_n<a_n<\left(\tfrac{3}{2}\overline{L}\right)b_n \qquad \text{for} \quad n\geqslant N.$$

The conclusion follows readily from this.

Corollary . *If $\sum a_n$ and $\sum b_n$ have nonnegative terms where $b_n>0$ for all n and*

$$0<\lim\frac{a_n}{b_n}<+\infty,$$

then $\sum a_n$ and $\sum b_n$ converge together or diverge together.

PROOF. Exercise.

PROB. 4.5 Prove: If $b_n>0$ for all n and $\sum a_n$ has nonnegative terms, then (a) $0\leqslant\overline{\lim}(a_n/b_n)<+\infty$ implies that $a_n=O(b_n)$ as $n\to+\infty$ and (b) $0<\underline{\lim}(a_n/b_n)\leqslant+\infty$ implies that $b_n=O(a_n)$ as $n\to+\infty$. In each case draw the appropriate conclusions about the relation between the convergence or divergence of $\sum a_n$ and that of $\sum b_n$.

Remark 4.2. If $b_n>0$ and $a_n\geqslant 0$ for all n and

$$\overline{\lim}\left(\frac{a_n}{b_n}\right)=+\infty \tag{4.11a}$$

or

$$\underline{\lim}\,\frac{a_n}{b_n}=0, \tag{4.11b}$$

then, in the first case, $a_n=O(b_n)$ as $n\to+\infty$ is false, and in the second case, $b_n=O(a_n)$ as $n\to+\infty$ is false. To see this, assume that $a_n=O(b_n)$ as $n\to+\infty$, so that there exist a positive integer N and a real $M>0$ such that $0\leqslant a_n\leqslant Mb_n$, or $a_n/b_n\leqslant M$ for $n\geqslant N$. This implies (explain) that $\overline{\lim}(a_n/b_n)<+\infty$. We must therefore conclude, under the hypothesis, that $\lim(a_n/b_n)=+\infty$ implies that $a_n=O(b_n)$, $n\to+\infty$, is false.

Similarly, if $b_n=O(a_n)$, $n\to+\infty$, there exists a positive integer N_1 and a real $M_1>0$ such that $a_n/b_n\geqslant M_1>0$ for $n\geqslant M_1$ (explain) so that

$\underline{\lim}(a_n/b_n) > 0$ (why?). It follows from the hypothesis that $\underline{\lim}(a_n/b_n) = 0$ implies that $b_n = O(a_n)$, $n \to +\infty$, is false.

Lemma 4.1. *If $\sum a_n$ and $\sum b_n$ have positive terms and there exists a positive integer N such that*

$$\frac{a_{n+1}}{a_n} \leqslant \frac{b_{n+1}}{b_n} \quad \text{for all} \quad n \geqslant N, \tag{4.12}$$

then $a_n = O(b_n)$ as $n \to +\infty$. Therefore, the convergence of $\sum b_n$ implies that of $\sum a_n$, and the divergence of $\sum a_n$ implies that of $\sum b_n$.

PROOF. From (4.12) we obtain

$$\frac{a_{n+1}}{b_{n+1}} \leqslant \frac{a_n}{b_n} \quad \text{for} \quad n \geqslant N.$$

This implies that the sequence $\langle a_n/b_n \rangle_{n \geqslant N}$ is monotonic decreasing. Since its terms are positive, it is also bounded from below. Accordingly, it converges. This implies that the sequence $\langle a_n/b_n \rangle$ converges (explain). We have

$$0 \leqslant \underline{\lim} \frac{a_n}{b_n} = \overline{\lim} \frac{a_n}{b_n} = \lim \frac{a_n}{b_n} < +\infty.$$

By Prob. 4.5(a), $a_n = O(b_n)$ as $n \to +\infty$. In view of Prob. 4.2, we conclude that the convergence of $\sum b_n$ implies that of $\sum a_n$ and the divergence of $\sum a_n$ implies that of $\sum b_n$.

PROB. 4.6. Test the following series for convergence or divergence.

(a) $\sum_{n=1}^{\infty} 1/(n^2 - n + 1)$,

(b) $\sum_{n=1}^{\infty} 1/\sqrt{2n-1}$,

(c) $\sum_{n=1}^{\infty} 1/\sqrt{n^2 - n + 1}$,

(d) $\sum_{n=1}^{\infty} n^{-1-(1/n)}$.

5. Ratio and Root Tests

Theorem 5.1 (Ratio Test). *If $\sum a_n$ is a series whose terms are all positive, then*

(a) $$\overline{\lim} \frac{a_{n+1}}{a_n} = \overline{L} < 1$$

implies that $\sum a_n$ converges, and

(b) $$\underline{\lim} \frac{a_{n+1}}{a_n} = \underline{L} > 1$$

implies that $\sum a_n$ diverges. If $\underline{L} \leqslant 1 \leqslant \overline{L}$, then the series may be convergent or divergent.

PROOF. Assume (a) holds, so that $\overline{L} < 1$. There exists a real q such that $\overline{L} < q < 1$. Write $\epsilon = q - \overline{L}$, so that $\epsilon > 0$ and $q = \overline{L} + \epsilon$. There exists a positive integer N such that

$$0 < \frac{a_{n+1}}{a_n} < \overline{L} + \epsilon = q < 1 \qquad \text{for} \quad n \geqslant N. \tag{5.1}$$

Since $0 < q < 1$, the geometric series $\sum_{n=1} q^n$ converges. Write $b_n = q^n$. From (5.1) we obtain

$$0 < \frac{a_{n+1}}{a_n} < q = \frac{b_{n+1}}{b_n} \qquad \text{for} \quad n \geqslant N.$$

This implies (Lemma 4.1) that $a_n = O(q^n)$ as $n \to +\infty$, and since $\sum b_n = \sum q^n$ converges, $\sum a_n$ converges.

Next assume $\underline{L} > 1$. Then a real q exists such that $1 < q < \underline{L}$. Put $\epsilon = \underline{L} - q$ so that $\epsilon > 0$ and $q = \underline{L} - \epsilon$. There exists a positive integer N such that

$$1 < q = L - \epsilon < \frac{a_{n+1}}{a_n} \qquad \text{if} \quad n \geqslant N.$$

This implies $a_n < a_{n+1}$ if $n \geqslant N$ so that $0 < a_N \leqslant a_n$ for $n \geqslant N$. It follows that $\lim a_n \neq 0$ (explain) and, hence, that $\sum a_n$ diverges.

We prove that the cases $\underline{L} \leqslant 1 \leqslant \overline{L}$ fail to distinguish between convergent and divergent series. Take $a_n = n^{-1}$ for all positive integers n. We have

$$\underline{\lim}\, \frac{a_{n+1}}{a_n} = \underline{\lim}\, \frac{(n+1)^{-1}}{n^{-1}} = \underline{\lim}\left(\frac{n}{n+1} \right) = \lim \frac{n}{n+1} = 1 = \overline{\lim}\, \frac{a_{n+1}}{a_n}$$

and $\sum a_n = \sum n^{-1}$ diverges. Next take $b_n = n^{-2}$ for all positive integers n. Here, too, it is clear that

$$\underline{\lim}\, \frac{b_{n+1}}{b_n} = \underline{\lim}\, \frac{(n+1)^{-2}}{n^{-2}} = \underline{\lim}\left(\frac{n}{n+1} \right)^2 = \lim\left(\frac{n}{n+1} \right)^2 = 1 = \overline{\lim}\, \frac{b_{n+1}}{b_n},$$

and $\sum b_n = \sum n^{-2}$ converges.

Corollary (Modified Ratio Test). *If $\sum a_n$ is a series whose terms are positive, then*

(a) $$\lim \frac{a_{n+1}}{a_n} = L < 1 \quad \textit{implies that } \sum a_n \textit{ converges}$$

and

(b) $$\lim \frac{a_{n+1}}{a_n} = L > 1 \quad \textit{implies that } \sum a_n \textit{ diverges.}$$

The case $\lim(a_{n+1}/a_n) = 1$ fails to distinguish between convergent and divergent series.

PROOF. Exercise.

EXAMPLE 5.1. For each $x \in \mathbb{R}$, consider the series

$$\sum_{n=0}^{\infty} \frac{x^n}{n!} = 1 + x + \frac{x^2}{2!} + \frac{x^3}{3!} + \cdots . \tag{5.2}$$

If $x = 0$, this series converges trivially. If $x \neq 0$, we apply the ratio test to the series

$$\sum_{n=0}^{\infty} \left| \frac{x^n}{n!} \right|. \tag{5.3}$$

We have:

$$\overline{\lim_{n\to+\infty}} \frac{|x^{n+1}/(n+1)!|}{|x^n/n!|} = \overline{\lim_{n\to+\infty}} \frac{|x|}{n+1} = |x| \lim_{n\to+\infty} \frac{1}{n+1} = 0 < 1.$$

This implies that (5.3) converges for $x \neq 0$. We conclude that (5.2) converges absolutely for all $x \in \mathbb{R}$ and, therefore, converges for all $x \in \mathbb{R}$. The series (5.2) is called the *exponential series*. We put:

$$\exp x = \sum_{n=0}^{\infty} \frac{x^n}{n!} \qquad \text{for} \quad x \in \mathbb{R}. \tag{5.4}$$

PROB. 5.1. Prove: If $x \in \mathbb{R}$, then $\lim_{n\to+\infty}(x^n/n!) = 0$.

PROB. 5.2. Prove: Each of the series

(a) $$\sum_{n=0}^{\infty} (-1)^n \frac{x^{2n+1}}{(2n+1)!} = x - \frac{x^3}{3!} + \frac{x^5}{5!} - \frac{x^7}{7!} + \cdots ,$$

(b) $$\sum_{n=0}^{\infty} (-1)^n \frac{x^{2n}}{(2n)!} = 1 - \frac{x^2}{2!} + \frac{x^4}{4!} - \frac{x^6}{6!} + \cdots$$

converges absolutely for each $x \in \mathbb{R}$.

PROB. 5.3. Prove: The series

$$\sum_{n=1}^{\infty} (-1)^{n+1} \frac{x^n}{n} = x - \frac{x^2}{2} + \frac{x^3}{3} - \frac{x^4}{4} + \cdots$$

converges for $-1 < x \leqslant 1$ and diverges for $|x| > 1$. Note also that the convergence is absolute for $-1 < x < 1$ and not for $x = 1$.

PROB. 5.4. Prove: If $b > 0$, then the series

$$x + \frac{a-b}{2!}x^2 + \frac{(a-b)(a-2b)}{3!}x^3 + \cdots$$

converges absolutely for $|x| < b^{-1}$ (Whittaker and Watson).

PROB. 5.5. Test the series $\sum_{n=1}^{\infty} n!/n^n$ for convergence.

Theorem 5.2 (Root Test). *If $\sum a_n$ is a series whose terms are nonnegative, then*

(a) $\overline{\lim} \sqrt[n]{a_n} < 1$ *implies that* $\sum a_n$ *converges and*
(b) $\overline{\lim} \sqrt[n]{a_n} > 1$ *implies that* $\sum b_n$ *diverges.*

The case $\overline{\lim}\ a_n = 1$ *fails to distinguish between convergent and divergent series.*

Proof. Assume that $\overline{\lim} \sqrt[n]{a_n} < 1$, so that a q exists such that $\overline{\lim} \sqrt[n]{a_n} < q < 1$. This implies that there is an integer N such that

$$\sqrt[n]{a_n} < q \qquad \text{if} \quad n \geqslant N,$$

i.e.,

$$0 \leqslant a_n < q^n \qquad \text{if} \quad n \geqslant N.$$

In turn this implies that $a_n = O(q^n)$ as $n \to +\infty$. Since $0 < q < 1$, $\sum q^n$ converges. The relation between a_n and q^n implies that $\sum a_n$ converges.

Now assume $\overline{\lim} \sqrt[n]{a_n} > 1$. This implies that

$$\sqrt[n]{a_n} > 1 \qquad \text{for infinitely many } n\text{'s}$$

We conclude that $\lim a_n \neq 0$, and hence that $\sum a_n$ diverges.

We consider the case $\overline{\lim} \sqrt[n]{a_n} = 1$. Consider the series (1) $\sum n^{-1}$ and (2) $\sum n^{-2}$. For each of these we have $\overline{\lim} \sqrt[n]{a_n} = 1$. Series (1) diverges and series (2) converges.

Corollary (Modified Root Test). *If $\sum a_n$ has nonnegative terms, then*

(a) $$\lim \sqrt[n]{a_n} < 1 \textit{ implies that } \sum a_n \textit{ converges}$$

and

(b) $$\lim \sqrt[n]{a_n} > 1 \textit{ implies that } \sum a_n \textit{ diverges.}$$

The case $\lim \sqrt[n]{a_n} = 1$ *fails to distinguish between convergence and divergence.*

Proof. Exercise.

Example 5.2. We apply the root test to the series

$$\sum_{n=1}^{\infty} \left(1 + \frac{1}{n}\right)^n x^n \tag{5.5}$$

and test for *absolute* convergence using the modified ratio test. We have

$$\lim_{n\to+\infty} \sqrt[n]{\left|\left(1+\frac{1}{n}\right)^n x^n\right|} = |x| \lim_{n\to+\infty}\left(1+\frac{1}{n}\right) = |x|.$$

Hence, if $|x| < 1$, then the series (5.5) converges absolutely. If $|x| \geqslant 1$, then

$$\lim_{n\to+\infty} |a_n| = \lim_{n\to+\infty}\left|\left(1+\frac{1}{n}\right)^n x^n\right| = \begin{cases} e & \text{if } |x| = 1 \\ +\infty & \text{if } |x| > 1. \end{cases}$$

Therefore, if $|x| \geqslant 1$, we have $\lim a_n \neq 0$. We conclude that (5.5) diverges for $|x| \geqslant 1$.

The next theorem relates the ratio and root tests.

Theorem 5.3. *If* $\sum a_n$ *has positive terms, then*

$$\underline{\lim}\, \frac{a_{n+1}}{a_n} \leqslant \underline{\lim}\, \sqrt[n]{a_n} \leqslant \overline{\lim}\, \sqrt[n]{a_n} \leqslant \overline{\lim}\, \frac{a_{n+1}}{a_n}. \tag{5.6}$$

Proof. Put

$$\underline{L}_1 = \underline{\lim}\, \frac{a_{n+1}}{a_n} \quad \text{and} \quad \underline{L}_2 = \underline{\lim}\, \sqrt[n]{a_n}. \tag{5.7}$$

We have $0 \leqslant \underline{L}_1$. If $L_1 = 0$, then $\underline{L}_1 = 0 \leqslant \underline{L}_2$, so that $\underline{L}_1 \leqslant \underline{L}_2$. If $0 < \underline{L}_1$, there exists a q such that $0 < q < \underline{L}_1$. This implies that a positive integer N exists such that

$$0 < q < \frac{a_{n+1}}{a_n} \quad \text{for} \quad n \geqslant N. \tag{5.8}$$

This implies

$$\begin{aligned} 0 < q < \frac{a_{N+1}}{a_N} & \qquad & 0 < a_N q < a_{N+1} \\ 0 < q < \frac{a_{N+2}}{a_{N+1}} & \qquad \text{so that} \qquad & 0 < a_N q^2 < a_{N+2} \\ \vdots \qquad & & \vdots \qquad \\ 0 < q < \frac{a_{N+m}}{a_{N+m-1}} & & 0 < a_N q^m < a_{N+m} \end{aligned}$$

for each positive integer m. Put $n = N + m$ so that $m = n - N$. Then

$$0 < a_N q^{n-N} < a_n \quad \text{if} \quad n > N.$$

This implies that

$$0 < \sqrt[n]{\frac{a_N}{q^N}}\, q < \sqrt[n]{a_n} \quad \text{for} \quad n \geqslant N.$$

Taking $\underline{\lim}$ of both sides we obtain

$$q \underline{\lim}_{n\to+\infty} \sqrt[n]{\frac{a_N}{q^N}} \leqslant \underline{\lim}_{n\to+\infty} \sqrt[n]{a_n} = \underline{L}_2. \tag{5.9}$$

Since here

$$\underline{\lim}_{n\to+\infty} \sqrt[n]{\frac{a_N}{q^N}} = \lim_{n\to+\infty} \sqrt[n]{\frac{a_N}{q^N}} = 1,$$

we obtain from this and (5.9) that, $0 < q \leqslant \underline{L}_2$. This proves: $q < \underline{L}_1$ implies $q \leqslant \underline{L}_2$ and, hence, that $\underline{L}_1 \leqslant \underline{L}_2$ in this case $(0 < \underline{L}_1)$ also.

Now let

$$\overline{L_2} = \overline{\lim}\sqrt[n]{a_n} \quad \text{and} \quad \overline{L_1} = \overline{\lim}\frac{a_{n+1}}{a_n}. \tag{5.10}$$

Thus far, $0 \leqslant \underline{L}_1 \leqslant \underline{L}_2 \leqslant \overline{L}_2$. If $\overline{L}_1 = +\infty$, since $\overline{L}_2 \leqslant +\infty = \overline{L}_1$, we have $\overline{L}_2 \leqslant \overline{L}_1$ in this case. If $\overline{L}_1 < +\infty$, there exists a real q such that $\overline{L}_1 < q$. Accordingly, there is a positive integer N_1 such that

$$0 < \frac{a_{n+1}}{a_n} < q \qquad \text{if} \quad n \geqslant N_1. \tag{5.11}$$

It follows from this that

$$\begin{aligned}
&0 < \frac{a_{N_1+1}}{a_{N_1}} < q & & 0 < a_{N_1+1} < q a_{N_1} \\
&0 < \frac{a_{N_1+2}}{a_{N_1+1}} < q, & \text{that is,} \quad & 0 < a_{N_1+2} < q^2 a_{N_1} \\
&\quad\vdots & & \quad\vdots \\
&0 < \frac{a_{N_1+m}}{a_{N_1+m-1}} < q & & 0 < a_{N_1+m} < q^m a_{N_1}
\end{aligned}$$

for each positive integer m. Putting $n = N_1 + m$, so that $m = n - N_1$, we have

$$0 < a_n < q^{n-N_1} a_{N_1} \qquad \text{if} \quad n \geqslant N. \tag{5.12}$$

Reasoning as we did before, we see from this that

$$\overline{\lim}\sqrt[n]{a_n} = \overline{L}_2 \leqslant q.$$

Thus, $\overline{L}_1 < q$ implies $\overline{L}_2 \leqslant q$ and we conclude that $\overline{L}_2 \leqslant \overline{L}_1$. This yields finally that $\underline{L}_1 \leqslant \underline{L}_2 \leqslant \overline{L}_2 \leqslant \overline{L}_1$ and completes the proof.

Remark 5.1. It follows from Theorem 5.3 that the root test is "at least as powerful" as the ratio test. Inequalities (5.6) inform us that whenever the ratio test detects convergence or divergence so does the root test (explain).

To see that the root test is "more powerful" than the ratio test, we present in Example 5.3 below a series for which the ratio test fails to detect convergence but for which the root test does detect convergence.

EXAMPLE 5.3.* Consider the series

$$\sum_{n=1}^{\infty} a_n = \frac{1}{2} + \frac{1}{3} + \frac{1}{2^2} + \frac{1}{3^2} + \frac{1}{2^3} + \frac{1}{3^3} + \cdots .$$

Here

$$a_n = \begin{cases} 2^{-(n+1)/2} & \text{if } n \text{ is odd} \\ 3^{-n/2} & \text{if } n \text{ is even,} \end{cases}$$

so

$$a_{n+1} = \begin{cases} 3^{-(n+1)/2} & \text{if } n \text{ is odd} \\ 2^{-(n+2)/2} & \text{if } n \text{ is even.} \end{cases}$$

Hence

$$\frac{a_{n+1}}{a_n} = \begin{cases} \dfrac{3^{-(n+1)/2}}{2^{-(n+1)/2}} & \text{if } n \text{ is odd} \\[2ex] \dfrac{2^{-(n+2)/2}}{3^{-n/2}} & \text{if } n \text{ is even.} \end{cases}$$

Consider the subsequences

$$\left\langle \frac{a_{2k}}{a_{2k-1}} \right\rangle \quad \text{and} \quad \left\langle \frac{a_{2k+1}}{a_{2k}} \right\rangle \qquad \text{of} \quad \left\langle \frac{a_{n+1}}{a_n} \right\rangle.$$

We have (Theorem III.4.5)

$$\varliminf_{n\to+\infty} \frac{a_{n+1}}{a_n} \leqslant \varliminf_{k\to+\infty} \frac{a_{2k}}{a_{2k-1}} = \lim_{k\to+\infty}\left(\frac{2}{3}\right)^k = 0$$

and

$$\varlimsup_{n\to+\infty} \frac{a_{n+1}}{a_n} \leqslant \varlimsup_{k\to+\infty} \frac{a_{2k+1}}{a_{2k}} = \lim_{k\to+\infty}\left(\frac{1}{2}\left(\frac{3}{2}\right)^k\right) = +\infty$$

so that $\varliminf(a_{n+1}/a_n) = 0 < 1 < +\infty = \varlimsup(a_{n+1}/a_n)$. The ratio test fails to detect convergence or divergence. We apply the root test. We have

$$\sqrt[n]{a_n} = (2^{-(n+1)/2})^{1/n} = 2^{-1/2}\cdot 2^{-1/2n} < 2^{-1/2} \qquad \text{if } n \text{ is odd}$$

and

$$\sqrt[n]{a_n} = (3^{-n/2})^{1/n} = 3^{-1/2} < 2^{-1/2} \qquad \text{if } n \text{ is even.}$$

Thus,

$$\sqrt[n]{a_n} < 2^{-1/2} \qquad \text{for all } n.$$

* John Randolph, *Basic and Abstract Analysis*, Academic Press, New York, p. 144.

This implies that $\overline{\lim}\sqrt[n]{a_n} \leqslant 2^{-1/2} < 1$. But then Theorem 5.2 yields convergence of the series.

Prob. 5.6. Prove: $\sum_{n=0}^{\infty}(1 - 1/(n+1))^{n^2}$ converges.

Prob. 5.7. Prove:

$$\lim_{n\to+\infty} \frac{\sqrt[n]{n!}}{n} = \frac{1}{e}$$

(Hint: cf. Prob. 5.5).

6. Kummer's and Raabe's Tests

If $\sum a_n$ has positive terms, then the ratio test fails to detect convergence in the cases $\underline{\lim}(a_{n+1}/a_n) \leqslant 1 \leqslant \overline{\lim}(a_{n+1}/a_n)$. The tests which follow can then be applied.

Theorem 6.1 (Kummer's Test). *If $\sum a_n$ has positive terms and* (a) *there exists* (1) *a real $c > 0$,* (2) *a sequence $\langle b_n \rangle$ whose terms are positive, and* (3) *a positive integer N such that*

$$\frac{a_n}{a_{n+1}} b_n - b_{n+1} \geqslant c \qquad \text{for} \quad n \geqslant N, \tag{6.1}$$

then $\sum a_n$ converges. (b) *If there exists a sequence $\langle b_n \rangle$ whose terms are positive such that $\sum b_n^{-1}$ diverges and such that for some positive integer N_1,*

$$\frac{a_n}{a_{n+1}} b_n - b_{n+1} \leqslant 0 \qquad \text{for} \quad n \geqslant N_1, \tag{6.2}$$

then $\sum a_n$ diverges.

Proof. We prove (a) first. By (6.1),

$$\begin{aligned}
a_N b_N \qquad &- a_{N+1} b_{N+1} \geqslant c a_{N+1} \\
a_{N+1} b_{N+1} \qquad &- a_{N+2} b_{N+2} \geqslant c a_{N+2} \\
&\vdots \\
a_{N+m-1} b_{N+m-1} \qquad &- a_{N+m} b_{N+m} \geqslant c a_{N+m} \qquad \text{for each positive integer } m.
\end{aligned}$$

Adding, we obtain

$$a_N b_N > a_N b_N - a_{N+m} b_{N+m} \geqslant c(a_{N+1} a_{N+2} + \cdots + a_{N+m}).$$

Hence

$$a_{N+1} + a_{N+2} + \cdots + a_{N+m} \leqslant \frac{1}{c} a_N b_N \qquad \text{for each positive integer } m.$$

We add $S_n = a_1 + \cdots + a_N$ to both sides to obtain

$$S_{N+m} \leqslant S_N + \frac{1}{c} a_N b_N \qquad \text{for each positive integer } m.$$

This implies that the sequence $\langle S_n \rangle$ of partial sums of the series $\sum a_n$ is bounded. Since the terms of $\sum a_n$ are positive, this implies that $\sum a_n$ converges.

We now prove (b). From (6.2) we obtain

$$\begin{aligned}
&a_{N_1}b_{N_1} &&- a_{N_1+1}b_{N_1+1} \leqslant 0 \\
&A_{N_1+1}b_{N_1+1} &&- a_{N_1+2}b_{N_1+2} \leqslant 0 \\
&\quad\vdots \\
&a_{N_1+m-1}b_{N_1+m-1} &&- a_{N_1+m}b_{N_1+m} \leqslant 0 \qquad \text{for each positive integer } m.
\end{aligned}$$

Adding, we have

$$a_{N_1}b_{N_1} - a_{N_1+m}b_{N_1+m} \leqslant 0 \qquad \text{for each positive integer } m.$$

Put $p = a_{N_1}b_{N_1}$ and $n = N_1 + m$. Then

$$a_n \geqslant \frac{p}{b_n} > 0 \qquad \text{for} \quad n > N_1 .$$

Thus,

$$\frac{1}{b_n} \leqslant \frac{1}{p} a_n \qquad \text{for} \quad n > N_1 .$$

Therefore $b_n^{-1} = O(a_n)$ as $n \to +\infty$. Since $\sum b_n^{-1}$ diverges, this implies that $\sum a_n$ also diverges.

EXAMPLE 6.1. Consider the series

$$\sum_{n=1}^{\infty} \frac{((2n)!)^3}{2^{6n}(n!)^6} . \tag{6.3}$$

Let us apply the modified ratio test. We have

$$\frac{a_{n+1}}{a_n} = \frac{\dfrac{((2n+2)!)^3}{2^{6n+6}((n+1)!)^6}}{\dfrac{((2n)!)^3}{2^{6n}(n!)^6}} = \frac{(2n+1)^3}{(2n+2)^3} \tag{6.4}$$

and, therefore,

$$\lim \frac{a_{n+1}}{a_n} = 1.$$

Thus, the ratio test fails to detect convergence. Let us apply Kummer's test. We choose the sequence $\langle b_n \rangle$, where $b_n = n$ for each positive integer n. Using (6.4), we have

$$\begin{aligned}
n\frac{a_n}{a_{n+1}} - (n+1) &= n\left(\frac{2n+2}{2n+1}\right)^3 - (n+1) \\
&= (n+1)\frac{4n^2+2n-1}{(2n+1)^3} > \frac{4n^2(n+1)}{(2n+2)^3} = \frac{1}{2}\,\frac{n^2}{(n+1)^2} .
\end{aligned}$$

Since

$$\frac{1}{2}\frac{n^2}{(n+1)^2} = \frac{1}{2}\frac{1}{(1+1/n)^2} \geqslant \frac{1}{2}\cdot\frac{1}{2^2} = \frac{1}{8} \qquad \text{for each positive integer } n,$$

we have

$$n\frac{a_n}{a_{n+1}} - (n+1) > \frac{1}{8} \qquad \text{for} \quad n \geqslant 1.$$

By Kummer's test, the series (6.4) converges.

EXAMPLE 6.2. Consider the series

$$\sum_{n=1}^{\infty} \frac{1\cdot 3\ldots(2n-1)}{2\cdot 4\ldots 2n} \tag{6.5}$$

and test it for convergence. We use the modified ratio test first. We have

$$\frac{a_{n+1}}{a_n} = \frac{2n+1}{2n+2} \to 1 \qquad \text{as} \quad n \to +\infty. \tag{6.6}$$

Hence, the ratio test fails. We now use Kummer's test. We take $b_n = n$ for each positive integer n and obtain

$$b_n\frac{a_n}{a_{n+1}} - b_{n+1} = n\frac{a_n}{a_{n+1}} - (n+1)$$

$$= n\left(\frac{2n+2}{2n+1}\right) - (n+1) = -\frac{n+1}{2n+1} < 0 \tag{6.7}$$

for $n \geqslant 1$. Since $\sum b_n^{-1}$ diverges, we see from (6.7) and Theorem 6.1, part (b) that $\sum b_n$ diverges.

Theorem 6.2 (Raabe's Test). *Let $\sum a_n$ have positive terms. If*

(a) $$\underline{\lim}\, n\left(\frac{a_n}{a_{n+1}} - 1\right) = \underline{L} > 1,$$

then $\sum a_n$ converges. If

(b) $$\overline{\lim}\, n\left(\frac{a_n}{a_{n+1}} - 1\right) = \overline{L} < 1,$$

then $\sum a_n$ diverges.

PROOF. Assume that (a) holds. There exists a real q such that $\underline{L} > q > 1$. Therefore, there exists a positive integer N such that

$$n\left(\frac{a_n}{a_{n+1}} - 1\right) > q > 1 \qquad \text{for} \quad n \geqslant N.$$

This implies that

$$n\frac{a_n}{a_{n+1}} - (n+1) > q - 1 > 0 \qquad \text{for} \quad n \geqslant N.$$

Since the sequence $\langle b_n \rangle$, where $b_n = n$ for all n, satisfies the hypothesis of part (a) of Theorem 6.1, $\sum a_n$ converges.

Now assume (b) so that a real q exists such that $0 < \bar{L} < q < 1$. Therefore, a positive integer N_1 exists such that

$$n\left(\frac{a_n}{a_{n+1}} - 1\right) < q \qquad \text{for} \quad n \geqslant N_1 .$$

This implies that

$$n\frac{a_n}{a_{n+1}} - (n+1) < q - 1 < 0 \qquad \text{for} \quad n \geqslant N_1 .$$

This time $\langle b_n \rangle$, where $b_n = n$ for each n, satisfies the hypothesis of part (b) of Theorem 6.1 since $\sum b_n^{-1} = \sum n^{-1}$ diverges. We conclude that $\sum a_n$ diverges.

Corollary (Modified Raabe's Test). *If $\sum a_n$ has positive terms, then*

(a) $$\lim_{n \to +\infty}\left(n\left(\frac{a_n}{a_{n+1}} - 1\right)\right) = L > 1 \quad \textit{implies that } \sum a_n \textit{ converges}$$

and

(b) $$\lim\left(n\left(\frac{a_n}{a_{n+1}} - 1\right)\right) = L < 1 \quad \textit{implies } \sum a_n \textit{ diverges.}$$

PROOF. Exercise.

Remark 6.1. Examining the proof of the last theorem, we see that Raabe's test is really a corollary of Kummer's test with the "testing" sequence $\langle b_n \rangle$, where $b_n = n$ for all n. We shall have occasion to use other testing sequences after we study *the natural logarithm*.

PROB. 6.1. We saw that the ratio test fails to detect convergence or divergence for the series $\sum n^{-2}$. Apply Raabe's test and show that it "works" for that series.

7. The Product of Infinite Series

In what follows, it will be convenient to use indices in our series whose ranges are the nonnegative integers. Let

$$A = \sum_{n=0}^{\infty} a_n \quad \text{and} \quad B = \sum_{n=0}^{\infty} b_n \tag{7.1}$$

be convergent series. We wish to obtain a series whose partial sums tend to their product AB. We first perform some "formal" calculations. The product AB should be given by

$$\begin{aligned} AB = \left(\sum_{n=0}^{\infty} a_n\right)\left(\sum_{k=0}^{\infty} b_k\right) &= a_0b_0 + a_0b_1 + a_0b_2 + a_0b_3 + \cdots \\ &+ a_1b_0 + a_1b_1 + a_1b_2 + a_1b_3 + \cdots \\ &+ a_2b_0 + a_2b_1 + a_2b_2 + a_2b_3 + \cdots \\ &+ a_3b_0 + a_3b_1 + a_3b_2 + a_3b_3 + \cdots \\ &+ \cdots + \cdots . \end{aligned} \tag{7.2}$$

We form the series $\sum_{n=0}^{\infty} c_n$, where

$$\begin{aligned} c_0 &= a_0b_0, \\ c_1 &= a_0b_1 + a_1b_0, \\ c_2 &= a_0b_2 + a_1b_1 + a_2b_0, \\ &\vdots \end{aligned} \tag{7.3}$$

The c's here are obtained by summing the terms marked by the lines in (7.2) and are then summed.

Def. 7.1. If $\sum a_n$ and $\sum b_n$ are infinite series, then their *product*, sometimes called their *Cauchy product*, is defined as $\sum_{n=0}^{\infty} c_n$, where for each n,

$$c_n = \sum_{k=0}^{n} a_k b_{n-k} = a_0b_n + a_1b_{n-1} + \cdots + a_nb_0 \tag{7.4}$$

(if each of the indices in the two series to be multiplied ranges over the *positive* integers, then we define $\sum_{n=1}^{\infty} c_n$ by putting

$$c_n = \sum_{k-1}^{n} a_k b_{n+1-k} = a_1b_n + a_2b_{n-1} + \cdots + a_nb_1 \tag{7.5}$$

for each positive integer n.) Note that

$$c_n = \sum_{k=0}^{n} a_k b_{n-k} = \sum_{k=0}^{n} a_{n-k} b_k \qquad \text{for each } n. \tag{7.6}$$

Theorem 7.1 (Mertens's Theorem). *Let $\sum a_n$ and $\sum b_n$ converge and let $\sum a_n$ converge absolutely. If $A = \sum_{n=0}^{\infty} a_n$ and $B = \sum_{n=0}^{\infty} b_n$, then*

$$\sum_{n=0}^{\infty} c_n = AB, \tag{7.7}$$

where $\sum_{n=0}^{\infty} c_n$ is the Cauchy product of $\sum a_n$ and $\sum b_n$.

PROOF.* Write the respective nth partial sums of $\sum a_n$, $\sum b_n$, and $\sum c_n$ as

$$A_n = \sum_{k=0}^{n} a_k, \quad B_n = \sum_{k=0}^{n} b_k, \quad \text{and} \quad \Gamma_n = \sum_{k=0}^{n} c_k. \tag{7.8}$$

* W. Rudin, *Principles of Mathematical Analysis*, Second Edition, McGraw-Hill, New York, p. 65.

We know that

$$\lim A_n = A, \quad \lim B_n = B, \quad \text{and} \quad \lim A_n B_n = AB. \tag{7.9}$$

We wish to prove that $\lim \Gamma_n = AB$. Now

$$A_n B_n = a_0 B_n + a_1 B_n + \cdots + a_n B_n \tag{7.10}$$

and

$$\begin{aligned} \Gamma_n &= c_0 + c_1 + \cdots + c_n \\ &= a_0 b_0 + (a_0 b_1 + a_1 b_0) + \cdots + (a_0 b_n + \cdots + a_n b_0) \\ &= a_0 B_n + a_1 B_{n-1} + \cdots + a_n B_0, \end{aligned} \tag{7.11}$$

so that

$$A_n B_n - \Gamma_n = a_1(B_n - B_{n-1}) + a_2(B_n - B_{n-2}) + \cdots + a_n(B_n - B_0). \tag{7.12}$$

Since $\sum a_n$ converges absolutely, the partial sums of $\sum |a_n|$ are bounded and an $M > 0$ exists such that

$$|a_1| + |a_2| + \cdots + |a_n| \leqslant |a_0| + |a_1| + |a_2| + \cdots + |a_n| \leqslant M \quad \text{for each } n. \tag{7.13}$$

Since $\langle B_n \rangle$ converges, it is a Cauchy sequence. Let $\epsilon > 0$ be given. There exists a nonnegative integer N such that

$$|B_m - B_n| < \frac{\epsilon}{2M} \quad \text{for} \quad n > m \geqslant N. \tag{7.14}$$

Using $n > m \geqslant N$, we have from (7.12),

$$\begin{aligned} A_n B_n - \Gamma_n = a_1(B_n - B_{n-1}) + \cdots + a_{n-m}(B_n - B_m) \\ + a_{n+1-m}(B_n - B_{m-1}) + \cdots + a_n(B_n - B_0). \end{aligned} \tag{7.15}$$

Therefore, fixing $m = N$ and taking $n > N = m$, we obtain for the sum of the first $n - m$ terms in (7.15) the inequality

$$\begin{aligned} |a_1(B_n - B_{n-1}) + \cdots + a_{n-N}(B_n - B_N)| &< (|a_1| + \cdots + |a_{n-N}|)\frac{\epsilon}{2M} \\ &\leqslant M\frac{\epsilon}{2M} = \frac{\epsilon}{2}. \end{aligned}$$

This and (7.15) imply

$$|A_n B_n - \Gamma_n| < \frac{\epsilon}{2} + |a_{n+1-N}(B_n - B_{N-1}) + \cdots + a_n(B_n - B_0)| \quad \text{for} \quad n > N. \tag{7.16}$$

Since N is fixed, there are N terms in the sum on the right inside the absolute value signs, regardless of $n > N$. Since $\sum a_n$ converges

$$\lim_{n \to +\infty} a_{n+1-N} = \lim_{n \to +\infty} a_{n+2-N} = \cdots = \lim_{n \to +\infty} a_n = 0.$$

Because $B_n \to B$ as $n \to +\infty$, we have for the second term on the right-hand side of (7.16)

$$\lim_{n\to+\infty} |a_{n+1-N}(B_n - B_{N-1}) + \cdots + a_n(B_n - B_0)| = 0.$$

Accordingly (7.16) yields

$$\overline{\lim_{n\to+\infty}} |A_nB_n - \Gamma_n| \leqslant \frac{\epsilon}{2} < \epsilon \qquad \text{for each} \quad \epsilon > 0.$$

This implies

$$\underline{\lim}(A_nB_n - \Gamma_n) = \overline{\lim}(A_nB_n - \Gamma_n) = 0$$

and, therefore, that

$$\lim_{n\to+\infty} (A_nB_n - \Gamma_n) = 0.$$

This, because of (7.9), implies

$$\begin{aligned}\lim \Gamma_n &= \lim\left[A_nB_n - (A_nB_n - \Gamma_n)\right] = \lim A_nB_n - \lim(A_nB_n - \Gamma_n)\\ &= AB - 0 = AB,\end{aligned}$$

completing the proof.

Example 7.1. We apply the last theorem to prove that the function $\exp : \mathbb{R} \to \mathbb{R}$ defined in Example 5.1 satisfies

$$\exp x \exp y = \exp(x + y) \qquad \text{for} \quad x \in \mathbb{R} \quad \text{and} \quad y \in \mathbb{R}. \tag{7.17}$$

By the definition of $\exp x$ given in Example 5.1,

$$\exp x = \sum_{n=0}^{\infty} \frac{x^n}{n!} \qquad \text{for} \quad x \in \mathbb{R}. \tag{7.18}$$

The series on the right converges absolutely for each $x \in \mathbb{R}$. By the last theorem,

$$\exp x \exp y = \left(\sum_{n=0}^{\infty} \frac{x^n}{n!}\right)\left(\sum_{m=0}^{\infty} \frac{y^m}{m!}\right) = \sum_{n=0}^{\infty} c_n, \tag{7.19}$$

where using $a_n = x^n/n!$ and $b_n = y^n/n!$ for each n, we have

$$\begin{aligned} c_n = \sum_{k=0}^{n} a_k b_{n-k} &= \sum_{k=0}^{n} \frac{x^k}{k!} \frac{y^{n-k}}{(n-k)!}\\ &= \frac{1}{n!} \sum_{k=0}^{n} \frac{n!}{k!(n-k)!} y^{n-k}x^k\\ &= \frac{1}{n!} \sum_{k=0}^{n} \binom{n}{k} y^{n-k}x^k\\ &= \frac{1}{n!}(y + x)^n. \end{aligned}$$

(This follows from the Binomial Theorem.) Substitution in (7.19) yields

$$\exp x \exp y = \sum_{n=0}^{\infty} \frac{(x+y)^n}{n!} = \exp(x + y) \qquad \text{for each} \quad x \in \mathbb{R}. \tag{7.20}$$

Theorem 7.2. *In addition to satisfying equality* (7.17), *the function* $\exp : \mathbb{R} \to \mathbb{R}$ *of Example* 5.1 *and the last example also has the following properties*:

(a) $\exp 0 = 1$,
(b) $\exp x > 0$ *for all* $x \in \mathbb{R}$,
(c) $\exp(-x) = 1/\exp(x)$ *for each* $x \in \mathbb{R}$,
(d) $1 + x \leqslant \exp x \leqslant 1 + x \exp x$ *for* $x \in \mathbb{R}$,
(e) $\exp x > 1$, *if* $x > 0$ *and* $0 < \exp x < 1$ *if* $x < 0$,
(f) $\exp 1 = e$,
(g) $x > y$ *implies* $\exp x > \exp y$.

PROOF. By the definition of $\exp 0$ and Example 5.1, (a) holds trivially. We prove (c). Apply (a) and (7.17) to obtain

$$1 = \exp 0 = \exp(x + (-x)) = \exp x \exp(-x). \tag{7.21}$$

This proves that $\exp x \neq 0$ for $x \in \mathbb{R}$ and (c) follows. To prove (b), note that

$$\exp x = \exp\left(\frac{x}{2} + \frac{x}{2}\right) = \left(\exp \frac{x}{2}\right)^2 > 0 \qquad \text{for} \quad x \in \mathbb{R}.$$

The strict inequality on the right follows from $\exp x \neq 0$ for $x \in \mathbb{R}$.

The proof of (d) is somewhat lengthy. For $x \in \mathbb{R}$,

$$\exp x = \sum_{n=0}^{\infty} \frac{x^n}{n!} = 1 + \sum_{n=1}^{\infty} \frac{x^n}{n!} = 1 + x \sum_{n=1}^{\infty} \frac{x^{n-1}}{n!}$$

from which we obtain

$$(\exp x) - 1 = x \sum_{n=1}^{\infty} \frac{x^{n-1}}{n!}. \tag{7.22}$$

For each positive integer n,

$$\frac{1}{n!} \leqslant \frac{1}{(n-1)!}$$

which implies

$$\frac{x^{n-1}}{n!} \leqslant \frac{x^{n-1}}{(n-1)!} \qquad \text{for} \quad x > 0, \quad n \geqslant 1.$$

This implies that

$$\sum_{n=1}^{\infty} \frac{x^{n-1}}{n!} \leqslant \sum_{n=1}^{\infty} \frac{x^{n-1}}{(n-1)!} = 1 + x + \frac{x^2}{2!} + \cdots = \exp x \qquad \text{for} \quad x > 0. \tag{7.23}$$

Because of (7.22) the inequality (7.23) yields

$$(\exp x) - 1 \leqslant x \exp x, \qquad \text{if} \quad x > 0. \tag{7.24}$$

Now, $x > 0$ implies that

$$\exp x = 1 + x + \frac{x^2}{2!} + \cdots \geqslant 1 + x + \frac{x^2}{2!} > 1 + x.$$

This and (7.24) yield

$$1 + x < \exp x \leqslant 1 + x \exp x, \qquad \text{if} \quad x > 0. \tag{7.25}$$

Next take $x < 0$ so that $-x > 0$. From (7.25) we obtain

$$1 - x < \exp(-x) \leqslant 1 - x\exp(-x) \qquad \text{if} \quad x < 0. \tag{7.26}$$

After multiplying through by $\exp x$, this implies that

$$(1 - x)\exp x < 1 \leqslant \exp x - x$$

and, after some elementary manipulations, that

$$1 + x \leqslant \exp x < 1 + x\exp x \qquad \text{if} \quad x < 0. \tag{7.27}$$

Thus far, we proved that (d) holds for $x \neq 0$. If $x = 0$, it is obvious that the equality in (d) holds. This completes the proof of (d).

To prove (e), we use (7.25) from which it follows that if $x > 0$, then $\exp x > 1$. This implies that if $x < 0$, then

$$\exp x = \frac{1}{\exp(-x)} < 1 \quad \text{since} \quad -x > 0.$$

This completes the proof of (e). (g) is an immediate consequence of (e), since

$$e^{x-y} > 1 \qquad \text{if} \quad x > y.$$

Multiplying both sides by $e^y > 0$, we obtain (g).

We prove (f). Recall from Section III.5 that

$$e = \lim_{n\to+\infty}\left(1 + \frac{1}{n}\right)^n \tag{7.28}$$

and by the inequality III.(5.4) that

$$\left(1 + \frac{1}{n}\right)^n \leqslant 1 + \sum_{k=1}^{n} \frac{1}{k!} \qquad \text{if } n \text{ is a positive integer.} \tag{7.29}$$

This implies

$$\left(1 + \frac{1}{n}\right)^n \leqslant 1 + \sum_{k=1}^{\infty} \frac{1}{k!} = \sum_{k=0}^{\infty} \frac{1}{k!} = \exp 1,$$

i.e.,

$$(1 + 1/n)^n \leqslant \exp 1 \qquad \text{for each positive integer } n.$$

Taking the limit on the left as $n \to +\infty$, we obtain

$$e = \lim_{n\to+\infty}\left(1 + \frac{1}{n}\right)^n \leqslant \exp 1, \tag{7.30}$$

so that $e \leqslant \exp 1$ holds.

We return to equality III.(5.2) and observe that if $m > n$ for the positive integers m and n, then

$$\begin{aligned}\left(1 + \frac{1}{m}\right)^m &= 1 + \sum_{k=1}^{m} \frac{1}{k!}\left(1 - \frac{1}{m}\right)\left(1 - \frac{2}{m}\right)\cdots\left(1 - \frac{k-1}{m}\right)\\ &> 1 + \sum_{k=1}^{n} \frac{1}{k!}\left(1 - \frac{1}{m}\right)\left(1 - \frac{2}{m}\right)\cdots\left(1 - \frac{k-1}{m}\right)\end{aligned}$$

and, since III.(5.9) holds,

$$e > 1 + \sum_{k=1}^{n} \frac{1}{k!}\left(1 - \frac{1}{m}\right) \cdots \left(1 - \frac{k-1}{m}\right) \quad \text{if} \quad m > n. \tag{7.31}$$

Here we fix n and note that the second sum on the right-hand side of this inequality consists of n terms each of which is a product of at most n factors. We take limits as $m \to +\infty$ in (7.31) and obtain

$$e \geqslant 1 + \sum_{k=1}^{n} \frac{1}{k!} = \sum_{k=0}^{n} \frac{1}{k!} \qquad \text{if } n \text{ is a positive integer.}$$

Here we take the limit on the right as $n \to +\infty$ to obtain

$$e \geqslant \sum^{\infty} \frac{1}{k!} = \exp 1.$$

This and the already proved inequality $e \leqslant \exp 1$ give us (f). With this, the proof of the theorem is complete.

Prob. 7.1. Prove: If r is rational, then $\exp r = e^r$ (Hint: first prove this for the case where r is a nonnegative integer, then for the case where r is an integer, etc.).

Prob. 7.2. Prove: If $|x| < 1$, then

$$\frac{1}{(1-x)^2} = \sum_{n=0}^{\infty} (n+1)x^n.$$

Observe that the series on the right converges for $|x| < 1$ and diverges for $|x| \geqslant 1$.

Prob. 7.3. Prove: $\sum_{n=0}^{\infty}(-1)^n(n+1)^{-1/2}$ converges and that the Cauchy product of the series with itself diverges. Reconcile this with Theorem 7.1.

8. The Sine and Cosine Functions

Def. 8.1. We define the sine and cosine functions by means of the following infinite series:

$$\sin x = \sum_{n=0}^{\infty} (-1)^n \frac{x^{2n+1}}{(2n+1)!} = x - \frac{x^3}{3!} + \frac{x^5}{5!} - \frac{x^7}{7!} + \cdots$$

$$\text{for each} \quad x \in \mathbb{R}. \tag{8.1}$$

$$\cos x = \sum_{n=0}^{\infty} (-1)^n \frac{x^{2n}}{(2n)!} = 1 - \frac{x^2}{2!} + \frac{x^4}{4!} - \frac{x^6}{6!} + \cdots. \tag{8.2}$$

Note that these series converge absolutely for each $x \in \mathbb{R}$ (Prob. 5.2). We must prove, of course, that the sine and cosine functions given by this definition possess all the properties of their intuitive namesakes. This will be done, in piecemeal fashion, in various parts of the book. But in this development, information about these functions is limited to the definitions above and their consequences.

PROB. 8.1. Prove: (a) $\sin 0 = 0$, (b) $\cos 0 = 1$, (c) $\sin(-x) = -\sin x$, and (d) $\cos(-x) = \cos x$ for $x \in \mathbb{R}$.

Theorem 8.1. *If x and y are real numbers, then*

(a) $$\sin(x+y) = \sin x \cos y + \cos x \sin y$$

and

(b) $$\cos(x+y) = \cos x \cos y - \sin x \sin y.$$

PROOF. By Prob. 8.1, parts (a) and (b), the theorem holds if $x = 0$ or $y = 0$. We, therefore, assume that $x \neq 0$ and $y \neq 0$. It follows that

$$\frac{\sin x}{x} = \sum_{k=0}^{\infty} (-1)^k \frac{x^{2k}}{(2k+1)!}, \tag{8.3}$$

where $x \neq 0$. By the ratio test we see that this series converges absolutely. By Theorem 7.1, we obtain

$$\frac{\sin x}{x} \cos y = \sum_{k=0}^{\infty} (-1)^k \frac{(x^2)^k}{(2k+1)!} \sum_{k=0}^{\infty} (-1)^k \frac{(y^2)^k}{(2k)!} = \sum_{n=0}^{\infty} C_n, \tag{8.4}$$

where for each n

$$\begin{aligned} C_n &= \sum_{k=0}^{n} (-1)^k \frac{(x^2)^k}{(2k+1)!} (-1)^{n-k} \frac{(y^2)^{(n-k)}}{(2(n-k))!} \\ &= (-1)^n \sum_{k=0}^{n} \frac{x^{2k} y^{2n-2k}}{(2k+1)!(2n-2k)!} \\ &= (-1)^n \sum_{k=0}^{n} \frac{x^{2k} y^{2n+1-(2k+1)}}{(2k+1)!(2n+1-(2k+1))!}. \end{aligned}$$

Multiplying and dividing by $(2n+1)!$ and then multiplying by x, we obtain

$$xC_n = \frac{(-1)^n}{(2n+1)!} \sum_{k=0}^{n} \frac{(2n+1)!}{(2k+1)!(2n+1-(2k+1))!} x^{2k+1} y^{2n+1-(2k+1)}$$

which can be written

$$xC_n = \frac{(-1)^n}{(2n+1)!} \sum_{k=0}^{n} \binom{2n+1}{2k+1} x^{2k+1} y^{2n+1-(2k+1)}. \tag{8.5}$$

In turn, this may be written

$$xC_n = \frac{(-1)}{(2n+1)!} \sum_{\substack{i=0,\\ i \text{ odd}}}^{2n+1} \binom{2n+1}{i} x^i y^{2n+1-i}. \tag{8.6}$$

Now we multiply both sides of (8.4) by x and obtain

$$\sin x \cos y = \sum_{n=0}^{\infty} xC_n . \tag{8.7}$$

We interchange x and y in the above to obtain

$$\cos x \sin y = \sum_{n=0}^{\infty} yD_n , \tag{8.8}$$

where each yD_n is obtained from the corresponding xC_n in (8.5) by interchanging x and y. Hence,

$$\begin{aligned} yD_n &= \frac{(-1)^n}{(2n+1)!} \sum_{k=0}^{n} \binom{2n+1}{2k+1} y^{2k+1} x^{2n+1-(2k+1)} \\ &= \frac{(-1)^n}{(2n+1)!} \sum_{k=0}^{n} \binom{2n+1}{2k+1} x^{2n-2k} y^{2k+1}. \end{aligned} \tag{8.9}$$

Now we put $j = n - k$ so that $k = n - j$ and $2k + 1 = 2n + 1 - 2j$ and note that for $k \in \{0, \ldots, n\}, j \in \{0, \ldots, n\}$. Then (8.9) becomes

$$yD_n = \frac{(-1)^n}{(2n+1)!} \sum_{j=0}^{n} \binom{2n+1}{2n+1-2j} x^{2j} y^{2n+1-2j}. \tag{8.10}$$

We observe that

$$\binom{2n+1}{2n+1-2j} = \binom{2n+1}{2j}$$

and substitute this in (8.10). It follows that

$$yD_n = \frac{+(-1)^n}{(2n+1)!} \sum_{j=0}^{n} \binom{2n+1}{2j} x^{2j} y^{2n+1-2j}.$$

The last expression may be rewritten

$$yD_n = \frac{(-1)^n}{(2n+1)!} \sum_{\substack{i=0,\\ i \text{ even}}}^{2n+1} \binom{2n+1}{i} x^i y^{2n+1-i}. \tag{8.11}$$

Adding (8.7) and (8.8) yields

$$\sin x \cos y + \cos x \sin y = \sum_{n=0}^{\infty} (xC_n + yD_n), \tag{8.12}$$

where for each n

$$\begin{aligned} xC_n + yD_n &= \frac{(-1)^n}{(2n+1)!}\left[\sum_{\substack{i=0,\\ i \text{ odd}}}^{2n+1}\binom{2n+1}{i}x^i y^{2n+1-i} + \sum_{\substack{i=0,\\ i \text{ even}}}^{2n+1}\binom{2n+1}{i}x^i y^{2n+1-i}\right] \\ &= \frac{(-1)^n}{(2n+1)!}\sum_{i=0}^{2n+1}\binom{2n+1}{i}x^i y^{2n+1-i}. \end{aligned}$$

By the binomial theorem we see that

$$xC_n + yD_n = \frac{(-1)^n}{(2n+1)!}(x+y)^{2n+1}. \tag{8.13}$$

Substituting this in (8.12) and using Def. 8.1 we obtain

$$\begin{aligned} \sin x \cos y + \cos x \sin y &= \sum_{n=0}^{\infty} \frac{(-1)^n}{(2n+1)!}(x+y)^{2n+1} \\ &= \sin(x+y). \end{aligned} \tag{8.14}$$

We prove (b) next. Calculations similar to the previous one yield

$$\cos x \cos y = \sum_{k=0}^{\infty}(-1)^k \frac{(x^2)^k}{(2k)!}\sum_{k=0}^{\infty}(-1)^k\frac{(y^2)^k}{(2k)!} = \sum_{n=0}^{\infty} E_n, \tag{8.15}$$

where for each n

$$E_n = \frac{(-1)^n}{(2n)!}\sum_{k=0}^{n}\binom{2n}{2k}x^{2k}y^{2n-2k}. \tag{8.16}$$

Note that $E_0 = 1$ and that (8.15) may be written

$$\cos x \cos y = 1 + \sum_{n=1}^{\infty} E_n. \tag{8.17}$$

Let $x \neq 0$ and $y \neq 0$. Using the series (8.3), we obtain from Theorem 7.1 the relation

$$\begin{aligned} \frac{\sin x}{x}\frac{\sin y}{y} &= \sum_{k=0}^{n}(-1)^k\frac{(x^2)^k}{(2k+1)!}\sum_{k=0}^{n}(-1)^k\frac{(y^2)^k}{(2k+1)!} \\ &= \sum_{n=0}^{\infty} C_n'' \end{aligned} \tag{8.18}$$

where for each n

$$C_n'' = \frac{(-1)^n}{(2n+2)!}\sum_{k=0}^{n}\binom{2n+2}{2k+1}x^{2k}y^{2n+1-(2k+1)}.$$

Multiplying both sides by xy, we obtain

$$xyC''_n = \frac{(-1)^n}{(2n+2)!} \sum_{k=0}^{n} \binom{2n+2}{2k+1} x^{2k+1} y^{2n+2-(2k+1)}.$$

Put $m = n + 1$. This m is a positive integer and

$$xyC''_{m-1} = \frac{(-1)^{m-1}}{(2m)!} \sum_{k=0}^{m-1} \binom{2m}{2k+1} x^{2k+1} y^{2m-(2k+1)}. \tag{8.19}$$

Because of this, (8.18) yields

$$-\sin x \sin y = \sum_{m=1}^{\infty} (-xyC''_{m-1}). \tag{8.20}$$

Replacing m in (8.19) by n yields

$$-xyC''_{n-1} = \frac{(-1)^{n-1}}{(2n)!} \sum_{k=0}^{n-1} \binom{2n}{2k+1} x^{2k+1} y^{2n-(2k+1)}. \tag{8.21}$$

Adding (8.17) and (8.20) yields

$$\cos x \cos y - \sin x \sin y = 1 + \sum_{n=1}^{\infty} (E_n - xyC''_{n-1}), \tag{8.22}$$

where for each positive integer n

$$E_n - xyC''_{n-1} = \frac{(-1)^n}{(2n)!} \sum_{k=0}^{n} \binom{2n}{2k} x^{2k} y^{2n-2k}$$

$$+ \sum_{k=0}^{n} \binom{2n}{2k+1} x^{2k+1} y^{2n-(2k+1)}. \tag{8.23}$$

Reasoning as we did in the first part of the proof, we see that

$$E_n - xyC''_{n-1} = \frac{(-1)^n}{(2n)!} \sum_{i=0}^{2n} \binom{2n}{i} x^i y^{2n-i} = \frac{(-1)^n}{(2n)!} (x+y)^{2n}. \tag{8.24}$$

Now using (8.22) and Def. 8.1, we obtain

$$\cos x \cos y - \sin x \sin y = 1 + \sum_{n=1}^{\infty} \frac{(-1)^n}{(2n)!} (x+y)^{2n}$$

$$= \sum_{n=0}^{\infty} \frac{(-1)^n}{(2n)!} (x+y)^{2n} = \cos(x+y).$$

The proof is now complete.

PROB. 8.2. Prove: If $x \in \mathbb{R}$ and $y \in \mathbb{R}$, then

(a) $\sin(x-y) = \sin x \cos y - \cos x \sin y$,
(b) $\cos(x-y) = \cos x \cos y + \sin x \sin y$,
(c) $\sin^2 x + \cos^2 x = 1$.

PROB. 8.3. Prove: If $x \in \mathbb{R}$, then $|\cos x| \leqslant 1$ and $|\sin x| \leqslant 1$.

PROB. 8.4. Prove: If $x \in \mathbb{R}$, then (a) $\sin 2x = 2\sin x \cos x$, and (b) $\cos 2x = \cos^2 x - \sin^2 x = 2\cos^2 x - 1 = 1 - 2\sin^2 x$, and (c) $\sin^2 x = (1 - \cos 2x)/2$ and $\cos^2 x = (1 + \cos 2x)/2$.

We proceed to obtain information about the sine and cosine functions from their definitions as infinite series.

Theorem 8.2. (a) *If* $0 < x \leqslant 1$, *then* $\sin x > 0$ *and if* $-1 \leqslant x < 0$, *then* $\sin x < 0$. (b) *If* $0 < |x| \leqslant 1$, *then* $|\sin x - x| < |x|^3/6$. (c) *If* $x \in \mathbb{R}$, *then* $|\sin x| \leqslant |x|$ *and we have*: *If* $0 < |x| \leqslant 1$, *then* $0 < (\sin x)/x < 1$.

PROOF. We begin with the series

$$\frac{\sin x}{x} = \sum_{n=0}^{\infty} (-1)^n \frac{x^{2n}}{(2n+1)!}, \tag{8.25}$$

where $x \neq 0$. Multiply both sides by -1 and obtain

$$-\frac{\sin x}{x} = \sum_{n=0}^{\infty} (-1)^{n+1} \frac{x^{2n}}{(2n+1)!} = -1 + \sum_{n=1}^{\infty} (-1)^{n+1} \frac{x^{2n}}{(2n+1)!}.$$

Hence,

$$1 - \frac{\sin x}{x} = \sum_{n=1}^{\infty} (-1)^{n+1} \frac{(x^2)^n}{(2n+1)!}. \tag{8.26}$$

We prove: If $0 < |x| \leqslant 1$, then the infinite series on the right of (8.26) is alternating. We begin by assuming $0 < |x| \leqslant 1$ so that $0 < x^2 \leqslant 1$. Therefore,

$$0 < (x^2)^{(n+1)} \leqslant (x^2)^n \quad \text{if} \quad 0 < |x| \leqslant 1 \quad \text{and } n \text{ is a positive integer.}$$

This implies

$$0 < \frac{x^{2n+2}}{(2n+3)!} \leqslant \frac{x^{2n}}{(2n+3)!} < \frac{x^{2n}}{(2n+1)!}$$

$$\text{if} \quad 0 < |x| \leqslant 1 \quad \text{and } n \text{ is a positive integer.}$$

This proves that if $0 < |x| \leqslant 1$, then the series in (8.26) is of the form $\sum_{n=1}^{\infty}(-1)^{n+1}a_n$, where $0 < a_{n+1} < a_n$ for each positive integer n. Moreover, as the reader can show by means of the ratio test, the above-mentioned series is absolutely convergent. This implies that

$$\lim_{n \to +\infty} \frac{(x^2)^n}{(2n+1)!} = \lim_{n \to +\infty} a_n = 0.$$

This completes the proof that the series on the right of (8.26) is alternating.

In the proof of Theorem 2.1 on alternating series we saw that the partial sums $\langle S_n \rangle$ of the alternating series were such that $S_2 < S < S_1$, where S is the sum of the series. Accordingly, we have for the series on the right of (8.26) that

$$S_2 = \frac{x^2}{6} - \frac{x^4}{24} < 1 - \frac{\sin x}{x} < S_1 = \frac{x^2}{6} \qquad \text{for} \quad 0 < |x| \leqslant 1. \quad (8.27)$$

Since $0 < x^2/4 \leqslant 1/4$ and $1 - x^2/4 \geqslant 3/4$ for $0 < |x| \leqslant 1$, it follows that

$$S_2 = \frac{x^2}{6} - \frac{x^2}{24} = \frac{x^6}{6}\left(1 - \frac{x^2}{4}\right) \geqslant \frac{x^2}{6} \cdot \frac{3}{4} \quad \text{and} \quad 0 < \frac{x^2}{6} \leqslant \frac{1}{6}$$

$$\text{if} \quad 0 < |x| \leqslant 1.$$

Therefore, by (8.27),

$$0 < \frac{x^2}{8} < 1 - \frac{\sin x}{x} < \frac{1}{6} < 1 \qquad \text{if} \quad 0 < |x| \leqslant 1.$$

This implies that

$$0 < 1 - \frac{\sin x}{x} < 1 \qquad \text{if} \quad 0 < |x| \leqslant 1. \qquad (8.28)$$

From this we conclude: (a) $\sin x > 0$ if $0 < x \leqslant 1$ and $\sin x < 0$ if $-1 \leqslant x < 0$. This proves (a). Returning to (8.27), we see that

$$\left|1 - \frac{\sin x}{x}\right| < \frac{x^2}{6} \qquad \text{if} \quad 0 < |x| \leqslant 1 \qquad (8.29)$$

so that

$$|\sin x - x| < \frac{|x|^3}{6} \qquad \text{if} \quad 0 < |x| \leqslant 1.$$

This proves (b).

To prove (c), we first use (8.28) to obtain

$$0 < \frac{\sin x}{x} < 1 \qquad \text{if} \quad 0 < |x| \leqslant 1. \qquad (8.30)$$

From this we conclude that $0 < |\sin x/x| < 1$ and, therefore, that $|\sin x| < |x|$ if $0 < |x| \leqslant 1$. If $|x| > 1$, we have from Prob. 8.3 that $|\sin x| \leqslant 1 < |x|$. Thus, $|\sin x| < |x|$ for $0 < |x|$. Since $\sin 0 = 0$ (Prob. 8.1), we obtain $|\sin x| \leqslant |x|$ for $x \in \mathbb{R}$. This proves (c) since we have already proved (8.30).

PROB. 8.5. Prove: $\cos x > 0$ if $|x| \leqslant 1$.

The remaining trigonometric functions are defined as follows:

Def. 8.2. The tangent and secant are defined as:

(a) $\tan x = \sin x/\cos x$ and
(b) $\sec x = 1/\cos x$ if $\cos x \neq 0$,

and the cosecant and cotangent as

(c) $\csc x = 1/\sin x$ and
(d) $\cot x = \cos x/\sin x$ if $\sin x \neq 0$.

PROB. 8.6. Prove:

(a) $1 + \tan^2 x = \sec^2 x$ if $\cos x \neq 0$,
(b) $\cot^2 x + 1 = \csc^2 x$ if $\sin x \neq 0$,
(c) $\tan(x + y) = ((\tan x + \tan y)/(1 - \tan x \tan y))$ if $\cos(x + y) \neq 0$, and $\cos x \cos y \neq 0$,
(d) $\cot(x + y) = ((\cot x \cot y - 1)/(\cot x + \cot y))$ if $\sin(x + y) \neq 0$,
(e) $\tan(2x) = ((2\tan x)/(1 - \tan^2 x))$ if $\cos 2x \neq 0$, and $\cos x \neq 0$,
(f) $\cot(2x) = ((\cot^2 x - 1)/(2\cot x))$ if $\sin 2x \neq 0$.

9. Rearrangements of Infinite Series and Absolute Convergence

We return to the discussion of absolute and conditional convergence.

As we saw,

$$\sum_{n=1}^{\infty} (-1)^{n+1}\frac{1}{n} = 1 - \frac{1}{2} + \frac{1}{3} - \frac{1}{4} + \cdots \tag{9.1}$$

converges conditionally. The sum is positive (why?). Let S be the sum of the series in (9.1). Then

$$0 < S = 1 - \tfrac{1}{2} + \tfrac{1}{3} - \tfrac{1}{4} + \tfrac{1}{5} - \tfrac{1}{6} + \tfrac{1}{7} - \tfrac{1}{8} + \tfrac{1}{9} - \tfrac{1}{10} + \tfrac{1}{11} - \tfrac{1}{12} + \cdots,$$
$$0 < \tfrac{1}{2}S = \tfrac{1}{2} - \tfrac{1}{4} + \tfrac{1}{6} - \tfrac{1}{8} + \tfrac{1}{10} - \tfrac{1}{12} + \cdots.$$

Adding, we have

$$0 < \tfrac{3}{2}S = 1 + \tfrac{1}{3} - \tfrac{1}{2} + \tfrac{1}{5} + \tfrac{1}{7} - \tfrac{1}{4} + \tfrac{1}{9} + \tfrac{1}{11} - \tfrac{1}{6} + \cdots.$$

This series has the same terms as the first one, but its terms are rearranged. Thus, rearranging the terms of a converging series may change its sum. We prove that this cannot occur with absolutely converging series (Theorem 9.1). We first make precise the notion of a *rearrangement* of a series (Def. 9.1).

Intuitively, a rearrangement of a series $\sum a_n$ is a series $\sum b_n$ such that each term of the first one occurs exactly once in the second and vice versa.

Def. 9.1. Let $f: \mathbb{Z}_+ \to \mathbb{Z}_+$ be a one-to-one correspondence on $\mathbb{Z}_+$ and $\langle a_n \rangle$ a *sequence*. If $\langle b_n \rangle$ is a sequence such that $b_n = a_{f(n)}$ for each $n \in \mathbb{Z}_+$, we call it a *rearrangement* of $\langle a_n \rangle$. The infinite series $\sum b_n$ is called a *rearrangement* of the infinite series $\sum a_n$ when $\langle b_n \rangle$ is a rearrangement of $\langle a_n \rangle$. The function f is called the *rearrangement function.*

For example, the nth term of the infinite series (9.1) is $(-1)^{n+1}(1/n)$. Define $f: \mathbb{Z}_+ \to \mathbb{Z}_+$ as follows: For $k \in \mathbb{Z}_+$:

(1) If $n = 3k - 2$, define $f(n) = f(3k - 2) = 4k - 3$,
(2) If $n = 3k$, define $f(n) = f(3k) = 2k$,
(3) If $n = 3k - 1$, define $f(n) = f(3k - 1) = 4k - 1$.

The sequence $\langle f(n)\rangle$ of positive integers is

$$\langle f(n)\rangle = \langle 1,3,2,5,7,4,9,11,6, \cdots \rangle.$$

Hence,

$$\langle b_n\rangle = \langle a_{f(n)}\rangle = \langle 1, \tfrac{1}{3}, -\tfrac{1}{2}, \tfrac{1}{5}, \tfrac{1}{7}, -\tfrac{1}{4}, \tfrac{1}{9}, \tfrac{1}{11}, -\tfrac{1}{6}, \ldots \rangle$$

and

$$\sum_{n=1}^{\infty} b_n = \sum_{n=1}^{\infty} a_{f(n)} = 1 + \tfrac{1}{3} - \tfrac{1}{2} + \tfrac{1}{5} + \tfrac{1}{7} - \tfrac{1}{4} + \tfrac{1}{9} + \tfrac{1}{11} - \tfrac{1}{6} + \cdots .$$

Remark 9.1. Note that if the sequence $\langle b_n\rangle = \langle a_{f(n)}\rangle$ is a rearrangement of $\langle a_n\rangle$, then the sequence $\langle a_n\rangle = \langle b_{f^{-1}(n)}\rangle$ is a rearrangement of $\langle b_n\rangle$.

Lemma 9.1. *If the sequence $\langle b_n\rangle$ is a rearrangement of the sequence $\langle a_n\rangle$, k is some positive integer, and f is the rearrangement function, then there exists a positive integer j such that*

$$\{a_1, a_2, \ldots, a_k\} \subseteq \{b_1, b_2, \ldots, b_n\} \qquad \textit{for} \quad n \geqslant j.$$

PROOF. Let n be an integer such that $n \geqslant j$, where

$$j = \max\{f^{-1}(1), f^{-1}(2), \ldots, f^{-1}(k)\}.$$

We have:

$$f^{-1}(i) \leqslant j \qquad \text{for} \quad 1 \leqslant i \leqslant k.$$

Hence,

$$\{f^{-1}(1), \ldots, f^{-1}(k)\} \subseteq \{1, \ldots, j\} \subseteq \{1, \ldots, n\}.$$

Since

$$a_1 = b_{f^{-1}(1)}, \quad a_2 = b_{f^{-1}(2)}, \ldots, \quad a_k = b_{f^{-1}(k)},$$

it follows that

$$\begin{aligned}\{a_1, a_2, \ldots, a_k\} &= \{b_{f^{-1}(1)}, b_{f^{-1}(2)}, \ldots, b_{f^{-1}(k)}\}\\ &\subseteq \{b_1, b_2, \ldots, b_j\} \subseteq \{b_1, b_2, \ldots, b_n\}\end{aligned}$$

for $n \geqslant j$.

Theorem 9.1. *If $\sum a_n$ converges absolutely, then all its rearrangements converge to its sum.*

PROOF. Let $\sum a_n$ converge absolutely and $S = \sum a_n$. Let $\sum b_n$ be some rearrangement of $\sum a_n$. This implies (Def. 9.1) that the sequence $\langle b_n\rangle$ is a rearrangement of the sequence $\langle a_n\rangle$. By Lemma 9.1, if k is a positive integer, then there exists a positive N such that

$$\{a_1, \ldots, a_k\} \subseteq \{b_1, \ldots, b_n\} \qquad \text{for} \quad n \geqslant N. \tag{9.2}$$

Now take $\epsilon > 0$. The series $\sum |a_n|$ converges by hypothesis. By the Cauchy criterion for series, there exists an N' such that

$$\sum_{i=k+1}^{j} |a_i| < \frac{\epsilon}{2} \qquad \text{if } j > k > N'. \tag{9.3}$$

The sequence $\langle S_n \rangle$ of partial sums of $\sum a_n$ satisfies

$$|S_j - S_k| = \left| \sum_{i=k+1}^{j} a_i \right| \leqslant \sum_{i=k+1}^{j} |a_i| < \frac{\epsilon}{2}. \tag{9.4}$$

This implies that

$$|S - S_k| = \lim_{j \to +\infty} |S_j - S_k| \leqslant \frac{\epsilon}{2} \qquad \text{for } k > N'. \tag{9.5}$$

Retain $k > N'$. We know that a positive integer N exists such that (9.2) holds. Let $\langle S'_n \rangle$ be the sequence of partial sums of the rearrangement $\sum b_n$ of $\sum a_n$ and consider

$$S'_n - S_k \qquad \text{for } k > N' \text{ and } n \geqslant N.$$

Note that $S'_n = \sum_{i=1}^{n} b_i$ and $S_k = \sum_{i=1}^{k} a_i$. Since $n \geqslant N$, (9.2) holds. If any terms remain at all in $S'_n - S_k$ after cancelling, they are a's whose indices are greater than k. Since $k > N'$, the sum of the absolute values of these terms is $\leqslant$ a sum of the form $\sum_{i=k+1}^{j} |a_i|$. This because of (9.3) is $< \epsilon/2$. By (9.4) this implies

$$|S'_n - S_k| < \frac{\epsilon}{2} \qquad \text{if } n \geqslant N \text{ and } k > N'.$$

This and (9.5) yield

$$|S'_n - S| \leqslant |S'_n - S_k| + |S_k - S| < \frac{\epsilon}{2} + \frac{\epsilon}{2} = \epsilon \qquad \text{for } n \geqslant N.$$

But then $\lim S'_n = S$. It follows that $\sum b_n$ converges to the sum S of $\sum a_n$. This completes the proof.

We recall that for $x \in \mathbb{R}$, we define x^+ and x^- as

$$x^+ = \max\{0, x\} \quad \text{and} \quad x^- = \min\{0, x\}. \tag{9.6}$$

By Prob. I.13.17,

$$x^+ = \frac{x + |x|}{2} \quad \text{and} \quad x^- = \frac{x - |x|}{2} \qquad \text{for } x \in \mathbb{R}$$

and, therefore,

$$x^+ + x^- = x \quad \text{and} \quad x^+ - x^- = |x| \qquad \text{for } x \in \mathbb{R}. \tag{9.7}$$

Theorem 9.2. *A real series $\sum a_n$ converges absolutely if and only if each of $\sum a_n^+$ and $\sum a_n^-$ converges. Moreover, if the latter two series converge, then*

$$\sum a_n = \sum a_n^+ + \sum a_n^- \tag{9.8a}$$

and

$$\sum |a_n| = \sum a_n^+ - \sum a_n^-. \tag{9.8b}$$

Proof. Assume that $\sum a_n^+$ and $\sum a_n^-$ converge. Then

$$\sum a_n^+ - \sum a_n^- = \sum (a_n^+ - a_n^-) = \sum |a_n|.$$

But then $\sum |a_n|$ converges and, therefore, $\sum a_n$ converges absolutely. We also have in this case that

$$\sum a_n^+ + \sum a_n^- = \sum (a_n^+ + a_n^-) = \sum a_n .$$

This proves (9.8).

Conversely, if $\sum a_n$ converges absolutely, $\sum |a_n|$ converges. From

$$2\sum a_n^+ = \sum (a_n + |a_n|) = \sum a_n + \sum |a_n|$$

and

$$2\sum a_n^- = \sum (a_n - |a_n|) = \sum a_n - \sum |a_n|,$$

it follows that $\sum a_n^+$ and $\sum a_n^-$ converge. This completes the proof.

Theorem 9.3. *If $\sum a_n$ converges conditionally, then $\sum a_n^+$ diverges to $+\infty$ and $\sum a_n^-$ diverges to $-\infty$.*

Proof. By hypothesis, $\sum a_n$ converges but $\sum |a_n|$ diverges. Taking partial sums, we have:

$$\sum_{k=1}^{n} a_k^+ = \frac{1}{2} \sum_{k=1}^{n} (a_k + |a_k|) = \frac{1}{2}\left(\sum_{k=1}^{n} a_k\right) + \frac{1}{2}\left(\sum_{k=1}^{n} |a_k|\right) \tag{9.9}$$

and

$$\sum_{k=1}^{n} a_k^- = \frac{1}{2} \sum_{k=1}^{n} (a_k - |a_k|) = \frac{1}{2}\left(\sum_{k=1}^{n} a_k\right) - \frac{1}{2}\left(\sum_{k=1}^{n} |a_k|\right). \tag{9.10}$$

Since $\sum |a_n|$ has positive terms and diverges, it diverges to $+\infty$. Thus (9.9) and (9.10) imply that

$$\sum_{k=1}^{\infty} a_k^+ = \frac{1}{2} S + \infty = +\infty \quad \text{and} \quad \sum_{k=1}^{\infty} a_k^- = \frac{1}{2} S - \infty = -\infty.$$

Prob. 9.1. Prove: (a) If $\sum a_n$ is a real series which is conditionally convergent, then it has infinitely many positive terms and infinitely many negative terms. (b) If $\sum p_n$ is the series whose terms are the positive terms of $\sum a_n$ (in the order they appear in this sum) and $\sum t_n$ is the series whose terms are the negative terms of $\sum a_n$ (in the order they appear in this sum), then $\sum p_n$ diverges to $+\infty$ and $\sum t_n$ diverges to $-\infty$.

Riemann proved the following remarkable theorem on conditionally converging series.

Theorem 9.4. *If $\sum a_n$ is a real conditionally converging series, then for any real number l, there exists a rearrangement $\sum b_n$ of $\sum a_n$ such that $\sum b_n = l$.*

PROOF. Let $\sum p_n$ and $\sum t_n$ be respectively the series of positive and negative terms of $\sum a_n$ where the terms of each are in the order in which they appear in $\sum a_n$. By Prob. 9.1 we have

$$\sum p_n = +\infty \tag{9.11a}$$

and

$$\sum t_n = -\infty. \tag{9.11b}$$

Let l be a given real number. There exists a positive integer m such that $\sum_{k=1}^{m} p_k > l$. Let $g(1)$ be the least such m. Then $1 \leq g(1)$. Since $\sum t_n = -\infty$, there exists a positive integer m such that

$$\sum_{k=g(1)+1}^{m} t_k < l - \sum_{k=1}^{g(1)} p_k .$$

Let $g(2)$ be the least such m. Then

$$\sum_{k=1}^{g(1)} p_k + \sum_{k=g(1)+1}^{g(2)} t_k < l, \tag{9.12}$$

where $g(1) < g(1) + 1 \leq g(2)$. If $g(1) + 1 = g(2)$, this implies

$$\sum_{k=1}^{g(1)} p_k + t_{g(2)} < l. \tag{9.13}$$

If $g(2) > g(1) + 1$, (9.12) implies, by the definition of $g(2)$, that

$$\sum_{k=1}^{g(1)} p_k + \sum_{k=g(1)+1}^{g(2)} t_k < l \leq \sum_{k=1}^{g(1)} p_k + \sum_{k=g(1)+1}^{g(2)-1} t_k . \tag{9.14}$$

Using the convention for sums according to which

$$\sum_{k=j+1}^{j} x_k = 0 \qquad \text{if } j \text{ is a nonnegative integer,}$$

we can say that (9.14) includes the case $g(2) = g(1) + 1$. The first $g(2)$ terms of $\sum a_n$ have now been given the arrangement $p_1, \ldots, p_{g(1)}, t_{g(1)+1}, \ldots, t_{g(2)}$. We write T_i for the ith partial sum of the rearrangement. We have

$$\left.\begin{aligned}
T_i &= \sum_{k=1}^{i} p_k && \text{if } \ 1 \leq i \leq g(1) \ \text{ and } \ T_{g(1)} = \sum_{k=1}^{g(1)} p_k, \\
T_i &= T_{g(1)} + \sum_{k=g(1)+1}^{i} t_k && \text{if } \ g(1) + 1 \leq i \leq g(2) \ \text{ so that} \\
T_{g(2)} &= T_{g(1)} + \sum_{k=g(1)+1}^{g(2)} t_k .
\end{aligned}\right\} \tag{9.15}$$

In terms of the T_i's, (9.14) can be written

$$T_{g(2)} < l \leq T_{g(2)-1} . \tag{9.16}$$

Since $T_{g(2)-1} - T_{g(2)} = -t_{g(2)}$, this implies that

$$0 < l - T_{g(2)} \leqslant -t_{g(2)} = |t_{g(2)}|$$

and hence that

$$|l - T_{g(2)}| \leqslant |t_{g(2)}|. \tag{9.17}$$

From the definition of $g(1)$, it follows that

$$T_{g(1)} > l. \tag{9.18}$$

We continue with the rearrangement and define $g(3)$ as the least positive integer m such that

$$l - T_{g(2)} < \sum_{k=g(2)+1}^{m} p_k$$

and obtain $g(1) < g(2) < g(3)$

$$T_{g(2)} + \sum_{k=g(2)+1}^{i} p_k \leqslant l < T_{g(2)} + \sum_{k=g(2)+1}^{g(3)} p_k \quad \text{for} \quad g(2) < i < g(3). \tag{9.19}$$

Now the first $g(3)$ terms of the arrangement are $p_1, \ldots, p_{g(1)}, t_{g(1)+1}, \ldots, t_{g(2)}, p_{g(2)+1}, \ldots, p_{g(3)}$. The corresponding partial terms T_i are given by (9.15) and

$$\begin{aligned} T_i &= T_{g(2)} + \sum_{k=g(2)+1}^{i} p_k \quad \text{if} \quad g(2) < i \leqslant g(3), \quad \text{so that} \\ T_{g(3)} &= T_{g(2)} + \sum_{k=g(2)+1}^{g(3)} p_k . \end{aligned} \tag{9.20}$$

This yields

$$T_{g(2)} \leqslant T_i \leqslant l < T_{g(3)} \quad \text{for} \quad g(2) < i < g(3) \tag{9.21}$$

and, in particular, that

$$T_{g(2)} \leqslant T_{g(3)-1} \leqslant l < T_{g(3)} . \tag{9.22}$$

This and (9.16) also imply $T_{g(2)} \leqslant T_i \leqslant l \leqslant T_{g(2)-1}$ for $g(2) < i < g(3)$. Accordingly, see (9.17), the partial sums T_i, $g(2) \leqslant i \leqslant g(3)$, satisfy

$$|l - T_i| \leqslant |l - T_{g(2)}| \leqslant |t_{g(2)}| \quad \text{for} \quad g(2) \leqslant i < g(3) \tag{9.23}$$

and

$$|l - T_{g(3)}| < T_{g(3)} - T_{g(3)-1} = p_{g(3)} = |p_{g(3)}|. \tag{9.24}$$

Continuing, we define $g(4)$ as the least positive integer m such that

$$T_{g(3)} + \sum_{k=g(3)+1}^{m} t_k < l. \tag{9.25}$$

We have $g(1) < g(2) < g(3) < g(4)$, the rearranged terms *after* the $g(3)$th being $t_{g(3)+1}, \ldots, t_{g(4)}$. The corresponding partial sums T_i are

$$T_i = T_{g(3)} + \sum_{k=g(3)+1}^{i} t_k \quad \text{for} \quad g(3) < i \leqslant g(4), \quad \text{so that}$$
$$T_{g(4)} = T_{g(3)} + \sum_{k=g(3)+1}^{g(4)} t_k . \tag{9.26}$$

This yields

$$T_{g(4)} < l \leqslant T_i \leqslant T_{g(3)} \quad \text{for} \quad g(3) < i < g(4), \tag{9.27}$$

and, in particular,

$$T_{g(4)} < l \leqslant T_{g(4)-1} \leqslant T_{g(3)} . \tag{9.28}$$

This and (9.22) also imply $T_{g(3)-1} \leqslant l \leqslant T_i \leqslant T_{g(3)}$ for $g(3) < i < g(4)$. Accordingly, see (9.24), the partial sums T_i, where $g(3) \leqslant i \leqslant g(4)$, satisfy the relations

$$|l - T_i| \leqslant |l - T_{g(3)}| \leqslant |p_{g(3)}| \quad \text{for} \quad g(3) \leqslant i < g(4) \tag{9.29}$$

and

$$|l - T_{g(4)}| < T_{g(4)-1} - T_{g(4)} = -t_{g(4)} = |t_{g(4)}|. \tag{9.30}$$

Proceeding inductively in this manner, we obtain a rearranged series with partial sums T_i which satisfy the relations

$$|l - T_i| \leqslant |t_{g(n)}| \quad \text{if} \quad g(n) \leqslant i < g(n+1) \quad \text{if } n \text{ is even} \tag{9.31}$$

and

$$|l - T_i| \leqslant |p_{g(n)}| \quad \text{if} \quad g(n) \leqslant i < g(n+1) \quad \text{if } n \text{ is odd.} \tag{9.32}$$

Here $g(n) < g(n+1)$ for all n.

Since our original series converges conditionally (by hypothesis), it follows that $p_{g(n)} \to 0$ and $t_{g(n)} \to 0$ as $n \to +\infty$. From the inequalities (9.31) and (9.32) we see that the rearrangement $\sum b_n$ of $\sum a_n$ is such that for its partial sum sequence $\langle T_n \rangle$ we have $T_n \to l$ as $n \to +\infty$. This completes the proof.

Corollary. *If all the rearrangements of a converging real series converge to its sum, then the series converges absolutely.*

PROOF. Let $\sum a_n$ be a converging series with sum S such that all its rearrangements also converge and have sum S. If $\sum a_n$ did not converge absolutely, then by Theorem 9.4, there would be two rearrangements of the series converging to different sums; this contradicts the assumption on $\sum a_n$. Therefore, $\sum a_n$ converges absolutely.

10. Real Exponents

Before beginning the next chapter it will be convenient to define real exponents, that is, to deal with p^x, where $p > 0$ and x is any real number. Our approach is to use the nonterminating decimal representation of x (cf. Example 3.1, Theorem III.10.4 and Remark 3.2).

Let

$$x = N + .d_1 d_2 \ldots \quad \text{and} \quad y = M + .\delta_1 \delta_2 \ldots \tag{10.1}$$

be nonterminating decimals. We have $x = \lim r_n$ and $y = \lim s_n$, where, for each n, r_n and s_n are the respective n truncations of the decimals of x and y in (10.1). By theorems on limits we have

$$x + y = \lim r_n + \lim s_n = \lim(r_n + s_n). \tag{10.2}$$

The sum $x + y$ is a real number and has a nonterminating decimal representation

$$x + y = P + .\Delta_1 \Delta_2 \ldots, \tag{10.3}$$

where P is an integer and Δ_k is a digit for each k. Let $\langle z_n \rangle$ be the sequence of truncations of this nonterminating decimal representation of $x + y$, so that

$$z_n = P + .\Delta_1 \Delta_2 \ldots \Delta_n \quad \text{for each } n.$$

Note that in general $r_n + s_n \neq z_n$. For example, consider $2 = 1.99 \ldots$ and $1 = .99 \ldots$. Let

$$r_n = 1 + \underbrace{.99 \ldots 9}_{n \text{ nines}} \quad \text{and} \quad s_n = \underbrace{.99 \ldots 9}_{n \text{ nines}} \quad \text{for each } n$$

be the respective nth truncations of the nonterminating decimal representations of 2 and 1. We have

$$r_n = 1 + \frac{10^n - 1}{10^n} \quad \text{and} \quad s_n = \frac{10^n - 1}{10^n},$$

so that

$$r_n + s_n = 3 - 2(10^{-n}) = 2 + .\overbrace{99 \ldots}^{n-1 \text{ nines}}98,$$

whereas the nonterminating decimal representation of $2 + 1 = 3$ is

$$2 + 1 = 3 = 2 + .99 \ldots.$$

The nth truncation of this representation is

$$z_n = 2 + .99 \ldots 9 = 2 + \frac{10^n - 1}{10^n} = 3 - \frac{1}{10^n}$$

and we see that here $r_n + s_n \neq z_n$ for any positive integer n. It is clear, however, that $\lim(r_n + s_n) = \lim z_n$ since the limit of each side of the equality is $x + y$. In general, we have

$$x + y = \lim(r_n + s_n) = N + M + \lim_{n \to +\infty} \sum_{k=1}^{n} (d_k + \delta_k) 10^{-k}. \tag{10.4}$$

We can now define p^x, where $p > 0$ and $x \in \mathbb{R}$. We first treat the case $p \geqslant 1$. Let $x = N + .d_1d_2 \ldots$, where the decimal on the right is nonterminating, and let $\langle r_n \rangle$ be the sequence of its truncations. We have

$$N < r_n < N + 1 \qquad \text{for all positive integers } n; \tag{10.5}$$

p^N, p^{r_n}, p^{N+1} are defined for rational exponents; and

$$p^N \leqslant p^{r_n} \leqslant p^{N+1} \qquad \text{for each } n. \tag{10.6}$$

It follows that the sequence $\langle p^{r_n} \rangle$ is bounded from above. Since it is also monotonically increasing (recall that $p \geqslant 1$ and the sequence $\langle r_n \rangle$ is increasing), this sequence converges. If $0 < p < 1$, then the sequence $\langle p^{r_n} \rangle$ is monotonic decreasing and

$$p^{N+1} < p^{r_n} < p^N \qquad \text{for each } n. \tag{10.7}$$

It follows that here too $\langle p^{r_n} \rangle$ converges. We now define:

Def. 10.1. If $p > 0$ and x are real numbers, then we define the function $E_p : \mathbb{R} \to \mathbb{R}$ as

$$E_p(x) = \lim_{n \to +\infty} p^{r_n}, \tag{10.8}$$

where for each positive integer n, r_n is the nth truncation of the nonterminating decimal representation of x. We also define

$$E_0(x) = 0 \qquad \text{if} \quad x > 0. \tag{10.9}$$

Remark 10.1. It follows from the remarks preceding this definition that the limit in (10.8) exists in $\mathbb{R}$. Therefore, $E_p(x)$ is defined for each $x \in \mathbb{R}$. As an example, we prove: If $p > 0$, then $E_p(0) = 1$. By Def. 10.1, we must use the nonterminating decimal representation of 0. This is,

$$0 = -1 + .99 \ldots . \tag{10.10}$$

The nth truncation of the decimal on the right is

$$r_n = -1 + \underbrace{.99 \ldots 9}_{n \text{ nines}} = -1 + \frac{10^n - 1}{10^n} = -10^{-n}. \tag{10.11}$$

Hence (Prob. III.8.5),

$$E_p(0) = \lim_{n \to +\infty} p^{-10^{-n}} = \lim_{n \to +\infty} (p^{-1})^{10^{-n}} = 1.$$

PROB. 10.1. Prove: $E_1(x) = 1$ for each $x \in \mathbb{R}$.

PROB. 10.2. Prove: If $p > 0$, then $E_p(1) = p$.

PROB. 10.3. Prove: If $p > 0$, then $E_p(x) > 0$ for $x \in \mathbb{R}$ (Hint: note (10.6) and (10.7)).

PROB. 10.4. Prove: If $p > 0$ and $q > 0$, then $E_p(x)E_q(x) = E_{pq}(x)$.

PROB. 10.5. Prove: If $p > 0$, then $E_{p^{-1}}(x) = E_p(x)^{-1}$.

Theorem 10.1. *If p, x and y are real numbers and $p > 0$, then*

$$E_p(x)E_p(y) = E_p(x+y). \tag{10.12}$$

PROOF. Let $x = N + .d_1d_2\ldots$ and $y = M + .\delta_1\delta_2\ldots$ be the nonterminating decimal representation of x and y and $\langle r_n \rangle$, $\langle s_n \rangle$, their respective sequences of truncations. Since $\langle p^{r_n} \rangle$ and $\langle p^{s_n} \rangle$ converge (see remarks preceding Def. 10.1), the definition of $E_p(x)$ and $E_p(y)$ implies that

$$\lim p^{r_n+s_n} = \lim p^{r_n}p^{s_n} = \lim p^{r_n} \lim p^{s_n} = E_p(x)E_p(y). \tag{10.13}$$

It remains to prove that

$$\lim p^{r_n+s_n} = E_p(x+y). \tag{10.14}$$

However, since the sequence $\langle z_n \rangle$ of truncations of the nonterminating decimal representation of $x+y$, in general, differs from the sequence $\langle r_n + s_n \rangle$ (see remarks in the opening paragraph), (10.14) is not obvious. We note that since $r_n < x$, $s_n < y$, we have $r_n + s_n < x + y$ for all n. Since $x + y = \lim z_n = \sup z_n$, it follows that

$$r_n + s_n < \sup z_n \qquad \text{for all } n. \tag{10.15}$$

Consequently, to each n there corresponds an m such that

$$r_n + s_n < z_m < \sup z_n = x + y. \tag{10.16}$$

For $p > 1$, this implies that

$$p^{r_n+s_n} < p^{z_m} \leqslant E_p(x+y) \qquad \text{for each } n. \tag{10.17}$$

Thus, for $p > 1$, the sequence $\langle p^{r_n+s_n} \rangle$ is bounded from above. Since it is also monotonic increasing (explain), it converges. From (10.13) and (10.17) we obtain

$$E_p(x)E_p(y) = \lim p^{r_n+s_n} \leqslant E_p(x+y) \qquad \text{for} \quad p > 1. \tag{10.18}$$

Now, for each n,

$$z_n < x + y = \lim(r_n + s_n) = \sup(r_n + s_n). \tag{10.19}$$

Hence, for each n, there exists an m such that

$$z_n < r_m + s_m.$$

It follows that if $p > 1$, then

$$p^{z_n} < p^{r_m+s_m} = p^{r_m}p^{s_m} \leqslant E_p(x)E_p(y) \qquad \text{for each } n.$$

This implies that for $p > 1$,

$$E_p(x+y) = \lim p^{z_n} \leqslant E_p(x)E_p(y). \tag{10.20}$$

This and (10.18) imply (10.12) for the case $p > 1$. If $p = 1$, then (10.12) holds trivially (Prob. 10.3).

As for the case $0 < p < 1$, we note that here $p^{-1} > 1$. Hence, by what was already proved

$$E_{p^{-1}}(x)E_{p^{-1}}(y) = E_{p^{-1}}(x+y) \qquad \text{if} \quad 0 < p < 1. \tag{10.21}$$

By Prob. 10.5, this yields

$$E_p^{-1}(x)E_p^{-1}(y) = E_p^{-1}(x+y)$$

and, therefore, taking reciprocals, we find that (10.12) holds in this case also. This completes the proof.

PROB. 10.6. Prove: If $p > 0$, then $E_p(-x) = E_p(x)^{-1} = E_{p^{-1}}(x)$.

PROB. 10.7. Prove: If $p > 1$, then $E_p(x-y) = E_p(x)/E_p(y)$.

PROB. 10.8. Prove: If $p > 1$, then (a) $x > 0$ implies $E_p(x) > 1$ and (b) $x < 0$ implies $0 < E_p(x) < 1$.

PROB. 10.9. Prove: If $0 < p < 1$, then (a) $x > 0$ implies that $0 < E_p(x) < 1$ and (b) $x < 0$ implies that $E_p(x) > 1$.

PROB. 10.10. Prove: (a) If $p > 1$, then the function E_p is strictly monotonically increasing; (b) if $0 < p < 1$, then the function E_p is strictly monotonically decreasing.

PROB. 10.11. Prove: If $p > 0$ and r is rational, then for each $x \in \mathbb{R}$, we have $E_p(rx) = (E_p(x))^r$. In particular, prove: If r is rational and $p > 0$, then $E_p(r) = p^r$ (Hint: first carry out the proof for the case where r is a nonnegative integer, then for the case when r is an integer, etc.).

Remark 10.2. We noted in the last problem that if $p > 0$ and r is rational, then $E_p(r) = p^r$. Thus, if $r = m/n$, where m and n are integers and $n > 0$, then $E_p(r) = E_p(m/n) = (\sqrt[n]{p}\,)^m$. Now let r have the nonterminating decimal representation $r = N + .d_1d_2 \ldots$ and let $\langle s_n \rangle$ be the sequence of truncations of the decimal representation of r. From $E_p(r) = p^r$ and Def. 10.1 it follows that

$$p^{m/n} = p^r = \lim_{k \to +\infty} p^{s_k}. \tag{10.22}$$

If x is irrational, we *define* $p^x = E_p(x)$. Thus, for x irrational, our interpretation of p^x is

$$p^x = \lim_{n \to +\infty} p^{r_n} = E_p(x), \tag{10.23}$$

where r_n is the nth truncation of the nonterminating decimal representation of x. By Theorem 10.1, we have

$$p^x p^y = p^{x+y} \qquad \text{for} \quad p > 0 \tag{10.24}$$

for real x and y.

Remark 10.3. We list all the results obtained thus far for the function E_p and express them in terms of the notation $E_p(x) = p^x$. We have: If $p > 0$ and $x = N + .d_1d_2 \ldots$, where the decimal on the right is nonterminating and $r_n = N + .d_1d_2 \ldots d_n$ is, for each n, the truncation of this decimal, then

(a) $p^x = \lim_{n\to+\infty} p^{r_n}$,
(b) $p^0 = 1, p^1 = p$,
(c) $0^x = 0$ if $x > 0$,
(d) $p > 0$ and $q > 0$ imply $(pq)^x = p^x q^x$,
(e) $(p^{-1})^x = 1/p^x = p^{-x}$,
(f) $p^x/p^y = p^{x-y}$,
(g) $p > 1$ and $x > 0$ imply $p^x > 1$; $p > 1$ and $x < 0$ imply $0 < p^x < 1$,
(h) $0 < p < 1$ and $x > 0$ imply $0 < p^x < 1$; $0 < p < 1$ and $x < 0$ imply $p^x > 1$.
(i) The result in Prob. 10.10 can be expressed as: $p > 1$ and $x_1 < x_2$ imply $0 < p^{x_1} < p^{x_2}$, $0 < p < 1$ and $x_1 < x_2$ imply $p^{x_1} > p^{x_2} > 0$.
(j) If r is rational, then $p^{rx} = (p^x)^r$. (In the next theorem this is extended to the case where r is any real number.)

Prob. 10.12. Prove: If $p > 0$, $q > 0$, then for $x \in \mathbb{R}$, $p^x/q^x = (p/q)^x$.

Prob. 10.13. Prove: If $0 \leqslant x_1 < x_2$ and $y > 0$, then $x_1^y < x_2^y$.

Theorem 10.2. *If p, x and y are real numbers and $p > 0$, then*

$$(p^x)^y = p^{xy}. \tag{10.25}$$

Proof. The conclusion is obvious if $x = 0$ or $y = 0$. Suppose that $x \neq 0$ and $y \neq 0$. We first consider the case (a) $p > 1$, $x > 0$. Let $y = N + .d_1d_2 \ldots$, where the decimal on the right is nonterminating and let $\langle r_n \rangle$ be its sequence of truncations. By Remark III.10.2

$$r_n < y \leqslant r_n + 10^{-n} \qquad \text{for each } n$$

and, hence,

$$xr_n < xy \leqslant xr_n + x10^{-n}.$$

It follows that

$$(p^x)^{r_n} = p^{xr_n} < p^{xy} \leqslant p^{xr_n + x10^{-n}} = p^{xr_n}p^{x10^{-n}}$$
$$= (p^x)^{r_n}(p^x)^{10^{-n}}. \tag{10.26}$$

We have

$$\lim_{n\to+\infty} (p^x)^{r_n} = (p^x)^y \tag{10.27}$$

and

$$\lim_{n\to+\infty} (p^x)^{r_n}(p^x)^{10^{-n}} = (p^x)^y 1 = (p^x)^y. \tag{10.28}$$

These two limit relations together with (10.26) and the sandwich theorem imply that

$$p^{xy} = \lim_{n \to +\infty} (p^x)^{r_n} = (p^x)^y$$

in case (a). In case (b) $p > 1$, $x < 0$, we have $-x > 0$. From what was proved in case (a), we obtain

$$(p^{-x})^y = p^{-xy}$$

which implies $(1/p^x)^y = 1/p^{xy}$, and, hence,

$$\frac{1}{(p^x)^y} = \frac{1}{p^{xy}} .$$

Thus, (10.25) holds for case (b) also. We leave the remaining cases (c) $0 < p < 1$, $x > 0$, $0 < p < 1$, $x < 0$ for the reader to prove.

Remark 10.4. The series

$$\zeta(p) = \sum_{n=1}^{\infty} \frac{1}{n^p}, \qquad p > 1 \tag{10.29}$$

(Example 3.2) was seen to converge for $p > 1$ and p rational. The reader can now use the properties of real exponents to demonstrate the convergence of this series for the case $p > 1$, where p is real.

CHAPTER V

Limit of Functions

1. Convex Set of Real Numbers

In Chapter III we dealt with limits of real sequences. These are real-valued functions whose domains are essentially $\mathbb{Z}_0$ or $\mathbb{Z}_1$. In this chapter we treat limits of real-valued functions of a real variable whose domains are not necessarily confined to $\mathbb{Z}_0$ or $\mathbb{Z}_1$. Of special interest are functions whose domains are intervals.

PROB. 1.1. Prove: The function $f: \mathbb{R} \to \mathbb{R}$, where

$$f(x) = \frac{x}{1 + |x|} \qquad \text{for} \quad x \in \mathbb{R}$$

is a one-to-one correspondence between $\mathbb{R}$ and the open interval $(-1; 1)$ whose inverse is g, where

$$g(y) = \frac{y}{1 - |y|} \qquad \text{for} \quad |y| < 1, \quad y \in \mathbb{R}.$$

Accordingly, $\mathbb{R}$ and the open interval $(-1; 1)$ are equipotent (Def. II.10.5).

PROB. 1.2. Prove: If a, b, c, d are real numbers such that $a < b$ and $c < d$, then the function $f: (a; b) \to \mathbb{R}$, where f is defined as

$$f(x) = c + \frac{d - c}{b - a}(x - a) \qquad \text{for} \quad x \in (a; b),$$

is a one-to-one correspondence between the intervals $(a; b)$ and $(c; d)$. Its inverse is $g:(c; d) \to (a; b)$, where

$$g(y) = a + \frac{b - a}{d - c}(y - c) \qquad \text{for} \quad y \in (c; d).$$

Consequently the intervals $(a; b)$ and $(c; d)$ are equipotent.

Prob. 1.3. Prove: If $a \in \mathbb{R}$, $b \in \mathbb{R}$ $(a < b)$, then the set $\mathbb{R}$ of real numbers and the interval $(a; b)$ are equipotent.

Intervals of real numbers can be characterized as *convex* sets of real numbers.

Def. 1.1. A subset $S \subseteq \mathbb{R}$ is called *convex* if and only if $x_1 \in S$, $x_2 \in S$, $x_1 < x_2$ imply $[x_1, x_2] \subseteq S$.

Prob. 1.4. Prove: If S is a family of convex subsets of $\mathbb{R}$, then its intersection $\bigcap S$ is convex.

Prob. 1.5. Prove: If I is an interval in $\mathbb{R}$, then I is convex.

Prob. 1.6. Prove: (a) The empty set and any singleton set in $\mathbb{R}$ are convex sets; (b) if S is a convex subset of $\mathbb{R}$ containing at least two members, then S is an interval.

Prob. 1.7. Prove: If a and b are real numbers such that $a < b$, then $x \in (a; b)$ if and only if there exists exactly one t such that $0 < t < 1$ and $x = (1 - t)a + tb$.

2. Some Real-Valued Functions of a Real Variable

In Chapter II we already defined some real-valued functions of a real variable. The notion of a *polynomial* on $\mathbb{R}$ was defined in Def. II.8.1.

Example 2.1 (Rational Functions). A *rational function* on $\mathbb{R}$ is a function R defined by the following means. Let P and Q be polynomials and Q not the zero polynomial,

$$R(x) = \frac{P(x)}{Q(x)} \quad \text{for} \quad x \in \mathbb{R} \quad \text{such that} \quad Q(x) \neq 0. \tag{2.1}$$

Prob. 2.1. Prove: (a) Every polynomial is a rational function, (b) the rational function R, where

$$R(x) = \frac{1}{x} \quad \text{for} \quad x \in \mathbb{R}, \quad x \neq 0$$

is not polynomial.

Example 2.2. If $p > 0$, the function $E_p : \mathbb{R} \to \mathbb{R}$ defined in Def. IV.10.1 is a useful function. In Remark IV.10.2 we introduced the notation

$$p^x = E_p(x) \quad \text{for} \quad x \in \mathbb{R}. \tag{2.2}$$

Of special importance is E_e, which we often write simply as E. Thus,

$$E(x) = e^x \qquad \text{for} \quad x \in \mathbb{R}. \tag{2.3}$$

By Prob. IV.7.1, we have:

$$\exp r = e^r = E(r) \qquad \text{for rational } r, \tag{2.4}$$

where $\exp x$ is defined in Example IV.5.1. Note, since $e > 2$, we have $e^x > 1$ for $x > 0$, and that $0 < e^x < 1$ for $x > 0$.

EXAMPLE 2.3 (Hyperbolic Sine and Cosine). The *hyperbolic sine* and *cosine functions*, abbreviated respectively as sinh and cosh, are defined as

$$\sinh x = \frac{e^x - e^{-x}}{2}, \qquad \cosh x = \frac{e^x + e^{-x}}{2} \qquad \text{for} \quad x \in \mathbb{R}. \tag{2.5}$$

Clearly,

$$\sinh 0 = 0 \quad \text{and} \quad \cosh 0 = 1. \tag{2.6}$$

PROB. 2.2. Prove: If $x \in \mathbb{R}$, then $\cosh x \geqslant 1$.

PROB. 2.3. Prove: (a) $\sinh(-x) = -\sinh x$, (b) $\cosh(-x) = \cosh x$.

PROB. 2.4. Prove: If $x > 0$, then $\sinh x > 0$ and if $x < 0$, then $\sinh x < 0$.

PROB. 2.5. Prove: If $x \in \mathbb{R}$, $y \in \mathbb{R}$, then

(a) $\cosh(x + y) = \cosh x \cosh y + \sinh x \sinh y$,
(b) $\sinh(x + y) = \sinh x \cosh y + \cosh x \sinh y$,
(c) $\cosh^2 x - \sinh^2 x = 1$.

PROB. 2.6. Prove: If n is an integer, then

$$(\cosh + \sinh x)^n = \cosh nx + \sinh nx \qquad \text{for} \quad x \in \mathbb{R}.$$

PROB. 2.7. Prove: If $x \in \mathbb{R}$, then

(a) $\sinh 2x = 2 \sinh x \cosh x$,
(b) $\cosh 2x = \cosh^2 x + \sinh^2 x = 2\cosh^2 x - 1 = 1 + 2\sinh^2 x$,
(c) $\sinh^2(x/2) = (\cosh x - 1)/2$ and $\cosh^2 x = (\cosh x + 1)/2$.

PROB. 2.8. We define *hyperbolic tangent*, *hyperbolic cosecant*, and *hyperbolic cotangent*, written respectively as $\tanh x$, csch, coth, by means of

$$\text{(a)} \quad \tanh x = \frac{\sinh x}{\cosh x}, \quad x \in \mathbb{R}, \qquad \text{(b)} \quad \operatorname{csch} x = \frac{1}{\sinh x} \quad \text{if} \quad x \neq 0. \tag{2.7}$$

$$\text{(c)} \quad \operatorname{sech} x = \frac{1}{\cosh x}, \quad x \in \mathbb{R}, \qquad \text{(d)} \quad \coth x = \frac{\cosh x}{\sinh x} \quad \text{if} \quad x \neq 0. \tag{2.8}$$

Prove: (a) $1 - \tanh^2 x = \operatorname{sech}^2 x$. (b) $\coth^2 x - 1 = \operatorname{csch}^2 x$ for $x \in \mathbb{R}$. (c) $\tanh 0 = 0$, $\operatorname{sech} 0 = 1$.

PROB. 2.9. Prove: (a) $-1 < \tanh x < 1$. (b) If $x > 0$, then $\coth x > 1$ and if $x < 0$, then $\coth x < -1$. (c) $0 < \operatorname{sech} x \leqslant 1$.

PROB. 2.10. Prove: The hyperbolic sine function is strictly monotonic increasing.

PROB. 2.11. Prove: The hyperbolic cosine function is strictly monotonic increasing on $[0, +\infty)$ and strictly decreasing on $(-\infty, 0]$.

Def. 2.1. A function $f: \mathbb{R} \to \mathbb{R}$ is called an *even* function if and only if

$$f(-x) = f(x) \qquad \text{for all} \quad x \in \mathbb{R}. \tag{2.9}$$

A function f is called an *odd* function if and only if

$$f(-x) = -f(x) \qquad \text{for all} \quad x \in \mathbb{R}. \tag{2.10}$$

Thus, the hyperbolic cosine function and the cosine function are examples of even functions, whereas the hyperbolic sine function and the sine function are examples of odd functions.

PROB. 2.12. Prove: The hyperbolic tangent and cotangent are odd functions.

Remark 2.1. Let $f: \mathbb{R} \to \mathbb{R}$ be a real-valued function of a real variable. The function g defined as

$$g(x) = \frac{f(x) + f(-x)}{2} \qquad \text{for each} \quad x \in \mathbb{R}$$

is even, and the function h defined as

$$h(x) = \frac{f(x) - f(-x)}{2} \qquad \text{for each} \quad x \in \mathbb{R}$$

is odd, as can be checked. Since $f(x) = g(x) + h(x)$ for each $x \in \mathbb{R}$ we see that any real-valued function of a real variable can be written as the sum of an even and an odd function.

PROB. 2.13. Prove: (a) The product of two even functions is even; (b) the product of two functions one of which is even and the other is odd is an odd function; (c) the product of two odd functions is an even function.

EXAMPLE 2.4 (Distance to the Nearest Integer). We define

$$\langle x \rangle^* = \text{the distance of } x \in \mathbb{R} \text{ to the integer nearest } x. \tag{2.11}$$

In the literature this is sometimes written as $\{x\}$. Since the latter notation conflicts with the notation $\{x\}$ = singleton set whose only member is x, we prefer to use the notation in (2.11). We have, $[x] \leqslant x < [x]+1$, where $[x]$ is the greatest integer $\leqslant x$, and that the respective distances of x from the integers $[x]$ and $[x]+1$ are $x-[x]$ and $[x]+1-x$. We have

$$\langle x\rangle^* = \min\{x-[x], [x]+1-x\} \quad \text{for each} \quad x \in \mathbb{R}. \tag{2.12}$$

We recall (Prob. I.13.17) that

$$\min\{a,b\} = \frac{a+b-|a-b|}{2} \quad \text{for} \quad a \in \mathbb{R}, \quad b \in \mathbb{R}. \tag{2.13}$$

This implies that in terms of the greatest integer function, we have (explain)

$$\langle x\rangle^* = \tfrac{1}{2} - |\tfrac{1}{2} - x + [x]| \quad \text{for} \quad x \in \mathbb{R}. \tag{2.14}$$

Prob. 2.14. Prove: $\langle [x]+\frac{1}{2}\rangle^* = \frac{1}{2}$ for $x \in \mathbb{R}$.

Prob. 2.15. Prove: $\langle x+1\rangle^* = \langle x\rangle^*$ for $x \in \mathbb{R}$.

Prob. 2.16. Prove: $0 \leqslant \langle x\rangle^* \leqslant \frac{1}{2}$ for $x \in \mathbb{R}$.

For the graph of the nearest integer function see Fig. 2.1. It is an example of a *periodic* function.

Def. 2.2. A function $f: \mathbb{R} \to \mathbb{R}$ is called *periodic* if there exists a real number $a \neq 0$ such that

$$f(x+a) = f(x) \quad \text{for each} \quad x \in \mathbb{R}. \tag{2.15}$$

The number a is called a *period* of the function. The least positive period is called *the fundamental period* or *the period* of the function.

Note that a constant function has any nonzero number as a period.

Prob. 2.17. Prove: If $f: \mathbb{R} \to \mathbb{R}$ is periodic and a is some period of f, then any nonzero integral multiple na of a is also a period of f. Thus, a periodic function always has some positive number as a period.

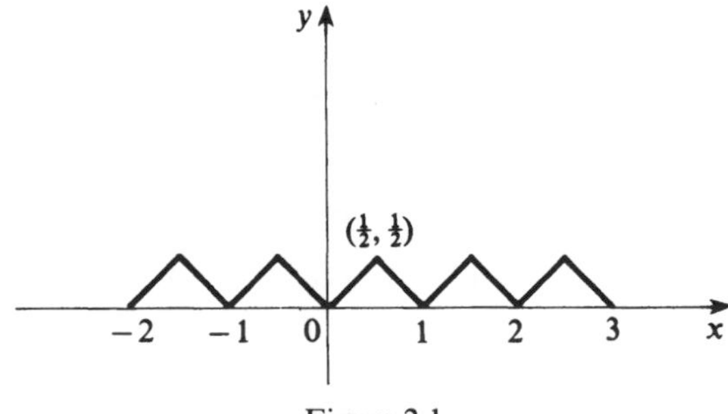

Figure 2.1

PROB. 2.18. Prove: (a) The period of the function $\langle\ \rangle^*$ is 1. (b) Also prove that it is an even function.

3. Neighborhood of a Point. Accumulation Point of a Set

We wish to extend the notion of limit to functions whose domains are subsets of $\mathbb{R}$. Before doing this it will be convenient to introduce the notion of a *neighborhood* of a point of $\mathbb{R}$ or of a point of $\mathbb{R}^*$.

Def. 3.1. Let $a \in \mathbb{R}$. By an *ϵ-neighborhood of a*, we mean the set $N(a,\epsilon)$ defined as

$$N(a,\epsilon) = \{x \in \mathbb{R} \mid |x - a| < \epsilon\}, \tag{3.1}$$

where ϵ is some positive real number. By *a deleted ϵ-neighborhood of a* we mean the set $N^*(a,\epsilon)$, where

$$N^*(a,\epsilon) = \{x \in \mathbb{R} \mid 0 < |x - a| < \epsilon\}. \tag{3.2}$$

We never use the notation $N(a,\epsilon)$ or $N^*(a,\epsilon)$ unless $\epsilon > 0$.

It is clear that

$$N(a,\epsilon) = (a - \epsilon; a + \epsilon). \tag{3.3}$$

$$N^*(a,\epsilon) = N(a,\epsilon) - \{a\}. \tag{3.4}$$

$$N^*(a,\epsilon) = (a - \epsilon; a) \cup (a; a + \epsilon). \tag{3.5}$$

PROB. 3.1. Prove: If a and b are real numbers such that $a < b$, then (see Fig. 3.1).

$$N\left(\frac{a+b}{2}, \frac{b-a}{2}\right) = (a; b). \tag{3.6}$$

PROB. 3.2. Prove: If $a \in \mathbb{R}$ and $\epsilon > 0$, then the set S, where

$$S = \left\{a \pm \frac{1}{n+1}\epsilon \,\middle|\, n \in \mathbb{Z}_+\right\}, \tag{3.7}$$

is a subset of $N^*(a,\epsilon)$. Thus, $N(a,\epsilon)$ is not empty and, as a matter of fact, contains infinitely many points.

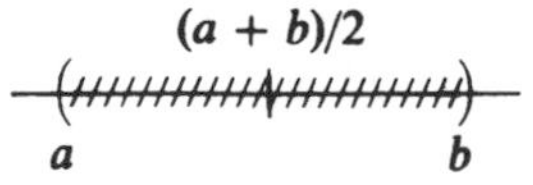

$(a + b)/2$

a b

Figure 3.1

Theorem 3.1. *If $x_0 \in \mathbb{R}$ and $N(x_0, \epsilon_1)$, $N(x_0, \epsilon_2)$ are ϵ-neighborhoods of x_0, then there exists an ϵ-neighborhood $N(x_0, \epsilon)$ of x_0 such that*

$$N(x_0, \epsilon) \subseteq N(x_0, \epsilon_1) \cap N(x_0, \epsilon_2). \tag{3.8}$$

PROOF. By hypothesis, $\epsilon_1 > 0$ and $\epsilon_2 > 0$ and each is in $\mathbb{R}$. Let

$$\epsilon = \min\{\epsilon_1, \epsilon_2\}.$$

We have $0 < \epsilon \leqslant \epsilon_1$ and $0 < \epsilon \leqslant \epsilon_2$. Use this ϵ. Assume $x \in N(x_0, \epsilon)$. We obtain

$$|x - x_0| < \epsilon \leqslant \epsilon_1 \quad \text{and} \quad |x - x_0| < \epsilon \leqslant \epsilon_2.$$

This implies that $x \in N(x_0, \epsilon_1) \cap N(x_0, \epsilon_2)$. This proves (3.8).

It is also useful to consider neighborhoods of $\pm\infty$.

Def. 3.2. By a *neighborhood of* $+\infty$, written $N(+\infty)$, we mean a subset of $\mathbb{R}^*$ of the form

$$N(+\infty) = \{x \in \mathbb{R}^* \mid x > B\}, \tag{3.9}$$

where B is some real number. Similarly, *a neighborhood of* $-\infty$, written as $N(-\infty)$, is a subset of $\mathbb{R}^*$ of the form

$$N(-\infty) = \{x \in \mathbb{R}^* \mid x < B\}, \tag{3.10}$$

where B is some real number. Clearly,

$$N(+\infty) = \{+\infty\} \cup (B; +\infty) \tag{3.11}$$

and

$$N(-\infty) = \{-\infty\} \cup (-\infty; B), \tag{3.12}$$

where $B \in \mathbb{R}$. Deleted neighborhoods of $\pm\infty$, written as $N^*(\pm\infty)$, are defined respectively as

$$N^*(+\infty) = \{x \in \mathbb{R} \mid x > B\} = (B; +\infty) \tag{3.13}$$

and

$$N^*(-\infty) = \{x \in \mathbb{R} \mid x < B\} = (-\infty; B), \tag{3.14}$$

where $B \in \mathbb{R}$. These are subsets of $\mathbb{R}$.

Def. 3.3. If $x_0 \in \mathbb{R}$, then by *an ϵ-neighborhood of x_0 from the right*, written as $N_+(x_0, \epsilon)$, we mean a set of the form

$$N_+(x_0, \epsilon) = [x_0, x_0 + \epsilon) = \{x \in \mathbb{R} \mid x_0 \leqslant x < x_0 + \epsilon\}. \tag{3.15}$$

Dually, by an *ϵ-neighborhood of x_0 from the left*, written as $N_-(x_0, \epsilon)$, we mean the set

$$N_-(x_0, \epsilon) = (x_0 - \epsilon, x_0] = \{x \in \mathbb{R} \mid x_0 - \epsilon < x \leqslant x_0\}. \tag{3.16}$$

These are called *one-sided* neighborhoods. Corresponding to them we have *deleted* one-sided neighborhoods $N_+^*(x_0,\epsilon)$ and $N_-^*(x_0,\epsilon)$, where

$$N_+^*(x_0,\epsilon) = (x_0\,;x_0+\epsilon) = \{x \in \mathbb{R} \mid x_0 < x < x_0 + \epsilon\} \tag{3.17}$$

and

$$N_-^*(x_0\epsilon) = (x_0-\epsilon;x_0) = \{x \in \mathbb{R} \mid x_0 - \epsilon < x < x_0\}. \tag{3.18}$$

Remark 3.1. A neighborhood of $+\infty$ can be viewed as a neighborhood of $+\infty$ from the left and a neighborhood of $-\infty$ as a neighborhood from the right of $-\infty$, respectively.

Prob. 3.3. Prove: If $L \in \mathbb{R}^*$ and $N_1^*(L)$ and $N_2^*(L)$ are deleted neighborhoods of L from the same side, then there exists a deleted neighborhood $N^*(L)$ of L, from that side, such that

$$N^*(L) \subseteq N_1^*(L) \cap N_2^*(L).$$

We will need the notion of an *accumulation point* of a set $S \subseteq \mathbb{R}$.

Def. 3.4. An *accumulation point, in* $\mathbb{R}$, of $S \subseteq \mathbb{R}$ is a point $x_0 \in \mathbb{R}$ such that each deleted ϵ-neighborhood $N^*(x_0,\epsilon)$ of x_0 contains points of S; that is, such that $N^*(x_0,\epsilon) \cap S \neq \emptyset$ for all $\epsilon > 0$. When there is no danger of confusion, an accumulation point $x_0 \in \mathbb{R}$ of $S \subseteq \mathbb{R}$ will be referred to simply as an *accumulation point* of S.

Example 3.1. Let

$$A = \{1, \tfrac{1}{2}, \tfrac{1}{3}, \ldots\} = \left\{ \frac{1}{n} \,\middle|\, n \text{ is a positive integer} \right\}.$$

The point 0 is an accumulation point of A. For, given $N^*(0,\epsilon)$, where $\epsilon > 0$, there exists an integer m such that $0 < 1/m < \epsilon$ and hence $1/m \in N^*(0,\epsilon)$. Since $1/m \in A$ also, it follows that

$$N^*(0,\epsilon) \cap A \neq \emptyset$$

for every $\epsilon > 0$. Note that $0 \notin A$.

Remark 3.2. The set of all accumulation points of a set $S \subseteq \mathbb{R}$ is called its *derived* set and is written as S'.

Prob. 3.4. Let $S = \{n + 1/m \mid n \text{ and } m \text{ are integers and } m > 0\}$. What is the derived set S' of S? Does $S' \subseteq S$ hold?

Prob. 3.5. Let $a \in \mathbb{R}$ and $b \in \mathbb{R}$. (a) Prove $(a;b)' = [a,b]$, (b) $[a,+\infty)' = [a,+\infty)$.

Remark 3.3. There exist subsets of $\mathbb{R}$ having no accumulation points. For example, let $x_0 \in \mathbb{R}$, then $A = \{x_0\}$ has no accumulation points (explain).

PROB. 3.6. Prove: The set $\mathbb{Z}$ of integers has no accumulation points.

Remark 3.4 (Extended Accumulation Points). When a set $S \subseteq \mathbb{R}$ is not bounded from above, then for each $B \in \mathbb{R}$, there exists $x_1 \in S$ such that $x_1 > B$. In terms of neighborhoods of $+\infty$, this states that each deleted neighborhood of $+\infty$ contains points of S. In this sense, although $+\infty$ is not an element of $\mathbb{R}$, it may be viewed as an accumulation point of S. We speak of $+\infty$ as an *extended accumulation* point of S. Similarly, if $S \subseteq \mathbb{R}$ is not bounded from below, $-\infty$ may be thought of as an accumulation point of S and is referred to as an *extended* accumulation point of S. Thus, $\pm\infty$ are called *extended accumulation* points of a set S, under appropriate conditions on S. When no reference is made at all as to whether the accumulation point we are dealing with is an extended one or not, we will mean an accumulation point *in* $\mathbb{R}$ referring to it as a *real* or *finite* accumulation point.

Remark 3.5 (Real, One-Sided Accumulation Points). If $S \subseteq \mathbb{R}$, then $x_0 \in \mathbb{R}$ is an accumulation point (in $\mathbb{R}$) of S *from the right* if and only if each deleted ϵ-neighborhood $(x_0; x_0 + \epsilon)$ of x_0 from the right contains points of S. Dually, $x_0 \in S$ is called an accumulation point of S *from the left*, if and only if each deleted ϵ-neighborhood $(x_0 - \epsilon, x_0)$ of x_0 from the left contains points of S. Such accumulation points are called *one-sided* accumulation points.

For the bounded open interval $(a; b)$, a is an accumulation point of $(a; b)$ from the right and b is an accumulation point of $(a; b)$ from the left.

Clearly, a one-sided accumulation point of a set is also an accumulation point of the set, but the converse does not hold.

Theorem 3.2. *If $S \subseteq \mathbb{R}$, then $x_0 \in \mathbb{R}$ is an accumulation point of S if and only if each ϵ-neighborhood $N(x_0, \epsilon)$ of x_0 contains infinitely many points of S.*

PROOF. Assume that each $N(x_0, \epsilon)$ of x_0 contains infinitely many points of S. Let $N^*(x_0, \epsilon')$ be some deleted ϵ-neighborhood of x_0. Now $\{x_0\} \cup N^*(x_0, \epsilon') = N(x_0, \epsilon')$ is an ϵ-neighborhood of x_0 and contains infinitely many points of S. This implies that $N^*(x_0, \epsilon') = N(x_0, \epsilon') - \{x_0\}$ contains points of S. Thus, each deleted ϵ-neighborhood $N^*(x_0, \epsilon')$ contains points of S. Therefore x_0 is an accumulation point of S.

Conversely, suppose x_0 is an accumulation point of S. Suppose that there exists some $N(x_0, \epsilon_1)$ of x_0 exists containing at most finitely many points of S. Since x_0 is an accumulation point of S, $N^*(x_0, \epsilon_1)$ contains some points of S. Since

$$N^*(x_0, \epsilon_1) \subset N(x_0, \epsilon_1),$$

we see that $N^*(x_0, \epsilon_1)$ contains at most finitely many such points. Take the set $N^*(x_0, \epsilon_1) \cap S$. This set is not empty and consists of points $x_1, \ldots, x_k$, where k is some positive integer. Consider the real numbers

$$d_1 = |x_1 - x_0|, \qquad d_2 = |x_2 - x_0|, \ldots, \qquad d_k = |x_k - x_0|.$$

Each d_j is a positive real number and the set $\{d_1, \ldots, d_k\}$ is a nonempty set of positive real numbers. Let

$$\epsilon' = \min\{d_1, \ldots, d_k\}.$$

We have $\epsilon' > 0$. Now, consider $N^*(x_0, \epsilon')$. If $x \in N^*(x_0, \epsilon')$, then

$$0 < |x - x_0| < \epsilon' \leqslant d_j = |x_j - x_0| \qquad \text{for } j \in \{1, 2, \ldots, k\}.$$

Thus, x differs from $x_1, \ldots, x_k$ and is not in $N^*(x_0, \epsilon') \cap S$. This implies

$$N^*(x_0, \epsilon') \cap S = \phi$$

and, hence, that the deleted ϵ-neighborhood $N^*(x_0, \epsilon')$ contains no points of S. This contradicts the assumption that x_0 is an accumulation point of S. Hence, we must conclude that each ϵ-neighborhood $N(x_0, \epsilon)$ of S contains infinitely many points of S. This completes the proof.

Theorem 3.3. *If $S \subseteq \mathbb{R}$, then x_0 is an accumulation point of S if and only if there exists a sequence $\langle x_n \rangle$ of distinct points of S such that* $\lim_{n \to +\infty} x_n = x_0$.

PROOF. Suppose first that there exists a sequence $\langle x_n \rangle$ of distinct points of S such that $\lim x_n = x_0$. Let $N(x_0, \epsilon)$ be some ϵ-neighborhood of x_0. There exists an N such that if $n > N$, then $|x_n - x_0| < \epsilon$. For at least one $n > N$, n_1 say, we have $x_{n_1} \in S$ and $x_{n_1} \neq x_0$ (otherwise $x_n = x_0$ holds for all $n > N$ and the points of $\langle x_n \rangle$ are not distinct). This implies that $x_{n_1} \in N^*(x_0, \epsilon) \cap S$. Thus, for each $\epsilon > 0$, we have $N^*(x_0, \epsilon) \cap S \neq \emptyset$. It follows that x_0 is an accumulation point of S.

Conversely, let x_0 be an accumulation point of S. There exists $x_1 \in S$ such that $x_1 \in N^*(x_0, 1)$. Take $\epsilon_2 = \frac{1}{2}$. $N(x_0, \frac{1}{2})$ contains infinitely many points of S. Hence, there exists $x_2 \in S$ such that $x_2 \neq x_0$, $x_2 \neq x_1$ and $x_2 \in N(x_0, \frac{1}{2})$. We have $x_2 \in N^*(x_0, \frac{1}{2})$ and $x_2 \neq x_1$. We continue this procedure inductively. If for some positive integer n there exist distinct $x_1, \ldots, x_n$ in S such that $x_j \in N^*(x_0, 1/j)$ for $j \in \{1, \ldots, n\}$, then take $\epsilon_{n+1} = 1/(n+1)$ and $N(x_0, 1/(n+1))$. There exists a point $x_{n+1} \in S$ differing from $x_0, x_1, \ldots, x_n$ such that $x_{n+1} \in N(x_0, 1/(n+1))$ (why?). We see that $x_{n+1} \in N^*(x_0, 1/(n+1))$ and that it differs from $x_1, \ldots, x_n$. The sequence $\langle x_n \rangle$ constructed in this manner consists of distinct elements of S. Since $|x_n - x_0| < 1/n$ for each n, it follows that $\lim x_n = x_0$. This completes the proof.

PROB. 3.7. Prove: If $x_0 \in \mathbb{R}$, then there exists a sequence r_n of rational numbers such that $r_n \to x_0$ as $n \to +\infty$ and there is a sequence $\langle c_n \rangle$ of irrational numbers such that $c_n \to x_0$ as $n \to +\infty$.

PROB. 3.8. Prove: If $S \subseteq T \subseteq \mathbb{R}$, then $S' \subseteq T'$.

PROB. 3.9. Suppose $S \subseteq \mathbb{R}$. Prove: (a) $+\infty$ is an *extended* accumulation point of S if and only if there exists a sequence $\langle x_n \rangle$ of distinct points of S such that $\lim x_n = +\infty$. (b) State and prove an analogous criterion for $-\infty$ to be an *extended* accumulation point of S.

4. Limits of Functions

We recall that a sequence $\langle x_n \rangle$ of real numbers converges with $L = \lim x_n$ if and only if for each $\epsilon > 0$, there exists an N such that if $n > N$, then $|x_n - L| < \epsilon$. We formulate this in terms of the notion of neighborhood.

A sequence $\langle x_n \rangle$ is a function whose domain is $\mathbb{Z}_+$. Let f be the function such that $f(n) = x_n$ for each $n \in \mathbb{Z}_+$. We have $\mathbb{Z}_+ = \mathfrak{D}(f)$. Since $\mathbb{Z}_+$ is not bounded from above, we can think of $+\infty$ as an (extended) accumulation point of $\mathbb{Z}_+ = \mathfrak{D}(f)$. The condition $n > N$ becomes $n \in (N; +\infty) \cap \mathfrak{D}(f)$, and the condition $|x_n - L| < \epsilon$ becomes $f(n) = x_n \in N(L, \epsilon)$, the last being an ϵ-neighborhood of L. Thus, we have the formulation: $\langle x_n \rangle$ converges with $L = \lim x_n$ if and only if for each ϵ-neighborhood $N(L, \epsilon)$ of L, there exists a deleted ϵ-neighborhood $N^*(+\infty) = (N; +\infty)$ of $+\infty$ such that

$$n \in N^*(+\infty) \cap \mathfrak{D}(f) = (N; +\infty) \cap \mathbb{Z}_+,$$

implies that

$$f(n) = x_n \in N(L, \epsilon).$$

Finally, this can be formulated as follows: $L = \lim_{n \to +\infty} x_n$ if and only if for each given ϵ-neighborhood $N(L, \epsilon)$ of L there exists a deleted neighborhood $N^*(+\infty)$ of $+\infty$ such that

$$f(N^*(+\infty) \cap \mathbb{Z}_+) \subseteq N(L, \epsilon).$$

Below, in Def. 4.1, we extend the definition of limit to functions that do not necessarily have $\mathbb{Z}_+$ as their domain of definition. This is general enough to accommodate the cases where the limit is $\pm\infty$.

Def. 4.1. Let f be a real-valued function of a real variable and $a \in \mathbb{R}^*$. We say that f *approaches* $L \in \mathbb{R}^*$ as x *approaches* a or that f has *limit* L *as* x *approaches* a and write

$$\lim_{x \to a} f(x) = L \tag{4.1}$$

if and only if: (1) a is an accumulation point of $\mathfrak{D}(f)$ (possibly in the extended sense) and (2) for each given neighborhood $N(L)$ of L there exists a deleted neighborhood $N_1^*(a)$ of a such that

$$f(N_1^*(a) \cap \mathfrak{D}(f)) \subseteq N(L). \tag{4.2}$$

Sometimes we also write $f(x) \to L$ as $x \to a$ for $\lim_{x \to a} f(x) = L$. If $L \in \mathbb{R}$, then we say that f has a *finite limit* as x *approaches* a.

In this definition we use deleted neighborhoods of a to ensure the independence of the limit of f as $x \to a$ from the value that f has at a (if it is defined there at all).

Remark 4.1. In Def. 4.1, there are three possibilities for a: $a \in \mathbb{R}$, $a = +\infty$, or $a = -\infty$. Similarly, there are three possibilities for L. Hence, there are altogether nine cases for the pair (a, L). We detail some of these cases below and leave the others for the reader.

Def. 4.2(a). Case of Def. 4.1 where $a \in \mathbb{R}$, $L \in \mathbb{R}$. In this case $f(x) \to L$ as $x \to a$ means: For each ϵ-neighborhood $N(L, \epsilon)$ of L, there exists a deleted δ-neighborhood $N^*(a, \delta)$ of a such that if $x \in N^*(a, \delta) \cap \mathfrak{D}(f)$, then $f(x) \in N(L, \epsilon)$. We translate this from the neighborhood terminology into the "language" of inequalities below:

Let $a \in \mathbb{R}$, $L \in \mathbb{R}$. We write $\lim_{x \to a} f(x) = L$ (where $a \in \mathbb{R}$, $L \in \mathbb{R}$) if and only if for each $\epsilon > 0$, there exists a $\delta > 0$ such that

$$x \in \mathfrak{D}(f) \quad \text{and} \quad 0 < |x - a| < \delta \tag{4.3}$$

imply

$$|f(x) - L| < \epsilon. \tag{4.4}$$

See Fig. 4.1.

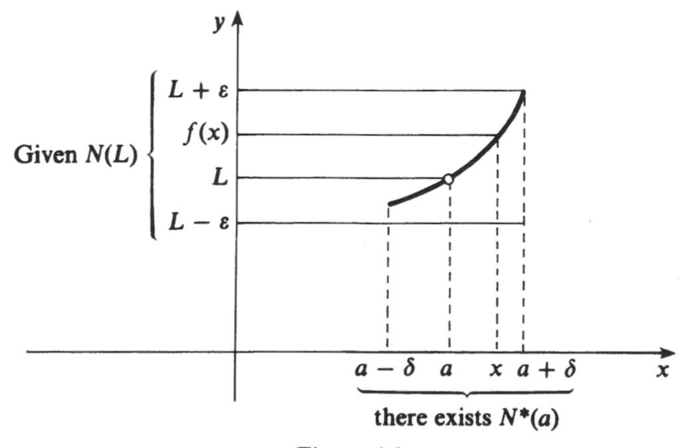

Figure 4.1

EXAMPLE 4.1. We prove:

$$\lim_{x\to 3}(2x^2 - x) = 15. \tag{4.5}$$

Here, $f(x) = 2x^2 - x$ and we take $\mathfrak{D}(f) = \mathbb{R}$. It is clear that 3 is an accumulation point of $\mathbb{R} = \mathfrak{D}(f)$. We do an analysis first. We wish to prove that if $\epsilon > 0$ is given, then there exists a $\delta > 0$ such that if $0 < |x - 3| < \delta$, then

$$|f(x) - 15| = |2x^2 - x - 15| = |(2x+5)(x-3)| = |2x+5||x-3| < \epsilon. \tag{4.6}$$

Given $\epsilon > 0$, we first take $\delta_1 = 1$ and x such that $0 < |x - 3| < \delta_1 = 1$, so that $x \neq 3$ and $-1 < x - 3 < 1$ or

$$x \neq 3 \quad \text{and} \quad 2 < x < 4. \tag{4.7}$$

These imply that $x \neq 3$ and

$$9 < 2x + 5 < 13. \tag{4.8}$$

Hence, by (4.6)

$$|f(x) - 15| = |2x+5||x-3| < 13|x-3|. \tag{4.9}$$

From this we see that if we take x such that $0 < |x - 3| < \epsilon/13$, then (4.6) will hold. Our analysis leads to the choice of a δ such that $0 < \delta \leqslant \min\{1, \epsilon/13\}$. We prove that such a δ "works." For x such that $0 < |x - 3| < \delta \leqslant \min\{1, \epsilon/13\}$, we have $0 < |x - 3| < 1$ and $0 < |x - 3| < \epsilon/13$. For such x, (4.8) holds and, hence, (4.9) holds. Therefore

$$|f(x) - 15| < 13|x-3| < 13\left(\frac{\epsilon}{13}\right) = \epsilon,$$

so that (4.6) holds. This proves (4.5).

EXAMPLE 4.2. We prove: If $a_0 \neq 0$ and n is a positive integer, then

$$\lim_{x\to 0}(a_0x^n + a_1x^{n-1} + \cdots + a_{n-1}x + a_n) = a_n. \tag{4.10}$$

We define $P: \mathbb{R} \to \mathbb{R}$ as $P(x) = a_0x^n + a_1x^{n-1} + \cdots + a_{n-1}x + a_n$ for $x \in \mathbb{R}$. The idea is to prove that $\lim_{x\to 0}P(x) = a_n$. We must prove: If $\epsilon > 0$ is given, then there exists a $\delta > 0$ such that if $0 < |x| = |x - 0| < \delta$, then

$$|a_0x^n + a_1x^{n-1} + \cdots + a_{n-1}x| = |P(x) - a_n| < \epsilon. \tag{4.11}$$

Let $\epsilon > 0$ be given. If $n = 1$, then $P(x) = a_0x + a_1$, so (4.11) becomes

$$|P(x) - a_1| = |a_0x| = |a_0||x| < \epsilon. \tag{4.12}$$

We take $0 < |x| < \epsilon/|a_0|$ and obtain

$$|P(x) - a_1| = |a_0||x| < |a_0|\frac{\epsilon}{|a_0|} = \epsilon.$$

As a matter of fact, this will hold for $0 < |x| < \delta$, where $0 < \delta \leqslant \epsilon/|a_0|$. Thus, (4.10) holds if $n = 1$. If $n \geqslant 2$, we first take x such that $0 < |x| < 1$.

For such x we have $|x|^n < |x|^{n-1} < \cdots < |x|$. Thus, $0 < |x| < 1$ implies that

$$|P(x) - a_n| = |a_0x^n + \cdots + a_{n-1}x| \leqslant |a_0||x|^n + \cdots + |a_n||x| \leqslant (|a_0| + \cdots + |a_n|)|x|. \tag{4.13}$$

Since $a_0 \neq 0$, we know that $|a_0| + \cdots + |a_n| > 0$. We take δ such that

$$0 < \delta \leqslant \min\left\{1, \frac{\epsilon}{|a_0| + \cdots + |a_n|}\right\} \tag{4.14}$$

and $0 < |x| < \delta$. For such x,

$$0 < |x| < 1 \tag{4.15a}$$

$$0 < |x| < \frac{\epsilon}{|a_0| + \cdots + |a_n|}. \tag{4.15b}$$

This and (4.13) imply that

$$|P(x) - a_n| < (|a_0| + \cdots + |a_n|)|x| < \epsilon.$$

We conclude that (4.10) holds also for $n \geqslant 2$ also.

Remark 4.2. As a special case of the result proved in the last example, we have: If n is a positive integer, then $\lim_{x\to 0} x^n = 0$. This could also be proved by noting that if $\alpha > 0$, then

$$\lim_{x\to 0} x^\alpha = 0. \tag{4.16}$$

(Recall that f, where $f(x) = x^\alpha$ and $\alpha \in \mathbb{R}$ is defined for $x > 0$.) Here $\mathfrak{D}(f) = (0; +\infty)$. Given $\epsilon > 0$, we take δ such that $0 < \delta \leqslant \epsilon^{1/\alpha}$ and $x \in (0; +\infty) = \mathfrak{D}(f)$ such that $0 < |x| < \delta$ and obtain $0 < x < \delta \leqslant \epsilon^{1/\alpha}$. This implies $0 < x^\alpha < (\epsilon^{1/\alpha})^\alpha = \epsilon$ and, hence, that

$$|f(x) - 0| = |x^\alpha - 0| = x^\alpha < \epsilon \qquad \text{for} \quad x \in (0; +\infty) \quad \text{and} \quad 0 < |x| < \delta.$$

Therefore, (4.16) follows.

PROB. 4.1. Let c be some real number and S some subset of $\mathbb{R}$ having a as an accumulation point (possibly an extended one). Let I_S be the identity function on S. Prove: $\lim_{x\to a} I_S(x) = c$.

Before proceeding further, we prove a theorem which can reduce some of the work involved in proving $\lim_{x\to a} f(x) = L$ when this is the case.

Theorem 4.1. *Let f and g have a common domain D and let $a \in \mathbb{R}^*$ be an accumulation point of D. If for some real number L there exists a deleted neighborhood $N_1^*(a)$ of a exists such that*

$$|f(x) - L| \leqslant |g(x)| \qquad \text{for all} \quad x \in N_1^*(a) \cap D \tag{4.17}$$

and $g(x) \to 0$ as $x \to a$, then

$$\lim_{x\to a} f(x) = L. \tag{4.18}$$

Proof. Assume $\epsilon > 0$. Since $g(x) \to 0$ as $x \to a$, there exists a deleted neighborhood $N_2^*(a)$ of a such that if $x \in N_2^*(a) \cap D$, then

$$|g(x)| = |g(x) - 0| < \epsilon. \tag{4.19}$$

There is a deleted neighborhood $N^*(a)$ of a such that $N^*(a) \subseteq N_1^*(a) \cap N_2^*(a)$. This implies $N^*(a) \subseteq N_1^*(a)$ and $N^*(a) \subseteq N_2^*(a)$. Therefore, if $x \in N^*(a) \cap D$, then both (4.17) and (4.19) hold so that

$$|f(x) - L| \leqslant |g(x)| < \epsilon.$$

This proves (4.18).

We illustrate the use of Theorem 4.1 by evaluating some special limits involving the sine and cosine functions.

Theorem 4.2. *The following limit statements hold*:

(a) $\lim_{x\to 0}\sin x = 0$
(b) $\lim_{x\to 0}((\sin x)/x) = 1$,
(c) $\lim_{x\to 0}\cos x = 1$,
(d) $\lim_{x\to 0}((1 - \cos x)/x) = 0$.

Proof. We use Theorem IV.8.2. By part (c) of that theorem, we have

$$|\sin x - 0| = |\sin x| \leqslant |x| \qquad \text{for all} \quad x \in \mathbb{R}. \tag{4.20}$$

We apply Theorem 4.1 with $f(x) = \sin x$, $g(x) = x$ for all $x \in \mathbb{R}$, with $a = L = 0$. Since $\lim_{x\to 0} g(x) = \lim_{x\to 0} x = 0$ (Remark 4.2 and (4.16)), we obtain (by Theorem 4.1)

$$\lim_{x\to 0} \sin x = 0.$$

This proves (a). To prove (b), we use Theorem IV.8.2, part (b) according to which we have

$$|\sin x - x| < \frac{|x|^3}{6} \qquad \text{for} \quad 0 < |x| < 1.$$

This implies that

$$\left|\frac{\sin x}{x} - 1\right| < \frac{x^2}{6} \qquad \text{if} \quad 0 < |x| < 1. \tag{4.21}$$

The reader can prove: $\lim_{x\to 0} x^2/6 = 0$. We can then apply Theorem 4.1 with

$$f(x) = \frac{\sin x}{x}, \qquad g(x) = \frac{x^2}{6}, \qquad a = 0, L = 1$$

and obtain from (4.21)

$$\lim_{x\to 0} \frac{\sin x}{x} = 1.$$

This proves (b). To prove (c), we use the equality $-\cos x = 2\sin^2(x/2)$ which holds for all $x \in \mathbb{R}$ (Prob. IV.8.4). We have

$$|\cos x - 1| = 2\sin^2\frac{x}{2} \leqslant 2\frac{x^2}{4} = \frac{x^2}{2} \qquad \text{for} \quad x \in \mathbb{R}. \tag{4.22}$$

After proving that $\lim_{x\to 0}x^2/2 = 0$, we can use Theorem 4.1 to prove $\lim_{x\to 0}\cos x = 1$. This proves (c).

By (4.22) we have

$$\left|\frac{1-\cos x}{x} - 0\right| = \left|\frac{1-\cos x}{x}\right| \leqslant \frac{|x|}{2} \qquad \text{for} \quad x \neq 0. \tag{4.23}$$

Since $\lim_{x\to 0}|x|/2 = 0$, this implies by Theorem 4.1 again that

$$\lim_{x\to 0}\frac{1-\cos x}{x} = 0.$$

This proves (d).

PROB. 4.2. Prove: $\lim_{x\to 1}(2x^2 - x) = 1$.

Def. 4.2(b). (Detailing Def. 4.1 for the case: $a = +\infty$, $L \in \mathbb{R}$.) Here

$$\lim_{x\to +\infty} f(x) = L, \qquad L \in \mathbb{R} \tag{4.24}$$

means: (1) The domain of f is not bounded from above ($+\infty$ is an (extended) accumulation point of $\mathfrak{D}(f)$) and (2) for each $\epsilon > 0$, there exists a real number X such that if

$$x \in \mathfrak{D}(f) \quad \text{and} \quad x > X, \tag{4.25}$$

then

$$|f(x) - L| < \epsilon. \tag{4.26}$$

(See Fig. 4.2.)

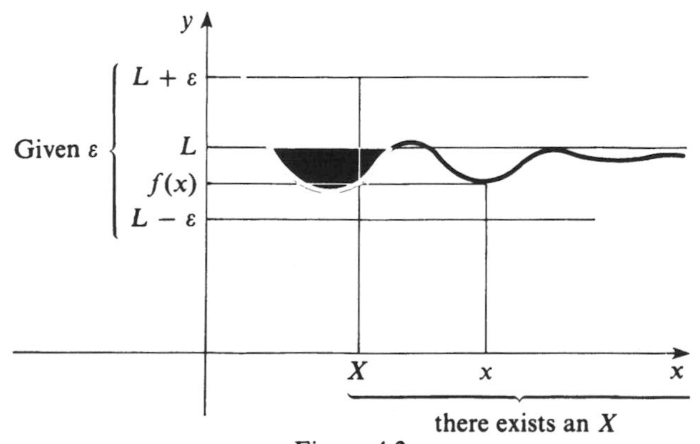

Figure 4.2

EXAMPLE 4.3. We prove that $\lim_{x\to+\infty}(2x^2-1)/(3x^2-x)=\frac{2}{3}$. Although nothing is said here about $\mathfrak{D}(f)$, we adhere to customary usage and assume that the domain of f is such that the operations involved in its definition yield a real-valued function of f. Accordingly,

$$\mathfrak{D}(f)=\{x\in\mathbb{R}\mid x\neq 0 \text{ and } x\neq\tfrac{1}{3}\}=\mathbb{R}-\{0,\tfrac{1}{3}\}.$$

This case is similar to that of limits of sequences except that here the domain contains real numbers that are not positive integers. In the case of sequences, the variable n is "discrete." Using Def. 4.2(b), we first make an analysis.

We must prove that if $\epsilon>0$ is given, there exists an X such that if $x\in\mathfrak{D}(f)$ and $x>X$ then

$$\left|f(x)-\frac{2}{3}\right|=\left|\frac{2x^2-1}{3x^2-x}-\frac{2}{3}\right|=\left|\frac{2x-3}{3(3x^2-x)}\right|<\epsilon.$$

Let $\epsilon>0$ be given. We first take $X_1=1$ and $x>1=X$. We then obtain $x\in\mathfrak{D}(f)$ and $x^2>x>1$. Adding $2x^2$ to both sides of $x^2>x$, we have

$$3x^2>2x^2+x,$$

so that

$$3x^2-x>2x^2>0.$$

This implies that if $x>1=X_1$, then

$$\left|f(x)-\frac{2}{3}\right|=\frac{|2x-3|}{3|3x^2-x|}<\frac{2|x|+3}{3(2x^2)}=\frac{2x+3}{6x^2}=\frac{1}{3x}+\frac{1}{2x^2}<\frac{1}{3x}+\frac{1}{2x}.$$

In short, if $x>1=X_1$, we have

$$\left|f(x)-\frac{2}{3}\right|<\frac{1}{3x}+\frac{1}{2x}=\frac{5}{6x}. \tag{4.27}$$

Here the right-hand side will be less than ϵ when $x>5/6\epsilon$. We, therefore, take $x>X\geqslant\max\{1,5/(6\epsilon)\}$. For such x we have $x\in\mathfrak{D}(f)$ and that both $x>1$, $x>5/6\epsilon$ hold. It follows that

$$\left|f(x)-\frac{2}{3}\right|<\frac{5}{6x}=\frac{5}{6}\cdot\frac{1}{x}<\frac{5}{6}\,\frac{6\epsilon}{5}=\epsilon.$$

But then

$$\lim_{x\to+\infty}f(x)=\tfrac{2}{3}.$$

PROB. 4.3. Prove:

(a) $\lim_{x\to+\infty}([x]/x)=1$,
(b) $\lim_{x\to+\infty}((\sin x)/x)=0$,
(c) $\lim_{x\to+\infty}((\cos x)/x)=0$,
(d) $\lim_{x\to+\infty}((3x^2-1)/(x^2+x+1))=3$.

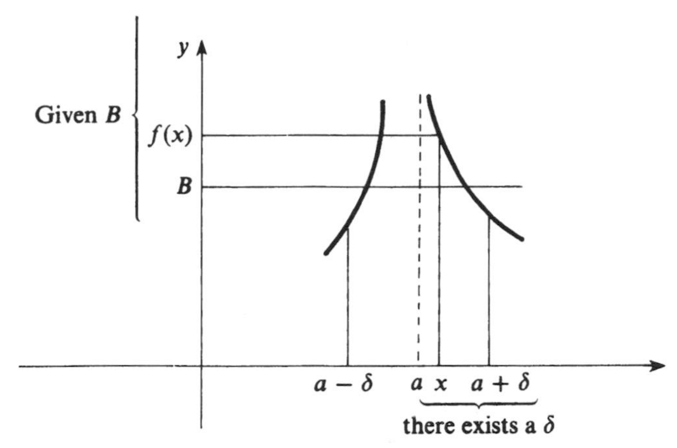

Figure 4.3

Def. 4.2(c). (Details of Def. 4.1 for the case $a \in \mathbb{R}$, $L = +\infty$.) $\lim_{x\to a} f(x) = +\infty$, where $a \in \mathbb{R}$. This means: (1) a is an accumulation point of $\mathfrak{D}(f)$ and (2) for each real B, there exists a $\delta > 0$ such that if $x \in \mathfrak{D}(f)$ and $0 < |x - a| < \delta$, then $f(x) > B$ (see Fig. 4.3).

EXAMPLE 4.4. Let $a \in \mathbb{R}$. We prove

$$\lim_{x\to a} \frac{1}{|x - a|} = +\infty. \tag{4.28}$$

Here $\mathfrak{D}(f) = \{x \in \mathbb{R} \mid x \neq a\} = \mathbb{R} - \{a\}$. Clearly, a is an accumulation point of $\mathfrak{D}(f)$ (explain). We prove that if B is given, then there exists a $\delta > 0$ such that if $0 < |x - a| < \delta$ and $x \in \mathfrak{D}(f)$, then

$$\frac{1}{|x - a|} = f(x) > B.$$

Given B, it suffices to take δ such that

$$0 < \delta \leqslant \frac{1}{1 + |B|}$$

and $0 < |x - a| < \delta$. This implies $x \in \mathfrak{D}(f)$ and

$$f(x) = \frac{1}{|x - a|} > \frac{1}{\delta} \geqslant 1 + |B| > |B| \geqslant B.$$

(4.28) now follows (explain).

Def. 4.2(d). (Details of Def. 4.1 for the case $a = -\infty$ and $L = -\infty$.) $\lim_{x\to -\infty} f(x) = -\infty$ means: (1) $\mathfrak{D}(f)$ is not bounded from below ($-\infty$ is an (extended) accumulation point of $\mathfrak{D}(f)$) and (2) if B is given, there exists an X such that if

$$x \in \mathfrak{D}(f) \quad \text{and} \quad x < X$$

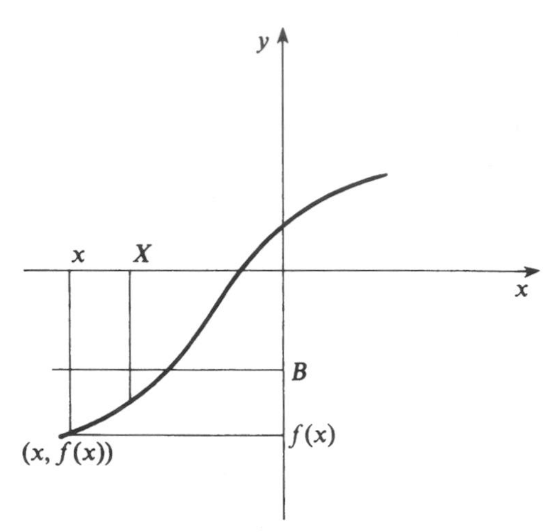

Figure 4.4

then

$$f(x) < B.$$

(See Fig. 4.4.)

PROB. 4.4. Prove: $\lim_{x\to-\infty} x^3 = -\infty$.

PROB. 4.5. Give a detailed definition, in terms of inequalities, of $\lim_{x\to a} f(x) = -\infty$, $a \in \mathbb{R}$, thus detailing Def. 4.1 for the case $a \in \mathbb{R}$, $L = -\infty$.

PROB. 4.6. Carry out the instructions in Prob. 4.5 for

(a) $\lim_{x\to+\infty} f(x) = +\infty$,
(b) $\lim_{x\to+\infty} f(x) = -\infty$,
(c) $\lim_{x\to-\infty} f(x) = L \in \mathbb{R}$,
(d) $\lim_{x\to-\infty} f(x) = +\infty$.

PROB. 4.7. Prove:

(a) $\lim_{x\to+\infty} x^3 = +\infty$,
(b) $\lim_{x\to+\infty} x^2 = +\infty$,
(c) $\lim_{x\to-\infty} x^2 = +\infty$.

PROB. 4.8. Prove: If n is a positive integer, then $\lim_{x\to+\infty} x^n = +\infty$.

PROB. 4.9. Prove: If n is an even positive integer, then $\lim_{x\to-\infty} x^n = +\infty$.

PROB. 4.10. Prove: If n is an odd positive integer, then $\lim_{x\to-\infty} x^n = -\infty$.

PROB. 4.11. Prove: If $\lim_{x\to a} f(x) = +\infty$, then $\lim_{x\to a}(-f(x)) = -\infty$ and if $\lim_{x\to a} f(x) = -\infty$, then $\lim_{x\to a}(-f(x)) = +\infty$.

PROB. 4.12. Assume $a \in \mathbb{R}$. Prove: $\lim_{x\to a}(-|x-a|^{-1}) = -\infty$.

We wish to give examples where $\lim_{x\to a} f(x)$ does not exist. The next theorem is useful in this connection.

Theorem 4.3. *Let $a \in \mathbb{R}^*$ and $L \in \mathbb{R}^*$ and a be a possibly extended accumulation point of $\mathfrak{D}(f)$, then* $\lim_{x\to a} f(a) = L$ *if and only if for each sequence $\langle x_n \rangle$ of elements of $\mathfrak{D}(f)$ such that $x_n \neq a$ for all n and* $\lim_{n\to+\infty} x_n = a$, *we have* $\lim_{n\to+\infty} f(x_n) = L$.

PROOF. First assume $\lim_{x\to a} f(x) = L$, so that a is an accumulation point of $\mathfrak{D}(f)$. Let $\langle x_n \rangle$ be a sequence of elements of $\mathfrak{D}(f)$ such that $x_n \neq a$ for all n and $\lim_{n\to+\infty} x_n = a$. (Such a sequence exists by Theorem 3.3 and Prob. 3.9.) Let $N(L)$ be some neighborhood of L. Since $\lim_{x\to a} f(x) = L$, there exists a deleted neighborhood $N^*(a)$ of a such that

$$f(N^*(a) \cap \mathfrak{D}(f)) \subseteq N(L). \tag{4.29}$$

Since $\lim_{a\to+\infty} x_n = a$, there exists an N such that if $n > N$, then $x_n \in N(a)$. Since $x_n \neq a$ for all n, we have $x_n \in N^*(a)$ for $n > N$. Since $x_n \in \mathfrak{D}(f)$ for all n, it follows that

$$x_n \in N^*(a) \cap \mathfrak{D}(f) \tag{4.30}$$

for $n > N$. By (4.29) this implies that

$$y_n = f(x_n) \in N(L)$$

for $n > N$. Thus, corresponding to each neighborhood $N(L)$ of L, there exists an N such that if $n > N$, then $f(x_n) \in N(L)$. This implies that $\lim_{x\to+\infty} f(x_n) = L$.

Now assume that $\lim_{x\to a} f(x) = L$ is false. This implies that some neighborhood $N(L)$ of L exists such that for each deleted neighborhood $N^*(a)$ of a we have

$$f(N^*(a) \cap \mathfrak{D}(f)) \nsubseteq N(L). \tag{4.31}$$

Thus, for each $N^*(a)$ there exists an x such that

$$x \in N^*(a) \quad \text{and} \quad x \in \mathfrak{D}(f) \quad \text{but} \quad f(x) \notin N(L). \tag{4.32}$$

Let

$$S = \{x \in \mathfrak{D}(f) \mid f(x) \notin N(L)\}. \tag{4.33}$$

Clearly, $S \subseteq \mathfrak{D}(f)$. Let $N^*(a)$ be any deleted neighborhood of a. By (4.32) and (4.33), $N^*(a)$ contains points of S. Hence a is an accumulation point of

S. Hence, there exists a sequence $\langle x_n \rangle$ of elements of S such that $x_n \neq a$ for all n and $\lim_{n \to +\infty} x_n = a$ (Theorem 3.3 and Prob. 3.9). For this sequence we have

$$x_n \in \mathfrak{D}(f) \quad \text{and} \quad f(x_n) \notin N(L) \qquad \text{for all } n. \tag{4.34}$$

This implies that $\lim_{n \to +\infty} f(x_n) = L$ is false (otherwise some N would exist that if $n > N$, then $x_n \in \mathfrak{D}(f)$ and $f(x_n) \in N(L)$, contradicting (4.34)). We, therefore, conclude that if for each sequence $\langle x_n \rangle$ of elements of $\mathfrak{D}(f)$ such that $x_n \neq a$ for all n and $\lim_{n \to a} x_n = a$ we have $\lim_{n \to +\infty} f(x_n) = L$, then $\lim_{x \to a} f(x) = L$. This completes the proof.

Example 4.5. Let $\langle x \rangle^*$ be the distance from x to the integer nearest x for $x \in \mathbb{R}$. We prove that $\lim_{x \to +\infty} \langle x \rangle$ does not exist. Construct the sequences $\langle x_n \rangle$ and $\langle x_n' \rangle$, where

$$x_n = n \quad \text{and} \quad x_n' = n + \tfrac{1}{2} \qquad \text{for each } n.$$

Clearly, $\lim_{n \to +\infty} x_n = \lim_{n \to +\infty} n = +\infty$ and $\lim_{n \to +\infty} x_n' = \lim_{n \to +\infty} (n + \frac{1}{2}) = +\infty$. However,

$$\langle x_n \rangle^* = \langle n \rangle^* = 0 \quad \text{and} \quad \langle x_n' \rangle^* = \langle n + \tfrac{1}{2} \rangle^* = \tfrac{1}{2}$$

for each n. It follows that

$$\lim_{n \to +\infty} \langle x_n \rangle^* = 0 \quad \text{and} \quad \lim_{n \to +\infty} \langle x_n' \rangle^* = \tfrac{1}{2}. \tag{4.35}$$

By Theorem 4.3, $\lim_{x \to +\infty} \langle x \rangle^*$ does not exist. If it did exist and had value L, then for each sequence $\langle z_n \rangle$ such that $\lim z_n = +\infty$ we would have $\lim \langle z_n \rangle^* = L$. This is not compatible with (4.35).

Example 4.6. We show that the condition $x_n \neq a$ for all n in Theorem 4.3 is needed. Let f be defined as

$$f(x) = \begin{cases} 1 & \text{if } x \neq 0 \\ 2 & \text{if } x = 0 \end{cases}$$

(see Fig. 4.5). One sees easily that here $\lim_{x \to 0} f(x) = 1$. Take the sequence

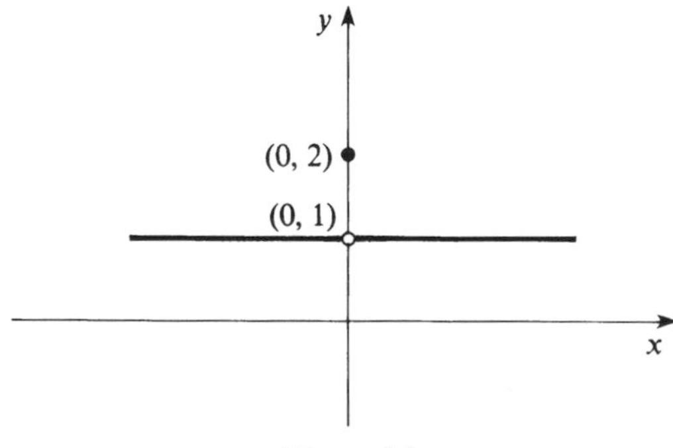

Figure 4.5

$\langle x_n \rangle$, where $x_n = 0$ for all n. Clearly, $\lim_{n\to+\infty} x_n = 0$, $x_n \in \mathbb{R} = \mathfrak{D}(f)$ for all n. Since

$$f(x_n) = f(0) = 2$$

for all n, we have

$$\lim_{n\to+\infty} f(x_n) = \lim_{n\to+\infty} 2 = 2 \neq 1 = \lim_{x\to 0} f(x).$$

Theorem 4.4. *Let $a \in \mathbb{R}^*$. If $\lim_{x\to a} f(x)$, exists, then it is unique.*

PROOF. Assume:

$$\lim_{x\to a} f(x) = L \quad \text{and} \quad \lim_{x\to a} f(x) = L'.$$

Suppose L and L' are in $\mathbb{R}$. Take any $\epsilon > 0$. There exist deleted neighborhoods $N_1^*(a)$ and $N_2^*(a)$ of a such that

$$|f(x) - L| < \frac{\epsilon}{2} \qquad \text{for} \quad x \in N_1^*(a) \cap \mathfrak{D}(f)$$

and

$$|f(x) - L'| < \frac{\epsilon}{2} \qquad \text{for} \quad x \in N_2^*(a) \cap \mathfrak{D}(f).$$

There exists a deleted neighborhood $N^*(a)$ of a such that $N^*(a) \subseteq N_1^*(a) \cap N_2^*(a)$. We have

$$\begin{aligned} N^*(a) \cap \mathfrak{D}(f) &\subseteq (N_1^*(a) \cap N_2^*(a)) \cap \mathfrak{D}(f) \\ &= (N_1^*(a) \cap \mathfrak{D}(f)) \cap (N_2^*(a) \cap \mathfrak{D}(f)). \end{aligned}$$

Let $x \in N^*(a) \cap \mathfrak{D}(f)$. We have $x \in N_1^*(a) \cap \mathfrak{D}(f)$ and $x \in N_2^*(a) \cap \mathfrak{D}(f)$, so that $|f(x) - L| < \epsilon/2$ and $|f(x) - L'| < \epsilon/2$. Hence,

$$|L' - L| \leqslant |L' - f(x)| + |L - f(x)| < \frac{\epsilon}{2} + \frac{\epsilon}{2} = \epsilon.$$

Thus, $|L' - L| < \epsilon$ for all $\epsilon > 0$. This implies that $|L' - L| \leqslant 0$ which yields $L = L'$.

Now assume that one of L or L' is $\pm\infty$ and the other is not. Say, $L = +\infty$ and $L' \neq +\infty$. Suppose, first that $L' = -\infty$. There exist deleted neighborhoods $N_1^*(a)$ and $N_2^*(a)$ of a such that $f(x) > 0$ for $x \in N_1^*(a) \cap \mathfrak{D}(f)$ and $f(x) < 0$ for $x \in N_2^*(a) \cap \mathfrak{D}(f)$. There exists a deleted neighborhood $N^*(a)$ of a such that $N^*(a) \subseteq N_1^*(a) \cap N_2(a)$. Let $x \in N^*(a) \cap \mathfrak{D}(f)$, so that $x \in N_1^*(a) \cap \mathfrak{D}(f)$ and $x \in N_2^*(a) \cap \mathfrak{D}(f)$. This implies that $f(x) > 0$ and $f(x) < 0$ which is impossible. Thus, $L' \neq -\infty$. Now assume $L' \in \mathbb{R}$. This time there exists a deleted neighborhood $N_3^*(a)$ of a such that $L' - 1 < f(x) < L' + 1$ for $x \in N_3^*(a) \cap \mathfrak{D}(f)$ and a deleted neighborhood $N_4^*(a)$ of a such that for $x \in N_4(a) \cap \mathfrak{D}(f)$, $f(x) > L' + 1$ (by hypothesis $L = +\infty$). As before, there exists a deleted neighborhood $N_5^*(a)$ of a such that $N_5^*(a) \subseteq N_3^*(a) \cap N_4^*(a)$. Let $x \in N_5^*(a) \cap \mathfrak{D}(f)$, so that $x \in N_3^*(a) \cap \mathfrak{D}(f)$ and $x \in N_4^*(a) \cap \mathfrak{D}(f)$. This implies $f(x) < L' + 1$ and $f(x) > L' + 1$

which is also impossible. Thus, $L' \in \mathbb{R}$ is also impossible. Since $L' \in \mathbb{R}^*$, we remain with the alternative $L' = +\infty = L$. Similarly, if $L = -\infty$, we can prove $L' = -\infty = L$. This completes the proof.

5. One-Sided Limits

Def. 5.1. Let f be a real-valued function of a real variable. Let $a \in \mathbb{R}$, $L \in \mathbb{R}^*$. We say that *f approaches L as x approaches a from the left*, or that *f has limit L as x approaches a from the left*, and write:

$$\lim_{x \to a-} f(x) = L \quad \text{or} \quad f(x) \to L \qquad \text{as} \quad x \to a- \tag{5.1}$$

if and only if: (1) a is an accumulation point from the left of $\mathfrak{D}(f)$ and (2) for each neighborhood $N(L)$ of L, there exists a deleted neighborhood $(a - \delta; a)$ of a from the left such that

$$f((a - \delta; a) \cap \mathfrak{D}((f)) \subseteq N(L). \tag{5.2}$$

Similarly, we say that *f approaches L as x approaches a from the right* or that *f has limit L as x approaches from the right* and write

$$\lim_{x \to a+} f(x) = L \quad \text{or} \quad f(x) \to L \qquad \text{as} \quad x \to a+ \tag{5.3}$$

if and only if (1) a is an accumulation point from the right of $\mathfrak{D}(f)$ and (2) for each neighborhood $N(L)$ of L there exists a deleted neighborhood $(a; a + \delta)$ of a from the right such that

$$f((a; a + \delta) \cap \mathfrak{D}(f)) \subseteq N(L). \tag{5.4}$$

(Note that in our definition, $a \in \mathbb{R}$. The cases $a = \pm\infty$ were considered in the last section.)

An alternate notation for one-sided limits is $f(a-)$ for $\lim_{x \to a-} f(x)$ and $f(a+)$ for $\lim_{x \to a+} f(x)$.

One-sided limits are related to the limits discussed in Section 4 by means of Theorem 5.1 below. Before stating this theorem we point out that by a *two-sided* accumulation point of a set $S \subseteq \mathbb{R}$, we mean one which is an accumulation point of S both from the right and from the left.

Theorem 5.1. *If f is a real-valued function of a real variable and $a \in \mathbb{R}$ is a two-sided accumulation point of $\mathfrak{D}(f)$ and $L \in \mathbb{R}^*$, then*

$$\lim_{x \to a} f(x) = L \tag{5.5}$$

if and only if both one-sided limits $f(a-)$ and $f(a+)$ exist and

$$f(a+) = L = f(a-). \tag{5.6}$$

Proof. Let a be a two-sided accumulation point of $\mathfrak{D}(f)$ and suppose that $f(x) \to L$ as $x \to a$. a is an accumulation point of $\mathfrak{D}(f)$ from the right and from the left. Let $N(L)$ be some neighborhood of L so that a deleted δ-neighborhood $N^*(a,\delta) = (a-\delta; a) \cup (a; a+\delta)$ of a exists such that

$$f(N^*(a,\delta) \cap \mathfrak{D}(f)) \subseteq N(L). \tag{5.7}$$

By properties of functions defined on sets, we obtain for the left-hand side of (5.7)

$$f(N^*(a,\delta) \cap \mathfrak{D}(f)) = f((a-\delta; a) \cap \mathfrak{D}(f)) \cup f((a; a+\delta) \cap \mathfrak{D}(f)).$$

This and (5.7) imply

$$f((a-\delta; a) \subseteq \mathfrak{D}(f)) \subseteq N(L), \tag{5.8a}$$

and

$$f((a; a+\delta) \subseteq \mathfrak{D}(f)) \subseteq N(L). \tag{5.8b}$$

Thus, for each $N(L)$, $(a-\delta; a)$ is a neighborhood of a from the left such that (5.8a) holds and $(a; a+\delta)$ is a δ-neighborhood from the right such that (5.8b) holds. Therefore,

$$f(a-) = L \quad \text{and} \quad f(a+) = L. \tag{5.9}$$

This proves (5.6).

Conversely, assume (5.6) holds. We know that a is an accumulation point of $\mathfrak{D}(f)$ from the right and from the left. Let $N(L)$ be some neighborhood of L. Because of (5.9), there exists a deleted δ-neighborhood $(a-\delta_1; a)$ of a from the left and a deleted δ-neighborhood $(a; a+\delta_2)$ of a from the right such that

$$f((a-\delta_1; a) \cap \mathfrak{D}(f)) \subseteq N(L) \quad \text{and} \quad f((a; a+\delta_2) \cap \mathfrak{D}(f)) \subseteq N(L). \tag{5.10}$$

By properties of sets of functions defined on sets, we obtain from (5.10) that

$$f((a-\delta_1; a) \cap \mathfrak{D}(f)) \cup f((a; a+\delta_2) \cap \mathfrak{D}(f)) \subseteq N(L),$$

and

$$f([(a-\delta_1; a) \cup (a; a+\delta_2)] \cap \mathfrak{D}(f)) \subseteq N(L). \tag{5.11}$$

Put $\delta = \min\{\delta_1, \delta_2\}$. Then $0 < \delta \leqslant \delta_1$ and $0 < \delta \leqslant \delta_2$. From this it is easily seen that

$$N^*(a;\delta) \subseteq (a-\delta_1; a) \cup (a; a+\delta_2).$$

This and (5.11) imply

$$f(N^*(a,\delta) \cap \mathfrak{D}(f)) \subseteq N(L). \tag{5.12}$$

We proved that for each neighborhood $N(L)$ of L, there exists a deleted δ-neighborhood of a such that (5.12) holds. We, therefore, conclude that $\lim_{x\to a} f(x) = L$. This completes the proof.

Remark 5.1. It follows from Theorem 5.1 that if a is a two-sided accumulation point of $\mathfrak{D}(f)$, then $\lim_{x\to a} f(x)$ does *not* exist if and only if either (1) one of $f(a-)$ or $f(a+)$ does not exist or (2) each of $f(a-)$ and $f(a+)$ exists but $f(a-) \neq f(a+)$.

EXAMPLE 5.1. We consider the signum function. Recall that

$$\operatorname{sig} x = \begin{cases} 1 & \text{if } x > 0 \\ 0 & \text{if } x = 0 \\ -1 & \text{if } x < 0. \end{cases}$$

We prove: $\lim \operatorname{sig}_{x\to 0-} x = -1$ and $\lim_{x\to 0+} \operatorname{sig} x = 1$ from which it will be seen that $\lim_{x\to 0} \operatorname{sig} x$ does not exist.

Given $\epsilon > 0$, take δ such that $0 < \delta \leqslant 1$ and $-\delta < x < 0$. For such x,

$$|\operatorname{sig} x - (-1)| = |(-1) + 1| = 0 < \epsilon.$$

This implies $\operatorname{sig} x \to -1$ as $x \to 0-$. Similarly, if $\epsilon > 0$ is given, take δ such that $0 < \delta \leqslant 1$ and $0 < x < \delta$. For such x,

$$|\operatorname{sig} x - 1| = |1 - 1| = 0 < \epsilon$$

and conclude $\operatorname{sig} x \to 1$ as $x \to 0+$.

EXAMPLE 5.2. We prove:

$$\lim_{x\to 0+} \frac{1}{x} = +\infty \quad \text{and} \quad \lim_{x\to 0-} \frac{1}{x} = -\infty$$

(see Fig. 5.1.) Here $f(x) = 1/x$ for $x \neq 0$, so $\mathfrak{D}(f) = \{x \in \mathbb{R} \mid x \neq 0\}$. Given B, take x such that $0 < x < \delta$, where

$$0 < \delta \leqslant \frac{1}{1 + |B|}.$$

This implies

$$\frac{1}{x} > \frac{1}{\delta} \geqslant 1 + |B| > |B| \geqslant B.$$

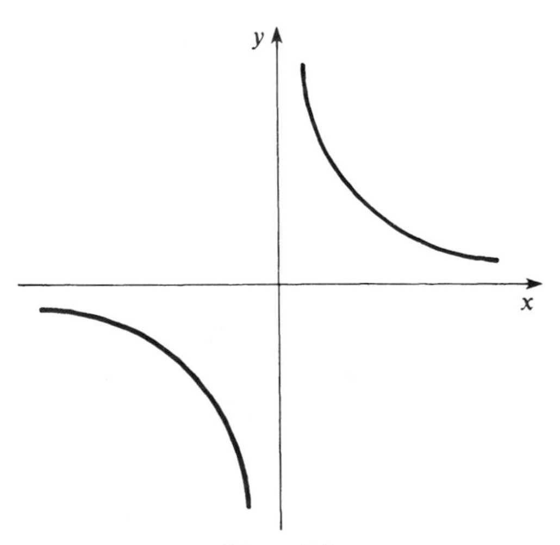

Figure 5.1

Hence, for each neighborhood $(B; +\infty]$ of $+\infty$ in $\mathbb{R}^*$, we have a deleted neighborhood $(0;\delta) = (0; 0 + \delta)$ of 0 from the right such that if $x \in (0;\delta) \cap \mathfrak{D}(f) = (0;\delta)$, then $f(x) \in (B; +\infty]$. Hence, $1/x \to +\infty$ as $x \to 0^+$. Next, given B, take x such that $-\delta = 0 - \delta < x < 0$, where

$$0 < \delta \leqslant \frac{1}{1 + |B|}.$$

Then

$$0 < -x < \frac{1}{1 + |B|}.$$

This implies that

$$\frac{1}{-x} > 1 + |B| > |B| \geqslant -B$$

and, hence, that

$$\frac{1}{x} < B \qquad \text{for} \quad x \in (-\delta; 0) \cap \mathfrak{D}(f) = (-\delta; 0).$$

We conclude from this $\lim_{x \to 0-} (1/x) = -\infty$.

Remark 5.2. A one-sided limit of a function f of a real variable can be viewed as an ordinary limit of the restriction of f to the set $(-\infty; a) \cap \mathfrak{D}(f)$ for the case $x \to a-$, and to the set $(a; +\infty) \cap \mathfrak{D}(f)$ for the case $x \to a+$. Because of this, many theorems true for $\lim_{x \to a} f(x)$ are also true for $f(a+)$ and $f(a-)$.

Example 5.3. We give an example where neither one-sided limit exists. Let

$$g(x) = \left\langle \frac{1}{x} \right\rangle^* \qquad \text{for} \quad x \neq 0.$$

(See, e.g., Fig. 5.2, where $\langle y \rangle^*$ is the distance from y to the integer nearest y.) We prove $\lim_{x \to 0+} g(x)$ does not exist. Let $\langle x_n \rangle$ and $\langle x_n' \rangle$ be sequences

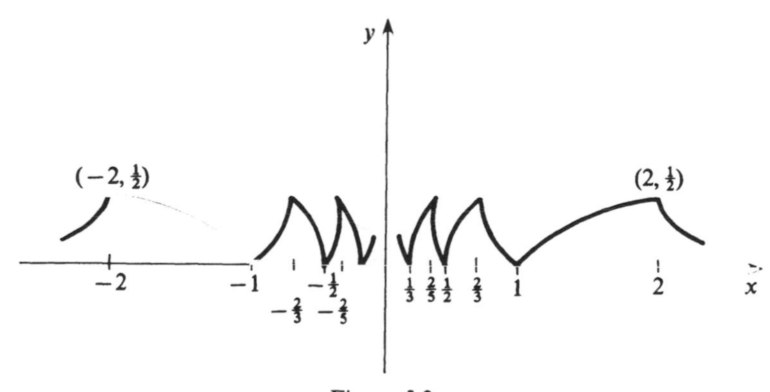

Figure 5.2

defined as

$$x_n = \frac{1}{n} \quad \text{and} \quad x_n' = \frac{2}{2n+1} \qquad \text{for each positive integer } n.$$

Then $x_n > 0$ and $x_n' > 0$ for all n, and

$$\lim_{n\to+\infty} x_n = 0 = \lim_{n\to+\infty} x_n' .$$

But $g(x_n) = g(1/n) = \langle n \rangle^* = 0$ and $g(x_n') = g(2/(2n+1)) = \langle n + \frac{1}{2} \rangle^* = \frac{1}{2}$ for each n. Hence,

$$\lim_{n\to+\infty} g(x_n) = 0 \quad \text{and} \quad \lim_{n\to+\infty} g(x_n') = \tfrac{1}{2} .$$

By Theorem 5.3 applied to the restriction of g to the interval $(0; +\infty)$ it follows that $\lim_{x\to 0+} g(x)$ does not exist. It is also easy to prove that $g(0-)$ does not exist either.

6. Theorems on Limits of Functions

Theorem 6.1. *Let $a \in \mathbb{R}^*$ and $L \in \mathbb{R}$. If $\lim_{x\to a} f(x) = L$, then some deleted neighborhood $N^*(a)$ of a exists such that f is bounded on the set $N^*(a) \cap \mathfrak{D}(f)$.*

Proof. Let $\epsilon = 1$. Since $f(x) \to L$ as $x \to a$, there exists a deleted neighborhood $N^*(a)$ of a such that

$$f(N^*(a) \cap \mathfrak{D}(f)) \subseteq N(L, 1) = (L-1; L+1).$$

This implies that if $x \in N^*(a) \cap \mathfrak{D}(f)$, then $L - 1 < f(x) < L + 1$ and the conclusion follows.

Theorem 6.2. *Let $a \in \mathbb{R}^*$ and $L > 0$. If $f(x) \to L$ as $x \to a$, then there exists a deleted neighborhood $N^*(a)$ of a such that $f(x) > 0$ for all $x \in N^*(a) \cap \mathfrak{D}(f)$; on the other hand, if $L < 0$, then there exists a deleted neighborhood $N_1^*(a)$ of a such that $f(x) < 0$ for all $x \in N_1^*(a) \cap \mathfrak{D}(f)$.*

Proof. We prove the first part and ask the reader to prove the second (Prob. 6.1). Suppose $L \in \mathbb{R}^*$ and $L > 0$. If $L = +\infty$, take $B = 0$. There exists a deleted neighborhood $N^*(a)$ of a such that if $x \in N^*(a) \cap \mathfrak{D}(f)$, then $f(x) > B = 0$. Thus, the conclusion of the first part holds in this case. If $0 < L < +\infty$, take $\epsilon = L/2$. There exists a deleted neighborhood $N^*(a)$ of a such that if $x \in N^*(a) \cap \mathfrak{D}(f)$, then

$$|f(x) - L| < \frac{L}{2} \quad \text{and hence} \quad -\frac{L}{2} < f(x) - L < \frac{L}{2} .$$

This implies that if $x \in N^*(a) \cap \mathfrak{D}(f)$, then $f(x) > L/2 > 0$. We see that the conclusion of the first part holds also for $0 < L < +\infty$.

Prob. 6.1. Complete the proof of Theorem 6.2 by proving the second part of that theorem.

Remark 6.1. Actually, we proved more than was stated in Theorem 6.2. We proved that if $f(x) \to L \in \mathbb{R}$, where $L > 0$, then some deleted neighborhood $N^*(a)$ exists such that f is "bounded away from 0" on $N^*(a) \cap \mathfrak{D}(f)$ in the sense that $f(x) > L/2 > 0$ for $x \in N^*(a) \cap \mathfrak{D}(f)$. If $L = +\infty$, then using $B = 1$, we find that some deleted neighborhood $N^*(a)$ of a exists such that $f(x) > B = 1$ for $x \in N^*(a) \cap \mathfrak{D}(f)$. Thus, in this case too there is a deleted $N^*(a)$ of a such that f is "bounded away from 0" on $N^*(a) \cap \mathfrak{D}(f)$.

Theorem 6.3. *If f and g are real-valued functions of a real variable having a common domain $\mathfrak{D}$ and*

$$\lim_{x \to a} f(x) = L, \qquad \lim_{x \to a} g(x) = M,$$

then

$$\lim_{x \to a} (f(x) + g(x)) = L + M = \lim_{x \to a} f(x) + \lim_{x \to a} g(x),$$

if $L + M$ is defined in $\mathbb{R}^*$.

Proof. Let $\langle x_n \rangle$ be a sequence of elements of $\mathfrak{D}$ such that $x_n \neq a$ for all n and $\lim_{n \to +\infty} x_n = a$ (such a sequence exists since a is an accumulation point of $\mathfrak{D}$). It follows (Theorem 4.3 and Prob. 3.9) that

$$\lim_{n \to +\infty} f(x_n) = L \quad \text{and} \quad \lim_{n \to +\infty} g(x_n) = M.$$

This implies that

$$\lim_{n \to +\infty} (f(x_n) + g(x_n)) = L + M \tag{6.1}$$

whenever $L + M$ is defined in $\mathbb{R}^*$. Since (6.1) holds for all sequences $\langle x_n \rangle$ of elements of $\mathfrak{D}$ such that $x_n \neq a$ for all n and $\lim x_n = a$, we have (Theorem 4.3) $\lim_{x \to a}(f(x) + g(x)) = L + M$ wherever $L + M$ is defined in $\mathbb{R}^*$.

Theorem 6.4. *If f and g are real-valued functions of a real variable having a common domain $\mathfrak{D}$, where $f(x) \to 0$ as $x \to a$ and some deleted neighborhood $N^*(a)$ of a exists such that g is bounded on $N^*(a) \cap \mathfrak{D}$, then*

$$\lim_{x \to a} f(x) g(x) = 0. \tag{6.2}$$

Proof. Although this theorem could be proved by using sequences as in the proof of Theorem 6.3, we proceed differently. Since g is bounded on $N^*(a) \cap \mathfrak{D}(f)$, there exists an $M > 0$ such that $|g(x)| \leqslant M$ for $x \in N^*(a) \cap \mathfrak{D}(f)$. Let $\epsilon > 0$ be given. Since $\lim_{x \to a} f(x) = 0$, there exists a deleted neighborhood $N_1^*(a)$ of a such that

$$|f(x)| < \frac{\epsilon}{M} \qquad \text{for} \quad x \in N_1^*(a) \cap \mathfrak{D}(f). \tag{6.3}$$

Also, there exists a deleted neighborhood $N_2^*(a)$ of a such that $N_2^*(a) \subseteq N_1^*(a) \cap N^*(a)$. Hence,

$$N_2^*(a) \cap \mathfrak{D}(f) \subseteq N_1^*(a) \cap N^*(a) \cap \mathfrak{D}(f)$$
$$\subseteq (N_1^*(a) \cap \mathfrak{D}(f)) \cap (N^*(a) \cap \mathfrak{D}(f)). \tag{6.4}$$

Assume that $x \in N_2^*(a) \cap \mathfrak{D}(f)$. From (6.4) we have $x \in N_1^*(a) \cap \mathfrak{D}(f)$ and $x \in N^*(a) \cap \mathfrak{D}(f)$. This implies that $|g(x)| \leqslant M$ and that (6.3) holds for all $x \in N_2^*(a) \cap \mathfrak{D}(f)$. Hence, for such x,

$$|f(x)g(x) - 0| = |f(x)g(x)| = |f(x)||g(x) < \frac{\epsilon}{M} M = \epsilon.$$

This yields $\lim_{x \to a} f(x)g(x) = 0$ and completes the proof.

Theorem 6.5. *Let $c \in \mathbb{R}$ and $L \in \mathbb{R}^*$. If $f(x) \to L$ as $x \to a$, where cL is defined, then*

$$\lim_{x \to a} (cf(x)) = cL = c \lim_{x \to a} f(x).$$

PROOF. Exercise.

Theorem 6.6. *Let a, L, and M be extended real numbers such that LM is defined in $\mathbb{R}^*$. If f and g are real-valued functions of a real variable having a common domain $\mathfrak{D}$ and*

$$\lim_{x \to a} f(x) = L, \qquad \lim_{x \to a} g(x) = M,$$

then

$$\lim_{x \to a} f(x)g(x) = LM = \lim_{x \to a} f(x) \lim_{x \to a} g(x).$$

PROOF. Exercise.

PROB. 6.2. Prove: If $f_1, \ldots, f_m$ are m functions with common domain $\mathfrak{D}$ and

$$\lim_{x \to a} f_i(x) = L_i \qquad \text{for each} \quad i \in \{1, \ldots, m\},$$

then

(a) $$\lim_{x \to a} (f_1(x) + \cdots + f_m(x)) = L_1 + \cdots + L_m$$

and

(b) $$\lim_{x \to a} (f_1(x) \ldots f_m(x)) = L_1 L_2 \ldots L_m$$

provided that the right-hand sides in (a) and (b) are defined in $\mathbb{R}^*$.

PROB. 6.3. Let P be a polynomial function in $\mathbb{R}$ and $x_0 \in \mathbb{R}$. Prove that

$$\lim_{x \to x_0} P(x) = P(x_0).$$

Theorem 6.7. *Let a, L, and M be extended real numbers, $M \neq 0$. If f and g are real-valued functions of a real variable with a common domain $\mathfrak{D}$, $g(x) \neq 0$ for $x \in \mathfrak{D}$ and $f(x) \to L$, $g(x) \to M$ as $x \to a$, then*

$$\lim_{x \to a} \frac{f(x)}{g(x)} = \frac{L}{M} = \frac{\lim_{x \to a} f(x)}{\lim_{x \to a} g(x)}$$

provided that L/M is defined in $\mathbb{R}^*$.

Proof. Exercise.

Prob. 6.4. Prove: If $f(x) \to L$ as $x \to a$, where a and L are in $\mathbb{R}$, then $|f(x)| \to |L|$ as $x \to a$.

Example 6.1. We prove that if $L \in \mathbb{R}$ and $L \neq 0$, then the converse of the result in Prob. 6.4 is false. Define f as

$$f(x) = \begin{cases} L & \text{if } x \text{ is rational} \\ -L & \text{if } x \text{ is irrational.} \end{cases}$$

We have $|f(x)| = |L|$ for all $x \in \mathbb{R}$. Hence, if $a \in \mathbb{R}$, then $|f(x)| \to |L|$ as $x \to a$. On the other hand, there exist sequences $\langle r_n \rangle$ and $\langle c_n \rangle$ such that r_n is rational and c_n is irrational for each n and $r_n \to a$, $c_n \to a$ as $n \to +\infty$. It follows that

$$\lim_{n \to +\infty} f(r_n) = L, \qquad \lim_{n \to +\infty} f(c_n) = -L.$$

Since $L \neq 0$, this implies that $\lim_{n \to +\infty} f(r_n) \neq \lim_{n \to +\infty} f(c_n)$, so that $\lim_{x \to a} f(x)$ does not exist.

Prob. 6.5. Prove: $f(x) \to 0$ as $x \to a$ if and only if $|f(x)| \to 0$ as $x \to a$.

Prob. 6.6. Prove: If $f(x) \leqslant L$ for all $x \in \mathfrak{D}(f)$ and $\lim_{x \to a} f(x)$ exists, then $\lim_{x \to a} f(x) \leqslant L$.

Prob. 6.7. Prove: If $f(x) \leqslant g(x)$ for all $x \in \mathfrak{D}$ and $\mathfrak{D}$ is a common domain of f and g, where both f and g have limits as $x \to a$, then $\lim_{x \to a} f(x) \leqslant \lim_{f \to a} g(x)$.

Prob. 6.8 (Sandwich Theorem for Functions). Prove: If f, g, and h are functions having a common domain $\mathfrak{D}$ and $f(x) \leqslant h(g) \leqslant g(x)$ for all $x \in \mathfrak{D}$ and $\lim_{x \to a} f(x) = \lim_{x \to a} g(x) = L$, then $\lim_{x \to a} h(x) = L$.

Prob. 6.9. Prove: If $a \in \mathbb{R}$, where $a \neq 0$, then

(a) $\lim_{x \to 0}((\sin ax)/x) = a$,
(b) $\lim_{a \to 0}((1 - \cos x)/x^2) = \frac{1}{2}$.

7. Some Special Limits

We first extend the validity of the inequalities in Theorem II.12.1. According to this theorem we have: If $x > 0$, $x \neq 1$, and r is rational, where $r < 0$ or $r > 1$, then

$$r(x-1) < x^r - 1 < rx^{r-1}(x-1). \tag{7.1}$$

Let y be a real number such that $y < 0$ or $y > 1$. Write y as a nonterminating decimal

$$y = N + .d_1 d_2 \ldots . \tag{7.2}$$

Now $y = \sup r_n$, where r_n is the sequence of truncations of $N + .d_1 d_2 \ldots$, i.e.,

$$r_n = N + 0.d_1 \ldots d_n \qquad \text{for all positive integers } n. \tag{7.3}$$

By the definition of x^y, we have (Def. IV.10.1, and Remark IV.10.2)

$$x^y = E_x(y) = \lim_{x \to +\infty} x^{r_n}. \tag{7.4}$$

By theorems on limits of sequences this implies that

$$\lim_{n \to +\infty} x^{r_n - 1} = \lim_{n \to +\infty} x^{-1} x^{r_n} = x^{-1} \lim_{n \to +\infty} x^{r_n} = x^{-1} x^y = x^{y-1} \qquad \text{if} \quad x > 0. \tag{7.5}$$

Since $y < 0$ or $y > 1$, we have, for the nth truncation r_n of y, $r_n < y < 0$ or $1 \leqslant r_n < y$. Using (7.1), this implies that

$$r_n(x-1) \leqslant x^{r_n} - 1 \leqslant r_n x^{r_n - 1}(x-1).$$

Taking limits as $n \to +\infty$ and using theorems on limits of sequences, we obtain

$$y(x-1) \leqslant x^y - 1 \leqslant yx^{y-1}(x-1). \tag{7.6}$$

Note, this holds trivially if $x = 1$ or $y = 0$. Thus, (7.6) holds for $y \leqslant 0$ or $y \geqslant 1$ and $x > 0$. At this stage we do not establish the strictness of the inequality in (7.6) for

$$y < 0 \quad \text{or} \quad y > 1 \quad \text{and} \quad x > 0, x \neq 1.$$

Reasoning as above and using the inequality

$$r(x-1) > x^r - 1 > rx^{r-1}(x-1) \qquad \text{if } r \text{ is rational and } 0 < r < 1, \tag{7.7}$$

we can prove: If $x > 0$ and $0 \leqslant y \leqslant 1$ where $y \in \mathbb{R}$, then

$$y(x-1) \geqslant x^y - 1 \geqslant x^{y-1}(x-1). \tag{7.8}$$

At this point we do not establish the strict inequality for $0 < y < 1$ and $x > 0$, $x \neq 1$. Summarizing we obtain the following theorem:

Theorem 7.1. *If x and y are real numbers, $x > 0$, then*

(a) $y(x-1) \leqslant x^y - 1 \leqslant yx^{y-1}(x-1)$ *if $y \leqslant 0$ or $y \geqslant 1$ and*
(b) $y(x-1) \geqslant x^y - 1 \geqslant yx^{y-1}$ *if $0 \leqslant y \leqslant 1$.*

PROB. 7.1. Prove: If a, b, and y are real numbers, where a and b are both positive, then

(a) $yb^{y-1}(a-b) \leqslant a^y - b^y \leqslant ya^{y-1}(a-b)$ if $y \leqslant 0$ or $y \geqslant 1$ and
(b) $ya^{y-1}(a-b) \leqslant a^y - b^y \leqslant yb^{y-1}(a-b)$ if $0 \leqslant y \leqslant 1$. (See Prob. II.12.6.)

Theorem 7.2. *If x_0 and y are in $\mathbb{R}$ and $x_0 > 0$, then*

$$\lim_{x \to x_0} x^y = x_0^y. \tag{7.9}$$

PROOF. x^y is defined for $x > 0$ and $y \in \mathbb{R}$. First take $y \geqslant 1$, $x > 0$, $x_0 > 0$. By Prob. 7.1, part (a),

$$yx_0^{y-1}(x - x_0) \leqslant x^y - x_0^y \leqslant yx^{y-1}(x - x_0). \tag{7.10}$$

Next take x such that $x_0 < x < x_0 + 1$. Since $y \geqslant 1$,

$$0 < x_0^{y-1} \leqslant x^{y-1} \leqslant (x_0 + 1)^{y-1}.$$

This and (7.10) imply that

$$0 < yx_0^{y-1}(x - x_0) \leqslant x^y - x_0^y \leqslant yx^{y-1}(x - x_0) \leqslant y(x_0 + 1)^{y-1}(x - x_0).$$

Hence, we have

$$0 < x^y - x_0^y \leqslant y(x_0 + 1)^{y-1}(x - x_0) \qquad \text{for} \quad y \geqslant 1 \quad \text{and} \quad x_0 < x < x_0 + 1.$$

This yields

$$\lim_{x \to x_0^+} x^y = x_0^y \qquad \text{for} \quad y \geqslant 1. \tag{7.11}$$

If $0 < x < x_0$, we use Prob. 7.1, part (a) again and obtain

$$yx^{y-1}(x_0 - x) \leqslant x_0^y - x^y \leqslant yx_0^{y-1}(x_0 - x) \qquad \text{for} \quad y \geqslant 1 \quad \text{and} \quad 0 < x < x_0.$$

This implies that

$$0 < x_0^y - x^y \leqslant yx_0^{y-1}(x_0 - x) \qquad \text{if} \quad y \geqslant 1, \quad 0 < x < x_0.$$

We conclude from this that

$$\lim_{x \to x_0-} x^y = x_0^y \qquad \text{for} \quad y \geqslant 1. \tag{7.12}$$

This and (7.11) yield the conclusion for $y \geqslant 1$.

If $y \leqslant 0$, we have $1 - y \geqslant 1$. Using what was just proved, we have

$$\lim_{x \to x_0} x^{1-y} = x_0^{1-y}.$$

This implies that

$$\lim_{x \to x_0} x^{-y} = \lim_{x \to x_0} x^{-1}(x^{1-y}) = \lim_{x \to x_0} x^{-1} \lim_{x \to x_0} x^{1-y} = x_0^{-1}x_0^{1-y} = x_0^{-y}.$$

From this it follows that

$$\lim_{x \to x_0} x^y = \lim_{x \to x_0} \frac{1}{x^{-y}} = \frac{1}{x_0^{-y}} = x_0^y.$$

Thus, the conclusion also holds if $y \leqslant 0$.

If $0 < y < 1$, then $1 + y > 1$. By what has already been proved,

$$\lim_{x \to x_0} x^{1+y} = x_0^{1+y}.$$

This yields

$$\lim_{x \to x_0} x^y = \lim_{x \to x_0} x^{-1}(x^{1+y}) = x_0^{-1} x_0^{1+y} = x_0^y \qquad \text{for} \quad 0 < y < 1.$$

This completes the proof.

PROB. 7.2. Prove: If $y \in \mathbb{R}$, then

$$\lim_{n \to +\infty} \left(1 + \frac{1}{n}\right)^{ny} = e^y.$$

Theorem 7.3. *If $z \in \mathbb{R}$, then*

$$1 + z \leqslant e^z \leqslant 1 + ze^z. \tag{7.13}$$

PROOF. The conclusion is trivially true if $z = 0$. Assume $z \neq 0$, taking $z < 0$ first. Using Theorem 7.1, part (a), with $x = 1 + 1/n$, $y = nz$, n a positive integer, we have

$$z = nz\left(1 + \frac{1}{n} - 1\right) \leqslant \left(1 + \frac{1}{n}\right)^{nz} - 1 \leqslant nz\left(1 + \frac{1}{n}\right)^{nz-1}\left(1 + \frac{1}{n} - 1\right)$$

so that

$$z \leqslant \left(1 + \frac{1}{n}\right)^{nz} - 1 \leqslant z\left(1 + \frac{1}{n}\right)^{nz-1}, \tag{7.14}$$

where $z < 0$ and n is a positive integer. Fix z and let $n \to +\infty$. By theorems on limits of sequences and Prob. 7.2,

$$z \leqslant e^z - 1 \leqslant ze^z, \tag{7.15}$$

where $z < 0$. Now assume that $z > 0$, so that $-z < 0$. By (7.15),

$$-z \leqslant e^{-z} - 1 \leqslant -ze^{-z} \quad \text{and, hence,} \quad ze^{-z} \leqslant 1 - e^{-z} \leqslant z.$$

Multiplying the second set of inequalities by e^z we obtain

$$z \leqslant e^z - 1 \leqslant ze^z \qquad \text{if} \quad z > 0.$$

This proves that (7.15) holds if $z > 0$. The proof is now complete.

Corollary 1. *The following hold*:

(a) $\lim_{x \to 0} e^x = 1$,
(b) $\lim_{x \to 0}((e^x - 1)/x) = 1$.

Proof. We prove (a) first. Assume that $x < 0$. By the theorem,

$$0 < -xe^x \leqslant 1 - e^x \leqslant -x = |x|,$$

so that

$$0 < 1 - e^x \leqslant |x| \qquad \text{if} \quad x < 0.$$

This implies

$$\lim_{x \to 0-} e^x = 1.$$

Now assume that $x > 0$ and use the theorem to obtain

$$0 < x \leqslant e^x - 1 \leqslant xe^x$$

from which we have, using $e^x e^{-x} = 1$, that

$$0 < xe^{-x} \leqslant 1 - e^{-x} \leqslant x = |x| \qquad \text{for} \quad x > 0.$$

It follows from this that

$$\lim_{x \to 0+} e^{-x} = 1$$

and, hence,

$$\lim_{x \to 0+} e^x = \lim_{x \to 0+} \frac{1}{e^{-x}} = 1.$$

This and (7.13) prove (a).

We prove (b). By the theorem,

$$1 \leqslant \frac{e^x - 1}{x} \leqslant e^x \qquad \text{for} \quad x > 0,$$

$$e^x \leqslant \frac{e^x - 1}{x} \leqslant 1 \qquad \text{for} \quad x < 0.$$

By part (a) and the Sandwich Theorem,

$$\lim_{x \to 0+} \frac{e^x - 1}{x} = 1 = \lim_{x \to 0-} \frac{e^x - 1}{x}.$$

This proves (b).

Prob. 7.3. Prove:

(a) $e^x \leqslant 1/(1 - x)$ for $x < 1$,
(b) $\lim_{x \to +\infty} e^x = +\infty$,
(c) $\lim_{x \to -\infty} e^x = 0$.

Prob. 7.4. Prove:

(a) $\lim_{x \to +\infty} \cosh x = +\infty = \lim_{x \to -\infty} \cosh x$,
(b) $\lim_{x \to +\infty} \sinh x = +\infty$ and $\lim_{x \to -\infty} \sinh x = -\infty$.

Prob. 7.5. Prove:

(a) $\lim_{x \to +\infty} \tanh x = 1$ and $\lim_{x \to -\infty} \tanh x = -1$,
(b) $\lim_{x \to +\infty} \operatorname{sech} x = 0 = \lim_{x \to -\infty} \operatorname{sech} x$.

PROB. 7.6. Prove:

(a) $\lim_{x\to 0+} \operatorname{csch} x = +\infty$ and $\lim_{x\to 0-} \operatorname{csch} x = -\infty$,
(b) $\lim_{x\to +\infty} \operatorname{csch} x = 0 = \lim_{x\to -\infty} \operatorname{csch} x$,
(c) $\lim_{x\to 0+} \coth x = +\infty$ and $\lim_{x\to 0-} \coth x = -\infty$,
(d) $\lim_{x\to +\infty} \coth x = 1$ and $\lim_{x\to -\infty} \coth x = -1$.

PROB. 7.7. Evaluate:

(a) $\lim_{x\to \pm\infty}(1/(1+e^{1/x}))$,
(b) $\lim_{x\to 0+}(1/(1+e^{1/x}))$ and $\lim_{x\to 0-}(1/(1+e^{1/x}))$.

Corollary 2 (of Theorem 7.3).

(a) $\lim_{x\to x_0} e^x = e^{x_0}$,
(b) $\lim_{x\to x_0}((e^x - e^{x_0})/(x-x_0)) = e^{x_0}$.

PROOF. Using Theorem 7.3, we have

$$\lim_{x\to x_0} e^x = \lim_{x\to x_0} e^{x_0}e^{x-x_0} = e^{x_0}\lim_{x\to x_0} e^{x-x_0} = e^{x_0}\cdot 1 = e^{x_0}$$

which proves (a), and

$$\lim_{x\to x_0}\frac{e^x - e^{x_0}}{x-x_0} = \lim_{x\to x_0} e^{x_0}\frac{e^{x-x_0}-1}{x-x_0} = e^{x_0}\lim_{x\to x_0}\frac{e^{x-x_0}-1}{x-x_0} = e^{x_0}\cdot 1 = e^{x_0}$$

which proves (b).

8. $P(x)$ as $x\to\pm\infty$, Where P Is a Polynomial on $\mathbb{R}$

Theorem 8.1. *Let $P:\mathbb{R}\to\mathbb{R}$ be a polynomial of degree $n \geqslant 1$,*

$$P(x) = a_0x^n + a_1x^{n-1} + \cdots + a_{n-1}x + a_n, \tag{8.1}$$

where $a_0 > 0$. Then

$$\lim_{x\to +\infty} P(x) = +\infty. \tag{8.2}$$

PROOF. Given a real B, let

$$M = \max\{|a_1|, |a_2|, \ldots, |a_{n-1}|, |a_n - B|\} \tag{8.3}$$

and

$$X \geqslant 1 + \frac{M}{a_0}. \tag{8.4}$$

Assume that $x > X$, so that $x > 1$ and $x - 1 > 0$. This implies that

$$x > 1 + \frac{M}{a_0}, \quad \text{and} \quad 1 > \frac{M}{a_0(x-1)} \geqslant 0. \tag{8.5}$$

Since $x^n > x^n - 1 > 0$, it follows that

$$x^n > \frac{M}{a_0} \frac{x^n - 1}{x - 1}.$$

Hence,

$$\begin{aligned} a_0 x^n &> M(x^{n-1} + \cdots + x + 1) \\ &= Mx^{n-1} + \cdots + Mx + M \\ &\geqslant |a_1|x^{n-1} + \cdots + |a_{n-1}|x + |a_n - B| \\ &= |a_1 x^{n-1} + \cdots + a_{n-1}x + a_n - B| \\ &\geqslant -(a_1 x^{n-1} + \cdots + a_{n-1}x + a_n - B). \end{aligned}$$

But then

$$P(x) = a_0 x^n + a_1 x^{n-1} + \cdots + a_{n-1}x + a_n > B$$

for $x > X \geqslant 1 + M/a_0$. The conclusion follows readily from this.

Prob. 8.1. If $a_0 < 0$ and n is a positive integer, then

$$\lim_{x \to +\infty} (a_0 x^n + a_1 x^{n-1} + \cdots + a_{n-1}x + a_n) = -\infty.$$

Prob. 8.2. Prove: Let n be a positive integer and $a_0, a_1, \ldots, a_n$ be $n + 1$ real numbers with $a_0 \neq 0$. Put

$$M = \max\{|a_1, \ldots, |a_n|\}.$$

Then for $x > 1 + M/|a_0|$ we have $a_0 x^n + \cdots + a_{n-1}x + a_n > 0$ or $a_0 x^n + \cdots + a_{n-1}x + a_n < 0$ according to whether $a_0 > 0$ or $a_0 < 0$.

Prob. 8.3. Prove: If $a_0 > 0$ and n is a positive integer, then

$$\lim_{x \to -\infty} (a_0 x^n + a_1 x^{n-1} + \cdots + a_{n-1}x + a_n) = +\infty \quad \text{or} \quad -\infty,$$

according to whether n is even or n is odd.

Prob. 8.4. Prove: (a) If n is a nonnegative integer and at least one of the $n + 1$ real numbers $a_0, a_1, \ldots, a_n$ is not zero, then there exists a real number x such that

$$a_0 x^n + a_1 x^{n-1} + \cdots + a_{n-1}x + a_n \neq 0.$$

(b) If n is a nonnegative integer and $a_0, a_1, \ldots, a_n$ are $n + 1$ real numbers such that

$$a_0 x^n + a_1 x^{n-1} + \cdots + a_{n-1}x + a_n = 0$$

for *all* $x \in \mathbb{R}$, then $a_0 = a_1 = \cdots = a_n = 0$.

PROB. 8.5. Prove: If m and n are integers such that $0 \leqslant m < n$ and

$$a_0x^n + a_1x^{n-1} + \cdots + a_{n-1}x + a_n$$
$$= b_0x^m + b_1x^{m-1} + \cdots + b_{m-1}x + b_m$$

for all $x \in \mathbb{R}$, then

$$a_0 = a_1 = \cdots = a_{n-m-1} = 0 \quad \text{and} \quad a_{n-m} = b_0,$$
$$a_{n-m+1} = b_1, \ldots, a_n = b_m.$$

9. Two Theorems on Limits of Functions. Cauchy Criterion for Functions

The theorem which follows is very useful. A special case of it is Theorem 4.3.

Theorem 9.1. *Let a, b, and L be in $\mathbb{R}^*$. If f and g are real-valued functions of a real variable and $\mathfrak{R}(g) \subseteq \mathfrak{D}(f)$, then*

$$\lim_{t\to a} g(t) = b, \quad \lim_{x\to b} f(x) = L, \quad \text{and} \quad g(t) \neq b \qquad \text{for all} \quad t \in \mathfrak{D}(f). \tag{9.1}$$

Then

$$\lim_{t\to a} f(g(t)) = L. \tag{9.2}$$

PROOF. Let $N(L)$ be a neighborhood of L. Since $f(x) \to L$ as $x \to b$, there exists a deleted neighborhood $N^*(b)$ of b such that if $x \in N^*(b) \cap \mathfrak{D}(f)$, then $f(x) \in N(L)$. Since $g(t) \to b$ as $t \to a$, there exists a deleted neighborhood $N^*(a)$ of a such that if $t \in N^*(a) \cap \mathfrak{D}(g)$, then $g(t) \in N(b)$. But $g(t) \neq b$ for all $t \in \mathfrak{D}(f)$. This implies that for $t \in N^*(a) \cap \mathfrak{D}(g)$, we have $g(t) \in N^*(b)$. Since $g(t) \in \mathfrak{D}(f)$ for all $t \in \mathfrak{D}(g)$, we see that $t \in N^*(a) \cap \mathfrak{D}(g)$ implies $g(t) \in N^*(b) \cap \mathfrak{D}(f)$ and, therefore, that $f(g(t)) \in N(L)$. In turn, this implies that (9.2) holds. The proof is now complete.

Theorem 9.2 (Cauchy Criterion for Limits of Functions). *Let $a \in \mathbb{R}^*$ and $L \in \mathbb{R}$. If a is an accumulation point of $\mathfrak{D}(f)$, then*

$$\lim_{x\to a} f(x) = L$$

if and only if for each $\epsilon > 0$ there exists a deleted neighborhood $N^*(a)$ of a such that if x' and x'' are in $N^*(a) \cap \mathfrak{D}(f)$, then

$$|f(x') - f(x'')| < \epsilon. \tag{9.3}$$

PROOF. Assume first $f(x) \to L \in \mathbb{R}$ as $x \to a \in \mathbb{R}^*$. Let $\epsilon > 0$ be given. There exists, therefore, a deleted neighborhood $N^*(a)$ of a such that

$$\text{if } x \in N^*(a) \cap \mathfrak{D}(f), \text{ then } |f(x) - L| < \frac{\epsilon}{2}. \tag{9.4}$$

Let $x' \in N^*(a) \cap \mathfrak{D}(f)$ and $x'' \in N^*(a) \cap \mathfrak{D}(f)$. We have

$$|f(x') - L| < \frac{\epsilon}{2} \quad \text{and} \quad |f(x'') - L| < \frac{\epsilon}{2}.$$

But then

$$|f(x') - f(x'')| \leqq |f(x') - L| + |f(x'') - L| < \frac{\epsilon}{2} + \frac{\epsilon}{2} = \epsilon.$$

Conversely, assume that for each $\epsilon > 0$ there exists a deleted neighborhood $N^*(a)$ of a such that if x' and x'' are in $N^*(a) \cap \mathfrak{D}(f)$ then (9.3) holds. Since a is an accumulation point of $\mathfrak{D}(f)$, there exists a sequence $\langle x_n \rangle$ of elements of $\mathfrak{D}(f)$ such that $x_n \neq a$ for all n and $\lim x_n = a$. We prove that the sequence $\langle f(x_n) \rangle$ is a Cauchy sequence. Let $\epsilon > 0$ be given, so that there exists a deleted neighborhood $N^*(a)$ of a such that if x' and x'' are in $N^*(a) \cap \mathfrak{D}(f)$, then (9.3) holds. Since $\lim x_n = a$, there exists an N such that if $n > N$, then $x_n \in N^*(a)$. Take $n > N$ and $m > N$. Then $x_n \in N^*(a) \cap \mathfrak{D}(f)$ and $x_m \in N^*(a) \cap \mathfrak{D}(f)$. Hence,

$$|f(x_n) - f(x_m)| < \epsilon.$$

Thus, $\langle f(x_n) \rangle$ is a Cauchy sequence of real numbers. As such it converges to some real limit L. We prove that $\lim_{x \to a} f(x) = L$. Let $\epsilon > 0$ be given. There exists a deleted neighborhood $N_1^*(a)$ such that if x' and x'' are in $N_1^*(a) \cap \mathfrak{D}(f)$, then

$$|f(x') - f(x'')| < \frac{\epsilon}{2}.$$

Take $x \in N_1^*(a) \cap \mathfrak{D}(f)$. Now an N_2 exists such that if $n > N_2$, then $x_n \in N_1^*(a) \cap \mathfrak{D}(f)$. We have

$$|f(x) - f(x_n)| < \frac{\epsilon}{2} \tag{9.5}$$

for $x \in N_1^*(a) \cap \mathfrak{D}(f)$ and $n > N_2$. Since $f(x_n) \to L$ as $n \to +\infty$, we have $\lim_{n \to +\infty} |f(x) - f(x_n)| = |f(x) - L|$. This and (9.5) imply

$$|f(x) - L| \leqslant \frac{\epsilon}{2} < \epsilon \qquad \text{for} \quad x \in N_1^*(a) \cap \mathfrak{D}(f). \tag{9.6}$$

We conclude from this that $\lim_{x \to a} f(x) = L$. The proof is now complete.

CHAPTER VI
Continuous Functions

1. Definitions

The notion of $\lim_{t\to a} f(x)$ was defined without reference to the value of f at a. Here we are interested in the value of f at a and its relation to the limit of f as x approaches a when this limit exists.

We recall that $\lim_{t\to a} f(x) = L$, where a and L are real numbers, means that, first of all, a is an accumulation point of $\mathfrak{D}(f)$, and, second, that for each ϵ-neighborhood $N(L,\epsilon)$ of L, there exists a deleted δ-neighborhood $N^*(a,\delta)$ of a such that if $x \in N^*(a) \cap \mathfrak{D}(f)$, then $f(x) \in N(L,\epsilon)$. This can be phrased more intuitively as: $f(x)$ is as close to L as we like provided that x is taken sufficiently close to a. Continuity of f at a can be phrased as: (1) f is defined at a and (2) any change in the value of f at a can be brought about by a sufficiently small change in the value of a. We state this with more precision in the following definition.

Def. 1.1. If f is a real-valued function of a real variable and $x_0 \in \mathbb{R}$, then f is called *continuous* at x_0 if and only if: (1) f is defined at x_0 and (2) for each ϵ-neighborhood $N(f(x_0),\epsilon)$ of $f(x_0)$, there exists a δ-neighborhood $N(x_0,\delta)$ of x_0 such that

$$f(N(x_0,\delta) \cap \mathfrak{D}(f)) \subseteq N(f(x_0),\epsilon). \tag{1.1}$$

This can also be written:

$$x \in N(x_0,\delta) \cap \mathfrak{D}(f) \quad \text{implies} \quad f(x) \in N(f(x_0),\epsilon) \tag{1.2}$$

or as

$$|x - x_0| < \delta \quad \text{and} \quad x \in \mathfrak{D}(f) \tag{1.3}$$

imply that

$$|f(x) - f(x_0)| < \epsilon. \tag{1.4}$$

We therefore rephrase Def. 1.1 as:

Def. 1.1′. f is continuous at x_0 if and only if: (1) $x_0 \in \mathfrak{D}(f)$ and (2) for each $\epsilon > 0$, there exists a $\delta > 0$ such that if

$$|x - x_0| < \delta \quad \text{and} \quad x \in \mathfrak{D}(f),$$

then

$$|f(x) - f(x_0)| < \epsilon.$$

If x_0 is not only in $\mathfrak{D}(f)$ but also an accumulation point of $\mathfrak{D}(f)$, then we may use limits to define continuity at x_0.

Def. 1.1″. If f is a real-valued function of a real variable and $x_0 \in \mathbb{R}$ is an accumulation point of $\mathfrak{D}(f)$, then f is continuous at x_0 if and only if: (1) $x_0 \in \mathfrak{D}(f)$ and (2)

$$\lim_{x \to x_0} f(x) = f(x_0). \tag{1.5}$$

Remark 1.1. By an *isolated* point of a set S we mean a point of S which is not an accumulation point of S. Def. 1.1″ cannot be used to define continuity of f at an isolated point x_0 of $\mathfrak{D}(f)$ since, in this case, $\lim_{x \to x_0} f(x)$ does not exist. We prove in Example 1.1 below that if x_0 is an isolated point of $\mathfrak{D}(f)$, then f is continuous at x_0.

PROB. 1.1. Prove: If $S \subseteq \mathbb{R}$, then x_0 is an isolated point of S if and only if there is a deleted δ-neighborhood $N^*(x_0, \delta)$ of x_0 such that $N^*(x_0, \delta) \cap S = \emptyset$. Accordingly, x_0 is an isolated point of S if and only if there exists a δ-neighborhood $N(x_0, \delta)$ of x_0 such that $N(x_0, \delta) \cap S = \{x_0\}$.

EXAMPLE 1.1. We prove that if x_0 is an isolated point of $\mathfrak{D}(f)$, where f is a real-valued function of a real variable, then f is continuous at x_0. Thus, let x_0 be an isolated point of $\mathfrak{D}(f)$. Then f is defined at x_0 and there is a δ-neighborhood $N(x_0, \delta)$ of x_0 such that $N(x_0, \delta) \cap \mathfrak{D}(f) = \{x_0\}$. Given $\epsilon > 0$, assume that $x \in \mathfrak{D}(f)$ and $|x - x_0| < \delta$. This implies that $x \in N(x_0, \delta) \cap \mathfrak{D}(f)$ and, hence, that $x = x_0$. Hence, $|f(x) - f(x_0)| = |f(x_0) - f(x_0)| = 0 < \epsilon$. According to Def. 1.1′, f is continuous at x_0.

Remark 1.2. Note that in defining continuity at x_0, we used δ-neighborhoods and not deleted δ-neighborhoods as in the definition of $\lim_{x \to x_0} f(x) = L$.

Def. 1.2. A function is called *continuous* if and only if it is continuous at each point of its domain. A function is called continuous on a *set* A if and only if it is continuous at each point of A.

For example, each polynomial function P on $\mathbb{R}$ is continuous since for each $x_0 \in \mathbb{R} = \mathfrak{D}(P)$ we have (Prob. V.6.3) $\lim_{x \to x_0} P(x) = P(x_0)$. Here the limit definition of continuity can be used, since each point of $\mathbb{R} = \mathfrak{D}(f)$ is an accumulation point of $\mathfrak{D}(f)$ (explain).

EXAMPLE 1.2. The function $E: \mathbb{R} \to \mathbb{R}$, where $E(x) = e^x$ for each $x \in \mathbb{R}$, is continuous since $e^x \to e^{x_0}$ as $x \to 0$ (see Corollary 2 of Theorem V.7.3).

EXAMPLE 1.3. The function $f: (0; +\infty) \to \mathbb{R}$, defined as $f(x) = x^y$ for $x > 0$, where y is some fixed real number is a continuous function. Its domain is $(0; +\infty)$. Each point of $(0; +\infty)$ is an accumulation point of $(0; +\infty)$ (prove this). Using limits, we have (Theorem V.7.2) $\lim_{x \to x_0} x^y = x_0^y$ for $x_0 > 0$.

EXAMPLE 1.4. The sine function is a continuous function. We have

$$\sin x - \sin x_0 = \sin\left(\frac{x + x_0}{2} + \frac{x - x_0}{2} \right) - \sin\left(\frac{x + x_0}{2} - \frac{x - x_0}{2} \right)$$
$$= 2 \cos \frac{x + x_0}{2} \sin \frac{x - x_0}{2}.$$

Since $|\cos(x + x_0)/2)| \leqslant 1$, it follows that

$$|\sin x - \sin x_0| \leqslant 2 \left|\sin \frac{x - x_0}{2}\right|. \tag{1.6}$$

But

$$\left|\sin \frac{x - x_0}{2}\right| \leqslant \frac{|x - x_0|}{2},$$

so that

$$|\sin x - \sin x_0| \leqslant |x - x_0| \qquad \text{for} \quad x \in \mathbb{R}, \quad x_0 \in \mathbb{R}. \tag{1.7}$$

From this it is clear that $\sin x \to \sin x_0$ as $x \to x_0$ and, hence, that sine is continuous for each $x_0 \in \mathbb{R} = \mathfrak{D}(\text{sine})$.

PROB. 1.2. Prove: The cosine function is continuous.

PROB. 1.3. Prove: The absolute value function is continuous.

PROB. 1.4. Prove: (a) If $x \geqslant 0$, then $|\sqrt{x} - \sqrt{x_0}| \leqslant \sqrt{|x - x_0|}$; (b) the function g: $[0, +\infty) \to \mathbb{R}$, defined as: $g(x) = \sqrt{x}$ for $x \in [0, +\infty)$, is a continuous function.

The following theorem should be compared with Theorem V.4.3. It gives a criterion for continuity at a point in terms of sequences.

Theorem 1.1. *f is continuous at x_0 if and only if for each sequence $\langle x_n \rangle$ of points of $\mathfrak{D}(f)$ such that $x_n \to x_0$ as $n \to +\infty$, we have $f(x_n) \to f(x_0)$ as $n \to +\infty$.*

PROOF. Suppose that f is continuous at x_0. Let $\langle x_n \rangle$ be a sequence of elements of $\mathfrak{D}(f)$ such that $x_n \to x_0$ as $n \to +\infty$. (Such sequences exist. For example, $x_n = x_0$ for all n is such a sequence.) Let $N(f(x_0), \epsilon)$ be a given ϵ-neighborhood of $f(x_0)$. By the continuity of f at x_0, there exists a δ-neighborhood $N(x_0, \delta)$ of x_0 such that

$$f(N(x_0, \delta) \cap \mathfrak{D}(f)) \subseteq N(f(x_0), \epsilon). \tag{1.8}$$

But since $x_n \to x_0$ as $n \to +\infty$, there exists an N such that if $n > N$, then $x_n \in N(x_0, \delta)$, and since $x_n \in \mathfrak{D}(f)$ for all n, we have: If $n > N$, then $x_n \in N(x_0, \delta) \cap \mathfrak{D}(f)$. By (1.8) this implies that

$$f(x_n) \in N(f(x_0), \epsilon) \qquad \text{for} \quad n > N.$$

Thus, for each $\epsilon > 0$, there exists an N such that

$$|f(x_n) - f(x_0)| < \epsilon \qquad \text{for} \quad n > N.$$

But then $f(x_n) \to f(x_0)$ as $n \to +\infty$.

Conversely, assume that for each sequence $\langle x_n \rangle$ of elements of $\mathfrak{D}(f)$ such that $x_n \to x_0$ as $n \to +\infty$, we have $f(x_n) \to f(x_0)$ as $n \to +\infty$. If x_0 is an isolated point of $\mathfrak{D}(f)$, then f is continuous at x_0. If x_0 is an accumulation point of $\mathfrak{D}(f)$, let $\langle x_n \rangle$ be a sequence of elements of $\mathfrak{D}(f)$ such that $x_n \neq x_0$ for all n and $x_n \to x_0$ as $n \to +\infty$. By the present hypothesis we have $f(x_n) \to f(x_0)$ as $n \to +\infty$. By Theorem V.4.3 we have $f(x) \to f(x_0)$ as $x \to x_0$, so that f is continuous at x_0 in this case also. This completes the proof.

The next theorem should be compared to Theorem V.9.1.

Theorem 1.2. *If g is a function such that $\mathfrak{R}(g) \subseteq \mathfrak{D}(f)$, where f is continuous at x_0 and $\lim_{t \to t_0} g(t) = x_0$ for some $t_0 \in \mathbb{R}^*$, then $\lim_{t \to t_0} f(g(t)) = f(x_0)$.*

PROOF. Exercise.

Theorem 1.3. (A Continuous Function of a Continuous Function is Continuous.) *Let g be continuous at $t_0 \in \mathbb{R}$, and $g(t_0) = x_0$. Let f be continuous at x_0, and let $\mathfrak{R}(g) \subseteq \mathfrak{D}(f)$. Then the composite function $f \circ g$ is continuous at t_0.*

PROOF. Exercise.

Remark 1.3. The criterion for continuity at a point in terms of sequences converging to the point given in Theorem 1.1 is called *sequential continuity*. Thus, Theorem 1.1 states that a function is continuous at a point if and only if it is sequentially continuous at the point. Theorem 1.1 is often used to demonstrate the *lack* of continuity at a point. We demonstrate this in the next example.

EXAMPLE 1.5. Consider the function $D:\mathbb{R}\to\mathbb{R}$ defined as:

$$D(x)=\begin{cases}1 & \text{if } x \text{ is rational}\\ 0 & \text{if } x \text{ is irrational.}\end{cases} \tag{1.9}$$

This function is called *Dirichlet's function*. This function is defined for all $x\in\mathbb{R}$ but is continuous at no $x\in\mathbb{R}$. We prove this.

Let r_0 be rational. Take a sequence $\langle c_n\rangle$ of irrational numbers such that $c_n\to r_0$ as $n\to+\infty$. (By Prob. V.3.7 such sequences exist.) We have $D(c_n)=0$ for all n, so that $\lim_{n\to+\infty}D(c_n)=0$. Since $D(r_0)=1$, we have $\lim_{n\to+\infty}D(c_n)\neq D(r_0)$ even though $\lim_{n\to+\infty}c_n=r_0$. By Theorem 1.1, D is not continuous at r_0. Now let c_0 be irrational and take a sequence $\langle r_n\rangle$ of rational numbers such that $r_n\to c_0$ and $n\to+\infty$. (By Prob. V.3.7 such sequences exist.) We have $D(r_n)=1$ for all n, so that $\lim_{n\to+\infty}D(r_n)=1$. Here, too, $r_n\to c_0$ as $n\to+\infty$ and $\lim_{n\to+\infty}D(r_n)\neq D(c_0)=0$. By Theorem 1.1, D is not continuous at c_0. Thus, D is not continuous at any real number.

We now apply Theorem 1.1 to prove an important identity. We recall that the function exp: $\mathbb{R}\to\mathbb{R}$ was defined in Example IV.5.1 as

$$\exp x=\sum_{n=0}^{\infty}\frac{x^n}{n!}\qquad\text{for}\quad x\in\mathbb{R}. \tag{1.10}$$

In Prob. IV.7.1 it was noted that

$$\exp r=e^r\qquad\text{if}\quad r\text{ is rational.} \tag{1.11}$$

In the next theorem we prove that this equality holds for *all* $r\in\mathbb{R}$.

Theorem 1.4. *If* $x\in\mathbb{R}$, *then*

$$\sum_{n=0}^{\infty}\frac{x^n}{n!}=\exp x=e^x. \tag{1.12}$$

PROOF. We already proved that the function E: $\mathbb{R}\to\mathbb{R}$ defined as

$$E(x)=e^x\qquad\text{for}\quad x\in\mathbb{R} \tag{1.13}$$

is continuous (see Example 1.2). We now prove that the function exp is a continuous. Observe that the function exp satisfies the following:

$$1+x\leqslant\exp x\leqslant 1+x\exp x\qquad\text{for}\quad x\in\mathbb{R}. \tag{1.14}$$

(See Theorem IV.7.2, part (d).) Now turn to Corollary 1 of Theorem V.7.3. There we proved that $\lim_{x\to0}e^x=1$. Examining the proof we find that all we used there was the inequality

$$1+z\leqslant e^z\leqslant 1+ze^z\qquad\text{for}\quad z\in\mathbb{R} \tag{1.15}$$

of Theorem V.7.3, and the fact that $e^ze^{-z}=1$. Since the function exp satisfies (1.14) which is similar to (1.15), and since exp also has the property: $\exp z\exp(-z)=1$ (see Theorem IV.7.2, part (c)), it is a simple matter to imitate the proof of Corollary 1 of Theorem V.7.3 and to prove that

$$\lim_{x\to 0} \exp x = 1. \tag{1.16}$$

We now note that exp also satisfies

$$\exp x \exp y = \exp(x+y). \tag{1.17}$$

(See formula (7.17) of Example IV.7.1.) This implies that

$$\begin{aligned}\lim_{x\to x_0} \exp x = \lim_{x\to x_0} (\exp x_0 \exp(x-x_0)) &= \exp x_0 \lim_{x\to x_0} \exp(x-x_0)\\ &= \exp x_0 \cdot 1\\ &= \exp x_0 .\end{aligned}$$

But then

$$\lim_{x\to x_0} \exp x = \exp x_0 . \tag{1.18}$$

This proves that the function exp is continuous at each $x \in \mathbb{R}$ and, therefore, that it is continuous.

Next we take any $x \in \mathbb{R}$ and a sequence $\langle r_n \rangle$ of rational numbers such that $r_n \to x$ as $n \to +\infty$. We have (1.11)

$$\exp r_n = e^{r_n} \qquad \text{for each } n.$$

This and the continuity of E and exp imply that

$$\exp x = \lim_{n\to+\infty} (\exp r_n) = \lim_{n\to+\infty} e^{r_n} = e^x,$$

so that

$$\exp x = e^x \qquad \text{for each} \quad x \in \mathbb{R},$$

and the proof is complete.

Prob. 1.5. The principle used in proving Theorem 1.4 is far-reaching. It can be stated as follows: Two functions continuous in $\mathbb{R}$ which have the same values for the rational numbers are identical. Prove the last statement.

Prob. 1.6. Prove:

(a) If $x > 0$ and n is a nonnegative integer, then

$$e^x > \frac{x^n}{n!} + \frac{x^{n+1}}{(n+1)!} .$$

(b) $\lim_{x\to+\infty}(e^x/x^n) = +\infty$.
(c) If $\alpha \in \mathbb{R}$, then $\lim_{x\to+\infty}(e^x x^{-\alpha}) = +\infty$.

Prob. 1.7. Prove:

$$\cosh x = \sum_{n=0}^{\infty} \frac{x^{2n}}{(2n)!} = 1 + \frac{x^2}{2!} + \frac{x^4}{4!} + \cdots ,$$

$$\sinh x = \sum_{n=0}^{\infty} \frac{x^{2n+1}}{(2n+1)!} = x + \frac{x^3}{3!} + \frac{x^5}{5!} + \cdots .$$

The Dirichlet function (Example 1.5) though defined for all of ℝ is continuous for no $x \in \mathbb{R}$. There are functions defined for all ℝ which are continuous at one point only.

Prob. 1.8. Prove: f, where

$$f(x) = \begin{cases} x^2 & \text{for } x \text{ rational} \\ 0 & x = 0, \end{cases}$$

is continuous at $x = 0$ only.

Prob. 1.9.

$$g(x) = \begin{cases} x & \text{if} \quad x \text{ is rational} \\ 1 - x & \text{if} \quad x \text{ is irrational,} \end{cases}$$

is continuous at $x = \frac{1}{2}$ only.

Prob. 1.10. Let f: $[0, 1]$aℝ be defined as

$$f(x) = \begin{cases} 1, & \text{if} \quad x = 1, \\ \dfrac{1}{q}, & \text{if} \quad 0 < x < 1, \quad \text{where } x \text{ is rational} \\ & \qquad x = p/q, p \text{ and } q \text{ being relatively prime positive integers,} \\ 1, & \text{if} \quad x = 0. \end{cases}$$

The figure (Fig. 1.1) is a poor attempt at portraying the graph of this function. Prove: If $x_0 \in [0, 1]$, then $\lim_{x \to x_0} f(x) = 0$. Conclude that f is continuous at x_0 if x_0 is irrational and discontinuous if x_0 is rational.

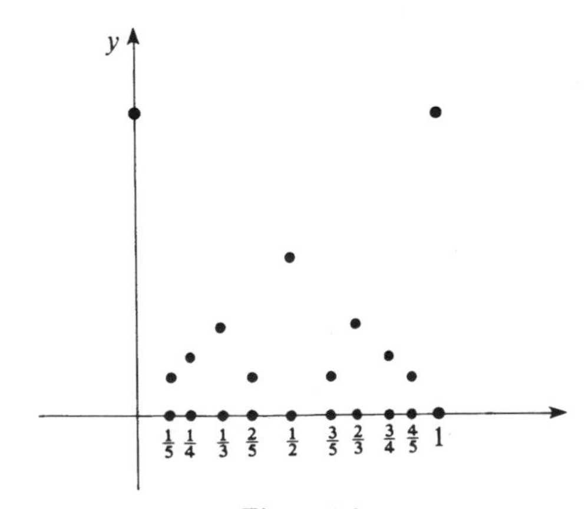

Figure 1.1

2. One-Sided Continuity. Points of Discontinuity.

Just as there are one-sided limits, we can speak of *one-sided continuity*.

Def. 2.1 (One-Sided Continuity). A real-valued function of a real variable is called continuous from the left at x_0 if and only if: (1) $x_0 \in \mathfrak{D}(f)$ and (2) for each ϵ-neighborhood $N(f(x_0),\epsilon)$ of $f(x_0)$, there exists a δ-neighborhood $N_-(x_0,\delta) = (x_0 - \delta, x_0]$ of x_0 from the left such that

$$f((x_0 - \delta, x_0] \cap \mathfrak{D}(f)) \subseteq N(f(x_0),\epsilon). \tag{2.1}$$

Similarly, f is called continuous from the *right at* x_0 if and only if: (1) $x_0 \in \mathfrak{D}(f)$ and (2) for each ϵ-neighborhood $N(f(x_0),\epsilon)$ of $f(x_0)$, there exists a δ-neighborhood $N_+(x_0,\delta) = [x_0, x_0 + \delta)$ of x_0 from the right such that

$$f((x_0, x_0 + \delta) \cap \mathfrak{D}(f)) \subseteq N(f(x_0,\epsilon)). \tag{2.2}$$

In terms of inequalities this is equivalent to:

Def. 2.1′. f is continuous from the *left at* x_0 if and only if: (1) $x_0 \in \mathfrak{D}(f)$ and (2) for each $\epsilon > 0$ there exists a $\delta > 0$ such that if $x_0 - \delta < x \leqslant x_0$ and $x \in \mathfrak{D}(f)$, then $|f(x) - f(x_0)| < \epsilon$. Similarly, f is continuous from the *right at* x_0 if and only if: (1) $x_0 \in \mathfrak{D}(f)$ and (2) for each $\epsilon > 0$ there exists a $\delta > 0$ such that if $x_0 \leqslant x < x_0 + \delta$ and $x \in \mathfrak{D}(f)$, then $|f(x) - f(x_0)| < \epsilon$.

Remark 2.1. If $x_0 \in \mathfrak{D}(f)$ but x_0 is not an accumulation point of $\mathfrak{D}(f)$ from one side, then f is continuous from that side at x_0. (See Example 1.1 where continuity of f at an isolated point of $\mathfrak{D}(f)$ is discussed.)

If x_0 is an accumulation point of $\mathfrak{D}(f)$ from one side, then the continuity of f from that side can be defined in terms of the limit of f as x approaches x_0 from that side.

Def. 2.3″. If x_0 is an accumulation point of $\mathfrak{D}(f)$ from the left, then f is continuous from the *left at* x_0 if and only if: (1) $x_0 \in \mathfrak{D}(f)$ and (2) $f(x_0 -) = f(x_0)$. Similarly, if x_0 is an accumulation point of $\mathfrak{D}(f)$ from the right, then f is continuous from the *right at* x_0 if and only if: (1) $x_0 \in \mathfrak{D}(f)$ and (2) $f(x_0 +) = f(x_0)$.

PROB. 2.1. Prove: f is continuous at x_0 if and only if f is continuous from the right and from the left at x_0.

Remark 2.2. A function f is continuous from the left at x_0 if and only if its restriction to $(-\infty; x_0] \cap \mathfrak{D}(f)$ is continuous at x_0. Similarly, f is continuous from the right at x_0 if and only if its restriction to $[x_0, +\infty) \cap \mathfrak{D}(f)$ is continuous at x_0. (See Remark V.5.2.)

EXAMPLE 2.1. We prove that the greatest integer function is continuous from the right but that it is not continuous from the left at each integer.

Assume $x_0 \in \mathbb{R}$. Given $\epsilon > 0$, take δ such that $0 < \delta \leqslant 1 + [x_0] - x_0$. (Note that $[x_0] \leqslant x_0 < [x_0] + 1$, so that $1 + [x_0] - x_0 > 0$.) Therefore, if $x_0 \leqslant x < x_0 + \delta$, then $x_0 \leqslant x < [x_0] + 1$ and

$$[x_0] \leqslant x_0 \leqslant x < [x_0] + 1$$

from which it follows that $[x] = [x_0]$ for such x. Hence,

$$|[x] - [x_0]| = |[x_0] - [x_0]| = 0 < \epsilon \qquad \text{for} \quad x_0 \leqslant x < x_0 + \delta,$$

and

$$\lim_{x \to x_0+} [x] = [x_0]. \tag{2.3}$$

On the other hand, if x_0 is an integer, then $x_0 = [x_0]$. Take δ_1 such that $0 < \delta_1 \leqslant 1$ and x such that $x_0 - \delta_1 < x < x_0$. For such x we have $[x_0] - 1 = x_0 - 1 < x < x_0 = [x_0]$, and therefore $[x] = [x_0] - 1 = x_0 - 1$, so that

$$\lim_{x \to x_0-} [x] = x_0 - 1 < x_0 = [x_0], \qquad \text{if} \quad x_0 \text{ is an integer.}$$

PROB. 2.2. Prove: If x_0 is not an integer, then $\lim_{x \to x_0-} [x] = [x_0]$.

A point $x_0 \in \mathbb{R}$ at which a function f is discontinuous is called a *point of discontinuity* of the function. According to the last example and the problem following it we may state: The greatest integer function is continuous for each noninteger x_0 and its points of discontinuity are the integers.

Let x_0 be an accumulation point of $\mathfrak{D}(f)$. If $\lim_{x \to x_0} f(x)$ exists and is finite but either f is not defined at x_0 or $\lim_{x \to x_0} f(x) \neq f(x_0)$, we call x_0 a *removable discontinuity* of f. This is to suggest that the discontinuity "can be removed" by redefining f appropriately at x_0. That is, a new function g can be defined where

$$g(x) = \begin{cases} f(x) & \text{for} \quad x \neq x_0 \\ \lim_{x \to x_0} f(x) & \text{for} \quad x = x_0. \end{cases} \tag{2.4}$$

This g is such that

$$\lim_{x \to x_0} g(x) = \lim_{x \to x_0} f(x) = g(x_0), \tag{2.5}$$

and is, therefore, continuous at x_0.

EXAMPLE 2.2. The function f, where

$$f(x) = \frac{x^2 - 4}{x - 2} \qquad \text{for} \quad x \neq 2,$$

has a discontinuity at $x = 2$ since it is not defined there. This discontinuity

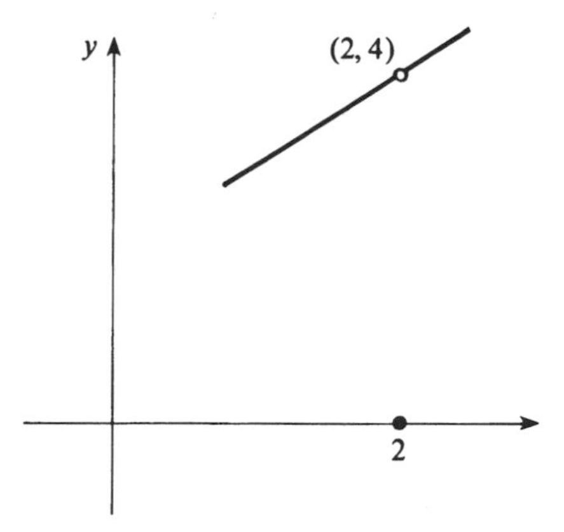

Figure 2.1

is removable since

$$\lim_{x\to 2} f(x) = \lim_{x\to 2} \frac{x^2-4}{x-2} = \lim_{x\to 2}(x+2) = 4$$

(see Fig. 2.1).

PROB. 2.3. Consider the function f, where

$$f(x) = \begin{cases} \dfrac{\sin x}{x} & \text{if } x \neq 0 \\ 2 & \text{if } x = 0. \end{cases}$$

(a) Show that f has a removable discontinuity at $x = 0$.
(b) Redefine f so that it is continuous at 0.

If x_0 is a two-sided accumulation point of $\mathfrak{D}(f)$ but at least one one-sided limit as $x \to x_0$ is infinite, we say that f has an *infinite discontinuity* at x_0.

EXAMPLE 2.3. Let f be defined as:

$$f(x) = \begin{cases} \dfrac{1}{x} & \text{if } x \neq 0 \\ 1 & \text{if } x = 0. \end{cases}$$

This function has an infinite discontinuity at $x = 0$. (See Fig. 2.2.) No matter how we define f at 0, the discontinuity there will persist (explain).

It may happen that x_0 is a two-sided accumulation point of $\mathfrak{D}(f)$ but $\lim_{x\to x_0} f(x)$ does not exist, even in the extended sense. One way in which this can occur is if both one-sided limits exist and are finite but $f(x_0-) \neq f(x_0+)$. Here $\lim_{x\to x_0} f(x)$ does not exist and f is discontinuous at x_0. This type of discontinuity is called a *jump discontinuity* and $f(x_0+) - f(x_0-)$ is called the *value of the jump*.

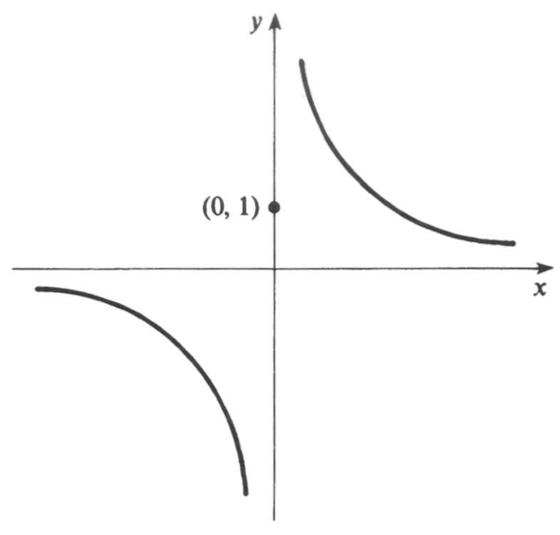

Figure 2.2

EXAMPLE 2.4. The signum function (cf. Example II.2.2 and Example V.5.1) has a jump discontinuity at $x = 0$, and the value of the jump is $\operatorname{sig}(1+) - \operatorname{sig}(1-) = 1 - (-1) = 2$.

If x_0 is a two-sided accumulation point of $\mathfrak{D}(f)$ and at least one of the one-sided limits of f as $x \to x_0$ does not exist, even in the extended case, then x_0 is yet another type of discontinuity of f.

EXAMPLE 2.5. Consider the function g, where

$$g(x) = \begin{cases} \left\langle \dfrac{1}{x} \right\rangle^* & \text{if} \quad x \neq 0 \\ 0 & \text{if} \quad x = 0. \end{cases}$$

(Recall $\langle z \rangle^*$ is the distance from z to the integer nearest z. It is shown in Example V.5.3 that neither of the limits $g(0-) = \langle 0- \rangle^*$, $g(0+) = \langle 0+ \rangle^*$ exist. Hence, $x = 0$ here is a discontinuity of g of the last type.

PROB. 2.4. Prove that $\lim_{x\to 0}(x\langle 1/x \rangle^*) = 0$.

3. Theorems on Local Continuity

The theorems which follow are similar to the ones in Section V.6 on limits. They concern themselves with the properties of functions which are continuous at a point x_0. We refer to such properties as *local* properties of continuous functions. On the other hand, properties of functions which are continuous on certain types of sets are called *global* properties, or properties *in the large*, of continuous functions.

Theorem 3.1. *If f is continuous at x_0, there exists a δ-neighborhood $N(x_0,\delta)$ of x_0 such that f is bounded on $N(x_0,\delta) \cap \mathfrak{D}(f)$.*

PROB. 3.1. Prove Theorem 3.1. (See the proof of Theorem V.6.1.)

Theorem 3.2. *Let f be continuous at x_0. If (a) $f(x_0) > 0$, then there exists a δ-neighborhood $N(x_0,\delta)$ of x_0 such that $f(x) > 0$ for $x \in N(x_0,\delta) \cap \mathfrak{D}(f)$. If (b) $f(x_0) < 0$, then there exists a δ-neighborhood $N(x_0,\delta)$ of x_0 such that $f(x) < 0$ for $x \in N(x_0,\delta) \cap \mathfrak{D}(f)$.*

PROB. 3.2. Prove Theorem 3.2. (See the proof of Theorem V.6.2.)

Theorem 3.3. *If f and g are functions having a common domain $\mathfrak{D}$ and each is continuous at x_0, then so are their sum $f+g$ and product fg.*

PROB. 3.3. Prove Theorem 3.3.

PROB. 3.4. Prove: If f is continuous at x_0 and c is some constant, then cf is continuous at x_0.

PROB. 3.5. Let $f_1, \ldots, f_n$ be n functions with a common domain $\mathfrak{D}$. If all the f_i are continuous at x_0, prove that (a) $f_1 + \cdots + f_n$ and (b) $f_1 f_2 \ldots f_n$ are continuous at x_0.

Def. 3.1. If f and g are functions with a common domain $\mathfrak{D}$, and g is not identically 0 on $\mathfrak{D}$, then we define f/g to be the function Q, where

$$Q(x) = \frac{f(x)}{g(x)} \qquad \text{for} \quad x \in \mathfrak{D} \quad \text{such that} \quad g(x) \neq 0.$$

EXAMPLE 3.1. Since a rational function in $\mathbb{R}$ is a quotient P/Q, where P and Q are polynomials and Q is not the zero polynomial, we know that the rational function R, where

$$R(x) = \frac{P(x)}{Q(x)} \qquad \text{for } x \text{ such that} \quad Q(x) \neq 0,$$

has a nonempty domain of definition (Prob. V.8.4). Polynomials are continuous functions. Hence, if $x_0 \in \mathbb{R}$ is such that $Q(x_0) \neq 0$, some δ-neighborhood $N(x_0,\delta)$ of x_0 exists such that $Q(x) \neq 0$ for $x \in N(x_0,\delta) \cap \mathbb{R} = N(x_0,\delta)$ (Theorem 3.2). Thus, there exists an open interval $(x_0 - \delta; x_0 + \delta)$ such that $Q(x) \neq 0$ for $x \in (x_0 - \delta; x_0 + \delta)$. This implies that x_0 is an accumulation point of $\mathfrak{D}(R)$ (explain) and

$$\lim_{x\to x_0} R(x) = \lim_{x\to x_0} \frac{P(x)}{Q(x)} = \frac{\lim_{x\to x_0} P(x)}{\lim_{x\to x_0} Q(x)} = \frac{P(x_0)}{Q(x_0)} = R(x_0).$$

This tells us that a rational function on $\mathbb{R}$ is continuous since it is continuous for each x_0 in its domain. (This is a special case of Theorem 3.4 below.) Note that each point of the domain of a rational function is one of its accumulation points.

Theorem 3.4. *If f and g are functions having a common domain $\mathfrak{D}$, where each is continuous at x_0 and $g(x_0) \neq 0$, then f/g is continuous at x_0.*

Prob. 3.6. Prove Theorem 3.4.

Prob. 3.7. Prove that the hyperbolic functions are continuous.

4. The Intermediate-Value Theorem

Here we encounter our first global property of continuous functions. It is a property of functions continuous on intervals.

Lemma 4.1. *If f is a real-valued function of a real variable which is continuous on an interval I, and if for some a and b in I we have $f(a) < 0 < f(b)$ or $f(b) < 0 < f(a)$, then there exists a c between a and b such that $f(c) = 0$.*

Proof. For the sake of definiteness let $a < b$ and $f(a) < 0 < f(b)$. Since I is an interval, it is a convex set of real numbers. Hence $[a,b] \subseteq I$, and f is continuous on $[a,b]$. We restrict f to the interval $[a,b]$. Since $f(a) < 0$, there exists a $\delta_1 > 0$ such that $f(x) < 0$ for $x \in [a, a+\delta_1) \cap [a,b]$ (Theorem 3.2), and since $f(b) > 0$, it follows that $a < a + \delta_1 \leqslant b$. Similarly, again by Theorem 3.2, there exists a $\delta_2 > 0$ such that for $x \in (b - \delta_2, b] \cap [a,b]$ we have $f(x) > 0$ and $a \leqslant b - \delta_2 < b$. Clearly, $a + \delta_1 \leqslant b - \delta_2$ (otherwise $a \leqslant b - \delta_2 < a + \delta_1 \leqslant b$ and for $b - \delta_2 < x < a + \delta_1$ we would have $f(x) > 0$ and $f(x) < 0$—an impossibility). Define the set S as

$$S = \{x \in [a,b] \mid f(x) > 0\}. \tag{4.1}$$

Since $S \subseteq [a,b]$, it is bounded from below, and since $f(b) > 0$, $S \neq \emptyset$. Hence, S has a real infimum. Let $c = \inf S$. Now $(b - \delta_2, b] \subseteq S$, and, therefore, $c = \inf S \leqslant \inf(b - \delta_2, b]$ (Prob. I.12.1). Since $\inf(b - \delta_2, b] = b - \delta_2$, we have $c \leqslant b - \delta_2 < b$. But $x \in [a, a + \delta_1)$ implies that $f(x) < 0$. In turn, this implies that $a + \delta_1$ is a lower bound for S (explain), and $a + \delta_1 \leqslant c$. Thus, $a < a + \delta_1 \leqslant c \leqslant b - \delta_2 < b$, and, therefore, $a < c < b$. Now observe that if $a \leqslant x < c$, then $x \notin S$. Since such an x is in $[a,b]$, it follows that $f(x) \leqslant 0$. In short, $a \leqslant x < c$ implies that $f(x) \leqslant 0$.

If $f(c) > 0$, then, because of the continuity of f at c, we know (Theorem 3.2) that there is an ϵ-neighborhood $(c - \epsilon; c + \epsilon)$ of c such that $x \in (c - \epsilon;$

$c+\epsilon) \cap [a,b]$ implies $f(x)>0$. Accordingly, if $x \in [a,b]$ and $c-\epsilon<x<c$, we have $f(x)>0$. By the last sentence of the last paragraph that is impossible. Hence, $f(c) \leqslant 0$. If $f(c)<0$, then there is an ϵ_1-neighborhood $(c-\epsilon_1; c+\epsilon_1)$ of c such that $x \in (c-\epsilon_1; c+\epsilon_1) \cap [a;b]$ implies $f(x)<0$. This and the last sentence of the last paragraph imply that if $x \in [a,b]$ and $a \leqslant x < c+\epsilon_1$, then $f(x) \leqslant 0$. Hence, $x \in S$ implies $x \geqslant c+\epsilon_1$. But then $c+\epsilon_1$ is a lower bound for S. This is impossible, for $c+\epsilon_1>c$, and, by definition, c is the *greatest* lower bound for S. Hence, $f(c) \geqslant 0$. Since we already showed that $f(c) \leqslant 0$ holds, we must have $f(c)=0$. This completes the proof.

Theorem 4.1 (The Intermediate-Value Theorem). *If f is continuous on an interval I and $f(a) \neq f(b)$ holds for some a and b in I, then for each y between $f(a)$ and $f(b)$, there exists a c between a and b such that $f(c)=y$.*

PROOF. We have

$$f(a)<y<f(b) \quad \text{or} \quad f(a)>y>f(b). \tag{4.2}$$

Define g on I as follows:

$$g(x)=f(x)-y \qquad \text{for} \quad x \in I. \tag{4.3}$$

This function g is continuous on I, and, because of (4.2), we have

$$g(a)<0<g(b) \quad \text{or} \quad g(a)>0>g(b). \tag{4.4}$$

Thus, g satisfies the hypothesis of Lemma 4.1. But then there exists a c between a and b such that $g(c)=0$. This implies that

$$f(c)-y=0,$$

i.e., $f(c)=y$, where c is between a and b. The proof is now complete.

This theorem is often phrased as follows.

Intermediate-Value Theorem. *A real-valued function of a real variable which is continuous on an interval assumes every value between any two of its values.*

Remark 4.1. The c of Theorem 4.1 is by no means unique. Theorem 4.1 merely asserts its existence.

When a real-valued function assumes every value between any two of its values we say that it has the *intermediate-value property*. Functions which are continuous on intervals have the intermediate-value property, but the converse is false. There exist functions defined on intervals having the intermediate-value property which are not continuous. We present an example.

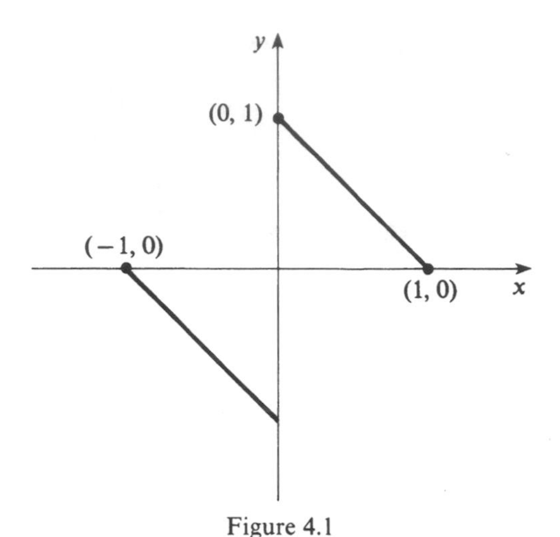

Figure 4.1

EXAMPLE 4.1. Let $f\colon [-1, 1] \to \mathbb{R}$ be defined as:

$$\begin{aligned} f(x) &= 1 - x \qquad &\text{if} \quad 0 \leqslant x \leqslant 1 \\ &= -1 - x \qquad &\text{if} \quad -1 \leqslant x < 0 \end{aligned}$$

(see Fig. 4.1). This function is defined on the interval $[-1, 1]$ has the intermediate-value property there but is not continuous (see Remark 4.3 and Probs. 4.6 and 4.7).

PROB. 4.1. Prove: A real-valued function has the intermediate-value property if and only if its range is a convex set (Def. V.1.1). Since a nonempty convex set of real numbers is either a point or an interval (Prob. V.1.6, part (b)), this result states that a real-valued function with nonempty domain has the intermediate-value property if and only if its range is either a point or an interval.

Remark 4.2. Because of the result stated in the last problem, alternate formulations of the intermediate-value theorem (Theorem 4.1) are: (a) a real-valued function of a real variable that is continuous on an interval I has a convex range $f(I)$; (b) if a real-valued function of a real variable is continuous on an interval I, then its range $f(I)$ is either a point or an interval.

PROB. 4.2. Prove: If P is a polynomial function in $\mathbb{R}$ of odd degree, then it has a real zero, i.e., a real r exists such that $P(r) = 0$.

EXAMPLE 4.2. Although we have already accumulated much information about the sine and cosine functions, we have no information about the zeros of these functions. We use the intermediate-value theorem to prove that the cosine function has real zeros.

We recall that if $|x| \leqslant 1$, then $\cos x > 0$ (Prob. IV.8.5). Thus, $\cos 1 > 0$. We prove that $\cos 2 < 0$. By definition,

$$\cos 2 = \sum_{n=0}^{\infty} (-1)^n \frac{2^{2n}}{(2n)!} = 1 - \frac{2^2}{2!} + \frac{2^4}{4!} - \frac{2^6}{6!} + \cdots$$

$$= -1 + \frac{2^4}{4!} - \frac{2^6}{6!} + \cdots . \tag{4.5}$$

Therefore,

$$1 + \cos 2 = \frac{2^4}{4!} - \frac{2^6}{6!} + \cdots . \tag{4.6}$$

We now prove that the series on the right is alternating. We write it in the form $\sum_{n=1}^{\infty}(-1)^{n+1}a_n$, where

$$a_n = \frac{2^{2n+2}}{(2n+2)!} \qquad \text{for each positive integer } n. \tag{4.7}$$

We have: (1) $a_n > 0$ for each n and (2) $a_n \to 0$ as $n \to +\infty$. The last holds because the series (4.5), and, therefore, the series (4.6), is converging. To complete the proof that the series in (4.6) is alternating we show that (3) $a_{n+1} < a_n$ for each n. Note that

$$0 < \frac{a_{n+1}}{a_n} = \frac{2^{2n+4}}{(2n+4)!} \Big/ \frac{2^{2n+2}}{(2n+2)!} = \frac{2}{(n+2)(2n+3)} < 1. \tag{4.8}$$

The inequality on the right is a consequence of

$$2n^2 + 7n + 6 > 2 \qquad \text{for} \quad n \geqslant 1.$$

Thus, (3) $0 < a_{n+1} < a_n$ for each n and the series in (4.6) is alternating. By Prob. IV.2.1,

$$0 < a_2 - a_3 + a_4 - a_5 + \cdots$$

Hence,

$$a_1 - a_2 + a_3 - a_4 + \cdots < a_1 = \frac{2^4}{4!} = \frac{2}{3} .$$

This and (4.6) yield

$$1 + \cos 2 < \tfrac{2}{3},$$

from which it follows that

$$\cos 2 < -\tfrac{1}{3} . \tag{4.8'}$$

Thus, $\cos 2 < 0 < \cos 1$. Since cosine is continuous on $\mathbb{R}$, the intermediate-value theorem tells that there exists a real c such that $1 < c < 2$ and $\cos c = 0$.

We now prove that the c such that $1 < c < 2$ and $\cos c = 0$ is unique. Since $\sin^2 c + \cos^2 c = 1$, $\sin^2 c = 1$ and, hence, $\sin c = \pm 1$. We prove that $\sin c = 1$. We do this by first proving that if $0 < x < \sqrt{6}$, then $\sin x > 0$. We use the definition of $\sin x$ and obtain, after reindexing,

$$\sin x = \sum_{n=1}^{\infty} (-1)^{n+1} \frac{x^{2n-1}}{(2n-1)!}. \tag{4.9}$$

We put

$$a_n = \frac{x^{2n-1}}{(2n-1)!} \qquad \text{for} \quad n \geqslant 1 \tag{4.10}$$

so that

$$\sin x = \sum_{n=1}^{\infty} (-1)^{n+1} a_n. \tag{4.11}$$

We take x such that $0 < x < \sqrt{6}$ and note that (1) $a_n > 0$ for all n and (2) $\lim_{n\to+\infty} a_n = 0$ (for (4.11) converges). We also have

$$0 < \frac{a_{n+1}}{a_n} = \frac{x^{2n+1}}{(2n+1)!} \Big/ \frac{x^{2n-1}}{(2n-1)!} = \frac{x^2}{(2n)(2n+1)} < \frac{6}{(2n)(2n+1)} \leqslant 1.$$

Here the last inequality follows from $(2n)(2n+1) \geqslant 6$ for $n \geqslant 1$. Thus, (3) $a_{n+1} < a_n$ for all n. We proved that if $0 < x < \sqrt{6}$, then the series (4.9) is alternating. Therefore,

$$0 < a_1 - a_2 + a_3 - a_4 + \cdots = \sin x \qquad \text{for} \quad 0 < x < \sqrt{6}. \tag{4.12}$$

Since $1 < c < 2 < \sqrt{6}$, it follows that $\sin c > 0$. This and $\sin c = \pm 1$ yield $\sin c = 1$. As a by-product we have

$$\sin x > 0 \qquad \text{if} \quad 0 < x \leqslant c. \tag{4.13}$$

Also

$$\sin(c - x) = \sin c \cos x - \cos c \sin x = \cos x,$$

i.e.,

$$\sin(c - x) = \cos x \qquad \text{for all} \quad x \in \mathbb{R}. \tag{4.14}$$

Now, if $0 \leqslant x < c$, then $0 < c - x \leqslant c$. It follows from this and (4.13) that $\sin(c - x) > 0$ and, hence, from (4.14), that $\cos x > 0$ for $0 \leqslant x < c$.

Finally, we show the uniqueness of c such that $1 < c < 2$ and $\cos c = 0$. To this end we take x such that $c < x \leqslant 2c$ so that $0 < x - c \leqslant c$. By (4.14), for such x, $-\cos x = -\sin(c - x) = \sin(x - c) > 0$. Thus,

$$\cos x < 0 \qquad \text{if} \quad c < x \leqslant 2c. \tag{4.15}$$

But $c < 2 < 2c$. It follows that if $c < x \leqslant 2$, then $\cos x < 0$. Since we also proved $\cos x > 0$ for $0 \leqslant x < c$, we have $\cos x \neq 0$ for $0 \leqslant x < c$ or $c < x \leqslant 2$. We conclude that the c such that $1 < c < 2$ and $\cos c = 0$ is unique. This c is the least positive x such that $\cos x = 0$.

Def. 4.1. We define $\pi = 2c$, where $1 < c < 2$ and $\cos c = 0$. Accordingly, $\cos(\pi/2) = 0$ and $2 < \pi < 4$.

We summarize our results in:

Theorem 4.2. *The following hold*:

(a) $2 < \pi < 4$ *and* $\cos(\pi/2) = 0$.
(b) π *is the only real number such that* (a) *holds*.
(c) $\sin(\pi/2) = 1$.
(d) *If* $0 < x \leqslant \pi/2$, *then* $\sin x > 0$.
(e) *If* $0 \leqslant x < \pi/2$, *then* $\cos x > 0$.
(f) *If* $x \in \mathbb{R}$, *then* $\sin(\pi/2 - x) = \cos x$.
(g) *If* $\pi/2 < x \leqslant \pi$, *then* $\cos x < 0$.
(h) $\pi/2$ *is the least positive* x *such that* $\cos x = 0$.

Prob. 4.3. Prove: (a) $\sin(-\pi/2) = -1$ and $\cos(-\pi/2) = 0$. (b) If $|x| < \pi/2$, then $\cos x > 0$. (c) If $x \in \mathbb{R}$, then $\cos(\pi/2 - x) = \sin x$. (d) If $\pi/2 \leqslant x < \pi$, then $\sin x > 0$. This, together with part (d) of Theorem 4.2, yields: If $0 < x < \pi$, then $\sin x > 0$. (e) If $-\pi < x < 0$, then $\sin x < 0$.

Prob. 4.4. Prove: (a) $\sin(\pm\pi) = 0$ and $\cos(+\pi) = -1$. (b) $\sin(3\pi/2) = -1$ and $\cos(\pm 3\pi/2) = 0$. (c) $\cos 2\pi = 1$ and $\sin 2\pi = 0$. (d) $\cos(x + 2\pi) = \cos x$ and $\sin(x + 2\pi) = \sin x$ for $x \in \mathbb{R}$, (e) $\sin(x \pm \pi) = -\sin x$ and $\cos(x \pm \pi) = -\cos x$ for $x \in \mathbb{R}$.

Prob. 4.5. Prove:

(a) $\sin(\pi/4) = 1/\sqrt{2} = \cos \pi/4$,
(b) $\cos(\pi/3) = \frac{1}{2} = \sin(\pi/6)$,
(c) $\cos(\pi/6) = \sqrt{3}/2 = \sin(\pi/3)$,
(d) $\cos(\pi/12) = (\sqrt{6} + \sqrt{2})/4 = \sqrt{2 + \sqrt{3}}/2$,
(e) $\sin(\pi/12) = (\sqrt{6} - \sqrt{2})/4 = \sqrt{2 - \sqrt{3}}/2$,
(f) $\tan(\pi/12) = 2 - \sqrt{3}$.

Remark 4.3. In connection with the intermediate value theorem, we introduce the following definition: A function will be said to have the *strong intermediate value property* on an interval I if it has the intermediate value property on every closed, bounded subinterval of I. For example, a function continuous on an interval I has the strong intermediate value property on I (why?). On the other hand, the function defined in Example 4.1, cited as an example of a function having the intermediate value property which is not continuous, does not have the strong intermediate value property (show this).

Prob. 4.6. (a) Let f be a function which has the *strong* intermediate value property on the interval $(a,b]$. Prove: If f is also strictly monotonically increasing on $(a; b)$, then f is strictly monotonically increasing on $(a, b]$. (b) More generally, let f be a function having the *strong* intermediate value property on an interval I which includes one of its endpoints, say c. Prove: If f is also strictly monotonic on the interval $J = I - \{c\}$, then f is strictly monotonic on I.

PROB. 4.7. Let f have the strong intermediate value property on an interval I. Prove: If f is also monotonic in the interior (the set of all interior points) of I, then f is continuous and monotonic on I.

5. The Natural Logarithm: Logs to Any Base

Let E: $\mathbb{R} \rightarrow$ be the function defined as $E(x) = e^x$ for $x \in \mathbb{R}$. We prove:

Theorem 5.1. *The range of E is* $(0; +\infty)$.

PROOF. We know that $E(x) = e^x > 0$ so that $\mathcal{R}(E) \subseteq (0; +\infty)$. We wish to prove that this subset relation can be reversed: that is, that $(0; +\infty) \subseteq \mathcal{R}(E)$. Assume $y \in (0; +\infty)$ so that $y > 0$. By Theorem V.7.3,

$$e^y \geqslant 1 + y > y > 0 \tag{5.1}$$

and

$$e^{1/y} \geqslant 1 + \frac{1}{y} > \frac{1}{y} > 0. \tag{5.2}$$

Taking reciprocals, we have

$$e^{-(1/y)} < y. \tag{5.3}$$

Thus, there exist real numbers $b = y$ and $a = -1/y$ such that

$$e^a < y < e^b. \tag{5.4}$$

Since E is continuous, (5.4) and the intermediate-value theorem imply that there is an x between a and b such that $E(x) = e^x = y$, and hence that $y \in \mathcal{R}(E)$. This proves that $(0; +\infty) \subseteq \mathcal{R}(E)$. This and the first sentence of the proof yield $\mathcal{R}(E) = (0; +\infty)$, as claimed.

Since $e > 1$, part (i) of Remark IV.10.3 shows that the function E: $\mathbb{R} \rightarrow \mathbb{R}$ such that $E(x) = e^x$ for $x \in \mathbb{R}$ is strictly monotonically increasing. Therefore, E has a strictly monotonic increasing inverse E^{-1} defined on its range $\mathcal{R}(E) = (0; +\infty)$, whose range is $\mathbb{R}$, the domain of E (Theorem II.11.1).

Def. 5.1. The inverse of the function E is called the *natural logarithm* function. If $x > 0$, then the unique y such that $E(y) = e^y = x$ is called the *natural logarithm* of x and is written

$$y = \ln x. \tag{5.5}$$

Theorem 5.2. *The domain of the natural logarithm function is* $(0; +\infty)$ *and its range is* $\mathbb{R}$. *We have*

$$e^{\ln x} = x \qquad \text{if} \quad x > 0 \tag{5.6}$$

and

$$\ln e^{y} = y \qquad \text{if} \quad y \in \mathbb{R}. \tag{5.7}$$

ln *is also a strictly monotonic increasing function.*

PROOF. The first statement is a consequence of the definition of the natural logarithm as the inverse of E. By properties of the inverse we have

$$e^{\ln x} = E\big(E^{-1}(x)\big) = x \qquad \text{for} \quad x \in (0; +\infty) = \mathfrak{R}(E) = \mathfrak{D}(\ln)$$

and

$$\ln e^{y} = E^{-1}(E(y)) = y \qquad \text{for} \quad y \in \mathbb{R} = \mathfrak{D}(E) = \mathfrak{R}(\ln).$$

This proves (5.6) and (5.7). The function ln is strictly monotonically increasing since it is the inverse of the strictly increasing function E.

Theorem 5.3. *We have*: (a) $\ln 1 = 0$, (b) $\ln e = 1$, *and* (c) $\ln x < 0$ *for* $0 < x < 1$ *and* $\ln x > 0$ *for* $x > 1$.

PROOF. $\ln 1 = 0$ follows from $e^{0} = 1$ and the definition of ln. Since $e^{1} = e$, $\ln e = 1$ for the same reason. In view of the strictly increasing character of ln,

$$0 < x < 1 \quad \text{implies that} \quad \ln x < \ln 1 = 0$$

and

$$x > 1 \quad \text{implies that} \quad \ln x > \ln 1 = 0.$$

Theorem 5.4. *If* $a > 0$ *and* $b > 0$, *then*

(a) $\ln(ab) = \ln a + \ln b$,
(b) $\ln(1/a) = -\ln a$,
(c) $\ln(a/b) = \ln a - \ln b$,
(d) $\ln a^{\alpha} = \alpha \ln a$ for $\alpha \in \mathbb{R}$.

PROOF. We prove (a). Let $u = \ln a$ and $v = \ln b$. Then

$$e^{u} = a \quad \text{and} \quad e^{v} = b$$

and, hence,

$$ab = e^{u}e^{v} = e^{u+v}.$$

By the definition of $\ln(ab)$, this implies that

$$\ln(ab) = u + v = \ln a + \ln b.$$

This proves (a). We prove (d) next. Since $a > 0$, we have

$$e^{\ln a} = a.$$

Hence, for $\alpha \in \mathbb{R}$,

$$e^{\alpha \ln a} = \left(e^{\ln a}\right)^{\alpha} = a^{\alpha}.$$

Taking the natural logarithm of the left- and right-hand sides, we obtain

$$\alpha \ln a = \ln e^{\alpha \ln a} = \ln a^{\alpha}.$$

This proves (d). We leave the proofs of (b) and (c) to the reader (Prob. 5.1).

PROB. 5.1. Complete the proof of the last theorem by proving parts (b) and (c).

Theorem 5.5. *We have*

(a)
$$\lim_{x \to +\infty} \ln x = +\infty$$

and

(b)
$$\lim_{x \to 0+} \ln x = -\infty.$$

PROOF. Take $B \in \mathbb{R}$ and $X \geqslant e^{B}$. Note that $e^{B} > 0$. If $x > X$, then $x > e^{B}$. Since ln is strictly increasing, this implies that

$$\ln x > \ln e^{B} = B.$$

Thus, for each B, there exists an X such that if $x > X$, then $\ln x > B$. This proves (a).

We prove (b). Given B, we have $e^{B} > 0$. Take δ such that $0 < \delta \leqslant e^{B}$ and $0 < x < \delta$. We have $0 < x < e^{B}$ and, therefore, $\ln x < \ln e^{B} < B$. Thus, for each $B \in \mathbb{R}$, there exists a $\delta > 0$ such that if $0 < x < \delta$, then $\ln x < B$. This proves (b).

Remark 5.1. It is important to note that if $p > 0$, then

$$p^{x} = \left(e^{\ln p}\right)^{x} = e^{x \ln p}. \tag{5.8}$$

PROB. 5.2. Prove:

(a) If $p > 1$, then $\lim_{x \to +\infty} p^{x} = +\infty$ and $\lim_{x \to -\infty} p^{x} = 0$;
(b) If $0 < p < 1$, then $\lim_{x \to +\infty} p^{x} = 0$ and $\lim_{x \to -\infty} p^{x} = +\infty$ (see Remark 5.1).

PROB. 5.3. Let $p > 0$ and $E_p\colon \mathbb{R} \to \mathbb{R}$ be the function $E_p(x) = p^{x}$ for $x \in \mathbb{R}$. Prove: E_p is continuous. Note that E_p is strictly monotonically increasing if $p > 1$ and strictly monotonically decreasing if $0 < p < 1$.

PROB. 5.4. Prove: $E_p\colon \mathbb{R} \to \mathbb{R}$, where $p > 0$, $p \neq 1$ has the range $(0; +\infty)$, and has an inverse E^{-1} that is strictly monotonically increasing for $p > 1$ and strictly monotonically decreasing for $0 < p < 1$.

Def. 5.2. If $p > 0$, $p \neq 1$, then the inverse E_p^{-1} of E_p is called the *logarithm to the base p*. If $x > 0$, then y such that

$$p^y = x \tag{5.9}$$

is called the *logarithm to the base p of x* and is written

$$y = \log_p x. \tag{5.10}$$

We have, as a special case,

$$\log_e x = \ln x \qquad \text{for} \quad x > 0. \tag{5.11}$$

Thus, the natural logarithm of $x > 0$ is the logarithm to base e of x.

Prob. 5.5. Prove: If $p > 0$, $p \neq 1$, then the domain of $\log_p$ is $(0; +\infty)$ and its range is $\mathbb{R}$. Also,

$$p^{\log_p x} = x \qquad \text{if} \quad x > 0$$

and

$$\log_p p^y = y \qquad \text{if} \quad y \in \mathbb{R}.$$

Prob. 5.6. Prove: If $p > 0$, $p \neq 1$, then (a) $\log_p 1 = 0$ and (b) $\log_p p = 1$.

Prob. 5.7. Prove: If $p > 0$, $p \neq 1$, then

$$\log_p x = \frac{\ln x}{\ln p} \qquad \text{for} \quad x > 0$$

and $(\log_p e)(\ln p) = 1$.

Prob. 5.8. Prove: If $p > 1$, then $\log_p$ is strictly increasing and

(a) $0 < x < 1$ implies $\log_p x < 0$,
(b) $x > 1$ implies $\log_p x > 0$.

Prob. 5.9. Prove: If $0 < p < 1$, then $\log_p$ is strictly decreasing and

(a) $0 < x < 1$ implies $\log_p x > 0$,
(b) $x > 1$ implies $\log_p x < 0$.

Prob. 5.10. Prove: If $p > 0$, $p \neq 1$, then $a > 0$ and $b > 0$ imply

(a) $\log_p(ab) = \log_p a + \log_p b$,
(b) $\log_p(1/a) = -\log_p a$,
(c) $\log_p(a/b) = \log_p a - \log_p b$,
(d) $\log_p a^\alpha = \alpha \log_p a$ if $\alpha \in \mathbb{R}$.

Prob. 5.11. Prove: If $p > 1$, then

(a)
$$\lim_{x \to +\infty} \log_p x = +\infty$$

and

(b) $$\lim_{x\to 0+} \log_p x = -\infty.$$

PROB. 5.12. Prove: If $0 < p < 1$, then

(a) $$\lim_{x\to +\infty} \log_p x = -\infty$$

and

(b) $$\lim_{x\to 0+} \log_p x = +\infty.$$

Theorem 5.6. *If $h > -1$, then*

$$\frac{h}{1+h} \leqslant \ln(1+h) \leqslant h. \tag{5.12}$$

PROOF. By Theorem V.7.3.,

$$1 + z \leqslant e^z \leqslant 1 + ze^z \qquad \text{for} \quad z \in \mathbb{R}. \tag{5.13}$$

Since $h > -1$ by hypothesis, we have $h + 1 > 0$. Let $z = \ln(1 + h)$ so that

$$e^z = 1 + h,$$

and, therefore,

$$h = e^z - 1.$$

Now use (5.13) in the form $z \leqslant e^z - 1 \leqslant ze^z$ to obtain

$$\ln(1+h) = z \leqslant e^z - 1 = h \leqslant ze^z = (\ln(1+h))(1+h),$$

i.e.,

$$\ln(1+h) \leqslant h \leqslant (1+h)\ln(1+h) \qquad \text{for} \quad h > -1. \tag{5.14}$$

If $h > 0$, then $\ln(1 + h) > 0$, so (5.14) yields

$$1 \leqslant \frac{h}{\ln(1+h)} \leqslant 1 + h.$$

Taking reciprocals, we obtain

$$\frac{1}{1+h} \leqslant \frac{\ln(1+h)}{h} \leqslant 1 \qquad \text{for} \quad h > 0,$$

from which (5.12) follows after multiplying by h. This establishes (5.12) for $h > 0$. If $-1 < h < 0$, then $0 < 1 + h < 1$ and $\ln(1 + h) < 0$. Multiplication by $1/\ln(1 + h)$ reverses the inequalities (in (5.14)) and we obtain

$$1 \geqslant \frac{h}{\ln(1+h)} \geqslant 1 + h > 0 \qquad \text{for} \quad -1 < h < 0.$$

Taking reciprocals, we obtain

$$1 \leqslant \frac{\ln(1+h)}{h} \leqslant \frac{1}{1+h} \qquad \text{for} \quad -1 < h < 0.$$

Inequality (5.12) now follows after we multiply the last inequalities by h ($h < 0$). Thus, (5.12) holds for $-1 < h < 0$ also. If $h = 0$, the conclusion holds trivially.

PROB. 5.13. Prove:

(a)
$$\lim_{h\to 0} \ln(1+h) = 0,$$

(b)
$$\lim_{h\to 0} \frac{\ln(1+h)}{h} = 1.$$

PROB. 5.14. Prove: $\lim_{h\to 0}(1+h)^{1/h} = e$.

Theorem 5.7. *If $x \neq 0$, then*

(a)
$$\lim_{h\to 0} \ln\left(1+\frac{h}{x}\right) = 0,$$

(b)
$$\lim_{h\to 0} \frac{1}{h}\ln\left(1+\frac{h}{x}\right) = \frac{1}{x}.$$

Let $x > 0$. If $h > -x$, then

(c)
$$\lim_{h\to 0} \ln(x+h) = \ln x,$$

(d)
$$\lim_{h\to 0} \frac{\ln(x+h) - \ln x}{h} = \frac{1}{x}.$$

PROOF. If $x > 0$, take $h > -x$ and obtain $h/x > -1$. If $x < 0$, take $h < -x$ and obtain again $h/x > -1$. By Theorem 5.6, if $h/x > -1$, then

$$\frac{h}{x+h} = \frac{h/x}{1+h/x} \leqslant \ln\left(1+\frac{h}{x}\right) \leqslant \frac{h}{x}. \tag{5.15}$$

Here $h/x > -1$; if $x > 0$ and $h > -x$, or $x < 0$ and $h < -x$. Letting $h \to 0$, we obtain (a).

If $h > 0$, we obtain from (5.15)

$$\frac{1}{x+h} \leqslant \frac{1}{h}\ln\left(1+\frac{h}{x}\right) \leqslant \frac{1}{x} \qquad \text{if} \quad \frac{h}{x} > -1, \tag{5.16}$$

from which it follows that

$$\lim_{h\to 0+} \frac{1}{h}\ln\left(1+\frac{h}{x}\right) = \frac{1}{x}. \tag{5.17}$$

If $h < 0$, then multiplication by $1/h$ reverses the inequalities (5.15), so

$$\frac{1}{x+h} \geqslant \frac{1}{h}\ln\left(1+\frac{h}{x}\right) \geqslant \frac{1}{x} \qquad \text{for} \quad \frac{h}{x} > -1. \tag{5.18}$$

It follows that

$$\lim_{h\to 0-} \frac{1}{h}\ln\left(1+\frac{h}{x}\right) = \frac{1}{x}. \tag{5.19}$$

Equations (5.17) and (5.19) yield (b) if $x \neq 0$.

If $x > 0$, then $h > -x$ implies $x + h > 0$, and we may write

$$\ln\left(1 + \frac{h}{x}\right) = \ln(x + h) - \ln x.$$

So (5.15) can be written

$$\frac{h}{x + h} \leqslant \ln(x + h) - \ln x \leqslant \frac{h}{x}.$$

This implies (c). Part (d) follows from (b) when $x > 0$ and $h > -x$ for then (b) can be written

$$\lim_{h \to 0} \frac{\ln(x + h) - \ln x}{h} = \frac{1}{x}.$$

Remark 5.2. It follows from part (c) of the last theorem that ln is a continuous function.

PROB. 5.15. Prove: If $x \in \mathbb{R}$, then

(a)
$$\lim_{h \to 0} (1 + hx)^{1/h} = e^x$$

and

(b)
$$\lim_{h \to +\infty} \left(1 + \frac{x}{h}\right)^h = e^x.$$

PROB. 5.16. Prove: If $x > 1$, then

$$0 < \ln x \leqslant 2(\sqrt{x} - 1).$$

PROB. 5.17. Prove: $\lim_{x \to +\infty}(\ln x / x) = 0$ (Hint: use Prob. 5.16).

PROB. 5.18. Assume $\alpha > 0$. Prove:

(a)
$$\lim_{x \to +\infty} x^\alpha = +\infty,$$

(b)
$$\lim_{x \to +\infty} \frac{\ln x^\alpha}{x^\alpha} = 0,$$

(c)
$$\lim_{x \to +\infty} \frac{\ln x}{x^\alpha} = 0.$$

PROB. 5.19. Prove:

(a) $\lim_{x \to 0+} (x \ln x) = 0$,
(b) If $\alpha > 0$, then $\lim_{x \to 0+} x^\alpha \ln x = 0$.

PROB. 5.20. Let $\langle b_n \rangle$ be a sequence of real numbers such that $b_n \to b \in \mathbb{R}$ as $n \to +\infty$. Prove:

(a) $n \ln(1 + b_n/n) \to b$ as $n \to +\infty$,
(b) $(1 + b_n/n)^n \to e^b$ as $n \to +\infty$,
(c) $(1 - b_n/n)^n \to e^{-b}$ as $n \to +\infty$.

Remark 5.3. It follows from Prob. 5.20 that if $\langle b_n \rangle$ is a sequence of real numbers such that $b_n \to b \in \mathbb{R}$, then for any positive integer k we have

$$\lim_{n\to+\infty} \binom{n}{k}\left(\frac{b_n}{n}\right)^k\left(1-\frac{b_n}{n}\right)^{n-k} = \frac{1}{k!}b^k e^{-b}. \tag{$*$}$$

We prove this. Note that

$$\begin{aligned}\binom{n}{k}\left(\frac{b_n}{n}\right)^k\left(1-\frac{b_n}{n}\right)^{n-k} &= \frac{n(n-1)\ldots(n-k+1)}{k!\,n^k}b_n^k\frac{(1-b_n/n)^n}{(1-b_n/n)^k}\\ &= \frac{1}{k!}\left(1-\frac{1}{n}\right)\left(1-\frac{2}{n}\right)\ldots\left(1-\frac{k-1}{n}\right)\frac{b_n^k}{(1-b_n/n)^k}\left(1-\frac{b_n}{n}\right)^n.\end{aligned}$$

Hence,

$$\binom{n}{k}\left(\frac{b_n}{n}\right)^k\left(1-\frac{b_n}{n}\right)^{n-k} = C_{k,n}\left(1-\frac{b_n}{n}\right)^n, \tag{5.19$'$}$$

where

$$C_{k,n} = \frac{1}{k!}\left(1-\frac{1}{n}\right)\left(1-\frac{2}{n}\right)\ldots\left(1-\frac{k-1}{n}\right)\frac{b_n^k}{(1-b_n/n)^k}. \tag{5.20}$$

Since k is fixed, we obtain

$$C_{k,n} \to \frac{b^k}{k!} \quad \text{as} \quad n\to+\infty. \tag{5.21}$$

This and Prob. 5.20, part (c), together with (5.19′), yield $(*)$.

The limit in (5.18) is useful in the theory of probability.

Prob. 5.21. Prove: If $b > 0$ and k is a positive integer and

$$p_k = \frac{1}{k!}b^k e^{-b} \qquad \text{for each positive integer } k,$$

then (1) $0 < p_k < 1$ for each k and (2) $\sum_{k=0}^{\infty} p_k = 1$.

Prob. 5.22. Prove: If $\langle a_n \rangle$ is a real sequence such that

(a) $$a_n > 0 \qquad \text{for all} \quad n$$

and

(b) $$a_n \to a \quad \text{as} \quad n\to+\infty, \qquad \text{where} \quad a>0,$$

then $\lim_{n\to+\infty}(a_1 a_2 \ldots a_n)^{1/n} = a$ (Hint: use the properties of natural logarithms and Example III.9.1).

PROB. 5.23. In Prob. IV.5.7 we asked the reader to prove

$$\lim_{n \to +\infty} \frac{\sqrt[n]{n!}}{n} = \frac{1}{e}.$$

As a hint we suggested that the reader use Prob. IV.5.5. This limit may also be evaluated by using the result cited in Prob. 5.22 above. Thus, we ask the reader to prove that

$$\lim_{n \to +\infty} \frac{n}{\sqrt[n]{n!}} = e$$

by constructing the sequence $\langle P_n \rangle$, where

$$P_n = \left(1 + \frac{1}{n}\right)^n \qquad \text{for each positive integer } n,$$

by observing that

$$P_1 P_2 \cdots P_n = \frac{(n+1)^n}{n!} = \frac{n^n}{n!}\left(1 + \frac{1}{n}\right)^n \qquad \text{for each } n,$$

and by using Prob. 5.22. It is also useful to examine Example III.6.3.

PROB. 5.24*. Prove: If $a > 0$, $x > 0$, then

$$e^x \geqslant \left(\frac{xe}{a}\right)^a, \qquad e^{-x} \leqslant \left(\frac{a}{ex}\right)^a$$

and

$$\ln x \leqslant \frac{a}{e} x^{1/a}, \qquad -\ln x \leqslant \frac{1}{eax^a}.$$

6. Bolzano–Weierstrass Theorem and Some Consequences

Def. 6.1. If $\langle x_n \rangle$ is a real sequence, then $c \in \mathbb{R}$ is called a *cluster point* of the sequence if and only if there exists a subsequence $\langle x_{n_k} \rangle$ of $\langle x_n \rangle$ which converges to c.

EXAMPLE 6.1. Let $x_n = (-1)^{n+1}$ for each positive integer n. We have $\langle x_n \rangle = \langle 1, -1, 1, -1, \dots \rangle$. The subsequence $\langle x_{2k-1} \rangle = \langle 1, 1, 1, \dots \rangle$ of odd-indexed terms converges to 1, while the subsequence $\langle x_{2k} \rangle = \langle -1, -1, -1, \dots \rangle$ of even-indexed terms converges to -1. Therefore, 1 and -1 are cluster points of the sequence.

*D. S. Mitrinovic, *Analytic Inequalities*, Springer-Verlag, New York, 1970, p. 266.

EXAMPLE 6.2. Let $x_n = n^{-1}$ if n is an odd positive integer and $x_n = 1 - n^{-1}$ if n is an even positive integer. We have

$$\langle x_n \rangle = \langle 1, \tfrac{1}{2}, \tfrac{1}{3}, \tfrac{3}{4}, \tfrac{1}{5}, \tfrac{5}{6}, \dots \rangle.$$

Since $\langle x_{2k-1} \rangle = \langle 1, \tfrac{1}{3}, \tfrac{1}{5}, \dots \rangle$, we have $x_{2k-1} \to 0$, as $k \to +\infty$, and since $\langle x_{2k} \rangle = \langle \tfrac{1}{2}, \tfrac{3}{4}, \tfrac{5}{6}, \dots \rangle$, we have $x_{2k} \to 1$, as $k \to +\infty$. Here 0 and 1 are cluster points of the sequence.

There exist real sequences having no cluster points.

EXAMPLE 6.3. Let $x_n = n$ for each positive integer so that $\langle x_n \rangle = \langle 1, 2, 3, \dots \rangle$. Here, no subsequence is bounded. Therefore, no subsequence converges. The sequence has no cluster points.

Remark 6.1. If a sequence $\langle x_n \rangle$ converges and has limit L, then every subsequence converges to L. This makes L the only cluster point of the sequence.

Theorem 6.1. *Every bounded (infinite) sequence of real numbers has a real cluster point.*

PROOF. Let $\langle x_n \rangle$ be a bounded sequence of real numbers. This implies that

$$-\infty < \underline{\lim}\, x_n \leq \overline{\lim}\, x_n < +\infty.$$

Let

$$\bar{L} = \overline{\lim}\, x_n \tag{6.1}$$

and $\epsilon = 1$. By Theorem III.6.4, part (c),

$$\bar{L} - 1 < x_n \qquad \text{for infinitely many } n\text{'s}. \tag{6.2}$$

By part (a) of the theorem just mentioned, there exists a positive integer N such that

$$x_n < \bar{L} + 1 \qquad \text{for} \quad n \geq N. \tag{6.3}$$

The set of n's for which (6.2) holds is an infinite set of positive integers and, thus, is not bounded. Hence, there exists a positive integer n_1 such that $n_1 > N$ and $\bar{L} - 1 < x_{n_1}$. It follows that

$$\bar{L} - 1 < x_{n_1} < \bar{L} + 1 \qquad \text{where} \quad n_1 > N. \tag{6.4}$$

Now, a positive integer N_2 exists such that

$$x_n < \bar{L} + 1 \qquad \text{for} \quad n \geq N_2. \tag{6.4'}$$

Let $N_2' = \max\{n_1, N_2\}$ and $N_2'' = N_2' + 1$. Again, since

$$\bar{L} - \tfrac{1}{2} < x_n \qquad \text{for infinitely many } n\text{'s}, \tag{6.5}$$

there exists among these n's one greater than N_2', n_2 say, and we have

$$\bar{L} - \tfrac{1}{2} < x_{n_2} < \bar{L} + \tfrac{1}{2} \qquad \text{where} \quad n_1 < n_2. \tag{6.6}$$

This procedure can be continued inductively to obtain a subsequence $\langle x_{n_k} \rangle$ of $\langle x_n \rangle$ such that

$$\bar{L} - \frac{1}{k} < x_{n_k} < \bar{L} + \frac{1}{k} \qquad \text{for each positive integer } k. \tag{6.7}$$

It is clear that $x_{n_k} \to \bar{L}$ as $k \to +\infty$. Thus, $\bar{L}$ is a cluster point of $\langle x_n \rangle$. We ask the reader to prove that $\underline{L} = \underline{\lim}\, x_n$ is also a cluster point of $\langle x_n \rangle$ (Prob. 6.1).

PROB. 6.1. Prove: If $\langle x_n \rangle$ is a bounded infinite sequence of real numbers, then $\underline{L} = \underline{\lim}\, x_n$ is a cluster point of $\langle x_n \rangle$.

PROB. 6.2. Prove: If $\langle x_n \rangle$ is a bounded infinite sequence of real numbers, then $\bar{L} = \overline{\lim}\, x_n$ and $\underline{L} = \underline{\lim}\, x_n$ are respectively the greatest and least cluster points of $\langle x_n \rangle$.

Remark 6.2. We call Theorem 6.1 the *Bolzano–Weierstrass Theorem for sequences*. The theorem which follows will be referred to as the *Bolzano–Weierstrass Theorem for sets*.

Theorem 6.2 (Bolzano–Weierstrass Theorem for Sets). *Every bounded infinite set of real numbers has at least one accumulation point.*

PROOF. Let S be a bounded infinite set of real numbers. We know that an infinite set contains a denumerable subset (Theorem II.10.3). Let $A \subseteq S$, where A is denumerable. Then there is a one-to-one correspondence f between $\mathbb{Z}_+$ and A. Let $x_n = f(n)$ for each $n \in \mathbb{Z}_+$. This gives us an infinite sequence $\langle x_n \rangle$ of distinct elements of S. Since S is bounded, so is $\langle x_n \rangle$. By the Bolzano–Weierstrass Theorem for sequences, the sequence $\langle x_n \rangle$ has a cluster point c. There exists a subsequence $\langle x_{n_k} \rangle$ of $\langle x_n \rangle$ such that $x_{n_k} \to c$ as $k \to +\infty$. The terms of this subsequence are all distinct (how do we know this?). Since there is a sequence of distinct elements of S converging to c, c is an accumuation point of S (Theorem V.3.3).

It is often useful to know whether or not an accumulation point of a set is also a member of the set. Sets containing *all* their accumulation points constitute an important class.

Def. 6.2. A set $S \subseteq \mathbb{R}$ is called *closed* (in $\mathbb{R}$) if and only if it contains all its accumulation points. Using the notation S' for the derived set of S (see Remark V.3.2), the definition may be phrased as: S is *closed* in $\mathbb{R}$ if and only if $S' \subseteq S$. Since all our sets will be in $\mathbb{R}$, we shall usually refer to a set simply as being *closed* or *not closed* without adding "in $\mathbb{R}$."

EXAMPLE 6.4. A bounded closed interval $[a,b]$ is a closed set. Let x_0 be an accumulation point of $[a,b]$. We prove that $x_0 \leqslant b$. Suppose that $x_0 > b$. Consider the neighborhood $N(x_0, x_0 - b)$ of x_0. Assume that $x \in N(x_0, x_0 - b)$ so that $|x - x_0| < x_0 - b$ and, hence, $b - x_0 < x - x_0 < x_0 - b$. The last inequalities imply that $b - x_0 < x - x_0$ and, hence, that $x > b$. This proves: If $x_0 > b$, then the neighborhood $N(x_0, x_0 - b)$ and, therefore, the deleted neighborhood $N^*(x_0, x_0 - b)$, of x_0 contains no points of $[a,b]$. It follows that if $x_0 > b$, then x_0 is not an accumulation point of $[a,b]$. Hence, if x_0 is an accumulation point of $[a,b]$, then $x_0 \leqslant b$. A similar indirect proof shows that if x_0 is an accumulation point of $[a,b]$, then $a \leqslant x_0$ (carry out this proof). It follows that if x_0 is an accumulation point of $[a,b]$, then $x_0 \in [a,b]$. Since the bounded closed interval $[a,b]$ contains all its accumulation points, it is a closed set.

EXAMPLE 6.5. The bounded interval $[a,b)$ is not a closed set since b is an accumulation point of $[a,b)$ which is not in $[a,b)$.

PROB. 6.3. Prove: $[a, +\infty)$ and $(-\infty, a]$, where $a \in \mathbb{R}$, are closed sets.

Remark 6.3. Clearly, $\mathbb{R}$ is closed. The empty set $\emptyset$ is also closed. For otherwise it would contain a point that is not an accumulation point of $\emptyset$. This is impossible, for $\emptyset$ has no elements. If $S' = \emptyset$, then S is closed, since then $S' = \emptyset \subseteq S$, so that $S' \subseteq S$. All finite subsets of $\mathbb{R}$ are closed since their derived sets are empty (prove this).

Theorem 6.3. *A set $S \subseteq \mathbb{R}$ is closed if and only if each converging sequence $\langle x_n \rangle$ of elements of S converges to a point of S.*

PROOF. Let S be a subset of $\mathbb{R}$ having the property that each converging sequence $\langle x_n \rangle$ of elements of S converges to a point of S. We prove that S is closed. Let x_0 be an accumulation point of S. There exists a sequence $\langle x_n \rangle$ of distinct points of S which converges to x_0 (Theorem V.3.3.). By the assumption on S, $x_0 \in S$. This proves that $S' \subseteq S$ and, hence, that S is closed.

Conversely, let S be a set that is closed in $\mathbb{R}$. Let $\langle x_n \rangle$ be a sequence of points of S which converges to some point c. Suppose c is not in S. By the present assumption on S, c is not an accumulation point of S and some deleted ϵ-neighborhood $N^*(c,\epsilon)$ exists containing no points of S. Since $c \notin S$ to begin with, this implies that $N(c,\epsilon) \cap S = \emptyset$. In turn, this implies that $x_n \notin N(c,\epsilon)$ for all n. Thus, $|x_n - c| \geqslant \epsilon > 0$ for all n and $\lim x_n \neq c$. This contradicts the assumption on the sequence $\langle x_n \rangle$. We must, therefore, conclude that $c \in S$. This proves that each converging sequence of elements of S converges to an element of S. This completes the proof.

Of special importance are sets which are bounded and closed.

Theorem 6.4. *A nonempty set of real numbers which is closed and bounded from above has a maximum.*

PROOF. Let $S \subseteq \mathbb{R}$ be nonempty, bounded from above and closed. This implies that S has a real supremum. Let $\mu = \sup S$. Let n be some positive integer so that $\mu - n^{-1} < \mu$. Corresponding to n, there is an $x_n \in S$ such that $\mu - n^{-1} < x_n \leqslant \mu$. This implies that

$$\mu = \lim x_n .$$

Thus, $\langle x_n \rangle$ is a sequence of elements of S which converges to μ. Since S is closed, we know (Theorem 6.3) that $\mu \in S$. Accordingly, μ is the maximum of S (explain).

PROB. 6.4. Prove: A nonempty set of real numbers that is closed and bounded from below has a minimum.

Remark 6.4. Combining the results of Theorem 6.4 and Prob. 6.4, we have:

Theorem 6.4′. *A nonempty set of real numbers that is closed and bounded has a maximum and a minimum.*

7. Open Sets in $\mathbb{R}$

Along with closed set in $\mathbb{R}$ we consider sets which are *open* in $\mathbb{R}$. Before defining the notion of an open set we introduce the notion of an interior point of a set $S \subseteq \mathbb{R}$.

Def. 7.1. If $S \subseteq \mathbb{R}$, then $x_0 \in S$ is called an *interior* point of S if and only if there is an ϵ-neighborhood $N(x_0, \epsilon)$ of x_0 such that $N(x_0, \epsilon) \subseteq S$.

EXAMPLE 7.1. A point of an interval I which is not an endpoint of I is an interior point of I (see Fig. 7.1). Although this seems intuitively clear, we prove it for the case of an interval $I = [a, +\infty)$, where $a \in \mathbb{R}$. Let x_0 be a point of I which differs from a. We have $x_0 > a$ so that $x_0 - a > 0$. Take the neighborhood $N(x_0, x_0 - a)$ of x_0. We prove that $N(x_0, x_0 - a) \subseteq I$. Let $x \in N(x_0, x_0 - a)$ so that $|x - x_0| < x_0 - a$ and, therefore, $a - x_0 < x - x_0 < x_0 - a$. This implies that $x > a$. But then $x \in I = [a, +\infty)$. We

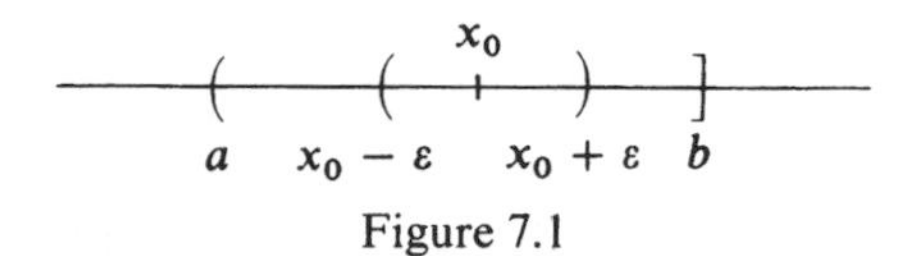

Figure 7.1

proved that $x \in N(x_0, x_0 - a)$ implies $x \in I$. This proves that $N(x_0, x_0 - a) \subseteq I$.

PROB. 7.1. Let $I = (-\infty; b)$ where $b \in \mathbb{R}$. Prove: If $x_0 \in I$ and $x_0 \neq b$, then x_0 is an interior point of I.

PROB. 7.2. Let $I = [a, b]$. Prove: If $a < x_0 < b$, then x_0 is an interior point of I.

PROB. 7.3. Prove: No endpoint of an interval is an interior point of an interval.

PROB. 7.4. Prove: If x_0 is an interior point of an interval I, then it is an accumulation point of I (Hint: see Prob. V.3.2).

PROB. 7.5. Prove: No finite subset of $\mathbb{R}$ has interior points.

PROB. 7.6. Prove: The set $\mathbb{Q}$ of rational numbers has no interior points.

Def. 7.2. A set $S \subseteq \mathbb{R}$ is called *open* if and only if each of its points is an interior point of S, that is, if and only if for each $x_0 \in S$ there exists some ϵ-neighborhood $N(x_0, \epsilon)$ of x_0 such that $N(x_0, \epsilon) \subseteq S$.

PROB. 7.7. Prove: An open interval is an open set.

Remark 7.1. The set of real numbers is open. The empty set is open (why?).

EXAMPLE 7.2. The bounded closed interval $I = (a, b]$ is not open. Indeed, $b \in I$ and b is not an interior point of I, that is, no ϵ-neighborhood $N(b, \epsilon)$ of b is contained in I. It is important to note that $(a, b]$ is not closed either. Thus, there exist sets that are neither open nor closed.

The next theorem relates open sets to closed ones.

Theorem 7.1. *A set $G \subseteq \mathbb{R}$ is open if and only if its complement $\mathbb{R} - G$ is closed. A set $F \subseteq \mathbb{R}$ is closed if and only if its complement $\mathbb{R} - F$ is open.*

PROOF. We prove the first statement. Assume that $G \subseteq \mathbb{R}$ is open. We prove that its complement $\mathbb{R} - G$ is closed. Let x_0 be an accumulation point of $\mathbb{R} - G$. If x_0 was in G, there would exist an ϵ-neighborhood $N(x_0, \epsilon)$ of x_0

such that $N(x_0, \epsilon) \subseteq G$. Since $N^*(x_0, \epsilon) \subseteq N(x_0, \epsilon)$, the deleted ϵ-neighborhood $N^*(x_0, \epsilon)$ is in G and contains no points of $\mathbb{R} - G$. This implies that x_0 is not an accumulation point of $\mathbb{R} - G$ and contradicts the assumption on x_0. We conclude that $x_0 \in \mathbb{R} - G$. Thus, $\mathbb{R} - G$ contains all its accumulation points and is, therefore, closed.

Conversely, assume that the complement $\mathbb{R} - G$ of G is closed. Then $\mathbb{R} - G$ contains all its accumulation points. Let $x_0 \in G$ so that x_0 is not in $\mathbb{R} - G$. Accordingly, there exists a deleted ϵ-neighborhood $N^*(x_0, \epsilon)$ of x_0 which contains no point of $\mathbb{R} - G$ and, consequently, is a subset of G. For this ϵ-neighborhood we have $N(x_0, \epsilon) = \{x_0\} \cup N^*(x_0, \epsilon) \subseteq G$. This means that if $x_0 \in G$, then x_0 is an interior point of G. Thus, each point of G is an interior point of G and G is open.

We ask the reader to prove the second statement (Prob. 7.8).

Prob. 7.8. Complete the proof of the last theorem by proving: A set F is closed if and only if its complement is open.

Theorem 7.2. *The intersection of two open sets is open.*

Proof. Let G_1 and G_2 be open sets. Consider their intersection $G_1 \cap G_2$. Let $x_0 \in G_1 \cap G_2$, so that $x_0 \in G_1$ and $x_0 \in G_2$. This implies that there exist ϵ-neighborhoods $N(x_0, \epsilon_1)$ and $N(x_0, \epsilon_2)$ of x_0 such that

$$N(x_0, \epsilon_1) \subseteq G_1 \quad \text{and} \quad N(x_0, \epsilon_2) \subseteq G_2.$$

But there exists an ϵ-neighborhood $N(x_0, \epsilon)$ of x_0 such that $N(x_0, \epsilon) \subseteq N(x_0, \epsilon_1) \cap N(x_0, \epsilon_2)$ (Theorem V.3.1). Hence,

$$N(x_0, \epsilon) \subseteq N(x_0, \epsilon_1) \cap N(x_0, \epsilon_2) \subseteq G_1 \cap G_2.$$

Thus, each point x_0 of $G_1 \cap G_2$ is an interior point of $G_1 \cap G_2$, and $G_1 \cap G_2$ is open.

Prob. 7.9. Prove: If F_1 and F_2 are closed sets, then their union is closed.

Theorem 7.3. *The intersection of any family of closed sets is closed.*

Proof. Let $\mathfrak{F}$ be a family of closed sets. We will prove that $\bigcap \mathfrak{F}$ is closed. Let x_0 be an accumulation point of $\bigcap \mathfrak{F}$. That is, let

$$x_0 \in \left(\bigcap \mathfrak{F}\right)'. \tag{7.1}$$

Let $F \in \mathfrak{F}$. By properties of intersections this implies that

$$\bigcap \mathfrak{F} \subseteq F. \tag{7.2}$$

But $S \subseteq T \subseteq \mathbb{R}$ implies $S' \subseteq T'$ (Prob. V.3.8) and, hence, by (7.2),

$$\left(\bigcap \mathfrak{F}\right)' \subseteq F' \subseteq F. \tag{7.3}$$

Here $F' \subseteq F$ holds because each $F \in \mathfrak{F}$ is closed. In view of (7.3) we see that $(\bigcap \mathfrak{F})'$ is a subset of every $F \in \mathfrak{F}$. Hence,

$$(\bigcap \mathfrak{F})' \subseteq \bigcap \mathfrak{F}.$$

This proves that $\bigcap \mathfrak{F}$ is closed.

Prob. 7.10. Prove: The union of any family of open sets is open.

Prob. 7.11. Prove: If f is a function which is continuous on $\mathbb{R}$ and $c \in \mathbb{R}$, then the sets

$$A = \{x \in \mathbb{R} \mid f(x) \geqslant c\}, \quad B = \{x \in \mathbb{R} \mid f(x) \leqslant c\}, \quad E = \{x \in \mathbb{R} \mid f(x) = c\}$$

are all closed and the sets

$$F = \{x \in \mathbb{R} \mid f(x) > c\}, \qquad G = \{x \in \mathbb{R} \mid x < c\}$$

are open.

8. Functions Continuous on Bounded Closed Sets

Prob. 8.1. Prove: A bounded closed interval $[a, b]$ is a bounded closed set.

Theorem 8.1. *A function continuous on a bounded closed set is bounded.*

Proof. Let f be continuous on the bounded closed set S. Suppose f is not bounded. This implies that corresponding to each positive integer n, there exists an $x_n \in S$ such that

$$|f(x_n)| > n. \tag{8.1}$$

This gives us a sequence $\langle x_n \rangle$ of elements of S for which (8.1) holds. Since S is bounded, $\langle x_n \rangle$ is a bounded sequence of elements of S. As such, it has a cluster point c. There exists a subsequence $\langle x_{n_k} \rangle$ of $\langle x_n \rangle$ such that $x_{n_k} \to c$ as $k \to +\infty$. Using (8.1), we obtain

$$|f(x_{n_k})| > n_k \geqslant k \qquad \text{for each positive integer } k. \tag{8.2}$$

This means that the sequence $\langle f(x_{n_k}) \rangle$ is not bounded. Since S is a closed set and $\langle x_{n_k} \rangle$ is a sequence of elements of S converging to c, we have $c \in S$. This implies that f is continuous at c and, therefore, sequentially continuous at c. It follows that $f(x_{n_k}) \to f(c)$ as $k \to +\infty$. Thus, $\langle f(x_{n_k}) \rangle$ converges. Consequently, it is a bounded sequence. This contradicts our previous conclusion, i.e., that $\langle f(x_{n_k}) \rangle$ is not bounded. Therefore, we conclude that f is bounded.

Prob. 8.2. The function f defined as

$$f(x) = \begin{cases} \dfrac{1}{x} & \text{if } 0 < x \leqslant 1 \\ 1 & \text{if } x = 0 \end{cases}$$

is defined on the bounded closed set [0, 1] and is not bounded. Reconcile this with Theorem 8.1.

PROB. 8.3. The function f, where

$$f(x) = \frac{1}{\sqrt{1 - x^2}} \qquad \text{for} \quad -1 < x < 1,$$

is continuous on the bounded set $(-1; 1)$ but is not bounded. Reconcile this with Theorem 8.1.

Theorem 8.2. *A function which is continuous on a bounded closed set has a range which is bounded and closed.*

PROOF. Let f be continuous on the bounded closed set S. By Theorem 8.1 we know that f and, hence, its range $\mathfrak{R}(f)$, is bounded. We prove that $\mathfrak{R}(f)$ is a closed set. Let $\langle y_n \rangle$ be a sequence of elements of $\mathfrak{R}(f)$ such that $y_n \to y$ as $n \to +\infty$. For each n there exists an $x_n \in S$ such that $y_n = f(x_n)$. The sequence $\langle x_n \rangle$ is a sequence of elements of S. Since S is bounded, the same is true of $\langle x_n \rangle$. Hence, $\langle x_n \rangle$ has a cluster point c and there exists a subsequence $\langle x_{n_k} \rangle$ of $\langle x_n \rangle$ which converges to c. Since S is closed, we have $c \in S$. f is continuous at c and $f(x_{n_k}) \to f(c)$ as $k \to +\infty$. But then $y_{n_k} \to f(c)$ as $k \to +\infty$. Since $y_n \to y$ as $n \to +\infty$ the subsequence $\langle y_{n_k} \rangle$ of $\langle y_n \rangle$ also converges to y. This yields $y = f(c)$ and, hence, $y \in \mathfrak{R}(f)$. We proved: If $\langle y_n \rangle$ is a sequence of elements of $\mathfrak{R}(f)$ converging to y, then $y \in \mathfrak{R}(f)$. This implies that $\mathfrak{R}(f)$ is closed (Theorem 6.3). Thus, $\mathfrak{R}(f)$ is closed and bounded, as claimed.

Corollary 1. *A real-valued function of a real variable which is continuous on a nonempty bounded closed set has a maximum and a minimum.*

PROOF. Let f be a function continuous on the bounded, nonempty, closed set S. By the hypothesis on f and by Theorem 8.2, the range $\mathfrak{R}(f)$ of f is nonempty closed and bounded. This implies (Theorem 6.4′) that $\mathfrak{R}(f)$ has a maximum and a minimum. The maximum and minimum of $\mathfrak{R}(f)$ are the maximum and minimum of f.

Corollary 2. *The range of a real-valued function of a real variable that is continuous on a bounded closed interval $[a, b]$ and not constant there is the bounded closed interval $[f(x_0), f(x_1)]$, where $f(x_0)$ and $f(x_1)$ are respectively the minimum and maximum of f.*

PROOF. By Corollary 1 above, f has a minimum $f(x_0), x_0 \in [a, b]$ and a maximum $f(x_1)$, $x_1 \in [a, b]$. Thus,

$$f(x_0) \leqslant f(x) \leqslant f(x_1) \qquad \text{for all} \quad x \in [a, b]. \tag{8.3}$$

If $f(x_0)=f(x_1)$, then $f(x)=f(x_0)$ for all $x\in[a,b]$, that is, f is constant on $[a,b]$. But this is precluded by the hypothesis, so that $f(x_0)<f(x_1)$. By (8.3) we have $\mathcal{R}(f)\subseteq[f(x_0),f(x_1)]$. By the intermediate-value property (recall that f is continuous on the bounded interval $[a,b]$, and so has the intermediate-value property there), we have $[f(x_0),f(x_1)]\subseteq\mathcal{R}(f)$. We conclude that $\mathcal{R}(f)=[f(x_0),f(x_1)]$.

Prob. 8.4. Let f be given by means of $f(x)=1/(1+x^2)$ for $x\in\mathbb{R}$. This function is continuous on the closed set $\mathbb{R}$ but its range $(0,1]$ (prove this is the range) is not closed. Reconcile this with Theorem 8.2.

Prob. 8.5. Let f be defined as $f(x)=x$ for $-1<x<1$. This f is continuous on the bounded interval $(-1;1)$ but has neither a maximum nor a minimum there. Reconcile this with Corollary 1 above.

Prob. 8.6. Define f as

$$f(x)=\begin{cases} x & \text{if} \quad -1<x<1 \\ 0 & \text{if} \quad x=\pm 1. \end{cases}$$

Here f is defined on the bounded closed interval $[-1,1]$ but it has neither a maximum nor a minimum there. Reconcile this with Corollary 1 above.

9. Monotonic Functions. Inverses of Functions

We first prove a lemma.

Lemma 9.1. *If I is an interval and a,b are accumulation points of I (possibly extended ones) such that $a<b$, then $(a;b)\subseteq I$.*

Proof. We first consider the case where a and b are real. Assume that $x\in(a;b)$ so that $a<x<b$, $b-x>0$, $x-a>0$. Since $b-a>0$, there exist positive integers n_1 and n_2 such that

$$\frac{b-a}{n_1}<x-a \quad \text{and} \quad \frac{b-a}{n_2}<b-x.$$

Hence,

$$a+\frac{b-a}{n_1}<x<b-\frac{b-a}{n_2}. \tag{9.1}$$

Since a and b are accumulation points of I, there exist points x_1 and x_2 of I such that $x_1\in N^*(a,(b-a)/n_1)$ and $x_2\in N^*(b,(b-a)/n_2)$ and, therefore,

$$x_1<a+\frac{b-a}{n_1} \quad \text{and} \quad b-\frac{b-a}{n_2}<x_2.$$

These and (9.1) imply that $x_1 < x < x_2$. Since I is a convex set, it follows that $x \in [x_1, x_2] \subseteq I$ and so $x \in I$. Thus, $(a;b) \subseteq I$ in this case.

Next we consider the case $a = -\infty$, $b \in \mathbb{R}$. Assume that $x \in (a;b) = (-\infty;b)$ so that $-\infty < x < b$. The deleted neighborhood $(-\infty;x)$ of $-\infty$ contains a point of I. Hence, there is $x_1 \in I$ such that $x_1 < x$. Let n_3 be a positive integer such that $n_3 > 1$. Then

$$0 < \frac{b-x}{n_3} < b - x \tag{9.2a}$$

and, therefore,

$$x < b - \frac{b-x}{n_3}. \tag{9.2b}$$

Now the deleted neighborhood $N^*(b,(b-x)/n_3)$ contains a point x_2 of I. This point is such that

$$b - \frac{b-x}{n_3} < x_2. \tag{9.3}$$

This and (9.2b) imply that $x < x_2$. Thus, there exist x_1 and x_2 in I such that $x_1 < x < x_2$. In view of the convexity of I, we see that $x \in I$. This proves that $(-\infty;b) \subseteq I$. We ask the reader to prove the theorem for the remaining cases (Prob. 9.1).

Prob. 9.1. Complete the proof of Lemma 9.1 by proving it for the cases $a \in \mathbb{R}$, $b = +\infty$ and $a = -\infty$, $b = +\infty$.

Theorem 9.1. *If f is monotonically increasing on the interval I and a, b are accumulation points of I (possibly extended ones) such that $a < b$, then $f(a+)$ and $f(b-)$ exist in $\mathbb{R}^*$ and $f(a+) \leqslant f(b-)$.* (1) *If $b \in I$, then $f(b-) \leqslant f(b)$.* (2) *If $b \notin I$, then $f(b-) \in \mathbb{R}$ or $f(b-) = +\infty$ according to whether f is bounded from above or not.* (3) *If $a \in I$, then $f(a) \leqslant f(a+)$.* (4) *If $a \notin I$, then $f(a+) \in \mathbb{R}$ or $f(a+) = -\infty$ according to whether f is bounded from below or not.*

Proof. By Lemma 9.1, we have $(a;b) \subseteq I$. Clearly, a is an accumulation point of I from the right, while b is an accumulation point from the left (these are possibly extended). First assume that f is bounded from above on I. Since $(a;b) \subseteq I$, this implies that f is bounded from above on $(a;b)$ so that f has a real supremum there. Write

$$L = \sup_{a<x<b} f(x). \tag{9.4}$$

Given $\epsilon > 0$, we have $L - \epsilon < L$, and there exists an x_1 such that $a < x_1 < b$ and

$$L - \epsilon < f(x_1) \leqslant L. \tag{9.5}$$

If $x_1 < x < b$, we have $a < x < b$ and $f(x_1) \leqslant f(x) \leqslant L$. This and (9.5) imply that

$$L - \epsilon < f(x) \leqslant L \qquad \text{for} \quad x_1 < x < b.$$

Hence: If $\epsilon > 0$ is given, there exists an $x_1 \in I$ such that

$$|f(x) - L| < \epsilon \qquad \text{if} \quad x_1 < x < b.$$

Whether $b = +\infty$ or $b \in \mathbb{R}$, this implies that

$$f(b-) = \lim_{x \to b-} f(x) = L \in \mathbb{R} \tag{9.6}$$

(explain). In case (1), where $b \in I$, we have $f(x) \leqslant f(b)$ for $a < x < b$. Therefore $f(b)$ is an upper bound for $f((a;b))$. This implies that $f(b-) = L \leqslant f(b)$. This proves (1). In case (2), if f is bounded from above, then (9.6) holds and so $f(b-) \in \mathbb{R}$. On the other hand, if, in case (2), f is not bounded from above, given B there exists an x' in I such that $f(x') > B$. Here $x' < b$ since b is a right endpoint (possibly extended) of I. Since f is monotonically increasing, it follows that if $x' < x < b$, then $f(x) \geqslant f(x') > B$. We conclude that

$$f(b-) = \lim_{x \to b-} f(x) = +\infty. \tag{9.7}$$

This proves the second part of case (2). Thus, cases (1) and (2) are dispensed with.

To deal with cases (3) and (4), we assume first that f is bounded from below I. Since $(a;b) \subseteq I$, this implies that f is bounded from below on $(a;b)$ so that f has a real infimum there. We put

$$l = \inf_{a<x<b} f(x). \tag{9.8}$$

Given $\epsilon > 0$, we note that $l < l + \epsilon$ and that an x_2 exists such that $a < x_2 < b$ and

$$l \leqslant f(x_2) < l + \epsilon. \tag{9.9}$$

If $a < x < x_2$ we have $a < x < b$ and $l \leqslant f(x) \leqslant f(x_2)$. These and (9.9) imply that

$$l \leqslant f(x) < l + \epsilon \qquad \text{for} \quad a < x < x_2.$$

Hence: If $\epsilon > 0$ is given, there exists an $x_2 \in I$ such that

$$|f(x) - l| < \epsilon \qquad \text{if} \quad a < x < x_2.$$

We conclude that

$$f(a+) = \lim_{x \to a+} f(x) = l \in \mathbb{R}. \tag{9.10}$$

The reader can now complete the proofs of cases (3) and (4). These proofs are analogous to those of cases (1) and (2).

It remains to prove that $f(a+) \leqslant f(b-)$. This is trivially true if $f(a+) = -\infty$ or $f(b-) = +\infty$. The cases $f(a+) \in \mathbb{R}$, $f(b-) \in \mathbb{R}$ are treated by comparing (9.4) and (9.8). These imply $l \leqslant L$ and, therefore, $f(a+) = l \leqslant L = f(b-)$.

Prob. 9.2. State and prove the dual of Theorem 9.1 for the case where f is monotonically decreasing on an interval I and a, b are extended accumulation points of I such that $a < b$.

Corollary (of Theorem 9.1). *If f is monotonically increasing on an interval I and x_0 is an interior point of I, then $f(x_0-)$ and $f(x_0+)$ are finite and $f(x_0-) \leqslant f(x_0) \leqslant f(x_0+)$.*

Proof. Since x_0 is assumed to be an interior point of I, there exist x_1 and x_2 in I such that $x_1 < x_0 < x_2$. Since, in this case, x_0 is a two-sided accumulation point of I, Theorem 9.1 implies that $f(x_0-) \leqslant f(x_0)$ and $f(x_0) \leqslant f(x_0+)$. This completes the proof.

Inverses of Functions

Theorem II.11.1 states that a strictly monotonic real-valued function of a real variable has a strictly monotonic inverse f^{-1}. Here we examine the effect of requiring that a strictly monotonic function is also continuous.

Example 9.1. Consider f, where

$$f(x) = \begin{cases} x+1 & \text{if } x < -1, \\ 0 & \text{if } x = 0, \\ x-1 & \text{if } x > 1. \end{cases}$$

(see Fig. 9.1). The domain of f is $\mathfrak{D}(f) = (-\infty; 1) \cup \{0\} \cup (1; +\infty)$. Since

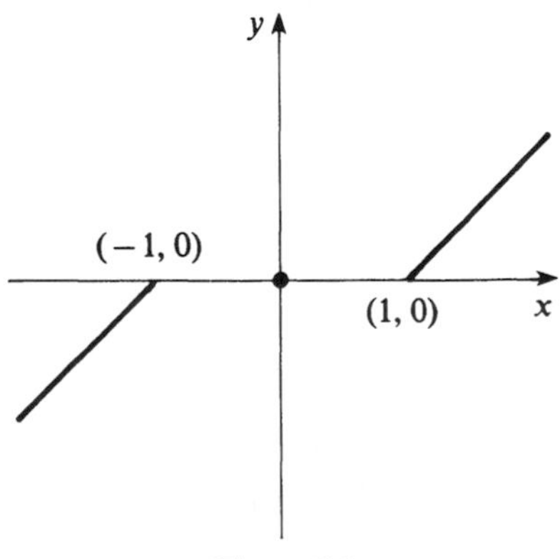

Figure 9.1

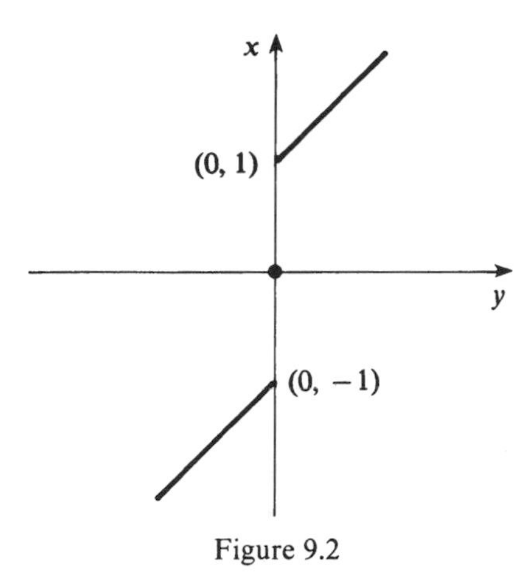

Figure 9.2

0 is an isolated point of $\mathfrak{D}(f)$, f is continuous at 0. f is also continuous also at $x \in \mathfrak{D}(f)$, $x \neq 0$. (Prove this.) Thus, f is a continuous function. The range of f is $\mathbb{R}$. Accordingly, f has an inverse f^{-1} defined as:

$$\begin{aligned} f^{-1}(y) &= y - 1 \quad &\text{if } y < 0, \\ &= 0 \quad &\text{if } y = 0, \\ &= y + 1 \quad &\text{if } y > 0 \end{aligned}$$

(see Fig. 9.2).

We note that f^{-1} is discontinuous, since it is not continuous at $y = 0$ (prove this). Thus, we can see that a function may be discontinuous but its inverse may be continuous (f^{-1} is discontinuous but its inverse f is continuous).

Theorem 9.2.* *If f is strictly monotonic on an interval I, then it has an inverse f^{-1} on $f(I)$ which is strictly monotonic and continuous.*

PROOF. We prove the theorem for the case where f is increasing. The proof for the decreasing case is similar.

Assume that f is strictly monotonically increasing on the interval I. Therefore, f has an inverse f^{-1} defined on $\mathfrak{R}(f) = f(I)$. Let $y_0 \in f(I)$. There exists $x_0 \in I$ such that $y_0 = f(x_0)$ and $x_0 = f^{-1}(y_0)$. Since $x_0 \in I$ and I is an interval, x_0 is either (1) an interior point or (2) a left endpoint of I, or (3) a right endpoint of I.

We examine case (1), where x_0 is an interior point of I. In this case, x_0 is a two-sided accumulation point of I. Given $\epsilon > 0$, there exist x_1 and x_2 in I such that $x_0 - \epsilon < x_1 < x_0 < x_2 < x_0 + \epsilon$. Put $y_1 = f(x_1)$ and $y_2 = f(x_2)$.

* Burril-Knudsen. *Real Variables*, Holt, Rhinehart, Winston, New York, p. 225.

We have

$$y_1 = f(x_1) < f(x_0) = y_0 < f(x_2) = y_2 \tag{9.11}$$

so that $y_1 < y_0 < y_2$. Therefore, $y_0 - y_1 > 0$, $y_2 - y_0 > 0$. Put $\delta_1 = y_0 - y_1$ and $\delta_2 = y_2 - y_0$. Then $y_1 = y_0 - \delta_1$ and $y_2 = y_0 + \delta_2$, where $\delta_1 > 0$ and $\delta_2 > 0$. Also put $\delta = \min\{\delta_1, \delta_2\}$ so that $\delta > 0$ and

$$y_1 = y_0 - \delta_1 \leqslant y_0 - \delta < y_0 < y_0 + \delta \leqslant y_0 + \delta_2 = y_2. \tag{9.12}$$

Take $y \in N(y_0, \delta)$ and $y \in f(I)$. There exists $x \in I$ such that $y = f(x)$, $x = f^{-1}(y)$, and

$$y_1 \leqslant y_0 - \delta < y < y_0 + \delta \leqslant y_2.$$

Accordingly, if $\epsilon > 0$ is given, we have

$$x_0 - \epsilon < x_1 = f^{-1}(y_1) < f^{-1}(y) < f^{-1}(y_2) = x_2 < x_0 + \epsilon$$

and, therefore,

$$|f^{-1}(y) - f^{-1}(y_0)| = |f^{-1}(y) - x_0| < \epsilon$$

$$\text{for} \quad y \in N(y_0, \delta) \quad \text{and} \quad y \in f(I). \tag{9.13}$$

It follows that f^{-1} is continuous at y_0 in case (1). We leave the proofs that f^{-1} is continuous at y_0 in cases (2) and (3) to the reader (Prob. 9.3). Thus, f^{-1} is continuous at each $y_0 \in f(I)$ and is, therefore, continuous on $f(I)$.

Prob. 9.3. Complete the proof of Theorem 9.2 by proving that the f^{-1} there is continuous at y_0, where $x_0 = f^{-1}(y_0)$ is either (2) a left endpoint of I or (3) a right endpoint of I.

Theorem 9.3. *If f is strictly monotonic and continuous on an interval I, then $f(I)$ is an interval, f^{-1} exists and is strictly monotonic and continuous on $f(I)$.*

Proof. The continuity of f on an interval I implies that its range $f(I)$ is either a point or an interval (Remark 4.2). However, since f is strictly monotonic on the interval I, $f(I)$ cannot be a point (explain). The existence, strict monotonicity, and continuity of the inverse is a consequence of Theorem 9.2.

Example 9.2. In Theorem 5.7 we saw that the natural logarithm is continuous (see Remark 5.2). Theorem 9.3 furnishes us with another proof of its continuity. The function ln is defined as the inverse of the function E, where $E(x) = e^x$ for all $x \in \mathbb{R}$. The latter is strictly monotonically increasing on the interval $\mathbb{R} = (-\infty; +\infty)$ and continuous there, so Theorem 9.3 guarantees the continuity and the strict monotonically increasing character of ln, the inverse of E on its range $\mathfrak{R}(E) = (0; +\infty)$.

Prob. 9.4. Prove: If f is one-to-one and continuous on an interval I, then it is strictly monotonic on I.

10. Inverses of the Hyperbolic Functions

We apply the results of Section 9 to the hyperbolic functions to obtain their inverses.

Inverse Hyperbolic Sine

The domain of sinh is $\mathbb{R}$. It is strictly monotonically increasing (Prob. V.2.10). By Prob. V.7.4 we also know that

$$\lim_{x\to-\infty}(\sinh x) = -\infty \quad \text{and} \quad \lim_{x\to+\infty}(\sinh x) = +\infty. \tag{10.1}$$

By Prob. 3.8, sinh is continuous. This, (10.1), and the intermediate-value theorem imply that the range $\mathfrak{R}(\sinh)$ of the hyperbolic sine is $\mathbb{R}$. Theorem 9.4 implies that sinh has a strictly monotonically increasing and continuous inverse defined on $\mathfrak{R}(\sinh)$. This inverse is written as $\sinh^{-1}$ or as Arcsinh, and we call it the *inverse hyperbolic sine* or *arc hyperbolic sine*. Given $x \in \mathbb{R}$ there exists a unique $y \in \mathbb{R}$ such that

$$x = \sinh y \quad \text{and} \quad y = \sinh^{-1}x. \tag{10.2}$$

We can obtain an explicit expression for $\sinh^{-1}x$ in terms of functions defined earlier. Thus, if x is given, then the y such that

$$x = \sinh y = \frac{e^y - e^{-y}}{2} = \frac{e^{2y}-1}{2e^y} \tag{10.3}$$

satisfies

$$e^{2y} - 2xe^y - 1 = 0 \quad \text{or} \quad (e^y)^2 - 2xe^y - 1 = 0.$$

This implies that

$$e^y = \frac{2x \pm \sqrt{4x^2+4}}{2} = x \pm \sqrt{x^2+1}\,. \tag{10.4}$$

Now $e^y > 0$ and

$$x - \sqrt{x^2+1} < 0 \quad \text{and} \quad x + \sqrt{x^2+1} > 0 \qquad \text{for} \quad x \in \mathbb{R} \tag{10.5}$$

(prove 10.5). Hence, y in (10.4) must satisfy

$$e^y = x + \sqrt{1+x^2} \qquad \text{for} \quad x \in \mathbb{R}.$$

This implies that

$$y = \ln\left(x + \sqrt{1+x^2}\right) \qquad \text{for} \quad x \in \mathbb{R}. \tag{10.6}$$

Prob. 10.1. Verify by direct substitution that

$$\sinh\left(\ln\left(x+\sqrt{1+x^2}\right)\right) = x \qquad \text{for} \quad x \in \mathbb{R}.$$

From the above it follows (explain) that

$$\sinh^{-1}x = \ln\left(x + \sqrt{1 + x^2}\right) \qquad \text{for} \quad x \in \mathbb{R}. \tag{10.7}$$

Inverse Hyperbolic Cosine

Since $\cosh(-x) = \cosh x$ for $x \in \mathbb{R}$, the hyperbolic cosine function is not one-to-one, so we define a *principal inverse hyperbolic cosine*.

The hyperbolic cosine is strictly monotonically increasing on $[0, +\infty)$ (Prob. V.2.11) and is continuous (Prob. 3.8). We also note that $\cosh 0 = 1$, and

$$\lim_{x \to +\infty} \cosh x = +\infty.$$

With the aid of the intermediate-value theorem we infer that cosh maps $[0, +\infty)$ onto $[1, +\infty)$. We restrict the hyperbolic cosine to $[0, +\infty)$ and conclude, using Theorem 9.3, that this restriction has a unique continuous and strictly monotonically increasing inverse defined on $[1, +\infty)$. We call the inverse of this restriction the *principal inverse hyperbolic cosine* or the *principal arc hyperbolic cosine*, and write it as $\cosh^{-1}$ or as Arccosh. Thus, if $x \geqslant 1$, then $\cosh^{-1}x$ is defined as the unique $y \geqslant 0$ such that

$$\frac{e^y + e^{-y}}{2} = \cosh y = x,$$

which implies that

$$e^{2y} - 2xe^y + 1 = 0.$$

This yields

$$e^y = x \pm \sqrt{x^2 - 1} \qquad \text{for} \quad x \geqslant 1. \tag{10.8}$$

Since $y \geqslant 0$, we know that $e^y \geqslant 1$. However, for $x \geqslant 1$ we have

$$\sqrt{x-1} - \sqrt{x+1} < 0 \quad \text{and} \quad \sqrt{x-1} + \sqrt{x+1} > 0. \tag{10.9}$$

We multiply each of the above inequalities by $\sqrt{x-1}$ and obtain $x - 1 - \sqrt{x^2-1} \leqslant 0$ and $x - 1 + \sqrt{x^2-1} \geqslant 0$, so that, if $x \geqslant 1$, then

$$x - \sqrt{x^2 - 1} \leqslant 1 \tag{10.10a}$$

and

$$x + \sqrt{x^2 - 1} \geqslant 1. \tag{10.10b}$$

Since $y \geqslant 0$, it follows that the x and y in (10.8) satisfy

$$e^y = x + \sqrt{x^2 - 1}\,, \tag{10.11}$$

where $y \geqslant 0$ and $x \geqslant 1$. Hence, $y = \ln(x + \sqrt{x^2-1}\,)$, that is,

$$y = \operatorname{Arccosh} x = \cosh^{-1}x = \ln\left(x + \sqrt{x^2 - 1}\,\right), \tag{10.12}$$

where $x \geqslant 1$.

Prob. 10.2. Recall the definition of sech by means of

$$\operatorname{sech} x = \frac{1}{\cosh x} \qquad \text{for} \quad x \in \mathbb{R}$$

in Prob. V.2.8. Prove that sech is strictly decreasing on $[0, +\infty)$. Prove: sech maps $[0, +\infty)$ onto $(0, 1]$. Define the *principal inverse hyperbolic secant* as the inverse of the restriction of sech to $[0, +\infty)$ and write it as Arcsech or as sech^{-1}. Prove: If $0 < x \leqslant 1$, then

$$\operatorname{Arcsech} x = \operatorname{sech}^{-1} x = \cosh^{-1} \frac{1}{x} = \ln\left(\frac{1 + \sqrt{1 - x^2}}{x} \right).$$

Prob. 10.3. In Prob. V.2.8 the hyperbolic tangent was defined as

$$\tanh x = \frac{\sinh x}{\cosh x} \qquad \text{for} \quad x \in \mathbb{R}.$$

Prove: tanh is strictly monotonically increasing and continuous and that its range is $(-1; 1)$. Define the *inverse hyperbolic tangent* or the *arc hyperbolic tangent* as the inverse of tanh. Prove: If $-1 < x < 1$, then

$$\operatorname{Arctanh} x = \tanh^{-1} x = \frac{1}{2} \ln \frac{1 + x}{1 - x}.$$

Prob. 10.4. Recall the definitions of the hyperbolic contangent and the hyperbolic cosecant:

$$\coth x = \frac{\cosh x}{\sinh x} \quad \text{and} \quad \operatorname{csch} x = \frac{1}{\sinh x} \qquad \text{for} \quad x \neq 0.$$

Prove that these are one-to-one functions. Using the obvious notation, prove that

$$\coth^{-1} x = \frac{1}{2} \ln \frac{x + 1}{x - 1} \qquad \text{for} \quad |x| > 1$$

and

$$\operatorname{csch}^{-1} x = \sinh^{-1} \frac{1}{x} = \ln\left(\frac{1}{x} + \frac{1}{|x|} \sqrt{1 + x^2} \right) \qquad \text{for} \quad x \neq 0.$$

11. Uniform Continuity

When we test a function for continuity on a set $A \subseteq \mathfrak{D}(f)$ we check to see if it is continuous for each $x_0 \in A$. Thus, given $\epsilon > 0$, we look for a $\delta > 0$ such that if $|x - x_0| < \delta$ and $x \in \mathfrak{D}(f)$, then $|f(x) - f(x_0)| < \epsilon$. The δ is not unique, for once one is found, any smaller one will do. The δ also depends on ϵ and will, in general, depend on x_0. For example, let f be given by

$$f(x) = \frac{1}{x} \qquad \text{for} \quad x > 0.$$

Here $\mathfrak{D}(f) = (0; +\infty)$. We prove that f is continuous at $x_0 > 0$. Suppose $\epsilon > 0$ is given. We wish to find a $\delta > 0$ such that if $|x - x_0| < \delta$ and $x > 0$, then

$$\left| \frac{1}{x} - \frac{1}{x_0} \right| < \epsilon.$$

We first take x such that $|x - x_0| < x_0/2$ so that

$$\frac{1}{2}x_0 = x_0 - \frac{x_0}{2} < x < x_0 + \frac{x_0}{2}.$$

Note that $x > x_0/2 > 0$ holds for such x and, hence, that

$$0 < \frac{1}{x} < \frac{2}{x_0}.$$

In short, if $|x - x_0| < x_0/2$, then $x > 0$ and

$$\left| \frac{1}{x} - \frac{1}{x_0} \right| = \frac{|x - x_0|}{|x||x_0|} = \frac{|x - x_0|}{x x_0} < \frac{2}{x_0^2}|x - x_0|.$$

Here the right-hand side—and therefore also the left-hand side—will be less than ϵ if $|x - x_0| < (x_0^2/2)\epsilon$. We, therefore, take δ such that

$$0 < \delta \leqslant \min\left\{ \frac{x_0}{2}, \frac{x_0^2}{2}\epsilon \right\}$$

and x such that $x > 0$ and $|x - x_0| < \delta$. This will imply that

$$|x - x_0| < \frac{x_0}{2} \quad \text{and} \quad |x - x_0| < \frac{x_0^2}{2}\epsilon,$$

and it will follow that for such x

$$\left| \frac{1}{x} - \frac{1}{x_0} \right| < \frac{2}{x_0^2} \frac{x_0^2}{2}\epsilon = \epsilon.$$

Hence, f is continuous at each $x_0 \in \mathfrak{D}(f)$.

However, for the same $\epsilon > 0$, the closer x_0 is to 0, the smaller the corresponding δ (see Fig. 11.1).

It is not always the case that δ depends on x_0. For example, let f be given by

$$f(x) = 2x + 3 \qquad \text{for each} \quad x \in \mathbb{R}.$$

If $x_0 \in \mathbb{R}$ and $\epsilon > 0$ are given, then taking x such that $|x - x_0| < \delta$, where δ is such that $0 < \delta \leqslant \epsilon/2$, yields

$$|f(x) - f(x_0)| = |2x + 3 - (2x_0 + 3)| = 2|x - x_0| < 2\frac{\epsilon}{2} = \epsilon.$$

Here, for a given $\epsilon > 0$, δ does not depend on x_0. (See Fig. 11.1(b).)

These distinctions are expressed by means of the notion of uniform continuity.

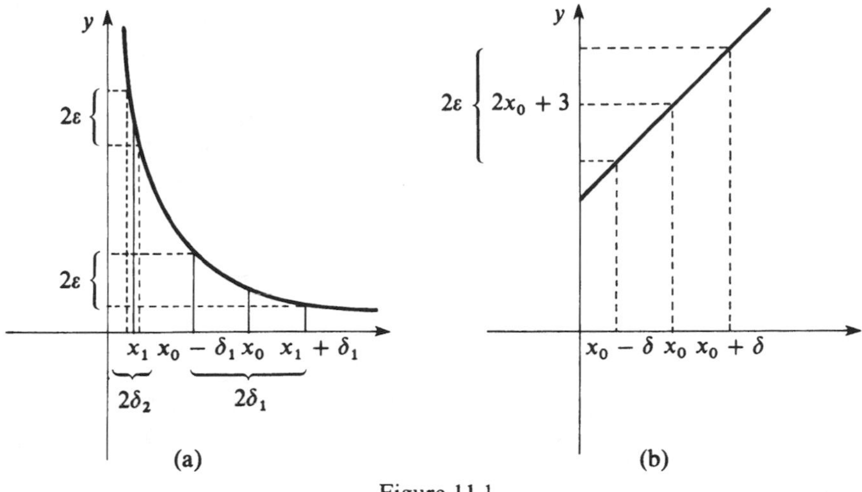

Figure 11.1

Def. 11.1. f is said to be *uniformly continuous* on a set A if and only if (1) $A \subseteq \mathfrak{D}(f)$ and (2) for each $\epsilon > 0$ there exists a $\delta > 0$ such that if $x_1 \in A$ and $x_2 \in A$ and $|x_1 - x_2| < \delta$, then $|f(x_1) - f(x_2)| < \epsilon$. (Here δ depends on ϵ only.) If f is uniformly continuous on $\mathfrak{D}(f)$, then we say that f is *uniformly continuous*.

Prob. 11.1. Prove: If f is uniformly continuous on a set A, then it is continuous on A.

Example 11.1. The sine function is uniformly continuous. Here $A = \mathfrak{D}(\text{sine}) = \mathbb{R}$. By Example 1.4, we have: If x_1 and x_2 are in $\mathbb{R}$, then

$$|\sin x_1 - \sin x_2| \leqslant |x_1 - x_2|.$$

Hence, if $\epsilon > 0$ is given, we take δ such that $0 < \delta \leqslant \epsilon$ to obtain: If $|x_1 - x_2| < \delta$, then

$$|\sin x_1 - \sin x_2| \leqslant |x_1 - x_2| < \delta \leqslant \epsilon.$$

Prob. 11.2. Prove: The cosine function is uniformly continuous.

Prob. 11.3. Prove: If $A = [1, +\infty)$, then the f such that $f(x) = 1/x$ for $x \in A$ is uniformly continuous on A.

Prob. 11.4. Prove: If $A = [0, 1]$ and f is given by $f(x) = x^2$ for $x \in A$, then f is uniformly continuous on A. (In Example 11.2 we see that if f is given by $f(x) = x^2$ and we take $A = \mathfrak{D}(f) = \mathbb{R}$, then f is not uniformly continuous.)

Prob. 11.5. Prove: If g is given by $\sqrt{x}$ for $x \in [0, +\infty)$, then g is uniformly continuous (cf. Prob. 1.4).

The next theorem gives a sequential criterion for uniform continuity. This will enable us, at least in some cases, to determine when a function is *not* uniformly continuous.

Theorem 11.1. *f is uniformly continuous on $A \subseteq \mathfrak{D}(f)$ if and only if for any two sequences $\langle x_n' \rangle$ and $\langle x_n'' \rangle$ of A such that $x_n' - x_n'' \to 0$ as $n \to +\infty$ we have $f(x_n') - f(x_n'') \to 0$ as $n \to +\infty$.*

PROOF. Assume that f is uniformly continuous on $A \subseteq \mathfrak{D}(f)$. Let $\langle x_n' \rangle$ and $\langle x_n'' \rangle$ be sequences of elements of A such that $x_n' - x_n'' \to 0$ as $n \to +\infty$. In view of the uniform continuity of f on A, given $\epsilon > 0$, there exists a $\delta > 0$ such that

$$|f(x_1) - f(x_2)| < \epsilon \quad \text{for} \quad x_1 \text{ and } x_2 \text{ in } A \text{ and } |x_1 - x_2| < \delta. \tag{11.1}$$

Since $x_n' - x_n'' \to 0$ as $n \to +\infty$, there exists an N such that

$$|x_n' - x_n''| < \delta \quad \text{for} \quad x_n', x_n'' \text{ in } A \text{ and } n > N.$$

Because of (11.1), this implies that

$$|f(x_n') - f(x_n'')| < \epsilon \quad \text{if} \quad n > N,$$

and we conclude that $f(x_n') - f(x_n'') \to 0$ as $n \to +\infty$.

Now assume that f is *not* uniformly continuous on $A \subseteq \mathfrak{D}(f)$. By Def. 11.2, this implies that some $\epsilon > 0$ exists such that for each $\delta > 0$ there exist x' and x'' in A such that

$$|x' - x''| < \delta \quad \text{and} \quad |f(x') - f(x'')| \geqslant \epsilon > 0. \tag{11.2}$$

It follows that, corresponding to each positive integer n, there exist x_n' and x_n'' in A such that

$$|x_n' - x_n''| < \frac{1}{n} \quad \text{and} \quad |f(x_n') - f(x_n'')| \geqslant \epsilon > 0. \tag{11.3}$$

Thus, $\langle x_n' \rangle$ and $\langle x_n'' \rangle$ are sequences of elements of A such that $x_n' - x_n'' \to 0$ as $n \to +\infty$, but $f(x_n') - f(x_n'') \to 0$ as $n \to +\infty$ is false. We conclude that if for any sequences $\langle x_n' \rangle$ and $\langle x_n'' \rangle$ of A such that $x_n' - x_n' \to 0$ we have $f(x_n') - f(x_n'') \to 0$, then f is necessarily uniformly continuous on A.

EXAMPLE 11.2. We prove that f, where $f(x) = x^2$ for $x \in \mathbb{R}$ is *not* uniformly continuous. Let $x_n' = \sqrt{n+1}$ and $x_n'' = +\sqrt{n}$ for each positive integer n. We have

$$x_n' - x_n'' = \sqrt{n+1} - \sqrt{n} = \frac{1}{\sqrt{n+1} + \sqrt{n}} \to 0$$

as $n \to +\infty$ and

$$|f(x_n') - f(x_n'')| = |x_n'^2 - x_n''^2| = |n + 1 - n| = 1$$

for all n so that $f(x_n') - f(x_n') \not\to 0$ as $n \to +\infty$. By Theorem 11.1, f is not uniformly continuous on $\mathbb{R} = \mathfrak{D}(f)$.

Prob. 11.6. Use Theorem 11.1 to prove that f, where $f(x) = 1/x$ for $x > 0$, is not uniformly continuous (Hint: consider the sequences $\langle x_n' \rangle$ and $\langle x_n'' \rangle$, where

$$x_n' = \frac{1}{n} \quad \text{and} \quad x_n'' = \frac{1}{n+1} \qquad \text{for each } n).$$

We show that there are several respects in which uniformly continuous functions are "better behaved" than continuous functions that are not uniformly continuous. We first show that there exists a function continuous on a set A which maps some Cauchy sequence of elements of A into a sequence which is not a Cauchy sequence.

Example 11.3. Let f: $(0, 1] \to \mathbb{R}$ be defined as

$$f(x) = \frac{1}{x} \qquad \text{for} \quad x \in (0, 1].$$

The sequence $\langle x_n \rangle$ given by $x_n = 1/n$ for each positive integer n converges to 0, and so is a Cauchy sequence of elements of $(0, 1]$. On the other hand, in spite of the continuity of f, the sequence $\langle f(x_n) \rangle$,

$$f(x_n) = \frac{1}{x_n} = n \qquad \text{for each positive integer } n,$$

is not bounded and diverges to $+\infty$, and so is not a Cauchy sequence.

A uniformly continuous function on a set A never behaves in this way. Specifically,

Theorem 11.2. *If f is uniformly continuous on a set $A \subseteq \mathfrak{D}(f)$ and $\langle x_n \rangle$ is a Cauchy sequence of elements of f, then the sequence $\langle f(x_n) \rangle$ of images of $\langle x_n \rangle$ is a Cauchy sequence.*

Proof. Let $\langle x_n \rangle$ be a Cauchy sequence of elements of A and $\langle f(x_n) \rangle$ the sequence of images of $\langle x_n \rangle$. Let $\epsilon > 0$ be given. Since f is uniformly continuous on A, there exists a $\delta > 0$ such that if x_1 and x_2 are in A and $|x_1 - x_2| < \delta$, then $|f(x_1) - f(x_2)| < \epsilon$. There exists an N such that if $m > N$ and $n > N$, then $|x_m - x_n| < \delta$. Hence, $|f(x_m) - f(x_n)| < \epsilon$ from $m > N$ and $n > N$. Clearly, $\langle f(x_n) \rangle$ is a Cauchy sequence.

Remark 11.1. The function f of the last example maps the bounded set $(0, 1]$ onto the unbounded set $[1, +\infty)$ even though it is continuous. This cannot occur when a function is uniformly continuous. In fact,

Theorem 11.3. *If f is uniformly continuous on a bounded set A, then f is bounded on A.*

Proof. Suppose that f is not bounded on A. This implies that corresponding to each positive integer n there exists an $x_n \in A$ such that

$$|f(x_n)| > n. \tag{11.4}$$

Since A is bounded and $x_n \in A$ for all n, the sequence $\langle x_n \rangle$ is bounded. By the Bolzano–Weierstrass Theorem for sequences, $\langle x_n \rangle$ has a cluster point c and there exists a subsequence $\langle x_{n_k} \rangle$ of $\langle x_n \rangle$ such that $x_{n_k} \to c$ as $k \to +\infty$. This implies that $\langle x_{n_k} \rangle$ is a Cauchy sequence of elements of A. By Theorem 11.2 it follows that $\langle f(x_{n_k}) \rangle$ is a Cauchy sequence. As such, $\langle f(x_{n_k}) \rangle$ is a bounded sequence. But this is impossible. In fact, since (see (11.3))

$$|f(x_{n_k})| > n_k \geqslant k \qquad \text{for each positive integer } k,$$

$\langle f(x_{n_k}) \rangle$ is not bounded. Because of this contradiction we conclude that f is bounded on A.

Theorem 11.4. *If f is uniformly continuous on a bounded interval of the form $(a; b]$, then $f(a+)$ exists and is finite.*

Proof. Given $\epsilon > 0$, there exists a $\delta > 0$ such that if x_1 and x_2 are in $(a, b]$ and

$$|x_1 - x_2| < \delta, \tag{11.5}$$

then

$$|f(x_1) - f(x_2)| < \epsilon. \tag{11.6}$$

Using the above δ, we take $x' \in N^*(a, \delta/2) \cap (a, b]$ and $x'' \in N^*(a, \delta/2) \cap (a, b]$. For such x' and x'', we have $x' \in (a, b]$, $x'' \in (a, b]$ and

$$|x' - x''| \leqslant |x' - a| + |x'' - a| < \frac{\delta}{2} + \frac{\delta}{2} = \delta$$

so that (11.5) and, therefore, also (11.6), hold with $x_1 = x'$, $x_2 = x''$. Thus, given $\epsilon > 0$, we have that $x' \in N^*(a, \delta/2) \cap (a, b]$ and $x'' \in N^*(a, \delta/2) \cap (a, b]$ imply that $|f(x') - f(x'')| < \epsilon$. By the Cauchy criterion for functions,

$$\lim_{x \to a} f(x) = L \qquad \text{for some} \quad L \in \mathbb{R}.$$

Since for all $x \in (a; b]$ we have $x > a$, this limit is one from the right as $x \to a$ and we have $f(a+) = L \in \mathbb{R}$, as claimed.

Prob. 11.7. Let f be uniformly continuous on a bounded interval of the form $[a, b)$. Prove: $f(b-)$ exists and is finite.

Theorem 11.5. *If f is continuous on a bounded closed set, then f is uniformly continuous there.*

Proof. Let F be a bounded closed subset of $\mathbb{R}$ and suppose f is continuous on F but not uniformly continuous there. The last implies that an $\epsilon > 0$ exists such that for each $\delta > 0$, there exist x_1 and x_2 in F such that

$$|x_1 - x_2| < \delta \quad \text{but} \quad |f(x_1) - f(x_2)| \geqslant \epsilon > 0.$$

This implies that corresponding to each positive integer n, there exist x_n' and

x_n'' such that

$$|x_n' - x_n''| < \frac{1}{n} \quad \text{but} \quad |f(x_n') - f(x_n'')| \geqslant \epsilon > 0. \tag{11.7}$$

Thus, $\langle x_n' \rangle$ is a sequence of elements of the bounded set F and is, therefore, a bounded sequence. As such, $\langle x_n' \rangle$ has a cluster point c and a subsequence $\langle x_{n_k}' \rangle$ of $\langle x_n' \rangle$ exists such that $x_{n_k}' \to c$ as $k \to +\infty$. Since F is closed, $c \in F$. We obtain

$$|x_{n_k}' - x_{n_k}''| < \frac{1}{n_k} \leqslant \frac{1}{k} \tag{11.8}$$

and

$$|f(x_{n_k}') - f(x_{n_k}'')| \geqslant \epsilon > 0 \tag{11.9}$$

for each positive integer k. By (11.7), we have $\lim_{k \to +\infty}(x_{n_k}' - x_{n_k}'') = 0$ and, therefore,

$$\begin{aligned} \lim_{k \to +\infty} x_{n_k}'' &= \lim_{k \to +\infty} (x_{n_k}'' - x_{n_k}' + x_{n_k}') \\ &= \lim_{k \to +\infty} (x_{n_k}'' - x_{n_k}') + \lim_{k \to +\infty} x_{n_k}' \\ &= 0 + c \\ &= c \end{aligned}$$

so that $x_{n_k}'' \to c$ as $k \to +\infty$. Since $c \in F$, f is continuous at c. Since $x_{n_k}'' \to c$, $x_{n_k}' \to c$ as $k \to +\infty$, it follows that

$$\lim_{k \to +\infty} f(x_{n_k}'') = f(c) = \lim_{k \to +\infty} f(x_{n_k}').$$

These yield $\lim_{k \to +\infty}(f(x_{n_k}'') - f(x_{n_k}')) = 0$. This is impossible by (11.9). Hence, f is necessarily uniformly continuous on F.

Corollary. *A function which is continuous on a bounded closed interval is uniformly continuous there.*

PROOF. Exercise.

PROB. 11.8. Prove: Each polynomial function on $\mathbb{R}$ is uniformly continuous on a bounded closed interval.

PROB. 11.9. Prove: f, where $f(x) = \sqrt{a^2 + x^2}$ for all $x \in \mathbb{R}$, $a \neq 0$, is uniformly continuous.

CHAPTER VII
Derivatives

1. The Derivative of a Function

Limits often arise from considering the derivative of a function at a point.

Def. 1.1. If f is a real-valued function of a real variable and a is an interior point of $\mathfrak{D}(f)$, then f is said to be *differentiable at* a if and only if

$$\lim_{x\to a} \frac{f(x)-f(a)}{x-a} \tag{1.1}$$

exists and is finite. When f is differentiable at a, the limit in (1.1) is called the *derivative of* f *at* a and will be written as $f'(a)$ or $(Df)(a)$. Thus,

$$(Df)(a) = f'(a) = \lim_{x\to a} \frac{f(x)-f(a)}{x-a}\,. \tag{1.2}$$

[We pause in our development to remind the reader that derivatives can be used to define the tangent line to the graph of a function. The slope of a line T' (see Fig. 1.1) joining $(a, f(a))$ and $(x, f(x))$, $x \neq a$, is

$$\frac{f(x)-f(a)}{x-a}\,.$$

If f is differentiable at a, then

$$f'(a) = \lim_{x\to a} \frac{f(x)-f(a)}{x-a}\,.$$

We define the tangent line T to the graph of f at $(a, f(a))$ as the line through $(a, f(a))$ having the slope $f'(a)$. The equation of the line T is

$$y = f(a) + f'(a)(x-a), \qquad x \in \mathbb{R}.] \tag{1.3}$$

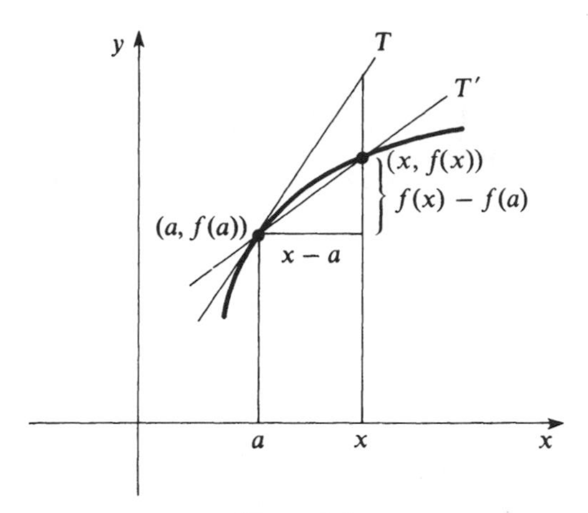

Figure 1.1

Def. 1.2. The *derivative* of f or its *derived function* is the function f' whose domain $\mathfrak{D}(f')$ is the set

$$\mathfrak{D}(f') = \{x \in \mathfrak{D}(f) \mid f \text{ is differentiable at } x\}$$

and whose value at x is $f'(x)$. We will also write the derivative of f as Df. Other notations for the derivative of a function will also be used.

Other Notation for Derivatives

The following is a slight variant of (1.2). Put $x - a = h$, then $x = a + h$ and

$$f'(a) = \lim_{h \to 0} \frac{f(a+h) - f(a)}{h}. \tag{1.3'}$$

The traditional notation introduced by Leibniz uses the notation y for the function f and dy/dx for its derivative. In this notation one often writes $x + \Delta x$ for a value different from x. Δx is called the *increment* or the *change* in x. The *increment in* y induced by the increment Δx in x is defined as

$$\Delta y = f(x + \Delta x) - f(x). \tag{1.4}$$

dy/dx is given by

$$\frac{dy}{dx} = \lim_{\Delta x \to 0} \frac{\Delta y}{\Delta x} = \lim_{\Delta x \to 0} \frac{f(x + \Delta x) - f(x)}{\Delta x}, \tag{1.5}$$

where $\Delta x \neq 0$.

Some books use

$$\frac{df}{dx} \quad \text{for} \quad f' \quad \text{and} \quad y' \quad \text{for} \quad \frac{dy}{dx}.$$

In the Leibniz notation the derivative of the function y at a is written as

$$\left(\frac{dy}{dx}\right)_{x=a} \quad \text{or as} \quad \left.\frac{dy}{dx}\right|_{x=a}.$$

A convenient notation we shall often use for the derivative of f at x is

$$\frac{df(x)}{dx} = f'(x). \tag{1.6}$$

EXAMPLE 1.1. From Corollary 1 of Theorem V.7.3 we know that

$$\lim_{h \to 0} \frac{e^h - 1}{h} = 1.$$

It follows that if $x \in \mathbb{R}$, then

$$\frac{de^x}{dx} = \lim_{h \to 0} \frac{e^{x+h} - e^x}{h} = \lim_{h \to 0} e^x \frac{e^h - 1}{h} = e^x \lim_{h \to 0} \frac{e^h - 1}{h} = e^x 1 = e^x,$$

so that

$$\frac{de^x}{dx} = e^x \qquad \text{for} \quad x \in \mathbb{R}. \tag{1.7}$$

EXAMPLE 1.2. We prove that

$$\frac{d \sin x}{dx} = \cos x \qquad \text{for} \quad x \in \mathbb{R}. \tag{1.8}$$

By Theorem V.4.2,

$$\lim_{h \to 0} \frac{\sin h}{h} = 1 \quad \text{and} \quad \lim_{h \to 0} \frac{\cosh - 1}{h} = 0.$$

If $h \neq 0$, then

$$\begin{aligned} \frac{\sin(x+h) - \sin x}{h} &= \frac{\sin x \cos h + \cos x \sin h - \sin x}{h} \\ &= \sin x \left(\frac{\cos h - 1}{h} \right) + \cos x \frac{\sin h}{h}. \end{aligned}$$

This implies that

$$\begin{aligned} \frac{d \sin x}{dx} &= \lim_{h \to 0} \frac{\sin(x+h) - \sin x}{h} = \lim_{h \to 0} \left(\sin x \frac{\cos h - 1}{h} + \cos x \frac{\sin h}{h} \right) \\ &= (\sin x)0 + (\cos x)(1) \\ &= \cos x, \end{aligned}$$

which proves (1.8).

PROB. 1.1. Prove that

$$\frac{d(\cos x)}{dx} = -\sin x.$$

Prob. 1.2. (a) Let c be the constant function whose value is c for all $x \in \mathbb{R}$. Prove:

$$\frac{dc}{dx} = 0.$$

(b) Let f be the identity function in $\mathbb{R}$, i.e., $f(x) = x$ for $x \in \mathbb{R}$. Prove: $d(x)/dx = 1$ for $x \in \mathbb{R}$.

In what follows it will be convenient to use the notion of a one-sided interior point of a subset of $\mathbb{R}$.

Def. 1.3. If $x_0 \in S \subseteq \mathbb{R}$, then we call x_0 an *interior point* of S *from the right* when some ϵ-neighborhood $N_+(x_0, \epsilon) = [x_0, x_0 + \epsilon)$ of x_0 from the right is contained in S. Dually, $x_0 \in S \subseteq \mathbb{R}$ will be called an *interior point* of S *from the left* when some ϵ-neighborhood $N_-(x_0, \epsilon) = (x_0 - \epsilon, x_0]$ of x_0 from the left is contained in S.

For example, let $[a, b]$ be some bounded closed interval. The point a is a left endpoint of $[a, b]$ and is an interior point of $[a, b]$ from the right. Also, b is an interior point of $[a, b]$ from the left. Certainly, an interior point of a set S is an interior point of S from the right and from the left. However, an interior point of a set from one side need not be an interior point of the set.

We now define one-sided derivatives at a point in the obvious way.

Def. 1.4. Let f be a real-valued function of a real variable and a an interior point of $\mathfrak{D}(f)$ from the right. If

$$\lim_{x \to a+} \frac{f(x) - f(a)}{x - a} \tag{1.9}$$

exists and is finite, then we say that f is *differentiable at a from the right*, call the limit the *derivative of f at a from the right*, and write

$$f_R'(a) = \lim_{x \to a+} \frac{f(x) - f(a)}{x - a}. \tag{1.10}$$

Similarly, if a is an interior part of $\mathfrak{D}(f)$ from the left and

$$\lim_{x \to a-} \frac{f(x) - f(a)}{x - a} \tag{1.11}$$

exists and is finite, then we say that f is *differentiable at a from the left*, call this limit the *derivative of f at a from the left*, and write

$$f_L'(a) = \lim_{x \to a-} \frac{f(x) - f(a)}{x - a}. \tag{1.12}$$

Prob. 1.3. Prove: If f is a real-valued function of a real variable and a is an interior point of $\mathfrak{D}(f)$, then (1) f is differentiable at a if and only if both

$f'_L(a)$ and $f'_R(a)$ exist, are finite, and equal, that is, $f'_L(a) = f'_R(a)$; (2) $f'(a) = f_L{}'(a) = f'_R(a)$ if and only if $f'_L(a) = f'_R(a)$.

EXAMPLE 1.3. We prove: If $\alpha \in \mathbb{R}$, then, for $x > 0$,

$$\frac{dx^{\alpha}}{dx} = \alpha x^{\alpha-1}. \tag{1.13}$$

We use Prob. V.7.1. By part (a) of that problem we have: If $\alpha \leqslant 0$ or $\alpha \geqslant 1$, and $x > 0$, $x + h > 0$, then

$$\begin{aligned}\alpha x^{\alpha-1}h = \alpha x^{\alpha-1}(x + h - x) \leqslant (x+h)^{\alpha} - x^{\alpha} &\leqslant \alpha(x+h)^{\alpha-1}(x + h - x)\\ &= \alpha(x+h)^{\alpha-1}h.\end{aligned}$$

This means that if $x > 0$, $x + h > 0$, $\alpha \leqslant 0$ or $\alpha \geqslant 1$, then

$$\alpha x^{\alpha-1}h \leqslant (x+h)^{\alpha} - x^{\alpha} \leqslant \alpha(x+h)^{\alpha-1}h. \tag{1.14}$$

It follows that

$$\begin{aligned}\alpha x^{\alpha-1} &\leqslant \frac{(x+h)^{\alpha-1} - x^{\alpha}}{h} \leqslant \alpha(x+h)^{\alpha-1} \qquad \text{if} \quad h > 0,\\ \alpha x^{\alpha-1} &\geqslant \frac{(x+h)^{\alpha-1} - x^{\alpha}}{h} \geqslant \alpha(x+h)^{\alpha-1} \qquad \text{if} \quad -x < h < 0.\end{aligned} \tag{1.15}$$

Since, for $x > 0$, $(x+h)^{\alpha-1} \to x^{\alpha-1}$ as $h \to 0$ (Theorem V.7.2), the first row of inequalities in (1.15) yields

$$\lim_{h\to 0+} \frac{(x+h)^{\alpha-1} - x^{\alpha}}{h} = \alpha x^{\alpha-1} \qquad \text{for} \quad x > 0, \quad \alpha \leqslant 0 \quad \text{or} \quad \alpha \geqslant 1,$$

and the second row of inequalities yields

$$\lim_{h\to 0-} \frac{(x+h)^{\alpha-1} - x^{\alpha}}{h} = \alpha x^{\alpha-1} \qquad \text{for} \quad x > 0, \quad \alpha \leqslant 0 \quad \text{or} \quad \alpha \geqslant 1.$$

Since the last two one-sided limits are equal, (1.13) holds for $\alpha \leqslant 0$ or $\alpha \geqslant 1$. To complete the proof of (1.13) for $0 < \alpha < 1$, all we need do is use part (b) of Prob. V.7.1 and reason as we did here (with appropriate modifications, of course).

EXAMPLE 1.4. Suppose $\alpha > 1$ and

$$f(x) = \begin{cases} x^{\alpha} & \text{if} \quad x > 0\\ 0 & \text{if} \quad x = 0.\end{cases} \tag{1.16}$$

Here f is defined on $[0, +\infty)$ and 0 is an interior point of $[0, +\infty)$ from the right. From the definition of $f'_R(0)$ it follows that if $x > 0$, then

$$\frac{f(x) - f(0)}{x - 0} = \frac{x^{\alpha}}{x} = x^{\alpha-1}. \tag{1.17}$$

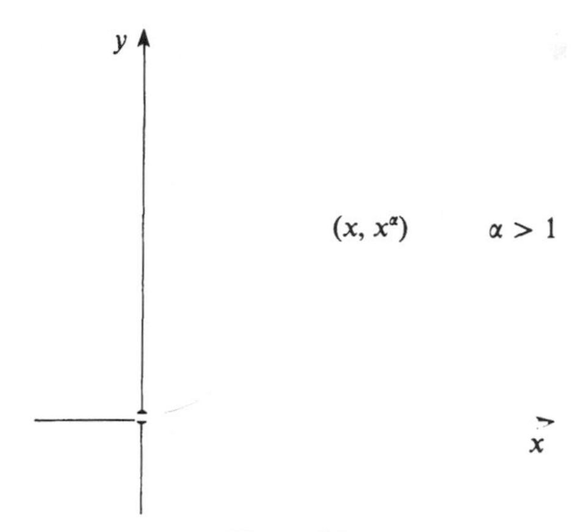

Figure 1.2

Since $\alpha - 1 > 0$, this implies that

$$f_R'(0) = \lim_{x \to 0+} \frac{f(x) - f(0)}{x - 0} = \lim_{x \to 0+} (x^{\alpha - 1}) = 0.$$

Since 0 is a left endpoint of the interval $[0, +\infty)$ on which f is defined, we say that f is differentiable at 0 and has the derivative $f'(0) = f_R'(0) = 0$ there (see Def. 1.5 below) even though it is just a one-sided derivative. We conclude that if $\alpha > 1$, then

$$f'(x) = \begin{cases} \alpha x^{\alpha - 1} & \text{if } x > 0 \\ 0 & \text{if } x = 0 \end{cases}$$

(see Fig. 1.2). Thus, for $\alpha > 1$, formula (1.13) retains its validity when $x = 0$.

PROB. 1.4. Prove: If $0 < \alpha < 1$ and f is defined as

$$f(x) = \begin{cases} x^\alpha & \text{if } x > 0 \\ 0 & \text{if } x = 0, \end{cases}$$

then f is not differentiable at 0 (see Fig. 1.3).

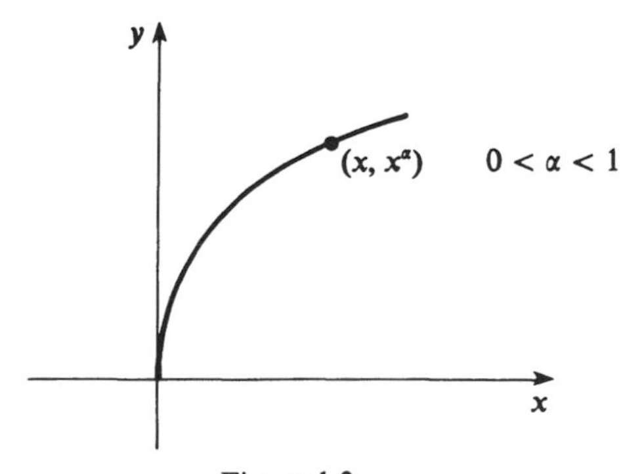

Figure 1.3

PROB. 1.5 (An Application of Formula (1.13)). Prove: (a) If $\alpha \in \mathbb{R}$, then

$$\lim_{n \to +\infty} \frac{(n+1)^{\alpha} - n^{\alpha}}{n^{\alpha-1}} = \alpha$$

and (b) use (a) and Theorem III.9.1 to prove that if $\alpha \geqslant 1$, then

$$\lim_{n \to +\infty} \frac{1^{\alpha-1} + 2^{\alpha-1} + \cdots + n^{\alpha-1}}{n^{\alpha}} = \frac{1}{\alpha}.$$

PROB. 1.6. Let $\alpha = m/n$, where m and n are integers and n is odd. Prove: (a) If $0 < n < m$, then formula (1.15) remains valid for all $x \in \mathbb{R}$; (b) If $m \leqslant n$, then formula (1.15) holds for all $x \in \mathbb{R}$, $x \neq 0$.

The notion of differentiability was defined at a point. We also speak of differentiability on a set.

Def. 1.5. When a function is differentiable at each point of its domain, then we say that it is *differentiable*. When the function is differentiable at each point of a set A, then we say that it is *differentiable on* A. If I is an interval of the form $(-\infty, a]$ or $[a, b]$ or $[a, +\infty)$, where a and b are in $\mathbb{R}$, then we say that f is *differentiable on* I if and only if it is differentiable at each interior point of I and differentiable from the appropriate side of an endpoint of I. The derivative of f at an endpoint of I is defined to be the appropriate one-sided derivative of f at that endpoint if the latter exists. For example, f is said to be differentiable on the bounded closed interval $[a, b]$ if and only if it is differentiable at each interior point of $[a, b]$ and differentiable at a from the right and at b from the left.

There exist functions defined on intervals which are differentiable at one point only (Prob. 1.7).

PROB. 1.7. Let f be defined as

$$f(x) = \begin{cases} x^2 & \text{if } x \text{ is rational} \\ 0 & \text{if } x \text{ is irrational.} \end{cases}$$

Prove that f is differentiable at $x = 0$ only.

2. Continuity and Differentiability. Extended Differentiability

Theorem 2.1. *If f is differentiable at a point, then it is continuous there.*

PROOF. Let f be differentiable at a. Then a is an interior point of $\mathfrak{D}(f)$ and there exists an ϵ-neighborhood $(a - \epsilon; a + \epsilon)$ such that $(a - \epsilon; a + \epsilon)$

$\subseteq \mathfrak{D}(f)$. Assume that $x \in (a - \epsilon; a + \epsilon)$ and $x \neq a$. Then

$$f(x) = f(a) + \frac{f(x) - f(a)}{x - a}(x - a),$$

and

$$\begin{aligned}\lim_{x \to a} f(x) &= \lim_{x \to a}\left(f(a) + \frac{f(x) - f(a)}{x - a}(x - a)\right)\\ &= f(a) + f'(a) \cdot 0\\ &= f(a).\end{aligned}$$

This establishes the conclusion.

The next example shows that the converse of Theorem 2.1 is false.

Example 2.1. The absolute value function is continuous on $\mathbb{R}$ and, in particular, at $x = 0$. However, it is not differentiable at $x = 0$. In fact, let $g(x) = |x|$ for $x \in \mathbb{R}$. Then

$$g_R'(0) = \lim_{x \to 0+} \frac{g(x) - g(0)}{x - 0} = \lim_{x \to 0+} \frac{|x| - |0|}{x} = \lim_{x \to 0+} \frac{x}{x} = \lim_{x \to 0+} 1 = 1$$

and

$$g_L'(0) = \lim_{x \to 0-} \frac{g(x) - g(0)}{x} = \lim_{x \to 0-} \frac{|x|}{x} = \lim_{x \to 0-} \frac{-x}{x} = -1.$$

Since $g_R'(0) = 1$ and $g_L'(0) = -1$, g is not differentiable at 0 even though it is continuous there. Later we shall give an example of a function which is continuous for all real x but is differentiable nowhere.

We note, however, that g is differentiable at $x \neq 0$. We first prove this for $x > 0$, i.e., we prove that for $x > 0$

$$g'(x) = \lim_{h \to 0} \frac{g(x + h) - g(x)}{h} = 1. \qquad (*)$$

Given $\epsilon > 0$, take h such that $0 < |h| < x$ so that $-x < h < x$ and $h \neq 0$ for such h. Since $h + x > 0$,

$$\left|\frac{|x + h| - |x|}{h} - 1\right| = \left|\frac{x + h - x}{h} - 1\right| = |1 - 1| = 0 < \epsilon \quad \text{for } 0 < |h| < x.$$

This proves $(*)$. We now prove: If $x < 0$, then

$$g'(x) = \lim_{h \to 0} \frac{g(x + h) - g(x)}{h} = -1. \qquad (**)$$

Given $\epsilon > 0$, take h such that $0 < |h| < -x$, so that $x < h < -x$ and $h \neq 0$. We have $x + h < 0$ and, therefore,

$$\left|\frac{|x + h| - |x|}{h} - (-1)\right| = \left|\frac{-x - h + x}{h} + 1\right| = |-1 + 1| = 0 < \epsilon$$

$$\text{for} \quad 0 < |h| < -x.$$

This proves $(**)$. Thus, the absolute value function is differentiable for all $x \neq 0$.

Def. 2.1. When x is an interior point of $\mathfrak{D}(f)$ and

$$\lim_{h \to 0} \frac{f(x+h) - f(x)}{h} = +\infty \quad \text{or} \quad -\infty,$$

then we will write, respectively, $f'(x) = +\infty$ or $f'(x) = -\infty$. We say that f is differentiable at x in the *extended sense* if and only if: either (a) f is differentiable at x (in the sense of Def. 1.1) or (b) f is continuous at x and $f'(x) = +\infty$ or $f'(x) = -\infty$. Similarly, if x is an interior point of $\mathfrak{D}(f)$ from one side, we call f differentiable at x in the *extended sense*, from that side, if and only if (c) f is differentiable from that side at x, or (d) f is continuous at x from that side and the derivative from that side at x is $+\infty$ or $-\infty$.

Example 2.2. It can happen that $f'(a) = +\infty$ or $f'(a) = -\infty$ and f is discontinuous at a. For example, let f be the signum function. We have

$$f(x) = \operatorname{sig} x \begin{cases} 1 & \text{if } x > 0 \\ 0 & \text{if } x = 0 \\ -1 & \text{if } x < 0. \end{cases}$$

Therefore,

$$f'_L(0) = \lim_{h \to 0-} \frac{\operatorname{sig}(h) - \operatorname{sig}(0)}{h} = \lim_{h \to 0-} \frac{-1}{h} = +\infty$$

and

$$f'_R(0) = \lim_{h \to 0+} \frac{\operatorname{sig}(h) - \operatorname{sig}(0)}{h} = \lim_{h \to 0+} \frac{1}{h} = +\infty.$$

Thus, $f'(0) = f'_R(0) = f'_L(0) = +\infty$. However, we do not call the signum function differentiable in the extended sense at 0 since it is not continuous there. (See Fig. 2.1.)

Example 2.3. Let m and n be odd integers such that $0 < m < n$. Let f be given by $f(x) = x^{m/n}$ for $x \in \mathbb{R}$. f is continuous for all $x \in \mathbb{R}$. At $x = 0$, we have

$$f'(0) = \lim_{x \to 0} \frac{x^{m/n} - 0^{m/n}}{x} = \lim_{x \to 0} \frac{1}{x^{(n-m)/n}} = +\infty.$$

The last equality holds since $n - m$ is an even positive integer and n is an odd positive integer (see Fig. 2.2). f is differentiable in the extended sense at $x = 0$ and is differentiable for all $x \neq 0$. Hence, f is differentiable in the extended sense.

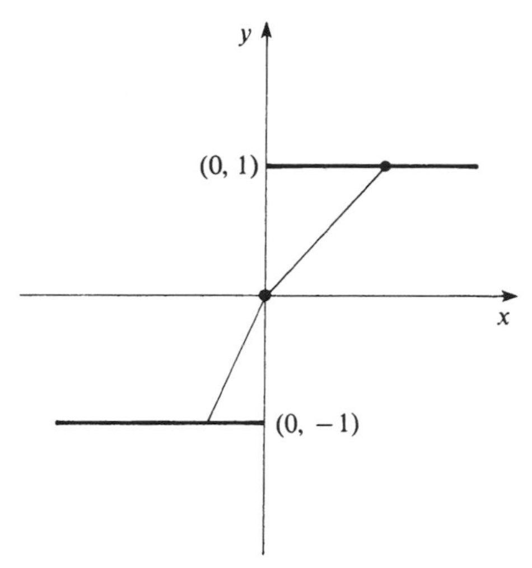

Figure 2.1

Example 2.4. Let h be given by $h(x) = x^{m/n}$ for $x \in \mathbb{R}$, where m and n are integers, m even and n odd and $0 < m < n$. Then

$$h'_R(0) = \lim_{x \to 0+} \frac{x^{m/n}}{x} = \lim_{x \to 0+} \frac{1}{x^{(n-m)/n}} = +\infty$$

and

$$h'_L(0) = \lim_{x \to 0-} \frac{1}{x^{(n-m)/n}} = -\infty.$$

Each of the equalities on the extreme left holds because $n - m$ and n are

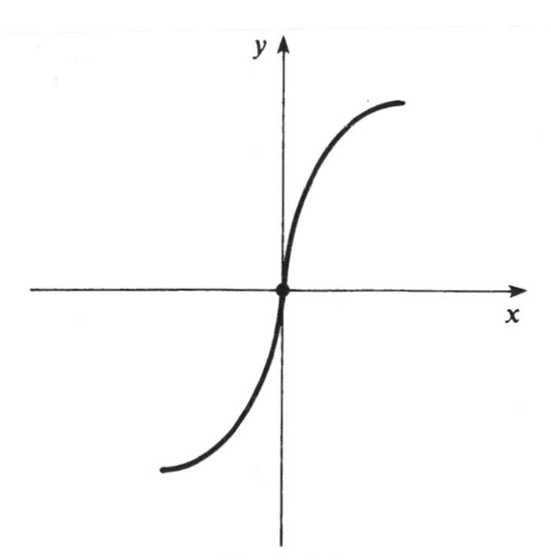

Figure 2.2

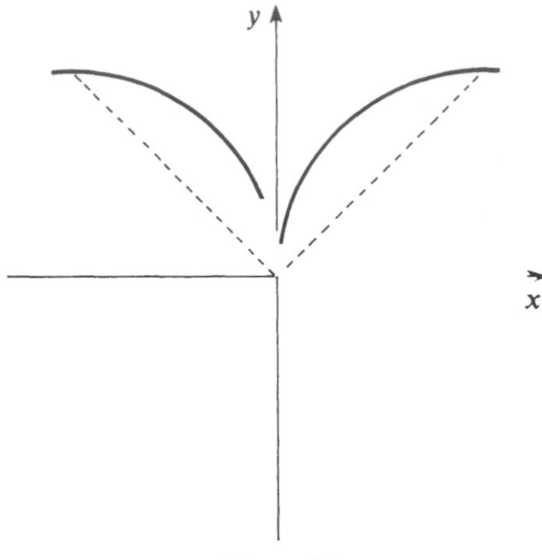

Figure 2.3

odd positive integers. Here $h'(0)$ does not exist, for $h'_L(0)$ and $h'_R(0)$ differ (see Fig. 2.3). h is continuous at 0, however, and so we have differentiability in the extended sense at $x = 0$ from two sides.

Terminology

If $f'(a) = +\infty$ or $f'(a) = -\infty$, then we say that $f'(a)$ exists even though we may not have differentiability in the extended sense at a.

3. Evaluating Derivatives. Chain Rule

PROB. 3.1. Prove: If u and v are functions having a common domain $\mathfrak{D}$ and each is differentiable at $x \in \mathfrak{D}$, then their sum $u + v$ and their difference $u - v$ are differentiable at x and

$$\frac{d(u(x) \pm v(x))}{dx} = \frac{du(x)}{dx} \pm \frac{dv(x)}{dx}.$$

PROB. 3.2. Prove: If n is a positive integer and $u_1, \ldots, u_n$ are functions having a common domain $\mathfrak{D}$ and each is differentiable at $x \in \mathfrak{D}$, then their sum $u_1 + u_2 + \cdots + u_n$ is differentiable at x and

$$\frac{d(u_1(x) + u_2(x) + \cdots + u_n(x))}{dx} = \frac{du_1(x)}{dx} + \frac{du_2(x)}{dx} + \cdots + \frac{du_n(x)}{dx}.$$

Prob. 3.3. Prove: If u is a function which is differentiable at x and c is some real number, then

$$\frac{d}{dx}(cf(x)) = c\frac{df(x)}{dx}.$$

Prob. 3.4. Prove:

$$\frac{d(\cosh x)}{dx} = \sinh x \quad \text{and} \quad \frac{d(\sinh x)}{dx} = \cosh x.$$

Prob. 3.5. Let n be some nonnegative integer and P a polynomial function on $\mathbb{R}$, where $P(x) = a_0x^n + a_1x^{n-1} + \cdots + a_{n-1}x + a_n$ for $x \in \mathbb{R}$. What is $P'(x)$?

Theorem 3.1. *If u and v are functions having a common domain $\mathfrak{D}$ and each is differentiable at $x \in \mathfrak{D}$, then their product is differentiable at x and*

$$\frac{du(x)v(x)}{dx} = u(x)\frac{dv(x)}{dx} + \frac{du(x)}{dx}v(x). \tag{3.1}$$

Proof. Note that since u is differentiable at x, it is continuous there so that

$$\lim_{h\to 0} u(x+h) = u(x). \tag{3.2}$$

Taking $h \neq 0$ and $x \in \mathfrak{D}$, $x + h \in \mathfrak{D}$, we obtain

$$\frac{u(x+h)v(x+h) - u(x)v(x)}{h} = u(x+h)\frac{v(x+h) - v(x)}{h} + \frac{u(x+h) - u(x)}{h}v(x). \tag{3.3}$$

Using (3.2), the fact that u and v are differentiable at x, and theorems on limits, we have

$$\begin{aligned}\frac{du(x)v(x)}{dx} &= \lim_{h\to 0}\left(u(x+h)\frac{v(x+h) - v(x)}{h} + \frac{u(x+h) - u(x)}{h}v(x)\right) \\ &= \lim_{h\to 0} u(x+h)\lim_{h\to 0}\frac{v(x+h) - v(x)}{h} \\ &\quad + \left(\lim_{h\to 0}\frac{u(x+h) - u(x)}{h}\right)v(x) \\ &= u(x)\frac{dv(x)}{dx} + \frac{du(x)}{dx}v(x),\end{aligned}$$

which proves the theorem.

PROB. 3.6. Let n be some positive integer and $u_1, u_2, \ldots, u_n$ functions having a common domain $\mathfrak{D}$. Prove: If $u_1, u_2, \ldots, u_n$ are all differentiable at x, then so is their product $u_1 u_2 \ldots u_n$, and

$$\begin{aligned}\frac{d}{dx}(u_1(x)u_2(x)\cdots u_n(x)) &= u_2(x)u_3(x)\cdots u_n(x)\frac{du_1(x)}{dx} \\ &\quad + u_1(x)u_3(x)\cdots u_n(x)\frac{du_2(x)}{dx} \\ &\quad + \cdots + u_1(x)u_2(x)\cdots u_{n-1}(x)\frac{du_n(x)}{dx}.\end{aligned}$$

Theorem 3.2. *If u and v are functions having a common domain $\mathfrak{D}$, both are differentiable at $x \in \mathfrak{D}$ and $v(x) \neq 0$, then their quotient u/v is differentiable at x and*

$$\frac{d}{dx}\left(\frac{u(x)}{v(x)}\right) = \frac{v(x)(du(x)/dx) - (dv(x)/dx)u(x)}{v^2(x)}. \tag{3.4}$$

PROOF. Since v is differentiable at x, it is necessarily continuous there. Since x is an interior point of $\mathfrak{D}$ and $v(x) \neq 0$, there exists a δ-neighborhood $N(x, \delta_1)$ of x such that $N(x, \delta_1) \subseteq \mathfrak{D}$. Because of the continuity of v at x and the fact that $v(x) \neq 0$, there exists a δ-neighborhood $N(x, \delta_2)$ of x such that $v(x+h) \neq 0$ for $x + h \in N(x, \delta_2) \cap \mathfrak{D}$. Also, there exists a δ-neighborhood $N(x, \delta)$ of x such that $N(x, \delta) \subseteq N(x, \delta_1) \cap N(x, \delta_2)$. Hence, $N(x, \delta) \subseteq N(x, \delta_1) \subseteq \mathfrak{D}$ and $N(x, \delta) \subseteq N(x, \delta_2)$ and $v(x + h) \neq 0$ for $x + h \in N(x, \delta)$. We now take $x + h \in N^*(x, \delta)$, so that $h \neq 0$ and $v(x + h) \neq 0$, and obtain

$$\begin{aligned}&\frac{1}{h}\left(\frac{u(x+h)}{v(x+h)} - \frac{u(x)}{v(x)}\right) \\ &\quad = \left[\left(\frac{u(x+h) - u(x)}{h}\right)v(x) - u(x)\frac{v(x+h) - v(x)}{h}\right] \\ &\quad\quad \times \frac{1}{v(x+h)v(x)}.\end{aligned} \tag{3.5}$$

The continuity of v at x implies that $v(x + h) \to v(x)$ as $h \to 0$. Using theorems on limits, (3.5) yields

$$\begin{aligned}\frac{d}{dx}\left(\frac{u(x)}{v(x)}\right) &= \lim_{h\to 0}\frac{1}{h}\left(\frac{u(x+h)}{v(x+h)} - \frac{u(x)}{v(x)}\right) \\ &= \frac{1}{v^2(x)}\left(v(x)\frac{du(x)}{dx} - \frac{dv(x)}{dx}u(x)\right).\end{aligned} \tag{3.6}$$

This completes the proof.

Prob. 3.7. Prove: If $\cos x \neq 0$, then

(a) $d\tan x/dx = \sec^2 x$ and
(b) $d\sec x/dx = \sec x \tan x$.

Prob. 3.8. Prove: If $\sin x \neq 0$,

(a) $d\cot x/dx = -\csc^2 x$ and
(b) $d\csc x/dx = -\csc x \cot x$.

Prob. 3.9. Prove: If $x \in \mathbb{R}$, then

(a) $d\tanh x/dx = \operatorname{sech}^2 x$,
(b) $d\operatorname{sech} x/dx = -\operatorname{sech} x \tanh x$.

Prob. 3.10. Prove: If $x \neq 0$, then

(a) $d\coth x/dx = -\operatorname{csch}^2 x$,
(b) $d\operatorname{csch} x/dx = -\operatorname{csch} x \coth x$.

Theorem 3.3 (The Chain Rule). *If f and g are real-valued functions of a real variable, $\mathfrak{R}(g) \subseteq \mathfrak{D}(f)$, g is differentiable at x and f at $u = g(x)$, then the composite function $f \circ g$ is differentiable at x and*

$$(f \circ g)'(x) = f'(g(x))g'(x). \tag{3.7}$$

Proof. The domain of $f \circ g$ is $\mathfrak{D}(g)$ and x is an interior point of $\mathfrak{D}(g) = \mathfrak{D}(f \circ g)$. There exists an ϵ-neighborhood $N(x, \delta_1)$ of x such that $N(x, \delta_1) \subseteq \mathfrak{D}(g)$. Take h such that $x + h \in \mathfrak{D}(g)$, $h \neq 0$, note that $g(x + h) \in \mathfrak{D}(f)$ and form

$$\frac{f(g(x+h)) - f(g(x))}{h}.$$

If $g(x + h) \neq g(x) = u$, then

$$\frac{f(g(x+h)) - f(g(x))}{h} = \frac{f(g(x+h)) - f(u)}{g(x+h) - u} \frac{g(x+h) - g(x)}{h}. \tag{3.8}$$

If $g(x + h) = g(x) = u$, then

$$\frac{f(g(x+h)) - f(g(x))}{h} = f'(u) \frac{g(x+h) - g(x)}{h} \tag{3.9}$$

since both sides are 0. Define the function v as follows:

$$v(x,h) = \begin{cases} \dfrac{f(g(x+h)) - f(u)}{g(x+h) - u} & \text{if} \quad g(x+h) \neq g(x) = u \\ f'(u) & \text{if} \quad g(x+h) = g(x) = u. \end{cases} \tag{3.10}$$

This, (3.8), and (3.9) yield

$$\frac{f(g(x+h)) - f(g(x))}{h} = v(x,h) \frac{g(x+h) - g(x)}{h}, \tag{3.11}$$

where $x + h \in N^*(x, \delta_1)$. We prove first that

$$\lim_{h \to 0} v(x,h) = f'(u). \tag{3.12}$$

Since f is differentiable at $u = g(x)$, there exists a $\delta_2 > 0$ such that if

$$u + k \in \mathfrak{D}(f) \quad \text{and} \quad 0 < |k| < \delta_2, \tag{3.13}$$

then

$$\left| \frac{f(u+k) - f(u)}{k} - f'(u) \right| < \epsilon. \tag{3.14}$$

But g is differentiable at x and, therefore, continuous there. Hence, a $\delta_3 > 0$ exists such that if

$$x + h \in \mathfrak{D}(g) = \mathfrak{D}(f \circ g) \quad \text{and} \quad |h| < \delta_3, \tag{3.15}$$

then

$$|g(x+h) - u| = |g(x+h) - g(x)| < \delta_2. \tag{3.16}$$

Take $0 < \delta \leqslant \min\{\delta_1, \delta_3\}$ so that $N(x,\delta) \subseteq N(x,\delta_1) \cap N(x,\delta_3) \subseteq \mathfrak{D}(g) = \mathfrak{D}(f \circ g)$. Put $k = g(x+h) - u$ so that $g(x+h) = u + k$, and take $x + h \in N^*(x,\delta)$. Accordingly, $x + h \in N(x,\delta_1)$, $x + h \in N(x,\delta_3)$, and $0 < |h| < \delta \leqslant \delta_3$. This implies that (3.15) holds and, therefore, that (3.16) holds. Thus, $u + k = g(x+h) \in N(u,\delta_2)$. If $g(x+h) = g(x) = u$, then

$$|v(x,h) - f'(u)| = |f'(u) - f'(u)| = 0 < \epsilon. \tag{3.17}$$

If $g(x+h) \neq g(x)$, then $k \neq 0$. Hence, $u + k \in N^*(u,\delta_2)$ and

$$0 < |g(x+h) - u| = |k| < \delta_2 \quad \text{and} \quad u + k = g(x+h) \in \mathfrak{D}(f). \tag{3.18}$$

Thus, (3.13) holds. But then (3.14) holds, and

$$|v(x,h) - f'(u)| = \left| \frac{f(g(x+h)) - f(u)}{g(x+h) - u} - f'(u) \right| < \epsilon. \tag{3.19}$$

Therefore, if $\epsilon > 0$ is given, then, for $0 < |h| < \delta$, one of (3.17) or (3.19) holds and

$$|v(x,h) - f'(u)| < \epsilon. \tag{3.20}$$

This proves (3.12). Returning to (3.11), we conclude that

$$\begin{aligned}
\lim_{h \to 0} \frac{f(g(x+h)) - f(g(x))}{h} &= \lim_{h \to 0} \left(v(x,h) \frac{g(x+h) - g(x)}{h} \right) \\
&= f'(u)\, g'(x) \\
&= f'(g(x))\, g'(x).
\end{aligned}$$

Thus, $f \circ g$ is differentiable at x and (3.7) holds.

Remark 3.1. If we put $y = f$ and $u = g$, then the chain rule can be written in the following easily remembered form:

$$\frac{dy}{dx} = \frac{dy(u)}{du}\frac{du}{dx} \tag{3.21}$$

or, more briefly, as

$$\frac{dy}{dx} = \frac{dy}{du} \cdot \frac{du}{dx}.$$

We can also write it as

$$\frac{d(y \circ u)}{dx} = \left(\frac{dy}{du} \circ u\right)\frac{du}{dx} \tag{3.22}$$

or as

$$\frac{df(g(x))}{dx} = \left(\frac{df}{du}\right)_{u=g(x)} \frac{dg(x)}{dx}.$$

EXAMPLE 3.1. If $p > 0$, then

$$p^x = e^{x \ln p}.$$

Hence, using the chain rule, we have

$$\frac{dp^x}{dx} = \frac{de^{x \ln p}}{dx} = e^{x \ln p}\frac{d(x \ln p)}{dx} = p^x \ln p.$$

PROB. 3.11. Let u be a function which is differentiable at x and $u(x) \neq 0$. Prove that $|u|$ is differentiable at x.

EXAMPLE 3.2. We claim that, for $x > 0$,

$$\frac{d \ln x}{dx} = \frac{1}{x}. \tag{3.23}$$

This follows from Theorem VI.5.7, part (d). Indeed, there we learned that, for $x > 0$,

$$\lim_{h \to 0} \frac{\ln(x+h) - \ln x}{h} = \frac{1}{x}.$$

If $x < 0$, then the chain rule implies that

$$\frac{d \ln(-x)}{dx} = \frac{1}{-x}\frac{d(-x)}{dx} = \frac{1}{-x}(-1) = \frac{1}{x}.$$

This and (3.23) can be summarized as

$$\frac{d \ln|x|}{dx} = \frac{1}{x} \quad \text{for} \quad x \neq 0. \tag{3.24}$$

Now let y be a differentiable function of x such that $y(x) \neq 0$. Then the chain rule and the above imply that

$$\frac{d\ln|y|}{dx} = \frac{1}{y}\frac{dy}{dx}. \tag{3.25}$$

EXAMPLE 3.3 (Logarithmic Differentiation). Let $a_1, a_2, \ldots, a_n; \alpha_1, \ldots, \alpha_n$ be real numbers where $a_1 < a_2 < \cdots < a_n$. Define y as

$$y(x) = (x-a_1)^{\alpha_1}(x-a_2)^{\alpha_2} \ldots (x-a_n)^{\alpha_n}, \tag{3.26}$$

where $x > a_n$. Since $x > a_n$, $y(x) > 0$. Hence, $\ln y(x)$ is defined. By properties of logarithms,

$$\ln y(x) = \alpha_1 \ln(x-a_1) + \alpha_2 \ln(x-a_2) + \cdots + \alpha_n \ln(x-a_n).$$

Taking derivatives of both sides, we have

$$\frac{1}{y}\frac{dy}{dx} = \frac{\alpha_1}{x-a_1} + \frac{\alpha_2}{x-a_2} + \cdots + \frac{\alpha_n}{x-a_n},$$

$$\begin{aligned}\frac{dy}{dx} &= y(x)\left(\frac{\alpha_1}{x-a_1} + \frac{\alpha_2}{x-a_2} + \cdots + \frac{\alpha_n}{x-a_n}\right)\\ &= \alpha_1(x-a_1)^{\alpha_1-1}(x-a_2)^{\alpha_2} \ldots (x-a_n)^{\alpha_n}\\ &\quad + \cdots + \alpha_n(x-a_1)^{\alpha_1} \ldots (x-a_{n-1})^{\alpha_n}(x-a_n)^{\alpha_n-1}.\end{aligned}$$

PROB. 3.12. Find dy/dx, where

(a) $y(x) = x^2\sin(1/x^2)$ if $x \neq 0$ and $y(0) = 0$
(b) $y = x\sqrt{1-x^2}$, $-1 \leqslant x \leqslant 1$,
(c) $y = \sqrt{(x-2)/(x-1)}$, where $x \geqslant 2$ or $x < 1$,
(d) $y = \ln(x^2-1)$, where $x^2 - 1 > 0$,
(e) $y = (x-1)(x-2)(x-3)(x-4)$.

PROB. 3.13. Prove: If $a > 0$, then

$$\frac{d\log_a x}{dx} = \frac{1}{x\ln a} = \frac{\log_a e}{x}, \qquad x > 0.$$

4. Higher-Order Derivatives

Def. 4.1(a). If f is a real-valued function of a real variable that is differentiable in some ϵ-neighborhood $N(x,\epsilon)$ of x and if its derivative f' is differentiable at x, then we say that f is *twice differentiable at* x and call the derivative of f' at x the *second derivative* of f at x. Some notations for the second derivative of f at x are

$$f''(x), \qquad (D^2f)(x), \qquad \frac{d^2f(x)}{dx^2}, \qquad \left.\frac{d^2f}{dx^2}\right|_x.$$

The function f'', whose domain $\mathfrak{D}(f'')$ is

$$\mathfrak{D}(f'') = \{x \in \mathfrak{D}(f) \mid f \text{ is twice differentiable at } x\},$$

is called the *second derivative* of f or the *derivative* of f *of the second order* (the first derivative of f is called the *derivative* of f *of the first order*). Some notations for it are

$$f'', \qquad D^2f, \qquad \frac{d^2f}{dx^2}.$$

We have

$$(f')' = f'', \qquad D(Df) = D^2f, \qquad \frac{d^2f}{dx^2} = \frac{d}{dx}\left(\frac{df}{dx}\right).$$

If f is differentiable on some ϵ-neighborhood of x and f' is differentiable in the extended sense at x, then we call f *twice differentiable in the extended sense* at x and the extended derivative of f' at x, the *second derivative of f in the extended sense* at x.

Remark 4.1. If f is defined on some open interval I, then we write $f^0 = f$, calling f itself the 0*th-order derivative* of f on I.

The nth-order derivative of a function can be defined by using complete induction.

Def. 4.1(b). Let f be a real-valued function of a real variable and x an interior point of $\mathfrak{D}(f)$, so that there exists an ϵ-neighborhood of x, $N(x,\epsilon)$, contained in $\mathfrak{D}(f)$. The 0th-order derivative of f is defined as $f^{(0)} = f$. Let n be some positive integer. If for all integers k such that $0 \leqslant k < n$, f is differentiable of order k for each point of $N(x,\epsilon)$, then f is said to be *differentiable of order* n at x if and only if $f^{(n-1)}$ is differentiable at x. In that case, the nth-order derivative of f at x is defined as

$$f^{(n)}(x) = (f^{(n-1)})'(x). \tag{4.1}$$

If $f^{(n-1)}$ is differentiable for each $x \in N(x,\epsilon)$, then we say that f is differentiable of order n on $N(x,\epsilon)$ and call the derivative of $f^{(n-1)}$ the *nth-order derivative* of f on $N(x,\epsilon)$. Thus,

$$f^{(n)} = (f^{(n-1)})'. \tag{4.2}$$

Alternate notations for the nth-order derivative of f at x and for the nth order derivative of f respectively are:

$$(D^{(n)}f)(x), \qquad \frac{d^nf(x)}{dx^n}$$

and

$$D^{(n)}f, \qquad \frac{d^n f}{dx^n}.$$

If n is a positive integer, then, by definition,

$$D^n f = D(D^{n-1}f) \quad \text{and} \quad \frac{d^n f}{dx^n} = \frac{d}{dx}\left(\frac{d^{n-1}f}{dx^{n-1}}\right). \tag{4.3}$$

If n is a nonnegative integer, then

$$f^{(n+1)} = (f^{(n)})'. \tag{4.4}$$

PROB. 4.1. Prove: If f is differentiable of order $n+1$, where n is some nonnegative integer, then

$$f^{(n+1)} = (f')^{(n)}.$$

PROB. 4.2. Prove: If m and n are nonnegative integers and f is differentiable of order $m+n$ on $N(x,\epsilon)$, then

$$f^{(m+n)} = \left(f^{(m)}\right)^{(n)}.$$

Remark 4.2. We exemplify the concept of higher-order *one-sided* derivatives by defining differentiability and derivative from the left at x, of order two. Let f be a real-valued function of a real variable, x an interior point of $\mathfrak{D}(f)$ from the left. Let f be differentiable from the left on some ϵ-neighborhood $N(x,\epsilon)$ of x from the left. Then f is called *differentiable* of order two at x from the left if and only if f'_L is differentiable from the left at x. We write the value of the derivative of f'_L at x from the left as $f''_L(x)$ and call it the *second derivative* of *f from the left* at x.

PROB. 4.3. Let f be defined as: $f(x) = |x|^3$ for each $x \in \mathbb{R}$. Prove: f is differentiable of order 2 on $\mathbb{R}$ but fails to be differentiable on $\mathbb{R}$ of order 3.

PROB. 4.4. Repeat the instructions of the last problem replacing f by the function g defined as $g(x) = x^{7/3}$ for $x \in \mathbb{R}$.

PROB. 4.5. Let P be a polynomial function on $\mathbb{R}$, where $P(x) = a_0x^n + a_1x^{n-1} + \cdots + a_{n-1}x + a_n$ for $x \in \mathbb{R}$. Prove: (a) $P^{(n)}(x) = n!\,a_0$ for $x \in \mathbb{R}$ and (b) $P^{(n+m)}(x) = 0$, where m is a positive integer.

A useful rule for calculating derivatives of higher order for the case of a function which is a product uv of functions u and v is *Leibniz's rule* stated in the next problem.

Prob. 4.6 (Leibniz's Rule). Prove: If u and v are differentiable functions of order n having a common domain $\mathfrak{D}$ and n is a positive integer, then

$$\frac{d^n(uv)}{dx^n} = \sum_{k=0}^{n} \binom{n}{k} \frac{d^{n-k}u}{dx^{n-k}} \frac{d^k v}{dx^k}$$

(Hint: use induction on n and the relation

$$\binom{n}{k-1} + \binom{n}{k} = \binom{n+1}{k},$$

where k is an integer such that $0 \leqslant k \leqslant n$).

Prob. 4.7. Using the same conditions on u and v as in Prob. 4.6 above, prove:

$$\frac{d}{dx}\left(\sum_{k=1}^{n} (-1)^{k+1} \frac{d^{n-k}u}{dx^{n-k}} \frac{d^{k-1}v}{dx^{k-1}} \right) = v\frac{d^n u}{dx^n} + (-1)^{n+1} u \frac{d^n v}{dx^n}.$$

An Application to Polynomials

We first define a special sequence p_n of polynomials as follows:

$$\begin{aligned} p_0(x) &= 1 \qquad \text{for all} \quad x \in \mathbb{R} \\ p_n(x) &= x^n \qquad \text{for} \quad x \in \mathbb{R}, \end{aligned} \tag{4.5}$$

where n is a positive integer. The 0th-order derivative of p_n is

$$p_n^{(0)}(x) = x^n. \tag{4.6}$$

If n and j are positive integers and $p_n^{(j)}$ is the jth-order derivative of p_n, then

$$\begin{aligned} & p_n^{(j)}(x) = n(n-1)\cdots(n-j+1)x^{n-j}, \qquad 1 \leqslant j \leqslant n, \\ & p_n^{(n)}(x) = n!, \\ & p_n^{(j)}(x) = 0 \qquad \text{if} \quad j > n. \end{aligned} \tag{4.7}$$

Using the notation for factorials of order n adopted in Section II.6 we recall that if n and j are non-negative integers, then

$$(n)_j = \begin{cases} 1 & \text{if} \quad j = 0 \\ n(n-1)\cdots(n-j+1) & \text{if} \quad 1 \leqslant j \leqslant n \\ 0 & \text{if} \quad j > n \end{cases} \tag{4.8}$$

and

$$\binom{n}{j} = \frac{(n)_j}{j!}. \tag{4.9}$$

Therefore,

$$n(n-1)\cdots(n-j+1)=j!\binom{n}{j}, \qquad \text{if} \quad 1 \leqslant j \leqslant n. \tag{4.10}$$

This and (4.7) yield

$$p_n^{(j)}(x)=j!\binom{n}{j}x^{n-j} \qquad \text{for} \quad x \in \mathbb{R} \quad \text{and} \quad 0 \leqslant j \leqslant n. \tag{4.11}$$

By the Binomial Theorem, if a and x are real numbers and n is a nonnegative integer, then

$$p_n(x)=x^n=(a+(x-a))^n=\sum_{j=0}^{n}\binom{n}{j}a^{n-j}(x-a)^j. \tag{4.12}$$

By (4.11), $p_n^{(j)}(a)=j!\binom{n}{j}a^{n-j}$. Substitution in (4.12) yields

$$x^n=p_n(x)=\sum_{j=0}^{n}\frac{p_n^{(j)}(a)}{j!}(x-a)^j. \tag{4.13}$$

This last formula is a special case of:

Theorem 4.1. *If P is a polynomial on $\mathbb{R}$, where*

$$P(x)=a_0x^n+a_1x^{n-1}+\cdots+a_{n-1}x+a_n=\sum_{k=0}^{n}a_{n-k}x^{n-k}$$

for x, a_n in $\mathbb{R}$,

then

$$P(x)=\sum_{k=0}^{n}\frac{p^{(k)}(a)}{k!}(x-a)^k. \tag{4.14}$$

Proof. Use the notation of the discussion preceding this theorem and put $p_n(x)=x^n$ for $x \in \mathbb{R}$ and $n \in \mathbb{Z}_0$. $P(x)$ may be written

$$P(x)=\sum_{k=0}^{n}a_{n-k}x^k=\sum_{k=0}^{n}a_{n-k}p_k(x). \tag{4.15}$$

Now use (4.13) and substitute in the above for $p_k(x)$ to obtain

$$P(x)=\sum_{k=0}^{n}a_{n-k}\left(\sum_{j=0}^{k}\frac{p_k^{(j)}(a)}{j!}(x-a)^j\right). \tag{4.16}$$

Note that $0 \leqslant j \leqslant k \leqslant n$. Fixing j, we sum with respect to k and then with respect to j. We conclude that

$$P(x)=\sum_{j=0}^{n}\frac{(x-a)^j}{j!}\sum_{k=j}^{n}a_{n-k}p_k^{(j)}(a). \tag{4.17}$$

We claim that $\sum_{j=0}^{n}a_{n-k}p_k^{(j)}(a)$ is the jth derivative of P at a. To see this, we take the jth derivative of P in (4.15),

$$P^{(j)}(x) = \sum_{k=0}^{n} a_{n-k} p_k^{(j)}(x). \tag{4.18}$$

If $j > k$, then $p_k^{(j)}(x) = 0$. Eq. (4.18) can be written

$$P^{(j)}(x) = \sum_{k=j}^{n} a_{n-k} p_k^{(j)}(x).$$

Hence

$$P^{(j)}(a) = \sum_{k=j}^{n} a_{n-k} p_k^{(j)}(a).$$

We substitute the second of these expressions in (4.17) and obtain

$$P(x) = \sum_{j=0}^{n} \frac{(x-a)^j}{j!} P^{(j)}(a) = \sum_{j=0}^{n} \frac{P^{(j)}(a)}{j!} (x-a)^j.$$

This completes the proof.

Remark 4.3. This theorem reveals a great deal about polynomial functions. We recall that if P is a polynomial of degree $n \geqslant 1$ and r is a real number such that $P(r) = 0$, then we call r a *real root* of the equation $P(x) = 0$. We also refer to r as a *real zero* of the polynomial P. Below we define roots of multiplicity k.

Def. 4.2. If P is a polynomial on $\mathbb{R}$ of degree $n \geqslant 1$, k is an integer such that $1 \leqslant k \leqslant n$, and r is a real number such that

$$\begin{aligned} &P(x) = (x-r)^k Q_{n-k}(x) \qquad \text{for all} \quad x \in \mathbb{R}, \\ &Q_{n-k}(r) \neq 0, \end{aligned} \tag{4.19}$$

where Q_{n-k} is a polynomial on $\mathbb{R}$ of degree $n-k$, then r is called a *zero of multiplicity* k of P or a *real root of multiplicity* k of the equation $P(x) = 0$. A zero of P of multiplicity 1 is called a *simple zero* of P or a *simple root* of $P(x) = 0$.

Corollary (of Theorem 4.1). *If P is a polynomial of degree $n \geqslant 1$, then r is a real zero of multiplicity k of P if and only if $r \in \mathbb{R}$,*

$$P(r) = P'(r) = \cdots = P^{(k-1)}(r) = 0 \quad \textit{and} \quad P^{(k)}(r) \neq 0. \tag{4.20}$$

PROOF. Let $P(x) = a_0 x^n + a_1 x^{n-1} + \cdots + a_{n-1} x + a_n$, where $a_0 \neq 0$. Suppose that (4.20) holds. By Theorem 4.1,

$$P(x) = P(r) + P'(r)(x-r) + \cdots + \frac{P^{(n)}(r)}{n!}(x-r)^n \qquad \text{for} \quad x \in \mathbb{R}. \tag{4.21}$$

Note that $P^{(n)}(r)/n! = a_0$ (Prob. 4.5). Using (4.20) in (4.21), we obtain

$$P(x) = \frac{P^{(k)}(r)}{k!}(x-r)^k + \cdots + \frac{P^{(n)}(r)}{n!}(x-r)^n$$
$$= (x-r)^k\left(\frac{P^{(k)}(r)}{k!} + \cdots + \frac{P^{(n)}(r)}{n!}(x-r)^{n-k}\right). \quad (4.22)$$

Put

$$Q_{n-k}(x) = \frac{P^{(k)}(r)}{k!} + \cdots + \frac{P^{(n)}(r)}{n!}(x-r)^{n-k}. \quad (4.23)$$

It is clear that Q_{n-k} is a polynomial of degree $n-k$. By (4.22) we have

$$P(x) = (x-r)^k Q_{n-k}(x) \qquad \text{for} \quad x \in \mathbb{R}.$$

Moreover, (4.23) implies that

$$Q_{n-k}(r) = \frac{P^{(k)}(r)}{k!} \neq 0.$$

Thus, r is a root of multiplicity k of P.

Conversely, let r be a zero of P of multiplicity k so that $P(x) = (x-r)^k \cdot Q_{n-k}(x)$ for all $x \in \mathbb{R}$, where $k \geqslant 1$ and $Q_{n-k}(r) \neq 0$. Take the ith derivative on both sides of the equality in the last sentence to obtain with the aid of Leibniz's rule (Prob. 4.6):

$$P^{(i)}(x) = \frac{d^i}{dx^i}\left((x-r)^k Q_{n-k}(x)\right)$$
$$= \sum_{j=0}^{i}\binom{i}{j}\frac{d^{i-j}Q_{n-k}(x)}{dx^{i-j}}\,\frac{d^j(x-r)^k}{dx^j}. \quad (4.24)$$

Now

$$\frac{d^j(x-r)^k}{dx^j} = \begin{cases}(x-r)^k & \text{if } j=0\\ k(k-1)\cdots(k-j+1)(x-r)^{k-j} & \text{if } 1 \leqslant j \leqslant k.\end{cases} \quad (4.25)$$

If $i < k$, then in (4.24), $0 \leqslant j \leqslant i < k$. It follows from (4.25) that

$$\left(\frac{d^j(x-r)^k}{dx^j}\right)_{x=r} = 0 \qquad \text{if} \quad i < k.$$

Therefore, (4.24) yields

$$P^{(j)}(r) = 0 \qquad \text{for} \quad 0 \leqslant i < k. \quad (4.26)$$

If $i = k$, however, then (4.24) becomes

$$P^{(k)}(x) = \sum_{j=0}^{k}\binom{k}{j}\frac{d^{k-j}Q_{n-k}(x)}{dx^{k-j}}\,\frac{d^j(x-r)^k}{dx^j}. \quad (4.27)$$

If we substitute r for x, then (4.25) implies that all the terms on the right vanish for $0 \leqslant j < k$ so that (4.27) yields, after substitution of r for x,

$$P^{(k)}(r) = \binom{k}{k} Q_{n-k}(r)k! = Q_{n-k}(r)k! \tag{4.28}$$

But, by hypothesis, r is a zero of multiplicity k of P. By definition, this implies that $Q_{n-k}(r) \neq 0$. By (4.28), this, in turn, implies that $P^{(k)}(r) \neq 0$. This and (4.26) imply (4.20), and the proof is complete.

PROB. 4.8. Prove: If P is a polynomial in $\mathbb{R}$ of degree $n \geqslant 1$ with leading coefficient a_0 and $r_1, r_2, \ldots, r_n$ are n distinct real zeros of P, then

$$P(x) = a_0(x - r_1)(x - r_2) \cdots (x - r_n) \qquad \text{for all} \quad x \in \mathbb{R}.$$

PROB. 4.9. Prove: If P is a polynomial in $\mathbb{R}$ of degree $n \geqslant 1$, then P has at most n distinct real zeros.

PROB. 4.10. Prove: If P and Q are polynomials in $\mathbb{R}$ each of degree not exceeding n (here, of course, n is some nonnegative integer) such that $P(x) = Q(x)$ holds for more than n distinct values of x, then $P = Q$, i.e., $P(x) = Q(x)$ holds for *all* values of x and both polynomials have the same coefficients.

Theorem 4.2 (Lagrange Interpolation Formula). *If n is a positive integer and $x_1, x_2, \ldots, x_n$ are n distinct real numbers, then for any real numbers $y_1, y_2, \ldots, y_n$ (not necessarily distinct) there exists a unique polynomial P of degree not exceeding $n-1$ such that*

$$P(x_1) = y_1, P(x_2) = y_2, \ldots, P(x_n) = y_n. \tag{4.29}$$

This polynomial is given by

$$P(x) = \frac{y_1 \prod_{k \neq 1}^{n}(x - x_k)}{\prod_{k \neq 1}^{n}(x_1 - x_k)} + \frac{y_2 \prod_{k \neq 1}^{n}(x - x_k)}{\prod_{k \neq 2}^{n}(x_2 - x_k)} + \cdots + \frac{y_n \prod_{k \neq n}^{n}(x - x_k)}{\prod_{k \neq n}^{n}(x_n - x_k)}, \tag{4.30}$$

where the symbol $\prod_{k \neq j}^{n}(x - x_k)$ for $1 \leqslant j \leqslant n$ is defined as follows: if $x \in \mathbb{R}$, then

$$\begin{aligned} \prod_{k \neq j}^{n}(x - x_k) &= \prod_{\substack{k=1 \\ k \neq j}}^{n}(x - x_k) \qquad \textit{for} \quad n > 1, \\ \prod_{k \neq 1}^{1}(x - x_k) &= 1 \qquad \textit{for} \quad n = 1. \end{aligned} \tag{4.31}$$

Before presenting a proof let us illustrate (4.30). Assume that $n = 3$, that x_1, x_2, x_3 are distinct, and that y_1, y_2, y_3 are given. We wish to find a polynomial P of degree not exceeding 2 such that $P(x_1) = y_1$, $P(x_2) = y_2$,

$P(x_3) = y_3$. In this case (4.30) becomes

$$P(x) = \frac{y_1(x - x_2)(x - x_3)}{(x_1 - x_2)(x_1 - x_3)} + \frac{y_2(x - x_1)(x - x_3)}{(x_2 - x_1)(x_2 - x_3)} + \frac{y_3(x - x_1)(x - x_2)}{(x_3 - x_1)(x_3 - x_2)}.$$

A simple check shows that P is a polynomial in x of degree not exceeding 2 such that $P(x_1) = y_1$, $P(x_2) = y_2$, $P(x_3) = y_3$.

We give a numerical example. We seek a polynomial P of degree not exceeding 2 such that $P(1) = 4$, $P(2) = 5$, $P(3) = 6$. Here $x_1 = 1$, $x_2 = 2$, $x_3 = 3$; $y_1 = 4$, $y_2 = 5$, $y_3 = 6$. According to (4.30), P is given by

$$P(x) = \frac{4(x-2)(x-3)}{(1-2)(1-3)} + \frac{5(x-1)(x-3)}{(2-1)(2-3)} + \frac{6(x-1)(x-2)}{(3-1)(3-2)}.$$

We simplify and obtain

$$P(x) = x + 3 \qquad \text{for} \quad x \in \mathbb{R}.$$

Note that P is of degree $1 < 2$ and that $P(1) = 1 + 3 = 4$, $P(2) = 2 + 3 = 5$, $P(3) = 3 + 3 = 6$.

Let us proceed with the proof of Theorem 4.2.

PROOF (of Theorem 4.2). If $y_1 = y_2 = \cdots = y_n = 0$, then (4.30) yields $P(x) = 0$ for all x, so P is the zero polynomial. In this case Eqs. (4.29) hold trivially, P has no degree, and, hence, its degree does not exceed $n - 1$ (explain). Suppose that at least one of $y_1, \ldots, y_n$ is not zero. If $n = 1$, there is only one y, i.e., y_1, and (4.30) becomes

$$P(x) = \frac{y_1 \prod_{k=1,k\neq 1}^{1}(x - x_k)}{\prod_{k=1,k\neq 1}^{1}(x_1 - x_k)} = \frac{y_1 \cdot 1}{1} = y_1 \neq 0 \qquad \text{for} \quad x \in \mathbb{R}.$$

This P has degree $0 = 1 - 1 = n - 1$, $P(x_1) = y_1$ holds and P satisfies conditions (4.29) with $n = 1$. If $n > 1$, then each term in (4.30) is a polynomial of degree not exceeding $n - 1$. This is so because it is a product of the constant

$$c_j = \frac{y_j}{\prod_{k=1,k\neq j}^{n}(x_j - x_k)}$$

and the polynomial P_j, where

$$P_j(x) = \prod_{\substack{k=1 \\ k\neq j}}^{n} (x - x_k) \qquad \text{for} \quad x \in \mathbb{R}.$$

P_j here has degree $n - 1$ for each j. Hence, the sum in (4.30) and, therefore, the polynomial P defined by that sum has degree not exceeding $n - 1$. Also, for each P_j in the sum on the right of (4.30), we have

$$P_j(x_j) = \prod_{\substack{k=1 \\ k\neq j}}^{n} (x_j - x_k),$$

$$P_j(x_i) = 0 \qquad \text{if} \quad i \neq j.$$

The equalities in the second line hold because $x_i \neq x_j$ and $x_i \in \{x_1, \ldots, x_n\}$. This causes all the terms in (4.30) except the jth to vanish. Hence,

$$P(x_j) = \frac{y_j P_j(x_j)}{P_j(x_j)} = y_j,$$

where $j \in \{1, \ldots, n\}$. Thus, the P in (4.30) is a polynomial of degree not exceeding $n-1$ which satisfies Eqs. (4.29).

As for the uniqueness of P, if Q also is a polynomial of degree not exceeding $n-1$ such that $Q(x_1) = y_1, \ldots, Q(x_n) = y_n$, then by Prob. 4.10 we have $P = Q$. This completes the proof.

Remark 4.4. There is another form of the Lagrange formula (4.30). Let $x_1, x_2, \ldots, x_n$ be distinct real numbers. Define the polynomial g as

$$g(x) = \prod_{k=1}^{n} (x - x_k) = (x - x_1)(x - x_2) \cdots (x - x_n) \qquad \text{for all} \quad x \in \mathbb{R}. \tag{4.32}$$

Using the notation adopted in (4.31), it is easy to see that

$$\begin{aligned}
\left(\frac{dg(x)}{dx}\right)_{x=x_1} &= g'(x_1) = \prod_{k \neq 1}^{n} (x_1 - x_k) = (x_1 - x_2) \cdots (x_1 - x_n), \\
\left(\frac{dg(x)}{dx}\right)_{x=x_2} &= g'(x_2) = \prod_{k \neq 2}^{n} (x_2 - x_k) = (x_2 - x_1)(x_2 - x_3) \cdots (x_2 - x_n), \\
&\vdots \\
\left(\frac{dg(x)}{dx}\right)_{x=x_n} &= g'(x_n) \\
&= \prod_{k \neq n}^{n} (x_n - x_k) = (x_n - x_1)(x_n - x_2) \cdots (x_n - x_{n-1}),
\end{aligned} \tag{4.33}$$

and that

$$\prod_{k \neq 1}^{n} (x - x_k) = \frac{g(x)}{x - x_1},$$

$$\prod_{k \neq 2}^{n} (x - x_k) = \frac{g(x)}{x - x_2}, \quad \ldots, \quad \prod_{k \neq n}^{n} (x - x_k) = \frac{g(x)}{x - x_n}, \tag{4.34}$$

where the first equality holds for $x \neq x_1$, the second for $x \neq x_2, \ldots$, and the last for $x \neq x_n$. Substituting (4.33) and (4.34) in (4.30) we obtain

$$P(x) = \frac{y_1 g(x)}{(x - x_1) g'(x_1)} + \frac{y_2 g(x)}{(x - x_2) g'(x_2)} + \cdots + \frac{y_n g(x)}{(x - x_n)(g'(x_n)} \tag{4.35}$$

for $x \neq x_1$, $x \neq x_2, \ldots, x \neq x_n$. This is the alternate form we spoke of. This also implies that

$$\frac{P(x)}{g(x)} = \frac{y_1}{(x - x_1)g'(x_1)} + \frac{y_2}{(x - x_2)g'(x_2)} + \cdots + \frac{y_n}{(x - x_n)g'(x_n)} \tag{4.36}$$

if x differs from $x_1, x_2, \ldots, x_n$. Note that $P(x_i) = y_i$ for $1 \leqslant i \leqslant n$. Upon substitution in (4.36), the latter can be written

$$\frac{P(x)}{g(x)} = \frac{P(x_1)}{(x - x_1)g'(x_1)} + \frac{P(x_2)}{(x - x_2)g'(x_2)} + \cdots + \frac{P(x_n)}{(x - x_n)g'(x_n)}, \tag{4.37}$$

where $x \neq x_i$ for $i \in \{1, \ldots, n\}$. In (4.37), P is a polynomial of degree not exceeding $n - 1$, and g is defined as $g(x) = \prod_{k=1}^{n}(x - x_k)$. The x_i's are distinct.

Prob. 4.11. Prove: If $x_1, \ldots, x_n$ are distinct and $g(x) = \prod_{k=1}^{n}(x - x_k)$, then

$$\frac{1}{g(x)} = \frac{1}{(x - x_1)g'(x_1)} + \frac{1}{(x - x_2)g'(x_2)} + \cdots + \frac{1}{(x - x_n)g'(x_n)}$$

for $x \in \mathbb{R}$, $x \neq x_1, x_2, \ldots, x_n$.

Prob. 4.12. Prove: If $x \notin \{-1, -2, \ldots, -n\}$ where n is some positive integer, then

$$\frac{1}{\prod_{k=0}^{n}(x + k)} = \frac{1}{x(x + 1)\cdots(x + n)} = \sum_{k=0}^{n}(-1)^k \frac{1}{k!\,(n - k)!\,(x + k)}$$

and

$$\frac{n!}{\prod_{k=0}^{n}(x + k)} = \sum_{k=0}^{n}(-1)^k \binom{n}{k} \frac{1}{x + k}$$

$$= \frac{1}{x} - n\frac{1}{x + 1} + \frac{n(n - 1)}{2!}\,\frac{1}{x + 2} + \cdots + (-1)^n \frac{1}{x + n}.$$

5. Mean-Value Theorems

We remind the reader that when we say that $f'(a)$ exists we mean that it may be infinite and f need not be differentiable in the extended sense.

Lemma 5.1. *Let f be a real-valued function of a real variable defined on an interval I and let $f(x_0)$ be a maximum of f on I. If x_0 is an interior point of*

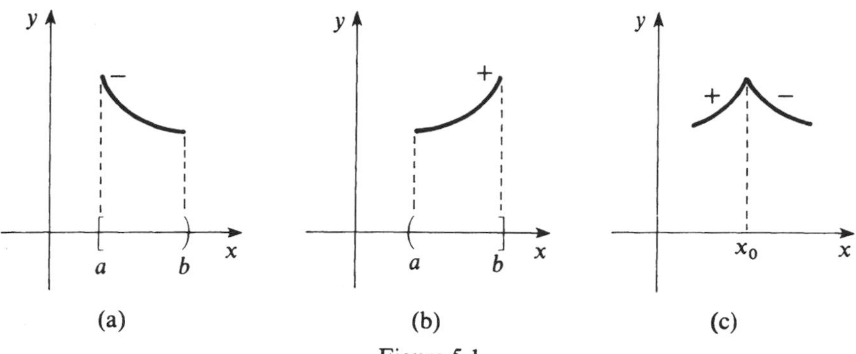

Figure 5.1

I from the right and $f'_R(x_0)$ exists, then $f'_R(x_0) \leqslant 0$. If x_0 is an interior point of I from the left and $f'_L(x_0)$ exists, then $f'_L(x_0) \geqslant 0$. If x_0 is an interior point of I and $f'_R(x_0)$ and $f'_L(x_0)$ exist, then

$$f'_L(x_0) \geqslant 0 \geqslant f'_R(x_0). \tag{5.1}$$

(*See* Fig. 5.1.)

Proof. Assume that x_0 is an interior point of I from the right. Take $x \in I$, $x > x_0$. Since $f(x_0)$ is the maximum of f, we have $f(x) \leqslant f(x_0)$. This implies that

$$\frac{f(x) - f(x_0)}{x - x_0} \leqslant 0 \quad \text{for} \quad x \in I, \quad x > x_0. \tag{5.2}$$

Hence,

$$f'_R(x_0) = \lim_{x \to x_0+} \frac{f(x) - f(x_0)}{x - x_0} \leqslant 0. \tag{5.3}$$

If x_0 is an interior point of I from the left, then we take $x \in I$, $x < x_0$ and obtain $f(x) \leqslant f(x_0)$. It follows that

$$\frac{f(x) - f(x_0)}{x - x_0} \geqslant 0 \quad \text{for} \quad x \in I, \quad x < x_0. \tag{5.4}$$

Hence,

$$f'_L(x_0) = \lim_{x \to x_0-} \frac{f(x) - f(x_0)}{x - x_0} \geqslant 0. \tag{5.5}$$

Now assume that x_0 is an interior point of I and that $f'_R(x_0)$ and $f'_L(x_0)$ exist. The above imply (5.1). This completes the proof.

Prob. 5.1. Let f be defined on an interval I and let $f(x_0)$ be a minimum of f on I. Prove: (a) If $f'_R(x_0)$ exists, then $f'_R(x_0) \geqslant 0$. (b) If $f'_L(x_0)$ exists, then $f'_L(x_0) \leqslant 0$. If x_0 is an interior point of I and $f'_L(x_0)$ and $f'_R(x_0)$ exist, then $f'_L(x_0) \leqslant 0 \leqslant f'_R(x_0)$.

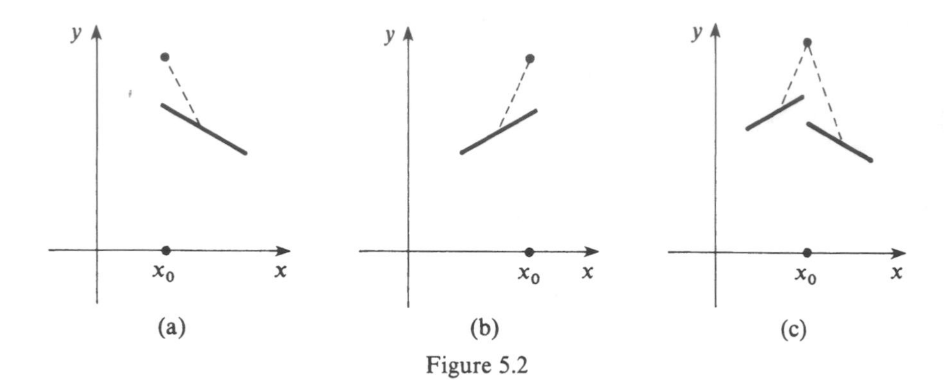

Figure 5.2

Def. 5.1. If f is defined on $S \neq \emptyset$ and $f(x_0)$ is a maximum or a minimum of f on S, then we call $f(x_0)$ an *extremum* of f on S.

PROB. 5.2. Let f be defined on an interval I and let x_0 be an interior point of I. Prove: If $f(x_0)$ is an extremum of on I and $f'(x_0)$ exists, then $f'(x_0) = 0$.

Remark 5.1. Lemma 5.1 includes cases such as those shown in Fig. 5.2 since, according to our terminology, existence of $f'_R(x_0)$ or $f'_L(x_0)$ includes the possibility that they are infinite. This allows for lack of one-sided differentiability even in the extended sense.

Theorem 5.1 (Rolle's Theorem with One-Sided Derivatives). *If f is a real-valued function of a real variable that is continuous on the closed interval $[a,b]$, differentiable (in the extended sense) from both sides at each interior point of $[a,b]$ and if $f(a) = f(b)$, then there exists an interior point x_0 of $[a,b]$ such that either*

$$f'_L(x_0) \geq 0 \geq f'_R(x_0) \tag{5.6a}$$

or

$$f'_L(x_0) \leq 0 \leq f'_R(x_0). \tag{5.6b}$$

PROOF. If f is constant on $[a,b]$, then it is differentiable at each x in $(a;b)$, and, moreover, $f'_L(x) = f'_R(x) = f'(x) = 0$ for all x in $(a;b)$, and the conclusion holds trivially. Now suppose f is not constant on $[a,b]$, so that a $c \in [a,b]$ exists such that $f(c) \neq f(a) = f(b)$. Suppose that $f(c) > f(a) = f(b)$. Since f is continuous on $[a,b]$ and the latter is a bounded closed set, f has a maximum $f(x_0)$ on $[a,b]$ so that $f(x_0) \geq f(c) > f(a) = f(b)$. This implies that x_0 is an interior point of $[a,b]$. Since $f'_L(x_0)$ and $f'_R(x_0)$ exist, Lemma 5.1 implies that (5.6a) holds. If $f(c) < f(a) = f(b)$, since we know that f has a minimum $f(x'_0)$ on $[a,b]$, it follows that $f(x'_0) \leq f(c) < f(a) = f(b)$. Thus, x'_0 is an interior point of $[a,b]$, and, by Prob. 5.1, we obtain

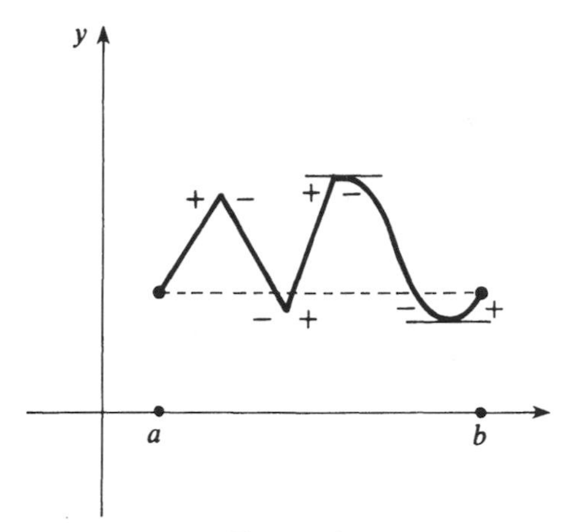

Figure 5.3

$f'_L(x'_0) \leqslant 0 \leqslant f'_R(x'_0)$. Thus, (5.6b) holds with x_0 replacing x'_0. This completes the proof. (See Fig. 5.3.)

Corollary (Rolle's Theorem). *If f is a real-valued function of a real variable which is continuous on the bounded closed interval $[a, b]$, differentiable in the extended case at each interior point of $[a, b]$ and $f(a) = f(b)$, then there exists an interior point x_0 of $[a, b]$ such that $f'(x_0) = 0$.*

PROOF. Exercise.

PROB. 5.3. Prove: (a) If f is continuous on an open interval $(a; b)$ where a and b are in $\mathbb{R}^*$ and $\lim_{x\to a+} f(x) = -\infty = \lim_{x\to b-} f(x)$, then f has a maximum on $(a; b)$. (b) If for the f in part (a) we also assume that it is differentiable in the extended sense on $(a; b)$, then there exists an x_0 in $(a; b)$ such that $f'(x_0) = 0$.

PROB. 5.4. Prove: (a) If f is continuous on an open interval $(a; b)$ where a and b are in $\mathbb{R}^*$ and $\lim_{x\to a+} f(x) = +\infty = \lim_{x\to b-} f(x)$, then f has a minimum on $(a; b)$. (b) If for the f in part (a) we also assume that it is differentiable in the extended sense on $(a; b)$, then there exists an x_0 in $(a; b)$ such that $f'(x_0) = 0$.

Theorem 5.2 (Mean-Value Theorem with One-Sided Derivatives). *If f is a real-valued function of a real variable that is continuous on the bounded closed interval $[a, b]$ and differentiable in the extended sense from both sides at each interior point of $[a, b]$, then there exists an interior point x_0 of $[a, b]$ such that either*

$$f'_L(x_0) \leqslant \frac{f(b) - f(a)}{b - a} \leqslant f'_R(x_0) \tag{5.7}$$

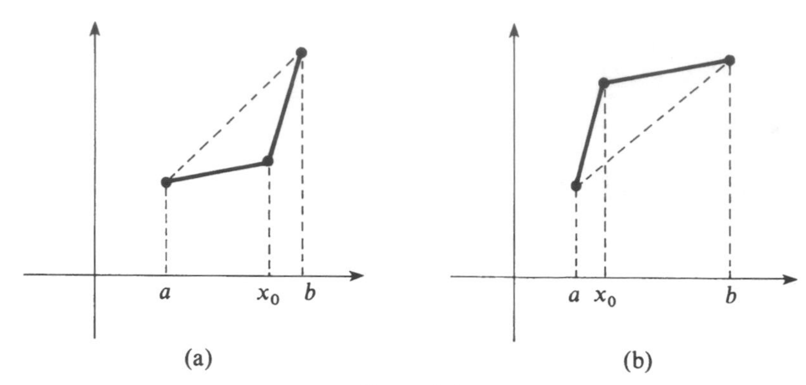

Figure 5.4

or

$$f'_L(x_0) \geqslant \frac{f(b)-f(a)}{b-a} \geqslant f'_R(x_0). \tag{5.8}$$

(*See* Fig. 5.4.)

PROOF. Construct the function g, where

$$g(x) = f(x) - f(a) - \frac{f(b)-f(a)}{b-a}(x-a) \qquad \text{for} \quad x \in [a,b]. \tag{5.9}$$

g is continuous on $[a,b]$ and $g(a) = g(b) = 0$. Let x be an interior point of $[a,b]$. Then

$$g'_L(x) = f'_L(x) - \frac{f(b)-f(a)}{b-a} \qquad \text{if} \quad f'_L(x) \in \mathbb{R}, \tag{5.10}$$

or $g'_L(x) = \pm\infty$ if $f'_L(x) = \pm\infty$. In the latter case, $g'_L(x) = \pm\infty = f'_L(x)$. Similarly, we obtain, from (5.9),

$$g'_R(x) = f'_R(x) - \frac{f(b)-f(a)}{b-a} \qquad \text{if} \quad f'_R(x) \text{ is finite} \tag{5.11}$$

or $g'_R(x) = \pm\infty$ if $f'_R(x) = \pm\infty$. In the latter case, $g'_R(x) = \pm\infty = f'_R(x)$. Thus, g is differentiable in the extended sense from each side at each interior point of $[a,b]$. We can, therefore, apply Theorem 5.1 to g and conclude that there exists an interior point x_0 of $[a,b]$ such that

$$g'_L(x_0) \leqslant 0 \leqslant g'_R(x_0) \tag{5.12a}$$

or

$$g'_L(x_0) \geqslant 0 \geqslant g'_R(x_0). \tag{5.12b}$$

Assume that (5.12a) holds. In this case, we know that $g'_L(x_0) \neq +\infty$ and $g'_R(x_0) \neq -\infty$ and, hence, that $f'_L(x_0) \neq +\infty$ and $f'_R(x_0) \neq -\infty$. Thus, either $f'_L(x_0) = -\infty$ or $f'_L(x_0) \in \mathbb{R}$ and $f'_R(x_0) = +\infty$ or $f'_R(x_0) \in \mathbb{R}$. Hence, there are four cases to consider: (1) $f'_L(x_0) = -\infty$, $f'_R(x_0) = +\infty$, (2) $f'_L(x_0) = -\infty$, $f'_R(x_0) \in \mathbb{R}$, (3) $f'_L(x_0) \in \mathbb{R}$, $f'_R(x_0) = +\infty$, (4) $f'_L(x_0) \in \mathbb{R}$,

$f'_R(x_0) \in \mathbb{R}$. If $f'_L(x_0) \in \mathbb{R}$, then (5.10) and (5.12a) imply that

$$f'_L(x_0) - \frac{f(b) - f(a)}{b - a} \leqslant 0$$

and, hence, that

$$f'_L(x_0) \leqslant \frac{f(b) - f(a)}{b - a}. \tag{5.13}$$

If $f'_R(x_0) \in \mathbb{R}$, then (5.11) and (5.12a) imply that

$$0 \leqslant f'_R(x_0) - \frac{f(b) - f(a)}{b - a}$$

and, hence, that

$$\frac{f(b) - f(a)}{b - a} \leqslant f'_R(x_0). \tag{5.14}$$

If $f'_L(x_0) \in \mathbb{R}$ *and* $f'_R(x_0) \in \mathbb{R}$, then the inequalities in (5.13) and (5.14) yield (5.7). Thus, (5.7) holds in case (4) above. By the properties of order in $\mathbb{R}^*$, (5.7) also holds in cases (1), (2) and (3). This proves the theorem if (5.12a) holds. We leave to the reader the proof that if (5.12b) holds, then (5.8) follows (Prob. 5.5).

Prob. 5.5. Complete the proof of the last theorem by proving that (5.12b) implies (5.8).

Remark 5.2 (Alternate Formulation of Theorem 5.2). Theorem 5.2 can be formulated as follows: If f is continuous on an interval I, differentiable in the extended sense from both sides at each interior point of I, and a, x are distinct points of I, then there exists a point x_0 between a and x such that

$$f(a) + f'_L(x_0)(x - a) \leqslant f(x) \leqslant f(a) + f'_R(x_0)(x - a) \tag{5.15}$$

or

$$f(a) + f'_L(x_0)(x - a) \geqslant f(x) \geqslant f(a) + f'_R(x_0)(x - a). \tag{5.16}$$

Prob. 5.6. Prove the formulation of Theorem 5.2 stated in the last remark.

Corollary (of Theorem 5.2) (The Mean-Value Theorem). *If f is continuous on the bounded closed interval $[a, b]$ and differentiable in the extended sense at each interior point of $[a, b]$, then there exists an interior point x_0 of $[a, b]$ such that*

$$f'(x_0) = \frac{f(b) - f(a)}{b - a}. \tag{5.17}$$

(*See* Fig. 5.5.)

Proof. Exercise.

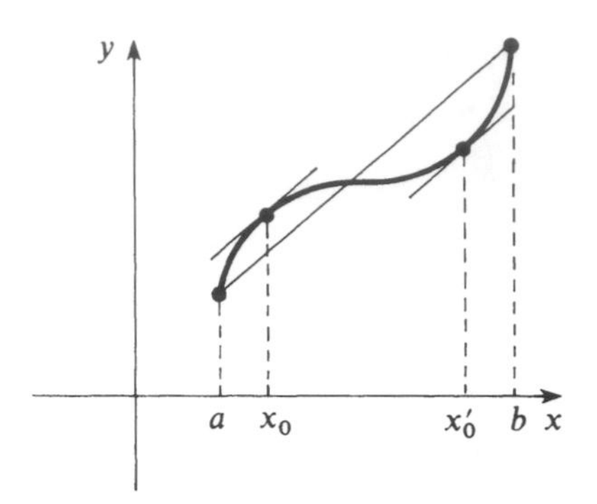

Figure 5.5

Remark 5.3 (Alternate Formulation of the Mean-Value Theorem). The corollary of Theorem 5.2 can be formulated as follows: If f is continuous on an interval I and differentiable in the extended sense at each interior point of I, and a, x are distinct points of I, then there exists a point x_0 between a and x such that

$$f(x) = f(a) + f'(x_0)(x - a). \tag{5.18}$$

PROB. 5.7. Prove the formulation of the corollary of Theorem 5.2 stated in Remark 5.3.

6. Some Consequences of the Mean-Value Theorems

Theorem 6.1. *Let f be continuous on an interval I and differentiable in the extended sense from both sides at each interior point of I.*

(a) *If $f_L'(x) \geqslant 0$ and $f_R'(x) \geqslant 0$ for each interior point x of I, then f is monotonically increasing on I; if these inequalities hold strictly, then f is strictly increasing on I.*
(b) *If $f_L'(x) \leqslant 0$ and $f_R'(x) \leqslant 0$ for each interior point x of I, then f is monotonically decreasing on I; if these inequalities hold strictly, then f is strictly decreasing on I.*

PROOF. We prove (a) and leave the proof of (b) to the reader (Prob. 6.1). Assume that (a) holds. Take x_1 and x_2 in I such that $x_1 < x_2$. By the formulation of Theorem 5.2 stated in Remark 5.2 we know that there exists an x_0 such that $x_1 < x_0 < x_2$ and

$$f(x_1) + f_L'(x_0)(x_2 - x_1) \leqslant f(x_2) \leqslant f(x_1) + f_R'(x_0)(x_2 - x_1) \tag{6.1}$$

or

$$f(x_1) + f'_L(x_0)(x_2 - x_1) \geqslant f(x_2) \geqslant f(x_1) + f'_R(x_0)(x_2 - x_1). \quad (6.2)$$

If (6.1) holds, then, since $f'_L(x_0) \geqslant 0$ and $x_2 - x_1 > 0$, it follows that $f'_L(x_0)(x_2 - x_1) \geqslant 0$ and, therefore that $f(x_1) \leqslant f(x_2)$. If (6.2) holds, then $f'_R(x_0)(x_2 - x_1) \geqslant 0$ and, therefore, $f(x_1) \leqslant f(x_2)$ holds in this case also. Thus, $x_1 < x_2$ implies that $f(x_1) \leqslant f(x_2)$ and, hence, f is monotonically increasing. If the inequalities in (a) are strict, then the same reasoning implies that if $x_1 < x_2$ in I, then $f(x_1) < f(x_2)$ so that f is strictly monotonically increasing.

PROB. 6.1. Complete the proof of the last theorem by proving part (b) there.

Corollary (of Theorem 6.1). *Let f be continuous on an interval I and differentiable in the extended sense at each interior point x of I. If (a) $f'(x) \geqslant 0$ for each interior point $x \in I$, then f is monotonic increasing on I; if the inequality in (a) is strict, then f is strictly monotonically increasing on I. If (b) $f'(x) \leqslant 0$ for each interior point $x \in I$, then f is monotonically decreasing on I. If in (b) the inequality is strict, then f is strictly monotonically decreasing.*

PROOF. Exercise.

Theorem 6.2. *If f is continuous on an interval I and differentiable in the extended sense at each interior point of I and $f'(x) = 0$ for each interior point $x \in I$, then f is constant in I.*

PROOF. Take $a \in I$ and $x \in I$ such that $x \neq a$. By the mean-value theorem, there exists an x_0 between a and x such that

$$f(x) = f(a) + f'(x_0)(x - a).$$

Since $f'(x_0) = 0$ by hypothesis, we obtain $f(x) = f(a)$. Thus, $f(x) = f(a)$ holds for all $x \in I$ and f is constant on I.

Theorem 6.3. *If f and g are continuous on an interval I, both are differentiable at each interior point of I and $f'(x) = g'(x)$ for each interior point $x \in I$, then a real c exists such that*

$$f(x) = g(x) + c \qquad \textit{for all} \quad x \in I. \quad (6.3)$$

PROOF. By hypothesis,

$$(f(x) - g(x))' = f'(x) - g'(x) = 0 \qquad \text{for each interior point } x \text{ of } I.$$

The function $f - g$ satisfies the hypothesis of Theorem 6.2 so it is a constant, c say, and, therefore, (6.3) holds.

Derivatives and Existence of Inverses

We now apply some of the above results to obtain criteria for the existence of inverses of functions.

Theorem 6.4. *If f is continuous on an interval I, the one-sided derivative f'_L exists in the extended sense at each interior point of I from the left, f'_R exists in the extended sense at each interior point of I from the right, a and b are points of I such that $a < b$, and*

$$f'_R(a) < 0 < f'_L(b) \quad \textit{or} \quad f'_R(a) > 0 > f'_L(b), \tag{6.4}$$

then there exists an x_0 such that $a < x_0 < b$ and

$$f'_L(x_0) \leqslant 0 \leqslant f'_R(x_0) \quad \textit{or} \quad f'_L(x_0) \geqslant 0 \geqslant f'_R(x_0). \tag{6.5}$$

Proof. Consider the interval $[a,b]$. We have $[a,b] \subseteq I$. Since f is continuous on $[a,b]$, it has a maximum and a minimum on $[a,b]$. Suppose the maximum and minimum both occur at the endpoints a and b of $[a,b]$. If both the maximum and the minimum occurred at the same point, then the function would be constant. This would imply that

$$f'_R(x) = 0 \quad \text{for} \quad a \leqslant x < b \quad \text{and} \quad f'_L(x) = 0 \quad \text{for} \quad a < x \leqslant b,$$

and, therefore, contradict the hypothesis. Hence, the maximum of f on $[a,b]$ would occur at a and the minimum at b or vice versa. If $f(a)$ is the maximum and $f(b)$ the minimum of f on $[a,b]$, then

$$f'_R(a) \leqslant 0 \quad \text{and} \quad f'_L(b) \leqslant 0.$$

If $f(a)$ is the minimum and $f(b)$ the maximum of f on $[a,b]$, then

$$f'_R(a) \geqslant 0 \quad \text{and} \quad f'_L(b) \geqslant 0.$$

In either case, (6.4) of the hypothesis is contradicted. Hence, the maximum or the minimum occur at some interior point x_0 of $[a,b]$ and one of the inequalities in (6.5) holds (Lemma 5.1 and Prob. 5.1). The conclusion therefore holds. This completes the proof.

Corollary. *If f is differentiable in the extended sense on an interval I and a, b are points of I such that $a < b$ and*

$$f'(a) < 0 < f'(b) \quad \textit{or} \quad f'(a) > 0 > f'(b), \tag{6.6}$$

then there exists an x_0 such that $a < x_0 < b$ and $f'(x_0) = 0$.

Proof. Since f is differentiable in the extended sense on I, it is continuous there. If x is an interior point, then $f'(x) = f'_R(x) = f'_L(x)$. If x is a left endpoint of I, then $f'(x) = f'_R(x)$. If x is a right endpoint of I, then $f'(x) = f'_L(x)$.

We can now apply the theorem to conclude that because of (6.6) an x_0 exists such that $a < x_0 < b$ and

$$f'(x_0) = f'_R(x_0) \leqslant 0 \leqslant f'_L(x_0) = f'(x_0)$$

or

$$f'(x_0) = f'_R(x_0) \geqslant 0 \geqslant f'_L(x_0) = f'(x_0).$$

Either of these implies that $f'(x_0) = 0$.

Prob. 6.2. Prove: If f is differentiable in the extended sense on an interval I and $f'(x) \neq 0$ for all $x \in I$, then either $f'(x) > 0$ for all $x \in I$ or $f'(x) < 0$ for all $x \in I$.

Theorem 6.5. *If f is continuous on an interval and $f'(x) \neq 0$ for each interior point $x \in I$, then $f(I)$ is an interval and f has a strictly monotonic and continuous inverse f^{-1} defined on $f(I)$. The inverse f^{-1} of f is also differentiable at each interior point of $f(I)$ and*

$$(f^{-1})'(y) = \frac{1}{f'(f^{-1}(y))} \quad \text{at each interior point } y \in f(I). \tag{6.6'}$$

Proof. The set of interior points of I is an open interval J and $f'(x) \neq 0$ there. Hence (Prob. 6.2), either $f'(x) > 0$ for all $x \in J$ or $f'(x) < 0$ for all $x \in J$. By the corollary of Theorem 6.1, f is strictly monotonic on I. Since f is also continuous on the interval I, $f(I)$ is an interval, and f has a strictly monotonic inverse f^{-1} on $f(I)$ which is continuous on $f(I)$ (Theorem VI.9.2).

We prove that f^{-1} is differentiable at each interior point of $f(I)$. Let y_0 be an interior point of $f(I)$. There exists an $x_0 \in I$ such that $y_0 = f(x_0)$ and $x_0 = f^{-1}(y_0)$. Take $y \in f(I)$ such that $y \neq y_0$. There exists $y \in f(I)$ such that $y = f(x)$ and $x = f^{-1}(y)$. Since f is continuous at y_0, we know that

$$\lim_{y \to y_0} f^{-1}(y) = f^{-1}(y_0) = x_0. \tag{6.7}$$

Since $y \neq y_0$ and f^{-1} is strictly monotonic and, therefore, one-to-one,

$$x = f^{-1}(y) \neq f^{-1}(y_0) = x_0. \tag{6.8}$$

Since f is differentiable at x_0,

$$\lim_{x \to x_0} \frac{f(x) - f(x_0)}{x - x_0} = f'(x_0). \tag{6.9}$$

Using theorems on limits and (6.7), (6.8) and (6.9), we see that

$$\lim_{y\to y_0} \frac{f(f^{-1}(y)) - f(x_0)}{f^{-1}(y) - x_0} = f'(x_0). \tag{6.10}$$

By a change of notation this becomes

$$\lim_{y\to y_0} \frac{y - y_0}{f^{-1}(y) - f^{-1}(y_0)} = f'(x_0). \tag{6.11}$$

Since $f'(x_0) \neq 0$, this implies that

$$\lim_{y\to y_0} \frac{f^{-1}(y) - f^{-1}(y_0)}{y - y_0} = \frac{1}{f'(x_0)}.$$

Hence,

$$(f^{-1})'(y_0) = \frac{1}{f'(x_0)} = \frac{1}{f'(f^{-1}(y_0))},$$

where y_0 is an interior point of $f(I)$.

Remark 6.1. The derivative of f^{-1} at y in Theorem 6.5 is

$$(f^{-1})'(y) = \frac{1}{f'(x)}, \tag{6.12}$$

where $x = f^{-1}(y)$. In the Leibniz notation y often serves as a notation for a function and dy/dx for its derivative. When y has a differentiable inverse, the relation between the derivative of y and its inverse is often written

$$\frac{dx}{dy} = \frac{1}{dy/dx}, \tag{6.13}$$

where $x = f^{-1}(y)$.

EXAMPLE 6.1. We apply Theorem 6.5 to obtain the derivative of the inverse hyperbolic sine. We have

$$y = \sinh^{-1}x \qquad \text{for} \quad x \in \mathbb{R} \tag{6.14}$$

and

$$x = \sinh y. \tag{6.15}$$

Since $\cosh y \geqslant 1$, it follows that

$$\frac{dy}{dx} = \frac{1}{dx/dy} = \frac{1}{\cosh y}. \tag{6.16}$$

We know that $\cosh^2 y - \sinh^2 y = 1$, so $\cosh y = \sqrt{1 + \sinh^2 y} = \sqrt{1 + x^2}$. This and (6.17) yield

$$\frac{d\sinh^{-1}x}{dx} = \frac{dy}{dx} = \frac{1}{\sqrt{1 + x^2}} \qquad \text{for} \quad x \in \mathbb{R}. \tag{6.17}$$

This result also follows from

$$\sinh^{-1}x = \ln\left(x + \sqrt{1 + x^2}\right) \qquad \text{for} \quad x \in \mathbb{R}.$$

EXAMPLE 6.2. We prove: If f is continuous on an interval I, $f'(x) \neq 0$ for each interior point x of I, and f is twice differentiable for such x, then f^{-1} is continuous on $f(I)$, twice differentiable for each interior point y of $f(I)$, and

$$(f^{-1})''(y) = -\frac{f''(x)}{(f'(x))^3}, \tag{6.18}$$

where $x = f^{-1}(y)$.

PROOF. From the hypothesis and Theorem 6.5 we see that f has a continuous inverse f^{-1} on $f(I)$, f^{-1} is differentiable at each interior point y of $f(I)$, and

$$(f^{-1})'(y) = \frac{1}{f'(x)}, \tag{6.19}$$

where $x = f^{-1}(y)$. By hypothesis, f' is differentiable at each interior point of I. Therefore, the composite $f' \circ f^{-1}$ is differentiable at each interior point of $f(I)$. Also,

$$(f' \circ f^{-1})(y) = f'(f^{-1}(y)) = f'(x) \neq 0 \tag{6.20}$$

for each interior point y of $f(I)$. But now we see that $(f^{-1})'$ is differentiable for each interior point y of $f(I)$ and, hence, that f^{-1} is twice differentiable for each interior point y of $f(I)$. Because of (6.20), the reciprocal of $f' \circ f^{-1}$ is differentiable at each interior point y of $f(I)$. Moreover, by the chain rule,

$$\begin{aligned}
(f^{-1})''(y) = ((f^{-1})'(y))' &= \left(\frac{1}{f'(f^{-1}(Y)}\right)' \\
&= -\frac{f''(f^{-1}(y))(f^{-1})'(y)}{(f'(f^{-1}(y)))^2} \\
&= -\frac{f''(x)}{(f'(x))^2}\frac{1}{f'(x)} = -\frac{f''(x)}{(f'(x))^3}.
\end{aligned}$$

Here it is perhaps more convenient to use the traditional Leibniz notation since it is less cumbersome. Beginning with

$$\frac{dx}{dy} = \frac{1}{dy/dx} = \left(\frac{dy}{dx}\right)^{-1} = (y')^{-1}$$

(here $(y')^{-1}$ is the *reciprocal* of y'), we obtain the second derivative of the inverse by means of the chain rule. We have

$$\frac{d^2x}{dy^2} = \frac{d}{dy}\left(\frac{dx}{dy}\right) = \frac{d}{dy}(y')^{-1} = -(y')^{-2}\frac{dy'}{dy} = -(y')^{2}\frac{dy'}{dx}\frac{dx}{dy}$$

$$= -(y')^{-2}y''\frac{1}{y'} = -\frac{y''}{(y')^3} = -\frac{d^2y/dx^2}{(dy/dx)^3},$$

where $x = f^{-1}(y)$. (Compare this with (6.18).)

Prob. 6.3. Prove: If f is continuous on an interval I, f is three times differentiable and $f'(x) \neq 0$ for each interior point x of I, then the inverse f^{-1} of f exists, is continuous on $f(I)$, is three times differentiable at each interior point of $f(I)$, and

$$(f^{-1})'''(y) = \frac{3(f''(x))^2 - f'(x)f'''(x)}{(f'(x))^5},$$

where $x = f^{-1}(y)$. This can also be written

$$\frac{d^3x}{dy^3} = \frac{3(d^2y/dx^2)^2 - (dy/dx)(d^3y/dx^3)}{(dy/dx)^5},$$

where $x = f^{-1}(y)$.

Prob. 6.4. Prove: If $x > 1$, then

(a) $d\cosh^{-1}x/dx = 1/\sqrt{x^2-1}$,
(b) $d\tanh^{-1}x/dx = 1/(1-x^2)$ for $-1 < x < 1$,
(c) $d\,\mathrm{sech}^{-1}x/dx = -1/x\sqrt{1-x^2}$ for $0 < x < 1$,
(d) $d\coth^{-1}x/dx = -1/(x^2-1)$ for $|x| > 1$,
(e) $d\,\mathrm{csch}^{-1}x/dx = -1/|x|\sqrt{1+x^2}$ for $x \neq 0$.

We saw that if $f'(x) \geqslant 0$ on an interval, then f is monotonically increasing, and if $f'(x) \leqslant 0$ is an interval, then f is monotonically decreasing, with strict inequalities on the interval implying strict monotonicity. We now consider possible converses of these propositions.

Theorem 6.6. *If f is monotonically increasing on an interval I and $f_L'(x)$ exists for some interior point x of I from the left, then $f_L'(x) \geqslant 0$. Similarly, if for some interior point x of I from the right $f_R'(x)$ exists, then $f_R'(x) \geqslant 0$. An analogous statement holds for a function which is monotonically decreasing on an interval I (Prob. 6.5).*

Proof. Assume that x is an interior point of I from the left. Then there exists a $\delta > 0$ such that $(x-\delta, x] \subseteq I$. For each h such that $-\delta < h < 0$, we have $x + h < x$ and $x + h \in I$ and, hence, $f(x+h) \leqslant f(x)$. Therefore, for all h such that $-\delta < h < 0$, we have

$$\frac{f(x+h)-f(x)}{h} \geqslant 0.$$

This implies that

$$f_L'(x) = \lim_{h \to 0-} \frac{f(x+h)-f(x)}{h} \geqslant 0.$$

Similarly, if x is an interior point of I from the right, there exists a $\delta_1 > 0$ such that $[x, x+\delta_1)$. Taking h such that $0 < h < \delta_1$, we have $x+h > x$, $x+h \in I$, $f(x+h) \geqslant f(x)$, and

$$\frac{f(x+h)-f(x)}{h} \geqslant 0,$$

so that

$$f_R'(x) = \lim_{h \to 0+} \frac{f(x+h)-f(x)}{h} \geqslant 0.$$

Prob. 6.5. Prove: If f is monotonically decreasing on an interval I and $f_L'(x)$ exists for some interior point x of I from the left, then $f_L'(x) \leqslant 0$. Similarly, if for some interior point x of I from the right $f_R'(x)$ exists, then $f_R'(x) \leqslant 0$.

Remark 6.2. In Theorem 6.6 the stronger assumption that f is *strictly* monotonically increasing does *not* yield the strict inequalities $f_L'(x) > 0$, $f_R'(x) > 0$. For example, the function f, where

$$f(x) = x^3 \qquad \text{for all} \quad x \in \mathbb{R},$$

is strictly monotonically increasing. Nevertheless,

$$f_R'(x) = f_L'(x) = f'(x) = 3x^2 \geqslant 0.$$

Here

$$f_L'(0) = f_R'(0) = f'(0) = 0.$$

7. Applications of the Mean-Value Theorem. Euler's Constant

Theorem V.7.1 states: If x and y are real numbers, $x > 0$, then

(a) $(x-1)y \leqslant x^y - 1 \leqslant yx^{y-1}(x-1)$ for $y \geqslant 1$ or $y \leqslant 0$ and
(b) $(x-1)y \geqslant x^y - 1 \geqslant yx^{y-1}(x-1)$ for $0 < y < 1$.

Here we improve these inequalities by stating conditions under which they are strict. We prove: If $x > 0$, $x \neq 1$ and $y > 1$ or $y < 0$, then the inequality in (a) is strict. We ask the reader to prove (Prob. 7.1): If $x > 0$, $x \neq 1$ and $0 < y < 1$, then the inequality in (b) is strict.

We assume that $y > 1$, $x > 0$. Let f be defined by $f(x) = x^y$ for $x > 0$. Assume $x \neq 1$. By the Mean-Value Theorem, we know that there exists an x_0 between x and 1 such that $f(x) = f(1) + f'(x_0)(x - 1)$. For our f this means that

$$x^y = 1 + yx_0^{y-1}(x - 1), \tag{7.1}$$

where $0 < x < x_0 < 1$ or $1 < x_0 < x$. We note that $y - 1 > 0$, so that

$$\begin{aligned} &(1) \qquad 0 < x^{y-1} < x_0^{y-1} < 1 \qquad \text{for} \quad 0 < x < x_0 < 1, \\ &(2) \qquad 1 < x_0^{y-1} < x^{y-1} \qquad \text{for} \quad 1 < x_0 < x. \end{aligned} \tag{7.2}$$

Using the fact that $y(x - 1) < 0$ in (1) above and $y(x - 1) > 0$ in (2), we obtain, after multiplying each set of inequalities by $y(x - 1)$,

$$y(x - 1) < yx_0^{y-1}(x - 1) < yx^{y-1}(x - 1) \qquad \text{for} \quad x > 0, \quad x \neq 1 \tag{7.3}$$

when $y > 1$.

We prove next that (7.3) also holds for $y < 0$. In this case, we have $1 - y > 1$. By hypothesis, $x > 0$ and $x \neq 1$, so there exists an x_0 between x and 1 such that (7.1) holds. This time we have $y - 1 < -1 < 0$, so that

$$\begin{aligned} &(1) \qquad 1 < x_0^{y-1} < x^{y-1} \qquad \text{for} \quad 0 < x < x_0 < 1, \\ &(2) \qquad 0 < x^{y-1} < x_0^{y-1} < 1 \qquad \text{for} \quad 1 < x_0 < x. \end{aligned} \tag{7.4}$$

But $y(x - 1) > 0$ in case (1) and $y(x - 1) < 0$ in case (2). Reasoning as we did in the previous paragraph, we have

$$y(x - 1) < yx_0^{y-1}(x - 1) < yx^{y-1}(x - 1) \qquad \text{for} \quad x > 0, \quad x \neq 1 \tag{7.5}$$

when $y < 0$. The strictness, under the stated conditions, of the inequality in (a) now follows from (7.1), (7.3), and (7.5).

Prob. 7.1. Prove:

$$y(x - 1) > x^y - 1 > yx^{y-1}(x - 1) \quad \text{for } x > 0, x \neq 1, \text{and } 0 < y < 1.$$

Prob. 7.2. Prove: If a and b are distinct positive reals, then

(a) $yb^{y-1}(a - b) < a^y - b^y < ya^{y-1}(a - b)$ for $y < 0$ or $y > 1$
(b) $yb^{y-1}(a - b) > a^y - b^y > ya^{y-1}(a - b)$ for $0 < y < 1$.

Prob. 7.3. Prove: If $x \in \mathbb{R}$, $x \neq 0$, then

$$1 + x < e^x < 1 + xe^x.$$

Prob. 7.4. Prove: If $x > -1$, $x \neq 0$, then

$$\frac{x}{x + 1} < \ln(1 + x) < x.$$

Euler–Mascheroni Constant

We use the inequality of the last problem to gain some information about the interesting sequence $\langle \gamma_n \rangle$, where

$$\gamma_n = 1 + \frac{1}{2} + \cdots + \frac{1}{n} - \ln n \qquad \text{for each positive integer } n. \tag{7.6}$$

Theorem 7.1. *Let γ_n be defined as in* (7.6). *Then*: (*a*) $0 < \gamma_n \leqslant 1$ *for each* n, *and* $0 < \gamma_n < 1$ *for* $n > 1$. (*b*) *The sequence* $\langle \gamma_n \rangle$ *is strictly monotonically decreasing. Hence,* $\langle \gamma_n \rangle$ *converges.*

PROOF. By Prob. 7.4, we have

$$\frac{1}{k+1} = \frac{1/k}{1+1/k} < \ln\left(1 + \frac{1}{k}\right) < \frac{1}{k} \qquad \text{for } k \text{ a positive integer.} \tag{7.7}$$

Hence,

$$\sum_{k=1}^{n} \frac{1}{k+1} < \sum_{k=1}^{n} \ln\left(1 + \frac{1}{k}\right) < \sum_{k=1}^{n} \frac{1}{k} \qquad \text{for each positive integer } n. \tag{7.8}$$

We note that $\ln(1 + 1/k) = \ln(k+1) - \ln k$ and, therefore, that

$$\begin{aligned} \sum_{k=1}^{n} \ln\left(1 + \frac{1}{k}\right) = \sum_{k=1}^{n} (\ln(k+1) - \ln k) &= \ln 2 - \ln 1 + \ln 3 - \ln 2 \\ &\quad + \cdots + \ln(n+1) - \ln n \\ &= \ln(n+1) - \ln 1 = \ln(n+1). \end{aligned}$$

By (7.8), this implies that

$$\sum_{k=1}^{n} \frac{1}{k+1} < \ln(n+1) < \sum_{k=1}^{n} \frac{1}{k}, \tag{7.9}$$

i.e., that

$$\frac{1}{2} + \frac{1}{3} + \cdots + \frac{1}{n+1} < \ln(n+1) < 1 + \frac{1}{2} + \frac{1}{3} + \cdots + \frac{1}{n} \tag{7.10}$$

holds for each positive integer n. We now add $-\ln n$ to the last inequality in (7.10) to obtain

$$0 < \ln\left(1 + \frac{1}{n}\right) = \ln(n+1) - \ln n < 1 + \frac{1}{2} + \frac{1}{3} + \cdots + \frac{1}{n} - \ln n = \gamma_n .$$

This proves that

$$0 < \ln\left(1 + \frac{1}{n}\right) < \gamma_n \qquad \text{for each positive integer } n. \tag{7.11}$$

Consider the first inequality in (7.10). By adding 1 on both sides we obtain

$$1 + \frac{1}{2} + \frac{1}{3} + \cdots + \frac{1}{n+1} < 1 + \ln(n+1). \tag{7.12}$$

This implies that

$$\gamma_{n+1} = 1 + \frac{1}{2} + \frac{1}{3} + \cdots + \frac{1}{n+1} - \ln(n+1) < 1 \quad \text{for each positive integer } n. \tag{7.13}$$

This and (7.11) yield

$$0 < \gamma_n < 1 \qquad \text{for each positive integer } n > 1. \tag{7.14}$$

Clearly, $\gamma_1 = 1 - \ln 1 = 1$. This and (7.14) prove (a). We prove (b). Note that if n is a positive integer, then

$$\gamma_{n+1} - \gamma_n = \frac{1}{n+1} - \ln(n+1) + \ln n = \frac{1}{n+1} - \ln\left(1 + \frac{1}{n}\right). \tag{7.15}$$

By (7.7), the right-hand side of (7.15) is negative and we have

$$\gamma_{n+1} - \gamma_n = \frac{1}{n+1} - \ln(n+1) < 0 \qquad \text{for each positive integer } n. \tag{7.16}$$

This implies that $\langle \gamma_n \rangle$ is a strictly decreasing sequence. This proves (b). The last statement holds because $\langle \gamma_n \rangle$ is bounded from below by 0 and is decreasing. Hence, $\langle \gamma_n \rangle$ converges. This completes the proof.

We define

$$\gamma = \lim_{n\to+\infty} \gamma_n = \lim_{n\to+\infty} \left(1 + \frac{1}{2} + \cdots + \frac{1}{n} - \ln n\right). \tag{7.17}$$

γ is called the *Euler–Mascheroni Constant*.* Euler evaluated γ to 16 places and Mascheroni to 32. However, an error was found in the 20th place. Later Gauss and Nicolai corrected the error. Correct to 10 places,

$$\gamma = 0.57722156649.$$

It is still not known whether or not γ is rational.

PROB. 7.5. Prove:

$$\lim_{n\to+\infty} \frac{1 + 1/2 + \cdots + 1/n}{\ln n} = 1.$$

Compare this with the result in Prob. III.9.2.

We present some further applications of the Mean-Value Theorem in problem form.

PROB. 7.6. Prove: (a) If f is continuous on $(a,b]$, $f'(x)$ exists for $a < x < b$, and $\lim_{x\to b-} f'(x) = k$, then $f'_L(b)$ exists and $f'_L(b) = k$. (b) If f is continuous on $[a,b)$, $f'(x)$ exists for $a < x < b$, and $\lim_{x\to a+} f'(x) = L$, then $f'_R(a)$ exists and $f'_R(a) = L$.

PROB. 7.7. Prove: If f is defined at some $a \in \mathbb{R}$ is differentiable in some ϵ-neighborhood of a and $\lim_{x\to a} f'(x) = B$, then f is differentiable at a and $f'(a) = B$.

* *Chrystal's Algebra*, Vol. 2, Chap. 25, Art. 13, Dover, New York.

PROB. 7.8. Prove: (a) If f has a bounded derivative on an interval at I, then it is uniformly continuous there. (b) If f has a bounded derivative on a bounded open interval $(a; b)$, then $f(a+)$ and $f(b-)$ exist and are finite. (See Theorem VI.11.4 and Prob. VI.11.7.)

Darboux's Theorem on the Values of the Derivative

We shall see later (Section VIII.5) that the derivative of a function need not be continuous. We prove, however, that the derivative of a function which is differentiable on an interval must have the intermediate-value property.

Theorem 7.2. *If g is a real-valued function of a real variable which is defined on an interval I and is the derivative of some function f defined on I, then g has the intermediate-value property.*

PROOF. This theorem is a consequence of the corollary of Theorem 6.4. By hypothesis

$$f'(x) = g(x) \qquad \text{for} \quad x \in I. \tag{$*$}$$

Suppose that $g(a) \neq g(b)$ for some a and b in I. Let μ be a number between $g(a)$ and $g(b)$. Consider the function h, where

$$h(x) = f(x) - \mu x \qquad \text{for} \quad x \in I.$$

Clearly,

$$h'(x) = f'(x) - \mu = g(x) - \mu \qquad \text{for} \quad x \in I$$

and, hence, $h'(a) = g(a) - \mu$ and $h'(b) = g(b) - \mu$. Since μ is between $g(a)$ and $g(b)$, either $g(a) < \mu < g(b)$ or $g(a) > \mu > g(b)$ so that either

$$h'(a) < 0 < h'(b) \quad \text{or} \quad h'(a) > 0 > h'(b).$$

By the corollary of Theorem 6.4, this implies that an x_0 exists between a and b such that $h'(x_0) = 0$, i.e., $g(x_0) - \mu = 0$ or, equivalently, that $g(x_0) = \mu$.

Remark 7.1. Let I be some interval. If $(*)$ holds on I, then it clearly holds on every closed, bounded subinterval of I. This and Theorem 7.2 imply that a function g which is the derivative of some function f on an interval has the *strong* intermediate value property on that interval (Remark VI.7.3).

An Application of the Corollary of Theorem 6.1.

The following is not a direct application of the Mean-Value Theorem, but is, instead, an application of the corollary of Theorem 6.1. We use the sign of the derivative of a function to gain information about its monotonicity.

We prove:

$$\frac{2}{2x+1} < \ln\left(1+\frac{1}{x}\right) \qquad \text{for} \quad x > 0. \tag{7.18}$$

Note first that this may be written

$$\frac{1}{x+\frac{1}{2}} < \ln\left(1+\frac{1}{x}\right) \qquad \text{for} \quad x > 0. \tag{7.19}$$

We already know that

$$\frac{1}{x+1} < \ln\left(1+\frac{1}{x}\right) \qquad \text{for} \quad x > 0. \tag{7.20}$$

Since

$$\frac{1}{x+1} < \frac{1}{x+\frac{1}{2}} \qquad \text{for} \quad x > 0,$$

(7.18) is an improvement on (7.20).

We prove (7.18). Define the function g, where

$$g(x) = \frac{2}{2x+1} - \ln\left(1+\frac{1}{x}\right) \qquad \text{for} \quad x > 0.$$

Now

$$g'(x) = \frac{1}{(2x+1)^2 x(x+1)} > 0 \qquad \text{for} \quad x > 0.$$

This implies that g is strictly increasing. It is easy to see that

$$\lim_{x \to +\infty} g(x) = 0.$$

Therefore,

$$0 = \lim_{x \to +\infty} g(x) = \sup_{x>0} g(x).$$

This implies that

$$g(x) \leqslant 0 \qquad \text{for} \quad x > 0.$$

Now take $x > 0$ and $x_1 > x > 0$. We have $g(x) < g(x_1) \leqslant 0$. Thus, $g(x) < 0$ for all $x > 0$ and so (7.18) holds. This inequality will be used later. A consequence of (7.18) is

$$1 < \left(n+\frac{1}{2}\right)\ln\left(1+\frac{1}{n}\right) \qquad \text{for each positive integer } n. \tag{7.21}$$

This implies that

$$e < \left(1+\frac{1}{n}\right)^{n+1/2} \qquad \text{for each positive integer } n. \tag{7.22}$$

Prob. 7.9. Prove: The sequence $\langle a_n \rangle$, where

$$a_n = \frac{n!\,e^n}{n^{n+1/2}} \qquad \text{for each positive integer } n,$$

is strictly monotonically decreasing and converges. Its limit will be evaluated later.

PROB. 7.10.* Prove: If $x > 0$, then

$$\ln\left(1 + \frac{1}{x}\right) < \frac{1}{\sqrt{x^2 + x}}.$$

PROB. 7.11. Prove: The function f, where

$$f(x) = \left(1 + \frac{1}{x}\right)^x \qquad \text{for} \quad x > 0,$$

is strictly monotonically increasing.

Another Application

We prove, using the Mean-Value Theorem, that

$$\frac{\tan x}{x} > 1 \qquad \text{for} \quad 0 < |x| < \frac{\pi}{2}. \tag{7.23}$$

If $0 < |x| < \pi.2$, then there is an x_0 between 0 and x such that

$$\frac{\tan x - \tan 0}{x} = \left(\frac{d \tan x}{dx}\right)\Bigg|_{x = x_0} = \sec^2 x_0.$$

Since $0 < |x_0| < \pi/2$, it follows that $0 < \cos x_0 < 1$ and, therefore, that $\sec^2 x_0 > 1$. This implies that

$$\frac{\tan x}{x} > 1 \qquad \text{for} \quad 0 < |x| < \frac{\pi}{2}. \tag{7.24}$$

We now prove Jordan's inequality† which states that

$$\frac{2}{\pi} \leqslant \frac{\sin x}{x} < 1 \qquad \text{for} \quad 0 < |x| \leqslant \frac{\pi}{2}. \tag{7.25}$$

For proof, consider f defined by

$$f(x) = \begin{cases} \dfrac{\sin x}{x} & \text{for} \quad 0 < x \leqslant \dfrac{\pi}{2} \\ 1 & \text{for} \quad x = 0. \end{cases} \tag{7.26}$$

This function is continuous on $[0, \pi/2]$ (explain). Note that

$$f'(x) = \frac{x \cos x - \sin x}{x^2} = \frac{\cos x}{x^2}(x - \tan x) \qquad \text{for} \quad 0 < x < \frac{\pi}{2}. \tag{7.27}$$

Since $\cos x > 0$ for $0 < x < \pi/2$, the last result and (7.24) show that $f'(x) < 0$ for $0 < x < \pi/2$. We conclude from this that f is strictly mono-

* Mitronovic, *Analytic Inequalities*, p. 273.

† *Ibid*, p. 33.

tonically decreasing on $[0,\pi/2]$. Therefore,

$$\frac{\sin x}{x} > \frac{\sin(\pi/2)}{\pi/2} = \frac{2}{\pi} \quad \text{for} \quad 0 < x < \frac{\pi}{2}. \tag{7.28}$$

Since (Theorem IV.8.2)

$$0 < \frac{\sin x}{x} < 1 \quad \text{for} \quad 0 < |x| \leqslant 1 < \frac{\pi}{2}, \tag{7.29}$$

we obtain the strict inequality

$$\frac{2}{\pi} < \frac{\sin x}{x} < 1 \quad \text{for} \quad 0 < x < \frac{\pi}{2}. \tag{7.30}$$

If $-\pi/2 < x < 0$, then $0 < -x < \pi/2$. Hence, by (7.30), we see that

$$\frac{2}{\pi} < \frac{\sin(-x)}{-x} < 1.$$

This and (7.30) imply that the strict inequality in (7.25) holds for $0 < |x| < \pi/2$ with equality holding trivially for $|x| = \pi/2$.

PROB. 7.12. Prove*: If $0 < x < \pi/2$, then

$$\cos x < \left(\frac{\sin x}{x}\right)^3$$

(Hint: consider f, where

$$f(x) = x - \sin x \cos^{-1/3} x \quad \text{if} \quad 0 \leqslant x < \pi/2, \tag{7.31}$$

then prove that f' is strictly decreasing on $[0,\pi/2)$ and $f'(x) < f'(0) = 0$).

PROB. 7.13. Prove*: If $a \leqslant 3$, then

$$\cos x < \left(\frac{\sin x}{x}\right)^a \quad \text{for} \quad 0 < x < \frac{\pi}{2}.$$

8. An Application of Rolle's Theorem to Legendre Polynomials

The *Legendre Polynomials* P_n for each nonnegative integer n are defined as follows:

$$P_0(x) = 1 \quad \text{for} \quad x \in \mathbb{R} \tag{8.1a}$$

$$P_n(x) = \frac{1}{2^n n!} \frac{d^n (x^2-1)^n}{dx^n} \quad \text{for} \quad x \in \mathbb{R} \quad \text{and } n \text{ a positive integer.} \tag{8.1b}$$

Formula (8.1b) is known as *Rodrigue's Formula*.

* *Ibid*, p. 238.

Using the Binomial Theorem, we obtain

$$(x^2-1)^n = \sum_{k=0}^{n} \binom{n}{k}(-1)^k x^{2(n-k)} = \sum_{k=0}^{n} (-1)^k \binom{n}{k} x^{2n-2k}. \tag{8.2}$$

Taking the nth derivative of both sides, we have

$$\frac{d^n(x^2-1)^n}{dx^n} = \sum_{k=0}^{n} (-1)^k \binom{n}{k} \frac{d^n(x^{2n-2k})}{dx^n}. \tag{8.3}$$

Here, the terms with $n > 2n-2k$, that is, with $k > n/2$, vanish. But if $0 \leqslant k \leqslant n/2$, then the corresponding term in (8.3) does not vanish. Since k is an integer, we have: If

$$0 \leqslant k \leqslant \left[\frac{n}{2}\right] = \text{greatest integer} \leqslant \frac{n}{2}, \tag{8.4}$$

then the terms in the sum on the right in (8.3) do not vanish. Note that

$$\left[\frac{n}{2}\right] = \begin{cases} \dfrac{n}{2} & \text{if } n \text{ is even} \\ \dfrac{n-1}{2} & \text{if } n \text{ is odd.} \end{cases} \tag{8.5}$$

Now,

$$\begin{aligned} \frac{d^n(x^{2n-2k})}{dx^n} &= (2n-2k)(2n-2k-1)\ldots(n-2k+1)x^{n-2k} \\ &= \frac{(2n-2k)(2n-2k-1)\ldots(n-2k+1)(n-2k)!\,x^{n-2k}}{(n-2k)!} \\ &= \frac{(2n-2k)!}{(n-2k)!}x^{n-2k}, \end{aligned}$$

where $0 \leqslant k \leqslant [n/2]$. Hence substitution into the sum on the right-hand side of (8.3) yields

$$\begin{aligned} \frac{d^n(x^2-1)^n}{dx^n} &= \sum_{k=0}^{[n/2]} (-1)^k \binom{n}{k} \frac{(2n-2k)!}{(n-2k)!} x^{n-2k} \\ &= \sum_{k=0}^{[n/2]} (-1)^k \frac{n!}{k!\,(n-k)!} \frac{(2n-2k)!}{(n-2k)!} x^{n-2k}. \end{aligned}$$

After multiplying both sides by $1/(2^n n!)$ we obtain

$$P_n(x) = \frac{1}{2^n n!} \frac{d^n(x^2-1)^n}{dx^n} = \sum_{k=0}^{[n/2]} (-1)^k \frac{(2n-2k)!}{2^n k!\,(n-k)!\,(n-2k)!} x^{n-2k}. \tag{8.6}$$

It follows that P_n is a polynomial of degree n.

Prob. 8.1. (a) Obtain the Legendre Polynomials P_1, P_2, P_3, and P_4. (b) Prove: If n is even, then $P_n(-x) = P_n(x)$ for $x \in \mathbb{R}$. (c) Show that the leading coefficient of P_n is

$$\frac{(2n-1)(2n-3)\ldots 3\cdot 1}{n!}.$$

We show that if j is an integer such that $0 \leqslant j < n$, then

$$\left.\frac{d^j(x^2-1)^n}{dx^j}\right|_{x=\pm 1} = 0. \tag{8.7}$$

Write $(x^2-1)^n = (x-1)^n(x+1)^n$ and use Leibniz's Rule (Prob. 4.6) to obtain the jth derivative of P_n. We have

$$\frac{d^j(x^2-1)^n}{dx^n} = \frac{d^j(x-1)^n(x+1)^n}{dx^j} = \sum_{k=0}^{j}\binom{j}{k}\frac{d^{j-k}(x-1)^n}{dx^{j-k}}\,\frac{d^k(x+1)^n}{dx^k}. \tag{8.8}$$

For $0 \leqslant j < n$, the index in the sum on the right satisfies $0 \leqslant k \leqslant j < n$, so that $0 \leqslant j-k \leqslant j < n$. Hence,

$$\begin{aligned}\frac{d^{j-k}(x-1)^n}{dx^{j-k}} &= n(n-1)\ldots(n-j+k+1)(x-1)^{n-j+k}\\ \frac{d^k(x+1)^n}{dx^k} &= n(n-1)\ldots(n-k+1)(x+1)^{n-k}.\end{aligned} \tag{8.9}$$

When $0 \leqslant k \leqslant j < n$, then the exponents in $(x-1)^{n-j+k}$ and $(x+1)^{n-k}$ are positive. This implies that when $x = \pm 1$, then the terms in the sum on the right in (8.8) all vanish. This proves (8.7).

Prob. 8.2. Prove: $P_n(1) = 1$ and $P_n(-1) = (-1)^n$.

We now come to the application of Rolle's Theorem mentioned in the heading of this section.

Theorem 8.1. *The Legendre Polynomial of degree $n \geqslant 1$ has exactly n zeros in the open interval* $(-1; 1)$.

Proof. Let v_n, where n is some positive integer, be defined by

$$v_n(x) = (x^2-1)^n \quad \text{for} \quad x \in \mathbb{R}. \tag{8.10}$$

Obviously, $v_n(1) = v_n(-1) = 0$. By Rolle's Theorem, the derivative v_n' has at least one zero, z_0 say, in $(-1; 1)$. Thus, z_1 exists such that $v_n'(z_1) = 0$, where $-1 < z_1 < 1$. When $n = 1$, we have

$$P_1(x) = \frac{1}{2}\,\frac{dv_1(x)}{dx} = \frac{1}{2}\,\frac{d(x^2-1)}{dx} = x.$$

We see that P_1 has exactly one zero in $(-1; 1)$. Now assume that $n > 1$. We saw in (8.7) that $v_n'(1) = 0 = v_n' = (-1)$. By Rolle's Theorem, there exist z_{12} and z_{22} with $-1 < z_{12} < z_1 < z_{22} < 1$ such that $v_n^{(2)}(z_{12}) = 0 = v^{(2)}(z_{22})$. These are distinct since they are separated by z_1. Thus, v_n'' has at least two zeros in $(-1; 1)$. Continuing up to $v_n^{(n-1)}$ we find that it has at least $n-1$ distinct zeros in $(-1; 1)$. We write these as $z_{1,n-1}, z_{2,n-1}, \ldots, z_{n-1,n-1}$. Then $-1 < z_{1,n-1} < \cdots < z_{n-1,n-1} < 1$. By (8.7),

$$v_n^{(n-1)}(-1) = 0 = v_n^{(n-1)}(1).$$

It follows that $v_n^{(n)}$ has n zeros $z_{1n}, z_{2n}, \ldots, z_{n,n}$ in $(-1; 1)$ (explain). Thus, P_n has at least n zeros in $(-1; 1)$. Since P_n is a polynomial of degree n, it cannot have more than n distinct zeros. Consequently, P_n has exactly n zeros in $(-1; 1)$.

PROB. 8.3. Let u be given by $u(x) = (x^2 - 1)^n$ for $x \in \mathbb{R}$, where n is some positive integer. Prove that

$$(x^2 - 1)\frac{du(x)}{dx} = 2nxu(x) \qquad \text{for} \quad x \subset \mathbb{R}. \tag{8.11}$$

Define y by means of

$$y(x) = \frac{1}{2^n n!}\frac{d^n(x^2-1)^n}{dx^n} = P_n(x) = \frac{1}{2^n n!}\frac{d^n(u(x))}{dx^n} \qquad \text{for} \quad x \in \mathbb{R}. \tag{8.12}$$

Take the $(n+1)$th derivative of both sides in (8.11) and show that y satisfies

$$(x^2 - 1)\frac{d^2y(x)}{dx^2} + 2x\frac{dy(x)}{dx} - n(n+1)y(x) = 0 \qquad \text{for} \quad x \in \mathbb{R}. \tag{8.13}$$

CHAPTER VIII
Convex Functions

1. Geometric Terminology

Let f be a real-valued function of a real variable and G be the graph of f. Thus,

$$G = \{(x, f(x)) \mid x \in \mathfrak{D}(f)\}.$$

The point (x, y), where $x \in \mathfrak{D}(f)$ is said to be *above* G if $y \geqslant f(x)$ and *below* G if $y \leqslant f(x)$. If $y > f(x)$, then we say that (x, y) is *strictly* above G and dually; when $y < f(x)$, then we say that (x, y) is *strictly* below G (see Fig. 1.1). Let (x_1, y_1), (x_2, y_2) and (x_3, y_3) be three points in $\mathbb{R}^{(2)} = \mathbb{R} \times \mathbb{R}$, such that $x_1 < x_2 < x_3$. Put

$$m_{13} = \frac{y_3 - y_1}{x_3 - x_1}, \qquad m_{12} = \frac{y_2 - y_1}{x_2 - x_1}, \qquad m_{23} = \frac{y_3 - y_2}{x_3 - x_2}.$$

Define the three functions f_{13}, f_{12}, f_{23} on $[x_1, x_3]$ as follows:

$$\begin{aligned} f_{13}(x) &= y_1 + m_{13}(x - x_1) \\ f_{12}(x) &= y_1 + m_{12}(x - x_1) \qquad \text{for} \quad x_1 \leqslant x \leqslant x_3 \\ f_{23}(x) &= y_2 + m_{23}(x - x_2) \end{aligned}$$

and let G_{13}, G_{12}, G_{23} be their respective graphs. It is easy to show that (x_2, y_2) is below G_{13} if and only if $m_{12} \leqslant m_{13}$ (see Fig. 1.2). To prove this, note first that

$$y_2 = y_1 + (y_2 - y_1) = y_1 + \frac{y_2 - y_1}{x_2 - x_1}(x_2 - x_1) = y_1 + m_{12}(x_2 - x_1).$$

Hence,

$$y_2 = y_1 + m_{12}(x_2 - x_1) \leqslant y_1 + m_{13}(x_2 - x_1) = f_{13}(x_2)$$

if and only if $m_{12} \leqslant m_{13}$.

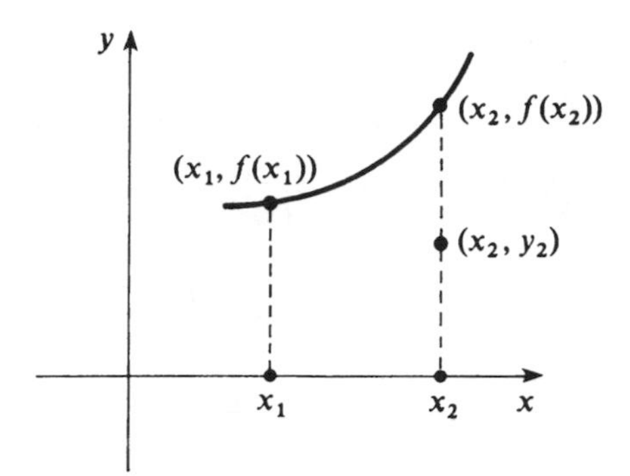

Figure 1.1

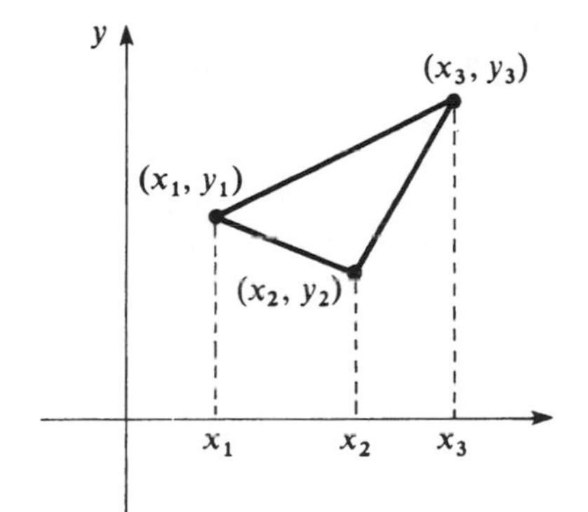

Figure 1.2

If $x_1 < x_2$, then the line l_{12} containing $P_1 = (x_1, y_1)$ and $P_2 = (x_2, y_2)$ is the graph h of f_{12}, where

$$f_{12}(x) = y_1 + \frac{y_2 - y_1}{x_2 - x_1}(x - x_1) \qquad \text{for all} \quad x \in \mathbb{R}.$$

The *line segment* joining $P_1 = (x_1, y_1)$ to $P_2 = (x_2, y_2)$ is the graph S_{12} of the restriction of f_{12} to the closed interval $[x_1, x_2]$ (see Fig. 1.3).

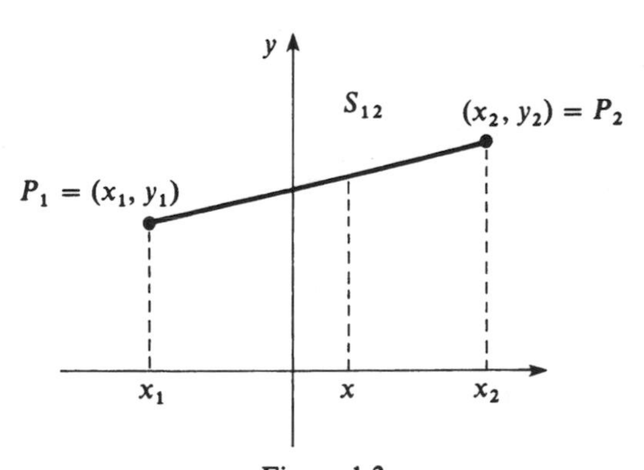

Figure 1.3

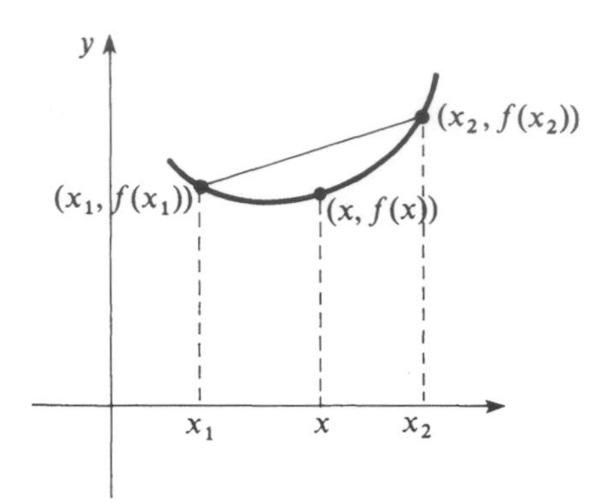

Figure 1.4

If $P_1 = (x_1, f(x_1))$ and $P_2 = (x_2, f(x_2))$ are distinct points of the graph G of a function f, so that $x_1 \neq x_2$, then we call the line segment joining P_1 to P_2 a *secant segment* of G.

Before "officially" defining the notion of a convex function, we phrase the definition using the geometric terminology just introduced. A *convex function* is a function defined on some interval I such that for any two points $P_1 = (x_1, f(x_1))$ and $P_2 = (x_2, f(x_2))$ of its graph, the graph of the restriction of f to the closed interval with endpoints x_1 and x_2 is below the secant segment joining P_1 to P_2 (see Fig. 1.4). We now give an analytic definition.

Def. 1.1. A real-valued function of a real variable which is defined on an interval I is called *convex* on I if and only if for x_1 and x_2 in I such that $x_1 < x_2$, we have

$$f(x) \leqslant f(x_1) + \frac{f(x_2) - f(x_1)}{x_2 - x_1}(x - x_1) \qquad \text{for all} \quad x \in [x_1, x_2]. \quad (1.1)$$

If $x_1 < x_2$ for x_1 and x_2 in I implies the strictness of the inequality (1.1), then we say that f is *strictly convex* on I.

Theorem 1.1. *A real-valued function of a real variable is convex on an interval I if and only if*

$$\begin{gathered} x_1 \in I, \quad x_2 \in I, \quad \textit{and} \quad 0 \leqslant t \leqslant 1 \quad \textit{imply} \\ f((1-t)x_1 + tx_2) \leqslant (1-t)f(x_1) + tf(x_2). \end{gathered} \qquad (1.2)$$

Proof. Let f be a function for which (1.2) holds. Take $x_1 \in I$, $x_2 \in I$, $x_1 < x_2$ and x such that $x_1 \leqslant x \leqslant x_2$. There exists a unique t such that

$$0 \leqslant t \leqslant 1 \quad \text{and} \quad x = (1-t)x_1 + tx_2 \qquad (1.3)$$

(Prob. V.1.7). This implies that

$$t = \frac{x - x_1}{x_2 - x_1}$$

and, since (1.2) holds, that

$$f(x) = f((1-t)x_1 + tx_2) \leqslant (1-t)f(x_1) + tf(x_2) = f(x_1) + t(f(x_2) - f(x_1))$$
$$= f(x_1) + \frac{f(x_2) - f(x_1)}{x_2 - x_1}(x - x_1).$$

Thus, the assumptions $x_1 \in I$, $x_2 \in I$, $x_1 < x_2$ imply (1.1) and, hence, that f is convex on I.

Conversely, assume that f is convex on I. Take $x_1 \in I$, $x_2 \in I$, and t such that $0 \leqslant t \leqslant 1$ and write

$$x = (1-t)x_1 + tx_2. \tag{1.4}$$

If $x_1 = x_2$, then (1.4) implies that $x = x_1 = x_2$ and $f(x) = f(x_1) = f(x_2)$. Therefore,

$$f(x) = (1-t)f(x) + tf(x) = (1-t)f(x_1) + tf(x_2).$$

In this case (1.2) is satisfied with equality holding there. Now consider the case $x_1 \neq x_2$, so that either $x_1 < x_2$ or $x_2 < x_1$. Solve for t in (1.4) to obtain

$$t = \frac{x - x_1}{x_2 - x_1}. \tag{1.5}$$

If $x_1 < x_2$, then (1.4) implies that $x_1 \leqslant x \leqslant x_2$. Since f is convex, we have

$$f((1-t)x_1 + tx_2) = f(x) \leqslant f(x_1) + \frac{f(x_2) - f(x_1)}{x_2 - x_1}(x - x_1)$$
$$= f(x_1) + t(f(x_2) - f(x_1)) = (1-t)f(x_1) + tf(x_2).$$

Thus, $x_1 < x_2$ in I implies (1.2). If $x_2 < x_1$, then (1.4) implies $x_2 \leqslant x \leqslant x_1$. This time the convexity of f on I yields

$$f(1-t)x_1 + tx_2) = f(x) \leqslant f(x_2) + \frac{f(x_1) - f(x_2)}{x_1 - x_2}(x - x_2)$$
$$= f(x_1) + \frac{f(x_2) - f(x_1)}{x_2 - x_1}(x - x_1) \tag{1.6}$$

(check this). Therefore, in view of (1.5), we have

$$f((1-t)x_1 + tx_2) \leqslant f(x_1) + \frac{f(x_2) - f(x_1)}{x_2 - x_1}(x - x_1)$$
$$= (f(x_1) + t(f(x_2) - f(x_1)) = (1-t)f(x_1) + tf(x_2),$$

as before. This completes the proof.

Remark 1.1. We note that $a < b$ implies that $a < (1-t)a + tb < b$, if and only if $0 < t < 1$. Hence, if x_1 and x_2 are in an interval I and $0 \leqslant t \leqslant 1$, then $(1-t)x_1 + tx_2 \in I$.

The condition for convexity obtained in Theorem 1.1 can be expressed somewhat differently.

Theorem 1.1′. *A real-valued function f of a real variable f is convex on an interval I if and only if for x_1 and x_2 in I, we have*

$$f(\alpha x_1 + \beta x_2) \leqslant \alpha f(x_1) + \beta f(x_2), \tag{1.7}$$

where $\alpha + \beta = 1$, $0 \leqslant \alpha$, $0 \leqslant \beta$.

EXAMPLE 1.1. The absolute value function is convex. We use Theorem 1.1′ to prove this: Take $\alpha \geqslant 0$, $\beta \geqslant 0$, $\alpha + \beta = 1$, then

$$|\alpha x_1 + \beta x_2| \leqslant |\alpha x_1| + |\beta x_1| = |\alpha||x_1| + |\beta||x_2| = \alpha|x_1| + \beta|x_2|.$$

Theorem 1.1′ furnishes us with an alternate definition of convexity.

Def. 1.2. If f is defined on an interval I, it is *convex* on I if and only if $x_1 \in I$, $x_2 \in I$ and $\alpha \geqslant 0$, $\beta \geqslant 0$, $\alpha + \beta = 1$ imply that

$$f(\alpha x_1 + \beta x_2) \leqslant \alpha f(x_1) + \beta f(x_2). \tag{1.8}$$

f is called *strictly* convex on I if and only if $x_1 \in I$, $x_2 \in I$, $x_1 \neq x_2$, $\alpha + \beta = 1$, $\alpha > 0$, $\beta > 0$ imply the strict inequality in (1.8).

Remark 1.2. Using Def. 1.2, we obtain: If f is strictly convex on I and there exist x_1, x_2 in I and α, β such that $\alpha > 0$, $\beta > 0$, $\alpha + \beta = 1$ and if

$$f(\alpha x_1 + \beta x_2) = \alpha f(x_1) + \beta f(x_2),$$

then $x_1 = x_2$.

EXAMPLE 1.2. The squaring function $(\ )^2$ is strictly convex on $\mathbb{R}$. For assume $x_1 \neq x_2$ in $\mathbb{R}$, so that $(x_1 - x_2)^2 > 0$. Clearly,

$$2x_1x_2 < x_1^2 + x_2^2. \tag{1.9}$$

If $\alpha > 0$, $\beta > 0$, $\alpha + \beta = 1$, then

$$2\alpha\beta x_1x_2 < \alpha\beta(x_1^2 + x_2^2). \tag{1.10}$$

This implies

$$\begin{aligned}
f(\alpha x_1 + \beta x_2) = (\alpha x_1 + \beta x_2)^2 &= \alpha^2x_1^2 + 2\alpha\beta x_1x_2 + \beta^2x_2^2 \\
&< \alpha^2x_1^2 + \alpha\beta(x_1^2 + x_2^2) + \beta^2x_2^2 \\
&= \alpha^2x_1^2 + \alpha\beta x_1^2 + \alpha\beta x_2^2 + \beta^2x_2^2 \\
&= \alpha(\alpha + \beta)x_1^2 + \beta(\alpha + \beta)x_2^2 \\
&= \alpha x_1^2 + \beta x_2^2 \\
&= \alpha f(x_1) + \beta f(x_2).
\end{aligned}$$

In short, we have proved that: If $x_1 \neq x_2$ and $\alpha > 0$, $\beta > 0$, $\alpha + \beta = 1$, then

$$(\alpha x_1 + \beta x_2)^2 < \alpha x_1^2 + \beta x_2^2.$$

An interesting special case of the above is

$$\left(\frac{x_1 + x_2}{2}\right)^2 < \frac{x_1^2 + x_2^2}{2}, \tag{1.11}$$

where $x_1 \neq x_2$.

PROB. 1.1. Prove: The absolute value function is not *strictly* convex on $\mathbb{R}$. (In Example 1.1, we saw that it is convex on $\mathbb{R}$.)

PROB. 1.2. A function f defined on an interval I is strictly convex if and only if: $x_1 \neq x_2$ and x is between x_1 and x_2 imply that

$$f(x) < f(x_1) + \frac{f(x_2) - f(x_1)}{x_2 - x_1}(x - x_1).$$

Concave Functions

The notion dual to that of convexity is that of concavity. It goes over into the former notion by reversing the sense of the inequalities in the definition.

Def. 1.3. Let I be an interval. We call f *concave* on I if and only if $x_1 \in I$, $x_2 \in I$, $x_1 < x < x_2$ imply

$$f(x) \geqslant f(x_1) + \frac{f(x_2) - f(x_1)}{x_2 - x_1}(x - x_1) \tag{1.12}$$

(see Fig. 1.5). When the inequality here is strict, f is called *strictly concave* on I.

The following theorem holds:

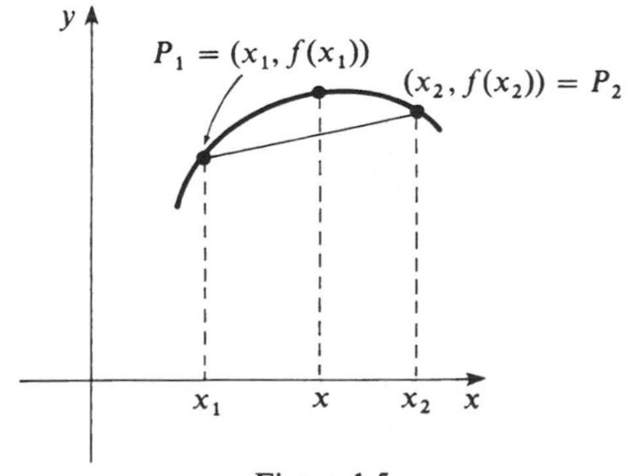

Figure 1.5

Theorem 1.2. *The function f defined on an interval I is concave if and only if $-f$ is convex.*

PROOF. Exercise.

Remark 1.3. Because of the last theorem, we deal in the sequel mainly with convex functions.

Theorem 1.3. *f is convex on an interval I if and only if for $x_1, x_2, \ldots, x_n$ in I, where n is some integer $n \geqslant 2$, we have*

$$f(\alpha_1 x_1 + \cdots + \alpha_n x_n) \leqslant \alpha_1 f(x_1) + \cdots + \alpha_n f(x_n)$$
$$\text{for} \quad \alpha_i \geqslant 0, \quad 1 \leqslant i \leqslant n, \quad \text{and} \quad \alpha_1 + \cdots + \alpha_n = 1. \tag{1.13}$$

PROOF. Let f be a function for which (1.13) holds for all integers $n \geqslant 2$. This implies, in particular, that (1.13) holds for $n = 2$. By Theorem 1.1′, f is convex on I.

Conversely, assume that f is convex on an interval I. We prove that (1.13) holds for each integer $n \geqslant 2$. If $n = 2$, it holds because of Theorem 1.1′. Assume that (1.13) holds for some integer $n \geqslant 2$. Take $x_1, \ldots, x_n, x_{n+1}$ in I and $\alpha_1, \ldots, \alpha_n, \alpha_{n+1}$ such that $\alpha_1 + \cdots + \alpha_{n+1} = 1$ and $\alpha_i \geqslant 0$ for $1 \leqslant i \leqslant n+1$. Write $\alpha = \alpha_1 + \cdots + \alpha_n$ and $\beta = 1 - \alpha$, so that $\beta = \alpha_{n+1}$. If $\alpha = 0$, then $\alpha_1 = \cdots = \alpha_n = 0$, and $\beta = \alpha_{n+1} = 1$. Hence,

$$f(\alpha_1 x_1 + \cdots + \alpha_{n+1} x_{n+1}) = f(x_{n+1}) \tag{1.14}$$

and

$$f(x_{n+1}) = \alpha_1 f(x_1) + \cdots + \alpha_{n+1} f(x_{n+1}), \tag{1.15}$$

from which (1.13) follows with the equality holding there for the integer $n + 1$. Now assume $\alpha > 0$. Write

$$m = \min\{x_1, \ldots, x_n\} \quad \text{and} \quad M = \text{Max}\{x_1, \ldots, x_n\} \tag{1.16}$$

and

$$y_1 = \frac{\alpha_1 x_1 + \cdots + \alpha_n x_n}{\alpha}. \tag{1.17}$$

Clearly,

$$m = \frac{m\alpha}{\alpha} = \frac{m(\alpha_1 + \cdots + \alpha_n)}{\alpha} \leqslant \frac{\alpha_1 x_1 + \cdots + \alpha_n x_n}{\alpha}$$
$$\leqslant \frac{(\alpha_1 + \cdots + \alpha_n)M}{\alpha} = M.$$

This implies that $m \leqslant y_1 \leqslant M$. Since each of m and M is one of $x_1, \ldots, x_n$, we know that m and M are in I. It follows that $y_1 \in I$. Put $y_2 = x_{n+1}$. Since f is convex and $\alpha + \beta = 1$, $\alpha > 0$, we know that

$$f(\alpha y_1 + \beta y_2) \leqslant \alpha f(y_1) + \beta f(y_2), \tag{1.18}$$

i.e.,

$$f(\alpha_1x_1 + \cdots + \alpha_nx_n + \alpha_{n+1}x_{n+1}) \leqslant \alpha f\left(\frac{\alpha_1x_1 + \cdots + \alpha_nx_n}{\alpha}\right) + \alpha_{n+1}f(x_{n+1}). \tag{1.19}$$

Since $1 = (\alpha_1 + \cdots + \alpha_n)/\alpha = (\alpha_1/\alpha) + \cdots + (\alpha_n/\alpha)$ and $0 \leqslant \alpha_i/\alpha$ for $i \in \{1, \ldots, n\}$, we have, by the induction hypothesis,

$$f\left(\frac{\alpha_1x_1 + \cdots + \alpha_nx_n}{\alpha}\right) \leqslant \frac{\alpha_1}{\alpha}f(x_1) + \cdots + \frac{\alpha_n}{\alpha}f(x_n)$$

and, therefore

$$\alpha f\left(\frac{\alpha_1x_1 + \cdots + \alpha_nx_n}{\alpha}\right) \leqslant \alpha_1f(x_1) + \cdots + \alpha_nf(x_n).$$

This and (1.19) imply, in this case ($\alpha \ni 0$) also, that

$$f(\alpha_1x_1 + \cdots + \alpha_nx_n + \alpha_{n+1}x_{n+1}) \leqslant \alpha_1f(x_1) + \cdots + \alpha_nf(x_n) + \alpha_{n+1}f(x_{n+1}).$$

The theorem holds by induction on n.

PROB. 1.3. Prove: If f is strictly convex on an interval I and $\alpha_1 + \cdots + \alpha_n = 1$, where $\alpha_i > 0$ for $i \in \{1, \ldots, n\}$, then the equality in (1.13) holds if and only if $x_1 = \cdots = x_n$.

2. Convexity and Differentiability

Lemma 2.1. *If f is convex on an interval I, x_0 is a point of I and g is defined as*

$$g(x) = \frac{f(x) - f(x_0)}{x - x_0} \quad \textit{for} \quad x \in I, \quad x \neq x_0, \tag{2.1}$$

then g is monotonically increasing. If f is strictly convex on I, then g is strictly monotonically increasing.

PROOF. Let x_0 be an interior point of I from the left. There exist points x_1 and x_2 in I such that $x_1 < x_2 < x_0$. Put

$$t = \frac{x_0 - x_2}{x_0 - x_1} = \frac{x_2 - x_0}{x_1 - x_0}. \tag{2.2}$$

Clearly, $0 < t < 1$ and $x_2 = x_0 + t(x_1 - x_0) = (1 - t)x_0 + tx_1$. From this and the convexity of f on I we obtain

$$f(x_2) = f((1 - t)x_0 + tx_1) \leqslant (1 - t)f(x_0) + tf(x_1).$$

This implies that

$$f(x_2) - f(x_0) \leqslant t(f(x_1) - f(x_0)) = \frac{x_2 - x_0}{x_1 - x_0}(f(x_1) - f(x_0)).$$

Divide the last inequality by $x_2 - x_0$. Since $x_2 - x_0 < 0$, we obtain

$$g(x_2) = \frac{f(x_2) - f(x_0)}{x_2 - x_1} \geqslant \frac{f(x_1) - f(x_0)}{x_1 - x_0} = g(x_1).$$

We proved that if $x_1 < x_2 < x_0$, then $g(x_1) \leqslant g(x_2)$. Note that the strict convexity of f would yield the strict inequality $g(x_1) < g(x_2)$ here.

Now let x_0 be an interior point of I from the right and let x_1 and x_2 be points of I such that $x_0 < x_1 < x_2$ and write

$$t = \frac{x_1 - x_0}{x_2 - x_0}, \tag{2.3}$$

so that $0 < t < 1$. We have $x_1 = (1 - t)x_0 + tx_2$. This time the convexity of f implies that

$$f(x_1) - f(x_0) \leqslant t(f(x_2) - f(x_0)) = \frac{x_1 - x_0}{x_2 - x_0}(f(x_2) - f(x_1)).$$

Since $x_1 - x_0 > 0$ here, dividing both sides in the above inequality by $x_1 - x_0$ yields

$$g(x_1) = \frac{f(x_1) - f(x_0)}{x_1 - x_0} \leqslant \frac{f(x_2) - f(x_0)}{x_2 - x_0} = g(x_2).$$

This proves that $x_0 < x_1 < x_2$ implies $g(x_1) \leqslant g(x_2)$.

Now assume that x_0 is an interior point of I and take points x_1 and x_2 of I, differing from x_0 such that $x_1 < x_2$. There are two cases: (1) $x_0 < x_1$ or (2) $x_1 < x_0$. In case (1), we have $x_0 < x_1 < x_2$. Earlier we proved that in this case $g(x_1) \leqslant g(x_2)$. If (2) holds, we have either (a) $x_1 < x_2 < x_0$ or (b) $x_1 < x_0 < x_2$. We proved above that in case (a), $g(x_1) \leqslant g(x_2)$. We confine our attention to (2)(b) so that $x_1 < x_0 < x_2$. Because f is convex, we know that

$$f(x_0) \leqslant f(x_1) + \frac{f(x_2) - f(x_1)}{x_2 - x_1}(x_0 - x_1).$$

Since $x_2 - x_1 > 0$, this implies that

$$(f(x_0) - f(x_1))(x_2 - x_1) \leqslant (f(x_2) - f(x_1))(x_0 - x_1). \tag{2.4}$$

But $f(x_2) - f(x_1) = f(x_2) - f(x_0) - (f(x_1) - f(x_0))$. Using this equality in the right-hand side of (2.4) yields

$$(f(x_0) - f(x_1))(x_2 - x_1) \leqslant (f(x_2) - f(x_0) - (f(x_1) - f(x_0))(x_0 - x_1).$$

By adding $(f(x_1) - f(x_0))(x_0 - x_1)$ to both sides of the last inequality we obtain, after some algebraic manipulation,

$$(f(x_0) - f(x_1))(x_2 - x_0) \leqslant (f(x_2) - f(x_0))(x_0 - x_1). \tag{2.5}$$

Upon dividing both sides of (2.5) by the positive number $(x_2 - x_0)(x_0 - x_1)$, we obtain

$$g(x_1) = \frac{f(x_0) - f(x_1)}{x_0 - x_1} \leqslant \frac{f(x_2) - f(x_0)}{x_2 - x_1} = g(x_2).$$

Thus, $x_1 < x_0 < x_2$ also implies $g(x_1) \leqslant g(x_2)$.

We summarize: (1) If x_0 is an interior point of I, either from the right or from the left, then g is monotonically increasing. (2) If x_0 is an interior point of I, then g is also monotonically increasing. Thus, $x_0 \in I$ implies that g is monotonically increasing. It is also seen that the strict convexity of f will yield the strict inequality $g(x_1) < g(x_2)$ when $x_1 < x_2$ in I where $x_1 \neq x_0$, $x_2 \neq x_0$. This completes the proof.

Theorem 2.1. *If f is convex on an interval I, then it is differentiable from both sides at each interior point x_0 of I and*

$$f'_L(x_0) \leqslant f'_R(x_0).$$

PROOF. The function g defined by

$$g(x) = \frac{f(x) - f(x_0)}{x - x_0} \qquad \text{for} \quad x \in I, \quad x \neq x_0$$

is monotonically increasing by the lemma. Since x_0 is an interior point of I, there exist x_1 and x_2 in I such that $x_1 < x_0 < x_2$. Hence,

$$g(x_1) = \frac{f(x_1) - f(x_0)}{x_1 - x_0} \leqslant \frac{f(x_2) - f(x_0)}{x_2 - x_0} = g(x_2).$$

Here, the left-hand side is bounded from above by $g(x_2)$ and the restriction of g to the set of points x of I such that $x < x_0$ is increasing. It follows that it has a finite limit as $x_1 \to x_0 -$. This implies that

$$f'_L(x_0) \leqslant g(x_2) \qquad \text{for} \quad x_0 < x_2. \tag{2.6}$$

For similar reasons, the restriction of g to the points x of I such that $x_0 < x$ is monotonically increasing and bounded from below by $f'_L(x_0)$. Accordingly, g has a finite limit as $x_2 \to x_0 +$. From this and (2.6) it follows that

$$f'_L(x_0) \leqslant f'_R(x_0),$$

as claimed.

Corollary. *If f is convex on an interval I, then it is continuous at each interior point of I.*

PROOF. Let x_0 be an interior point of I. By the theorem, the convexity of f on I implies that f is differentiable from both sides at x_0. This implies that f is continuous from both sides at x_0 and consequently that f is continuous at x_0.

Theorem 2.2. *If f is convex on an interval I, then f'_R and f'_L are monotonically increasing on the interior (the set of all interior points) of I. Also, if x_1 and x_2 are interior points of I such that $x_1 < x_2$, then*

$$f'_R(x_1) \leqslant f'_L(x_2). \tag{2.7}$$

Moreover, if f is strictly convex, then this inequality is strict and f'_L and f'_R are strictly monotonically increasing in the interior of I.

PROOF. Take x such that $x_1 < x < x_2$. By Lemma 2.1, we have

$$\frac{f(x_1) - f(x)}{x_1 - x} \leqslant \frac{f(x_2) - f(x)}{x_2 - x}. \tag{2.8}$$

By Theorem 2.1 and Lemma 2.1, f is differentiable from the left at x_2 and from the right at x_1

$$f'_L(x_2) = \lim_{x \to x_2 -} \frac{f(x) - f(x_2)}{x - x_2} = \lim_{x \to x_2 -} \frac{f(x_2) - f(x)}{x_2 - x}$$

$$= \sup_{x_1 < x < x_2} \frac{f(x) - f(x_2)}{x - x_2},$$

$$f'_R(x_1) = \lim_{x \to x_1 +} \frac{f(x) - f(x_1)}{x - x_1} = \lim_{x \to x_1 +} \frac{f(x_1) - f(x)}{x_1 - x_1}$$

$$= \inf_{x_1 < x < x_2} \frac{f(x_1) - f(x)}{x_1 - x}$$

hold. Hence,

$$f'_R(x_1) \leqslant \frac{f(x_1) - f(x)}{x_1 - x} \leqslant \frac{f(x_2) - f(x)}{x_2 - x} \leqslant f'_L(x_2). \tag{2.9}$$

This proves (2.7).

We now use Theorem 2.1 and (2.7) to obtain

$$f'_L(x_1) \leqslant f'_R(x_1) \leqslant f'_L(x_2) \leqslant f'_R(x_2), \tag{2.10}$$

implying

$$f'_L(x_1) \leqslant f'_L(x_2) \tag{2.11a}$$

and

$$f'_R(x_1) \leqslant f'_R(x_2) \tag{2.11b}$$

for interior points x_1 and x_2 of I such that $x_1 < x_2$. Thus, f'_L and f'_R are monotonically increasing functions in the interior of I.

Now suppose that f is strictly convex. Assume $x_1 < x_2$ and $x_1 < x < x_2$ where x_1 and x_2 are interior points of I. By Lemma 2.1, the inequality (2.8) is strict, so (2.9) becomes

$$f'_R(x_1) \leqslant \frac{f(x_1) - f(x)}{x_1 - x} < \frac{f(x_2) - f(x)}{x_2 - x} \leqslant f'_L(x_2).$$

Accordingly, in this case, (2.7) is a strict inequality and (2.10) becomes

$$f'_L(x_1) \leqslant f'_R(x_1) < f'_L(x_2) \leqslant f'_R(x_2). \tag{2.12}$$

This implies the strictness of the inequalities (2.11). This completes the proof.

Corollary 1. *If f is convex on an interval I and differentiable in the interior of I, then f' is monotonically increasing there. If f is strictly convex on an interval I and differentiable in the interior of I, then f' is strictly increasing there.*

PROOF. Exercise.

Corollary 2. *If f is convex on an interval I and twice differentiable in its interior, then $f''(x) \geqslant 0$ for each interior point x of I.*

PROOF. Exercise.

Thus far, the conditions we obtained for convexity were necessary ones. We now obtain sufficient conditions.

Theorem 2.3.* *If f is continuous on an interval I, differentiable from both sides at each interior point of I,*

$$f'_L(x) \leqslant f'_R(x) \qquad \text{for each interior point } x \text{ of } I, \tag{2.13}$$

and

$$f'_R(x_1) \leqslant f'_L(x_2) \qquad \text{if } x_1 \text{ and } x_2 \text{ are interior points of } I, \text{ such that } \quad x_1 < x_2, \tag{2.14}$$

then the one-sided derivatives of f are monotonically increasing in the interior of I and f is convex on I.

PROOF. Take interior points x_1 and x_2 of I such that $x_1 < x_2$ and obtain

$$f'_L(x_1) \leqslant f'_R(x_1) \leqslant f'_L(x_2) \leqslant f'_R(x_2). \tag{*}$$

This implies that

(a) $f'_L(x_1) \leqslant f'_L(x_2)$ and
(b) $f'_R(x_1) \leqslant f'_R(x_2)$. (**)

Hence, f'_L and f'_R are monotonically increasing in the interior of I.

We prove now that our conditions imply that f is convex on I. To do this we turn to Theorem VII.5.2 (the Mean-Value Theorem with one-sided derivatives). Take points x_1, x, x_2 of I such that $x_1 < x < x_2$. Applying Theorem VII.5.2 to f and the interval $[x_1, x]$, we know that there exists an

*E. Artin, *The Gamma Function*, Holt, Rinehart, Winston, New York, 1964, pp. 2–4.

x_0 such that $x_1 < x_0 < x$ and

$$f'_L(x_0) \leqslant \frac{f(x) - f(x_1)}{x - x_1} \leqslant f'_R(x_0) \quad \text{or} \quad f'_L(x_0) \geqslant \frac{f(x) - f(x_1)}{x - x_1} \geqslant f'_R(x_0).$$

Since x_0 is an interior point of I, the hypothesis implies that the first of these possibilities holds, and we have

$$f'_L(x_0) \leqslant \frac{f(x) - f(x_1)}{x - x_1} \leqslant f'_R(x_0), \tag{2.15}$$

where $x_1 < x_0 < x_2$. Using the same reasoning on the interval $[x, x_2]$ we see that there exists an x'_0 such that

$$f'_L(x'_0) \leqslant \frac{f(x_2) - f(x)}{x_2 - x} \leqslant f'_R(x'_0), \tag{2.16a}$$

where $x < x'_0 < x_2$. Since $x_1 < x_0 < x < x'_0 < x_2$, we have $x_0 < x'_0$. By the hypothesis we obtain

$$f'_R(x_0) \leqslant f'_L(x'_0). \tag{2.16b}$$

This (2.15) and (2.16) imply that

$$\frac{f(x) - f(x_1)}{x - x_1} \leqslant \frac{f(x_2) - f(x)}{x_2 - x}. \tag{2.17}$$

Since $x - x_1 > 0$ and $x_2 - x > 0$, this implies that

$$(x_2 - x)(f(x) - f(x_1)) \leqslant (x - x_1)(f(x_2) - f(x)). \tag{2.18}$$

We replace $f(x_2) - f(x)$ with $f(x_2) - f(x_1) - (f(x) - f(x_1))$ in (2.18) and obtain

$$(x_2 - x)(f(x) - f(x_1)) \leqslant (x - x_1)(f(x_2) - f(x_1)) - (x - x_1)(f(x) - f(x_1)).$$

It follows that

$$(x_2 - x_1)(f(x) - f(x_1)) \leqslant (x - x_1)(f(x_2) - f(x_1)). \tag{2.19}$$

Since $x_2 - x_1 > 0$, this implies

$$f(x) \leqslant f(x_1) + \frac{f(x_2) - f(x_1)}{x_2 - x_1}(x - x_1) \tag{2.20}$$

for $x_1 < x < x_2$. Hence, f is convex on I.

Corollary. *If f is continuous on an interval I, differentiable at each interior point of I, and f' is monotonically increasing in the interior of I, then f is convex on I.*

Proof. We prove that our f satisfies the hypothesis of the theorem. First of all, f is continuous on I. Second, since f is differentiable at each interior point of I, the one-sided derivatives of f exist for such points. Third, we have, because f is differentiable in the interior of I, that $f'_L(x) = f'_R(x)$ for each interior point $x \in I$ and, hence, that (2.13) holds for such points. Finally, since f' is monotonic increasing in the interior of I, we have: If x_1 and x_2 are interior points of I such that $x_1 < x_2$, then

$$f_R'(x_1) = f'(x_1) \leqslant f'(x_2) \leqslant f_L'(x_2).$$

It follows from this and the theorem that f is convex on I.

Combining this corollary with Corollary 1 of Theorem 2.2 we have:

Theorem 2.4. *A function which is continuous on an interval I and differentiable in the interior of I is convex on I if and only if its derivative is monotonically increasing in the interior of I.*

PROB. 2.1. Prove: If f is continuous on an interval I and twice differentiable in the interior of I, then f is convex on I if and only if $f''(x) \geqslant 0$ for each interior point x of I.

We state some theorems concerning strict convexity.

Theorem 2.5. *If f is defined on an interval I and satisfies the hypothesis of Theorem 2.3 with the proviso that inequality (2.14) holds strictly, then the one-sided derivatives f_L' and f_R' are strictly monotonically increasing in the interior of I and f is strictly convex on I.*

PROOF. Repeat the proof of Theorem 2.3 but use the strict inequality permitted by the present hypothesis to obtain, instead of (*): If x_1 and x_2 are interior points of I such that $x_1 < x_2$, then

$$f_L'(x_1) \leqslant f_R'(x_1) < f_L'(x_2) \leqslant f_R'(x_2)$$

so that

$$f_L'(x_1) < f_L'(x_2) \quad \text{and} \quad f_R'(x_1) < f_R'(x_2)$$

and, hence, that f_L' and f_R' are strictly increasing in the interior of I.

Now continue the proof of Theorem 2.3, using points x_1, x, and x_2 of I such that $x_1 < x < x_2$. It will follow from the reasoning used there that there exist x_0 and x_0' such that $x_1 < x_0 < x < x_0' < x_2$ and such that (2.16b) holds. The present hypothesis justifies the strict inequality $f_R'(x_0) < f_L'(x_0)$. Now (2.17) will be replaced by the strict inequality

$$\frac{f(x) - f(x_1)}{x - x_1} < \frac{f(x_2) - f(x)}{x_2 - x}. \tag{2.21}$$

It follows that (2.20) can be replaced by the strict inequality

$$f(x) < f(x_1) + \frac{f(x_2) - f(x_1)}{x_2 - x_1}(x - x_1) \qquad \text{for} \quad x_1 < x < x_2.$$

The strict convexity of f on I follows.

Corollary. *If f is continuous on an interval I and differentiable in its interior and f' is strictly increasing there, then f is strictly convex on I.*

PROOF. Exercise.

This corollary and Corollary 1 of Theorem 2.2 yield:

Theorem 2.6. *If f is continuous on an interval I and differentiable in its interior, then f is strictly convex on I if and only if f' is strictly increasing in the interior of I.*

PROB. 2.2 (cf. Prob. 2.1). Prove: If f is continuous on an interval I, twice differentiable in the interior of I, and $f''(x) > 0$ holds for each x in the interior of I, then f is strictly convex on I.

Remark 2.1. Note the result in the last problem is not in "if and only if" form. We give an example of a function which is twice differentiable in the interior of an interval and is strictly convex and continuous there for which the weaker $f''(x) \geqslant 0$ holds but cannot be strengthened to a strict inequality. We take $(\ )^4$ on $\mathbb{R}$ and prove it is strictly convex. Take x_1 and x_2 such that $x_1 \neq x_2$ and α, β such that $\alpha > 0$, $\beta > 0$ and $\alpha + \beta = 1$. Since $(\ )^2$ is strictly convex on $\mathbb{R}$ (Example 1.2), we have

$$(\alpha x_1 + \beta x_2)^2 < \alpha x_1^2 + \beta x_2^2.$$

Squaring again yields

$$(\alpha x_1 + \beta x_2)^4 < (\alpha x_1^2 + \beta x_2^2)^2 \leqslant \alpha x_1^4 + \beta x_2^4.$$

Accordingly f, where $f(x) = x^4$, $x \in \mathbb{R}$ is strictly convex on $\mathbb{R}$. We have, however, $f''(x) = 12x^2 \geqslant 0$, $x \in \mathbb{R}$. Since $f''(0) = 0$, this inequality cannot be strengthened.

There are corresponding results on concavity. We state these in problem form.

PROB. 2.3. Prove: If f is continuous on an interval I and differentiable in the interior of I, then f is concave on I if and only if f' is monotonically decreasing.

PROB. 2.4. Prove: If f is continuous on an interval I and twice differentiable in the interior of I, then f is concave on I if and only if $f''(x) \leqslant 0$ holds for each interior point x of I.

PROB. 2.5. Prove: If f is continuous on an interval I and twice differentiable in the interior of I and $f''(x) < 0$ for x in the interior of I, then f is *strictly* concave on I.

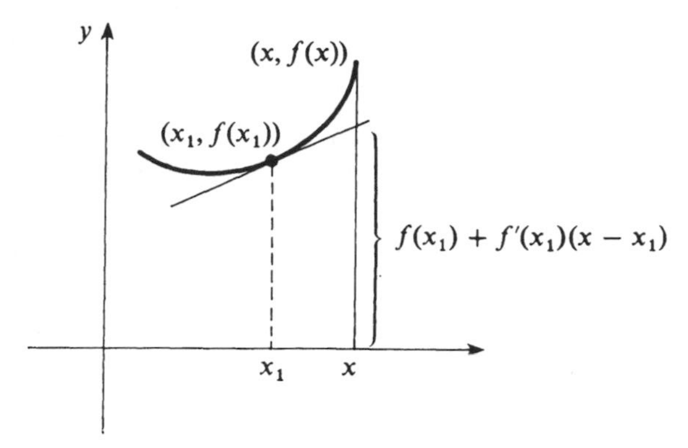

Figure 2.1

We present still another characterization of convexity for differentiable functions.

Def. 2.1. If f is differentiable on an interval I, then we say that its *graph lies above its tangents* if and only if, for each $x_1 \in I$, we have

$$f(x) \geqslant f(x_1) + f'(x_1)(x - x_1) \qquad \text{for each} \quad x \in I \tag{2.22}$$

(see Fig. 2.1). If at some x_1 in I the inequality above holds, we say that *the graph lies above its tangent at* x_1. When the inequality (2.22) is strict for each x_1 in I we say that the graph of f lies *strictly* above its tangents. If for each x_1 in I_1 the sense of the inequality (2.22) is reversed, then we say that the *graph of f lies below* its tangents. If x_1 is an endpoint of I, then $f'(x_1)$ is the appropriate one-sided derivative of f at x_1.

Theorem 2.7. *If f is continuous on an interval I and differentiable in the interior of I, then it is convex on I if and only if its graph lies above all its tangents at all interior points of I.*

PROOF. We first assume that f is convex on I. By hypothesis, f is continuous on I and differentiable in the interior of I. By Theorem 2.4, f' is monotonically increasing on I. Let x_1 be an interior point of I and x a point of I such that $x > x_1$. By the Mean-Value Theorem, there exists a point x_0 such that

$$f(x) = f(x_1) + f'(x_0)(x - x_1), \tag{2.23}$$

where $x_1 < x_0 < x$. But $f'(x_1) \leqslant f'(x_0)$ and $x - x_1 > 0$ and, hence,

$$f'(x_1)(x - x_1) \leqslant f'(x_0)(x - x_1).$$

This implies, using (2.3), that

$$f(x) \geqslant f(x_1) + f'(x_1)(x - x_1). \tag{2.24}$$

Now take x in I such that $x < x_1$. Again there exists an x_0' such that

$$f(x) = f(x_1) + f'(x_0')(x - x_1), \tag{2.25}$$

where $x < x_0' < x_1$. This time, we have $f'(x_0') \leqslant f'(x_1)$ and $x - x_0 < 0$, so that $f'(x_0')(x - x_1) \geqslant f'(x_1)(x - x_1)$. This and (2.25) imply (2.24) in this case also. Also note that for $x = x_1$, (2.24) holds trivially. This establishes (2.24) for all $x \in I$ for each x_1 in the interior of I and, hence, that the graph of f lies above all its tangents at all interior points of I.

Conversely, let f be continuous on I and differentiable in the interior of I and suppose the graph of f lies above all its tangents at all interior points of I. Let x_1 and x_2 be interior points of I such that $x_1 < x_2$. We have

$$f(x_2) \geqslant f(x_1) + f'(x_1)(x_2 - x_1) \tag{2.26a}$$

and

$$f(x_1) \geqslant f(x_2) + f'(x_2)(x_1 - x_2). \tag{2.26b}$$

Since $x_2 - x_1 > 0$ and $x_1 - x_2 < 0$, it follows from these two inequalities that

$$f'(x_1) \leqslant \frac{f(x_2) - f(x_1)}{x_2 - x_1} \leqslant f'(x_2)$$

and that $f'(x_1) \leqslant f'(x_2)$. This proves that f' is monotonically increasing in the interior of I. By Theorem (2.4), f is convex on I. The proof is complete.

PROB. 2.6. Prove: If f is continuous on an interval I and twice differentiable in the interior of I, then the graph of f lies above all its tangents at all interior points of I if and only if $f''(x) \geqslant 0$ for each interior point x of I.

PROB. 2.7. Prove: If f is continuous on an interval I and differentiable in the interior of I, then f is strictly convex on I if and only if the graph of f lies strictly above all its tangents at all interior points of I.

PROB. 2.8. Prove: If f is continuous on an interval I and differentiable in the interior of I, and $f''(x) > 0$ for each interior point x of I, then the graph of f lies strictly above all its tangents at all interior points of I.

PROB. 2.9. Let f be differentiable in an interval I. Prove that the graph of f lies below its tangents on I, if and only if $-f$ lies above its tangents on I.

PROB. 2.10. Prove: If f is continuous on I and differentiable in its interior, then (a) f is concave on I if and only if its graph lies below all its tangents at all interior points of I. (b) f is strictly concave on I if and only if the graph of f lies strictly below all its tangents at all interior points of I.

PROB. 2.11. Let f be continuous on an interval I and twice differentiable in the interior of I. Prove: (a) the graph of f lies below all its tangents at all interior points of I if and only if $f''(x) \leqslant 0$ for each x in the interior of I. (b) If $f''(x) < 0$ for each x in the interior of I, f lies strictly below all its tangents at all interior points of I.

We shall see the importance of the notion of convexity when we study the gamma function. The relation between convexity, concavity, and the location of maxima and minima of functions will be examined in Chapter IX. Meanwhile, in the next three problems, we cite some results on convex functions needed for our later work.

PROB. 2.12. Prove: If f and g are convex functions on an interval I, then so is $f+g$.

PROB. 2.13. Prove: If f is a function which is convex on an interval I and c is a *positive* constant, then cf is convex on I.

PROB. 2.14. Prove: If $\langle f_n \rangle$ is a sequence of functions convex on an interval I and

$$\lim_{n \to +\infty} f_n(x) = f(x) \qquad \text{for each} \quad x \in I,$$

then f is convex on I.

An Application to an Inequality Involving ln

Consider the function f where

$$f(x) = x \ln x \qquad \text{for} \quad x > 0. \tag{2.27}$$

We have,

$$f'(x) = 1 + \ln x \quad \text{and} \quad f''(x) = \frac{1}{x} \qquad \text{for} \quad x > 0.$$

Thus, f is strictly convex. Also if $x_1, \ldots, x_n > 0$ and $\alpha_1 + \cdots + \alpha_n = 1$, where $\alpha_i > 0$, $i \in \{1, \ldots, n\}$, then (Theorem 1.3)

$$f(\alpha_1 x_1 + \cdots + \alpha_n x_n) \leqslant \alpha_1 f(x_1) + \cdots + \alpha_n f(x_n).$$

Hence, if $x_1, \ldots, x_n$ are positive and $\alpha_1 + \cdots + \alpha_n = 1$, where $\alpha_i > 0$, $i \in \{1, \ldots, n\}$, then

$$(\alpha_1 x_1 + \cdots + \alpha_n x_n) \ln(\alpha_1 x_1 + \cdots + \alpha_n x_n)$$
$$\leqslant \alpha_1 x_1 \ln x_1 + \cdots + \alpha_n x_n \ln x_n. \tag{2.28}$$

Here the equality holds if and only if $x_1 = x_2 = \cdots = x_n$.

PROB. 2.15. Prove: If $u > 0$, $v > 0$, then

(a) $((u+v)/2)^k \leqslant (u^k + v^k)/2$ if $k > 1$ and
(b) $((u+v)/2)^k \geqslant (u^k + v^k)/2$ if $0 < k < 1$.

3. Inflection Points

Def. 3.1. If f is differentiable on an interval I and x_0 is an interior point of I, then the point $(x_0, f(x_0))$ is called an *inflection* point of the graph of f if and only if there exists a δ-neighborhood $N(x_0, \delta)$ of x_0 contained in I such that either f is strictly convex on $(x_0 - \delta, x_0)$ and strictly concave on $(x_0, x_0 + \delta)$ or f is strictly concave on $(x_0 - \delta, x_0)$ and strictly concave on $(x_0, x_0 + \delta)$ (see Fig. 3.1).

Theorem 3.1. *If f is differentiable on an interval I and $(x_0, f(x_0))$ is an inflection point of the graph of f and f is twice differentiable at x_0, then $f''(x_0) = 0$.*

PROOF. Use the definition of inflection point to obtain a δ-neighborhood $N(x_0, \delta)$ in I such that either (1) f is strictly convex on $(x_0 - \delta, x_0)$ and strictly concave on $(x_0, x_0 + \delta)$ or (2) f is strictly concave on $(x_0 - \delta, x_0)$ and strictly convex on $(x_0, x_0 + \delta)$. For the sake of definiteness suppose that (1) is the case. This implies that f' is strictly increasing on $(x_0 - \delta, x_0)$ and strictly decreasing on $(x_0, x_0 + \delta)$. Since f' is a derivative on the interval I, it has the *strong* intermediate value property on I (Remark VII.7.1). Therefore, f' will have the strong intermediate value property on $(x_0 - \delta, x_0]$ and on $[x_0, x_0 + \delta)$. Since f' is also strictly increasing on $(x_0 - \delta; x_0)$ and decreasing on $(x_0; x_0 + \delta)$, it follows from Prob. VI.4.6 that f' is strictly increasing on $(x_0 - \delta, x_0]$ and strictly decreasing on $[x_0, x_0 + \delta)$. Accordingly, if $x_0 - \delta < x < x_0$, then $f'(x) < f'(x_0)$ and $x - x_0 < 0$, so that

$$\frac{f'(x) - f'(x_0)}{x - x_0} > 0 \tag{3.1a}$$

and, hence,

$$f''(x_0) = \lim_{x \to x_0-} \frac{f'(x) - f'(x_0)}{x - x_0} \geqslant 0; \tag{3.1b}$$

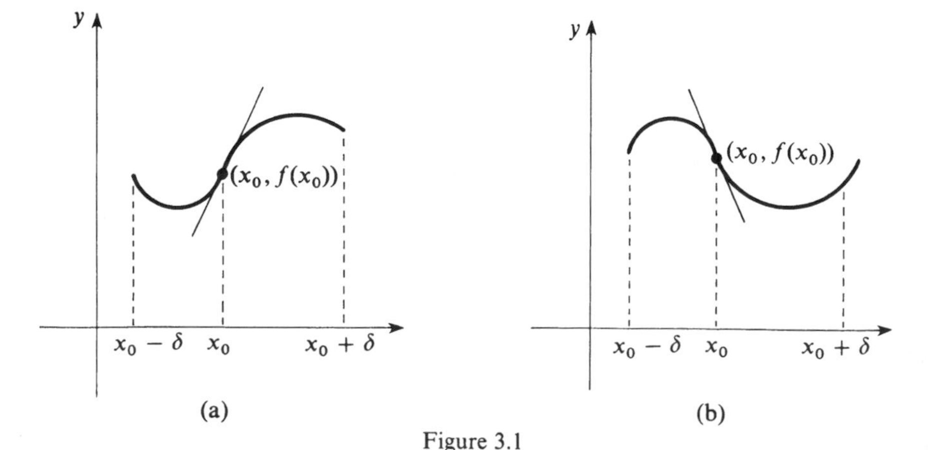

Figure 3.1

and if $x_0 < x < x_0 + \delta$, we have $f'(x) < f'(x_0)$ and $x - x_0 > 0$, so that

$$\frac{f'(x) - f'(x_0)}{x - x_0} < 0 \tag{3.2a}$$

and, hence,

$$f''(x_0) = \lim_{x \to x_0+} \frac{f'(x) - f'(x_0)}{x - x_0} \leqslant 0. \tag{3.2b}$$

Thus, we have $f''(x_0) \geqslant 0$ and $f''(x_0) \leqslant 0$, so that $f''(x_0) = 0$.

Remark 3.1. Under the hypothesis of the last theorem we find that $f''(x_0) = 0$ is a necessary condition for $(x_0, f(x_0))$ to be an inflection point of the graph of f. This condition is, however, not sufficient. For example, let f be defined as $f(x) = x^4$ for $x \in \mathbb{R}$. We have $f''(x) = 12x^2$, $f''(0) = 0$. It is easy to see that $(0, f(0)) = (0, 0)$ is not an inflection point of the graph of f. Since $f'(x) = 4x^3$, f' is strictly increasing on $\mathbb{R}$ and, therefore, f is strictly convex on every δ-neighborhood $N(0, \delta) = (-\delta; \delta)$ of 0. Accordingly, $(0, f(0)) = (0, 0)$ is not an inflection point of the graph of f. Below, we give a sufficient condition for a point to be an inflection point.

Theorem 3.2. *If f' is continuous in some δ-neighborhood $N(x_0, \delta)$ of x_0 and either* (1) *$f''(x) > 0$ for $x_0 - \delta < x < x_0$ and $f''(x) < 0$ for $x_0 < x < x_0 + \delta$ or* (2) *$f''(x) < 0$ for $x_0 - \delta < x < x_0$ and $f''(x) > 0$ for $x_0 < x < x_0 + \delta$, then $(x_0, f(x_0))$ is an inflection point of the graph of f. (Note, the hypothesis does not require f to be twice differentiable at x_0).*

Proof. We consider an f such that f' is continuous on $N(x_0, \delta)$ and for which (1) holds. (In case (2) holds, the proof is similar.) Since $f''(x) > 0$ for $x_0 - \delta < x < x_0$, and $f''(x) < 0$ for $x_0 < x < x_0 + \delta$, f' is strictly increasing on $(x_0 - \delta; x_0)$ and strictly decreasing on $(x_0; x_0 + \delta)$. It follows that f is strictly convex on $(x_0 - \delta; x_0)$ and strictly concave on $(x_0; x_0 + \delta)$ and that $(x_0, f(x_0))$ is an inflection point of the graph of f.

Corollary. *If f is twice differentiable in some δ-neighborhood $N(x_0, \delta)$ of x_0 and either* (1) *$f''(x) > 0$ for $x_0 - \delta < x < x_0$ and $f''(x) < 0$ for $x_0 < x < x_0 + \delta$, or* (2) *$f''(x) < 0$ for $x_0 - \delta < x < x_0$ and $f''(x) > 0$ for $x_0 < x < x_1 + \delta$, then $f''(x_0) = 0$.*

Proof. Exercise.

Prob. 3.1. Find the inflection points, if any, for (1) $f(x) = e^{-x^2}$, $x \in \mathbb{R}$ and (2) $g(x) = (1 + x^2)^{-1}$, $x \in \mathbb{R}$.

Prob. 3.2. Let f be given by $f(x) = x^{5/3}$, $x \in \mathbb{R}$. Show that $(0, f(0)) = (0,0)$ is an inflection point of the graph of f and $f''(0)$ does not exist.

PROB. 3.3. Let g be given by $g(x) = x^{1/3}$, $x \in \mathbb{R}$. Show that $(0, f(0)) = (0, 0)$ is an inflection point of the graph of g, $g'(0) = +\infty$, and $g''(0)$ does not exist.

4. The Trigonometric Functions

The reader is probably familiar with the properties of the trigonometric functions noted in this section. We derive these here in order to illustrate the theory and also to make this book self-contained.

The sine and cosine functions were defined in Section IV.8 and some of their properties were noted there. Some limits involving the sine and cosine were obtained in Theorem V.4.2, and in Example VI.1.4 we noted that these functions are continuous on $\mathbb{R}$. The number π was defined in Def. VI.4.1 as $\pi = 2c$, where $2 < c < 4$ and $\cos c = 0$. More information about sine and cosine is contained in Theorem VI.4.2 and Probs. VI.4.3–4.5. In Example VII.1.2 and Prob. VII.1.1 we saw that $d(\cos x)/dx = -\sin x$ and $d(\sin x)/dx = \cos x$. We note further properties of these functions.

We have $\sin(\pi/2) = 1$ and $\sin(-\pi/2) = -1$. Since the sine is continuous for all $x \in \mathbb{R}$, this implies that sine maps $[-\pi/2, \pi/2]$ onto $[-1, 1]$. We recall that

$$\frac{d \sin x}{dx} = \cos x \tag{4.1a}$$

and

$$\frac{d(\cos x)}{dx} = -\sin x \quad \text{for} \quad x \in \mathbb{R}, \tag{4.1b}$$

so that

$$\frac{d^2 \sin x}{dx^2} = -\sin x \tag{4.2a}$$

and

$$\frac{d^2 \cos x}{dx^2} = -\cos x \quad \text{for} \quad x \in \mathbb{R}. \tag{4.2b}$$

From (4.1a) and $\cos x > 0$ for $-\pi/2 < x < \pi/2$, we obtain that sine is strictly increasing on $[-\pi/2, \pi/2]$ and, hence, that the restriction of sine to $[-\pi/2, \pi/2]$ is one-to-one. Since $\cos x < 0$ for $\pi/2 < x < \pi$ (Theorem VI.4.2, part (g)), (4.1a) implies that sine is strictly decreasing on $[\pi/2, \pi]$. From $\cos(-x) = \cos x$, we see that $\cos x < 0$ for $-\pi < x < -\pi/2$, so (4.2a) implies that sine is decreasing on $[-\pi, -\pi/2]$. We now use (4.2a) and the fact that $\sin x < 0$ for $-\pi < x < 0$ and $\sin x > 0$ for $0 < x < \pi$ to obtain that sine is convex on $[-\pi, 0]$ and concave on $[0, \pi]$, and that $(0, \sin 0) = (0, 0)$ is an inflection point of the graph of sine. Fig. 4.1(a) is a sketch of the graph of sine on $[-\pi, \pi]$.

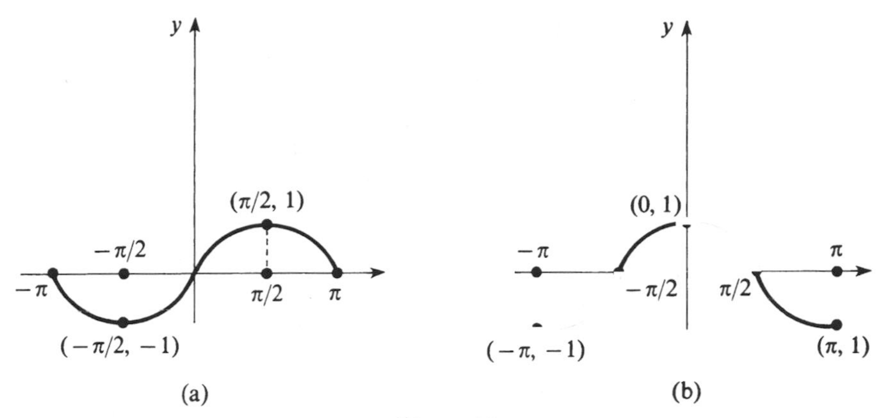

Figure 4.1

We now turn to cosine. We have $\cos 0 = 1$ and $\cos\pi = -1$. Since cosine is continuous for all $x \in \mathbb{R}$, it maps the interval $[0,\pi]$ onto the interval $[-1,1]$. From (4.1b) and $\sin x > 0$ for $0 < x < \pi$, it follows that cosine is strictly decreasing on $[0,\pi]$ and, hence, that the restriction of cosine to $[0,\pi]$ is one-to-one. It is also easy to see that cosine is concave on $[-\pi/2,\pi/2]$ and convex on the intervals $[-\pi,-\pi/2]$, $[\pi/2,\pi]$, and that the points $(-\pi/2,\cos(-\pi/2)) = (-\pi/2,0)$ and $(\pi/2,\cos\pi/2) = (\pi/2,0)$ are inflection points of the graph of cosine (prove these statements). See Fig. 4.1(b) for a sketch of cosine on $[-\pi,\pi]$.

Prob. 4.1. Prove: If x and y are real numbers such that $x^2 + y^2 = 1$, then there exists exactly one t such that $-\pi < t \leqslant \pi$ with $x = \cos t$ and $y = \sin t$.

Prob. 4.2. Prove: If n is an integer, then $\cos(n\pi) = (-1)^n$ and $\sin(n\pi) = 0$.

Prob. 4.3. Prove: If n is an integer, then for each $x \in \mathbb{R}$ (a) $\sin(x + 2n\pi) = \sin x$ and (b) $\cos(x + 2n\pi) = \cos x$.

It follows from Prob. 4.3 that if n is a nonzero integer, then $2n\pi$ is a period for the sine and cosine. In the theorem which follows we prove that these functions have no other periods.

Theorem 4.1. *(a) If* $\sin p = 0$, *then* $p = n\pi$, *where n is an integer. (b) If* $\cos p = 1$, *then* $p = 2k\pi$, *where k is an integer. (c) If* $\cos p = -1$, *then* $p = (2k+1)\pi$, *where k is an integer. (d) If* $\sin(x+p) = \sin x$ *for all* $x \in \mathbb{R}$, *then* $p = 2n\pi$, *where n is an integer, and similarly for cosine.*

Proof. We first prove (a). Assume $\sin p = 0$ for some $p \in \mathbb{R}$. There exists an integer n such that $n \leqslant p/\pi < n+1$. This implies that $0 \leqslant p - n\pi < \pi$. But

$$\sin(p - n\pi) = \sin p\cos(n\pi) - \cos p\sin n\pi = (-1)^n\sin p = 0,$$

that is, $\sin(p - n\pi) = 0$. If $0 < p - n\pi$, we would have $0 < p - n\pi < \pi$, from which it would follow that $\sin(p - n\pi) > 0$. This contradiction leads to the conclusion that $0 = p - n\pi$ and, hence, $p = n\pi$, n an integer.

We prove (b). Assume that $\cos p = 1$ for some $p \in \mathbb{R}$. There exists an integer n such that $n \leqslant p/\pi < n + 1$, which implies that $0 \leqslant p - n\pi < \pi$. Since cosine is strictly decreasing on $[0, \pi]$, we have

$$1 = \cos 0 \geqslant \cos(p - n\pi) > \cos\pi = -1.$$

If n were odd, we would have

$$\cos(p - n\pi) = (-1)^n \cos p = (-1)^n = -1$$

which contradicts the previous inequality. Thus, n is an even integer, $n = 2k$. It follows that $0 \leqslant p - 2k\pi < \pi$, where k is an integer. If $0 < p - 2k\pi$, we would have $0 < p - 2k\pi < \pi$ and

$$\cos(p - 2k\pi) = \cos p = 1 = \cos 0.$$

This is impossible since cosine is one-to-one $[0, \pi]$. Hence, $0 = p - 2k\pi$, or $p = 2k\pi$, where k is an integer.

We prove (c). Assume that $\cos p = -1$ for some $p \in \mathbb{R}$. This implies that

$$\cos(p - \pi) = \cos p \cos\pi = (-1)(-1) = 1.$$

In view of (b), this implies that $p - \pi = 2k\pi$, where k is an integer. Hence, $p = (2k + 1)\pi$, where k is an integer.

We prove the part of (d) involving sine. Assume that $\sin(x + p) = \sin x$ for all $x \in \mathbb{R}$. This implies, in particular, that $\sin(\pi/2 + p) = \sin(\pi/2) = 1$. Since $\cos p = \sin(\pi/2 + p)$, it follows that $\cos p = \sin(\pi/2 + p) = 1$. In view of (c), $p = 2n\pi$, where n is an integer. We leave the proof of the part of (d) involving cosine to the reader (Prob. 4.4).

Corollary. *The fundamental period of the sine function is* 2π.

PROOF. Exercise.

PROB. 4.4. Prove: If $\cos(x + p) = \cos x$ for all $x \in \mathbb{R}$, then $p = 2n\pi$, where n is an integer and that the fundamental period of cosine is 2π.

The graphs of sine and cosine are drawn in Fig. 4.2.

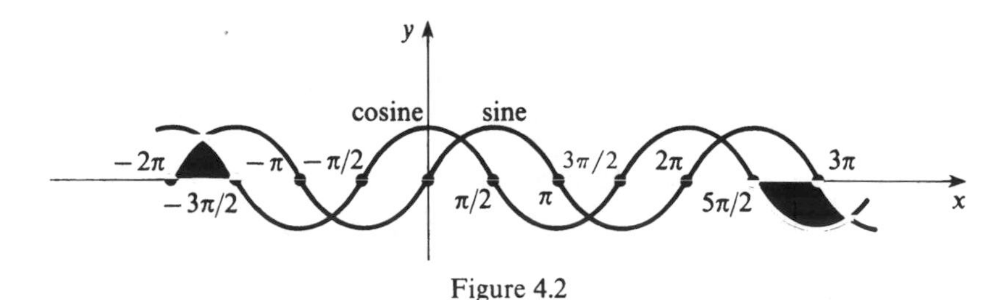

Figure 4.2

Prob. 4.5. Prove: $\sin x = 1$ if and only if $x = (4n+1)(\pi/2)$, where n is an integer, and $\sin x = -1$ if and only if $x = (4n-1)(\pi/2)$, where n is an integer.

Prob. 4.6. Prove: $\cos x = 0$ if and only if $x = (2n+1)(\pi/2)$, where n is an integer.

Prob. 4.7. Prove: $\tan x = 0$ if and only if $x = n\pi$, where n is an integer, and that $\tan x > 0$ if $n\pi < x < (2n+1)(\pi/2)$, and $\tan x < 0$ if $(2n-1)(\pi/2) < x < n\pi$, where n is an integer.

Prob. 4.8. Note that the domain of tangent is $\{x \in \mathbb{R} \mid x \neq (2n+1)(\pi/2)$, where n is an integer$\}$. Show tangent is periodic with fundamental period π. Also show that

$$\lim_{x \to ((2n+1)(\pi/2))-} \tan x = +\infty \quad \text{and} \quad \lim_{x \to ((2n+1)(\pi/2))+} \tan x = -\infty.$$

Prob. 4.9. Prove that the tangent function is strictly increasing on the intervals $((2n-1)(\pi/2); (2n+1)(\pi/2))$, n an integer, and that in each of these intervals it assumes each real number exactly once.

Prob. 4.10. Prove: Tangent is strictly concave on $((2n-1)(\pi/2), n\pi]$ and strictly convex on $[n\pi, (2n+1)(\pi/2))$, n an integer.

The graph of tangent appears along with the graph of secant in Fig. 4.3. The latter is sketched using heavy broken lines.

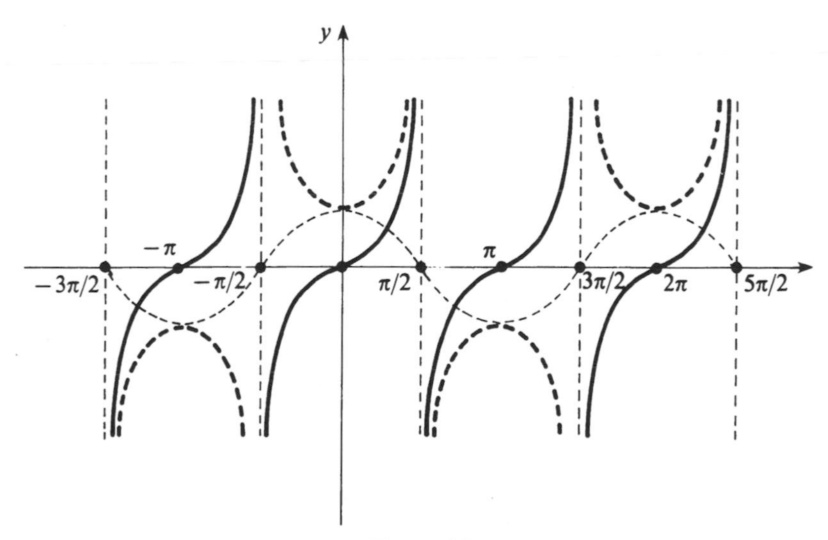

Figure 4.3

Prob. 4.11. Note that the domain of the secant function is $\{x \in \mathbb{R} \mid x \neq (2n+1)(\pi/2), n \text{ an integer}\}$. Prove:

(a) $\lim_{x\to(\pi/2)-} \sec x = +\infty = \lim_{x\to(-\pi/2)+} \sec x$,
(b) $\lim_{x\to(\pi/2)+} (\sec x) = -\infty = \lim_{x\to(-\pi/2)-} (\sec x)$,
(c) $\sec(n\pi) = (-1)^n$, n an integer,
(d) $\sec x \geqslant 1$ for $-\pi/2 < x < \pi/2$, and that $\sec x \leqslant -1$ for $x \in (-3\pi/2; -\pi/2) \cup (\pi/2; 3\pi/2)$.
(e) secant is periodic with fundamental period 2π.

Prob. 4.12. Prove: Secant is convex on $(-\pi/2; \pi/2)$ and concave on $(-3\pi/2; -\pi/2)$ and on $(\pi/2; 3\pi/2)$.

Prob. 4.13. Note the domain of contangent is $\{x \in \mathbb{R} \mid x \neq n\pi, n \text{ an integer}\}$. Prove: (a) $\cot x = 0$ if and only if $x = (2n+1)(\pi/2)$ n an integer. (b) $\cot x \geqslant 0$ if and only if $n\pi < x < (2n+1)(\pi/2)$, and $\cot x < 0$ if and only if $(2n-1)(\pi/2) < x < n\pi$, n is an integer.

Prob. 4.14. Prove: Cotangent is periodic with fundamental period π. Also prove that for an integer n

$$\lim_{x\to(n\pi)+} \cot x = +\infty \quad \text{and} \quad \lim_{x\to(n\pi)-} \cot x = -\infty.$$

Prob. 4.15. Prove: Cotangent is strictly decreasing on $(n\pi; (n+1)\pi)$, n an integer, and that on each of these intervals it assumes each real number exactly once.

Prob. 4.16. Prove: Cotangent is strictly convex on $(n\pi, (2n+1)(\pi/2)]$ and strictly concave on $[(2n-1)(\pi/2), n\pi)$.

The cotangent function is graphed in Fig. 4.4 together with cosecant. The graph of the latter is drawn with heavy broken lines.

Prob. 4.17. Note that the domain of cosecant is $\{x \in \mathbb{R} \mid x \neq n\pi, n \text{ an integer}\}$. Prove:

(a) $\lim_{x\to 0+} \csc x = +\infty = \lim_{x\to\pi-} \csc x$,
(b) $\lim_{x\to 0-} \csc x = -\infty = \lim_{x\to\pi+} \csc x$,
(c) $\csc((4n+1)(\pi/2)) = 1$, $\csc((4n-1)(\pi/2) = -1$,
(d) $\csc x \geqslant 1$ for $0 < x < \pi$, and $\csc x \leqslant -1$ for $-\pi < x < 0$,
(e) cosecant is periodic with fundamental period 2π.

Prob. 4.18. Prove: Cosecant is convex on $(0; \pi)$ and concave on $(-\pi; 0)$.

Prob. 4.19. Prove: $\lim_{x\to\pm\infty} \sin x$ and $\lim_{x\to\pm\infty} \cos x$ do not exist.

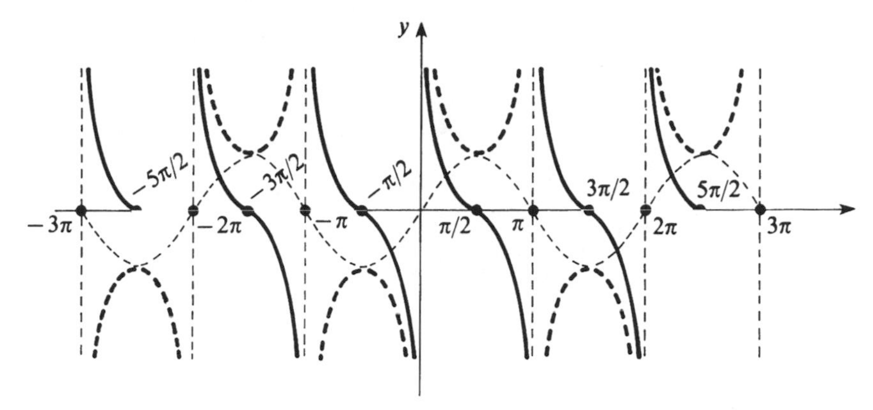

Figure 4.4

PROB. 4.20. Show that $\sin 18° = (\sqrt{5} - 1)/4$. Then show that if $\tan x = \cos x$, then $\sin x = 2\sin 18°$.

5. Some Remarks on Differentiability

We first examine the function f, where

$$f(x) = \sin\frac{1}{x}, \qquad x \neq 0.$$

We show that

$$\lim_{x\to 0+} \sin\frac{1}{x} \quad \text{does not exist.}$$

Consider the sequences $\langle x_n \rangle$ and $\langle x_n' \rangle$, where

$$x_n = \frac{1}{(4n+1)(\pi/2)} \quad \text{and} \quad x_n' = \frac{1}{n\pi} \qquad \text{for each positive integer } n.$$

Clearly,

$$\lim_{n\to+\infty} x_n = 0 = \lim_{n\to+\infty} x_n', \tag{5.1}$$

$$\lim_{n\to+\infty} \sin\frac{1}{x_n} = \lim_{n\to+\infty} \sin(4n+1)(\pi/2) = \lim_{n\to+\infty} 1 = 1, \tag{5.2}$$

and

$$\lim_{n\to+\infty} \sin\frac{1}{x_n'} = \lim_{n\to+\infty} \sin(n\pi) = \lim_{n\to+\infty} 0 = 0. \tag{5.3}$$

Comparing this limit with (5.2), we see that $\lim_{x\to 0+} \sin(1/x)$ does not exist

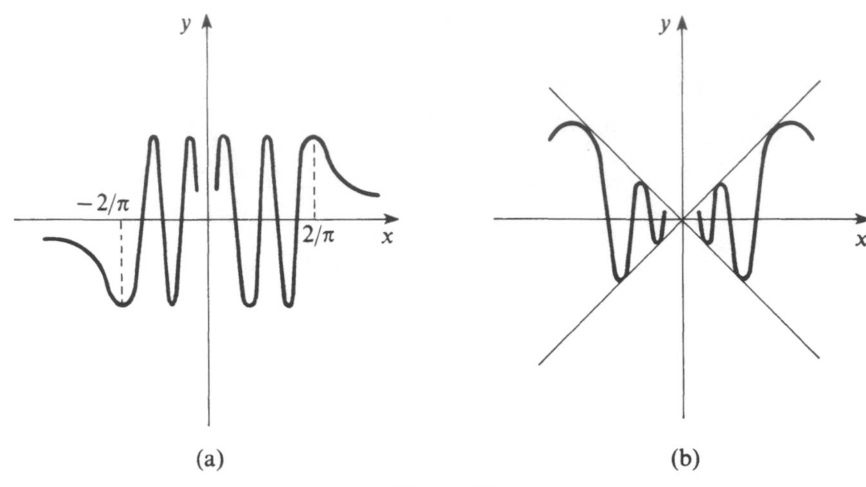

Figure 5.1

(explain). Using similar sequences we see that

$$\lim_{x\to 0-} \sin\frac{1}{x}$$

does not exist (see Fig. 5.1(a)) either.

We now consider the function g, where

$$g(x) = \begin{cases} x\sin\dfrac{1}{x} & \text{for} \quad x \neq 0, \\ 0 & \text{for} \quad x = 0 \end{cases}$$

(see Fig. 5.1(b)). Since $\lim_{x\to 0} x = 0$ and $\sin(1/x)$ is bounded, it follows that

$$\lim_{x\to 0} g(x) = \lim_{x\to 0}\left(x\sin\frac{1}{x}\right) = 0 = g(0). \tag{5.4}$$

This proves that g is continuous at 0. In fact, g is continuous for all $x \in \mathbb{R}$. However,

$$\lim_{x\to 0}\frac{g(x)-g(0)}{x} = \lim_{x\to 0}\sin\frac{1}{x} \quad \text{does not exist.}$$

Hence, g is not differentiable at 0. Of course, g is differentiable for all $x \neq 0$. This gives us another example of a function which is continuous at a point but not differentiable there (see Example VII.2.1).

Finally we examine h, where

$$h(x) = \begin{cases} x^2\sin\dfrac{1}{x} & \text{for} \quad x \neq 0 \\ 0 & \text{for} \quad x = 0 \end{cases} \tag{5.5}$$

(see Fig. 5.2). Since

$$h'(0) = \lim_{x\to 0}\frac{h(x)-h(0)}{x} = \lim_{x\to 0}\left(x\sin\frac{1}{x}\right) = 0, \tag{5.6}$$

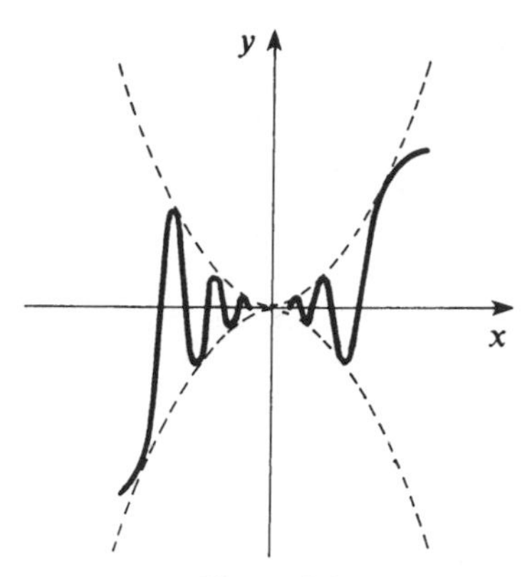

Figure 5.2

h is differentiable at 0 and $h'(0) = 0$. For $x \neq 0$, since h is a product of differentiable functions, it is differentiable. Thus,

$$h'(x) = \begin{cases} -\cos\dfrac{1}{x} + 2x\sin\dfrac{1}{x} & \text{for} \quad x \neq 0 \\ 0 & \text{for} \quad x = 0. \end{cases}$$

Thus, h is differentiable for all $x \in \mathbb{R}$. Note, however, that the derivative h' of h is not continuous at 0 (explain). The derivative of a function need not be continuous. If the derivative of a function is continuous, then we say that the function is *continuously differentiable*.

PROB. 5.1. Define f by

$$f(x) = \begin{cases} x^2\sin\dfrac{1}{x^2} & \text{for} \quad x \neq 0 \\ 0 & \text{for} \quad x = 0. \end{cases}$$

Prove that f is differentiable everywhere but its derivative is not continuous at 0 and is not bounded in any interval containing 0 as an interior point.

PROB. 5.2. Prove: The function g, where

$$g(x) = \begin{cases} x + 2x^2\sin\dfrac{1}{x} & \text{for} \quad x \neq 0 \\ 0 & \text{for} \quad x = 0, \end{cases}$$

is differentiable for all x and that $g'(0) = 1 > 0$. Also prove that even though $g'(0) > 0$, there exists no δ-neighborhood $N(0, \delta)$ of 0 in which g is monotonically increasing (see the next problem).

PROB. 5.3. Prove: If f is defined in some δ-neighborhood $N(a, \delta)$ of $a \in \mathbb{R}$, and $f'(a) > 0$, then there exists some ϵ-neighborhood $N(a, \epsilon) = (a - \epsilon; a + \epsilon)$ such that $f(x) < f(a)$ for $a - \epsilon < x < a$ and $f(x) > f(a)$ for $a < x < a + \epsilon$.

Remark 5.1. From the example given in Prob. 5.2 we see that $f'(a) > 0$, for some $a \in \mathfrak{D}(f)$ does not imply that f is monotonically increasing in some neighborhood of a. The result contained in the last problem is the most information we can gain from $f'(a) > 0$ at some $a \in \mathfrak{D}(f)$. To conclude from the sign of f' that f is monotonically increasing in some ϵ-neighborhood of a, we need to know that $f'(x) > 0$ for all x in this neighborhood. However, if f is continuously differentiable in some interval I and $f'(a) > 0$ for some interior point a of I, then some ϵ-neighborhood of a exists such that $f'(x) > 0$ for x in this neighborhood. In turn, this implies that f is strictly monotonically increasing there.

6. Inverses of Trigonometric Functions. Tschebyscheff Polynomials

The trigonometric functions are not one-to-one. However, upon restricting them to intervals on which they are one-to-one and which they map onto their full range, these restrictions do have inverses.

Inverse Sine

We restrict the sine to $[-\pi/2, \pi/2]$. This restriction maps $[-\pi/2, \pi/2]$ onto $[-1, 1]$ and is one-to-one. The inverse of this restriction is defined as the *principal inverse sine* or *Arcsin*. We write this function as $\sin^{-1}$ or as Arcsin. Note, that for $-1 \leqslant x \leqslant 1$, there is exactly one y with $\sin y = x$ such that $-\pi/2 \leqslant y \leqslant \pi/2$.

Def. 6.1. If $-1 \leqslant x \leqslant 1$, then $\operatorname{Arcsin} x$ or $\sin^{-1} x$ is defined as the unique y such that $\sin y = x$ and $-\pi/2 \leqslant y \leqslant \pi/2$. This y is also called the *principal value* of the inverse sine of x.

Thus,

$$\sin^{-1} 0 = \operatorname{Arcsin} 0 = 0,$$

since $\sin 0 = 0$ and $0 \in [-\pi/2, \pi/2]$ and also

$$\sin^{-1} 1 = \operatorname{Arcsin} 1 = \frac{\pi}{2} \quad \text{and} \quad \sin^{-1}(-1) = \operatorname{Arcsin}(-1) = -\frac{\pi}{2}$$

since

$$\sin \frac{\pi}{2} = 1, \qquad \sin\left(-\frac{\pi}{2}\right) = -1$$

and $\pi/2 \in [-\pi/2, \pi/2]$, $-\pi/2 \in [-\pi/2, \pi/2]$.

If $y = \operatorname{Arcsin} x$, where $-1 \leqslant x \leqslant 1$, then $\sin y = x$ and $-\pi/2 \leqslant y \leqslant \pi/2$, so

$$\frac{\pi}{2} \leqslant \operatorname{Arcsin} x \leqslant \frac{\pi}{2} \qquad \text{for} \quad -1 \leqslant x \leqslant 1. \tag{6.1}$$

(See Fig. 6.1.)

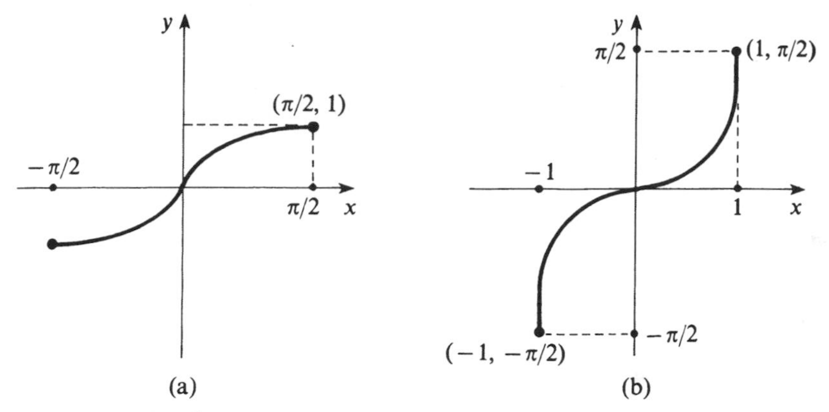

Figure 6.1. (a) Graph of restriction of sine to $[-\pi/2, \pi/2]$. (b) Graph of Arcsin.

By definition,

$$\sin(\operatorname{Arcsin} x) = x \qquad \text{if} \quad -1 \leqslant x \leqslant 1, \tag{6.2}$$

$$\operatorname{Arcsin}(\sin y) = y \qquad \text{if} \quad -\frac{\pi}{2} \leqslant y \leqslant \frac{\pi}{2}\,. \tag{6.3}$$

PROB. 6.1. Prove: If $-1 \leqslant x \leqslant 1$, then $\operatorname{Arcsin}(-1) = -\operatorname{Arcsin} x$.

PROB. 6.2. Prove: If $(2n-1)(\pi/2) \leqslant y \leqslant (2n+1)(\pi/2)$, where n is an integer, then $\operatorname{Arcsin}(\sin y) = (-1)^n(y - n\pi)$.

Note that if $-1 \leqslant x \leqslant 1$, then

$$\cos(\operatorname{Arcsin} x) = \sqrt{1 - x^2}\,. \tag{6.4}$$

In fact $y = \operatorname{Arcsin} x$ implies that $-\pi/2 \leqslant y \leqslant \pi/2$ and $x = \sin y$. For such y, $\cos y \geqslant 0$. Hence,

$$\cos(\operatorname{Arcsin} x) = \cos y = \sqrt{1 - \sin^2 y} = \sqrt{1 - x^2}\,.$$

PROB. 6.3. Verify: If $-1 < x < 1$, then

(a) $\tan(\operatorname{Arcsin} x) = x/\sqrt{1 - x^2}$,
(b) $\cot(\operatorname{Arcsin} x) = \sqrt{1 - x^2}/x$ if $x \neq 0$,
(c) $\sec(\operatorname{Arcsin} x) = 1/\sqrt{1 - x^2}$,
(d) $\csc(\operatorname{Arcsin} x) = 1/x$ for $x \neq 0$.

Remark 6.1. Since Arcsine is the inverse of a function which is strictly monotonically increasing and continuous on an interval, it is continuous and strictly monotonically increasing. Moreover (Theorem VII.6.5), since

$$\frac{d \sin y}{dy} = \cos y > 0 \qquad \text{for} \quad -\frac{\pi}{2} < y < \frac{\pi}{2}\,,$$

it follows that the restriction of the sine function to $[-\pi/2, \pi/2]$ is differentiable in the interior $(-1; 1)$ of $[1, 1]$ and

$$\frac{d(\operatorname{Arcsin} x)}{dx} = \frac{d\sin^{-1}x}{dx} = \frac{1}{(d\sin y/dy)|_{y=\sin^{-1}x}} = \frac{1}{\sqrt{1-x^2}}, \tag{6.5}$$

where $-1 < x < 1$.

Inverse of Cosine

The cosine function is strictly decreasing on $[0, \pi]$ and maps $[0, \pi]$ onto $[-1, 1]$. We define its *principal inverse*, Arccosine or $\cos^{-1}$, as the inverse of the restriction of cosine to $[0, \pi]$ (see Fig. 6.2).

Def. 6.2. If $-1 \leqslant x \leqslant 1$, then $\operatorname{Arccos} x$ or $\cos^{-1}x$ is defined as the y such that $\cos y = x$ and $0 \leqslant y \leqslant \pi$. Accordingly,

$$0 \leqslant \operatorname{Arccos} x \leqslant \pi. \tag{6.6}$$

It follows that

$$-\frac{\pi}{2} \leqslant \frac{\pi}{2} - \operatorname{Arccos} x \leqslant \frac{\pi}{2}. \tag{6.7}$$

Put $y = \operatorname{Arccos} x$. Then $\cos y = x$. Hence,

$$\sin\left(\frac{\pi}{2} - \operatorname{Arccos} x\right) = \sin\left(\frac{\pi}{2} - y\right) = \cos y = x$$

for $-1 \leqslant x \leqslant 1$. Since (6.7) holds, we obtain from this that

$$\frac{\pi}{2} - \operatorname{Arccos} x = \operatorname{Arcsin} x \qquad \text{for} \quad -1 \leqslant x \leqslant 1. \tag{6.8}$$

We also observe that

$$\cos(\operatorname{Arccos} x) = x \qquad \text{for} \quad -1 \leqslant x \leqslant 1 \tag{6.9}$$

and

$$\operatorname{Arccos}(\cos y) = y \qquad \text{for} \quad 0 \leqslant y \leqslant \pi. \tag{6.10}$$

Arccos x is called the *principal value* of the inverse cosine of x.

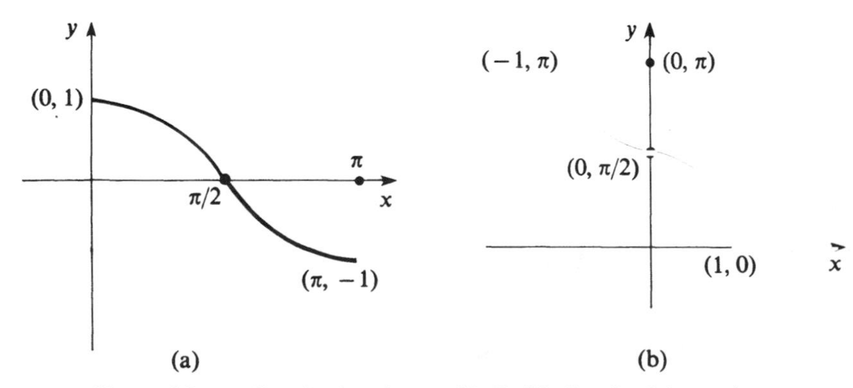

Figure 6.2. (a) Graph of cosine on [0, 1]. (b) Graph of Arccosine.

Prob. 6.4. Show that Arccosine is strictly decreasing and continuous on $[-1, 1]$, differentiable on $(-1, 1)$ and that

$$\frac{d\operatorname{Arccos} x}{dx} = -\frac{1}{\sqrt{1-x^2}} \quad \text{if} \quad -1 < x < 1.$$

Prob. 6.5. Show (a) $\operatorname{Arccos} 0 = \pi/2$, $\operatorname{Arccos} 1 = 0$, $\operatorname{Arccos}(-1) = \pi$, and that $\operatorname{Arccos}(-x) = \pi/2 + \operatorname{Arcsin} x$ for $-1 \leqslant x \leqslant 1$.

Inverse Tangent

The tangent function is strictly monotonically increasing on $(-\pi/2; \pi/2)$ and it maps this interval onto $\mathbb{R}$. The *principal* inverse of the tangent or the *Arctangent* is defined as the inverse of the restriction of the tangent to the interval $(-\pi/2; \pi/2)$ and is written as Arctan or as $\tan^{-1}$ (see Fig. 6.3).

Def. 6.3. If $x \in \mathbb{R}$, then $\operatorname{Arctan} x$ or $\tan^{-1} x$ is defined as the y such that $\tan y = x$ and $-\pi/2 < y < \pi/2$.

Thus,

$$-\frac{\pi}{2} < \operatorname{Arctan} x < \frac{\pi}{2}. \tag{6.11}$$

Also,

$$\tan(\operatorname{Arctan} x) = x \quad \text{for} \quad x \in \mathbb{R} \tag{6.12}$$

and

$$\operatorname{Arctan}(\tan y) = y \quad \text{for} \quad -\frac{\pi}{2} < y < \frac{\pi}{2}. \tag{6.13}$$

Prob. 6.6. (a) $\operatorname{Arctan} 0 = 0$, (b) $\operatorname{Arctan}(-x) = -\operatorname{Arctan} x$, (c) $\operatorname{Arctan} 1 = \pi/4$ and $\operatorname{Arctan}(-1) = -\pi/4$.

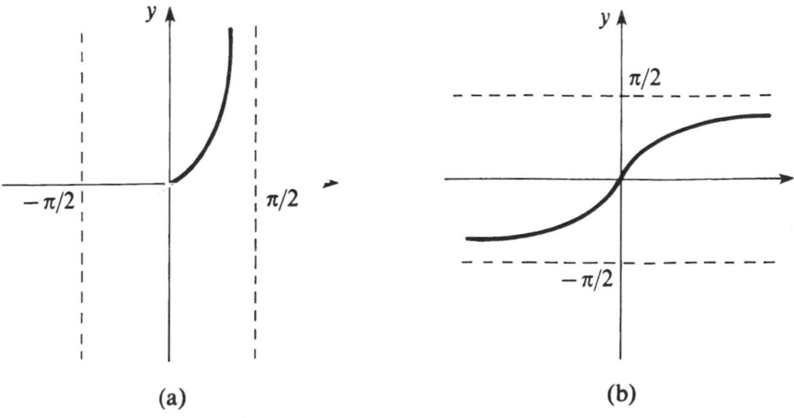

Figure 6.3. (a) Graph of the restriction of tangent to $(-\pi/2, \pi/2)$. (b) Graph of Arctan.

Prob. 6.7. Prove:

$$\lim_{x\to+\infty}(\text{Arctan}\,x)=\frac{\pi}{2}\quad\text{and}\quad\lim_{x\to-\infty}(\text{Arctan}\,x)=-\frac{\pi}{2}.$$

Prob. 6.8. Verify: If $x \in \mathbb{R}$, then

(a) $\cos(\text{Arctan}\,x) = 1/\sqrt{1+x^2}$,
(b) $\sin(\text{Arctan}\,x) = x/\sqrt{1+x^2}$,
(c) $\sec(\text{Arctan}\,x) = \sqrt{1+x^2}$.

Prob. 6.9. Verify: if $x \neq 0$, then

(a) $\cot(\text{Arctan}\,x) = 1/x$ and
(b) $\csc x = \sqrt{1+x^2}\,/x$.

Prob. 6.10. Prove: Arctan is differentiable with

$$\frac{d(\text{Arctan}\,x)}{dx}=\frac{1}{1+x^2}\qquad\text{for}\quad x\in\mathbb{R},$$

and that it is strictly monotonically increasing.

Inverses of the Remaining Trigonometric Functions

We define

(a) $\text{Arccot}\,x = \pi/2 - \text{Arctan}\,x$ for $x \in \mathbb{R}$,
(b) $\text{Arcsec}\,x = \text{Arccos}(1/x)$ for $|x| \geqslant 1$,
(c) $\text{Arccsc}\,x = \text{Arcsin}(1/x)$ for $|x| \geqslant 1$.

Prob. 6.11. Prove: (a) $0 < \text{Arccot}\,x < \pi$. (b) $\cot(\text{Arccot}\,x) = x$ for $x \in \mathbb{R}$. (c) $\text{Arccot}(\tan y) = \pi/2 - y$ for $|y| < \pi/2$.

Prob. 6.12. Prove: (a) If $x \geqslant 1$, then $0 \leqslant \text{Arcsec}\,x < \pi/2$ and, also, if $x \leqslant -1$, then $\pi/2 < \text{Arcsec}\,x \leqslant \pi$. (b) If $|x| \geqslant 1$, then $\sec(\text{Arcsec}\,x) = x$. (c) $\text{Arcsec}(\sec y) = y$ for $y \in [0,\pi/2) \cup (\pi/2,\pi]$.

Prob. 6.13. Prove:

(a) $\lim_{x\to+\infty}(\text{Arcsec}\,x) = \pi/2$ and
(b) $\lim_{x\to-\infty}(\text{Arcsec}\,x) = \pi/2$.

Prob. 6.14. Prove:

(a) $\lim_{x\to+\infty}\text{Arccot}\,x = 0$ and
(b) $\lim_{x\to-\infty}\text{Arccot}\,x = \pi$.

Prob. 6.15. Prove: (a) $0 < \operatorname{Arccsc} x \leqslant \pi/2$ for $x \geqslant 1$ and $-\pi/2 \leqslant \operatorname{Arccsc} x < 0$ for $x \leqslant -1$. (b) If $|x| \geqslant 1$, then $\csc(\operatorname{Arccsc} x) = x$. (c) If $y \in [-\pi/2, 0) \cup (0, \pi/2)$, then $\operatorname{Arccsc}(\csc y) = y$.

Prob. 6.16. Prove:

(a) $\lim_{x \to +\infty}(\operatorname{Arccsc} x) = 0 = \lim_{x \to -\infty}(\operatorname{Arccsc} x)$.

Prob. 6.17. Prove:

(a) $d(\operatorname{Arccot} x)/dx = -1/(1 + x^2)$ for $x \in \mathbb{R}$,

(b) $d(\operatorname{Arcsec} x)/dx = 1/|x|\sqrt{x^2 - 1}$ for $|x| > 1$,

(c) $d(\operatorname{Arc}(\csc x))/dx = -1/|x|\sqrt{x^2 - 1}$ for $|x| > 1$.

Prob. 6.18. Let $E(x) = e^x$ and $g(x) = E(-x) = e^{-x}$ for $x \in \mathbb{R}$. Sketch the graph of E and g.

Prob. 6.19. Sketch the graph of the natural logarithm function.

Prob. 6.20. Sketch the graphs of the hyperbolic functions.

Prob. 6.21. Sketch the graphs of the inverse hyperbolic functions.

Prob. 6.22. Note $\cos 2\theta = \cos^2\theta - \sin^2\theta = 2\cos^2\theta - 1$. This expresses $\cos 2\theta$ as a polynomial of degree 2 in $\cos\theta$. (a) Show that if n is a positive integer, then

$$\cos(n + 1)\theta = 2\cos n\theta \cos\theta - \cos(n - 1)\theta. \tag{6.14}$$

This last is a *recurrence* relation, which expresses the cosine of a positive integral multiple of θ in terms of the cosine of smaller positive multiples of θ. (b) Obtain $\cos 3\theta$ as a polynomial of degree 3 in $\cos\theta$. (c) Prove: If n is a nonnegative integer, then $\cos n\theta$ can be expressed as a polynomial of degree n in $\cos\theta$.

Prob. 6.23. The *Tschebyscheff polynomial* T_n of degree n, n an integer, is defined as: $T_0(x) = 1$ and

$$T_n(x) = \frac{1}{2^{n-1}}\cos(n \operatorname{Arccos} x) \qquad \text{for} \quad -1 \leqslant x \leqslant 1. \tag{6.15}$$

We write for each $n \in \mathbb{Z}_+$,

$$\tilde{T}_n(x) = 2^{n-1}T_n(x) = \cos(n \operatorname{Arccos} x), \qquad -1 \leqslant x \leqslant 1. \tag{6.16}$$

We have, for example, $\tilde{T}_1(x) = \cos(\operatorname{Arccos} x) = x$ for $-1 \leqslant x \leqslant 1$. Put $\theta = \operatorname{Arccos} x$ and use Eq. (6.14) of Prob. 6.22 to prove that

$$\tilde{T}_{n+1}(x) = 2x\tilde{T}_n(x) - \tilde{T}_{n-1}(x), \qquad -1 \leqslant x \leqslant 1,$$

for each positive integer n, and show that $\tilde{T}_2(x) = 2x^2 - 1$, $\tilde{T}_3(x) = 4x^3 - 3x$, $\tilde{T}_4(x) = 8x^4 - 8x^2 + 1$, $\tilde{T}_5(x) = 16x^5 - 20x^3 + 5x$. Show, for each positive integer n,

$$\tilde{T}_n(x) = 2^{n-1}x^n + \text{terms of lower degree in } x.$$

Thus show that $\tilde{T}_n$ and hence T_n has degree n and that T_n has leading coefficient 1.

PROB. 6.24. Using the notation and terminology of Prob. 6.23, show that $\tilde{T}_n(1) = 1$, $\tilde{T}_n(-1) = (-1)^n$ and $T_n(1) = 2^{-(n-1)}$, $T_n(-1) = (-1)^n 2^{-(n-1)}$.

Theorem 6.1. *The nth* $(n \geqslant 1)$ *Tschebyscheff polynomial* T_n *(see* Probs. 6.23–6.24) *has exactly n zeros* $x_1, \ldots, x_n$ *in the open interval* $(-1; 1)$. *These are given by*

$$x_i = \cos\left(\frac{2k-1}{2n}\pi\right) \quad \textit{for} \quad k \in \{1, \ldots, n\}. \tag{6.17}$$

PROOF. The θ_k given by

$$\theta_k = \frac{2k-1}{2n}\pi \qquad \text{for} \quad k \in \{1, \ldots, n\}$$

are distinct real numbers in the interval $(0; \pi)$. Cosine is strictly decreasing on $(0; \pi)$, so it maps this interval in a one-to-one way onto the interval $(-1; 1)$. It follows from this that $x_1, \ldots, x_n$ are distinct numbers in $(-1; 1)$. We have

$$\begin{aligned} \tilde{T}(x_k) &= \cos\left(n \operatorname{Arccos}\left(\cos \frac{2k-1}{2n}\pi\right)\right) \\ &= \cos\left(n \frac{2k-1}{2n}\pi\right) = \cos\left(\frac{2k-1}{2}\pi\right) = 0 \end{aligned}$$

for $k \in \{1, \ldots, n\}$. This proves that $x_1, \ldots, x_n$ constitute n zeros of $\tilde{T}_n$ and hence of T_n (see (6.15) and (6.16)). Since $\tilde{T}_n$ and therefore T_n are both polynomials of degree n, neither can have zeros other than these. It follows that T_n has exactly n zeros, all of them being in $(-1; 1)$.

Theorem 6.2. *The nth* $(n \geqslant 1)$ *Tschebyscheff polynomial* T_n *has extreme values at the* $n + 1$ *points*

$$z_k = \cos \frac{k}{n}\pi, \qquad \textit{where} \quad k \in \{0, 1, \ldots, n\} \tag{6.18}$$

and no others. These z_k*'s all lie in the closed interval* $[-1, 1]$ *and we have*

$$T_n(z_k) = \frac{(-1)^k}{2^{n-1}} \qquad \textit{for} \quad k \in \{0, 1, \ldots, n\}. \tag{6.19}$$

PROOF. It is clear from (6.18) that $-1 \leqslant z_k \leqslant 1$ for each $k \in \{0, 1, \ldots, n\}$. In particular we note that $z_0 = 1$ and $z_n = -1$. Reasoning as we did in the proof of Theorem 6.1, we see that the z_k's are all distinct and are therefore $n+1$ in number. Since cosine is strictly decreasing on $[0,\pi]$ we see that $i<j$ in $\{0,1,\ldots,n\}$ implies $z_i > z_j$. We also note that by the definition of T_n,

$$-\frac{1}{2^{n-1}} \leqslant T_n(x) \leqslant \frac{1}{2^{n-1}} \qquad \text{for} \quad -1 \leqslant x \leqslant 1. \tag{6.20}$$

Also,

$$\begin{aligned}T(z_k) &= \frac{1}{2^{n-1}}\tilde{T}_n(z_k) = \frac{1}{2^{n-1}}\cos\left(n \operatorname{Arccos}\left(\cos\frac{k}{n}\pi\right)\right)\\ &= \frac{1}{2^{n-1}}\cos\left(n\frac{k}{n}\pi\right) = \frac{1}{2^{n-1}}\cos(k\pi) = \frac{(-1)^k}{2^{n-1}}\end{aligned}$$

for $k \in \{0,1,\ldots,n\}$. This proves (6.19). It follows from this and (6.20) that T_n has extreme values which are equal to $(-1)^k 2^{-(n-1)}$ at the points z_k.

We check to see that T_n has no other extreme values. We evaluate T_n' at the points $z_1, z_2, \ldots, z_{n-1}$ in the interior of $[-1,1]$. We have

$$\begin{aligned}T_n'(x) &= \frac{1}{2^{n-1}}(-\sin(n \operatorname{Arccos}(\cos x)))\left(-\frac{1}{\sqrt{1-x^2}}\right)\\ &= \frac{n\sin(n \operatorname{Arccos} x)}{2^{n-1}\sqrt{1-x^2}}\end{aligned}$$

and

$$T'(z_k) = \frac{n\sin\left(n\operatorname{Arccos}\left(\cos\frac{k}{n}\pi\right)\right)}{2^{n-1}\sqrt{1-\cos\frac{2k}{n}\pi}} = \frac{n\sin\left(n\frac{k}{n}\pi\right)}{2^{n-1}\sqrt{1-\cos\frac{2k}{n}\pi}} = 0$$

for $k \in \{1,\ldots,n-1\}$. In short, $T_n'(z_k)=0$ for $k \in \{1,\ldots,n-1\}$. Now T_n is a polynomial of degree n, so that T_n' is a polynomial of degree $n-1$. Consequently T_n has no extreme values other than the z_k's in the interior of $[-1,1]$ (explain). At the endpoints $z_0 = 1$ and $z_n = 1$ of $[-1,1]$ we already saw that T_n assumes extreme values. Therefore T_n has $n+1$ extreme values at $z_0, z_1, \ldots, z_n$, these values being alternately $2^{-(n-1)}$ and $-2^{-(n-1)}$ beginning with $T_n(z_0) = 2^{-(n-1)}$. The last extreme value of T_n in $[-1,1]$ occurs at $z_n = -1$ and is equal to $(-1)^n 2^{-(n-1)}$.

Theorem 6.3. *Let T_n $(n \geqslant 1)$ be the nth Tschebyscheff polynomial. For all real polynomials P of degree n with leading coefficient* 1, *we have*

$$\frac{1}{2^{n-1}} = \max_{-1 \leqslant x \leqslant 1} |T_n(x)| \leqslant \max_{-1 \leqslant x \leqslant 1} |P(x)|. \tag{6.21}$$

Proof. The proof is indirect. Assume that for some real polynomial P we have

$$\max_{-1 \leqslant x \leqslant 1} |P(x)| < 2^{-(n-1)},$$

so that

$$-\frac{1}{2^{n-1}} < P(x) < \frac{1}{2^{n-1}} \qquad \text{for} \quad -1 \leqslant x \leqslant 1 \tag{6.22}$$

and hence

$$-\frac{1}{2^{n-1}} < P(z_k) < \frac{1}{2^{n-1}} \qquad \text{for} \quad z_k = \cos\left(\frac{k}{n}\pi\right)$$

where $k \in \{0, 1, \ldots, n\}$. By Theorem 6.2,

$$T_n(z_k) = (-1)^k \frac{1}{2^{n-1}} \qquad \text{for} \quad k \in \{0, 1, \ldots, n\}.$$

Therefore

$$T(z_k) - P(z_k) > 0 \qquad \text{if} \quad k \text{ is even} \tag{6.23}$$

and

$$T(z_k) - P(z_k) < 0 \qquad \text{if} \quad k \text{ is odd.} \tag{6.24}$$

Therefore, beginning with $T_n(z_0) - P(z_0) > 0$, $T(z_k) - P(z_k)$ alternates successively in sign n times as $k \in \{0, 1, \ldots, n\}$ ranges from 0 through to n. By the intermediate value theorem the polynomial $T_n - P$ takes on the value 0, n times. Since T_n and P are polynomials of degree n, both having leading coefficients 1, we see that $T_n - P$ is a polynomial of degree $n - 1$. Since $T_n - P$ has n zeros, it follows that $T_n(x) = P_n(x)$ for $-1 \leqslant x \leqslant 1$. This yields

$$\frac{1}{2^{n-1}} = T_n(z_k) = P_n(z_k) \qquad \text{if} \quad k \text{ is even}$$

and

$$-\frac{1}{2^{n-1}} = T_n(z_k) = P_n(z_k) \qquad \text{if} \quad k \text{ is odd.}$$

This contradicts (6.22). We therefore conclude that (6.21) holds and the proof is complete.

7. Log Convexity

Def. 7.1. If f is a positive-valued function on an interval I, then it is called *log convex* on I if and only if the function $g: I \to \mathbb{R}$ such that $g(x) = \ln f(x)$, $x \in I$, is convex on I.

For example, if $f(x) = 1/x$, $x > 0$, then f is log convex on $I = (0; +\infty)$ since g where

$$g(x) = \ln \frac{1}{x} = -\ln x \qquad \text{for} \quad x > 0$$

is convex on I. Note that g is convex on $(0, +\infty)$ since $g''(x) = 1/x^2 > 0$, $x > 0$.

Prob. 7.1. Prove: If f_1 and f_2 are log convex on an interval I, then their product $f_1 f_2$ is log convex on I.

Prob. 7.2. Prove: If $\langle f_n \rangle$ is a sequence of log convex functions on an interval I and

$$\lim_{x \to +\infty} f_n(x) = f(x) > 0 \qquad \text{for all} \quad x \in I,$$

then f is log convex on I.

Prob. 7.3. Let f be continuous and positive on an interval I. Prove: If f is twice differentiable at each interior point of I, then f is log convex on I if and only if $f(x)f''(x) - (f'(x))^2 \geqslant 0$ for all $x \in I$.

Prob. 7.4. Prove: If $f(x) > 0$ for x in some interval I and f is log convex on I, then f is convex on I.

Prob. 7.5. First prove: (a) If $a_1, a_2, b_1, b_2, c_1, c_2$ are real numbers such that

$$a_1 > 0, \quad a_2 > 0 \quad \text{and} \quad a_1c_1 - b_1^2 \geqslant 0, \quad a_2c_2 - b_2^2 \geqslant 0,$$

then $(a_1 + a_2)(c_1 + c_2) - (b_1 + b_2)^2 \geqslant 0$. Next prove: (b) If f and g are positive and continuous on an interval I, twice differentiable in the interior of I, and both are log convex on I, then their sum $f + g$ is log convex on I (Hint: use part (a) and Prob. 7.3.)

Prob. 7.6. Let g_n be defined by means of

$$g_n(x) = \frac{n^x n!}{x(x+1) \cdots (x+n)}, \qquad x > 0$$

for each positive integer n. Prove that g_n is log convex.

CHAPTER IX

L'Hôpital's Rule—Taylor's Theorem

1. Cauchy's Mean-Value Theorem

Theorem 1.1 (Cauchy's Mean-Value Theorem). *If f and g are real-valued functions of a real variable, both continuous on the bounded closed interval $[a,b]$, differentiable in the extended sense on $(a;b)$ with $g'(x) \neq 0$ for $x \in (a;b)$, having derivatives which are not simultaneously infinite, then* (1) $g(a) \neq g(b)$; (2) *there exists an $x_0 \in (a;b)$ such that*

$$\frac{f(b)-f(a)}{g(b)-g(a)} = \frac{f'(x_0)}{g'(x_0)}; \tag{1.1}$$

(3) *if $f(a) \neq f(b)$, then at the x_0 in* (1.1), *$f'(x_0)$ and $g'(x_0)$ are both finite.*

PROOF. If $g(a) = g(b)$ were to hold, there would exist a c such that $a < c < b$ and $g'(c) = 0$. This is ruled out by the hypothesis on g. Thus, $g(a) \neq g(b)$, proving (1).

We prove (2). If $f(a) = f(b)$, then there exists an x_0 such that $f'(x_0) = 0$. Hence, (1.1) holds in this case, since

$$\frac{f'(x_0)}{g'(x_0)} = 0 = \frac{f(b)-f(a)}{g(b)-g(a)}.$$

If $f(a) \neq f(b)$, consider the function F defined as

$$F(x) = f(x)[g(b)-g(a)] - g(x)[f(b)-f(a)] \quad \text{for} \quad x \in [a,b]. \tag{1.2}$$

Calculation shows that

$$F(a) = f(a)g(b) - g(a)f(b) = F(b). \tag{1.3}$$

Also, F has the same continuity and differentiability properties on $[a,b]$ as f and g. We can now use Rolle's Theorem to obtain an $x_0 \in (a;b)$ with $F'(x_0) = 0$. We now prove that $f'(x_0)$ and $g'(x_0)$ are both finite. By hypothesis at least one of $f'(x_0)$ or $g'(x_0)$ is finite. Using (1.2) we have, for $h \neq 0$ and $x_0 + h \in (a;b)$, that

$$\frac{F(x_0+h)-F(x_0)}{h} = \frac{f(x_0+b)-f(x_0)}{h}(g(b)-g(a)) - \frac{g(x_0+b)-g(x_0)}{h}(f(b)-f(a)). \tag{1.4}$$

Taking the limit as $h \to 0$, we obtain $F'(x_0) = 0$ on the left and therefore on the right. Suppose that $f'(x_0)$ or $g'(x_0)$ is $\pm\infty$. Since the other is necessarily finite, the limit as $h \to 0$ of the right-hand side of (1.4) is infinite and we have a contradiction. It follows that $f'(x_0)$ and $g'(x_0)$ are both finite. This proves (3). Since

$$0 = F'(x_0) = f'(x_0)[g(b)-g(a)] - g'(x_0)[f(b)-f(a)]$$

as is seen by differentiating (1.2), it follows that

$$\frac{f'(x_0)}{g'(x_0)} = \frac{f(b)-f(a)}{g(b)-g(a)},$$

which proves (2).

Theorem 1.2 (L'Hôpital's Rule for the Indeterminate Form $0/0$). *Let f and g be functions of which we assume that they are differentiable in the extended sense on the interval $(a;b)$, that f' and g' are not simultaneously infinite on $(a;b)$, that and $g'(x) \neq 0$ for $x \in (a;b)$, and that*

$$f(a+) = 0 = g(a+) \tag{1.5a}$$

and

$$\lim_{x \to a+} \frac{f'(x)}{g'(x)} = L, \tag{1.5b}$$

then

$$\lim_{x \to a+} \frac{f(x)}{g(x)} = L. \tag{1.6}$$

(*Here, we allow a, b and L to be in $\mathbb{R}^*$.*)

PROOF. Suppose $L = +\infty$. Let $B \in \mathbb{R}$. Because of (1.5b) there exists a deleted neighborhood $N^*(a)$ of a from the right such that for $x \in N^*(a) \cap (a;b)$ we have

$$\frac{f'(x)}{g'(x)} > B + 1. \tag{1.7}$$

Now $N^*(a) \cap (a;b)$ is a deleted neighborhood $N_1^*(a)$ of a from the right contained in $(a;b)$ and in $N^*(a)$. Assume that $x \in N_1^*(a)$ so that $a < x < b$ and $x \in N^*(a)$. There exists x_1 such that $a < x_1 < x$. By Theorem 1.1, there exists an x_0 such that

$$\frac{f(x) - f(x_1)}{g(x) - g(x_1)} = \frac{f'(x_0)}{g'(x_0)}, \tag{1.8}$$

where $a < x_1 < x_0 < x$. Since x_0 is necessarily in $N_1^*(a)$ (explain), we have $x_0 \in N^*(a)$. Hence,

$$\frac{f(x) - f(x_1)}{g(x) - g(x_1)} = \frac{f'(x_0)}{g'(x_0)} > B + 1.$$

Thus, $a < x_1 < x \in N_1^*(a)$ implies that

$$\frac{f(x) - f(x_1)}{g(x) - g(x_1)} > B + 1.$$

Taking limits as $x_1 \to a+$, we have

$$\frac{f(x)}{g(x)} = \lim_{x_1 \to a+} \frac{f(x) - f(x_1)}{g(x) - g(x_1)} \geqslant B + 1 > B$$

for $x \in N_1^*(a)$. Thus,

$$\frac{f(x)}{g(x)} > B \qquad \text{for} \quad x \in N_1^*(a).$$

We conclude that

$$\lim_{x \to +\infty} \frac{f(x)}{g(x)} = +\infty = L.$$

Thus, (1.6) holds if $L = +\infty$. If $L = -\infty$, a similar proof yields (1.6).

Next consider the case $L \in \mathbb{R}$. Let $\epsilon > 0$ be given. There exists a deleted neighborhood $N_2^*(a)$ of a from the right such that if $x \in N_2^*(a) \cap (a;b)$, then

$$L - \frac{\epsilon}{2} < \frac{f'(x)}{g'(x)} < L + \frac{\epsilon}{2}. \tag{1.9}$$

As before, $N_2^*(a) \cap (a;b)$ is a deleted neighborhood $N_3^*(a)$ of a from the right contained in $N_2^*(a)$ and in $(a;b)$. Assume that $x \in N_3^*(a)$ so that $a < x < b$ and $x \in N_2^*(a)$ and take x_1 such that $a < x_1 < x$. We obtain an x_0 such that

$$\frac{f(x) - f(x_1)}{g(x) - g(x_1)} = \frac{f'(x_0)}{g'(x_0)}, \tag{1.10}$$

where $a < x_1 < x_0 < x$. Since $x_0 \in N_3^*(a) \subseteq N^*(a)$, it follows from (1.10) and (1.9) that

$$L - \frac{\epsilon}{2} < \frac{f(x) - f(x_1)}{g(x) - g(x_1)} < L + \frac{\epsilon}{2} \qquad \text{for} \quad a < x_1 < x \in N_3^*(a).$$

Taking limits as $x_1 \to a+$ yields

$$L - \epsilon < L - \frac{\epsilon}{2} \leqslant \frac{f(x)}{g(x)} \leqslant L + \frac{\epsilon}{2} < L + \epsilon \qquad \text{for} \quad x \in N_3^*(a)$$

and, hence,

$$\left| \frac{f(x)}{g(x)} - L \right| < \epsilon \qquad \text{for} \quad x \in N_3^*(a).$$

We conclude from this that

$$\lim_{x \to a+} \frac{f(x)}{g(x)} = L$$

in this case also. This completes the proof.

Theorem 1.3 (L'Hôpital's Rule for the Case L/∞). *If f and g are differentiable in the extended sense on the interval $(a;b)$, with derivatives not simultaneously infinite there, $g'(x) \neq 0$ for $x \in (a;b)$ and*

$$\lim_{x \to a+} g(x) = \pm\infty, \tag{1.11a}$$

$$\lim_{x \to a+} \frac{f'(x)}{g'(x)} = L, \tag{1.11b}$$

then

$$\lim_{x \to a+} \frac{f(x)}{g(x)} = L. \tag{1.12}$$

(*Here, too, we permit a, b, and L to be in* $\mathbb{R}^*$.)

Proof. Suppose $L \in \mathbb{R}$. Let $\epsilon > 0$ be given. There exists a deleted neighborhood $N_1^*(a)$ of a from the right contained in $(a;b)$ such that

$$\left| \frac{f'(x)}{g'(x)} - L \right| < \frac{\epsilon}{3} \qquad \text{for} \quad x \in N_1^*(a). \tag{1.13}$$

Fix x' in $N_1^*(a)$. The interval $(a;x')$ is a deleted neighborhood of a from the right contained in $N_1^*(a)$ and (1.13) holds for all x such that $a < x < x'$. Given x such that $a < x < x'$, there exists an x_0 such that

$$\frac{f(x) - f(x')}{g(x) - g(x')} = \frac{f'(x_0)}{g'(x_0)}, \tag{1.14}$$

where $a < x < x_0 < x'$. Since x_0 is necessarily in $N_1^*(a)$, (1.13) holds for $x = x_0$. Now (1.14) and (1.13) imply that

$$\left|\frac{f(x) - f(x')}{g(x) - g(x')} - L\right| < \frac{\epsilon}{3} \qquad \text{for} \quad a < x < x' \in N_1^*(a). \tag{1.15}$$

This may be written

$$\left|\frac{f(x)/g(x) - f(x')/g(x)}{1 - g(x')/g(x)} - L\right| < \frac{\epsilon}{3} \qquad \text{for} \quad a < x < x' \in N_1^*(a)$$

so that

$$\left|\frac{f(x)}{g(x)} - L - \frac{f(x') - Lg(x')}{g(x)}\right| < \frac{\epsilon}{3}\left|1 - \frac{g(x')}{g(x)}\right| \tag{1.16}$$

for $a < x < x' < N_1^*(a)$. In turn, this implies that

$$\left|\frac{f(x)}{g(x)} - L\right| < \left|1 - \frac{g(x')}{g(x)}\right|\frac{\epsilon}{3} + \left|\frac{f(x') - Lg(x')}{g(x)}\right| \tag{1.17}$$

for $a < x < x' \in N_1(a)$. Since $\lim_{x\to a+} g(x) = \pm\infty$, $\lim_{x\to a+}|g(x)| = +\infty$ holds, so a deleted neighborhood $N_2^*(a)$ of a from the right exists in $(a;b)$ such that

$$|g(x)| > \max\left\{2|g(x')|, \frac{2}{\epsilon}|f(x') - Lg(x')|\right\}$$

for $x \in N_2^*(a)$. Thus,

$$|g(x)| > 2|g(x')| \quad \text{and} \quad |g(x)| > \frac{2}{\epsilon}|f(x') - Lg(x')| \tag{1.18}$$

and, hence,

$$\left|\frac{g(x')}{g(x)}\right| < \frac{1}{2} \quad \text{and} \quad \left|\frac{f(x') - Lg(x')}{g(x)}\right| < \frac{\epsilon}{2} \tag{1.19}$$

for $x \in N_2^*(a)$. Note that

$$\left|1 - \frac{g(x')}{g(x)}\right| \leqslant 1 + \left|\frac{g(x')}{g(x)}\right| \leqslant \frac{3}{2}.$$

Take $x \in (a;x') \cap N_2^*(a)$. For such x, (1.17) and (1.19) hold. Therefore,

$$\left|\frac{f(x)}{g(x)} - L\right| < \frac{3}{2}\,\frac{\epsilon}{3} + \frac{\epsilon}{2} < \epsilon.$$

Thus, for each $\epsilon > 0$,

$$\left|\frac{f(x)}{g(x)} - L\right| < \epsilon \qquad \text{for} \quad x \in (a;x') \cap N_2^*(a).$$

Since $(a;x') \cap N_2^*(a)$ is a neighborhood $N_3^*(a)$ of a from the right in $(a;b)$, we conclude that (1.12) holds for the case $L \in \mathbb{R}$. We ask the reader to complete the proof for the cases $L = \pm\infty$ (Prob. 1.1).

PROB. 1.1. Complete the proof of Theorem 1.3 by showing that the conclusion there holds for the cases $L = \pm\infty$.

Remark 1.1. The last two theorems are stated in terms of the limit as $x \to a+$. It is clear that, with appropriate modifications of the hypotheses and conclusions, these theorems can be formulated in terms of $x \to a-$ and also in terms of the two-sided limit $x \to a$.

EXAMPLE 1.1. We evaluate

$$\lim_{\alpha \to 0+} \frac{\ln(1+x)}{x^{\alpha}},$$

where $\alpha \in \mathbb{R}$. If $\alpha = 0$, L'Hôpital's rule, whether in the form of Theorem 1.2 or of Theorem 1.3, does not apply since $\lim_{x\to 0}\ln(1+x) = 0$ and $\lim_{x\to 0+} x^{\alpha} = \lim_{x\to 0+} x^{0} = 1$. However, the limit can be evaluated as follows:

$$\lim_{x\to 0+} \frac{\ln(1+x)}{x^{\alpha}} = \lim_{x\to 0+} \frac{\ln(1+x)}{x^{0}} = \lim_{x\to 0+} \ln(1+x) = 0.$$

Here we do not have an indeterminate case (explain).

Similarly, we do not have an indeterminate case if $\alpha < 0$. In fact, in that case, $\lim_{x\to 0+} \ln(1+x) = 0$ and $\lim_{x\to 0+} x^{\alpha} = +\infty$. However, L'Hôpital's rule in the form of Theorem 1.2 does apply since here $\lim_{x\to a+} g(x) = \lim_{x\to 0+} x^{\alpha} = +\infty$. Using it, we obtain

$$\lim_{x\to 0+} \frac{\ln(1+x)}{x^{\alpha}} = \lim_{x\to 0+} \frac{1/(1+x)}{\alpha x^{\alpha-1}} = \lim_{x\to 0+} \frac{x^{1-\alpha}}{(1+x)\alpha} = 0$$

(since $1-\alpha > 0$). The reader can check that this limit can be evaluated without using L'Hôpital's rule.

Finally, if $\alpha > 0$, we have the indeterminate case $0/0$. By L'Hopital's rule, we obtain

$$\lim_{x\to 0+} \frac{\ln(1+x)}{x^{\alpha}} = \lim_{x\to 0+} \frac{1}{\alpha x^{\alpha-1}(1+x)} = \begin{cases} 0 & \text{if } 0<\alpha<1 \\ 1 & \text{if } \alpha = 1 \\ +\infty & \text{if } \alpha > 1. \end{cases}$$

In summary

$$\lim_{x\to 0+} \frac{\ln(1+x)}{x^{\alpha}} = \begin{cases} 0 & \text{if } \alpha<1 \\ 1 & \text{if } \alpha = 1 \\ +\infty & \text{if } \alpha > 1. \end{cases}$$

PROB. 1.2. Note that in

$$\lim_{x\to +\infty} \frac{\sin x}{x} = 0,$$

the limit can be evaluated without the use of L'Hôpital's rule. Explain why the rule cannot be used here.

EXAMPLE 1.2. Consider

$$\lim_{x\to+\infty}\left[(x+1)^{\alpha}-x^{\alpha}\right], \qquad \alpha>0.$$

Since $\alpha>0$ implies $\lim_{x\to+\infty}(1+x)^{\alpha}=+\infty=\lim_{x\to+\infty}x^{\alpha}$, we describe this situation as being indeterminate of the form $(+\infty)-(+\infty)$. We write

$$\lim_{x\to+\infty}\left[(x+1)^{\alpha}-x^{\alpha}\right]=\lim_{x\to+\infty}x^{\alpha}\left[\left(1+\frac{1}{x}\right)^{\alpha}-1\right].$$

Here

$$\lim_{x\to+\infty}x^{\alpha}=+\infty \quad\text{and}\quad \lim_{x\to+\infty}\left[\left(1+\frac{1}{x}\right)^{\alpha}-1\right]=0$$

so we have the indeterminate form $(+\infty)0$. This can be converted to the form $0/0$ by writing

$$(x+1)^{\alpha}-x^{\alpha}=\frac{(1+1/x)^{\alpha}-1}{1/x^{\alpha}}=\frac{(1+1/x)^{\alpha}-1}{x^{-\alpha}}.$$

L'Hôpital's rule can now be applied to obtain

$$\begin{aligned}\lim_{x\to+\infty}\left[(x+1)^{\alpha}-x^{\alpha}\right]&=\lim_{x\to+\infty}\frac{(1+1/x)^{\alpha}-1}{x^{-\alpha}}\\&=\lim_{x\to+\infty}\frac{\alpha(1+1/x)^{\alpha-1}(-1/x^{2})}{-\alpha x^{-\alpha-1}}\\&=\lim_{x\to+\infty}\frac{(1+1/x)^{\alpha-1}}{x^{1-\alpha}}=\begin{cases}+\infty & \text{if } \alpha>1\\ 1 & \text{if } \alpha=1\\ 0 & \text{if } 0<\alpha<1.\end{cases}\end{aligned}$$

State what occurs if $\alpha\leqslant 0$.

EXAMPLE 1.3. Other indeterminate forms are: 0^0, $0^{\pm\infty}$, $(+\infty)^0$, $1^{\pm\infty}$. An example of the form 0^0 is

$$\lim_{x\to0+}x^{x}.$$

To evaluate this limit by L'Hôpital's rule, we write

$$y=x^{x} \qquad \text{for} \quad x>0$$

and take the natural logarithm of both sides to obtain

$$\ln y=x\ln x=\frac{\ln x}{1/x} \qquad \text{as} \quad x\to0+.$$

Applying L'Hôpital's rule gives

$$\lim_{x\to0+}\ln y=\lim_{x\to0+}\left(\frac{\ln x}{1/x}\right)=\lim_{x\to0+}\frac{1/x}{-1/x^{2}}=0$$

so that we have

$$\lim_{x\to 0+} x^x = \lim_{x\to 0+} y = \lim_{x\to 0+} e^{\ln y} = e^0 = 1.$$

The remaining forms are treated in the problems.

Remark 1.2. To evaluate certain limits it may be necessary to apply L'Hôpital's rule several times. For example,

$$\lim_{x\to+\infty} \frac{e^x}{x^2} = \lim_{x\to+\infty} \frac{e^x}{2x} = \lim_{x\to+\infty} \frac{e^x}{2} = +\infty.$$

An important limit is

$$\lim_{x\to+\infty} \frac{e^x}{x^\alpha} = +\infty \qquad \alpha \in \mathbb{R}. \tag{1.20}$$

For $\alpha \leqslant 0$, this is not indeterminate since

$$\lim_{x\to+\infty} \frac{e^x}{x^\alpha} = \begin{cases} \lim\limits_{x\to+\infty} e^x = +\infty & \text{if } \alpha = 0 \\ \lim\limits_{x\to+\infty} x^{-\alpha} e^x = +\infty & \text{if } \alpha < 0 \end{cases}$$

If $\alpha > 0$, then the limit in (1.20) is indeterminate of the form $+\infty/+\infty$. If $0 < \alpha < 1$, then by L'Hôpital's rule, we have

$$\lim_{x\to+\infty} \frac{e^x}{x^\alpha} = \lim_{x\to+\infty} \left(\frac{e^x}{\alpha x^{\alpha-1}}\right) = \lim_{x\to+\infty} \left(\frac{1}{\alpha} x^{1-\alpha} e^x\right) = +\infty \tag{1.21}$$

(since $0 < 1 - \alpha < 1$). If $\alpha = 1$, then L'Hôpital's rule again gives the limit $+\infty$. If $\alpha > 1$, put $n = [\alpha]$, so that $1 \leqslant n \leqslant \alpha$. In this case,

$$\frac{d^n x^\alpha}{dx^n} = \alpha(\alpha - 1) \ldots (\alpha - n + 1) x^{\alpha - n}.$$

If $\alpha \geqslant n$, then the limit in (1.20) is obtained from

$$\lim_{x\to+\infty} \frac{e^x}{x^\alpha} = \lim_{n\to+\infty} \frac{e^x}{\alpha x^{\alpha-1}} = \cdots = \lim_{x\to+\infty} \frac{e^x}{\alpha(\alpha-1)\cdots(\alpha-n+1)x^{\alpha-n}}. \tag{1.22}$$

If $\alpha = n$, then $x^{\alpha-n} = 1$, so the limit on the right is $+\infty$. If $\alpha > n$, then we have $n = [\alpha] < \alpha < [\alpha] + 1 = n + 1$, so $0 < \alpha - n < 1$. We saw in (1.21) that, in this case, the limit in the right-hand side of (1.22) is $+\infty$.

Prob. 1.3. Evaluate

(a) $\lim_{x\to 0}(1 + x^2)^{1/x}$,
(b) $\lim_{x\to a}((x^r - a^r)/(x^s - a^s))$, where $\alpha > 0$ and r, s are real numbers,
(c) $\lim_{x\to\pi/2}(\tan x - \sec x)$,
(d) $\lim_{x\to+\infty}(\ln(x + 1)/\ln x)$,

(e) $\lim_{x \to +\infty}(\ln(\ln(1+x))/\ln(\ln x))$,
(f) $\lim_{x \to 0+} x^{(x^x)}$,
(g) $\lim_{x \to +\infty}(\cos(1/x))^x$,
(h) $\lim_{x \to 0}((a^x - b^x)/x)$,
(i) $\lim_{x \to +\infty}(x/e^x)$, $\alpha \in \mathbb{R}$,
(j) $\lim_{x \to +\infty}(1/x^2)e^{-(1/x)}$,
(k) $\lim_{x \to +\infty}(\cot(1/x) - 1/x)$,
(l) $\lim_{x \to 0+} \ln(1-x)\ln(x)$,
(m) $\lim_{x \to 0+} (\cos x)^{1/x^2}$

PROB. 1.4. Note that $\lim_{x \to 1} x^2/(x^2+1) = \frac{1}{2}$. Explain why L'Hôpital's rule fails here.

PROB. 1.5. Prove: If $\lim_{x \to +\infty} f(x) = +\infty$ and $\lim_{x \to +\infty} f'(x) = L$, then $\lim_{x \to +\infty}(f(x)/x) = L$.

2. An Application to Means and Sums of Order t

The arithmetic and harmonic means of x_1 and x_2 are defined respectively as

$$\text{(a)} \qquad \frac{x_1 + x_2}{2}$$

and

$$\text{(b)} \qquad \frac{1}{\frac{1}{2}(1/x_1 + 1/x_2)} = \frac{2x_1x_2}{x_1 + x_2},$$

where $x_1 > 0$, $x_2 > 0$. These can be treated using the notion of "mean of order t of x_1 and x_2 with weights α_1, α_2." We write this as $M_t(x_1, x_2; \alpha_1, \alpha_2)$ and define it as

$$M_t(x_1, x_2; \alpha_1, \alpha_2) = \left(\alpha_1 x_1^t + \alpha_2 x_2^t\right)^{1/t}, \tag{2.1}$$

where $\alpha_1 + \alpha_2 = 1$, $\alpha_1 > 0$, $\alpha_2 > 0$ and $x_1 > 0$, $x_2 > 0$.

Using "weights" $\alpha_1 = \alpha_2 = \frac{1}{2}$, we see that the arithmetic and harmonic means of x_1 and x_2 are seen to occur when $t = 1$ and $t = -1$ are used in $M_t(x_1, x_2; \frac{1}{2}, \frac{1}{2})$. Thus,

$$M_1(x_1, x_2; \tfrac{1}{2}, \tfrac{1}{2}) = \frac{x_1 + x_2}{2}$$

and

$$M_{-1}(x_1, x_2; \tfrac{1}{2}, \tfrac{1}{2}) = \left(\tfrac{1}{2}x_1^{-1} + \tfrac{1}{2}x_2^{-1}\right)^{-1} = \frac{2x_1x_2}{x_1 + x_2}.$$

This notion can be generalized to apply to n positive real numbers.

Def. 2.1. If $x_1, x_2, \ldots, x_n$ are positive real numbers, their mean of order $t \neq 0$, with *weights* $\alpha_1, \alpha_2, \ldots, \alpha_n$, written as $M_t(x_1, \ldots, x_n; \alpha_1, \ldots, \alpha_n)$, is defined as

$$M_t(x_1, \ldots, x_n; \alpha_1, \ldots, \alpha_n) = \left(\alpha_1 x_1^t + \cdots + \alpha_n x_n^t\right)^{1/t}, \tag{2.2}$$

where

$$\alpha_i > 0 \quad \text{for} \quad i \in \{1, \ldots, n\} \quad \text{and} \quad \alpha_1 + \cdots + \alpha_n = 1. \tag{2.3}$$

Thus, the mean of order 1, with weights $\alpha_1 = \cdots = \alpha_n = 1/n$, of the positive numbers $x_1, \ldots, x_n$ is their arithmetic mean

$$M_1\left(x_1, \ldots, x_n; \frac{1}{n}, \ldots, \frac{1}{n}\right) = \frac{x_1 + \cdots + x_n}{n}, \tag{2.4}$$

and the mean of order -1, with the same weights, is their harmonic mean, i.e.,

$$M_{-1}\left(x_1, \ldots, x_n; \frac{1}{n}, \ldots, \frac{1}{n}\right) = \left(\frac{1}{n} x_1^{-1} + \cdots + \frac{1}{n} x_n^{-1}\right)^{-1}$$
$$= \frac{n}{1/x_1 + \cdots + 1/x_n}. \tag{2.5}$$

It is natural to ask if the geometric mean is among the means of order t. It turns out that the answer to this question is found by taking $\lim_{t \to 0} M_t$. We prove this.

Beginning with (2.2) we take the natural logarithm of both sides in (2.2) to obtain

$$\ln M_t = \frac{1}{t} \ln\left(\alpha_1 x_1^t + \cdots + \alpha_n x_n^t\right). \tag{2.6}$$

Using

$$\frac{dx_i^t}{dt} = x_i^t \ln x_i \qquad \text{for} \quad i \in \{1, \ldots, n\}, \tag{2.7}$$

we obtain by L'Hôpital's rule:

$$\lim_{t \to 0} \ln M_t = \lim_{t \to 0} \frac{\alpha_1 x_1^t \ln x_1 + \cdots + \alpha_n x_n^t \ln x_n}{\alpha_1 x_1^t + \cdots + \alpha_n x_n^t}$$
$$= \frac{\alpha_1 \ln x_1 + \cdots + \alpha_n \ln x_n}{\alpha_1 + \cdots + \alpha_n} = \ln\left(x_1^{\alpha_1} \ldots x_n^{\alpha_n}\right).$$

Hence,

$$\lim_{t \to 0} M_t = \lim_{t \to 0} e^{\ln M_t} = e^{\ln(x_1^{\alpha_1} \ldots x_n^{\alpha_n})} = x_1^{\alpha_1} \ldots x_n^{\alpha_n}. \tag{2.8}$$

Because of this we extend the definition of M_t by defining

$$M_0(x_1, \ldots, x_n; \alpha_1, \ldots, \alpha_n) = x_1^{\alpha_1} \ldots x_n^{\alpha_n}. \tag{2.9}$$

We have now $M_t \to M_0$ as $t \to 0$. Using weights $\alpha_1 = \cdots = \alpha_n = 1/n$, we obtain

$$M_0\left(x_1, \ldots, x_n; \frac{1}{n}, \ldots, \frac{1}{n}\right) = x_1^{1/n} \ldots x_n^{1/n} = \sqrt[n]{x_1 \ldots x_n}, \quad (2.10)$$

which is the geometric mean of the positive numbers $x_1, \ldots, x_n$.

We also prove: If $x_1, \ldots, x_n$ are positive real numbers, then

$$\lim_{t\to+\infty} M_t(x_1, \ldots, x_n; \alpha_1, \ldots, \alpha_n) = \max\{x_1, \ldots, x_n\} \quad (2.11^*)$$

and

$$\lim_{t\to-\infty} M_t(x_1, \ldots, x_n; \alpha_1, \ldots, \alpha_n) = \min\{x_1, \ldots, x_n\}. \quad (2.12^*)$$

To prove (2.11), we write $x_k = \max\{x_1, \ldots, x_n\}$ and obtain

$$\alpha_k^{1/t} x_k \leqslant \left(\alpha_1 x_1^t + \cdots + \alpha_n x_n^t\right)^{1/t} \leqslant x_k. \quad (2.13)$$

Noting $\lim_{t\to+\infty} \alpha_k^{1/t} = 1$, we obtain (2.11) from (2.13) by letting $t \to +\infty$. We prove (2.12) next.

Note first that

$$M_{-t}(x_1, \ldots, x_n; \alpha_1, \ldots, \alpha_n) = \frac{1}{M_t(1/x_1, \ldots, 1/x_n; \alpha_1, \ldots, \alpha_n)} \quad (2.14)$$

and that

$$\max\left\{\frac{1}{x_1}, \ldots, \frac{1}{x_n}\right\} = \frac{1}{\min\{x_1, \ldots, x_n\}}. \quad (2.15)$$

Since

$$\begin{aligned}\lim_{t\to+\infty} M_{-t} &= \lim_{t\to+\infty} \frac{1}{M_t(1/x_1, \ldots, 1/x_n; \alpha_1, \ldots, \alpha_n)} \\ &= \frac{1}{\lim_{t\to+\infty} M_t(1/x_1, \ldots, 1/x_n; \alpha_1, \ldots, \alpha_n)} \\ &= \frac{1}{\max\{1/x_1, \ldots, 1/x_n\}},\end{aligned}$$

it follows that

$$\lim_{t\to-\infty} M_t = \lim_{t\to+\infty} M_{-t} = \frac{1}{\max\{1/x_1, \ldots, 1/x_n\}} = \min\{x_1, \ldots, x_n\}$$

which proves (2.12). We define $M_{-\infty} = \min\{x_1, \ldots, x_n\}$ and $M_{+\infty} = \max\{x_1, \ldots, x_n\}$.

PROB. 2.1. Prove: (a) M_t is a monotonically increasing function (as function of t, for fixed $x_1, \ldots, x_n; \alpha_1, \ldots, \alpha_n$). (b) M_t is *strictly* monotonically

* Beckenback and Bellman, *Inequalities*, Ergebmisse D. Mathematik, Band 30, 1965, p. 16.

increasing if and only if $x_1, \ldots, x_n$ are distinct (Hint: see inequality VIII.2.28).

Remark 2.1. The result cited in the last problem constitutes another proof of the arithmetic–geometric inequality (Corollary of Theorem II.12.3) since it states that $M_0(x_1, \ldots, x_n; 1/n, \ldots, 1/n) \leqslant M_1(x_1, \ldots, x_n; 1/n, \ldots, 1/n)$. This is another way of saying that the geometric mean of $x_1, \ldots, x_n$ does not exceed the arithmetic mean. It also places that inequality in a broader context since it states that $M_{-\infty} \leqslant M_{-1} \leqslant M_0 \leqslant M_1 \leqslant M_{+\infty}$ and relates the minimum, harmonic, geometric–arithmetic, and maximum of positive $x_1, \ldots, x_n$. It also states that $-1 < t < 1$ implies that $M_{-1} \leqslant M_t \leqslant M_1$.

PROB. 2.2. Prove: If n is a positive integer, then for nonnegative numbers $x_1, x_2, \ldots, x_n$, we have

$$\sqrt[n]{x_1 \ldots x_n} \leqslant \left(\frac{\sqrt[n]{x_1} + \cdots + \sqrt[n]{x_n}}{n} \right)^n \leqslant \frac{x_1 + \cdots + x_n}{n}.$$

Sums of Order t

Def. 2.2. If $x_1, \ldots, x_n$ are positive, then their *sum* $S_t(x_1, \ldots, x_n)$ *of order* $t \neq 0$ is defined as

$$S_t(x_1, \ldots, x_n) = \left(x_1^t + \cdots + x_n^t\right)^{1/t}. \tag{2.16}$$

The sum of order t behaves somewhat differently than the mean of order t. For example, we have for $n \geqslant 2$,

$$S_{0+}(x_1, \ldots, x_n) = +\infty \tag{2.17a}$$

and

$$S_{0-}(x_1, \ldots, x_n) = 0, \tag{2.17b}$$

whereas, for $\alpha_i > 0$, $\alpha_1 + \cdots + \alpha_n = 1$, we have

$$\lim_{t \to 0} M_t(x_1, \ldots, x_n; \alpha_1, \ldots, \alpha_n) = x_1^{\alpha_1} \ldots x_n^{\alpha_n}. \tag{2.18}$$

We prove (2.17). Take the natural logarithm of both sides in (2.16) to obtain

$$\ln S_t = \frac{\ln\left(x_1^t + \cdots + x_n^t\right)}{t}, \qquad n \geqslant 2, \tag{2.19}$$

$$\lim_{t \to 0} \ln\left(x_1^t + \cdots + x_n^t\right) = \ln n > 0, \qquad n \geqslant 2.$$

It follows from this and (2.19) that

$$\lim_{t \to 0+} \ln S_t = +\infty \quad \text{and} \quad \lim_{t \to 0-} \ln S_t = -\infty, \qquad n \geqslant 2$$

and, therefore, that

$$\lim_{t\to 0+} S_t = \lim_{t\to 0+} e^{\ln S_t} = +\infty \quad \text{and} \quad \lim_{t\to 0-} S_t = \lim_{t\to 0-} e^{\ln S_t} = 0,$$

proving (2.17). (Note that if $n = 1$, then $S_t(x_1) = x_1$ for all t.)

Prob. 2.3. Prove: If $x_1, \ldots, x_n$ are positive, then

(a) $$\lim_{t\to+\infty} S_t(x_1, \ldots, x_n) = \max\{x_1, \ldots, x_n\},$$

(b) $$\lim_{t\to-\infty} S_t(x_1, \ldots, x_n) = \min\{x_1, \ldots, x_n\}.$$

Remark 2.2. Concerning S_t, there is the well-known inequality of *Jensen* which states that: If $0 < t_1 < t_2$ or $t_1 < t_2 < 0$, then

$$S_{t_1}(x_1, \ldots, x_n) \geqslant S_{t_2}(x_1, \ldots, x_n). \tag{2.20}$$

To see this, note

$$\ln S_t = \frac{\ln(x_1^t + \cdots + x_n^t)}{t}, \qquad t \neq 0.$$

Taking the derivative of both sides here with respect to t, we obtain, after some easy calculations,

$$\frac{t^2}{S_t}\frac{dS_t}{dt} = \ln\left[\left(\frac{x_1^t}{x_1^t + \cdots + x_n^t}\right)^{x_1^t} \cdots \left(\frac{x_n^t}{x_1^t + \cdots + x_n^t}\right)^{x_n^t}\right]. \tag{2.21}$$

But we have

$$0 < \left(\frac{x_i^t}{x_1^t + \cdots + x_n^t}\right)^{x_i^t} \leqslant 1 \qquad \text{for} \quad i \in \{1, \ldots, n\}$$

(explain). It follows that the product inside the square bracket on the right-hand side of (2.21) is $\leqslant 1$ so that its natural logarithm is $\leqslant 0$. Thus, (2.21) implies that

$$\frac{dS_t}{dt} \leqslant 0 \qquad \text{for} \quad t \neq 0 \tag{2.22}$$

it follows that S_t viewed as a function of t for fixed $x_1, \ldots, x_n$, is monotonic decreasing on $(0; +\infty)$ and also on $(-\infty; 0)$.

Prob. 2.4. Prove: If $0 < t \leqslant 1$ and $x_1 \geqslant 0, x_2 \geqslant 0, \ldots, x_n \geqslant 0$, then

$$(x_1 + \cdots + x_n)^t \leqslant (x_1^t + \cdots + x_n^t).$$

Prob. 2.5. Prove: If $x_1 \geqslant 0$, $x_2 \geqslant 0$, and $0 < t \leqslant 1$, then

$$|x_1^t - x_2^t| \leqslant |x_1 - x_2|.$$

PROB. 2.6. Prove: If $0 < t \leqslant 1$, then the function given by $f(x) = x^t$ for $0 \leqslant x$ is uniformly continuous.

3. The O–o Notation for Functions

We introduced the "big O" and "little o" notation for sequences in Defs. IV.4.1 and IV.4.2. Here we extend these notions to functions which are not necessarily sequences.

Def. 3.1. If f and g are real-valued functions defined on a set $D \neq \emptyset$ and an $A > 0$ exists such that

$$|f(x)| \leqslant A|g(x)| \qquad \text{for all} \quad x \in D, \tag{3.1}$$

then we write

$$f(x) = O(g(x)) \qquad \text{for} \quad x \in D. \tag{3.2}$$

Thus, we write $\sin x = O(x)$, $x \in \mathbb{R}$, since $|\sin x| \leqslant |x|$ for $x \in \mathbb{R}$.

PROB. 3.1. Prove: $|e^x - 1| \leqslant e|x|$ for $|x| \leqslant 1$. Consequently $e^x - 1 = O(x)$, $x \in [-1, 1]$.

Def. 3.2. Let f and g be real-valued functions defined on a set D and a an accumulation point, possibly extended, of D. If there exists a deleted neighborhood $N^*(a)$ such that

$$f(x) = O(g(x)) \qquad \text{for} \quad x \in N^*(a) \cap D,$$

then we say that f is *big O* of g as $x \to a$ and write

$$f(x) = O(g(x)) \qquad \text{as} \quad x \to a. \tag{3.3}$$

For example, if $a = +\infty$, then $f(x) = O(g(x))$, $x \to a$, means that there exists $A > 0$ and X such that

$$|f(x)| = A|g(x)| \qquad \text{for} \quad x \in D \quad \text{and} \quad x > X. \tag{3.4}$$

This is similar to the use of big O for sequences (cf. Def. IV.4.1).

On the other hand, if $a \in \mathbb{R}$, then $f(x) = O(g(x))$ as $x \to a$ means that there exist $A > 0$ and $\epsilon > 0$ such that

$$|f(x)| \leqslant A|g(x)| \qquad \text{for} \quad x \in \mathfrak{D}(f) \quad \text{and} \quad 0 < |x - a| < \epsilon. \tag{3.5}$$

Def. 3.3. If the functions f, g, and h have a common domain D and

$$f(x) - h(x) = O(g(x)) \qquad \text{as} \quad x \to a,$$

then we write

$$f(x) = h(x) + O(g(x)) \qquad \text{as} \quad x \to a. \tag{3.6}$$

PROB. 3.2. Prove: If f and g are defined on D and $g(x) \neq 0$ for $x \in D - \{a\}$ and

$$\lim_{t \to a} \frac{f(x)}{g(x)} = L \in \mathbb{R},$$

then $f(x) = O(g(x))$ as $x \to a$.

PROB. 3.3. Prove: (a) $\sin x = O(x)$ as $x \to 0$; (b) $\sin x = O(x)$ as $x \to +\infty$; (c) $\cos x = 1 + O(x^2)$ as $x \to 0$.

PROB. 3.4. Prove:

$$\frac{1}{x+1} = \frac{1}{x} + O\left(\frac{1}{x^2}\right) \qquad \text{as} \quad x \to +\infty.$$

PROB. 3.5. Prove: If $f(x) = O(g(x))$ and $g(x) = O(h(x))$ as $x \to a$, then $f(x) = O(h(x))$ as $x \to a$.

PROB. 3.6. Prove: If

$$\lim_{x \to a} \frac{f(x)}{g(x)} = L,$$

where $0 < L < +\infty$, then

$$f(x) = O(g(x)) \quad \text{and} \quad g(x) = O(f(x)) \qquad \text{as} \quad x \to a.$$

Def. 3.4. If f and g have a common domain D and both $f(x) = O(g(x))$, $g(x) = O(g(x))$ hold as $x \to a$, then we write $f(x) \asymp g(x)$ as $x \to a$, and say that the f and g are of the *same* order of magnitude as $x \to a$.

PROB. 3.7. Prove: $f(x) \asymp g(x)$ as $x \to a$ if and only if positive constants A and B exist such that for some deleted neighborhood $N^*(a)$ of a we have

$$B|g(x)| \leqslant f(x) \leqslant A|g(x)| \qquad \text{for} \quad x \in D \cap N^*(a).$$

Asymptotic Equivalence

Def. 3.5. When f and g have a common domain D, a is an accumulation point, possibly extended, of D, $g(x) \neq 0$ for $x \in D - \{a\}$, and

$$\lim_{x \to a} \frac{f(x)}{g(x)} = 1, \tag{3.7}$$

then we say that f and g are *asymptotically equivalent* as $x \to a$ and write

$$f(x) \sim g(x) \qquad \text{as} \quad x \to a. \tag{3.8}$$

For example, (a) $\sin x \sim x$ as $x \to 0$; (b) $\ln(1+x) \sim x$ as $x \to 0$; (c) $x+1 \sim x$ as $x \to +\infty$.

PROB. 3.8. Prove: (a) $\cosh x \sim e^x/2$ as $x \to +\infty$ and (b) $\sinh x \sim e^x/2$ as $x \to +\infty$.

Remark 3.1. Clearly $f(x) \sim g(x)$ as $x \to a$ implies $f(x) \asymp g(x)$ as $x \to a$.

PROB. 3.7′. Prove: If f, g, and h have a common domain $\mathfrak{D}$ and f, g, h are nonzero in D, then:

(a) $f(x) \sim f(x)$ as $x \to a$,
(b) $f(x) \sim g(x)$ as $x \to a$ implies $g(x) \sim f(x)$ as $x \to a$,
(c) $f(x) \sim h(x)$ and $h(x) \sim g(x)$ as $x \to a$ imply $f(x) \sim g(x)$ as $x \to a$.

PROB. 3.8′. Prove: If $f_1(x) = O(g_1(x))$ and $f_2(x) = O(g_2(x))$ as $x \to a$, then;

(a) $f_1(x)f_2(x) = O(g_1(x)g_2(x))$ as $x \to a$,
(b) $f_1(x) + f_2(x) = O(|g_1(x)| + |g_2(x)|)$ as $x \to a$.

Little o as $x \to a$

Def. 3.6. If f and g are defined on a common domain D, a is an accumulation point of D, $g(x) \neq 0$ for $x \in D - \{a\}$, and

$$\lim_{x \to a} \frac{f(x)}{g(x)} = 0, \tag{3.9}$$

then we say that f is "little o" of g as $x \to a$ and write

$$f(x) = o(g(x)) \qquad \text{as} \quad x \to a. \tag{3.10}$$

If $f(x) - h(x) = o(g(x))$ as $x \to a$, then we write

$$f(x) = h(x) + o(g(x)) \qquad \text{as} \quad x \to a. \tag{3.11}$$

For example,

$$\cos x = 1 + o(x) \quad \text{and} \quad \sin^2 x = o(x) \qquad \text{as} \quad x \to 0.$$

The notion of "little o" for functions as $x \to a$ is similar to the one for sequences where we spoke of "little o" as $n \to +\infty$.

Theorem 3.1. *If f_1, f_2; g_1, g_2 are all defined on a common domain D, $g_1(x) > 0$ and $g_2(x) > 0$ for $x \in D - \{a\}$, and*

$$f_1(x) = o(g_1(x)), \qquad f_2(x) = o(g_2(x)) \qquad \textit{as} \quad x \to a, \tag{3.12}$$

then

(a) $$f_1(x) + f_2(x) = o(g_1(x) + g_2(x)) \qquad \textit{as} \quad x \to a,$$

(b) $$f_1(x)f_2(x) = o(g_1(x)g_2(x)) \qquad \textit{as} \quad x \to a.$$

Proof. We prove (a) and ask the reader to prove (b) in Prob. 3.9. Given $\epsilon > 0$, we use (3.12) to establish the existence of some deleted neighborhood $N^*(a)$ of s such that

$$-\epsilon g_1(x) < f_1(x) < \epsilon g_1(x)$$

and

$$-\epsilon g_2(x) < f_2(x) < \epsilon g_2(x)$$

for $x \in N^*(a) \cap D$. Adding these inequalities we see that

$$-\epsilon < \frac{f_1(x) + f_2(x)}{g_1(x) + g_2(x)} < \epsilon \qquad \text{for} \quad x \in N^*(a) \cap D.$$

It follows that

$$\lim_{x \to a} \frac{f_1(x) + f_2(x)}{g_1(x) + g_2(x)} = 0$$

and, hence, that (a) holds.

Prob. 3.9. Complete the proof of the last theorem by proving its part (b).

Prob. 3.10. Prove: If $f_1(x) = O(g(x))$ and $f_2(x) = o(g(x))$, each as $x \to a$, then

(a) $$f_1(x) + f_2(x) = O(g(x)) \qquad \text{as} \quad x \to a$$

(b) $$f_1(x) f_2(x) = o(g(x)) \qquad \text{as} \quad x \to a.$$

4. Taylor's Theorem of Order n

Let f be defined in some ϵ-neighborhood $N(a, \epsilon)$ of $a \in \mathbb{R}$ and continuous at a, so that

$$\lim_{h \to 0} f(a + h) = f(a). \tag{4.1}$$

This implies that $\lim_{h \to 0}(f(a + h) - f(a)) = 0$ or, in the little o notation, that

$$f(a + h) = f(a) + o(1) \qquad \text{as} \quad h \to 0. \tag{4.2}$$

Now, in addition, we assume that f is differentiable at a so that

$$\lim_{h \to 0} \frac{f(a + h) - f(a)}{h} = f'(a). \tag{4.3}$$

Using the big O notation we can write this as

$$f(a + h) = f(a) + O(h) \qquad \text{as} \quad h \to 0. \tag{4.4}$$

Equation (4.3) can also be written

$$\lim_{h\to 0}\frac{f(a+h)-(f(a)+f'(a)h)}{h}=\lim_{h\to 0}\left(\frac{f(a+h)-f(a)}{h}-f'(a)\right)=0. \tag{4.5}$$

This implies in the little o notation that

$$f(a+h)-(f(a)+f'(a)h)=o(h) \quad \text{as} \quad h\to 0$$

and, hence, that

$$f(a+h)=f(a)+f'(a)h+o(h) \quad \text{as} \quad h\to 0. \tag{4.6}$$

The above result is based on the assumption that f is defined in the ϵ-neighborhood $N(a,\epsilon)$ of a and differentiable at a. Now assume that f is differentiable in $N(a,\epsilon)$ and twice differentiable at a. We obtain by L'Hôpital's rule

$$\begin{aligned}\lim_{h\to 0}\frac{f(a+h)-(f(a)+f'(a)h)}{h^2}&=\lim_{h\to 0}\frac{f'(a+h)-f'(a)}{2h}\\&=\frac{1}{2}\lim\frac{f'(a+h)-f'(a)}{h}\\&=\frac{1}{2}f''(a).\end{aligned} \tag{4.7}$$

This implies that

$$f(a+h)-(f(a)+f'(a)h)=O(h^2) \quad \text{as} \quad h\to 0$$

and, hence, that

$$f(a+h)=f(a)+f'(a)h+O(h^2) \quad \text{as} \quad h\to 0. \tag{4.8}$$

Equality (4.7) implies that

$$\lim_{h\to 0}\frac{f(a+h)-\left(f(a)+f'(a)h+(f''(a)/2!)h^2\right)}{h^2}=0,$$

from which it follows that

$$f(a+h)=f(a)+f'(a)h+\frac{f''(a)}{2!}h^2+o(h^2) \quad \text{as} \quad h\to 0. \tag{4.9}$$

PROB. 4.1. Prove: If f is twice differentiable in some ϵ-neighborhood $N(a,\epsilon)$ of a and three times differentiable at a, then

$$f(a+h)=f(a)+f'(a)h+\frac{f''(a)}{2!}h^2+O(h^3) \quad \text{as} \quad h\to 0$$

and

$$f(a+h)=f(a)+f'(a)h+\frac{f''(a)}{2!}h^2+\frac{f'''(a)}{3!}h^3+o(h^3) \quad \text{as} \quad h\to 0.$$

Theorem 4.1. *If f is differentiable of order n at a, where n is some positive integer, then*

$$f(a+h)=f(a)+f'(a)h+\frac{f''(a)}{2!}h^2+\cdots+\frac{f^{(n)}(a)}{n!}h^n+o(h^n) \quad \textit{as} \quad h\to 0. \tag{4.10}$$

PROOF. We use induction on n. We already proved (4.10) for $n=1,2$. Suppose (4.10) holds for some positive integer n. Assume that f is differentiable of order $n+1$ at a. This implies that f' is differentiable of order n at a. By the induction hypothesis,

$$f'(a+h)=f'(a)+f''(a)h+\cdots+\frac{f^{(n+1)}(a)}{n!}h^n+o(h^n) \quad \text{as} \quad h\to 0, \tag{4.11}$$

i.e.,

$$\lim_{h\to 0}\frac{f'(a+h)-\left(f'(a)+f''(a)h+\cdots+\left(f^{(n+1)}(a)/n!\right)h^n\right)}{h^n}=0. \tag{4.12}$$

Hence, by L'Hôpital's rule and (4.12) we have

$$\lim_{h\to 0}\left[f(a+h)-\left(f(a)+f'(a)h+\cdots+\frac{f^{(n+1)}(a)}{(n+1)!}h^{n+1}\right)\right]\Big/h^{n+1}$$

$$=\lim_{h\to 0}\left(f'(a+h)-\left(f'(a)+f''(a)h+\cdots+\frac{f^{(n+1)}(a)}{n!}h^n\right)\right)\Big/(n+1)h^n$$

$$=0.$$

It follows that

$$f(a+h)=f(a)+f'(a)h+\frac{f''(a)}{2!}h+\cdots+\frac{f^{(n+1)}(a)}{(n+1)!}h^{n+1}+o(h^{n+1}) \quad \text{as} \quad h\to 0. \tag{4.13}$$

The conclusion follows by applying the principle of mathematical induction.

PROB. 4.2. Prove: If f is differentiable of order $n+1$ at a, where n is some nonnegative integer, then

$$f(a+h)=f(a)+f'(a)h+\cdots+\frac{f^{(n)}(a)}{n!}h^n+O(h^{n+1}) \quad \text{as} \quad h\to 0.$$

Remark 4.1. Let f be differentiable of order n at a, where n is some positive integer. Put

$$u_n(a,h) = \frac{\left(f(a+h) - \sum_{k=0}^{n}\left(f^{(k)}(a)/k!\right)h^k\right)}{h^n}. \tag{4.14}$$

Theorem 4.1 states that $\lim_{h\to 0} u_n(a,h) = 0$. From (4.14) it follows that

$$f(a+h) = \sum_{k=0}^{n} \frac{f^{(k)}(a)}{k!} + u_n(a,h)h^n, \tag{4.15}$$

where $u_n(a,h) \to 0$ as $h \to 0$. Formula (4.15) is called *Taylor's formula* of order n. The polynomial in h,

$$\sum_{k=0}^{n} \frac{f^{(k)}(a)}{k!} h^k = f(a) + f'(a)h + \cdots + \frac{f^{(n)}(a)}{n!} h^n, \tag{4.16}$$

is called the *Taylor Polynomial of order n at a*, while the term $u_n(a,h)h^n$ in Taylor's formula of order n is called the *remainder* in Taylor's formula of order n.

Remark 4.2. Put $x = a + h$ and order n and

$$v_n(a,x) = u_n(a, x-a).$$

Formula (4.15) then becomes

$$f(x) = \sum_{k=0}^{n} \frac{f^{(k)}(a)}{k!}(x-a)^k + v_n(a,x)(x-a)^n, \tag{4.17}$$

where $v_n(a,x) \to 0$ as $x \to a$. In this notation the remainder $R_{n+1}(a,x)$ and the Taylor polynomial are, respectively,

$$R_{n+1}(a,x) = v_n(a,x)(x-a)^n \tag{4.18a}$$

and

$$\sum_{k=0}^{n} \frac{f^{(k)}(a)}{k!}(x-a)^k. \tag{4.18b}$$

Calculation shows that Taylor's Polynomial of order n for f has the same value and derivatives, up to order n, that f has.

Remark 4.3. If P is a polynomial of degree $n \geqslant 1$ (cf. Theorem VII.4.1)

$$P(x) = \sum_{k=0}^{n} \frac{P^{(k)}(a)}{k!}(x-a)^k$$

$$= P(a) + P'(a)(x-a) + \cdots + \frac{P^{(n)}(a)}{n!}(x-a)^n$$

and, therefore, the Taylor Polynomial of order n is identical with P itself, the remainder being identically 0. In general, this is not the case. However under the appropriate conditions on f, the remainder can be given several forms. We deduce them in the next theorem and its corollary. We begin with what is called *Schlömlich's form* for the remainder.

Theorem 4.2. *Let n be an integer, $n \geqslant 0$ and p be some real number $p > -1$. If f and its derivatives up to and including order n are continuous on an interval I and $f^{(n+1)}(x)$ exists at least for x in the interior of I, then, for distinct a and x in I, there exists a c between a and x such that*

$$f(x) = \sum_{k=0}^{n} \frac{f^{(k)}(a)}{k!}(x-a)^k + \frac{f^{(n+1)}(c)}{(p+1)n!}(x-c)^{n-p}(x-a)^{p+1}. \quad (4.19)$$

PROOF. Fix a and x in I, $x \neq a$ and construct F, where

$$F(t) = f(x) - f(t) - \sum_{k=1}^{n} \frac{f^{(k)}(t)}{k!}(x-t)^k$$
$$- \frac{(x-t)^{p+1}}{(x-a)^{p+1}}\left(f(x) - \sum_{k=0}^{n} \frac{f^{(k)}(a)}{k!}(x-a)^k\right) \quad (4.20)$$

for $t \in I$. Note that

$$F(a) = 0 = F(x). \quad (4.21)$$

We can, therefore, apply Rolle's Theorem to obtain a c between a and x such that $F'(c) = 0$. Taking the derivative $f'(t)$ in (4.20), we have

$$F'(t) = -f'(t) - \sum_{k=1}^{n}\left(\frac{f^{(k+1)}(t)}{k!}(x-t)^k - \frac{f^{(k)}(t)}{(k-1)!}(x-t)^{k-1}\right)$$
$$+ \frac{(p+1)(x-t)^p}{(x-a)^{p+1}}\left(f(x) - \sum_{k=0}^{n} \frac{f^{(k)}(a)}{k!}(x-a)^k\right). \quad (4.22)$$

We have a "collapsing" sum

$$f'(t) + \sum_{k=1}^{n}\left(\frac{f^{(k+1)}(t)}{k!}(x-t)^k - \frac{f^{(k)}(t)}{(k-1)!}(x-t)^{k-1}\right)$$
$$= \frac{f^{(n+1)}(t)}{n!}(x-t)^n.$$

Putting this in (4.22) we obtain

$$F'(t) = -\frac{f^{(n+1)}(t)}{n!}(x-t)^n$$
$$+ \frac{(p+1)(x-t)^p}{(x-a)^{p+1}}\left(f(x) - \sum_{k=0}^{n} \frac{f^{(k)}(a)}{k!}(x-a)^k\right). \quad (4.23)$$

Since $F'(c) = 0$, this formula yields

$$0 = -\frac{f^{(n+1)}(c)}{n!}(x-c)^n + \frac{(p+1)(x-c)^p}{(x-a)^{p+1}}\left(f(x) - \sum_{k=0}^{n} \frac{f^{(k)}(a)}{k!}(x-a)^k\right),$$

from which (4.19) follows.

Remark 4.4. The remainder $R_{n+1}(a, x)$ in Theorem 4.2 is

$$R_{n+1}(a,x) = \frac{f^{(n+1)}(c)}{(p+1)n!}(x-c)^{n-p}(x-a)^{p+1}, \qquad p > -1. \quad (4.24)$$

This is known as *Schlömlich's form* of the remainder in Taylor's Theorem of order n. When $p = n$, we have

$$R_{n+1}(a,x) = \frac{f^{(n+1)}(c)}{(n+1)!}(x-a)^{n+1}, \quad (4.25)$$

where c is between a and x. This is known as the *Lagrange form* of the remainder in Taylor's Theorem of order n. When $p = 0$, we obtain from (4.24)

$$R_{n+1}(a,x) = \frac{f^{(n+1)}(c)}{n!}(x-c)^n(x-a). \quad (4.26)$$

This is known as *Cauchy's form* of the remainder in Taylor's Theorem of order n. This proves the corollary below.

Corollary (of Theorem 4.2). *If f and its derivatives up to and including order n are continuous on an interval I and $f^{(n+1)}(x)$ exists at least for x in the interior of I, where n is some nonnegative integer, then for distinct a and x in I, there exists a c between a and x such that*

$$f(x) = \sum_{k=0}^{n} \frac{f^{(k)}(a)}{k!}(x-a)^k + \frac{f^{(n+1)}(c)}{(n+1)!}(x-a)^{n+1}. \quad (4.27)$$

Remark 4.5. Using $n = 0$ in the last corollary we obtain a form of the Mean-Value Theorem. This is why we call the last corollary the *extended mean-value theorem.*

Remark 4.6. In Theorem 4.2, let us use the fact that c is between a and x and write

$$\theta = \frac{c-a}{x-a}, \qquad x \neq a,$$

so that $0 < |\theta| < 1$ and $c = a + \theta(x - a)$. Writing $h = x - a$, we obtain

$x - c = x - a - (c - a) = (1 - \theta)(x - a)$. The Schlömlich form of the remainder becomes

$$R_{n+1}(a,x) = \frac{(1-\theta)^{n-p}}{(p+1)n!}(x-a)^{n+1} f^{(n+1)}(a + \theta(x-a)) \quad (4.28)$$

and the Lagrange form becomes

$$R_{n+1}(a,x) = \frac{f^{(n+1)}(a+\theta(x-a))}{(n+1)!}(x-a)^{n+1}. \quad (4.29)$$

In both cases $0 < |\theta| < 1$.

5. Taylor and Maclaurin Series

When f has derivatives of all orders at $a \in \mathbb{R}$, then the series

$$\sum_{n=0}^{\infty} \frac{f^{(n)}(a)}{n!}(x-a)^n = f(a) + f'(a)(x-a) + \frac{f^{(2)}(a)}{2!}(x-a)^2 + \cdots \quad (5.1)$$

is called the *Taylor series* of f at a. When $a = 0$, this series has the form

$$\sum_{n=0}^{\infty} \frac{f^{(n)}(0)}{n!} x^n = f(0) + f'(0)x + \frac{f^{(n)}(0)}{2!}x^2 + \cdots . \quad (5.2)$$

This series is called the *Maclaurin series* of f. Thus, the Maclaurin series of f is its Taylor series at $a = 0$.

When f has a Taylor series at a, several questions arise. (1) Does it converge? Of course, at $x = a$ the Taylor series of f at a converges trivially and reduces to $f(a)$. The convergence question becomes significant when $x \neq a$. (2) Does the Taylor series (5.1) converge to $f(x)$? If for some x, the Taylor series of f at a converges to $f(x)$, then we say that it represents $f(x)$ for that x. For example, when the Taylor series of f at a exists, then it always represents $f(a)$. Whether or not the Taylor series of f at a represents $f(x)$ for some $x \neq a$ can be decided by examining the remainder $R_{n+1}(a,x)$ in Taylor's formula of order n.

Let f have derivatives of all orders at a. Then, for each positive integer n, we have

$$f(x) = \sum_{k=0}^{n} \frac{f^{(k)}(a)}{k!}(x-a)^k + R_{n+1}(a,x). \quad (5.3)$$

There are three possibilities. If $\lim_{n\to+\infty} R_n(a,x) = 0$, then

$$f(x) = \lim_{n\to+\infty} \sum_{k=0}^{n} \frac{f^{(k)}(a)}{k!}(x-a)^k = \sum_{k=0}^{\infty} f^k \frac{(a)}{k!}(x-a)^k \quad (5.4)$$

and the Taylor series of f at a converges and represents $f(x)$. If $\lim_{n\to+\infty} R_n(a,x)$ does not exist or is infinite, then the Taylor series of f at x

does not converge. If $\lim_{n\to+\infty} f_n(a,x)$ exists and is finite but is not equal to 0, then the Taylor series of f at a converges but does not represent $f(x)$. We give some examples.

EXAMPLE 5.1. In Example IV.5.1 we defined

$$\exp x = \sum_{n=0}^{\infty} \frac{x^n}{n!} \qquad \text{for each} \quad x \in \mathbb{R}. \tag{5.5}$$

Using the ratio test, we found that the series on the right converges. In Theorem VI.1.4 we proved

$$e^x = \exp x = \sum_{n=0}^{\infty} \frac{x^n}{n!} \qquad \text{for} \quad x \in \mathbb{R}. \tag{5.6}$$

Simple calculations show that the series on the right is really the Maclaurin series for the function E defined as $E(x) = e^x$ for $x \in \mathbb{R}$. It follows that the Maclaurin series (the Taylor series for E at 0) for E represents $E(x)$. Here we establish this in another way by proving, more generally, that if $a \in \mathbb{R}$, then the Taylor series for E at a represents $E(x)$ for $x \in \mathbb{R}$. We do this by estimating the remainder $R_{n+1}(a,x)$ for E in Taylor's Theorem of order n. For each nonnegative integer n

$$E^{(n)}(x) = e^x, \qquad E^{(n)}(a) = e^a. \tag{5.7}$$

Assume that $x \neq a$. Using the Lagrange form of the remainder (Corollary of Theorem 4.2) we see that there exists a c between a and x such that

$$\begin{aligned} E(x) &= \sum_{k=0}^{n} \frac{E^{(k)}(a)}{k!}(x-a)^k + \frac{E^{(n+1)}(c)}{(n+1)!}(x-a)^{n+1} \\ &= \sum_{k=0}^{n} \frac{e^a}{k!}(x-a)^k + \frac{e^c}{(n+1)!}(x-a)^{n+1}. \end{aligned} \tag{5.8}$$

Here

$$R_{n+1}(a,x) = \frac{e^c}{(n+1)!}(x-a)^{n+1}, \tag{5.9}$$

where c is between a and x. If $a < x$, then $a < c < x$ and $0 < c - a < x - a$ so that

$$e^c = e^{a+(c-a)} = e^a e^{c-a} < e^a e^{x-a} = e^a e^{(x-a)}. \tag{5.10}$$

If $x < a$, then $x < c < a$ and $x - a < c - a < 0$ so that

$$e^c = e^{a+(c-a)} = e^a e^{c-a} < e^a e^0 = e^a < e^a e^{-(x-a)}. \tag{5.11}$$

This and (5.9) show that

$$|R_{n+1}(a,x)| = \frac{e^c}{(n+1)!}|x-a|^{n+1} < \frac{e^{|x-a|}}{(n+1)!}|x-a|^{n+1}.$$

But (Prob. IV.5.1)

$$\lim_{n\to+\infty} \frac{|x-a|^{n+1}}{(n+1)!} = 0$$

which implies that $\lim_{n\to+\infty} R_{n+1}(a,x) = 0$. Using this, (5.8) and (5.9) we have: If $x \in \mathbb{R}$, $a \in \mathbb{R}$, then

$$E(x) = e^x = \sum_{k=0}^{\infty} \frac{e^a}{k!}(x-a)^k. \tag{5.12}$$

On the right we have the Taylor series for E. Thus, (5.12) states that the Taylor series for E at a represent $E(x) = e^x$ for each $x \in \mathbb{R}$. We also have from (5.12) that

$$E(x) = e^x = e^a \sum_{k=0}^{n} \frac{(x-a)^k}{k!} = e^a \exp(x-a).$$

Putting $a = 0$, we have again

$$E(x) = \exp x \qquad \text{for} \quad x \in \mathbb{R}. \tag{5.13}$$

This constitutes another proof that $E = \exp$.

Prob. 5.1. Prove: The Maclaurin series for the sine and cosine functions are the series by means of which these functions were defined in Def. IV.8.1.

Prob. 5.2. Obtain the Maclaurin series expansion for the hyperbolic sine and cosine functions and prove these series represent them for all $x \in \mathbb{R}$. (See Example V.2.3.)

Example 5.2. Consider the function f, where

$$f(x) = \ln(1+x), \qquad x > -1. \tag{5.14}$$

We have $f(0) = 0$ and

$$f'(x) = \frac{1}{1+x}, \qquad f'(0) = 1,$$

$$f''(x) = -\frac{1}{(1+x)^2}, \qquad f''(0) = -1.$$

As a matter of fact, if n is a positive integer, then

$$f^{(n)}(x) = (-1)^{n+1}\frac{(n-1)!}{(1+x)^n} \quad \text{and} \quad f^{(n)}(0) = (-1)^{n+1}(n-1)!. \tag{5.15}$$

Therefore, the Maclaurin series for f is

$$f(0) = \sum_{n=1}^{\infty} \frac{f^{(n)}(0)}{n!}x^n = \sum_{n=1}^{\infty} (-1)^{n+1}\frac{(n-1)!}{n!}x^n$$

$$= \sum_{n=0}^{\infty} (-1)^{n+1}\frac{x^n}{n} = x - \frac{x^2}{2} + \frac{x^3}{3} - \frac{x^4}{4} + \cdots. \tag{5.16}$$

We examine the convergence of this series. The series obviously converges if $x = 0$. If $x \neq 0$, the ratio test yields

$$\frac{|a_{n+1}|}{|a_n|} = \left| \frac{x^{n+1}/(n+1)!}{x^n/n!} \right| = \frac{n}{n+1}|x| \underset{n}{\rightarrow} |x|. \tag{5.17}$$

Therefore, the series converges if $|x| < 1$ and diverges for $|x| > 1$. At $x = 1$, the Maclaurin series for f is

$$1 - \tfrac{1}{2} + \tfrac{1}{3} - \tfrac{1}{4} + \cdots$$

which converges since it is an alternating series. For $x = -1$, we have

$$-1 - \tfrac{1}{2} - \tfrac{1}{3} - \tfrac{1}{4} - \cdots .$$

which diverges. Thus, the Maclaurin series for f converges for $x \in (-1, 1]$ and diverges for $(-\infty, -1] \cup (1; +\infty)$. We now inquire whether or not the Maclaurin series for f represents $f(x)$ for $-1 < x \leqslant 1$. We know that $f(0) = 0$ and that

$$f^{(n+1)}(x) = (-1)^n \frac{n!}{(1 + x)^{n+1}} .$$

Using the Corollary of Theorem 4.2, we find that if $x \neq 0$, then for a non-negative integer n there exists a c between 0 and x such that

$$\ln(1 + x) = x - \frac{x^2}{2} + \cdots + (-1)^n \frac{x^{n+1}}{(n+1)(1+c)^{n+1}} . \tag{5.18}$$

The remainder $R_{n+1}(0, x)$ is

$$R_{n+1}(0, x) = (-1)^n \frac{x^{n+1}}{(n+1)(1+c)^{n+1}} . \tag{5.19}$$

If $0 < x \leqslant 1$, then $0 < c < x \leqslant 1$ and

$$|R_{n+1}(0, x)| = \frac{|x|^{n+1}}{(1+c)^{n+1}(n+1)} \leqslant \frac{|x|^{n+1}}{n+1} \leqslant \frac{1}{n+1} .$$

This inequality is valid also for $x = 0$. Thus,

$$\lim_{n \to +\infty} |R_{n+1}(0, x)| \leqslant \lim_{n \to +\infty} \frac{1}{n+1} = 0 \qquad \text{if} \quad 0 \leqslant x \leqslant 1.$$

It follows that

$$\ln(1 + x) = x - \frac{x^2}{2} + \frac{x^3}{3} - \frac{x^4}{4} + \cdots \qquad \text{for} \quad 0 \leqslant x \leqslant 1. \tag{5.20}$$

The series represents $\ln(1 + x)$ for $0 \leqslant x \leqslant 1$. In particular,

$$\ln 2 = \ln(1 + 1) = 1 - \tfrac{1}{2} + \tfrac{1}{3} - \tfrac{1}{4} + \cdots . \tag{5.21}$$

We now turn to the case $-1 < x < 0$. Here we resort to Cauchy's form for the remainder (4.26) to obtain

$$\ln(1 + x) = x - \frac{x^2}{2} + \cdots + (-1)^{n+1} \frac{x^n}{n} + R^*_{n+1}(0, x), \tag{5.22}$$

where

$$R^*_n(0, x) = \frac{f^{(n+1)}(c)}{n!}(x - c)^n x = (-1)^n \frac{(x - c)^n x}{(1+c)^{n+1}} , \tag{5.23}$$

where $-1 < x < c < 0$. We prove:

$$\left|\frac{x-c}{1+c}\right| < |x| \tag{5.24a}$$

and

$$\left|\frac{x}{1+c}\right| < \frac{|x|}{1+x}. \tag{5.24b}$$

Indeed, (5.24b) follows because we have $0 < 1+x < 1+c < 1$ and, therefore,

$$\left|\frac{x}{1+c}\right| = \frac{|x|}{1+c} < \frac{|x|}{1+x}.$$

Inequality (5.24a) holds because $0 < -c < -x < 1$ and $c < 0$, which implies that $-cx > c$, so that

$$(-x)(1+c) = -x - cx > c - x > 0, \tag{5.25}$$

which implies that

$$\left|\frac{x-c}{1+c}\right| = \frac{|x-c|}{1+c} = \frac{c-x}{1+c} < (-x)\frac{(1+c)}{1+c} = -x = |x|.$$

Thus, (5.24a) and (5.24b) hold for $-1 < x < c < 0$. Returning to (5.23), we conclude that

$$|R_n^*(0,x)| = \frac{|x-c|^n|x|}{(1+c)^{n+1}} = \left|\frac{x-c}{1+c}\right|^n \frac{|x|}{1+c} < |x|^n \frac{|x|}{1+x} = \frac{|x|^{n+1}}{1+x}.$$

Since $|x| < 1$ here this implies $R_n^*(0,x) \to 0$ as $n \to +\infty$. We conclude from this and (5.22) that

$$\ln(1+x) = x - \frac{x^2}{2} + \cdots$$

holds also if $-1 < x < 0$. Thus,

$$\ln(1+x) = \sum_{n=1}^{\infty} (-1)^{n+1} \frac{x^n}{n} \qquad \text{for} \quad -1 < x \leqslant 1. \tag{5.26}$$

Prob. 5.3. Prove: If $0 < x \leqslant 2$, then

$$\ln x = (x-1) - \frac{(x-1)^2}{2} + \frac{(x-1)^3}{3} - \frac{(x-1)^4}{4} + \cdots.$$

Prob. 5.4. Prove: If $a > 0$ and $0 < x \leqslant 2a$, then

$$\ln x = \ln a + \frac{x-a}{a} - \frac{(x-a)^2}{2a^2} + \frac{(x-a)^3}{3a^3} - \frac{(x-a)^4}{4a^4} + \cdots.$$

Prob. 5.5. Use

$$\operatorname{Arctanh} x = \frac{1}{2}\ln\frac{1+x}{1-x} \qquad \text{if} \quad |x| < 1$$

to obtain

$$\operatorname{Arctanh} x = x + \frac{x^3}{3} + \frac{x^5}{5} + \cdots \qquad \text{for} \quad |x| < 1.$$

EXAMPLE 5.3. Our examples illustrate a case where the Taylor series of f at a converges and represents $f(x)$, and where the Taylor series for f at a diverges for some x ($\ln(1+x)$ for $x \in (-\infty, -1] \cup (1; +\infty)$). We exhibit now a function f which has derivatives of all orders at a whose Taylor series at a converges for $x \neq a$ but does not represent $f(x)$ for $x \neq a$.

Let f be defined as

$$f(x) = \begin{cases} e^{-(1/x^2)} & \text{if} \quad x \neq 0 \\ 0 & \text{if} \quad x = 0. \end{cases} \tag{5.27}$$

We have

$$f'(0) = \lim_{x \to 0} \frac{e^{-(1/x^2)}}{x} = \lim_{x \to 0} \frac{1/x}{e^{(1/x^2)}} = 0,$$

so that

$$f'(x) = \begin{cases} \dfrac{2}{x^3} e^{-(1/x^2)} & \text{for} \quad x \neq 0 \\ 0 & \text{for} \quad x = 0. \end{cases}$$

We can prove by induction on n that if $x \neq 0$, then

$$f^{(n)}(x) = \frac{e^{-(1/x^2)} P_n(x)}{x^{3n}},$$

where P_n is a polynomial of degree $< 3n$ and, hence,

$$f^{(n)}(x) = e^{-(1/x^2)} Q_n\left(\frac{1}{x}\right) \qquad \text{for} \quad x \neq 0, \tag{5.28}$$

where Q_n is a polynomial depending on n. We prove that $f^{(n)}(0) = 0$ for all positive integers n by using induction on n. We have $f'(0) = 0 = f(0)$. If $f^{(n)}(0)$ for some positive integer n, then using (5.28), we obtain

$$\begin{aligned} f^{(n+1)}(0) &= \lim_{x \to 0} \frac{e^{-(1/x^2)} Q_n(1/x)}{x} \\ &= \lim_{x \to 0} \left(\frac{1}{x} Q_n\left(\frac{1}{x}\right) \right) \Big/ e^{(1/x^2)} \\ &= 0. \end{aligned}$$

The last limit is 0. In fact, we can write

$$\frac{1}{x} Q_n\left(\frac{1}{x}\right) = R_n\left(\frac{1}{x}\right),$$

where R_n is some polynomial, so that

$$R_n\left(\frac{1}{x}\right) = b_0 + b_1 \frac{1}{x} + \cdots + b_m \frac{1}{x^m}.$$

Here m is some positive integer and $b_1, \ldots, b_m$ are constants. Hence,

$$f^{(n+1)}(0) = \lim_{x \to 0} \frac{(1/x)Q_n(1/x)}{e^{(1/x^2)}} = \lim_{x \to 0} \frac{R_n(1/x)}{e^{(1/x^2)}} = 0.$$

This proves that $f^{(n)}(0) = 0$ for each positive integer n. Constructing the Taylor series $Tf(x)$ at 0 for f we obtain

$$Tf(x) = \sum_{n=0}^{\infty} a_n x^n = \sum_{n=0}^{\infty} 0x^n = 0 \qquad \text{for} \quad x \neq 0.$$

The series has only zeros for its coefficients, and thus converges to 0 for all $x \in \mathbb{R}$. Since $f(x) = e^{-(1/x^2)} > 0$ for $x \neq 0$, the Taylor series for f at c cannot represent $f(x)$ for $x \neq 0$.

6. The Binomial Series

We suggest that at this point the reader review the material on factorials in Section II.6.

PROB. 6.1. Prove: If k is a positive integer and $\alpha < 0$, then

$$(-1)^k \binom{\alpha}{k} > \frac{-\alpha}{k} > 0.$$

Lemma 6.1. *If $\alpha \in \mathbb{R}$ and n is a positive integer, then*

$$\lim_{n \to +\infty} \left| \binom{\alpha}{n} \right| = \begin{cases} +\infty & \text{for} \quad \alpha < -1 \\ 1 & \text{for} \quad \alpha = -1 \\ 0 & \text{for} \quad \alpha > -1. \end{cases} \tag{6.1}$$

PROOF. If $\alpha < -1$, then $-(\alpha + 1) > 0$ and

$$\left| \binom{\alpha}{n} \right| = \left| \frac{\alpha(\alpha - 1) \cdots (\alpha - n + 1)}{n!} \right| = \frac{|\alpha + 1 - 1||\alpha + 1 - 2| \cdots |\alpha + 1 - n|}{n!}$$

$$= (1 - (\alpha + 1))\left(1 - \frac{\alpha + 1}{2}\right) \cdots \left(1 - \frac{\alpha + 1}{n}\right)$$

$$\geqslant 1 - (\alpha + 1)\left(1 + \frac{1}{2} + \cdots + \frac{1}{n}\right).$$

The last inequality is a consequence of Prob. II.12.14; it implies that

$$\lim_{n \to +\infty} \left| \binom{\alpha}{n} \right| = +\infty \qquad \text{for} \quad \alpha < -1. \tag{6.2}$$

If $\alpha = -1$, then

$$\left| \binom{\alpha}{n} \right| = \left| \binom{-1}{n} \right| = \left| \frac{(-1)(-1-1) \cdots (-1-n+1)}{n!} \right| = \left| \frac{1 \cdot 2 \cdots n}{n!} \right| = 1.$$

Therefore,

$$\lim_{n\to+\infty}\left|\binom{\alpha}{n}\right| = 1 \qquad \text{for} \quad \alpha = -1.$$

If $\alpha > -1$, then we consider the following cases: either (a) $-1 < \alpha < 0$ or (b) $0 \leqslant \alpha$. In case (a) we have $0 < 1 + \alpha < 1$ and also

$$|\alpha - 1| = 1 - \alpha, \qquad |\alpha - 2| = 2 - \alpha, \ldots, \qquad |\alpha - n + 1| = n - 1 - \alpha,$$

so that

$$\begin{aligned}\left|\binom{\alpha}{n}\right| &= \frac{|\alpha|\,|\alpha-1|\cdots|\alpha-n+1|}{n!} = \frac{(-\alpha)(1-\alpha)\cdots(n-1-\alpha)}{1\cdot 2\cdots n}\\ &= \frac{(1-(\alpha+1))(2-(\alpha+1))\cdots(n-(\alpha+1))}{1\cdot 2\cdots n}\\ &= \left(1-\frac{\alpha+1}{1}\right)\left(1-\frac{\alpha+1}{2}\right)\cdots\left(1-\frac{\alpha+1}{n}\right).\end{aligned}$$

Since $0 < (\alpha+1)/n \leqslant \alpha + 1 < 1$, the above and Prob. II.12.15(c) imply that: If $-1 < \alpha < 0$, then

$$\begin{aligned}\left|\binom{\alpha}{n}\right| &= \left(1-\frac{\alpha+1}{1}\right)\left(1-\frac{\alpha+1}{2}\right)\cdots\left(1-\frac{\alpha+1}{n}\right)\\ &< \frac{1}{1+(\alpha+1)(1+1/2+\cdots+1/n)}.\end{aligned} \tag{6.3}$$

This time we have: If $-1 < \alpha < 0$, then $|\binom{\alpha}{n}| \to 0$ as $n \to +\infty$. In case (b), $\alpha \geqslant 0$, we prove

$$\binom{\alpha}{n} = O\left(\frac{1}{n}\right) \qquad \text{as} \quad n \to +\infty. \tag{6.4}$$

If α is a nonnegative integer, then for sufficiently large n, we have $\alpha < n$, so $\binom{\alpha}{n} = 0$. Thus, (6.4) holds in this case. If $0 < \alpha < 1$, then

$$\left|\binom{\alpha}{n}\right| = \alpha\frac{|\alpha-1||\alpha-2|\cdots|\alpha-n+1|}{n!} < \alpha\frac{1\cdot 2\cdots(n-1)}{n!} = \frac{\alpha}{n}. \tag{6.5}$$

If $1 < \alpha$ and α is not an integer, then there exists a positive integer m such that $m < \alpha < m + 1$. Fix m and take $n > m + 1$, where n is an integer. We have $n - 1 \geqslant m + 1$ and

$$\begin{aligned}\left|\binom{\alpha}{n}\right| &= \left|\frac{\alpha(\alpha-1)\cdots(\alpha-m)}{1\cdot 2\cdots(m+1)}\cdot\frac{(\alpha-m-1)\cdots(\alpha-n+1)}{(m+2)(m+3)\cdots n}\right|\\ &= \left|\binom{\alpha}{m+1}\right|\frac{(m+1-\alpha)\cdots(n-1-\alpha)}{(m+2)(m+3)\cdots n}\\ &\leqslant \left|\binom{\alpha}{m+1}\right|\frac{(m+1)(m+2)\cdots(n-1)}{(m+2)(m+3)\cdots n}\\ &= \left|\binom{\alpha}{m+1}\right|\frac{m+1}{n}.\end{aligned}$$

This can be written

$$\left|\binom{\alpha}{n}\right| \leqslant \left|\binom{\alpha}{[\alpha]+1}\right| \frac{[\alpha]+1}{n} \qquad \text{for} \quad 1 < \alpha, \quad \alpha \notin \mathbb{Z}. \tag{5.6}$$

Our study of cases for (b) $\alpha \geqslant 0$ shows that (6.4) holds if $\alpha \geqslant 0$. This together with case (a) $-1 < \alpha < 0$, proves that if $\alpha > -1$, then

$$\left|\binom{\alpha}{n}\right| \to 0 \qquad \text{as} \quad n \to +\infty.$$

and completes the proof.

PROB. 6.2. Prove:

$$\lim_{n\to+\infty} \left|\binom{\alpha+n}{n}\right| = \begin{cases} +\infty & \text{for} \quad \alpha > 0 \\ 1 & \text{for} \quad \alpha = 0 \\ 0 & \text{for} \quad \alpha < 0 \end{cases}$$

(Hint: see Prob. II.6.14 and Lemma 6.1).

Remark 6.1. The result in Prob. II.6.13 states that

$$\sum_{k=0}^{n} \binom{\alpha+k-1}{k} = \binom{\alpha+n}{n} \qquad \text{if} \quad n \in \mathbb{Z}_0, \quad \alpha \in \mathbb{R}. \tag{6.7}$$

Using the result in Prob. II.6.14, i.e.,

$$\binom{-\alpha}{k} = (-1)^k \binom{\alpha+k-1}{k}, \qquad \alpha \in \mathbb{R}, \quad k \in \mathbb{Z}_0,$$

(6.7) can be written

$$\sum_{k=0}^{n} (-1)^k \binom{-\alpha}{k} = \binom{n+\alpha}{n}, \qquad \alpha \in \mathbb{R}, \quad n \in \mathbb{Z}_0.$$

In turn, this implies that

$$\sum_{k=0}^{n} (-1)^k \binom{\alpha}{k} = \binom{n-\alpha}{n}, \qquad \alpha \in \mathbb{R}, \quad n \in \mathbb{Z}_0. \tag{6.8}$$

From this and Prob. 6.2 we conclude that

$$\begin{aligned} \sum_{k=0}^{\infty} (-1)^k \binom{\alpha}{k} &= \lim_{n\to+\infty} \sum_{k=0}^{n} (-1)^k \binom{\alpha}{k} \\ &= \lim_{n\to+\infty} \binom{n-\alpha}{n} = \begin{cases} +\infty & \text{if} \quad \alpha < 0 \\ 0 & \text{if} \quad \alpha > 0. \end{cases} \end{aligned} \tag{6.9}$$

Binomial Series

We consider the function g given by

$$g(x) = (1+x)^{\alpha} \qquad \text{for} \quad x > -1, \quad \alpha \in \mathbb{R}. \tag{6.9'}$$

If $\alpha = n$, where n is a non-negative integer, the Binomial Theorem applies and we have

$$g(x) = (1+x)^{\alpha} = (1+x)^n = \sum_{k=0}^{n} \binom{n}{k} x^k. \tag{6.10}$$

Since $\binom{n}{k} = 0$ if $k > n$ we can think of this as an infinite series, whose terms after the nth are equal to 0. This series converges for all $x \in \mathbb{R}$ since it "collapses" to a finite sum. We will obtain the Maclaurin series for g, when α is not a non-negative integer. We have

$$\begin{aligned} g(x) &= (1+x)^{\alpha} & g(0) &= 1, \\ g'(x) &= \alpha(1+x)^{\alpha-1}, & g'(0) &= \alpha, \\ g''(x) &= \alpha(\alpha-1)(1+x)^{\alpha-2}, & g''(0) &= \alpha(\alpha-1). \end{aligned}$$

Quite generally, if n is a positive integer, then

$$\begin{aligned} g^{(n)}(x) &= \alpha(\alpha-1)\ldots(\alpha-n+1)(1+x)^{\alpha-n}, \\ g^{(n)}(0) &= \alpha(\alpha-1)\ldots(\alpha-n+1). \end{aligned} \tag{6.11}$$

These may be written as

$$g^{(n)}(x) = (\alpha)_n(1+x)^{\alpha-n}, \qquad g^{(n)}(0) = (\alpha)_n. \tag{6.12}$$

The Maclaurin series for g is

$$\sum_{n=0}^{\infty} \frac{g^{(n)}(0)}{n!} x^n = \sum_{n=0}^{\infty} \frac{(\alpha)_n}{n!} x^n = \sum_{n=0}^{\infty} \binom{\alpha}{n} x^n. \tag{6.13}$$

This series is called the *Binomial* series. To investigate its convergence we apply the ratio test to obtain for $x \neq 0$, $\alpha \notin \mathbb{Z}$,

$$\frac{|a_{n+1}|}{|a_n|} = \frac{\left|\binom{\alpha}{n+1}\right|}{\left|\binom{\alpha}{n}\right|} |x| = \left|\frac{\alpha-n}{n+1}\right| |x|.$$

Since

$$\lim_{n\to+\infty} \left|\frac{\alpha-n}{n+1}\right| = 1,$$

we have

$$\lim_{n\to+\infty} \frac{|a_{n+1}|}{|a_n|} = |x|.$$

If $\alpha \in \mathbb{R}$, $\alpha \notin \mathbb{Z}$, then series (6.13) converges absolutely if $|x| < 1$ and diverges if $|x| > 1$. The ratio test fails if $|x| = 1$. Let $\alpha > 0$. If $x = -1$ then by (6.13) and (6.9),

$$\sum_{n=0}^{\infty} (-1)^n \binom{\alpha}{n} = 0 = (1 + (-1))^{\alpha}. \tag{6.14}$$

Thus, the binomial series represents $g(-1)$ for $\alpha > 0$. If $x = 1$, then the binomial series becomes

$$\sum_{n=0}^{\infty} \binom{\alpha}{n}. \tag{6.15}$$

This series diverges if $\alpha \leqslant -1$ (see Lemma 6.1). The case $\alpha > -1$ will be considered later.

We now consider the representation of $g(x)$ by its Maclaurin series for $x \neq -1$ and first treat the case $0 < x \leqslant 1$. Using (6.12), we note that

$$g^{(n+1)}(x) = (\alpha)_{n+1}(1+x)^{\alpha-n-1}. \tag{6.16}$$

By the Extended Mean-Value Theorem (Corollary of Theorem 4.2), there exists a c such that $0 < c < x \leqslant 1$ and

$$\begin{aligned}(1+x)^{\alpha} &= \sum_{k=0}^{n} \frac{g^{(k)}(0)}{k!} x^k + \frac{g^{(n+1)}(c)}{(n+1)!} x^{n+1} \\ &= \sum_{k=0}^{n} \binom{\alpha}{k} x^k + \frac{(\alpha)_{n+1}}{(n+1)!}(1+c)^{\alpha-n-1} x^{n+1} \\ &= \sum_{k=0}^{n} \binom{\alpha}{k} x^k + \binom{\alpha}{n+1} \frac{x^{n+1}}{(1+c)^{n+1-\alpha}}.\end{aligned}$$

Thus,

$$(1+x)^{\alpha} = \sum_{k=0}^{n} \binom{\alpha}{k} x^k + \binom{\alpha}{n+1} \frac{x^{n+1}}{(1+c)^{n+1-\alpha}}, \tag{6.17}$$

where $0 < c < x \leqslant 1$. For sufficiently large n, we have $n+1-\alpha > 0$. Estimating the remainder, we see that for such n

$$|R_{n+1}(0,x)| = \left|\binom{\alpha}{n+1}\right| \frac{x^{n+1}}{(1+c)^{n+1-\alpha}} \leqslant \left|\binom{\alpha}{n+1}\right| x^{n+1} \qquad \text{for} \quad 0 < x \leqslant 1. \tag{6.18}$$

We consider two cases: (a) $0 < x < 1$, (b) $x = 1$. In case (a) the series converges, so that

$$\lim_{n \to +\infty} \binom{\alpha}{n+1} x^{n+1} = 0 \qquad \text{for all} \quad \alpha \in \mathbb{R}. \tag{6.19}$$

In case (b), Lemma 6.1 implies that

$$\lim_{n \to +\infty} \binom{\alpha}{n+1} = 0 \qquad \text{if} \quad \alpha > -1. \tag{6.20}$$

Accordingly, we have

$$\lim_{n \to +\infty} R_{n+1}(0,x) = 0 \qquad \text{if} \quad \begin{cases} 0 < x < 1 & \text{for all} \quad \alpha \in \mathbb{R} \\ x = 1 & \alpha > -1. \end{cases} \tag{6.21}$$

(We recall that the series (6.15) diverges for $\alpha \leqslant -1$.) Thus, we have

$$(1+x)^{\alpha} = \sum_{n=0}^{\infty} \binom{\alpha}{n} x^n \qquad \text{for} \quad 0 < x < 1, \quad \alpha \in \mathbb{R} \tag{6.22}$$

and

$$2^\alpha = (1+1)^\alpha = \sum_{n=0}^{\infty} \binom{\alpha}{n} \quad \text{for} \quad \alpha > -1. \tag{6.23}$$

We now consider what happens to the remainder as $n \to +\infty$ if $-1 < x < 0$. We use Cauchy's form (4.26) for the remainder. In this case, there exists a c such that $-1 < x < c < 0$ and

$$\begin{aligned} R_{n+1}^*(0,x) &= \frac{g^{(n+1)}(c)}{n!}(x-c)^n x \\ &= \frac{(\alpha)_{n+1}}{n!}(1+c)^{\alpha-n-1}(x-c)^n x \\ &= \frac{(\alpha)_{n+1}}{n!}(1+c)^{\alpha-1}\left(\frac{x-c}{1+c}\right)^n x. \end{aligned}$$

We proved earlier (see (5.24)) that if $-1 < x < c < 0$, then

$$\left|\frac{x-c}{1+c}\right| < |x|.$$

Accordingly,

$$\begin{aligned} |R_{n+1}^*(0,x)| &= \frac{|(\alpha)_{n+1}|}{n!}(1+c)^{\alpha-1}\left|\frac{x-c}{1+c}\right|^n |x| \\ &< \frac{|(\alpha)_{n+1}|}{n!}(1+c)^{\alpha-1}|x|^{n+1}. \end{aligned} \tag{6.24}$$

Since $0 < 1 + x < 1 + c < 1$, we have

$$\begin{aligned} (1+c)^{\alpha-1} &< 1 + c < 1 && \text{if} \quad \alpha > 1, \\ (1+c)^{\alpha-1} &= \frac{1}{(1+c)^{1-\alpha}} < \frac{1}{(1+x)^{1-\alpha}} && \text{if} \quad \alpha < 1. \end{aligned} \tag{6.25}$$

(For $\alpha = 1$, we are in the case where α is a positive integer and the series surely represents $g(x)$ then.) This and (6.24) imply that

$$\begin{aligned} |R_{n+1}^*(0,x)| &< \frac{|(\alpha)_{n+1}|}{n!}|x|^{n+1} && \text{if} \quad \alpha > 1, \\ &< \frac{|(\alpha)_{n+1}|}{n!}\frac{|x|^{n+1}}{(1+x)^{1-\alpha}} && \text{if} \quad \alpha < 1. \end{aligned} \tag{6.26}$$

We saw that $\sum_{n=0}^{\infty}\binom{\alpha}{n}x^n$ converges for $|x| < 1$, $\alpha \in \mathbb{R}$, so we know that

$$\lim_{n\to+\infty}\binom{\alpha}{n}x^n = 0 \quad \text{if} \quad -1 < x < 0 \quad \text{and} \quad \alpha \in \mathbb{R}. \tag{6.27}$$

Since

$$\frac{|(\alpha)_{n+1}|}{n!}|x|^{n+1} = |x|\left|\alpha\frac{(\alpha-1)\ldots(\alpha-n)}{n!}x^n\right| = |\alpha||x|\left|\binom{\alpha-1}{n}x^n\right|,$$

we conclude from this and (6.27) that

$$\frac{|(\alpha)_{n+1}|}{n!}|x|^{n+1} \to 0 \quad \text{as} \quad n \to +\infty.$$

This implies that $R_{n+1}^*(0,x) \to 0$ if $-1 < x < 0$ and $\alpha \in \mathbb{R}$. This and (6.21) imply that

$$(1+x)^\alpha = \sum_{n=0}^{\infty} \binom{\alpha}{n} x^n \quad \text{if} \quad |x| < 1 \quad \text{and} \quad \alpha \in \mathbb{R}. \tag{6.28}$$

We summarize these results in

Theorem 6.1. *Equality* (6.28) *holds if* (*a*) $x \in \mathbb{R}$, $\alpha \in \mathbb{Z}_0$; (*b*) *for* $|x| < 1$ *and* $\alpha \in \mathbb{R}$; (*c*) $x = -1$ *if* $\alpha > 0$; (*d*) *for* $x = 1$ *and* $\alpha > -1$. *In all other cases the binomial series diverges.*

PROB. 6.3. Prove: If $|x| \leqslant 1$, then

(a)
$$\sqrt{1-x} = 1 - \frac{x}{2} - \sum_{k=2}^{\infty} \frac{1 \cdot 3 \cdots (2k-3)}{2^k k!} x^k.$$

(b)
$$\frac{1}{2} + \sum_{k=2}^{\infty} \frac{1 \cdot 3 \cdots (2k-3)}{2^k k!} = 1.$$

PROB. 6.4. Prove:

(a) If $|x| < 1$, $\alpha \in \mathbb{R}$, then

$$\frac{1}{(1-x)^\alpha} = (1-x)^{-\alpha} = 1 + \alpha x + \frac{\alpha(\alpha+1)}{2!}x^2 + \cdots$$
$$= \sum_{n=0}^{\infty} \binom{\alpha+n-1}{n} x^n.$$

(b) If $\alpha < 1$, then

$$\frac{1}{2^\alpha} = 2^{-\alpha} = 1 - \alpha + \frac{\alpha(\alpha+1)}{2!} - \frac{\alpha(\alpha+1)(\alpha+3)}{3!} + \cdots$$
$$= \sum_{n=0}^{\infty} (-1)^n \binom{\alpha+n-1}{n}.$$

7. Tests for Maxima and Minima

In Section VII.5 we obtained necessary conditions for a value $f(x_0)$ of f be a maximum or a minimum of f. Here we seek sufficient conditions.

We introduce the notion of a local maximum or minimum of a function.

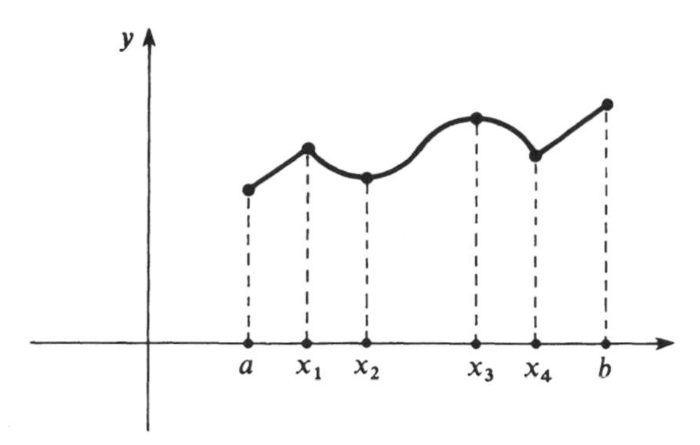

Figure 7.1

Def. 7.1. Let f be a real-valued function of a real variable defined on a set $D \neq \emptyset$. We call $f(x_0)$ a *local maximum* of f if some δ-neighborhood $N(x_0, \delta)$ exists such that $f(x_0)$ is a maximum of the restriction of f to $N(x_0, \delta) \cap D$, that is, if $f(x) \leqslant f(x_0)$ for $x \in N(x_0, \delta) \cap D$. When we have the strict inequality $f(x) < f(x_0)$ for $x \in N^*(x_0, \delta) \cap D$, then we say that $f(x_0)$ is a *strict* local maximum of f. Dually, $f(x_0)$ is called a *local* minimum of f if some δ-neighborhood $N(x_0, \delta)$ of x_0 exists such that $f(x_0)$ is a minimum of the restriction of f to $N(x_0, \delta) \cap D$, i.e., if $f(x) \geqslant f(x_0)$ holds for $x \in N(x_0, \delta) \cap D$. When we have the strict inequality $f(x) > f(x_0)$ for $x \in N^*(x_0, \delta) \cap D$, then we call $f(x_0)$ a strict *local* minimum of f. Local maxima or minima will be called local *extrema*. An extremum on D itself will be called an *absolute* extremum. Thus, we speak of an *absolute maximum or minimum* at x_0 if $f(x_0)$ is respectively a maximum or minimum of f on D. (See Fig. 7.1.)

In Fig. 7.1, $f(a)$, $f(x_2)$, and $f(x_4)$ are local minima of f, $f(x_1)$, $f(x_3)$, and $f(b)$ are local maxima of f, $f(a)$ is an absolute minimum, and $f(b)$ is an absolute maximum of f on D.

Theorem 7.1. *Let f be continuous on an interval I, and let x_0 be an interior point of I. If $f'(x) \geqslant 0$ for $x < x_0$, $x \in I$, and $f'(x) \leqslant 0$ for $x > x_0$, $x \in I$, then $f(x_0)$ is a maximum of f on I. (See* Fig. 7.2(a).)

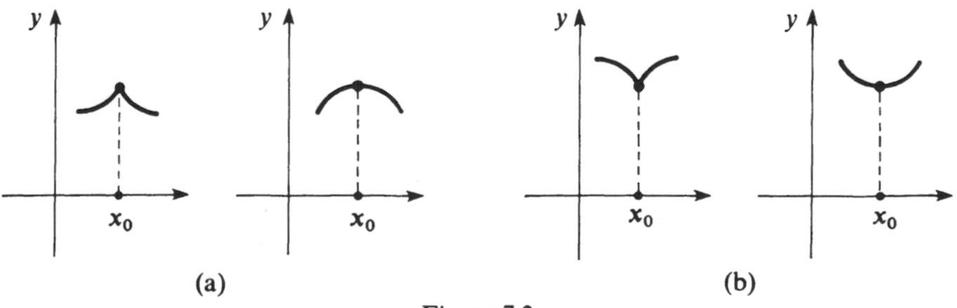

Figure 7.2

PROOF. Assume $x \in I$ and $x \neq x_0$. By the Mean-Value Theorem, there exists a c between x and x_0 such that

$$f(x) = f(x_0) + f'(c)(x - x_0). \tag{7.1}$$

If $x < x_0$, $x \in I$, then $x < c < x_0$, so that $x - x_0 < 0$ and $f'(c) \geqslant 0$. This implies that $f'(c)(x - x_0) \leqslant 0$. Therefore, (7.1) implies that $f(x) \leqslant f(x_0)$ holds for $x \in I$, $x < x_0$. If $x > x_0$, we have $x_0 < c < x$. In this case, $x - x_0 > 0$ and $f'(c) \leqslant 0$, so that again $f'(c)(x - x_0) \leqslant 0$. By (7.1), $f(x) \leqslant f(x_0)$ for $x \in I$, $x > x_0$. Thus, $f(x) \leqslant f(x_0)$ for all $x \in I$, and $f(x_0)$ is a maximum of f on I.

Theorem 7.2. *Let f be continuous on an interval I. If $f'(x) \leqslant 0$ for $x < x_0$, $x \in I$ and $f'(x) \geqslant 0$ for $x > x_0$, $x \in I$, then $f(x_0)$ is the minimum of f on I.* (*See* Fig. 7.2(b).)

PROB. 7.1. Prove the last theorem.

Remark 7.1. We refer to Theorems 7.1 and 7.2 as the *first derivative test* for an extremum. Note that in these theorems we did not assume that f was differentiable at x_0. Hence, the test can be used even if $f'(x_0)$ does not exist. If, in addition to the hypotheses in the theorems mentioned, $f'(x_0)$ exists, then we necessarily have $f'(x_0) = 0$. For, in each case, the hypothesis implies, that $f(x_0)$ is an extremum of f and since x_0 is an interior point of I, it follows from Prob. VII.5.2 that $f'(x_0) = 0$. Sometimes the first derivative test for an extremum is given the form of Prob. 7.2 below.

PROB. 7.2. Prove: If x_0 is an interior point of an interval I on which the function f is differentiable and $f'(x_0) = 0$, then (a) $f'(x) \geqslant 0$ for $x < x_0$, $x \in I$, $f'(x) \geqslant 0$ for $x > x_0$, $x \in I$ imply that $f(x_0)$ is a maximum of f on I, while (b) $f'(x) \leqslant 0$ for $x < x_0$, $x \in I$ and $f'(x) \geqslant 0$ for $x > x_0$, $x \in I$ imply that $f(x_0)$ is a minimum of f on I.

Remark 7.2. The example of f, where $f(x) = x^3$ for $x \in \mathbb{R}$, shows that $f'(x_0) = 0$ can occur without $f(x_0)$ being an extremum of f. We have, in this case, $f'(0) = 0$, yet for $\epsilon > 0$, $f(\epsilon) = \epsilon^3 > 0 = f(0)$ and $f(-\epsilon) = -\epsilon^3 < 0 = f(0)$, and $f(0)$ is not an extremum of f.

PROB. 7.3. Prove: If f is continuous on a bounded closed interval $[a, b]$ and differentiable in its interior and $f'(x) \neq 0$ for x in the interior of I, then the extrema of f occur on the "boundary of $[a, b]$; that is, at the endpoints of $[a, b]$.

Knowing whether f is concave or convex on an interval is helpful in dealing with questions about the extrema of f.

Theorem 7.3. *Let f be differentiable in the interior of an interval I and $f'(x_0) = 0$ for some interior point of I. The following hold: (a) if f is concave on I, then $f(x_0)$ is a maximum of f on I, and (b) if f is convex on I, then $f(x_0)$ is a minimum of f on I.*

PROOF. Assume that (a) f is concave on I. The graph of f then lies below its tangents at the points in the interior of I (Prob. VIII.2.10). Hence, for x in I, we have

$$f(x) \leqslant f(x_0) + f'(x_0)(x - x_0).$$

Since $f'(x_0) = 0$ by hypothesis, it follows that

$$f(x) \leqslant f(x_0) \qquad \text{for each point } x \text{ of } I. \tag{7.2}$$

This proves that $f(x_0)$ is a maximum of f on I. As for what happens if f is convex on I, we leave this case to the reader (Prob. 7.4).

PROB. 7.4. Complete the proof of the last theorem by proving part (b).

PROB. 7.5. Let f be differentiable in the interior of an interval I and $f'(x_0) = 0$ for some interior point x_0 of I. Prove: (a) If f is strictly concave on I, then $f(x_0)$ is a *strict* maximum of f on I and (b) if f is strictly convex on I, then $f(x_0)$ is a *strict* minimum of f on I.

PROB. 7.6. Let f be continuous on an interval I, twice differentiable in the interior of I and $f'(x_0) = 0$ for some interior point $x_0 \in I$. Prove: (a) if $f''(x) \leqslant 0$ for x in the interior of I, then $f(x_0)$ is a maximum of f on I and (b) if $f''(x) \geqslant 0$, for x in the interior of I, then f is a minimum of f on I. Also prove that the strict inequalities in (a) and (b) imply that $f(x_0)$ is a strict maximum in case (a) and a strict minimum in case (b).

Remark 7.3. The second derivative test described in the last problem is one for absolute extrema. This is why the condition $f''(x) \leqslant 0$ (or $f''(x) \geqslant 0$) is assumed for all x in the interior of the interval I. For local extrema it suffices to prescribe $f''(x_0) < 0$ (or $f''(x_0) > 0$) at an interior point x_0 of f where $f'(x_0) = 0$. This is seen in the next theorem.

Theorem 7.4. *If f is continuous on an interval I, differentiable in the interior of I, and $f'(x_0) = 0$ for some interior point x_0 of I, then $f''(x_0) < 0$ implies that $f(x_0)$ is a local maximum of f, while $f''(x_0) > 0$ implies that $f(x_0)$ is a local minimum of f.*

PROOF.* Since $f'(x_0) = 0$, we have, by the definition of $f''(x_0)$,

$$\lim_{x \to x_0} \frac{f'(x)}{x - x_0} = \lim_{x \to x_0} \frac{f'(x) - f'(x_0)}{x - x_0} = f''(x_0). \tag{7.3}$$

* J. Olmsted, *Advanced Calculus*, Appleton-Century-Crofts, New York, 1961.

Assume that $f''(x_0) < 0$. This and (7.3) imply that a deleted δ-neighborhood $N^*(x_0, \delta)$ of x_0 exists such that

$$\frac{f'(x)}{x - x_0} < 0 \qquad \text{for} \quad x \in N^*(x_0, \delta) \cap I. \tag{7.4}$$

From this we obtain: $f'(x) > 0$ for $x_0 - \delta < x < x_0$, $x \in I$ and $f'(x) < 0$ for $x_0 < x < x_0 + \delta$, $x \in I$. The set $J = N(x_0, \delta) \cap I$ is an interval. Apply Theorem 7.1 to J and obtain that $f(x_0)$ is a maximum of f on J and, hence, that $f(x_0)$ is a local maximum of f. The case $f''(x_0) > 0$ can be treated analogously.

Remark 7.4. If $f'(x_0) = 0 = f''(x_0)$ at some interior point x_0 of interval I, then $f(x_0)$ may or may not be an extremum of f. The function f, where $f(x) = x^3$ for $x \in \mathbb{R}$, is such that $f'(0) = 0 = f''(0)$, and here $f(0)$ is neither a maximum nor a minimum of f. On the other hand, the function g, where $g(x) = x^4$, for $x \in \mathbb{R}$ also has the property $g'(0) = 0 = g''(0)$ but $g(0)$ is a minimum of f. The next theorem is an extension of Theorem 7.4 and can be used, under the appropriate conditions, when $f'(x_0) = 0 = f''(x_0)$.

Theorem 7.5 (Extension of Theorem 7.4). *Let n be a positive integer and f a function such that $f, f', \ldots, f^{(n)}$ are continuous on an interval I. Let x_0 be an interior point of I. Let $f'(x_0) = f''(x_0) = \cdots = f^{(n)}(x_0) = 0$, but $f^{(n+1)}(x_0) \neq 0$. If n is odd (so that $n + 1$ is even), then $f(x_0)$ is a local maximum or local minimum of f according to whether $f^{(n+1)}(x_0) < 0$ or $f^{(n+1)}(x_0) > 0$. If n is even (so that $n + 1$ is odd), then $f(x_0)$ is neither a local maximum nor a local minimum of f.*

PROOF.* Suppose that $x \in I$, $x \neq x_0$. There exists a c between x and x_0 such that

$$\begin{aligned} f(x) &= f(x_0) + f'(x_0)(x - x_0) + \cdots \\ &\quad + \frac{f^{(n-1)}(x_0)}{(n-1)!}(x - x_0)^{n-1} + \frac{f^{(n)}(c)}{(n)!}(x - x_0)^n \\ &= f(x_0) + \frac{f^{(n)}(c)}{(n)!}(x - x_0)^n. \end{aligned} \tag{7.5}$$

Assume that $f^{(n+1)}(x_0) < 0$. Since $f^{(n)}(x_0) = 0$, we have

$$\lim_{x \to x_0} \frac{f^{(n)}(x)}{x - x_0} = \lim_{x \to x_0} \frac{f^{(n)}(x) - f^{(n)}(x_0)}{x - x_0} = f^{(n+1)}(x_0) < 0. \tag{7.6}$$

Hence, there exists a deleted δ-neighborhood $N^*(x_0, \delta)$ of x_0 such that

$$\frac{f^{(n)}(x)}{x - x_0} < 0 \qquad \text{for} \quad x \in N^*(x_0, \delta) \cap I. \tag{7.7}$$

*J. Olmsted, *Advanced Calculus*, Appleton-Century-Crofts, New York, 1961.

This implies that $f^{(n)}(x) > 0$ for $x_0 - \delta < x < x_0$, $x \in I$ and $f^{(n)}(x) < 0$ for $x_0 < x < x_0 + \delta$, $x \in I$. In the first case we have $x_0 - \delta < x < c < x_0$ and, therefore, $f^{(n)}(c) > 0$. In the second case we have $x_0 < c < x < x_0 + \delta$ and, therefore, $f^{(n)}(c) > 0$. Now suppose n is odd. In the first case, $x - x_0 < 0$, so that $(x - x_0)^n < 0$, yielding $f^{(n)}(c)(x - x_0)^n/n! < 0$. Using (7.5), we obtain $f(x) < f(x_0)$, if $x_0 - \delta < x < x_0$. In the second case, we have $x - x_0 > 0$, $(x - x_0)^n > 0$, so that once more, $f^{(n)}(c)(x - x_0)^n/n! < 0$. Using (7.5) we have $f(x) < f(x_0)$ if $x_0 < x < x_0 + \delta$. Thus, if n is odd, then $f(x) \leqslant f(x_0)$ for $x \in N(x_0, \delta) \cap I$ and, therefore, that $f(x_0)$ is a local maximum of f. Next assume that n is even. Then, $(x - x_0)^n > 0$ for $x \neq x_0$. We now obtain, reasoning as above, that $f^{(n)}(c)(x - x_0)^n/n! > 0$ if $x_0 - \delta < x < x_0$ so that $f(x) > f(x_0)$ by (7.5). On the other hand, if $x_0 < x < x_0 + \delta$, we now have $f^{(n)}(c)(x - x_0)^n/n! < 0$ and $f(x) < f(x_0)$. Thus, if n is odd, $f(x_0)$ is neither a local maximum nor a local minimum of f.

The case $f^{(n+1)}(x_0) > 0$ can be treated analogously.

Prob. 7.7. Let a_1, a_2, a_3 be real numbers such that $a_1 < a_2 < a_3$. (a) Locate the local maximum and minimum of f, where $f(x) = (x - a_1)(x - a_2)(x - a_3)$. (b) Also locate the inflection point of its graph. How does the graph look?

Prob. 7.8. Assume that $a_1 < a_2$. Locate the local maximum and minimum of f and g, where $f(x) = (x - a_1)(x - a_2)^2$, $x \in \mathbb{R}$, $g(x) = (x - a_1)^2(x - a_2)$, $x \in \mathbb{R}$. Also locate the points of inflection of the graphs of f and g.

Prob. 7.9. Prove: If $f(x) = x^3 + ax + b$ for $x \in \mathbb{R}$, then the inequalities $27b^2 + 4a^3 \lesseqgtr 0$ determine the number and multiplicity of the real zeros of f. State which inequality corresponds to which possibility for the zeros of f.

Prob. 7.10. Let $x > 0$, $x \neq e$. Prove: (a) $\ln x/x < 1/e$, (b)* $x^e < e^x$.

8. The Gamma Function

We return to Example 5.2 and consider f, where

$$f(x) = \ln(1 + x), \qquad x > -1. \tag{8.1}$$

We take $x > 0$. By (5.18), there exists a c such that

$$\ln(1 + x) = x - \frac{x^2}{2(1 + c)^2}, \tag{8.2}$$

*J. T. Varner, Comparing a^b and b^a using elementary calculus, *The Two-Year College Mathematics Journal*, 7(1976), p. 46.

where $0 < c < x$. We have

$$0 < \frac{x^2}{2(1+c)^2} = x - \ln(1+x). \tag{8.3}$$

Since $c > 0$, it follows that

$$\frac{x^2}{2(1+c)^2} < \frac{x^2}{2}.$$

This and (8.3) imply that

$$0 < x - \ln(1+x) < \frac{x^2}{2} \quad \text{if} \quad x > 0. \tag{8.4}$$

Replace x in this inequality by x/k, where k is a positive integer. Then

$$0 < \frac{x}{k} - \ln\left(1 + \frac{x}{k}\right) < \frac{x^2}{2k^2}, \qquad x > 0 \quad \text{and} \quad k \in \mathbb{Z}_+. \tag{8.5}$$

For each positive integer n, we take $k \in \{1, 2, \ldots, n\}$ in (8.5) and sum. We arrive at

$$0 < \left(\sum_{k=1}^{n} \frac{1}{k}\right)x - \ln(1+x)\left(x + \frac{x}{2}\right) \ldots \left(1 + \frac{x}{n}\right) < \frac{x^2}{2} \sum_{k=1}^{n} \frac{1}{k^2}. \tag{8.6}$$

We write for each n,

$$P_n = (1+x)\left(1 + \frac{x}{2}\right) \cdots \left(1 + \frac{x}{n}\right), \qquad u_n = \left(\sum_{k=1}^{n} \frac{1}{k}\right)x - \ln P_n. \tag{8.7}$$

Now (8.6) can be written as follows:

$$0 < u_n < \frac{x^2}{2} \sum_{k=1}^{n} \frac{1}{k^2}. \tag{8.8}$$

Since $\zeta(2) = \sum_{k=1}^{\infty} k^{-2}$ converges and has positive terms, (8.8) yields

$$0 < u_n < \frac{x^2}{2} \zeta(2), \tag{8.9}$$

implying that $\langle u_n \rangle$ is, for each $x > 0$, a bounded sequence. We now prove that $\langle u_n \rangle$ is monotonically increasing for each $x > 0$. Note first that

$$u_{n+1} - u_n = \frac{1}{n+1} x - \ln P_{n+1} + \ln P_n = \frac{1}{n+1} x - \ln \frac{P_{n+1}}{P_n}.$$

This implies that

$$u_{n+1} - u_n = \frac{x}{n+1} - \ln\left(1 + \frac{x}{n+1}\right), \tag{8.10}$$

where $x > 0$, $n \in \mathbb{Z}_+$. Since $\ln(1+z) < z$ for $z > 0$, this implies that $u_{n+1} - u_n > 0$ and, hence, that $\langle u_n \rangle$ is strictly monotonically increasing for $x > 0$. We already saw that $\langle u_n \rangle$ is bounded (see (8.9)). We conclude: If $x > 0$, then $\langle u_n \rangle$ converges.

We return to the second equality in (8.7) and subtract and add $x \ln n = \ln n^x$ to obtain

$$u_n = \left(\sum_{k=1}^{n} \frac{1}{k}\right)x - \ln P_n = \left(\sum_{k=1}^{n} \frac{1}{k} - \ln n\right)x + \ln n^x - \ln P_n$$

or

$$\frac{u_n}{x} = \sum_{k=1}^{n} \frac{1}{k} - \ln n + \frac{1}{x} \ln \frac{n^x}{P_n}. \tag{8.11}$$

We recall that $\langle \gamma_n \rangle$, where $\gamma_n = \sum_{k=1}^{n} k^{-1} - \ln n$ for each n, converges to the Euler–Mascheronic constant γ (see formula VII(7.17)). Thus, (8.11) becomes

$$\frac{u_n}{x} = \gamma_n + \frac{1}{x} \ln \frac{n^x}{P_n}, \tag{8.12}$$

and we conclude that from this that $\ln(n^x/P_n)$ converges as $n \to +\infty$ for $x > 0$. This implies that the sequence $\langle g_n(x) \rangle$, where

$$g_n(x) = \frac{n^x}{xP_n}, \tag{8.13}$$

converges for $x > 0$. Next, we note that

$$g_n(x) = \frac{n^x}{xP_n} = \frac{n^x}{x\prod_{k=1}^{n}(1 + x/k)} = \frac{n^x n!}{x(x+1)\cdots(x+n)} \tag{8.14}$$

for each positive integer n and $x \notin \{0, -1, -2, \ldots, -n\}$. We proved:

Lemma 8.1. *If $x > 0$, then the sequence $\langle g_n(x) \rangle$, where $g_n(x)$ is defined in* (8.14), *converges.*

Remark 8.1. In Theorem 8.1 below we shall prove that $\langle g_n(x) \rangle$ converges for all x other than the nonpositive integers, i.e., for $x \in \mathbb{R} - (\mathbb{Z}_- \cup \{0\})$.

Def. 8.1. We define $\Gamma(x)$ as

$$\Gamma(x) = \lim_{n\to+\infty} g_n(x) = \lim_{n\to+\infty} \frac{n^x n!}{x(x+1)\cdots(x+n)} \tag{8.15}$$

for all x for which the limit on the right exists and is finite. The function Γ defined in this manner is called the *Gamma Function*. Thus far we know that it is well-defined for $x > 0$.

PROB. 8.1. Prove:

$$\Gamma(k) = (k-1)!, \tag{8.16}$$

where k is a positive integer.

Theorem 8.1. (*a*) *The Gamma Function is positive and log-convex on the interval* $(0; +\infty)$. (*b*) *The domain of the Gamma Function is* $\mathfrak{D}(\Gamma) = \mathbb{R} -$

$(\mathbb{Z}_- \cup \{0\})$. (*c*) *We have for the Gamma Function*

$$\Gamma(x+1) = x\Gamma(x) \qquad \textit{for} \quad x \in \mathfrak{D}(\Gamma). \tag{8.17}$$

Proof. We first prove $\Gamma(x) > 0$ for $x > 0$. Turn to the discussion preceding Lemma 8.1 at the beginning of this section. By (8.12) and (8.13)

$$g_n(x) = \frac{n^x}{xP_n} = \frac{1}{x} e^{u_n - x\gamma_n}, \qquad x > 0.$$

Writing $u = \lim_{n\to+\infty} u_n$, we obtain from this and (8.15) that

$$\Gamma(x) = \frac{1}{x} e^{u - x\gamma} > 0 \qquad \text{for} \quad x > 0.$$

Next we prove that Γ is log-convex on $(0; +\infty)$. We have from (8.14) and Prob. VIII.7.16 that each g_n, where

$$g_n(x) = \frac{n^x n!}{x(x+1)\cdots(x+n)},$$

is log-convex on $(0; +\infty)$. By Prob. VIII.7.2, since

$$\Gamma(x) = \lim_{n\to+\infty} g_n(x) \qquad \text{for} \quad x > 0,$$

and $\Gamma(x) > 0$ for $x > 0$, we see that Γ is log-convex on $(0; +\infty)$.

We prove (c). From the definition of Γ in (8.15) we see that $\Gamma(x)$ is *not* defined if $x \in \mathbb{Z}_- \cup \{0\}$ and, hence, that $\mathfrak{D}(\Gamma) \subseteq \mathbb{R} - (\mathbb{Z}_- \cup \{0\})$. Take $x \in \mathfrak{D}(\Gamma)$, so that $x \notin \{0, -1, -2, \dots\} = \mathbb{Z}_- \cup \{0\}$. But

$$\frac{n^{x+1} n!}{(x+1)(x+2)\cdots(x+n+1)} = x \frac{n^x n!}{x(x+1)\cdots(x+n)} \frac{n}{(x+n+1)}. \tag{8.18}$$

Taking limits as $n \to +\infty$, we see that the right-hand side converges to $x\Gamma(x)$. It follows from (8.18), since the limit of its left-hand side is $\Gamma(x+1)$, that (8.17) holds. This proves (c). It also follows from this that $x \in \mathfrak{D}(\Gamma)$ implies that $x+1 \in \mathfrak{D}(\Gamma)$.

Now take $x \in \mathfrak{D}(\Gamma) - \{1\}$, so that $x \notin \mathbb{Z}_- \cup \{0, 1\}$ and the limit in (8.15) exists and is finite. Since

$$\frac{n^{x-1} n!}{(x-1)x(x+1)\cdots(x+n-1)} = \frac{1}{x-1} \frac{n^x n!}{x(x+1)\cdots(x+n)} \frac{x+n}{n} \tag{8.19}$$

for such x and the right-hand side here converges to the limit $\Gamma(x)/(x-1)$ as $n \to +\infty$, the left-hand side converges to $\Gamma(x)/(x-1)$. But the left-hand side, when it converges, is equal to $\Gamma(x-1)$. This proves that

$$\Gamma(x-1) = \frac{\Gamma(x)}{x-1} \qquad \text{if} \quad x \in \mathfrak{D}(\Gamma) - \{1\} \tag{8.20}$$

and that $x \in \mathfrak{D}(\Gamma) - \{1\}$ implies that $x - 1 \in \mathfrak{D}(\Gamma)$.

We prove that $\mathbb{R} - \{\mathbb{Z}_- \cup \{0\}\} \subseteq \mathfrak{D}(\Gamma)$. We already know that $(0; +\infty) \subseteq \mathfrak{D}(\Gamma)$ (Lemma 8.1). We prove: If n is a positive integer, then $\Gamma(x)$ is defined for $x \in (-n; -n+1)$, and

$$\Gamma(x) = \frac{\Gamma(x+n)}{x(x+1)\cdots(x+n-1)}. \tag{8.21}$$

We use induction on n. Take $n = 1$ and $x \in (-1;0) = (-1; -1+1)$ so that $-1 < x < 0$ and $0 < x+1 < 1$. By what was proved in the last paragraph, we have $x = x+1-1 \in \mathfrak{D}(\Gamma)$ and

$$\Gamma(x) = \frac{\Gamma(x+1)}{x}. \tag{8.22}$$

Thus, our statement holds for $n = 1$. Assume that our statement holds for some positive integer n. Take $-n-1 < x < -n$, so that $-n < x+1 < -n+1$ and, hence, that $x+1 \in \mathfrak{D}(\Gamma)$ and

$$\Gamma(x) = \frac{\Gamma(x+1+n)}{(x+1)(x+2)\cdots(x+n)} = \frac{\Gamma(x+n+1)}{(x+1)(x+2)\cdots(x+n)}. \tag{8.23}$$

By what was proved in the last paragraph, we have $x = x+1-1 \in \mathfrak{D}(\Gamma)$ and

$$\Gamma(x) = \frac{\Gamma(x+1)}{x}$$

and, therefore, by (8.23) that

$$\Gamma(x) = \frac{\Gamma(x+n+1)}{x(x+1)\cdots(x+n)} \qquad \text{for} \quad x \in (-n-1; -n).$$

By induction our statement holds for all positive integers n. Since $(-n; -n+1) \subseteq \mathfrak{D}(\Gamma)$ for each positive integer n and $(0; +\infty) \subseteq \mathfrak{D}(\Gamma)$, we have $\mathbb{R} - \{\mathbb{Z}_- \cup \{0\}\} \subseteq \mathfrak{D}(\Gamma)$. But $\mathfrak{D}(\Gamma) \subseteq \mathbb{R} - \{\mathbb{Z}_- \cup \{0\}\}$ holds. Hence, $\mathfrak{D}(\Gamma) = \mathbb{R} - \{\mathbb{Z}_- \cup \{0\}\}$. This completes the proof.

PROB. 8.2. Prove that (8.21) holds for $x \in \mathfrak{D}(\Gamma)$ for each positive integer n.

Corollary 1 (of Theorem 8.1). *The Gamma Function is continuous.*

PROOF. The Gamma Function is log-convex on $(0; +\infty)$ and, hence, convex there (Prob. VIII.7.4). A function which is convex on an interval I is continuous in the interior of I (Corollary of Theorem VIII.2.1). Hence, Γ is continuous for $x > 0$.

If $-n < x < -n+1$, where n is a positive integer, then $0 < x+n < 1$. The function h, where $h(x) = \Gamma(x+n)$, is continuous for $-n < x < -n+1$. By (8.21), Γ is continuous for $-n < x < -n+1$. Thus, Γ is also continuous on $(-n; -n+1)$. Hence, Γ is continuous on $\mathbb{R} - (\mathbb{Z}_- \cup \{0\})$, its domain, and is, therefore, continuous.

Corollary 2 (of Theorem 9.1). *If n is a positive integer, then $\Gamma(x)$ is positive or negative for $x \in (-n; -n+1)$ according to whether n is even or odd.*

PROOF. Assume $x \in (-n; -n+1)$ so that $0 < x+n < 1$ and $\Gamma(x+n) > 0$. In the product $x(x+1)\cdots(x+n-1)$ there are n negative factors, so this product is >0 or <0 according to whether n is even or odd. By (8.21) it follows that $\Gamma(x)$ is positive or negative according to whether n is even or odd.

Theorem 8.2. *If $x \in \mathfrak{D}(\Gamma)$ and n is a positive integer, then*

$$\Gamma(x+n) \sim n^x \Gamma(n) \qquad \text{as} \quad n \to +\infty \tag{8.24}$$

PROOF. By (8.21) and Prob. 8.2, we have

$$\frac{\Gamma(x)}{\Gamma(x+n)} = \frac{1}{x(x+1)\cdots(x+n)}(x+n).$$

Hence,

$$\frac{\Gamma(x)n^x\Gamma(n)}{\Gamma(x+n)} = \frac{\Gamma(x)n^x(n-1)!}{\Gamma(x+n)} = \frac{n^x n!}{x(x+1)\cdots(x+n)} \frac{x+n}{n}.$$

Here the right-hand side approaches $\Gamma(x)$ as $n \to +\infty$, yielding

$$\lim_{n\to+\infty} \frac{\Gamma(x)n^x\Gamma(n)}{\Gamma(x+n)} = \Gamma(x).$$

This implies that

$$\lim_{n\to+\infty} \frac{n^x\Gamma(n)}{\Gamma(x+n)} = 1$$

from which (8.24) follows.

Corollary. *We have*

$$\lim_{x\to+\infty} \Gamma(x) = +\infty. \tag{8.25}$$

PROOF. By the theorem, we know that

$$\lim_{n\to+\infty} \frac{n^x\Gamma(n)}{\Gamma(x+n)} = 1$$

Take $y > 0$. There exists a positive integer N_1 such that if $n \geqslant N_1$, then

$$-\frac{1}{2} < \frac{n^y\Gamma(n)}{\Gamma(y+n)} - 1 < \frac{1}{2},$$

which implies that

$$\tfrac{2}{3} n^y \Gamma(n) < \Gamma(n+y) \qquad \text{for} \quad n \geqslant N_1. \tag{8.26}$$

Let $N = \max\{N_1, 2\}$ so that $N \geqslant N_1$ and $N \geqslant 2$, and

$$\tfrac{2}{3}N^y\Gamma(N) < \Gamma(N+y). \tag{8.27}$$

Taking $x > N$, we have $x - N > 0$. By (8.26) this implies that

$$\tfrac{2}{3}N^{x-N} \leqslant \tfrac{2}{3}N^{x-N}\Gamma(N) < \Gamma(N + x - N) = \Gamma(x),$$

i.e., that

$$\tfrac{2}{3}N^{x-N} < \Gamma(x) \qquad \text{for} \quad x > N \geqslant 2. \tag{8.28}$$

Since

$$\lim_{x \to +\infty} N^{x-N} = +\infty \qquad \text{for} \quad N > 1,$$

it follows from this and (8.28) that $\lim_{x\to+\infty}\Gamma(x) = +\infty$, proving the theorem.

For a sketch of the Gamma Function see Fig. X.11.1.

PROB. 8.3. Prove: (a) $\Gamma(0+) = +\infty$, (b) $\Gamma(0-) = -\infty$, (c) $\Gamma((-n+1)-) = -\infty = \Gamma((-n)+)$ if n is an odd positive integer, (d) $\Gamma((-n+1)-) = +\infty = \Gamma((-n)+)$ if n is an even positive integer.

9. Log-Convexity and the Functional Equation for Γ

Equation (8.17) is called the *functional* equation for the Gamma Function. This equation does not determine a unique function. For if f satisfies

$$f(x+1) = xf(x), \tag{9.1}$$

then g, where $g(x) = f(x)\sin(2\pi x)$ also satisfies it. In fact,

$$\begin{aligned} g(x+1) &= f(x+1)\sin[2\pi(x+1)] = xf(x)\sin(2\pi x + 2\pi) \\ &= xf(x)\sin(2\pi x) = xg(x). \end{aligned}$$

However, a unique solution of (9.1) is obtained when further conditions are imposed on the solution of (9.1).

Theorem 9.1.* *If f is positive for $x > 0$, log-convex, $f(1) = 1$, and also satisfies* (9.1), *then $f(x) = \Gamma(x)$ for $x > 0$.*

PROOF. Using (9.1) and induction on n, it is easy to show that if n is a positive integer, then

$$f(n) = (n-1)!, \tag{9.2a}$$

$$f(x) = \frac{f(x+n)}{x(x+1)\cdots(x+n-1)} \qquad \text{for} \quad x > 0. \tag{9.2b}$$

This can be done for f in the same way it was done earlier for Γ.

* E. Artin, *The Gamma Function*, Holt, Rinehart, Winston, New York, 1964.

We now take x such that $0 < x \leqslant 1$ and $n \geqslant 2$, so that

$$-1 + n < n < x + n \leqslant n + 1. \tag{9.3}$$

Consider the function g, defined as

$$g(x) = \frac{\ln f(x) - \ln f(n)}{x - n}, \qquad x > 0, \quad x \neq n. \tag{9.4}$$

This function is monotonically increasing for fixed $n \geqslant 2$ since f is log-convex and therefore convex on $(0; +\infty)$. This and (9.3) imply that

$$\begin{aligned}\frac{\ln f(n-1) - \ln f(n)}{-1} &\leqslant \frac{\ln f(x+n) - \ln f(n)}{x} \\ &\leqslant \frac{\ln f(n+1) - \ln f(n)}{1}.\end{aligned}$$

By (9.2a), this implies that

$$\begin{aligned}x(\ln(n-1)! - \ln(n-2)!) &\leqslant \ln f(x+n) - \ln(n-1)! \\ &\leqslant x(\ln n! - \ln(n-1)!).\end{aligned}$$

Using properties of the natural logarithm, this implies that

$$x\ln(n-1) + \ln(n-1)! \leqslant \ln f(x+n) \leqslant x\ln n + \ln(n-1)!$$

This is equivalent to

$$\ln\left[(n-1)^x(n-1)!\right] \leqslant \ln f(x+n) \leqslant \ln\left[n^x(n-1)!\right] \tag{9.5}$$

or

$$(n-1)^x(n-1)! \leqslant f(x+n) \leqslant n^x(n-1)!. \tag{9.6}$$

Using (9.2b), this yields

$$\frac{(n-1)^x(n-1)!}{x(x+1)\cdots(x+n-1)} \leqslant f(x) \leqslant \frac{n^x(n-1)!}{x(x+1)\cdots(x+n-1)}, \tag{9.7}$$

where $n \geqslant 2$, $0 < x \leqslant 1$. Here the inequality on the left holds for all integers n such that $n - 1 \geqslant 1$, so it yields, after replacing $n - 1$ by n,

$$\frac{n^x n!}{x(x+1)\cdots(x+n)} \leqslant f(x) \qquad \text{for} \quad n \geqslant 1, \quad 0 < x \leqslant 1. \tag{9.8}$$

This, using the inequality on the right in (9.7), implies that

$$\frac{n^x n!}{x(x+1)\cdots(x+n)} \leqslant f(x) \leqslant \frac{n^x n!}{x(x+1)\cdots(x+n)} \frac{x+n}{n} \tag{9.9}$$

for $n \geqslant 2$, $0 < x \leqslant 1$. Take limits as $n \to +\infty$. Then

$$\begin{aligned}\Gamma(x) &= \lim_{n\to+\infty} \frac{n^x n!}{x(x+1)\cdots(x+n)} \\ &= \lim_{n\to+\infty}\left(\frac{n^x n!}{x(x+1)\cdots(x+n)} \frac{x+n}{n}\right).\end{aligned}$$

From this and (9.9) we conclude using the Sandwich Theorem that

$$\Gamma(x) = f(x) \qquad \text{for} \quad 0 < x \leqslant 1. \tag{9.10}$$

We now prove this equality is valid for all $x > 0$. To establish this, we note, by the induction on n, we can prove (do this) that

$$f(x+n) = \Gamma(x+n) \qquad \text{for} \quad 0 < x \leqslant 1, \quad n \in \mathbb{Z}_0. \tag{9.11}$$

We already observed that f and Γ have the same values for the positive integers. Take $x > 1$, where x is not an integer. Let $n = [x]$. We have $n < x < n + 1$, which implies that $0 < x - n < 1$. Let $y = x - n$ so that $0 < y < 1$. By (9.11), we have

$$f(x) = f(x - n + n) = f(y + n) = \Gamma(y + n) = \Gamma(x)$$

for $x > 1$ and x not an integer. In sum, $f(x) = \Gamma(x)$ for all $x > 0$, as claimed.

PROB. 9.1. Prove: If $t > 0$, then the function f, defined as $f(x) = \Gamma(tx)$ for $x > 0$, is log-convex. Also prove that g, where $g(x) = \Gamma(x + 1)$ for $x > 0$, is log-convex.

PROB. 9.2. Prove: If f is defined on $S = \mathbb{R} - (\mathbb{Z}_- \cup \{0\})$ and log-convex on $(0; +\infty)$, where $f(1) = 1$, and $f(x + 1) = xf(x)$ for each $x \in S$, then $f(x) = \Gamma(x)$ for all $x \in S$.

PROB. 9.3.* Let p be a positive integer. Write

$$a_p = p\Gamma\left(\frac{1}{p}\right)\Gamma\left(\frac{2}{p}\right)\cdots\Gamma\left(\frac{p}{p}\right),$$

$$h(x) = p^x\Gamma\left(\frac{x}{p}\right)\Gamma\left(\frac{x+1}{p}\right)\cdots\Gamma\left(\frac{x+p-1}{p}\right),$$

where $x > 0$. Prove: The function H, where

$$H(x) = \frac{h(x)}{a_p},$$

where $x > 0$, is identical with the Gamma Function. Thus, prove that

$$p^x\Gamma\left(\frac{x}{p}\right)\Gamma\left(\frac{x+1}{p}\right)\cdots\Gamma\left(\frac{x+p-1}{p}\right) = a_p\Gamma(x)$$

if p is a positive integer and $x > 0$ (Hint: prove that H satisfies the hypothesis of Theorem 9.1).

PROB. 9.4. If k and p are integers such that $1 \leqslant k \leqslant p$, we have, by

*E. Artin, *The Gamma Function*, Holt, Rinehart, Winston, New York, 1964.

definition of $\Gamma(k/p)$,

$$\Gamma\left(\frac{k}{p}\right) = \lim_{n\to+\infty} \frac{n^{k/p} n!}{(k/p)(k/p+1)\cdots(k/p+n)}$$

$$= \lim_{n\to+\infty} \frac{n^{k/p} n!\, p^{n+1}}{k(k+p)\cdots(k+np)}$$

for $k \in \{1, \ldots, p\}$. Obtain:

$$a_p = p\Gamma\left(\frac{1}{p}\right)\cdots\Gamma\left(\frac{p}{p}\right) = p \lim_{n\to+\infty} \frac{n^{(p+1)/2}(n!)^p p^{np+p}}{(np+p)!}.$$

Prob. 9.5. Note that

$$\lim_{n\to+\infty}\left(1+\frac{1}{np}\right)\left(1+\frac{2}{np}\right)\cdots\left(1+\frac{p}{np}\right) = 1$$

or

$$\lim_{n\to+\infty} \frac{(np+p)!}{(np)!\,(np)^p} = 1.$$

Multiply a_p by 1, where 1 is in the form of the last limit, and obtain

$$a_p = a_p \cdot 1 = a_p \lim_{n\to+\infty} \frac{(np+p)!}{(np)!\,(np)^p}.$$

Use this and the results cited in Prob. 9.4 to conclude that

$$a_p = p \lim_{n\to+\infty} \frac{n^{(p+1)/2}(n!)^p p^{np+p}}{(np+p)!} \cdot \lim_{n\to+\infty} \frac{(np+p)!}{(np)!\,(np)^p}.$$

Prove:

$$a_p = p \lim_{n\to+\infty} \frac{(n!)^p p^{np}}{(np)!\, n^{(p-1)/2}}.$$

Prob. 9.6. Prove: (a) If $x > 0$, then

$$\Gamma(x) = \frac{2^{x-1}\Gamma(x/2)\Gamma((x+1)/2)}{\Gamma(1/2)}.$$

(See Prob. 9.3.)

Prob. 9.7.* Prove:

$$\lim_{n\to+\infty} \frac{x(x+1)(x+2)\cdots(x+2n-1)}{1\cdot 3\cdot 5\cdots(2n-1)(2x)(2x+2)\cdots(2x+2n-2)} = 2^{x-1}.$$

* Bromwich, *Infinite Series*, 2nd ed., MacMillan, New York, 1942, p. 115.

CHAPTER X

The Complex Numbers. Trigonometric Sums. Infinite Products

1. Introduction

In order to solve the equation

$$ax^2 + bx + c = 0, \tag{1.1}$$

where a, b, c are real numbers and $a \neq 0$, for $x \in \mathbb{R}$, we use the identity

$$ax^2 + bx + c = a\left[\left(x + \frac{b}{2a}\right)^2 + \frac{4ac - b^2}{4a^2}\right], \tag{1.2}$$

obtained by "completing the square." A real number x satisfying (1.1) must satisfy

$$\left(x + \frac{b}{2a}\right)^2 = \frac{b^2 - 4ac}{4a^2}. \tag{1.3}$$

This implies that

$$b^2 - 4ac \geqslant 0. \tag{1.4}$$

When $b^2 - 4ac < 0$, no $x \in \mathbb{R}$ exists satisfying (1.1). To obtain a system of numbers in which (1.1) can be solved we extend the system $\mathbb{R}$ to the complex number system. In this number system there will be a z such that $z^2 = -p$, where $p > 0$.

2. The Complex Number System

Def. 2.1. A *complex number* is an ordered pair (α, β) of real numbers α and β. The set of complex numbers will be denoted by $\mathbb{C}$.

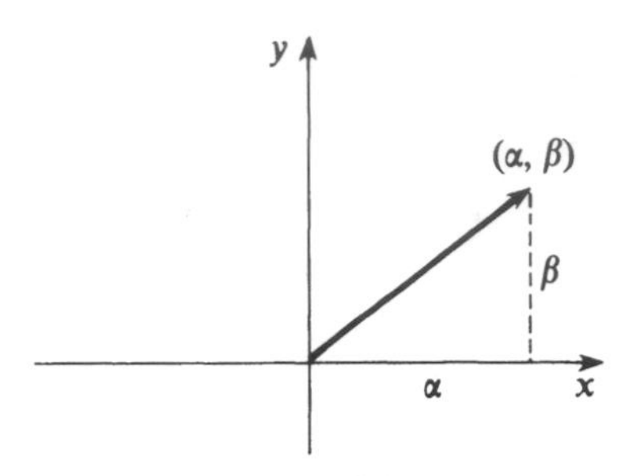

Figure 2.1

Complex numbers may be visualized by taking a plane p together with a rectangular coordinate system in it. In this way there is assigned to each point P in p an ordered pair (α, β) of real numbers α and β called the *coordinates* of P and conversely each ordered pair (α, β) of real numbers determines a unique point P having α and β as coordinates (see Fig. 2.1). We shall often speak of the point $a = (\alpha, \beta)$ rather than use the more precise "the point P represented by $a = (\alpha, \beta)$." It is also convenient to view $a = (\alpha, \beta)$ as the *arrow* or *vector* from the origin $0 = (0,0)$ to $a = (\alpha, \beta)$.

By properties of ordered pairs we have: If $a_1 = (\alpha_1, \beta_1)$ and $a_2 = (\alpha_2, \beta_2)$ are complex numbers, then

$$a_1 = a_2 \quad \text{if and only if} \quad \alpha_1 = \alpha_2 \quad \text{and} \quad \beta_1 = \beta_2. \tag{2.1}$$

PROB. 2.1. Prove: $(\alpha, \beta) \neq (0,0)$ if and only if $\alpha^2 + \beta^2 > 0$.

We define addition of complex numbers.

Def. 2.2. If $a_1 = (\alpha_1, \beta_1)$ and $a_2 = (\alpha_2, \beta_2)$ are complex numbers, we define the *sum* $a_1 + a_2$ as

$$a_1 + a_2 = (\alpha_1, \beta_1) + (\alpha_2, \beta_2) = (\alpha_1 + \alpha_2, \beta_1 + \beta_2) \tag{2.2}$$

(see Fig. 2.1′).

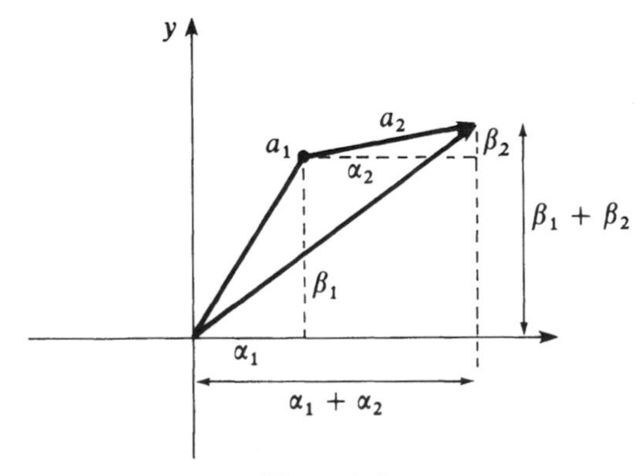

Figure 2.1′

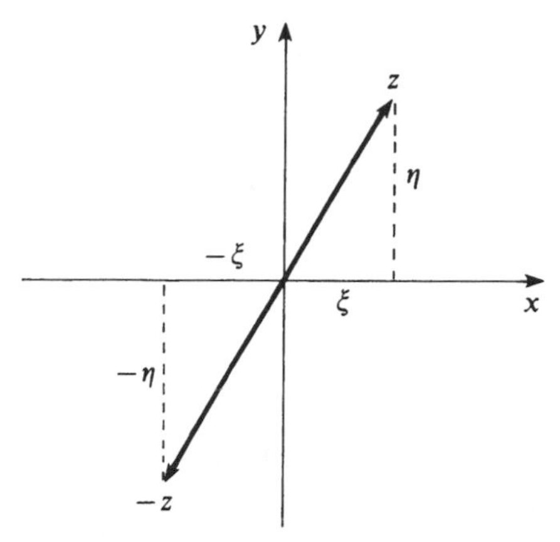

Figure 2.2

For example, let $a_1 = (-1, 3)$ and $a_2 = (-2, -4)$, then $a_1 + a_2 = (-1, 3) + (-2, -4) = (-3, -1)$.

Since for any complex number $z = (\xi, \eta)$, we have $z + (0, 0) = (\xi, \eta) + (0, 0) = (\xi + 0, \eta + 0) = (\xi, \eta) = z$, the complex number $(0, 0)$ is the *additive identity* in $\mathbb{C}$.

Def. 2.3. The *negative* or *additive inverse* of $z = (\xi, \eta)$ is defined as $-z = -(\xi, \eta) = (-\xi, -\eta)$ (see Fig. 2.2).

For example, $-(-2, 3) = (2, -3)$. In particular,

$$-(1, 0) = (-1, 0) \quad \text{and} \quad -(0, 0) = (-0, -0) = (0, 0). \tag{2.3}$$

PROB. 2.2. Prove: If $z \in \mathbb{C}$, then $z + (-z) = (0, 0)$.

PROB. 2.3. Prove: If a_1, a_2, and a_3 are complex numbers, then

(1) $(a_1 + a_2) + a_3 = a_1 + (a_2 + a_3)$ and
(2) $a_1 + a_2 = a_2 + a_1$.

PROB. 2.4. Prove: If $z \in \mathbb{C}$, then $-(-z) = z$.

Def. 2.4. If a_1 and a_2 are complex numbers, then define $a_1 - a_2$ by means of the equality $a_1 - a_2 = a_1 + (-a_2)$.

PROB. 2.5. Prove: If $a_1 = (\alpha_1, \beta_1)$ and $a_2 = (\alpha_2, \beta_2)$, then $a_1 - a_2 = (\alpha_1 - \alpha_2, \beta_1 - \beta_2)$.

PROB. 2.6. Prove: If a_1 and a_2 are complex numbers, then (1) $a_2 + (a_1 - a_2) = a_1$ and (2) $-(a_1 - a_2) = a_2 - a_1$.

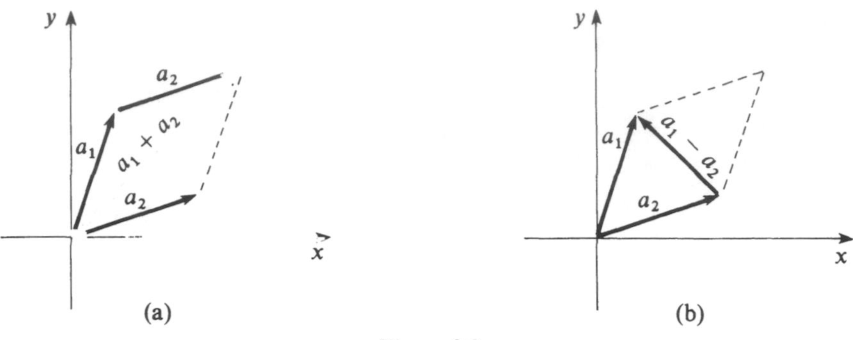

Figure 2.3

In Figs. 2.3(a) and 2.3(b) we give graphic representation of $a_1 + a_2$ and $a_1 - a_2$, respectively. $a_1 - a_2$ is the z such that $a_2 + z = a_1$.

Next, we define the product a_1a_2 of complex numbers a_1 and a_2.

Def. 2.5. If $a_1 = (\alpha_1, \beta_1)$ and $a_2 = (\alpha_2, \beta_2)$ are complex numbers, then we define their *product* a_1a_2 as

$$a_1a_2 = (\alpha_1, \beta_1)(\alpha_2, \beta_2) = (\alpha_1\alpha_2 - \beta_1\beta_2, \alpha_1\beta_2 + \alpha_2\beta_1). \tag{2.4}$$

For example, let $a_1 = (2, 3)$ and $a_2 = (4, 5)$. Using the definition of a_1a_2 we see that

$$a_1a_2 = (2,3)(4,5) = (2 \cdot 4 - 3 \cdot 5, 2 \cdot 5 + 3 \cdot 4) = (-7, 22).$$

We see easily that $(1, 0)$ is the *multiplicative identity* in $\mathbb{C}$. In fact, if $z = (\xi, \eta) \in \mathbb{C}$, then

$$z(1,0) = (\xi,\eta)(1,0) = (\xi \cdot 1 - \eta \cdot 0, \xi \cdot 0 + \eta \cdot 1) = (\xi, \eta) = z. \tag{2.5}$$

PROB. 2.7. Prove: If a_1, a_2 and a_3 are complex numbers, then

(1) $(a_1a_2)a_3 = a_1(a_2a_3)$,
(2) $a_1a_2 = a_2a_1$,
(3) $a_1(a_2 + a_3) = a_1a_2 + a_1a_3$.

PROB. 2.8. Prove: If one of the complex numbers a or b is 0, then $ab = 0$.

PROB. 2.9. Prove: If a, b, and c are complex numbers, then

(1) $a(-b) = -(ab)$,
(2) $(-a)(-b) = ab$,
(3) $a(b - c) = ab - ac$.

Theorem 2.1. *If $a \in \mathbb{C}$, $a \neq (0,0)$, then there exists a $z \in \mathbb{C}$ such that $az = (1,0)$.*

Proof. Let $a = (\alpha, \beta)$. Since $(\alpha, \beta) = a \neq (0,0)$, we have $\alpha^2 + \beta^2 > 0$ (Prob. 2.1). Take z as

$$z = \left(\frac{\alpha}{\alpha^2 + \beta^2}, \frac{-\beta}{\alpha^2 + \beta^2} \right). \tag{2.6}$$

We have

$$\begin{aligned} az &= (\alpha, \beta)\left(\frac{\alpha}{\alpha^2 + \beta^2}, \frac{-\beta}{\alpha^2 + \beta^2} \right) \\ &= \left(\alpha \frac{\alpha}{\alpha^2 + \beta^2} - \beta \frac{-\beta}{\alpha^2 + \beta^2}, \alpha \frac{-\beta}{\alpha^2 + \beta^2} + \beta \frac{\alpha}{\alpha^2 + \beta^2} \right) \\ &= \left(\frac{\alpha^2 + \beta^2}{\alpha^2 + \beta^2}, 0 \right) = (1,0), \end{aligned}$$

as claimed.

Def. 2.6. If $a = (\alpha, \beta) \neq (0,0)$, then the z defined as in (2.6) is written as a^{-1}, so that

$$aa^{-1} = (1,0) = a^{-1}a. \tag{2.7}$$

a^{-1} is called the *reciprocal* or *multiplicative inverse* of a.

Prob. 2.10. Prove: If a, b, and c are complex numbers, $a \neq 0$, then $ab = ac$ implies $b = c$.

Def. 2.7. If a and b are complex numbers, $a \neq 0$, then we define

$$\frac{b}{a} = ba^{-1}. \tag{2.8}$$

Clearly,

$$\frac{(1,0)}{a} = a^{-1}. \tag{2.9}$$

Prob. 2.11. Prove: If a and b are complex numbers, $a \neq 0$, then $a(b/a) = b$.

Prob. 2.12. Prove: If a and b are complex numbers, where $a \neq (0,0)$, then

$$\frac{a}{b} = a \frac{(1,0)}{b}.$$

We next examine the subset $\mathbb{C}_{\mathbb{R}}$ of $\mathbb{C}$ where

$$\mathbb{C}_{\mathbb{R}} = \{(\alpha, 0) \mid \alpha \notin \mathbb{R}\}. \tag{2.10}$$

PROB. 2.12$'$. Prove: If $(\alpha, 0) \in \mathbb{C}_{\mathbb{R}}$ and $(\beta, 0) \in \mathbb{C}_{\mathbb{R}}$, then

(a) $(\alpha, 0) + (\beta, 0) = (\alpha + \beta, 0)$,
(b) $(\alpha, 0)(\beta, 0) = (\alpha\beta, 0)$,
(c) $(\alpha, 0) - (\beta, 0) = (\alpha - \beta, 0)$,
(d) $(\alpha, 0)/(\beta, 0) = (\alpha/\beta, 0)$ if $\beta \neq 0$.

We define the "less than" relation in $\mathbb{C}_{\mathbb{R}}$ as follows:

$$(\alpha, 0) < (\beta, 0) \qquad \text{if and only if} \quad \alpha < \beta. \tag{2.11}$$

When $(\alpha, 0) < (\beta, 0)$, we write $(\beta, 0) > (\alpha, 0)$.

Remark 2.1. The subset $\mathbb{C}_{\mathbb{R}}$ of $\mathbb{C}$ defined in (2.10), under the operations of addition and multiplication in $\mathbb{C}$ and the ordering relation just defined, is easily seen to be a complete ordered field.

We next define the mapping $f: \mathbb{R} \to \mathbb{C}_{\mathbb{R}}$ by means of

$$f(\alpha) = (\alpha, 0) \qquad \text{for each} \quad \alpha \in \mathbb{R}. \tag{2.12}$$

One sees readily that this f is a one-to-one correspondence between $\mathbb{R}$ and $\mathbb{C}_{\mathbb{R}}$ such that

(1) $f(\alpha + \beta) = f(\alpha) + f(\beta)$,
(2) $f(\alpha\beta) = f(\alpha)f(\beta)$,
(3) $f(\alpha - \beta) = f(\alpha) - f(\beta)$,
(4) $f(\alpha/\beta) = f(\alpha)/f(\beta)$ if $\beta \neq 0$,

and that

$$\alpha < \beta \qquad \text{if and only if} \quad f(\alpha) < f(\beta).$$

It follows that the mapping f enables us to obtain for each theorem in $\mathbb{R}$ a duplicate theorem in $\mathbb{C}_{\mathbb{R}}$. $\mathbb{C}_{\mathbb{R}}$ is, thus, seen to be a copy of $\mathbb{R}$. Each $(\alpha, 0)$ in $\mathbb{C}_{\mathbb{R}}$ can be viewed as simply another notation for α in $\mathbb{R}$. f is called an *isomorphism* between $\mathbb{R}$ and $\mathbb{C}_{\mathbb{R}}$. $\mathbb{R}$ and $\mathbb{C}$ are called *isomorphic* ordered fields.

Agreement 1. If $\alpha \in \mathbb{R}$, then we write $\alpha = (\alpha, 0)$. We discard the system $\mathbb{R}$ of real numbers, replace it with the system $\mathbb{C}_{\mathbb{R}}$, and reserve the term "real number" for an element of $\mathbb{C}_{\mathbb{R}}$. We write $\mathbb{R} = \mathbb{C}_{\mathbb{R}}$. As examples we have

$$1 = (1, 0), \qquad 0 = (0, 0), \qquad -1 = (-1, 0). \tag{2.13}$$

Note, that if $\lambda \in \mathbb{R}$ and $(\alpha, \beta) \in \mathbb{C}$, then

$$\lambda(\alpha, \beta) = (\lambda\alpha, \lambda\beta). \tag{2.14}$$

This follows from

$$\lambda(\alpha, \beta) = (\lambda, 0)(\alpha, \beta) = (\lambda\alpha - 0\beta, \lambda\beta + 0\alpha) = (\lambda\alpha, \lambda\beta).$$

PROB. 2.13. Prove: If λ and μ are elements of $\mathbb{R}$ $(=\mathbb{C}_{\mathbb{R}})$ and $z \in \mathbb{C}$, $w \in \mathbb{C}$, then

(a) $(\lambda\mu)z = \lambda(\mu z)$,
(b) $\lambda(z + w) = \lambda z + \lambda w$,
(c) $(\lambda + \mu)w = \lambda w + \mu w$,
(d) $1z = z$.

Finally, we write $i = (0, 1)$. Then

$$i^2 = (0,1)(0,1) = (0 \cdot 0 - 1 \cdot 1, 0 \cdot 1 + 1 \cdot 0) = (-1,0) = -1$$

so that

$$i^2 = -1. \tag{2.15}$$

We also have

$$(\alpha, \beta) = \alpha + \beta i \qquad \text{for each } (\alpha, \beta) \text{ in } \mathbb{C}. \tag{2.16}$$

This follows from

$$(\alpha, \beta) = (\alpha, 0) + (0, \beta) = \alpha(1,0) + \beta(0,1) = \alpha 1 + \beta i = \alpha + \beta i.$$

Agreement 2. We agree to write each (α, β) in $\mathbb{C}$ as $\alpha + \beta i$. In this notation

$$\mathbb{C} = \{\alpha + \beta i \mid \alpha \in \mathbb{R} \text{ and } \beta \in \mathbb{R}\}. \tag{2.17}$$

Remark 2.2. The procedure just described is called an *embedding* of the real numbers in the system $\mathbb{C}$ of complex numbers. The complex number $a = \alpha + \beta i$ is called *imaginary* if $\beta \neq 0$ and *real* if $\beta = 0$. If $\alpha = 0$, then a is called *pure* imaginary. Also, α is called the *real part* of a and β its *imaginary* part. If a plane p is equipped with a rectangular coordinate system and the ordered pair (α, β) associated with a point P is written as $\alpha + \beta i$, then the latter is called the *complex coordinate* of P. The x-axis is then called the *real axis* and y-axis the *imaginary axis* (Fig. 2.4).

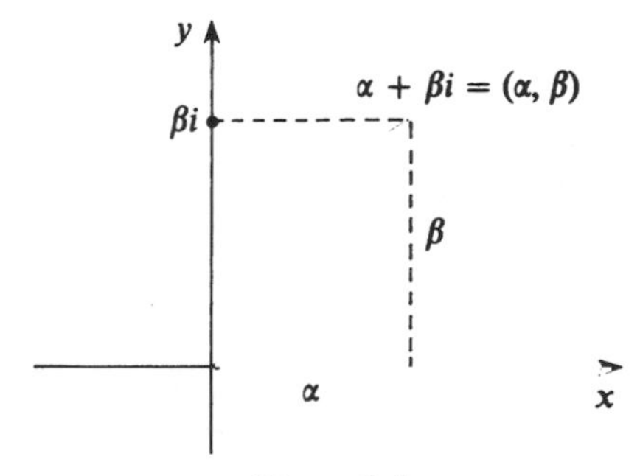

Figure 2.4

PROB. 2.14. Prove: If $\alpha_1, \alpha_2, \beta_1, \beta_2$ are real numbers, then

(1) $(\alpha_1 + \beta_1 i) \pm (\alpha_2 + \beta_2 i) = (\alpha_1 \pm \alpha_2) + (\beta_1 \pm \beta_2)i$,
(2) $(\alpha_1 + \beta_1 i)(\alpha_2 + \beta_2 i) = (\alpha_1\alpha_2 - \beta_1\beta_2) + (\alpha_1\beta_2 + \alpha_2\beta_1)i$,
(3) $\alpha_1 + \beta_1 i = \alpha_2 + \beta_2 i$ if and only if $\alpha_1 = \alpha_2$ and $\beta_1 = \beta_2$.

The product $(\alpha_1 + \beta_1 i)(\alpha_2 + \beta_2 i)$ can be obtained by "multiplying out" and using the relation $i^2 = -1$. Thus,

$$\begin{aligned}(\alpha_1 + \beta_1 i)(\alpha_2 + \beta_2 i) &= (\alpha_1 + \beta_1 i)\alpha_2 + (\alpha_1 + \beta_1 i)\beta_2 i \\ &= \alpha_1\alpha_2 + \alpha_2\beta_1 i + \alpha_1\beta_2 i - \beta_1\beta_2 \\ &= \alpha_1\alpha_2 - \beta_1\beta_2 + (\alpha_1\beta_2 + \alpha_2\beta_1)i.\end{aligned}$$

Def. 2.8. If $a = \alpha + \beta i \in \mathbb{C}$, we define the *modulus* $|a|$ of a as follows:

$$|a| = |\alpha + \beta i| = \sqrt{\alpha^2 + \beta^2}\,. \tag{2.18}$$

For example, $|3 + 4i| = \sqrt{3^2 + 4^2} = 5$. Note that

$$|i| = |0 + 1i| = \sqrt{0^2 + 1^2} = 1. \tag{2.19}$$

Remark 2.3. If $a = \alpha$, where α is real, then $|a| = \sqrt{\alpha^2 + 0^2} = \sqrt{\alpha^2} = |\alpha|$. Thus, the modulus of a real number agrees with its absolute value. The modulus of a complex number is, therefore, an extension of the notion of the absolute value of a real number to the complex numbers. We also refer to the modulus of a complex number as its *absolute value*.

Def. 2.9. We define the *conjugate* $\bar{a}$ of $a = \alpha + \beta i$ as

$$\bar{a} = \overline{\alpha + \beta i} = \alpha - \beta i. \tag{2.20}$$

For example, $\overline{2 + 3i} = 2 - 3i$. In particular, we have

$$\bar{i} = -i. \tag{2.21}$$

We prove that

$$a\bar{a} = |a|^2 \qquad \text{for} \quad a \in \mathbb{C}. \tag{2.22}$$

Assume that $a = \alpha + \beta i$. Then

$$a\bar{a} = (\alpha + \beta i)(\alpha - \beta i) = \alpha^2 - (-\beta^2) + 0i = \alpha^2 + \beta^2 = |\alpha + \beta i|^2 = |a|^2.$$

It follows that

$$|a| = \sqrt{a\bar{a}} \qquad \text{for} \quad a \in \mathbb{C}. \tag{2.23}$$

PROB. 2.15. Prove: If $a \in \mathbb{C}$, then (1) $|a| \geqslant 0$ and (2) $|a| = 0$ if and only if $a = 0$. Thus, prove: $|a| > 0$ if and only if $a \neq 0$.

Prob. 2.16. Prove: If $a \in \mathbb{C}$ and $b \in \mathbb{C}$, then

(1) $\overline{a+b} = \bar{a} + \bar{b}$,
(2) $\overline{ab} = \bar{a}\bar{b}$,
(3) $\overline{(a/b)} = \bar{a}/\bar{b}$ if $b \neq 0$.

Prob. 2.17. Prove: If $a = \alpha + \beta i$, then

$$\alpha = \frac{a + \bar{a}}{2} \quad \text{and} \quad \beta = \frac{a - \bar{a}}{2i}.$$

Prob. 2.18. Prove: (1) a is real if and only if $\bar{a} = a$ and (2) $\bar{a} = -a$ if and only if a is pure imaginary.

Remark 2.4. The reciprocal of $a \neq 0$ is a number z such that $az = 1$. Multiplying both sides by $\bar{a}$, we have $a\bar{a}z = \bar{a}$. Since $a\bar{a} > 0$, we obtain

$$a^{-1} = z = \frac{\bar{a}}{a\bar{a}} = \frac{\bar{a}}{|a|^2}.$$

This agrees with (2.5). For, writing $a = (\alpha, \beta) \neq 0$, we have

$$a^{-1} = \frac{\bar{a}}{|a|^2} = \frac{\alpha - \beta i}{\alpha^2 + \beta^2} = \frac{\alpha}{\alpha^2 + \beta^2} - \frac{\beta}{\alpha^2 + \beta^2} i = \left(\frac{\alpha}{\alpha^2 + \beta^2}, \frac{-\beta}{\alpha^2 + \beta^2} \right).$$

As an example, then

$$\frac{1}{i} = \frac{\bar{i}}{i\bar{i}} = \frac{-i}{1} = -i. \tag{2.24}$$

Again, if $a \neq 0$, then

$$\frac{b}{a} = b\frac{1}{a} = b\frac{\bar{a}}{|a|^2} = \frac{b\bar{a}}{|a|^2}. \tag{2.25}$$

The conjugate $\bar{a}$ of a is symmetric to a with respect to the x-axis (Fig. 2.5).

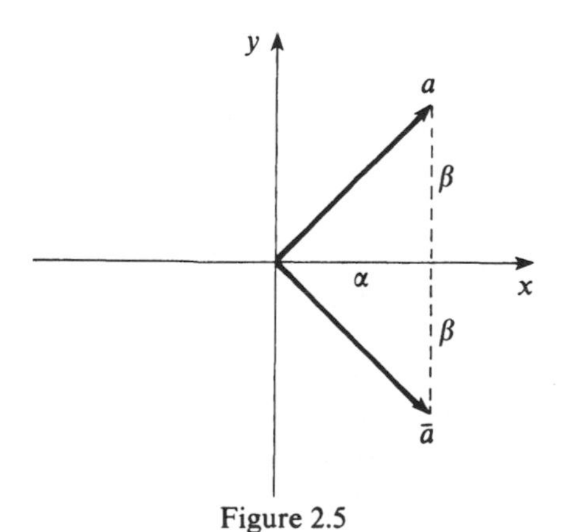

Figure 2.5

PROB. 2.19. Prove: $(a+\bar{a})^2 \geqslant 0$ and $(a-\bar{a})^2 \leqslant 0$ (see Prob. 2.17).

If α and β are real and $a = \alpha + \beta i$, then we write

$$\alpha = \operatorname{Re} a \quad \text{and} \quad \beta = \operatorname{Im} a \tag{2.26}$$

for the real and imaginary parts of a, respectively. We have (Prob. 2.17)

$$\operatorname{Re} a = \frac{a+\bar{a}}{2} \quad \text{and} \quad \operatorname{Im} a = \frac{a-\bar{a}}{2i}. \tag{2.27}$$

PROB. 2.20. Prove: If a_1 and a_2 are complex numbers, then $a_1\bar{a}_2 + \bar{a}_1 a_2$ is real and $a_1\bar{a}_2 - \bar{a}_1 a_2$ is pure imaginary.

Lemma 2.1. *If a_1 and a_2 are complex numbers, then*

$$|a_1\bar{a}_2 + \bar{a}_1 a_2| \leqslant 2|a_1||a_2|. \tag{2.28}$$

Equality holds if and only if $a_1\bar{a}_2 - \bar{a}_1 a_2 = 0$.

PROOF. Observe that $|a_1|^2|a_2|^2 = a_1\bar{a}_1 a_2\bar{a}_2 = (a_1a_2)(\bar{a}_1\bar{a}_2)$. Hence,

$$(a_1\bar{a}_2 + \bar{a}_1 a_2)^2 - 4|a_1|^2|a_2|^2 = (a_1\bar{a}_2 - \bar{a}_1 a_2)^2. \tag{2.29}$$

By Prob. 2.20, $a_1\bar{a}_2 + \bar{a}_1 a_2$ is real and $a_1\bar{a}_2 - \bar{a}_1 a_2$ is pure imaginary. Therefore $(a_1\bar{a}_2 - \bar{a}_1 a_2)^2 \leqslant 0$ and

$$(a_1\bar{a}_2 + \bar{a}_1 a_2)^2 \leqslant 4|a_1|^2|a_2|^2. \tag{2.30}$$

This implies (2.28).

Examination of (2.29) shows that its left-hand side is equal to 0 if and only if $a_1\bar{a}_2 - \bar{a}_1 a_2 = 0$. It follows that

$$(a_1\bar{a}_2 + \bar{a}_1 a_2)^2 = 4|a_1|^2|a_2|^2 \qquad \text{if and only if} \quad a_1\bar{a}_2 - \bar{a}_1 a_2 = 0.$$

Both sides are real and nonnegative. Taking square roots, we conclude that

$$|a_1\bar{a}_2 + \bar{a}_1 a_2| = 2|a_1||a_2| \qquad \text{if and only if} \quad a_1\bar{a}_2 - \bar{a}_1 a_2 = 0.$$

This completes the proof.

Theorem 2.2. *If a_1 and a_2 are complex numbers, then*

(1) $|-a_1| = |a_1|$ *and* $|\bar{a}_1| = |a_1|$,
(2) $|a_1a_2| = |a_1||a_2|$,
(3) $|a_1^2| = |a_1|^2$,
(4) *if* $a_2 \neq 0$, *then* $|a_1/a_2| = |a_1|/|a_2|$,
(5) $|a_1 + a_2| \leqslant |a_1| + |a_2|$.

PROOF. We ask the reader to prove parts (1) through (4), (Prob. 2.21) and we prove only part (5) here. We have

$$|a_1 + a_2|^2 = (a_1 + a_2)(\bar{a}_1 + \bar{a}_2) = |a_1|^2 + |a_2|^2 + a_1\bar{a}_2 + \bar{a}_1 a_2.$$

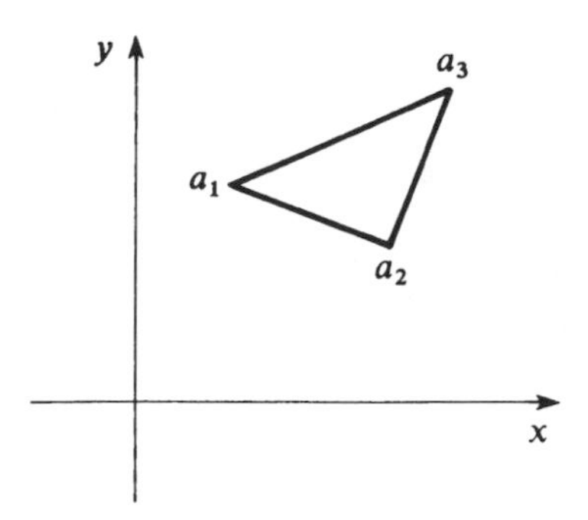

Figure 2.6

By Lemma 2.1, $a_1\bar{a}_2 + \bar{a}_1 a_2 \leqslant 2|a_1||a_2|$. This and the above imply that

$$|a_1 + a_2|^2 \leqslant |a_1|^2 + |a_2|^2 + 2|a_1||a_2| = (|a_1| + |a_2|)^2.$$

Now (1) follows after taking square roots of both sides.

Prob. 2.21. Complete the proof of the last theorem by proving parts (1), (2), (3), and (4).

Prob. 2.22. Prove: If a_1, a_2, and a_3 are complex numbers, then (1) $|a_1 - a_2| \geqslant 0$; (2) $|a_1 - a_2| = 0$ if and only if $a_1 = a_2$; (3) $|a_1 - a_2| = |a_2 - a_1|$; (4) $|a_3 - a_1| \leqslant |a_2 - a_1| + |a_3 - a_2|$; (5) $||a_3 - a_1| - |a_2 - a_1|| \leqslant |a_3 - a_2|$.

Def. 2.10. The distance $d(a_1, a_2)$ between complex numbers a_1 and a_2 is defined as

$$d(a_1, a_2) = |a_1 - a_2|. \tag{2.31}$$

Remark 2.5. By this definition and Prob. 2.22 we have: If a_1, a_2, and a_3 are complex numbers, then (1) $d(a_1, a_3) \geqslant 0$; (2) $d(a_1, a_2) = 0$ if and only if $a_1 = a_2$; (3) $d(a_1, a_2) = d(a_2, a_1)$; (4) $d(a_1, a_3) \leqslant d(a_1, a_2) + d(a_2, a_3)$. The last inequality is known as the *triangle inequality* (see Fig. 2.6).

3. Polar Form of a Complex Number

We saw that if a and b are complex numbers, then $|ba| = |b||a|$. In particular, we have: If $b = \lambda$, where λ is real and $\lambda \geqslant 0$, then $|\lambda a| = |\lambda||a| = \lambda|a|$. From this we obtain: If $z \in \mathbb{C}$, $z \neq 0$, then the u such that

$$u = \frac{z}{|z|} \tag{3.1}$$

has modulus 1, that is, $|u| = 1$. This is seen from

$$|u| = \left|\frac{z}{|z|}\right| = \left|\frac{1}{|z|}z\right| = \left|\frac{1}{|z|}\right||z| = \frac{1}{|z|}|z| = 1.$$

Def. 3.1. A complex number whose modulus is 1 is called a *direction*. It is also referred to as a *unit vector*.

If $z \neq 0$, then the u given by

$$u = \frac{z}{|z|}$$

is called the *direction of* z. We assign no direction to $z = 0$.

PROB. 3.1. Prove: If u is a direction, then so are $-u$, $\bar{u}$, and u^{-1}.

PROB. 3.2. Prove: If u and v are directions, then so are uv and uv^{-1}.

Def. 3.2. If u is a direction, then we refer to u and $-u$ as *opposite* directions, each being called the direction *opposite* to the other (see Fig. 3.1).

As examples of directions we have

$$u_1 = \frac{1}{\sqrt{2}} + \frac{1}{\sqrt{2}}i, \qquad u_2 = -\frac{1}{2} + \frac{\sqrt{3}}{2}i.$$

Note also that $-1, 1, i$, and $-i$ are directions.

PROB. 3.3. Prove: If $z \neq 0$, then z and $-z$ have opposite directions.

By Def. 3.1 each $z \neq 0$ has a unique direction and

$$z = |z|u, \tag{3.2}$$

where u is the direction of z.

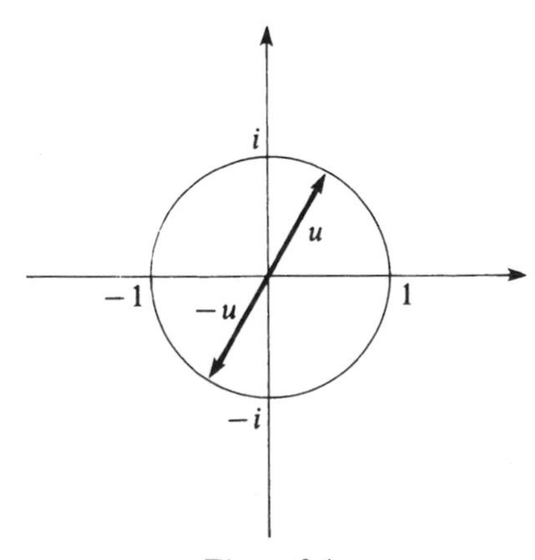

Figure 3.1

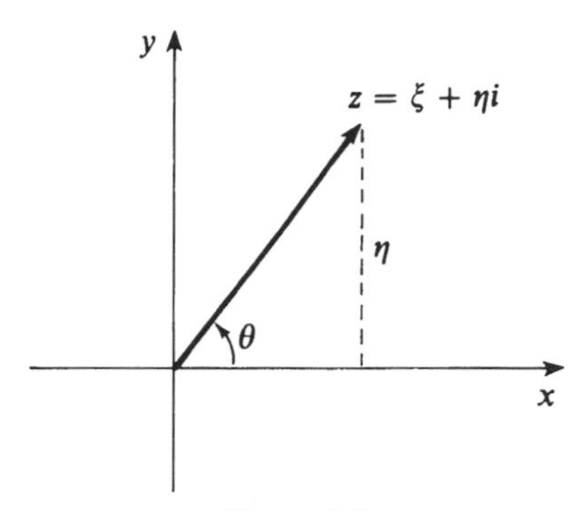

Figure 3.2

PROB. 3.4. Prove: If z and w are nonzero complex numbers, then $z = w$ if and only if z and w have the same modulus and direction.

Let $u = \alpha + \beta i$ be the direction of $z \neq 0$. By (3.2)

$$z = |z|u = |z|(\alpha + \beta i), \tag{3.3}$$

where $\alpha^2 + \beta^2 = |u|^2 = 1$. By Prob. VIII.4.1 there exists exactly one θ such that

$$\alpha = \cos\theta \quad \text{and} \quad \beta = \sin\theta, \tag{3.4}$$

where $-\pi < \theta \leqslant \pi$ so that

$$z = |z|(\cos\theta + i\sin\theta), \tag{3.5}$$

where $-\pi < \theta \leqslant \pi$. We visualize this in Fig. 3.2, where $z = \xi + \eta i = |z|(\cos\theta + i\sin\theta)$. The θ in (3.5) is called the *principal argument* of z and is written as $\operatorname{Arg} z$. Clearly,

$$-\pi < \operatorname{Arg} z \leqslant \pi. \tag{3.6}$$

For each integer n, write $\theta_n = \theta + 2n\pi$, where $\theta = \operatorname{Arg} z$. We have

$$|z|(\cos\theta_n + i\sin\theta_n) = |z|(\cos\theta + i\sin\theta) = z.$$

Each θ_n is called *an argument* of z and is written as $\theta_n = \arg_n z$. Note that $\arg_0 z = \operatorname{Arg} z$. In $z = |z|(\cos\theta + i\sin\theta)$, where θ is an argument of z, we call the right-hand side the *polar form* of z. $r = |z|$ and θ are called *polar coordinates* of z. By $\arg z$ (as distinguished from $\operatorname{Arg} z$) we mean some argument of z. The polar coordinates of $z = 0$ are defined as $r = 0$ and θ, where θ is any real number.

For example, let $z = 1 + i$. Then

$$z = 1 + i = \sqrt{2}\left(\frac{1}{\sqrt{2}} + \frac{1}{\sqrt{2}}i\right).$$

Since $\cos(\pi/4) = 1/\sqrt{2} = \sin(\pi/4)$, we have

$$z = \sqrt{2}\left(\cos\frac{\pi}{4} + i\sin\frac{\pi}{4}\right).$$

Also

$$i = \cos 0 + i \sin 0, \qquad -1 = \cos \pi + i \sin \pi$$
$$i = \cos \frac{\pi}{2} + i \sin \frac{\pi}{2}, \qquad -i = \cos\left(-\frac{\pi}{2}\right) + i \sin\left(-\frac{\pi}{2}\right). \tag{3.7}$$

PROB. 3.5. Find polar forms of: (1) $1-i$, (2) $-1-i$, (3) $-1+i$, (4) $1+\sqrt{3}\,i$, $-1+\sqrt{3}\,i$, $-1-\sqrt{3}\,i$, (5) $1-\sqrt{3}\,i$.

Theorem 3.1. *Nonzero complex numbers z and z' are equal if and only if they have the same modulus and their arguments differ by an integral multiple of 2π. If principal arguments are used, then we have: z and z' are equal if and only if they have the same modulus and their principal arguments are equal.*

PROOF. Write z and z' in polar form

$$z = |z|(\cos\theta + i \sin\theta), \qquad z' = |z'|(\cos\theta' + i\sin\theta').$$

If $|z| = |z'|$ and $\theta' = \theta + 2n\pi$, n an integer, then it is clear that $z = z'$.

Conversely, assume $z = z'$. Clearly, $|z| = |z'|$. It then follows that $\cos\theta + i\sin\theta = \cos\theta' + i\sin\theta'$ so that $\cos\theta = \cos\theta'$ and $\sin\theta = \sin\theta'$. Since

$$1 = \cos^2\theta + \sin^2\theta = \cos\theta\cos\theta' + \sin\theta\sin\theta' = \cos(\theta' - \theta) \tag{3.8}$$

and

$$0 = \sin\theta\cos\theta - \cos\theta\sin\theta = \sin\theta\cos\theta' - \cos\theta\sin\theta' = \sin(\theta' - \theta), \tag{3.9}$$

it follows from (3.9) that $\theta - \theta' = k\pi$, where k is some integer. By (3.8),

$$1 = \cos(\theta' - \theta) = \cos(k\pi) = (-1)^k.$$

This implies that k is even and that $k = 2n$ for some integer n. Hence, $\theta - \theta' = k\pi = 2n\pi$, that is, $\theta = \theta' + 2n\pi$. Thus, θ and θ' differ by an integral multiple of 2π, as contended.

If θ and θ' are principal arguments, then $-\pi < \theta \leqslant \pi$ and $-\pi < \theta' \leqslant \pi$ so that $|\theta - \theta'| < 2\pi$. Since $\theta - \theta' = 2n\pi$, this implies that $|2n\pi| < 2\pi$, or $|n| < 1$. Because n is an integer, this implies that $n = 0$ and, therefore, that $\theta = \theta'$.

Translations and Vectors

If a is a complex number, then by a *translation of the complex plane through* a we mean a function $T: \mathbb{C} \to \mathbb{C}$ defined as follows:

$$T(z) = z + a \qquad \text{for each} \quad z \in \mathbb{C}. \tag{3.10}$$

Each translation is called a *vector*.

If $a = 0$, then $T(z) = z$ for all $z \in \mathbb{C}$. T is, therefore, the identity mapping on $\mathbb{C}$. Then the vector is called the *zero* vector. If $z_1 \in \mathbb{C}$ and $z_2 = T(z_1)$, then T sends z_1 into z_2, and $z_2 - z_1 = a$. Thus, an ordered pair (z_1, z_2) of points z_1 and z_2 is created which we write as $\overrightarrow{z_1z_2}$. We call $\overrightarrow{z_1z_2}$ an

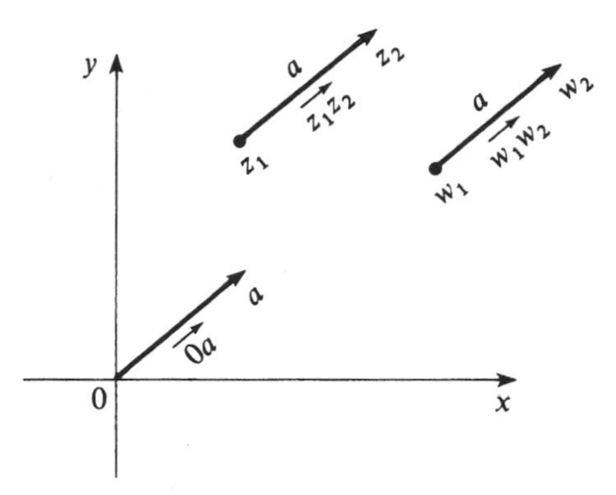

Figure 3.3

arrow with *initial* point z_1 and *terminal* point z_2. We say that it *represents* the vector a and write $\overrightarrow{z_1z_2} = a$. We abuse terminology and call $\overrightarrow{z_1z_2}$ the *vector* with initial point z_1 and terminal point z_2. Clearly

$$\overrightarrow{z_1z_2} = a \qquad \text{if and only if} \quad z_2 - z_1 = a. \tag{3.11}$$

The modulus $|a|$ of a is called the *magnitude* of the vector a. If $a \neq 0$, then the direction of a is termed the *direction* of the vector a. We call vectors $\overrightarrow{z_1z_2}$ and $\overrightarrow{w_1w_2}$ *equal* and write $\overrightarrow{z_1z_2} = \overrightarrow{w_1w_2}$ if and only if they represent the same vector. Thus,

$$\overrightarrow{z_1z_2} = \overrightarrow{w_1w_2} \qquad \text{if and only if} \quad z_2 - z_1 = w_2 - w_1. \tag{3.12}$$

Since $T(0) = 0 + a = a$, the vector $\overrightarrow{0a}$ represents a and we have $\overrightarrow{0a} = a$. $\overrightarrow{0a}$ is called the *position vector* of the point a. In the above spirit we have

$$\overrightarrow{z_1z_2} = z_2 - z_1. \tag{3.13}$$

(See Fig. 3.3.)

Prob. 3.6. Prove: Nonzero vectors $\overrightarrow{z_1z_2}$ and $\overrightarrow{w_1w_2}$ are equal if and only if they have the same direction and magnitude.

By the *sum* of the vectors a and b we mean the composite of the translation a followed by the translation b. Therefore, the function S: $\mathbb{C} \to \mathbb{C}$ defined as

$$S(z) = (z + a) + b \qquad \text{for each} \quad z \in \mathbb{C} \tag{3.14}$$

is the sum of a and b. Since $(z + a) + b = z + (a + b)$ for each $z \in \mathbb{C}$, we see that the sum of the vectors a and b is the vector $a + b$. We also have: If $a = \overrightarrow{z_1z_2}$ and $b = \overrightarrow{z_2z_3}$, then

$$a + b = \overrightarrow{z_1z_2} + \overrightarrow{z_2z_3} = z_2 - z_1 + z_3 - z_2 = z_3 - z_2 = \overrightarrow{z_2z_3} \tag{3.15}$$

(see Fig. 3.4). The additive inverse of the vector a is the vector $-a$. The latter is the additive inverse of the translation a because $a + (-a) = 0$ and 0 is the "identity" translation. Since $\overrightarrow{z_1z_2} + \overrightarrow{z_2z_1} = \overrightarrow{z_1z_1} = z_1 - z_1 = 0$, we see that $-(\overrightarrow{z_1z_2}) = \overrightarrow{z_2z_1}$.

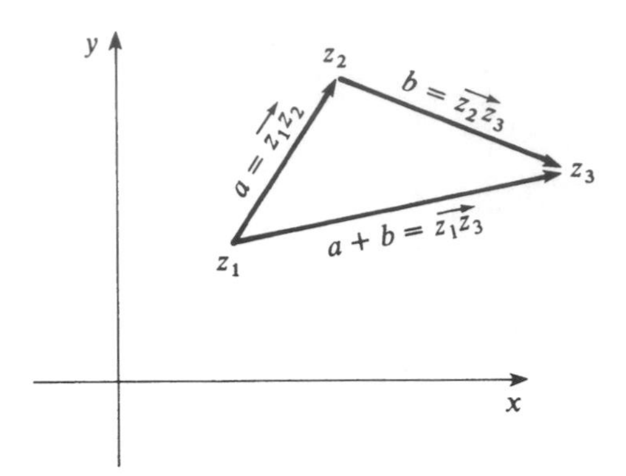

Figure 3.4

Prob. 3.7. Prove: If a, b, and c are vectors in the complex plane, then (1) $a + b = b + a$ and (2) $(a + b) + c = a + (b + c)$ (see Fig. 3.5).

We define the *difference* $a - b$ of vectors a and b as $a - b = a + (-b)$. It is easy to see that $b + (a - b) = a$. Thus, $a - b$ is the c such that $b + c = a$. We have: If $a = \overrightarrow{z_1z_2}$ and $b = \overrightarrow{z_1z_3}$, then

$$\begin{aligned}\overrightarrow{z_1z_2} - \overrightarrow{z_1z_3} = a - b = a + (-b) = \overrightarrow{z_1z_2} + (-\overrightarrow{z_1z_3}) = \overrightarrow{z_1z_2} + \overrightarrow{z_3z_1} \\ = \overrightarrow{z_3z_1} + \overrightarrow{z_1z_2} = \overrightarrow{z_3z_2}\end{aligned}$$

(see Fig. 3.6).

Prob. 3.8. Prove: $\overrightarrow{z_1z_2} + \overrightarrow{z_2z_3} + \overrightarrow{z_3z_4} = 0$ if and only if $z_1 = z_4$.

Let a be a nonzero vector in $\mathbb{C}$ and λ a nonzero real number, so that $\lambda a \neq 0$. The direction of λa is the u such that

$$u = \frac{\lambda a}{|\lambda a|} = \frac{\lambda a}{|\lambda||a|} = \frac{\lambda}{|\lambda|}\frac{a}{|a|} = \begin{cases} \dfrac{a}{|a|} & \text{if } \lambda > 0 \\ -\dfrac{a}{|a|} & \text{if } \lambda < 0. \end{cases}$$

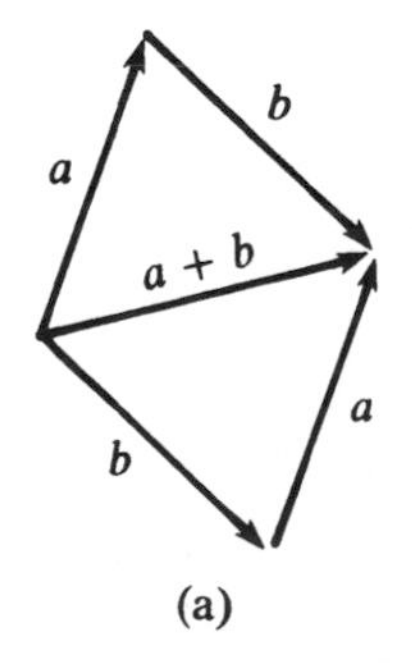

(a)

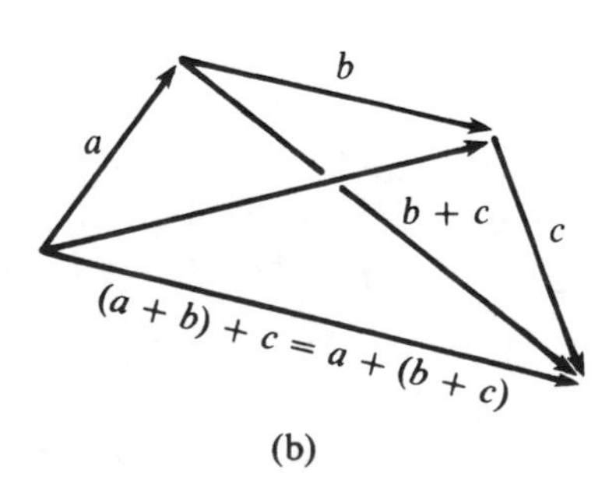

(b)

Figure 3.5

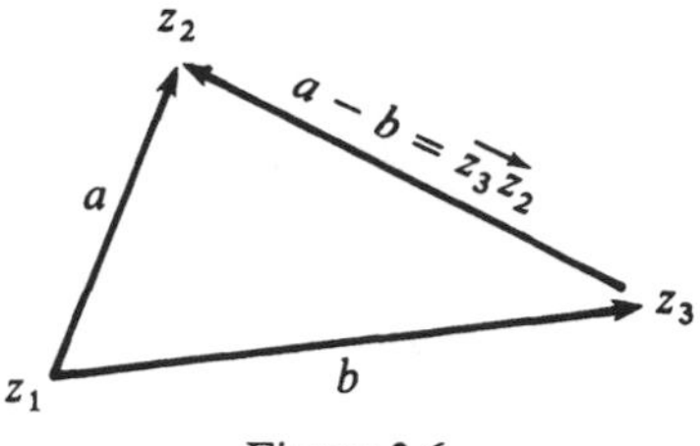

Figure 3.6

Thus, the direction of λa is equal to the direction of a if $\lambda > 0$, and equal to the direction opposite a if $\lambda < 0$. As to the magnitude of λa, we have $|\lambda a| = |\lambda||a|$ (see Fig. 3.7).

Theorem 3.2. *If $z_1 = r_1(\cos\theta_1 + i\sin\theta_1)$ and $z_2 = r_2(\cos\theta_2 + i\sin\theta_2)$, where $r_1 = |z_1|$, $r_2 = |z_2|$, and θ_1, θ_2 are arguments of z_1 and z_2, respectively, then*

(1) $z_1 z_2 = r_1 r_2(\cos(\theta_1 + \theta_2) + i\sin(\theta_1 + \theta_2))$,
(2) $\bar{z}_1 = r_1(\cos\theta_1 - i\sin\theta_1)$,
(3) $z_1/z_2 = (r_1/r_2)(\cos(\theta_1 - \theta_2) + i\sin(\theta_1 - \theta_2)$.

PROOF. We prove (1). We have

$$\begin{aligned} z_1 z_2 &= r_1\left[r_1(\cos\theta_1 + i\sin\theta_1)\right]\left[r_2(\cos\theta_2 + i\sin\theta_2)\right] \\ &= r_1 r_2(\cos\theta_1 + i\sin\theta_1)(\cos\theta_2 + i\sin\theta_2) \\ &= r_1 r_2\left[\cos\theta_1\cos\theta_2 - \sin\theta_1\sin\theta_2 + i(\sin\theta_1\cos\theta_2 + \cos\theta_1\sin\theta_2)\right] \\ &= r_1 r_2\left[\cos(\theta_1 + \theta_2) + i\sin(\theta_1 + \theta_2)\right], \end{aligned}$$

which proves (1). We ask the reader to prove parts (2) and (3) (Prob. 3.9).

PROB. 3.9. Complete the proof of Theorem 3.2 by proving parts (2) and (3).

From the last theorem we see that if z_1 and z_2 are nonzero complex numbers, then the modulus of their product is the product of their moduli

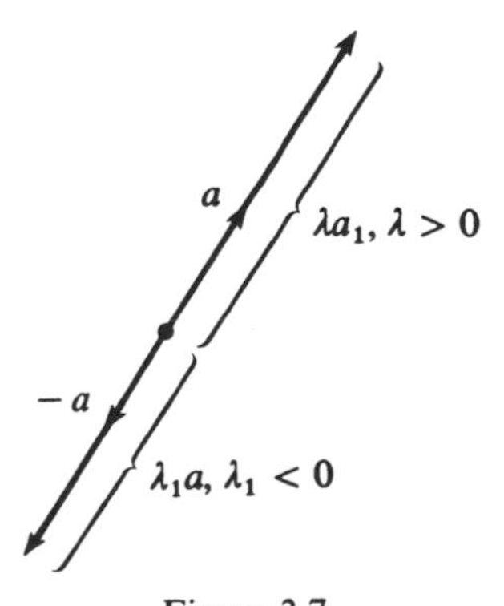

Figure 3.7

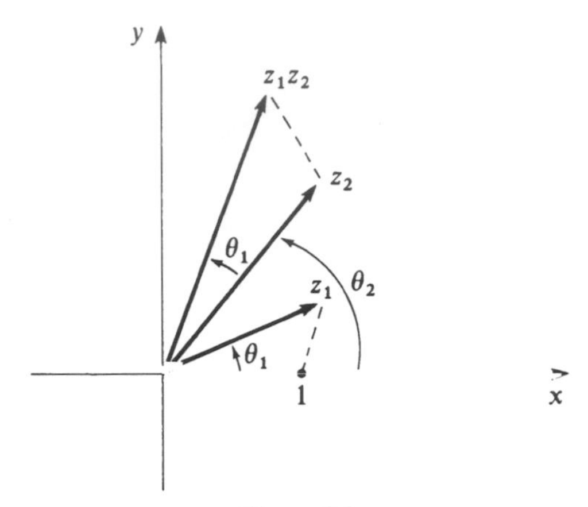

Figure 3.8

and that an argument of their *product* is the *sum* of their arguments (Fig. 3.8). On the other hand, the quotient z_1/z_2 has as its modulus the quotient $|z_1|/|z_2|$ of their moduli and one of its arguments is the difference $\arg z_1 - \arg z_2$.

4. The Exponential Function on $\mathbb{C}$

To define z^n for $z \in \mathbb{C}$, where n is a nonnegative integer, we use induction on n and define

(1) $$z^0 = 1$$

(2) $$z^{n+1} = z^n z \qquad \text{for each nonnegative integer } n. \tag{4.1}$$

We then define z^n, where $z \neq 0$ and n is a negative integer, as

$$z^n = (z^{-1})^{(-n)}. \tag{4.2}$$

We can then prove that the following relations hold: If $z \neq 0$ and m, n are integers, then

(a) $z^m z^n = z^{m+n}$,
(b) $(z^m)^n = z^{mn}$,
(c) $z^m/z^n = z^{m-n}$,
(d) if also $w \neq 0$, then $(wz)^n = w^n z^n$.

PROB. 4.1. Prove: If $z \in \mathbb{C}$ and n is a nonnegative integer, then

$$(\bar{z})^n = \overline{z^n}. \tag{4.3}$$

If $z \neq 0$, then (4.3) holds for all integers n.

PROB. 4.2 (De Moivre's Theorem). Prove: If $\theta \in \mathbb{R}$ and n is an integer, then

$$(\cos\theta + i\sin\theta)^n = \cos(n\theta) + i\sin(n\theta). \tag{4.4}$$

We illustrate the usefulness of (4.4) for $n = 2$ and $n = 3$. We have

$$(\cos\theta + i\sin\theta)^2 = \cos(2\theta) + i\sin(2\theta).$$

Squaring on the left yields

$$\cos^2\theta - \sin^2\theta + 2i\sin\theta\cos\theta = \cos 2\theta + i\sin 2\theta.$$

Equating real and imaginary parts, we arrive at the identities

$$\cos 2\theta = \cos^2\theta - \sin^2\theta \quad \text{and} \quad \sin 2\theta = 2\sin\theta\cos\theta. \tag{4.5}$$

Again, we have

$$(\cos\theta + i\sin\theta)^3 = \cos 3\theta + i\sin 3\theta.$$

So cubing on the left yields

$$\cos^3\theta + 3i\cos^2\theta\sin\theta - 3\cos\theta\sin^2\theta - i\sin^3\theta = \cos 3\theta + i\sin 3\theta.$$

Equating real and imaginary parts we have

$$\cos 3\theta = \cos^3\theta - 3\cos\theta\sin^2\theta \quad \text{and} \quad \sin 3\theta = 3\cos^2\theta\sin\theta - \sin^3\theta. \tag{4.6}$$

We now define $E: \mathbb{C} \to \mathbb{C}$ as

$$E(z) = E(\xi + \eta i) = e^{\xi}(\cos\eta + i\sin\eta) \qquad \text{for each} \quad z = \xi + \eta i \in \mathbb{C}. \tag{4.7}$$

This definition yields

$$E(\xi) = E(\xi + 0i) = e^{\xi}(\cos 0 + i\sin 0) = e^{\xi}1 = e^{\xi} \qquad \text{for} \quad \xi \in \mathbb{R},$$

so that

$$E(\xi) = e^{\xi} \qquad \text{for} \quad \xi \in \mathbb{R}. \tag{4.8}$$

Thus, the restriction of E to $\mathbb{R}$ is the exponential function on $\mathbb{R}$. E is, therefore, an extension of the exponential function on $\mathbb{R}$ to the set $\mathbb{C}$ of complex numbers. In particular,

$$E(0) = 1 \quad \text{and} \quad E(1) = e. \tag{4.9}$$

We define

$$e^z = E(z) \qquad \text{for each} \quad z \in \mathbb{C}. \tag{4.10}$$

Theorem 4.1. *If z_1 and z_2 are complex numbers, then*

(1) $e^{z_1+z_2} = e^{z_1}e^{z_2}$,
(2) $e^{z_1} \neq 0$,
(3) $e^{-z_1} = 1/e^{z_1}$,
(4) $e^{z_2}/e^{z_1} = e^{z_2-z_1}$.

Proof. Let $z_1 = \xi_1 + \eta_1 i$ and $z_2 = \xi_2 + \eta_2 i$. We have

$$\begin{aligned} e^{z_1}e^{z_2} = E(z_1)E(z_2) &= e^{\xi_1}(\cos\eta_1 + i\sin\eta_1)e^{\xi_2}(\cos\eta_2 + i\sin\eta_2) \\ &= e^{\xi_1+\xi_2}(\cos\eta_1 + i\sin\eta_1)(\cos\eta_2 + i\sin\eta_2) \\ &= e^{\xi_1+\xi_2}(\cos(\eta_1+\eta_2) + i\sin(\eta_1+\eta_2)) \\ &= E((\xi_1+\xi_2) + (\eta_1+\eta_2)i) \\ &= E(z_1 + z_2) \\ &= e^{z_1+z_2}. \end{aligned}$$

This proves (1). It follows that

$$e^{z_1}e^{-z_1} = e^{z_1+(-z_1)} = e^0 = 1 \tag{4.11}$$

and, hence, that $e^{z_1} \neq 0$ for $z_1 \in \mathbb{C}$. This proves (2). Part (3) follows from (4.11). Part (4) holds because

$$\frac{e^{z_2}}{e^{z_1}} = e^{z_2}\frac{1}{e^{z_1}} = e^{z_2}e^{-z_1} = e^{z_2+(-z_1)} = e^{z_2-z_1}.$$

This completes the proof.

Theorem 4.2. *If $z \in \mathbb{C}$ and n is an integer, then*

$$e^{z+2n\pi i} = e^z. \tag{4.12}$$

Moreover, if

$$e^{z+a} = e^z \tag{4.13}$$

for some $z \in \mathbb{C}$, then $a = 2n\pi i$, where n is some integer.

Proof. Equation (4.12) holds because

$$\begin{aligned} e^{z+2n\pi i} &= e^z e^{zn\pi i} = e^z(\cos 2n\pi + i\sin 2n\pi) \\ &= e^z \cdot 1 = e^z. \end{aligned}$$

We prove the second part of the theorem. Assume that z and a in $\mathbb{C}$ exist such that

$$e^{z+a} = e^z.$$

This implies that

$$e^a = 1. \tag{4.14}$$

Let $a = \alpha + \beta i$, $\alpha \in \mathbb{R}$, $\beta \in \mathbb{R}$. This and (4.14) imply that

$$e^{\alpha+\beta i} = e^\alpha(\cos\beta + i\sin\beta) = 1 = 1(\cos 0 + i\sin 0).$$

By Theorem 3.1, we obtain from this

$$e^\alpha = 1 \quad \text{and} \quad \beta = 2n\pi,$$

where n is an integer. But α real and $e^\alpha = 1$ imply that $\alpha = 0$. Hence, we have $a = \alpha + \beta i = 2n\pi i$, where n is an integer. This completes the proof.

Theorem 4.3 (Euler's Formula). *If $y \in \mathbb{R}$, then*

$$e^{yi} = \cos y + i \sin y \tag{4.15}$$

and

$$\cos y = \frac{e^{yi} + e^{-yi}}{2} \tag{4.16a}$$

and

$$\sin y = \frac{e^{yi} - e^{-yi}}{2i}. \tag{4.16b}$$

Proof. Since $y \in \mathbb{R}$, we have

$$e^{yi} = E(yi) = E(0 + yi) = e^0(\cos y + i \sin y) = \cos y + i \sin y.$$

This proves (4.7). To prove (4.16) note that

$$e^{-yi} = \cos y - i \sin y. \tag{4.17}$$

This and (4.15) imply (4.16a) and (4.16b).

Euler's formula (4.15) enables us to write each nonzero complex number as

$$z = re^{i\theta}, \tag{4.18}$$

where $r = |z|$ and θ is some argument of z.

Prob. 4.3. Prove: (a) If $z = |z|e^{i\theta}$, $w = |w|e^{i\psi}$, where $z \neq 0$, $w \neq 0$, $\theta = \operatorname{Arg} z$, and $\psi = \operatorname{Arg} w$, then $z = w$ if and only if $|z| = |w|$ and $\theta = \psi$. If, however, θ and ψ are not necessarily principal arguments, then $\theta = \psi + 2n\pi$, where n is some integer. (b) Prove: If u is a direction, then there is exactly one θ such that $\theta = \operatorname{Arg} z$ and $u = e^{i\theta}$.

Prob. 4.4. Prove: If $z \in \mathbb{C}$, then for any integer n, we have

$$(e^z)^n = e^{nz}.$$

Prob. 4.5. Prove: If $z \in \mathbb{C}$, then $\overline{e^z} = e^{\bar{z}}$.

Polynomials on $\mathbb{C}$

Polynomial functions on $\mathbb{C}$ can be defined in the same way polynomials on $\mathbb{R}$ were defined (see Def. II.8.1). Thus, a polynomial on $\mathbb{C}$ is a function $f: \mathbb{C} \to \mathbb{C}$ defined by means of

$$f(z) = a_0 z^n + a_1 z^{n-1} + \cdots + a_{n-1}x + a_n \qquad \text{for each} \quad z \in \mathbb{C}, \tag{4.19}$$

where n is a nonnegative integer and $a_0, a_1, \ldots, a_n$ are complex numbers.

The latter are called the *coefficients* of the polynomial. The notions of *degree* of a polynomial, a *polynomial equation* and its *root*, the *zero* of a polynomial all carry over from $\mathbb{R}$ to $\mathbb{C}$ in the obvious manner.

PROB. 4.6. Prove: If $a_0, a_1, \ldots, a_n$ are real and P is a polynomial on $\mathbb{C}$ of degree $n \geqslant 1$ which has r as a zero, then $\bar{r}$ is also a zero of P.

Remark 4.1. The following are results holding in $\mathbb{R}$ which can be extended to $\mathbb{C}$. Let n be a positive integer and a, b complex numbers. Then

$$a^n - b^n = (a-b)\sum_{k=0}^{n-1} a^{n-1-k}b^k \tag{4.20}$$

and (the binomial theorem on $\mathbb{C}$)

$$(a+b)^n = \sum_{k=0}^{n} \binom{n}{k} a^{n-k}b^k. \tag{4.21}$$

Theorem 4.4 (Remainder Theorem). *If P is a polynomial on $\mathbb{C}$ of degree $n \geqslant 1$, then*

$$P(z) = (z-a)Q(z) + P(a) \qquad \text{for each} \quad z \in \mathbb{C}, \tag{4.22}$$

where Q is a polynomial of degree $n-1$.

PROOF. Assume that

$$P(z) = a_0 z^n + a_1 z^{n-1} + \cdots + a_{n-1}z + a_n = \sum_{k=0}^{n} a_{n-k}z^k,$$

where $a_0 \neq 0$. Then

$$\begin{aligned}
P(z) - P(a) &= \sum_{k=0}^{n} a_{n-k}z^k - \sum_{k=0}^{n} a_{n-k}a^k \\
&= \sum_{k=1}^{n} a_{n-k}(z^k - a^k) \\
&= \sum_{k=1}^{n} a_{n-k}\left[(z-a)\sum_{j=0}^{k-1} z^j a^{k-1-j}\right] \qquad \text{(see (4.20))} \\
&= (z-a)\sum_{k=1}^{n} a_{n-k}\sum_{j=0}^{k-1} z^j a^{k-(1+j)} \\
&= (z-a)\sum_{k=1}^{n} a_{n-k}\sum_{m=1}^{k} z^{m-1}a^{k-m}
\end{aligned}$$

(reindex by writing $m = j+1$, so that $1 \leqslant m \leqslant k$ and $j = m-1$).

Note that $1 \leqslant m \leqslant k \leqslant n$. Interchange the order of summation by fixing m and summing first with respect to k, where $m \leqslant k \leqslant n$ and then with respect to m to obtain

$$P(z) - P(a) = (z-a)\sum_{m=1}^{n} a_{n-k}\sum_{k=m}^{n} z^{m-1}a^{k-m}$$
$$= (z-a)\sum_{m=1}^{n} z^{m-1}\sum_{k=m}^{n} a_{m-k}a^{k-m}. \tag{4.23}$$

Now write $A_m = \sum_{k=m}^{n} a_{n-k}a^{k-m}$ and note that

$$A_n = \sum_{k=n}^{n} a_{n-k}a^{k-n} = a_0 .$$

Then (4.23) becomes

$$P(z) - P(a) = (z-a)\big(A_1 + A_2 z + \cdots + A_{n-1}z^{n-2} + a_0 z^{n-1}\big). \tag{4.24}$$

Next write

$$Q(z) = A_1 + A_2 z + \cdots + A_{n-1}z^{n-2} + a_0 z^{n-1}. \tag{4.25}$$

Q is a polynomial of degree $n-1$. This and (4.24) lead to

$$P(z) = (z-a)\,Q(z) + P(a), \tag{4.26}$$

where Q is a polynomial of degree $n-1$.

PROB. 4.7 (Factor Theorem). Prove: If P is a polynomial on $\mathbb{C}$ of degree $n \geqslant 1$, then $a \in \mathbb{C}$ is a zero of P if and only if $P(z) = (z-a)Q(z)$ for all $z \in \mathbb{C}$, where $Q(z)$ is a polynomial of degree $n-1$. If a_0 is the leading coefficient of P, then it is the leading coefficient of Q.

PROB. 4.8. Prove: If P is a polynomial on $\mathbb{C}$ of degree $n \geqslant 1$ and a is a zero of P, then there exists an integer k such that $1 \leqslant k \leqslant n$, and $P(z) = (z-a)^k Q_{n-k}(z)$ for all $z \in \mathbb{C}$, where Q_{n-k} is a polynomial of degree $n-k$ with $Q_{n-k}(a) \neq 0$ (Hint: try induction on n).

Remark 4.2. If P is a polynomial on $\mathbb{C}$ of degree $n \geqslant 1$ having a as one of its zeros, then the k of Prob. 4.8 is called the *multiplicity* of a. If $k = 1$, then a is called a *simple zero* of P.

PROB. 4.9. Prove: If P is a polynomial on $\mathbb{C}$ of degree $n \geqslant 1$ having a_0 as its leading coefficient and $r_1, \ldots, r_n$ are n distinct zeros of P, then

$$P(z) = a_0(z-r_1)(z-r_0)\ldots(z-r_n) \qquad \text{for all} \quad z \in \mathbb{C}.$$

PROB. 4.10. Prove: If P is a polynomial on $\mathbb{C}$ of degree $n \geqslant 1$, then P has at most n distinct real zeros.

PROB. 4.11. Prove: If P and Q are polynomials each of degree not exceeding some nonnegative integer n such that $P(z) = Q(z)$ holds for more than n distinct values of z, then $P = Q$ and both polynomials have the same coefficients.

Prob. 4.12. Extend Theorem VII.4.2 (the Lagrange interpolation formula) to the complex numbers.

Remark 4.3. Let $z_1, z_2, \ldots, z_n$ be n zeros of a polynomial of degree $n \geqslant 1$, where

$$P(z) = a_0 z^n + a_1 z^{n-1} + \cdots + a_{n-1} z + a_n, \tag{4.27}$$

so that

$$P(z) = a_0(z - z_1)(z - z_2) \ldots (z - z_n). \tag{4.28}$$

Multiply on the right and obtain

$$P(z) = a_0\left(z^n - \sigma_1 z^{n-1} + \sigma_2 z^{n-2} + \cdots + (-1)^n \sigma_n\right), \tag{4.29}$$

where

$$\begin{aligned}
\sigma_1 &= z_1 + z_2 + \cdots + z_n, \\
\sigma_2 &= z_1 z_2 + \cdots + z_1 z_n + z_2 z_3 + \cdots + z_2 z_n + \cdots + z_{n-1} z_n \\
&= \text{sum of the products of } z_1, \ldots, z_n \text{ taken two at a time} \\
\sigma_3 &= z_1 z_2 z_3 + \cdots + z_{n-1} z_{n-2} z_n \\
&= \text{sum of the products of } z_1, \ldots, z_n \text{ taken three at a time} \\
&\vdots \\
\sigma_n &= z_1 z_2 \ldots z_n.
\end{aligned} \tag{4.30}$$

These σ's are known as the *elementary symmetric functions* of $z_1, z_2, \ldots, z_n$. Since (4.29) holds for all z, we can equate coefficients to conclude that

$$\sigma_1 = -\frac{a_1}{a_0}, \qquad \sigma_2 = \frac{a_2}{a_0}, \qquad \sigma_3 = -\frac{a_3}{a_0}, \ldots, \qquad \sigma_n = (-1)^n \frac{a_n}{a_0}. \tag{4.31}$$

A polynomial is called *monic* if its leading coefficient a_0 is equal to 1. Let M be a monic polynomial and

$$M(z) = z^n + p_1 z^{n-1} + \cdots + p_{n-1} z + a_n \tag{4.32}$$

and let its zeros be $z_1, \ldots, z_n$. We then have, in view of (4.31),

$$\sigma_1 = -p_1, \qquad \sigma_2 = p_2, \ldots, \qquad \sigma_n = (-1)^n p_n. \tag{4.33}$$

5. nth Roots of a Complex Number. Trigonometric Functions on $\mathbb{C}$

Theorem 5.1. *If $a \in \mathbb{C}$, where $a \neq 0$, and n is a positive integer ($n \geqslant 2$). Then there exist exactly n complex numbers $z_0, z_1, \ldots, z_{n-1}$ such that*

$$z_k^n = a \qquad \text{for each} \quad k \in \{0, \ldots, n-1\}. \tag{5.1}$$

PROOF. Write a in polar form as $a = \rho e^{i\alpha}$, $\alpha = \operatorname{Arg} a$. Suppose z exists such that $z^n = a$. Write z in polar form as $z = re^{i\theta}$, where θ is some argument of z. Then

$$r^n e^{n\theta i} = \left(re^{i\theta}\right)^n = z^n = u = \rho e^{i\alpha}.$$

This implies that

$$r = \rho^{1/n} \quad \text{and} \quad n\theta = \alpha + 2\pi m, \tag{5.2}$$

where m is an integer. Put

$$\theta_m = \frac{\alpha}{n} + \frac{2\pi m}{n} \tag{5.3a}$$

and

$$z_m = \rho^{1/n} e^{i\theta_m}. \tag{5.3b}$$

This proves: If $z^n = a$, then for some integer m,

$$z = z_m = \rho^{1/n} e^{i\theta_m},$$

where θ_m is defined in (5.3). It is a simple matter to verify that $z_m^n = a$ for each integer m.

We now prove that any z with $z^n = a$ is of the form $z_k = \rho^{1/n} e^{i\theta_k}$, where $k \in \{0, 1, \ldots, n-1\}$. This will prove that there are at most n z's with $z^n = a$. Since m and n are integers and $n > 0$, there exist integers q and k such that $m = nq + k$, where $0 \leqslant k < n$. Thus,

$$\theta_m = \frac{\alpha}{n} + \frac{(nq + k)2\pi}{n} = \theta_k + 2q\pi, \tag{5.4}$$

where $k \in \{0, 1, \ldots, n-1\}$ and

$$z = \rho^{1/n} e^{i\theta_m} = \rho^{1/n} e^{i(\theta_k + 2q\pi)} = \rho^{1/n} e^{i\theta_k} e^{2q\pi i} = \rho^{1/n} e^{i\theta_k} = z_k, \tag{5.5}$$

where $k \in \{0, 1, \ldots, n-1\}$.

Finally, we prove that $z_0, z_1, \ldots, z_{n-1}$ are all distinct. Assume that k_1 and k_2 are distinct elements of $\{0, 1, \ldots, n-1\}$. There is no loss of generality if we take $k_1 < k_2$. We have $0 < k_2 - k_1 \leqslant k_2 < n$ where k_1 and k_2 are integers. Assume that $z_{k_1} = z_{k_2}$ so that

$$e^{i\theta_{k_1}} = e^{i\theta_{k_2}} \quad \text{and, therefore,} \quad e^{i(\theta_{k_2} - \theta_{k_1})} = 1$$

i.e.,

$$e^{i((k_2 - k_1)/n)2\pi} = 1 = e^{0i}. \tag{5.6}$$

This implies that $0 < ((k_2 - k_1)/n)2\pi = 2m\pi$, where m is an integer. Here m is necessarily a positive integer and we have $k_2 - k_1 = mn \geqslant n$. This contradicts the earlier $k_2 - k_1 < n$. Thus, $k_1 \neq k_2$ in $\{0, 1, \ldots, n-1\}$ implies $z_{k_1} \neq z_{k_2}$. In other words, $z_0, z_1, \ldots, z_{n-1}$ are all distinct. This together with what was proved above implies that $z_0, z_1, \ldots, z_{n-1}$ constitute n distinct complex numbers whose nth power is equal to a. Moreover, these numbers are given by

$$z_k = \rho^{1/n} e^{i(\alpha/n + 2k\pi/n)},$$

where $k \in \{0, 1, \ldots, n-1\}$. This completes the proof.

Def. 5.1. If $a \in \mathbb{C}$ and n is a positive integer, then a $z \in \mathbb{C}$ such that $z^n = a$ is called an nth *root* of a. When $a = 1$, then a z such that $z^n = 1$ and $z \neq 1$ is called an nth *root of unity*.

Remark 5.1. In Theorem 5.1 we proved: Let $a \neq 0$ in $\mathbb{C}$. Put $a = \rho e^{i\alpha}$, where $\alpha = \operatorname{Arg} a$. If n is an integer such that $n \geqslant 2$, then a has exactly n distinct roots, $z_0, z_1, \ldots, z_{n-1}$, where for each $k \in \{0, 1, \ldots, n-1\}$

$$z_k = \rho^{1/n} e^{i\theta_k} = \rho^{1/n} e^{i(\alpha/n + 2k\pi/n)}. \tag{5.7}$$

Writing

$$z_0 = \rho^{1/n} e^{i\alpha/n}$$

we have

$$z_1 = z_0 e^{(2\pi/n)i}, \qquad z_2 = z_0 e^{(4\pi/n)i}, \ldots, \qquad z_{n-1} = z_0 e^{((n-1)/n)2\pi i}. \tag{5.8}$$

If $a = 1$, then $\alpha = 0$ and $z_0 = 1$. In this case (5.8) becomes

$$z_1 = e^{(2\pi/n)i}, \qquad z_2 = e^{(4\pi/n)i}, \ldots, \qquad z_{n-1} = e^{((n-1)/n)2\pi i}. \tag{5.9}$$

We see that if $n \geqslant 2$, then 1 has $n-1$ roots of unity. We usually write $\omega = z_1 = e^{(2\pi/n)i}$. In this case the nth roots of unity can be written

$$z_1 = \omega, \qquad z_2 = \omega^2, \ldots, \qquad z_{n-1} = \omega^{n-1}. \tag{5.10}$$

We saw in the proof of the last theorem that these are all distinct. It follows that all the nth roots of $a = \rho e^{i\alpha}$, where $\rho > 0$, $\alpha = \operatorname{Arg} a$, can be written

$$z_k = z_0 \omega^k, \qquad k \in \{0, 1, \ldots, n-1\}, \tag{5.11}$$

where

$$z_0 = \rho^{1/n} e^{\alpha/n} \quad \text{and} \quad \omega = e^{(2\pi/n)i}.$$

Remark 5.2. If n is an integer, $n \geqslant 2$, and ω is an nth root of unity, so that $\omega^n = 1$, $\omega \neq 1$, then

$$(\omega - 1)(\omega^{n-1} + \cdots + \omega + 1) = \omega^n - 1 = 0.$$

The fact that $\omega \neq 1$ implies that

$$1 + \omega + \omega^2 + \cdots + \omega^{n-1} = 0, \tag{5.12}$$

where ω is an nth root of unity and $n \geqslant 2$. Since each of the numbers $\omega, \omega^2, \ldots, \omega^{n-1}$ is an nth root of unity, it satisfies the equation

$$1 + z + z^2 + \cdots + z^{n-1} = 0. \tag{5.13}$$

By factor theorem, this implies that

$$1 + z + z^2 + \cdots + z^{n-1} = (z - \omega)(z - \omega^2) \ldots (z - \omega^{n-1}) \tag{5.14}$$

and, hence, that

$$z^n - 1 = (z - 1)(z - \omega) \ldots (z - \omega^{n-1}) \qquad \text{for each} \quad z \in \mathbb{C}. \tag{5.15}$$

It follows that if a and b are complex numbers and n is a positive integer $\geqslant 2$, then

$$a^n - b^n = (a - b)(a - \omega b)(a - \omega^2 b) \ldots (a - \omega^{n-1} b), \tag{5.16}$$

where $\omega = e^{(2\pi/n)i}$.

As an example we compute the cube roots of unity. We must solve $z^3 = 1$, so we search for a θ such that

$$e^{3\theta i} = 1 \tag{5.17}$$

and obtain

$$3\theta = 2k\pi,$$

where k is an integer. Distinct roots of (5.17) are obtained from $\theta_0 = 0$, $\theta_1 = \frac{2}{3}\pi$, $\theta_2 = \frac{4}{3}\pi$. These yield

$$z_0 = 1, \qquad z_1 = e^{(2\pi/3)i}, \qquad z_2 = e^{(4\pi/3)i}. \tag{5.18}$$

Graphically, we have Fig. 5.1 and

$$e^{(2\pi/3)i} = \cos\left(\frac{2\pi}{3}\right) + i\sin\left(\frac{2\pi}{3}\right) = -\frac{1}{2} + \frac{\sqrt{3}}{2}i,$$

$$e^{(4\pi/3)i} = \cos\left(\frac{4\pi}{3}\right) + i\sin\left(\frac{4\pi}{3}\right) = -\frac{1}{2} - \frac{\sqrt{3}}{2}i.$$

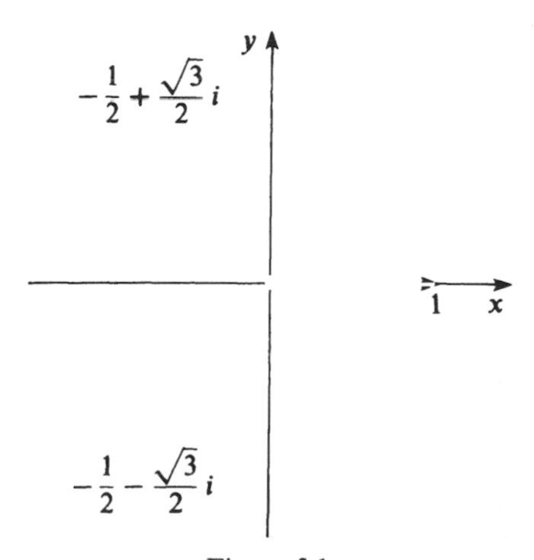

Figure 5.1

The three roots of $z^3 = 1$ are $z_0 = 1$, $\omega = -1/2 + (\sqrt{3}/2)i$, and $\omega^2 = -1/2 - (\sqrt{3}/2)i$.

PROB. 5.1. Solve: $z^3 = 1 + i$.

PROB. 5.2. Prove: If $\omega = e^{(2\pi/n)i}$, where n is an integer, $n \geqslant 2$, then

$$(1-\omega)(1-\omega^2)\ldots(1-\omega^{n-1}) = n.$$

PROB. 5.3. Prove: If α is an nth root of unity, where $n \geqslant 2$, then any integral power α^m of α is also an nth root of unity.

PROB. 5.4. Prove: (a) If m and n are positive integers and (m,n) is their greatest common divisor, then any solution z of both $z^m = 1$ and $z^n = 1$ is also a solution of $z^{(m,n)} = 1$. (b) If m and n are relatively prime, then the only solution z common to $z^n = 1$ and $z^m = 1$ is $z = 1$.

Hyperbolic and Trigonometric Functions on $\mathbb{C}$

By Theorem 4.3, Euler's Formula implies that

$$\cos y = \frac{e^{yi} + e^{-yi}}{2} \quad \text{and} \quad \sin y = \frac{e^{iy} - e^{-iy}}{2i} \qquad \text{for} \quad y \in \mathbb{R}. \tag{5.19}$$

We use these to extend the definition of the sine and cosine functions from $\mathbb{R}$ to $\mathbb{C}$. We define

$$\cos z = \frac{e^{iz} + e^{-iz}}{2} \quad \text{and} \quad \sin z = \frac{e^{iz} - e^{-iz}}{2i} \qquad \text{for each} \quad z \in \mathbb{C}. \tag{5.20}$$

PROB. 5.5. Prove: (a) $e^{iz} = \cos z + i \sin z$ for $z \in \mathbb{C}$ and (b) $\cos^2 z + \sin^2 z = 1$ for all $z \in \mathbb{C}$. (c) $\overline{\cos z} = \cos \bar{z}$ and $\overline{\sin z} = \sin \bar{z}$ for $z \in \mathbb{C}$.

In a similar manner we extend the definition of hyperbolic sine and cosine to $\mathbb{C}$ by defining

$$\cosh z = \frac{e^z + e^{-z}}{2} \tag{5.21a}$$

and

$$\sinh z = \frac{e^z - e^{-z}}{2} \qquad \text{for} \quad z \in \mathbb{C}. \tag{5.21b}$$

It follows from (5.20) and (5.21) that

$$\begin{aligned} \cos z &= \cosh iz, \\ \cos iz &= \cosh z \end{aligned} \tag{5.22a}$$

and

$$\begin{aligned} i \sin z &= \sinh iz, \\ \sin iz &= i \sinh z. \end{aligned} \tag{5.22b}$$

PROB. 5.6. Prove: $\cosh^2 z - \sinh^2 z = 1$ for $z \in \mathbb{C}$.

PROB. 5.7. Prove: If a and b are *complex* numbers, then

(1) $\cos(a + b) = \cos a \cos b - \sin a \sin b$,
(2) $\cosh(a + b) = \cosh a \cosh b + \sinh a \sinh b$,
(3) $\sin(a + b) = \sin a \cos b + \cos a \sin b$,
(4) $\sinh(a + b) = \sinh a \cosh b + \cosh a \sinh b$.

PROB. 5.8. Prove: If $z \in \mathbb{C}$, then (1) $\cos(-z) = \cos z$, (2) $\cosh(-z) = \cosh z$, (3) $\sin(-z) = -\sin z$, (4) $\sinh(-z) = -\sinh z$.

Using (5.22) and Prob. 5.7, we have for $a = \alpha + \beta i \in \mathbb{C}$,

$$\begin{aligned} \cos a &= \cos(\alpha + \beta i) = \cos\alpha \cos \beta i - \sin\alpha \sin \beta i \\ &= \cos\alpha \cosh \beta - i \sin\alpha \sinh \beta, \end{aligned} \tag{5.23}$$

$$\begin{aligned} \sin a &= \sin(\alpha + \beta i) = \sin\alpha \cos \beta i + \cos\alpha \sin \beta i \\ &= \sin\alpha \cosh \beta + i \cos\alpha \sinh \beta, \end{aligned} \tag{5.24}$$

$$\cosh a = \cosh(\alpha + \beta i) = \cosh\alpha \cos \beta + i \sinh\alpha \sin \beta, \tag{5.25}$$

$$\sinh a = \sinh(\alpha + \beta i) = \sinh\alpha \cos \beta + i \cosh\alpha \sin \beta. \tag{5.26}$$

It follows from the last four identities that if $a = \alpha + i\beta \in \mathbb{C}$, then

$$|\cos a|^2 = |\cos(\alpha + \beta i)|^2 = \cos^2\alpha \cosh^2\beta + \sin^2\alpha \sinh^2\beta, \tag{5.27}$$

$$|\sin a|^2 = |\sin(\alpha + \beta i)|^2 = \sin^2\alpha \cosh^2\beta + \cos^2\alpha \sinh^2\beta, \tag{5.28}$$

$$|\cosh a|^2 = |\cosh(\alpha + \beta i)|^2 = \cosh^2\alpha \cos^2\beta + \sinh^2\alpha \sin^2\beta, \tag{5.29}$$

$$|\sinh a|^2 = |\sin(\alpha + \beta i)|^2 = \sinh^2\alpha \cos^2\beta + \cosh^2\alpha \sin^2\beta. \tag{5.30}$$

PROB. 5.9. Prove: If $a = \alpha + \beta i \in \mathbb{C}$, then

(1) $|\cos a| = |\cos(\alpha + \beta i)| = \sqrt{\cos^2\alpha + \sinh^2\beta}$,

(2) $|\sin a| = |\sin(\alpha + \beta i)| = \sqrt{\sin^2\alpha + \sinh^2\beta}$,

(3) $|\cosh a| = |\cosh(\alpha + \beta i)| = \sqrt{\cos^2\beta + \sinh^2\alpha}$,

(4) $|\sinh a| = |\sinh(\alpha + \beta i)| = \sqrt{\sin^2\beta + \sinh^2\alpha}$.

PROB. 5.10. Prove: (1) If $a = \alpha + \beta i$ is imaginary, then $\cos a \neq 0$ and $\sin a \neq 0$; (2) $\cos a = 0$ if and only if $a = (2n + 1)\pi/2$, where n is an integer; (3) $\sin a = 0$ if and only if $a = n\pi$, where n is an integer.

PROB. 5.11. Prove: (1) If a is not pure imaginary, then $\cosh a \neq 0$ and $\sinh a \neq 0$; (2) $\cosh a = 0$ if and only if $a = (2n + 1)(\pi/2)i$, where n is an integer; (3) $\sinh a = 0$ if and only if $a = n\pi i$, where n is an integer.

PROB. 5.12. Prove: (1) The sine and cosine functions on $\mathbb{C}$ are periodic with fundamental period 2π; (2) the hyperbolic sine and cosine functions sinh and cosh are periodic with period $2\pi i$.

The tan, cotangent, secant, and cosecant and similarly the hyperbolic counterparts of these, such as tanh, coth, sech, and cosh can be defined on $\mathbb{C}$ in the usual way. For example:

$$\tan z = \frac{\sin z}{\cos z}, \quad \tanh z = \frac{\sinh z}{\cosh z}, \quad \text{etc.}$$

6. Evaluation of Certain Trigonometric Sums

We evaluate the sum

$$\sum_{k=1}^{n} \sin k\theta, \qquad \theta \in \mathbb{R}, \tag{6.1}$$

where n is a positive integer. Note that

$$\sin k\theta = \frac{e^{ik\theta} - e^{-ik\theta}}{2i} = \frac{(e^{i\theta})^k - (e^{-i\theta})^k}{2i}, \tag{6.2}$$

where k is an integer. Write $e^{i\theta} = u$, so that $e^{-i\theta} = \bar{u}$. Hence, (6.1) becomes

$$\begin{aligned}\sum_{k=1}^{n} \sin k\theta &= \sum_{k=0}^{n} \sin k\theta = \frac{1}{2i} \sum_{k=0}^{n} (u^k - \bar{u}^k) \\ &= \frac{1}{2i}\left(\sum_{k=0}^{n} u^k - \sum_{k=0}^{n} \bar{u}^k\right) = \frac{1}{2i}\left(\frac{1 - u^{n+1}}{1 - u} - \frac{1 - \bar{u}^{n+1}}{1 - \bar{u}}\right) \\ &= \frac{1}{2i} \frac{u - \bar{u} + \bar{u}^{n+1} - u^{n+1} + u\bar{u}(u^n - \bar{u}^n)}{(1 - u)(1 - \bar{u})}.\end{aligned}$$

In the last expression on the right, we have (a) $u\bar{u} = 1$, (b) $u - \bar{u} = 2i \sin\theta$, (c) $u + \bar{u} = 2\cos\theta$, (d) $(1 - u)(1 - \bar{u}) = 1 - (u + \bar{u}) + u\bar{u} = 2 - (u + \bar{u}) = 2 - 2\cos\theta$, (e) $u^n - \bar{u}^n = 2i \sin n\theta$, (f) $\bar{u}^{n+1} - u^{n+1} = -2i\sin(n + 1)\theta$.

Substituting these in the expression following (6.2), we obtain

$$\sum_{k=1}^{n} \sin k\theta = \frac{\sin\theta - \sin(n+1)\theta + \sin n\theta}{2 - 2\cos\theta}, \tag{6.3}$$

where $\cos\theta \neq 1$.

PROB. 6.1. Prove: If n is a positive integer and $\theta \in \mathbb{R}$, then

$$\sum_{k=1}^{n} \sin k\theta = \frac{\sin(n\theta/2)\sin((n+1)/2)\theta}{\sin(\theta/2)}, \qquad \text{if } \theta \text{ is not an integral multiple of } 2\pi.$$

PROB. 6.2. Prove: If n is a positive integer and $\theta \in \mathbb{R}$, then

$$\sum_{k=1}^{n} \cos k\theta = \frac{\sin(n\theta/2)\cos((n+1)/2)\theta}{\sin(\theta/2)} \qquad \text{if } \theta \text{ is not an integral multiple of } 2\pi.$$

PROB. 6.3. Prove: If n is a positive integer and θ is not an integral multiple of π, then

(a) $\sum_{k=1}^{n}\sin(2k-1)\theta = \sin^2 n\theta/\sin\theta$ and
(b) $\sum_{k=1}^{n}\cos(2k-1)\theta = \sin 2n\theta/2\sin\theta$.

Def. 6.1. A function T of the form

$$T(\theta) = A + \sum_{k=1}^{n} (a_k\cos k\theta + b_k\sin k\theta), \tag{6.4}$$

where θ and A, $a_1, \ldots, a_n$; $b_1, b_2, \ldots, b_n$ are real, is called *a trigonometric polynomial of order n on* $\mathbb{R}$. (Here n is some nonnegative integer.) T is called *even* if $b_1 = b_2 = \cdots = b_n = 0$, and *odd* if $A = a_1 = a_2 = \cdots = a_n = 0$.

Theorem 6.1. *If n is a nonnegative integer and* $\theta \in \mathbb{R}$, *then*

$$2^{n-1}\cos^n\theta = \sum_{k=0}^{(n-1)/2} \binom{n}{k}\cos(n-2k)\theta \qquad \text{if } n \text{ is odd} \tag{6.5}$$

and

$$2^{n-1}\cos^n\theta = \frac{1}{2}\left(\binom{n}{\frac{n}{2}}\right) + \sum_{k=0}^{n/2-1} \binom{n}{k}\cos(n-2k)\theta \qquad \text{if } n \text{ is even.} \tag{6.6}$$

PROOF. Write $u = e^{i\theta}$, so that $\bar{u} = e^{-i\theta}$ and $2\cos\theta = u + \bar{u}$. Therefore,

$$2^n\cos^n\theta = (u+\bar{u})^n = \sum_{k=0}^{n} \binom{n}{k} u^{n-k}\bar{u}^k. \tag{6.7}$$

The sum on the right has $n+1$ terms. Assume that n is odd. We obtain from (6.7)

$$2^n\cos^n\theta = \sum_{k=0}^{(n-1)/2} \binom{n}{k} u^{n-k}\bar{u}^k + \sum_{k=(n+1)/2}^{n} \binom{n}{k} u^{n-k}\bar{u}^k. \tag{6.8}$$

Noting that

$$\binom{n}{k} = \binom{n}{n-k},$$

we can write the second sum on the right in (6.8) as

$$\sum_{k=(n+1)/2}^{n} \binom{n}{k} u^{n-k}\bar{u}^k = \sum_{k=(n+1)/2}^{n} \binom{n}{n-k} u^{n-k}\bar{u}^k. \tag{6.9}$$

Reindex by putting $j = n - k$. For $k \in \{(n+1)/2, \ldots, n\}$, we have $j \in \{0, \ldots, (n-1)/2\}$. Accordingly, we obtain from (6.9)

$$\sum_{k=(n+1)/2}^{n} \binom{n}{n-k} u^{n-k}\bar{u}^k = \sum_{j=0}^{(n-1)/2} \binom{n}{j} u^j\bar{u}^{n-j}. \tag{6.10}$$

Here we may replace the "dummy" variable j by k and obtain from (6.10)

$$\sum_{k=(n+1)/2}^{n} \binom{n}{n-k} u^{n-k}\bar{u}^k = \sum_{k=0}^{(n-1)/2} \binom{n}{k} \bar{u}^{n-k}u^k. \tag{6.11}$$

Using this in (6.8), the latter becomes

$$2^n\cos^n\theta = \sum_{k=0}^{(n-1)/2} \binom{n}{k} (u^{n-k}\bar{u}^k + \bar{u}^{n-k}u^k). \tag{6.12}$$

Now note that $u\bar{u} = 1$, so that

$$u^{n-k}\bar{u}^k = u^{n-2k}u^k\bar{u}^k = u^{n-2k} \quad \text{and similarly} \quad \bar{u}^{n-k}u^k = \bar{u}^{n-2k}.$$

Substituting these in (6.12) yields

$$2^n\cos^n\theta = \sum_{k=0}^{(n-1)/2} \binom{n}{k} (u^{n-2k} + \bar{u}^{n-2k}). \tag{6.13}$$

Since

$$u^{n-2k} + \bar{u}^{n-2k} = e^{i(n-2k)\theta} + e^{-i(n-2k)\theta} = 2\cos(n-2k)\theta,$$

it follows from (6.13) that (6.5) holds.

Next assume that n is even. In (6.7), the sum on the right consists of an odd number of terms, and we have from there

$$2^n\cos^n\theta = \sum_{k=0}^{n/2-1} \binom{n}{k} u^{n-k}\bar{u}^k + \left(\binom{n}{\frac{n}{2}}\right) u^{n/2}\bar{u}^{n/2} + \sum_{k=n/2+1}^{n} \binom{n}{k} u^{n-k}\bar{u}^k. \tag{6.14}$$

Reasoning as we did in the first case and noting that $u^{n/2}\bar{u}^{n/2} = 1$, we obtain

$$2^n\cos^n\theta = \sum_{k=0}^{n/2-1}\binom{n}{k}(u^{n-2k}+\bar{u}^{n-2k}) + \binom{n}{\frac{n}{2}}$$

$$= \binom{n}{\frac{n}{2}} + \sum_{k=0}^{n/2-1}\binom{n}{k}2\cos(n-2k)\theta.$$

Now (6.6) follows.

PROB. 6.4. Prove:

(a) If n is an odd positive integer, then

$$2^{n-1}(-1)^{(n-1)/2}\sin^n\theta = \sum_{k=0}^{(n-1)/2}(-1)^k\binom{n}{k}\sin(n-2k)\theta;$$

(b) if n is an even nonnegative integer, then

$$2^{n-1}(-1)^{n/2}\sin^n\theta = \frac{1}{2}\binom{n}{\frac{n}{2}}(-1)^{n/2} + \sum_{k=0}^{n/2-1}(-1)^k\binom{n}{k}\cos(n-2k)\theta.$$

Theorem 6.2. *If n is a positive integer, then, for $\theta \in \mathbb{R}$, we have*

$$\sin n\theta = \sum_{k=1, k\text{ odd}}^{n}(-1)^{(k-1)/2}\binom{n}{k}\cos^{n-k}\theta\sin^k\theta \qquad \text{if } n \text{ is odd} \quad (6.15)$$

and

$$\sin n\theta = \sum_{k=1, k\text{ odd}}^{n-1}(-1)^{(k-1)/2}\binom{n}{k}\cos^{n-k}\theta\sin^k\theta \qquad \text{if } n \text{ is even.} \quad (6.16)$$

PROOF. We have

$$2i\sin n\theta = \cos n\theta + i\sin n\theta - (\cos n\theta - i\sin n\theta)$$

$$= (\cos\theta + i\sin\theta)^n - (\cos\theta - i\sin\theta)^n$$

$$= \sum_{k=0}^{n}\binom{n}{k}\cos^{n-k}\theta(i^k\sin^k\theta) - \sum_{k=0}^{n}\binom{n}{k}\cos^{n-k}\theta((-i)^k\sin^k\theta)$$

$$= \sum_{k=0}^{n}\binom{n}{k}(\cos^{n-k}\theta\sin^k\theta)(i^k - (-i)^k).$$

Since $i^k - (-i)^k = 0$ if k is even, the only terms appearing in the last sum are those for which k is odd. Hence,

$$2i\sin n\theta = \sum_{k=1, k\text{ odd}}^{n} 2i^k\cos^{n-k}\theta\sin^k\theta. \quad (6.17)$$

We divide both sides by $2i$ to obtain from this

$$\sin n\theta = \sum_{k=1,k\text{ odd}}^{n} \binom{n}{k} i^{k-1}\cos^{n-k}\theta \sin^k\theta. \tag{6.18}$$

In this sum, $k-1$ is even. Hence, $i^{k-1} = (i^2)^{(k-1)/2} = (-1)^{(k-1)/2}$, and (6.18) becomes

$$\sin n\theta = \sum_{k=1,k\text{ odd}}^{n} (-1)^{(k-1)/2}\binom{n}{k}\cos^{n-k}\theta \sin^k\theta \tag{6.19}$$

if n is odd and

$$\sin n\theta = \sum_{k=1,k\text{ odd}}^{n-1} (-1)^{(k-1)/2}\binom{n}{k}\cos^{n-k}\theta \sin^k\theta$$

if n is even.

Corollary. *If n is an odd positive integer and $\theta \in \mathbb{R}$, then $\sin n\theta$ can be written as a polynomial of degree n in $x = \sin\theta$ of the form*

$$\sin n\theta = A_1x + A_3x^3 + \cdots + A_nx^n, \tag{6.20}$$

where $A_1, A_3, \ldots, A_n$ are constants not depending on θ.

Proof. Since n is odd, the theorem yields

$$\begin{aligned}\sin n\theta &= n\cos^{n-1}\theta\sin\theta - \binom{n}{3}\cos^{n-3}\theta\sin^3\theta + \cdots + (-1)^{(n-1)/2}\sin^n\theta \\ &= \sin\theta\left[\cos^{n-1}\theta - \binom{n}{3}\cos^{n-3}\theta\sin^2\theta + \cdots + (-1)^{(n-1)/2}\sin^{n-1}\theta\right].\end{aligned} \tag{6.21}$$

In the second factor on the right the exponents $n-1, n-3, \ldots, 0$ are all even and (6.21) can be written

$$\begin{aligned}\sin n\theta = \sin\theta\Big[n(1-\sin^2\theta)^{(n-1)/2} &- \binom{n}{3}(1-\sin^2\theta)^{(n-3)/2}\sin^2\theta \\ &+ \cdots + (-1)^{(n-1)/2}\sin^{n-1}\theta\Big].\end{aligned} \tag{6.22}$$

On the right here the exponents $(n-1)/2, (n-3)/2, \ldots$ are all nonnegative integers. We expand

$$(1-\sin^2\theta)^{(n-1)/2}, \qquad (1-\sin^2\theta)^{(n-3)/2}, \ldots \tag{6.23}$$

by the binomial theorem and find that the factor in square brackets on the right in (6.22) has the form $A_1 + A_3x^2 + \cdots + A_nx^{n-1}$, where $x = \sin\theta$ and $A_1 = n$. Thus, (6.22) has the form

$$\sin n\theta = x(A_1 + A_3x^2 + \cdots + A_nx^{n-1}) = A_1x + A_3x^3 + \cdots + A_nx^n, \tag{6.24}$$

where $x = \sin\theta$ and $A_1 = n$. We examine the term $t = A_n x^n$. We have for this term (from (6.22), after using the binomial theorem in (6.23)),

$$\begin{aligned} t &= \sin\theta\Big(n(-1)^{(n-1)/2}(\sin^2\theta)^{(n-1)/2} - \binom{n}{3}(-1)^{(n-3)/2}(\sin^2\theta)^{(n-3)/2}\sin^2\theta \\ &\qquad + \cdots + (-1)^{(n-1)/2}\sin^{n-1}\theta\Big) \\ &= \sin\theta\Big((-1)^{(n-1)/2}n\sin^{n-1}\theta + (-1)^{(n-1)/2}\binom{n}{3}\sin^{n-1}\theta \\ &\qquad + \cdots + (-1)^{(n-1)/2}\sin^{n-1}\theta\Big) \\ &= (-1)^{(n-1)/2}\Big(n + \binom{n}{3} + \cdots + 1\Big)\sin^n\theta. \end{aligned}$$

Thus,

$$A_n = (-1)^{(n-1)/2}\Big(n + \binom{n}{3} + \cdots + 1\Big) \neq 0,$$

and the polynomial in (6.24) in $x = \sin\theta$ is of degree n, as claimed.

PROB. 6.5. Prove: If n is an odd positive integer, then $\cos n\theta$ can be written as a polynomial of degree n in $x = \cos\theta$ of the form

$$\cos n\theta = B_1 x + B_3 x^3 + \cdots + B_n x^n.$$

The following result will be needed later (Section 9).

Theorem 6.3. *If n is an odd positive integer and t is real, then*

$$\sin t = n \sin\frac{t}{n}\prod_{k=1}^{(n-1)/2}\left[1 - \frac{\sin^2(t/n)}{\sin^2(k/n)\pi}\right]. \tag{6.25}$$

PROOF. We begin with the last corollary. From (6.20) and the fact that s n $n\theta$ vanishes for $n\theta = k\pi$, where k is an integer, we see that the right-hand side of (6.20) vanishes for x_k, where

$$x_k = \sin\Big(k\frac{\pi}{n}\Big), \qquad k = 0, \pm 1, \pm 2, \ldots . \tag{6.26}$$

Since n is odd, $\pm(n-1)/2$ are integers. We know that the sine function is one-to-one on the interval $[-\pi/2, \pi/2]$. Therefore, if k is an integer such that $|k| \leqslant (n-1)/2$, so that

$$\left|k\frac{\pi}{n}\right| \leqslant \frac{n-1}{n}\,\frac{\pi}{2} < \frac{\pi}{2}, \tag{6.27}$$

the x_k's in (6.26) are distinct zeros of the polynomial on the right of (6.20). But $|k| \leqslant (n-1)/2$ holds for $n-1$ distinct nonzero k's and for $k = 0$ and, therefore, for n of the k's. We conclude that the polynomial on the right-hand side of (6.20) has n distinct zeros x_k, where $|k| \leqslant (n-1)/2$. This polynomial is of degree n in x and has no other zeros. By the factor theorem, we have for this polynomial

$$\sin n\theta = A_n x\left(x - \sin\frac{\pi}{n}\right)\left(x + \sin\frac{\pi}{n}\right)\dots$$
$$\cdot\left(x - \sin\frac{n-1}{2}\frac{\pi}{n}\right)\left(x + \sin\frac{n-1}{2}\frac{\pi}{n}\right)$$
$$= A_n x\left(x^2 - \sin^2\frac{\pi}{n}\right)\left(x^2 - \sin^2\frac{2\pi}{n}\right)\dots$$
$$\cdot\left(x^2 - \sin^2\frac{n-1}{2}\frac{\pi}{n}\right)$$
$$= A_n \sin\theta\left(\sin^2\theta - \sin^2\frac{\pi}{n}\right)\left(\sin^2\theta - \sin^2\frac{2\pi}{n}\right)\dots$$
$$\cdot\left(\sin^2\theta - \sin^2\frac{n-1}{2}\frac{\pi}{n}\right). \tag{6.28}$$

We multiply and divide the right side of the above by

$$\sin^2\frac{\pi}{n}\sin^2 2\frac{\pi}{n}\dots\sin^2\frac{n-1}{2}\frac{\pi}{n},$$

and then rearrange its factors to obtain from (6.28)

$$\sin n\theta = B\sin\theta\left(1 - \frac{\sin^2\theta}{\sin^2(\pi/n)}\right)\left(1 - \frac{\sin^2\theta}{\sin^2(2/n)\pi}\right)\dots$$
$$\cdot\left(1 - \frac{\sin^2\theta}{\sin^2((n-1)/2)\pi/n}\right), \tag{6.29}$$

where B is some constant depending only on n. We have

$$\lim_{\theta\to 0}\frac{\sin n\theta}{\sin\theta} = \lim_{\theta\to 0} n\frac{(\sin n\theta)/n\theta}{(\sin\theta)/\theta} = n \tag{6.30}$$

and

$$\lim_{\theta\to 0}\left(1 - \frac{\sin^2\theta}{\sin^2(\pi/n)}\right)\left(1 - \frac{\sin^2}{\sin^2(2\pi/n)}\right)\dots$$
$$\cdot\left(1 - \frac{\sin^2\theta}{\sin^2((n-1)/2)(\pi/n)}\right) = 1. \tag{6.31}$$

These relations and (6.29) imply that $B = n$. Thus, (6.29) can be written

$$\sin n\theta = n\sin\theta\left(1 - \frac{\sin^2\theta}{\sin^2(\pi/n)}\right)\left(1 - \frac{\sin^2\theta}{\sin^2(2\pi/n)}\right)\dots$$
$$\cdot\left(1 - \frac{\sin^2\theta}{\sin^2((n-1)/2)(\pi/n)}\right). \tag{6.32}$$

Equation (6.25) is obtained by putting $t = n\theta$.

Theorem 6.4.* *If m is a positive integer, then*

$$\sum_{k=1}^{m} \cot^2 \frac{k\pi}{2m+1} = \frac{m(2m-1)}{3}. \tag{6.33}$$

PROOF. Use Theorem 6.2. Let $n = 2m + 1$, where m is a positive integer and $\theta \neq \nu\pi$, where ν is an integer. By (6.15), we have

$$\begin{aligned}\sin(2m+1)\theta &= (2m+1)\cos^{2m}\theta \sin\theta - \binom{2m+1}{3}\cos^{2m-2}\theta \sin^3\theta \\ &\quad + \cdots + (-1)^m \sin^{2m+1}\theta \\ &= \sin^{2m+1}\theta \Big[(2m+1)\cot^{2m}\theta - \binom{2m+1}{3}\cot^{2m-2}\theta \\ &\quad + \cdots + (-1)^m \Big]. \end{aligned} \tag{6.34}$$

The factor in square brackets on the right is a polynomial $P(x)$ of degree m in $x = \cot^2\theta$, and we may write (6.34) as

$$\sin(2m+1)\theta = (\sin^{2m+1}\theta)P(x). \tag{6.35}$$

The left-hand side vanishes for $\theta = k\pi/(2m+1)$, $k \in \{1, \ldots, m\}$. Clearly,

$$0 < \frac{k\pi}{2m+1} < \frac{\pi}{2} \qquad \text{for} \quad k \in \{1, \ldots, m\}.$$

Therefore, we know that the x_k's given by

$$x_k = \cot^2\theta \frac{k\pi}{2m+1}, \qquad k \in \{1, \ldots, m\}, \tag{6.36}$$

are all distinct. Since $\sin^{2m+1}\theta$ does not vanish on $(0; \pi/2)$, it is not equal to 0 at any of the x_k's above. These x_k's are m distinct zeros of the mth degree polynomial P and therefore constitute *all* the zeros of P. Their sum is $\sigma_1 = x_1 + \cdots + x_m$. Hence (cf. (4.29)), σ_1 is the negative of the coefficient of x^{m-1} divided by the leading coefficient $2m + 1$ of P. Thus,

$$\begin{aligned}\sum_{k=1}^{m} \cot^2 \frac{k\pi}{2m+1} &= \sum_{k=1}^{m} x_k = \sigma_1 = \binom{2m+1}{3}(2m+1)^{-1} \\ &= \frac{(2m+1)2m(2m-1)}{3!\,(2m+1)} \\ &= \frac{m(2m-1)}{3}. \end{aligned}$$

This proves the theorem.

*L. B. W. Jolley, *Summation of Series*, formula (451), Dover, New York, 1961.

This theorem enables one to evaluate $\zeta(2)$, where ζ is the zeta function, i.e.,

$$\zeta(2)=\sum_{n=1}^{\infty}\frac{1}{n^2}=1+\frac{1}{2^2}+\frac{1}{3^2}+\cdots.$$

The proof of the next theorem is due to Apostol.

Theorem 6.5. *We have*

$$\frac{\pi^2}{6}=\sum_{n=1}^{\infty}\frac{1}{n^2}=\zeta(2). \tag{6.37}$$

Proof. Let $0<x<\pi/2$, so that $0<\sin x<x<\tan x$. It follows that $0<\cot x<1/x<\csc x$ and

$$\cot^2 x<1/x^2<\csc^2 x=1+\cot^2 x; \tag{6.38}$$

substitute $x=k\pi/(2m+1)$, $k\in\{1,\ldots,m\}$ and sum to obtain

$$\sum_{k=1}^{m}\cot^2\frac{k\pi}{2m+1}<\frac{(2m+1)^2}{\pi^2}+\frac{(2m+1)^2}{2^2\pi^2}+\cdots+\frac{(2m+1)^2}{m^2\pi^2}$$
$$<m+\sum_{k=1}^{m}\cot^2\frac{k\pi}{2m+1}.$$

Use Theorem 6.4 to conclude that

$$\frac{m(2m-1)}{3}<\frac{(2m+1)^2}{\pi^2}\left(1+\frac{1}{2^2}+\cdots+\frac{1}{m^2}\right)<m+\frac{m(2m-1)}{3}$$

and, therefore, that

$$\frac{m(2m-1)}{3(2m+1)^2}\pi^2<1+\frac{1}{2^2}+\cdots+\frac{1}{m^2}<\frac{\pi^2}{3(2m+1)^2}(2m^2+2m). \tag{6.39}$$

Now let $m\to+\infty$ and note

$$\lim_{m\to+\infty}\frac{m(2m-1)}{3(2m+1)^2}\pi^2=\lim_{m\to+\infty}\frac{\pi^2}{3(2m+1)^2}(2m^2+2m)=\frac{\pi^2}{6}.$$

By the sandwich theorem, this and (6.39) imply that

$$\frac{\pi^2}{6}=\lim_{m\to+\infty}\sum_{k=1}^{m}\frac{1}{k^2}=\frac{1}{1^2}+\frac{1}{2^2}+\frac{1}{3^2}+\cdots.$$

Prob. 6.6. Prove

$$\sum_{n=1}^{\infty}\frac{1}{n^2(n+1)}=\frac{\pi^2}{6}-1.$$

7. Convergence and Divergence of Infinite Products

Finite products such as

$$\prod_{k=1}^{n} a_k \quad \text{and} \quad \prod_{k=0}^{n} a_k$$

were defined in Chapter II. Here we define infinite products of real numbers.

Def. 7.1. Let $\langle u_n \rangle$ be an infinite sequence of real numbers. The sequence $\langle P_n \rangle$, where

$$P_n = \prod_{k=1}^{n} u_k \qquad \text{for each } n, \tag{7.1}$$

is called an *infinite product* of terms of the sequence $\langle u_n \rangle$. The nth term of $\langle P_n \rangle$ is called the nth *partial product* of the infinite product. We write this infinite product as

$$\prod_{n=1}^{\infty} u_n . \tag{7.2}$$

(This is analogous to the procedure we used in defining infinite series.) Next we wish to assign a number to the infinite product. We do not proceed in full analogy with infinite series. If we were to do so, an infinite product would be called convergent if the sequence of its partial products converges. It would then follow that if one of the terms of $\langle u_n \rangle$ is equal to 0, say $u_j = 0$, then all the partial products P_n for $n > j$ would equal 0. Therefore, the partial product sequence would converge to 0, regardless of what the terms u_n for $n > j$ are like. The partial product sequence of the subsequence $\langle u_{j+1}, u_{j+2}, \ldots \rangle$ might not converge even though the sequence of partial products of the sequence $\langle u_1, u_2, \ldots \rangle$ converges. We, therefore, restrict the sequences $\langle u_n \rangle$ whose infinite products we form. We form infinite products of sequences $\langle u_n \rangle$ for which there exists a positive integer N such that $u_n \neq 0$ for $n > N$. Then we define the convergence of the infinite product

$$\prod_{k=N+1}^{\infty} u_k = u_{N+1} u_{N+2} \cdots \tag{7.3}$$

and form the sequence $\langle P'_{N+1}, P'_{N+2}, \ldots \rangle$, where

$$P'_n = \prod_{k=N+1}^{n} u_k \qquad \text{for} \quad n \geqslant N+1.$$

The P'_n's are the partial products of (7.3). Here too we impose a restriction and call (7.3) *convergent* when the sequence of its partial products converges to a nonzero limit. If we were to allow $\lim P'_n = 0$ and write

$$\prod_{k=N+1}^{\infty} u_k = \lim_{n\to+\infty} \prod_{k=N+1}^{n} u_k = \lim_{n\to+\infty} P'_n,$$

we would have

$$\prod_{k=N+1}^{\infty} u_k = 0$$

even though $u_{N+1}, u_{N+2}, \ldots$ are all different from 0. This behavior of infinite products would differ from that of finite products. The latter have the property that

$$\prod_{K=N+1}^{n} u_k = 0$$

implies one of $u_1, \ldots, u_n$ is zero.

Next we give an "official" definition of convergence for infinite products.

Def. 7.2. The infinite product $\prod_{n=1}^{\infty} u_n$ will be called *convergent* if (1) there exists a positive integer N such that $u_n \neq 0$ for $n > N$ and (2)

$$\lim_{n\to+\infty} \prod_{k=N+1}^{n} u_k \tag{7.4}$$

is a nonzero real number. Writing P' for the number in (7.4), we define the *product* P of $u_1, u_2, \ldots$ as

$$P = u_1 u_2 \ldots u_N P'. \tag{7.5}$$

In this case, we write

$$\prod_{k=N+1}^{\infty} u_k = P' = \lim_{n\to+\infty} \prod_{k=N+1}^{\infty} u_k$$

and

$$\prod_{k=1}^{\infty} u_k = P = u_1 u_2 \ldots u_N \prod_{k=N+1}^{\infty} u_k. \tag{7.6}$$

If for some N the limit in (7.4) does not exist, or is infinite, or equal to 0, we then say that the infinite product $\prod_{k=1}^{\infty} u_k$ *diverges*.

In analogy with infinite series, the symbol $\prod_{k=1}^{\infty} u_k$ is given two meanings. Indeed, on the one hand, it represents the sequence of partial products of $\langle u_n \rangle$ and, on the other hand, it is the number defined by (7.5). It should be clear how to define the infinite product

$$\prod_{k=0}^{\infty} u_k.$$

Example 7.1. Consider

$$\prod_{n=1}^{\infty}\left(1-\frac{1}{(n+1)^2}\right)=\left(1-\frac{1}{2^2}\right)\left(1-\frac{1}{3^2}\right)\cdots.$$

The nth partial product of this infinite product is

$$P_n=\prod_{k=1}^{n}\left(1-\frac{1}{(k+1)^2}\right)=\left(1-\frac{1}{2^2}\right)\left(1-\frac{1}{3^2}\right)\cdots\left(1-\frac{1}{(n+1)^2}\right).$$

Since

$$\begin{aligned}P_n&=\left(1-\frac{1}{2^2}\right)\left(1-\frac{1}{3^2}\right)\cdots\left(1-\frac{1}{(n+1)^2}\right)\\&=\left(1-\frac{1}{2}\right)\left(1-\frac{1}{3}\right)\cdots\left(1-\frac{1}{n+1}\right)\\&\quad\cdot\left(1+\frac{1}{2}\right)\left(1+\frac{1}{3}\right)\cdots\left(1+\frac{1}{n+1}\right)\\&=\frac{1}{2}\cdot\frac{2}{3}\cdots\frac{n}{n+1}\cdot\frac{3}{2}\cdot\frac{4}{3}\cdots\frac{n+2}{n+1}\\&=\frac{1}{2}\,\frac{n+2}{n+1},\end{aligned}$$

we have

$$P=\lim P_n=\lim\left(\frac{1}{2}\,\frac{n+2}{n+1}\right)=\frac{1}{2}.$$

Prob. 7.1. Prove:

$$\prod_{n=2}^{\infty}\left(1-\frac{2}{n(n+1)}\right)=\frac{1}{3}.$$

Example 7.2. The infinite product

$$\prod_{n=1}^{\infty}\left(1+\frac{1}{n}\right)$$

diverges. We have for the nth partial product

$$\begin{aligned}P_n&=\prod_{k=1}^{n}\left(1+\frac{1}{k}\right)=(1+1)\left(1+\frac{1}{2}\right)\cdots\left(1+\frac{1}{n}\right)\\&=2\cdot\frac{3}{2}\cdots\frac{n+1}{n}=n+1\to+\infty\qquad\text{as}\quad n\to 0.\end{aligned}$$

Example 7.3. The infinite product

$$\prod_{n=1}^{\infty}\left(1-\frac{1}{n+1}\right)$$

diverges to 0. For each n,

$$P_n = \left(1 - \frac{1}{2}\right)\left(1 - \frac{1}{3}\right) \cdots \left(1 - \frac{1}{n+1}\right)$$
$$= \frac{1}{2} \cdot \frac{2}{3} \cdots \frac{n}{n+1} = \frac{1}{n+1} \to 0.$$

EXAMPLE 7.4 (The Gamma Function as an Infinite Product). We recall that

$$\Gamma(x) = \lim_{n \to +\infty} \frac{n^x n!}{x(x+1)(x+2) \ldots (x+n)}$$

if $x \neq 0$ or x is not a negative integer.

For each positive integer n, define

$$g_n(x) = \frac{n^x n!}{x(x+1)(x+2) \ldots (x+n)}.$$

We have

$$g_n(x) = \frac{n^x}{x(1+x)(1+x/2) \ldots (1+x/n)} = \frac{e^{x \ln n}}{x(1+x) \ldots (1+x/n)}$$
$$= \frac{1}{x} e^{x[\ln n - (1+1/2+ \cdots +1/n)]} \frac{e^x}{(1+x)} \frac{e^{x/2}}{(1+x/2)} \cdots \frac{e^{x/n}}{1+x/n}$$
$$= \frac{1}{x} e^{-\gamma_n x} \prod_{k=1}^{n} \frac{e^{x/k}}{(1+x/k)},$$

where

$$\gamma_n = 1 + \frac{1}{2} + \cdots + \frac{1}{n} - \ln n.$$

The sequence $\langle \gamma_n \rangle$ is dealt with in Theorem VII.7.1. There we saw that it converges. Its limit γ is the Euler–Mascheroni constant. We have

$$e^{-\gamma x} \frac{1}{x} \prod_{k=1}^{\infty} \frac{e^{x/k}}{(1+x/k)} = \lim_{n \to +\infty} e^{-\gamma_n x} \prod_{k=1}^{n} \frac{e^{x/k}}{1+x/k}$$
$$= \lim_{n \to +\infty} g_n(x) = \Gamma(x),$$

i.e.,

$$\Gamma(x) = e^{-\gamma x} \frac{1}{x} \prod_{k=1}^{\infty} \frac{e^{x/k}}{(1+x/k)}. \qquad (*)$$

Next we state a necessary condition for the convergence of an infinite product.

Theorem 7.1. *If $\prod_{n=1}^{\infty} u_n$ converges, then $u_n \to 1$ as $n \to +\infty$.*

Proof. Since the product converges, there exists a positive integer N such that if $n > N$, then $u_n \neq 0$ and

$$P' = \lim_{n\to+\infty} \prod_{k=N+1}^{n} u_k,$$

where P' is some nonzero real number. This implies that

$$\lim_{n\to+\infty} u_n = \lim_{n\to+\infty} \frac{\prod_{k=N+1}^{n} u_k}{\prod_{k=N+1}^{n-1} u_k} = \frac{\lim_{n\to+\infty}\prod_{k=N+1}^{n} u_k}{\lim_{n\to+\infty}\prod_{k=N+1}^{n-1} u_k} = \frac{P'}{P'} = 1$$

and completes the proof.

It is customary to write $u_n = 1 + a_n$ for each n so that

$$\prod_{n=1}^{\infty} u_n = \prod_{n=1}^{\infty} (1 + a_n).$$

In this notation Theorem 7.1 takes the form:

Theorem 7.1′. *If $\prod_{n=1}^{\infty}(1 + a_n)$ converges, then $a_n \to 0$ as $n \to +\infty$.*

Remark 7.1. The criterion for convergence given in the last theorem is necessary but not sufficient. Consider the infinite product of Example 7.2, i.e., $\prod_{n=1}^{\infty}(1 + 1/n)$. This diverges, but $a_n = 1/n \to 0$ as $n \to +\infty$. Note also that the infinite product

$$\prod_{n=1}^{\infty}\left(1 - \frac{1}{n+1}\right)$$

of Example 7.3 diverges to 0 even though $a_n = 1/(n+1) \to 0$ as $n \to +\infty$. The divergence of both products mentioned here, the first to $+\infty$ and the second to 0, can also be deduced from the inequalities

$$1 + \frac{1}{1} + \frac{1}{2} + \cdots + \frac{1}{n} \leqslant \left(1 + \frac{1}{1}\right)\left(1 + \frac{1}{2}\right)\cdots\left(1 + \frac{1}{n}\right) \tag{7.7}$$

and

$$\left(1 - \frac{1}{2}\right)\left(1 - \frac{1}{3}\right)\cdots\left(1 - \frac{1}{n+1}\right) < \frac{1}{1 + 1/2 + \cdots + 1/(n+1)} \tag{7.8}$$

(see Prob. II.12.15), and the divergence of the harmonic series to $+\infty$.

Prob. 7.2. Prove: If $x \neq 0$, $x \in \mathbb{R}$, then

$$\prod_{n=1}^{\infty}\left(1 + \frac{x}{n}\right)$$

diverges (see Prob. IX.6.2).

PROB. 7.3. Prove:

$$\prod_{n=1}^{\infty}\left(1-(-1)^n\frac{1}{n}\right)=1.$$

Remark 7.2. Consider the infinite products

$$\prod_{n=1}^{\infty}(1+a_n) \tag{7.9a}$$

and

$$\prod_{n=1}^{\infty}(1-a_n), \tag{7.9b}$$

where $0 \leqslant a_n < 1$ for all n. Let their respective partial product sequences be $\langle P_n\rangle$ and $\langle Q_n\rangle$. The first sequence is monotonically increasing and the second is monotonically decreasing. We also see that $P_n > 0$ and $Q_n > 0$ for all n. It therefore follows that if (7.9a) diverges, then it diverges to $+\infty$ and if (7.9b) diverges, it diverges to 0.

Theorem 7.2. *If* $0 \leqslant a_n < 1$ *for all* n, *then the infinite product* (7.9a) *converges if and only if* $\langle P_n\rangle$ *is bounded, whereas the product* (7.9b) *converges if and only if* $\langle Q_n\rangle$ *is bounded from below by some positive number.*

PROOF. Exercise.

Theorem 7.3. *Given the infinite products in* (7.9), *the convergence of*

$$\sum a_n \tag{7.10}$$

is a necessary and sufficient condition for (7.9a) *to converge provided that* $a_n \geqslant 0$ *for all* n, *and is a necessary and sufficient condition for* (7.9b) *to converge, provided that* $0 \leqslant a_n < 1$ *for all* n.

PROOF. Assume that $0 \leqslant a_n$ for all n. Consider (7.9a). Assume that (7.10) converges. Let $S = \sum a_n$, $\langle S_n\rangle$ be its sequence of partial *sums* and $\langle P_n\rangle$ its sequence of partial products. We know that $\ln(1+x) \leqslant x$ for $x > -1$. Hence,

$$\ln \prod_{k=1}^{n}(1+a_k) = \sum_{k=1}^{n}\ln(1+a_k) \leqslant \sum_{k=1}^{n} a_k = S_n \leqslant S \qquad \text{for all } n.$$

This implies that

$$P_n = \prod_{k=1}^{n}(1+a_k) \leqslant e^S \qquad \text{for all } n. \tag{7.11}$$

Thus, the monotonically increasing sequence $\langle P_n\rangle$ (see Remark 7.2) is bounded and, therefore, converges to a finite limit P. It is also clear that $P \geqslant 1$. Thus, (7.9a) converges. Conversely, assume that the infinite product (7.9a) converges. We have $P_n \leqslant P$ for all n, where P is a value of the infinite product (explain). By Prob. II.12.14 we have

$$1 + \sum_{k=1}^{n} a_k \leqslant \prod_{k=1}^{n} (1 + a_k) = P_n \leqslant P \qquad \text{for all } n.$$

This tells us that $\langle S_n \rangle$ is bounded. Since $a_n \geqslant 0$ for all n, we know that $\sum a_n$ converges.

We now concern ourselves with (7.9b). Assume that $0 \leqslant a_n < 1$ for all n and that $\sum a_n$ converges. Write S for its sum and S_n for the nth partial sum for each n. There exists a positive integer N such that

$$0 \leqslant \sum_{k=N+1}^{\infty} a_k = S - S_N < 1.$$

Put $R_N = \sum_{k=N+1}^{\infty} a_k$ so that

$$0 \leqslant \sum_{k=N+1}^{n} a_k \leqslant R_N < 1 \qquad \text{if} \quad n > N. \tag{7.12}$$

Hence,

$$0 < 1 - R_N \leqslant 1 - \sum_{k=N+1}^{n} a_k \qquad \text{for} \quad n > N. \tag{7.13}$$

By Prob. II.12.15 we have

$$1 - \sum_{k=N+1}^{n} a_k \leqslant \prod_{k=N+1}^{n} (1 - a_k) \qquad \text{for} \quad n > N.$$

Therefore,

$$0 < 1 - R_N \leqslant \prod_{k=N+1}^{n} (1 - a_k) \qquad \text{for} \quad n > N. \tag{7.14}$$

Thus, the sequence $\langle P_n' \rangle$ of partial sums of

$$\sum_{n=N+1}^{\infty} (1 - a_n) \tag{7.15}$$

is bounded from below by the positive number $1 - R_N$. By Theorem 7.2 the infinite product (7.15) converges. It follows from Def. 7.2 that (7.9b) converges.

Now return to the condition $0 \leqslant a_n < 1$ for all n and assume that (7.9b) converges. Let $\langle Q_m \rangle$ be its sequence of partial products. By Theorem 7.2 the sequence $\langle Q_n \rangle$ is bounded from below by a positive number, say, B. We have $0 < B \leqslant Q_n$ for all n. We now use Prob. II.12.15, part (b) and obtain

$$0 < B \leqslant Q_n = \prod_{k=1}^{n} (1 - a_k) < \frac{1}{\prod_{k=1}^{n}(1 + a_k)} \qquad \text{for all } n.$$

This implies that

$$\prod_{k=1}^{n} (1 + a_k) < \frac{1}{B} \qquad \text{for all } n.$$

This and Theorem 7.2, part (a) imply that (7.9a) converges. This and the assumption that $0 \leqslant a_n < 1$ for all n guarantee, by the first part of this theorem, that $\sum a_n$ converges. The proof is now complete.

EXAMPLE 7.5. The infinite product

$$f(x) = \prod_{n=1}^{\infty}\left(1 - \frac{x^2}{n^2}\right) \tag{7.16}$$

converges for each $x \in \mathbb{R}$. This is clear if $|x| < 1$, for then

$$0 \leqslant \frac{x^2}{n^2} < \frac{1}{n^2} \qquad \text{for each positive integer } n \tag{7.17}$$

and

$$\sum_{n=1}^{\infty} \frac{x^2}{n^2} = x^2 \sum_{n=1}^{\infty} \frac{1}{n^2}$$

converges. Using the second part of the last theorem we obtain from this that the infinite product (7.16) converges. If $|x| \geqslant 1$, let $N = [|x|]$ so that $1 \leqslant N \leqslant |x| < N + 1$. For $n \geqslant N + 1$, we have $0 \leqslant x^2/n^2 < 1$ and that $\sum_{k=N+1}^{\infty} x^2/k^2$ converges. Thus,

$$\prod_{n=N+1}^{\infty}\left(1 - \frac{x^2}{n^2}\right)$$

converges. This yields the convergence of (7.16) also for the case $|x| \geqslant 1$.

PROB. 7.4. Prove:

$$\prod_{n=1}^{\infty}\left(1 + \frac{1}{n^{\alpha}}\right)$$

converges if $\alpha > 1$, and diverges to $+\infty$ if $\alpha \leqslant 1$. (See Remark 7.1 for the case $\alpha = 1$.)

PROB. 7.5. Prove:

$$\prod_{n=1}^{\infty}\left(1 - \frac{1}{n^{\alpha}}\right)$$

converges for $\alpha > 1$ and diverges for $\alpha \leqslant 1$.

Theorem 7.4. *The infinite product*

$$\prod_{n=1}^{\infty}(1 + a_n)$$

(here a_n is not necessarily positive for all n) converges, if and only if for each $\epsilon > 0$ there exists a positive integer N such that if $m > n > N$, then

$$\left|\prod_{k=n+1}^{m}(1 + a_k) - 1\right| < \epsilon. \tag{7.18}$$

PROOF. We prove sufficiency first. Suppose that for each $\epsilon > 0$ there exists a positive integer N such that if $m > n > N$, then (7.18) holds. Let $\epsilon > 0$ be given. There exists a positive integer N such that if $m > n > N_1$, then

$$\left|\prod_{k=n+1}^{m}(1+a_k)-1\right|<\frac{\epsilon}{3}. \tag{7.19}$$

There also exists a positive integer N_2 such that if $m > n > N_2$, then

$$\left|\prod_{k=n+1}^{m}(1+a_k)-1\right|<\frac{1}{2}. \tag{7.20}$$

Let $N = \max\{N_1, N_2\}$. By (7.20), we have

$$\left|\prod_{k=N+1}^{m}(1+a_k)-1\right|<\frac{1}{2} \qquad \text{for} \quad m \geqslant N+1 \tag{7.21}$$

so that

$$0<\frac{1}{2}<\prod_{k=N+1}^{m}(1+a_k)<\frac{3}{2} \qquad \text{for} \quad m \geqslant N+1. \tag{7.22}$$

We consider the sequence $\langle P_n' \rangle$ of partial products of

$$\prod_{k=N+1}^{\infty}(1+a_k). \tag{7.23}$$

We know that $N \geqslant N_1$ and $N \geqslant N_2$. Take $m > n > N$ so that $m > n > N_1$ and $m > n > N_2$. We have, first of all, that (7.19) holds and, therefore, that

$$\left|\frac{P_m'}{P_n'}-1\right|<\frac{\epsilon}{3}.$$

This, the fact that $m \geqslant N+1$, and (7.22) imply that

$$|P_m'-P_n'|<P_n'\frac{\epsilon}{3}<\frac{3}{2}\,\frac{\epsilon}{3}<\epsilon \qquad \text{for} \quad m>n>N.$$

Thus, $\langle P_n' \rangle$ is a Cauchy sequence and converges. But (7.22) also implies that $\lim P_n' > 0$. Thus, (7.23) converges by Def. 7.2. We conclude that $\prod_{n=1}^{\infty}(1+a_n)$ converges.

We prove the necessity next. Assume that $\prod_{n=1}^{\infty}(1+a_n)$ converges. Then there exists a positive integer N such that if $n > N$, then $1 + a_n \neq 0$ and that (7.23) converges to a nonzero number. It is easy to prove for the partial products P_n' of (7.23) that some $M > 0$ exists such that

$$|P_n'| > M \qquad \text{for all } n. \tag{7.24}$$

Since $\langle P_n' \rangle$ is necessarily a Cauchy sequence of real numbers, there exists a positive integer N_1 such that

$$|P_m'-P_n'|<M\epsilon \qquad \text{for} \quad m>n>N_1. \tag{7.25}$$

Take $N_2 = \max\{N, N_1\}$ and $m > n > N_2$. Thus, (7.24) and (7.25) hold and

$$\left|\prod_{k=n+1}^{m}(1+a_k)-1\right|=\left|\frac{P_m'}{P_n'}-1\right|<\frac{M}{|P_n'|}\epsilon<\epsilon$$

for $m > n > N_2$. This completes the proof.

8. Absolute Convergence of Infinite Products

Def. 8.1. When the infinite product

$$\prod_{n=1}^{\infty}(1+|a_n|) \tag{8.1}$$

converges, we say that the infinite product $\prod_{n=1}^{\infty}(1+a_n)$ is *absolutely convergent*.

Theorem 8.1. *The infinite product $\prod(1+a_n)$ converges absolutely if and only if $\sum a_n$ converges absolutely.*

PROOF. Exercise.

PROB. 8.1. Prove:

$$\prod_{n=1}^{\infty}(1+x^n) \quad \text{and} \quad \prod_{n=1}^{\infty}(1-x^n)$$

converge absolutely if $|x|<1$.

Theorem 8.2. *If an infinite product converges absolutely, it converges.*

PROOF. Assume that i and j are integers such that $j>i\geqslant 1$. We obtain, after "multiplying out,"

$$\begin{aligned}\left|\prod_{k=i}^{j}(1+a_k)-1\right| &= |a_i+a_{i+1}+\cdots+a_j+\cdots+a_ia_{i+1}\ldots a_j| \\ &\leqslant |a_i|+|a_{i+1}|+\cdots+|a_j|+\cdots+|a_i||a_{i+1}|\ldots|a_j| \\ &= \prod_{k=i}^{j}(1+|a_k|)-1=\left|\prod_{k=i}^{j}(1+|a_k|)-1\right|.\end{aligned}$$

Now assume $\prod(1+a_k)$ converges absolutely. Let $\epsilon>0$ be given. There exists a positive integer N such that if $m>n>N$, then

$$\left|\prod_{k=n+1}^{m}(1+a_k)-1\right|\leqslant\left|\prod_{k=n+1}^{m}(1+|a_k|)-1\right|<\epsilon.$$

This implies (by Theorem 7.4) that $\prod(1+a_k)$ converges.

As in the case of infinite series, the absolute convergence of an infinite product implies that all its rearrangements converge to it.

We defined the notion of a rearrangement of an infinite series $\sum a_n$. The notion of a rearrangement of an infinite product is similar. Intuitively, a rearrangement of an infinite product $\prod(1+a_n)$ is an infinite product $\prod(1+b_n)$ such that each factor of the first product occurs exactly once as a factor in the second and vice versa. This definition can be made more precise as follows: Let $f: \mathbb{Z}_+ \to \mathbb{Z}_+$ be a one-to-one correspondence on $\mathbb{Z}_+$ and $\langle a_n \rangle$ a sequence. If $\langle b_n \rangle$ is a sequence such that $b_n = a_{f(n)}$ for each $n \in \mathbb{Z}_+$, then we call it a *rearrangement* of $\langle a_n \rangle$. The infinite product $\prod(1 + b_n)$ is called a *rearrangement* of $\prod(1 + a_n)$ if $\langle b_n \rangle$ is a rearrangement of $\langle a_n \rangle$. The function f is called the *rearrangement function*. Remark IV.9.1 and Lemma IV.9.1 apply to rearrangements of infinite products. Below we state an infinite product analogue of Theorem IV.9.1.

Theorem 8.3. *If $\prod_{n=1}^{\infty}(1 + a_n)$ converges absolutely, then all its rearrangements converge to its product P.*

Proof. We first consider the case where $1 + a_n \neq 0$ for all n and let

$$P = \prod_{n=1}^{\infty} (1 + a_n), \tag{8.2}$$

where the product on the right converges absolutely. We then assume that $\prod_{n=1}^{\infty}(1 + b_n)$ is some rearrangement of our product. Thus, $1 + b_n \neq 0$ for all n. We have

$$P = \lim P_n, \tag{8.3}$$

where $\langle P_n \rangle$ is the sequence of partial products of $\prod(1 + a_n)$. We shall prove that

$$P = \lim_{n \to +\infty} Q_n, \tag{8.4}$$

where $\langle Q_n \rangle$ is the sequence of partial products of $\prod(1 + b_n)$. Let $\langle \overline{P}_n \rangle$ be the sequence of partial products of $\prod(1 + |a_n|)$. The latter product converges by hypothesis. Let $\epsilon > 0$ be given. There exists a positive integer N_1 such that if $j > k > N_1$, then

$$\left| \frac{\overline{P}_j}{\overline{P}_k} - 1 \right| = \left| \prod_{n=k+1}^{j} (1 + |a_n|) - 1 \right| < \frac{\epsilon}{2}, \tag{8.5}$$

By properties of the absolute value, this implies that

$$\left| \frac{P_j}{P_k} - 1 \right| = \left| \prod_{n=k+1}^{j} (1 + a_n) - 1 \right| \leqslant \left| \prod_{n=k+1}^{j} (1 + |a_n|) - 1 \right| < \frac{\epsilon}{2}. \tag{8.6}$$

Now we let $j \to +\infty$ and obtain

$$\left| \frac{P}{P_k} - 1 \right| = \lim_{j \to +\infty} \left| \frac{P_j}{P_k} - 1 \right| \leqslant \frac{\epsilon}{2} \qquad \text{for} \quad k > N_1. \tag{8.7}$$

Fix such a k. By Lemma IV.9.1, there exists a positive integer N_2 such that

$$\{a_1, \ldots, a_k\} \leqslant \{b_1, b_2, \ldots, b_n\} \qquad \text{for} \quad n > N_2. \tag{8.8}$$

Assume that $n > \max\{N_2, k\}$ so that $n > N_2$ and $n > k$, and consider

$$\frac{Q_n}{P_k} - 1 \qquad \text{for such } n.$$

Since $n > N_2$, (8.8) holds. If any factors of $\prod(1 + a_n)$ remain at all after cancelling, they have indices greater than k. Hence, by (8.6) we have

$$\left|\frac{Q_n}{P_k} - 1\right| < \frac{\epsilon}{2} \qquad \text{for} \quad n > \max\{N_2, k\}. \tag{8.9}$$

This implies that

$$\left|\frac{Q_n}{P_k} - \frac{P}{P_k}\right| \leqslant \left|\frac{Q_n}{P_k} - 1\right| + \left|\frac{P}{P_k} - 1\right| < \frac{\epsilon}{2} + \frac{\epsilon}{2} = \epsilon \tag{8.10}$$

for $n > \max\{N_2, k\}$. From this it follows that

$$\lim_{n \to +\infty} \frac{Q_n}{P_k} = \frac{P}{P_k}$$

and, hence, that

$$\lim_{n \to +\infty} Q_n = P.$$

This completes the proof of (8.4) in the present case.

We turn to the case where finitely many $1 + a_n$'s are equal to 0, so that a positive integer N exists such that if $n > N$, then $1 + a_n \neq 0$. By Def. 7.2, the product P is defined as

$$P = (1 + a_1) \ldots (1 + a_N)P',$$

where

$$P' = \lim_{n \to +\infty} \prod_{k=N+1}^{n} (1 + a_k).$$

There exists a positive integer N' such that

$$\{a_1, a_2, \ldots, a_N\} \subseteq \{b_1, b_2, \ldots, b_n\} \qquad \text{for} \quad n > N'.$$

For such n, the factors of Q_n include those of

$$P_N = (1 + a_1) \ldots (1 + a_N).$$

We omit the factors of P_N from the rearrangement $\prod_{n=1}^{\infty}(1 + b_n)$ of $\prod_{n=1}^{\infty}(1 + a_n)$, and obtain thereby an infinite product which is a rearrangement of

$$P' = \prod_{n=N+1}^{\infty} (1 + a_n).$$

We now apply the first part of the proof and ascertain that the infinite product of the rearrangement with the factors of P_N deleted has the same infinite product P'. Hence,

$$\prod_{n=1}^{\infty} (1 + b_n) = P_N P' = \prod_{n=1}^{\infty} (1 + a_n).$$

EXAMPLE 8.1.* Consider the infinite products

$$Q_0 = \prod_{n=1}^{\infty} (1 - q^{2n}), \qquad Q_1 = \prod_{n=1}^{\infty} (1 + q^{2n}),$$
$$Q_2 = \prod_{n=1}^{\infty} (1 + q^{2n-1}), \qquad Q_3 = \prod_{n=1}^{\infty} (1 - q^{2n-1}), \tag{8.11}$$

where $|q| < 1$. Since the series

$$\sum_{n=1}^{\infty} q^{2n} \quad \text{and} \quad \sum_{n=1}^{\infty} q^{2n-1}, \tag{8.12}$$

where $|q| < 1$ are absolutely convergent, so are all the products in (8.11) (Theorem 8.1).

We have

$$\begin{aligned} Q_3 Q_0 &= \lim_{n\to+\infty} \prod_{k=1}^{n} (1 - q^{2k-1}) \lim_{n\to+\infty} \prod_{k=1}^{n} (1 - q^{2k}) \\ &= \lim_{n\to+\infty} \left(\prod_{k=1}^{n} (1 - q^{2k-1}) \prod_{k=1}^{n} (1 - q^{2k}) \right). \end{aligned} \tag{8.13}$$

But

$$\begin{aligned} \prod_{k=1}^{n} (1 - q^{2k-1}) \prod_{k=1}^{n} (1 - q^{2k}) &= (1 - q)(1 - q^3) \ldots \\ &\quad (1 - q^{2n-1})(1 - q^2)(1 - q^4) \ldots (1 - q^{2n}) \\ &= (1 - q)(1 - q^2)(1 - q^3) \ldots \\ &\quad (1 - q^{2n-1})(1 - q^{2n}) \\ &= \prod_{k=1}^{2n} (1 - q^k) \end{aligned}$$

and, hence,

$$Q_3 Q_0 = \lim_{n\to+\infty} \prod_{k=1}^{n} (1 - q^{2k-1})(1 - q^{2k}) = \lim_{n\to+\infty} \prod_{k=1}^{2n} (1 - q^k). \tag{8.14}$$

Since the sequence $\langle \prod_{k=1}^{2n} (1 - q^k) \rangle$ is a subsequence of the sequence $\langle \prod_{k=1}^{n} (1 - q^k) \rangle$, it has the same limit as the latter. Hence, we conclude from (8.14) that if $|q| < 1$, then

$$Q_3 Q_0 = \lim_{n\to+\infty} \prod_{k=1}^{2n} (1 - q^k) = \lim_{n\to+\infty} \prod_{k=1}^{n} (1 - q^k) = \prod_{k=1}^{\infty} (1 - q^k). \tag{8.15}$$

The last infinite product converges absolutely when $|q| < 1$ since

$$\sum_{k=1}^{\infty} q^k$$

* Harris Hancock, *Lectures on the Theory of Elliptic Functions*, Dover, New York, 1958, p. 396.

does. Similarly,

$$Q_2 Q_1 = \prod_{k=1}^{\infty} (1 + q^k) \qquad \text{if} \quad |q| < 1.$$

PROB. 8.2. Prove: If Q_0, Q_1, Q_2, and Q_3 are defined as in Example 2.1, then

$$Q_0 Q_1 Q_2 Q_3 = Q_0 \quad \text{and} \quad Q_1 Q_2 Q_3 = 1.$$

PROB. 8.3. Prove: If $|q| < 1$, then

$$\prod_{n=1}^{\infty} (1 + q^n) = \frac{1}{\prod_{n=1}^{\infty}(1 - q^{2n-1})}.$$

PROB. 8.4.* Prove: If $|q| < 1$, then

$$(1 + q + q^2 + \cdots + q^9)(1 + q^{10} + q^{20} + \cdots + q^{90}) \cdots$$
$$(1 + q^{100} + q^{200} + \cdots + q^{900}) \cdots = \frac{1}{1 - q}.$$

9. Sine and Cosine as Infinite Products. Wallis' Product. Stirling's Formula

Theorem 9.1. *If* $t \in \mathbb{R}$, *then*

$$\sin t = t \prod_{k=1}^{\infty} \left(1 - \frac{t^2}{k^2\pi^2}\right) \tag{9.1a}$$

and

$$\cos t = \prod_{k=1}^{\infty} \left(1 - \frac{4t^2}{\pi^2(2k-1)^2}\right). \tag{9.1b}$$

PROOF. We prove (9.1a) first. Begin with Theorem 6.3. That theorem states: If n is an odd positive integer and $t \in \mathbb{R}$, then

$$\sin t = n \sin \frac{t}{n} \prod_{k=1}^{(n-1)/2} \left[1 - \frac{\sin^2(t/n)}{\sin^2(k/n)\pi}\right]. \tag{9.2}$$

We fix $t \neq 0$ and take a positive integer m such that $m > \max\{|t|/2, t^2/4\}$, so that $0 < |t|/2 < m$ and $m > t^2/4$. We then take an odd integer n such that $n > \max\{2m + 1, 2|t|/\pi\}$. This implies that $n > 2m + 1$ and n

* Polya–Szego, *Aufgaben und Lehrsätze aus der Analysis*, Vol. I, Dover, 1945, Prob. 18.

$> 2|t|/\pi$, and, therefore, that

$$0 < m < \frac{n-1}{2} \tag{9.3a}$$

and

$$0 < \frac{|t|}{n} < \frac{\pi}{2}. \tag{9.3b}$$

Because of the second inequality, we have $\sin(t/n) \neq 0$. From (9.2) we obtain

$$\frac{\sin t}{n\sin(t/n)} = \prod_{k=1}^{(n-1)/2}\left[1 - \frac{\sin^2(t/n)}{\sin^2(k/n)\pi}\right], \tag{9.4}$$

where n is odd. For the left-hand side we have

$$\lim_{n\to+\infty}\frac{\sin t}{n\sin(t/n)} = \lim_{n\to+\infty}\frac{(\sin t)/t}{(\sin t/n)(t/n)^{-1}} = \frac{\sin t}{t}. \tag{9.5}$$

This implies, using (9.4), that

$$\lim_{n\to+\infty}\prod_{k=1}^{(n-1)/2}\left[1 - \frac{\sin^2(t/n)}{\sin^2(k/n)\pi}\right] = \frac{\sin t}{t}. \tag{9.6}$$

We examine the right-hand side of (9.4) and recall (9.3a). We have

$$\prod_{k=1}^{(n-1)/2}\left[1 - \frac{\sin^2(t/n)}{\sin^2(k/n)\pi}\right]$$
$$= \prod_{k=1}^{m}\left[1 - \frac{\sin^2(t/n)}{\sin^2(k/n)\pi}\right]\prod_{k=m+1}^{(n-1)/2}\left[1 - \frac{\sin^2(t/n)}{\sin^2(k/n)\pi}\right]. \tag{9.7}$$

Note that

$$0 < \sin^2\frac{t}{n} < \frac{t^2}{n^2}. \tag{9.8}$$

We also note that, since $1 \leqslant k \leqslant (n-1)/2$, as is the case for the index k in the products on either side of (9.7), we have

$$0 < k\frac{\pi}{n} \leqslant \frac{n-1}{n}\frac{\pi}{2} < \frac{\pi}{2}. \tag{9.9}$$

By Jordan's inequality (VII.7.25), we obtain from (9.9)

$$\frac{\sin k(\pi/n)}{k(\pi/n)} \geqslant \frac{2}{\pi} \quad \text{for} \quad 0 < k \leqslant \frac{n-1}{2}. \tag{9.10}$$

This and (9.8) imply that

$$0 < \frac{\sin^2(t/n)}{\sin^2(k/n)\pi} < \frac{t^2}{4k^2} \quad \text{for} \quad 0 < k \leqslant \frac{n-1}{2}. \tag{9.11}$$

However, for the index k in the second factor on the right of (9.7) we have

$0 < m < k$ and, therefore, $1/4k^2 < 1/4m^2$. But, to begin with, we have $0 < |t|/2 < m$. Hence, we have

$$\frac{t^2}{4k^2} < \frac{t^2}{4m^2} < 1 \quad \text{if} \quad m < k \leqslant \frac{n-1}{2}. \tag{9.12}$$

This and (9.11) imply that

$$0 < \frac{\sin^2(t/n)}{\sin^2(k/n)\pi} < 1 \quad \text{if} \quad m < k \leqslant \frac{n-1}{2}. \tag{9.13}$$

We write the second factor on the right-hand side of (9.7) as R_m. Because of (9.13) it follows that

$$0 < R_m = \prod_{k=m+1}^{(n-1)/2} \left[1 - \frac{\sin^2(t/n)}{\sin^2 k(\pi/n)}\right] < 1. \tag{9.14}$$

Next we use Prob. II.12.15 to obtain, since (9.13) holds,

$$1 > R_m > 1 - \left[\frac{\sin^2(t/n)}{\sin^2(m+1)(\pi/n)} + \frac{\sin^2(t/n)}{\sin^2(m+2)(\pi/n)} + \cdots + \frac{\sin^2(t/n)}{\sin^2((n-1)/2)(\pi/n)}\right]. \tag{9.15}$$

Now use (9.11) to obtain from (9.15), after writing $j = (n-1)/2$,

$$1 > R_m > 1 - \frac{t^2}{4}\left(\frac{1}{(m+1)^2} + \frac{1}{(m+2)^2} + \cdots + \frac{1}{j^2}\right). \tag{9.16}$$

Note that

$$\frac{1}{(m+1)^2} + \frac{1}{(m+2)^2} + \cdots + \frac{1}{j^2}$$
$$< \frac{1}{m(m+1)} + \frac{1}{(m+1)(m+2)} + \cdots + \frac{1}{(j-1)j}$$
$$= \frac{1}{m} - \frac{1}{m+1} + \frac{1}{m+1} - \frac{1}{m+2} + \cdots + \frac{1}{j-1} - \frac{1}{j}$$
$$= \frac{1}{m} - \frac{1}{j} < \frac{1}{m}.$$

This, (9.12), and (9.16) imply, since $m > t^2/4$, that

$$1 > R_m > 1 - \frac{t^2}{4m} > 0. \tag{9.17}$$

This, by the definition of R_m in (9.14) and by (9.4) and (9.7) implies that

$$0 < 1 - \frac{t^2}{4m} < \frac{\sin t}{n\sin(t/n)} \frac{1}{\prod_{k=1}^{m}(1 - \sin^2(t/n)/\sin^2(k\pi/n))} = R_m < 1. \tag{9.18}$$

Now fix m and let $n \to +\infty$. Since

$$\frac{\sin^2(t/n)}{\sin^2(k\pi/n)} = \frac{t^2}{k^2\pi}\left[\frac{\sin(t/n)(t/n)^{-1}}{\sin(k\pi/n)(k\pi/n)^{-1}}\right]^2 \to \frac{t^2}{k^2\pi^2} \qquad \text{as} \quad n \to +\infty,$$

we have

$$\lim_{n\to+\infty} \prod_{k=1}^{m}\left[1 - \frac{\sin^2(t/n)}{\sin^2(k\pi/n)}\right] = \prod_{k=1}^{m}\left(1 - \frac{t^2}{k^2\pi^2}\right). \tag{9.19}$$

Take limits as $n \to +\infty$ in (9.18), use (9.19) and (9.5), and obtain

$$0 < 1 - \frac{t^2}{4m} \leqslant \frac{\sin t}{t} \frac{1}{\prod_{k=1}^{m}(1 - t^2/k^2\pi^2)} \leqslant 1. \tag{9.20}$$

Now let $m \to +\infty$ and obtain from (9.20)

$$1 = \frac{\sin t}{t} \cdot \frac{1}{\prod_{k=1}^{\infty}(1 - t^2/k^2\pi^2)} \qquad \text{for} \quad t \neq 0.$$

This proves (9.1a).

Next we prove (9.1b). Use the identity $\sin 2t = 2\sin t\cos t$ and the appropriate product formula for the sine on each side to arrive at

$$2t\prod_{k=1}^{\infty}\left(1 - \frac{4t^2}{k^2\pi^2}\right) = 2t\prod_{k=1}^{\infty}\left(1 - \frac{t^2}{k^2\pi^2}\right)\cos t. \tag{9.21}$$

Now

$$\begin{aligned}\prod_{k=1}^{\infty}\left(1 - \frac{4t^2}{k^2\pi^2}\right) &= \lim_{n\to+\infty}\prod_{k=1}^{n}\left(1 - \frac{4t^2}{k^2\pi^2}\right)\\ &= \lim_{n\to+\infty}\prod_{k=1}^{2n}\left(1 - \frac{4t^2}{k^2\pi^2}\right)\end{aligned} \tag{9.22}$$

and

$$\begin{aligned}\lim_{n\to+\infty}\prod_{k=1}^{2n}\left(1 - \frac{4t^2}{k^2\pi^2}\right) &= \lim_{n\to+\infty}\prod_{\substack{k=2\\ k \text{ even}}}^{2n}\left(1 - \frac{4t^2}{k^2\pi^2}\right)\prod_{\substack{k=1\\ k \text{ odd}}}^{2n-1}\left(1 - \frac{4t^2}{k^2\pi^2}\right)\\ &= \lim_{n\to+\infty}\prod_{j=1}^{n}\left(1 - \frac{4t^2}{(2j)^2\pi^2}\right)\prod_{j=1}^{n}\left(1 - \frac{4t^2}{(2j-1)^2\pi^2}\right)\\ &= \lim_{n\to+\infty}\prod_{j=1}^{n}\left(1 - \frac{t^2}{j^2\pi^2}\right)\prod_{j=1}^{n}\left(1 - \frac{4t^2}{(2j-1)\pi^2}\right).\end{aligned} \tag{9.23}$$

Since the product

$$\prod_{j=1}^{\infty}\left(1 - \frac{4t^2}{(2j-1)^2\pi^2}\right) \tag{9.24}$$

converges (why?), we obtain from (9.23)

$$\lim_{n\to+\infty}\prod_{k=1}^{2n}\left(1-\frac{4t^2}{k^2\pi^2}\right)=\lim_{n\to+\infty}\prod_{j=1}^{n}\left(1-\frac{t^2}{j^2\pi^2}\right)\lim_{n\to+\infty}\prod_{j=1}^{n}\left(1-\frac{4t^2}{(2j-1)^2\pi^2}\right)$$
$$=\prod_{k=1}^{\infty}\left(1-\frac{t^2}{k^2\pi^2}\right)\prod_{k=1}^{\infty}\left(1-\frac{4t^2}{(2k-1)^2\pi^2}\right).$$

By (9.22), this implies that

$$\prod_{k=1}^{\infty}\left(1-\frac{4t^2}{k^2\pi^2}\right)=\prod_{k=1}^{\infty}\left(1-\frac{t^2}{k^2\pi^2}\right)\prod_{k=1}^{\infty}\left(1-\frac{4t^2}{(2k-1)^2\pi^2}\right). \tag{9.25}$$

Substitute in (9.21) to obtain

$$2t\prod_{k=1}^{\infty}\left(1-\frac{t^2}{k^2\pi^2}\right)\prod_{k=1}^{\infty}\left(1-\frac{4t^2}{(2k-1)^2\pi^2}\right)=2t\prod_{k=1}^{\infty}\left(1-\frac{t^2}{k^2\pi^2}\right)\cos t.$$

This implies that

$$\cos t=\prod_{k=1}^{\infty}\left(1-\frac{4t^2}{(2k-1)^2\pi^2}\right).$$

This proves (9.1b).

Corollary 1. *If $x\in\mathbb{R}$, then*

$$\sin\pi x=\pi x\prod_{k=1}^{\infty}\left(1-\frac{x^2}{k^2}\right) \tag{9.26a}$$

and

$$\cos\pi x=\prod_{k=1}^{\infty}\left(1-\frac{4x^2}{(4k-1)^2}\right). \tag{9.26b}$$

PROOF. Obvious.

Remark 9.1. The infinite product expansions for $\sin t$ and $\cos t$ become comprehensible intuitively by making an analogy with the case of a polynomial $P(x)=a_0x^n+a_1x^n+\cdots+a_n$ of degree $n\geqslant 1$ having zeros $r_1,r_2,\ldots,r_n$.

P can be expressed as a product

$$P(x)=a_0(x-r_1)(x-r_2)\ldots(x-r_n).$$

$\sin t$ is expressed in (9.1a) as a product of factors

$$t,\qquad 1-\frac{t^2}{\pi^2},\qquad 1-\frac{t^2}{4\pi^2},\qquad 1-\frac{t^2}{9\pi^2},\ldots$$

which vanish at the zeros $0,\pm\pi,\pm 2\pi,\pm 3\pi,\ldots$ of $\sin t$. Similarly $\cos t$ is expressed in (9.1b) as a product of factors

$$1-\frac{4t^2}{\pi^2},\qquad 1-\frac{4t^2}{9\pi^2},\qquad 1-\frac{4t^2}{25\pi^2},\dots$$

which vanish at the zeros $\pm\pi/2, \pm 3\pi/2, \pm 5\pi/2, \dots$ of $\cos t$. This was how Euler discovered these formulas.

Corollary 2 (Wallis Product). *We have*

$$\frac{\pi}{2}=\prod_{k=1}^{\infty}\left(\frac{2k}{2k-1}\right)\left(\frac{2k}{2k+1}\right)=\left(\frac{2}{1}\cdot\frac{2}{3}\right)\left(\frac{4}{3}\cdot\frac{4}{5}\right)\left(\frac{6}{5}\cdot\frac{6}{7}\right)\cdots. \tag{9.27}$$

PROOF. Substitute $1/2$ for x in (9.26a) and obtain

$$1=\frac{\pi}{2}\prod_{k=1}^{\infty}\left(1-\frac{1}{4k^2}\right)=\frac{\pi}{2}\prod_{k=1}^{\infty}\left(1-\frac{1}{2k}\right)\left(1+\frac{1}{2k}\right)$$

which implies (9.27).

10. Some Special Limits. Stirling's Formula

We prove

$$\sqrt{\pi}=\lim_{n\to+\infty}\left(\frac{(n!)^2 2^{2n}}{(2n)!}\cdot\frac{1}{\sqrt{n}}\right). \tag{10.1}$$

By Wallis' Product, we have

$$\frac{\pi}{2}=\lim_{n\to+\infty}\prod_{k=1}^{n}\left(\frac{2k}{2k-1}\cdot\frac{2k}{2k+1}\right)=\lim_{n\to+\infty}\frac{2^2\cdot 4^2\cdots(2n)^2}{3^2\cdot 5^2\cdots(2n-1)^2(2n+1)}. \tag{10.2}$$

Since

$$\lim_{n\to+\infty}\frac{2n}{2n+1}=1,$$

(10.2) implies that

$$\frac{\pi}{2}=\lim_{n\to+\infty}\left[\frac{2^2\cdot 4^2\cdots(2n-2)^2}{3^2\cdot 5^2\cdots(2n-1)^2}2n\right].$$

Taking square roots, we have

$$\sqrt{\frac{\pi}{2}}=\lim_{n\to+\infty}\left(\frac{2\cdot 4\cdots(2n-2)}{3\cdot 5\cdots(2n-1)}\sqrt{2n}\right). \tag{10.3}$$

We multiply the numerator and denominator inside the limit by $2 \cdot 4 \cdots 2n-2$ and obtain

$$\sqrt{\frac{\pi}{2}} = \lim_{n\to+\infty} \frac{2^2 \cdot 4^2 \cdots (2n-2)^2}{(2n-1)!} \sqrt{2n}$$

$$= \lim_{n\to+\infty} \frac{2^2 \cdot 4^2 \cdots (2n-2)^2 (2n)^2}{(2n)!} \frac{\sqrt{2n}}{2n}$$

$$= \lim_{n\to+\infty} \left(\frac{2^{2n}(n!)^2}{(2n)!} \frac{1}{\sqrt{n}} \frac{1}{\sqrt{2}} \right).$$

This yields, after multiplying both sides by $\sqrt{2}$,

$$\sqrt{\pi} = \lim_{n\to+\infty} \left(\frac{2^{2n}(n!)^2}{(2n)!} \frac{1}{\sqrt{n}} \right)$$

and proves (10.1).

We are now in a position to evaluate the number a_2 in Prob. IX.9.3. There we offered the definition: If p is a positive integer, then

$$a_p = p\Gamma\left(\frac{1}{p}\right)\Gamma\left(\frac{2}{p}\right) \cdots \Gamma\left(\frac{p}{p}\right), \tag{10.4}$$

where Γ is the gamma function. This formula yields

$$a_2 = 2\Gamma\left(\tfrac{1}{2}\right)\Gamma\left(\tfrac{2}{2}\right) = 2\Gamma\left(\tfrac{1}{2}\right). \tag{10.5}$$

In Prob. IX.9.5 we have the result

$$a_p = p \lim_{n\to+\infty} \frac{(n!)^p p^{np}}{(np)!\, n^{(p-1)/2}}. \tag{10.6}$$

This implies that

$$a_2 = 2 \lim_{n\to+\infty} \frac{(n!)^2 2^{2n}}{(2n)!\, n^{1/2}}. \tag{10.7}$$

This and (10.1) yield

$$a_2 = 2\sqrt{\pi}. \tag{10.8}$$

Using (10.5) we conclude from this that

$$2\sqrt{\pi} = a_2 = 2\Gamma\left(\tfrac{1}{2}\right)$$

and obtain

$$\Gamma\left(\tfrac{1}{2}\right) = \sqrt{\pi}. \tag{10.9}$$

PROB. 10.1. Evaluate: $\Gamma\left(\frac{3}{2}\right)$ and $\Gamma\left(-\frac{1}{2}\right)$.

PROB. 10.2. Prove: If n is a positive integer, then

$$\Gamma(n+\tfrac{1}{2}) = \frac{1\cdot 3\cdot 5\cdots(2n-1)}{2^n}\sqrt{\pi}.$$

Theorem 10.1. *If x is not an integer, then*

$$\Gamma(x)\Gamma(1-x) = \frac{\pi}{\sin \pi x}. \tag{10.10}$$

PROOF. By hypothesis, we have $x \in \mathfrak{D}(\Gamma)$. Hence, $1-x \in \mathfrak{D}(\Gamma)$. We recall that

$$\Gamma(x) = \lim_{n\to+\infty} g_n(x),$$

where

$$g_n(x) = \frac{n^x n!}{x(x+1)\dots(x+n)} = \frac{n^x}{x(1+x)(1+x/2)\dots(1+x/n)}. \tag{10.11}$$

Substituting $1-x$ for x, we have

$$\begin{aligned} g_n(1-x) &= \frac{n^{1-x}n!}{(1-x)(2-x)(3-x)\dots(n+1-x)} \\ &= \frac{n^{1-x}}{(1-x)(1-x/2)\dots(1-x/n)}\,\frac{1}{n+1-x}. \end{aligned} \tag{10.12}$$

From (10.11) and (10.12) we conclude that

$$\begin{aligned} &g_n(x)\,g_n(1-x) \\ &\quad= \frac{n^x n^{1-x}}{x(1+x)(1+x/2)\dots(1+x/n)(1-x)(1-x/2)\dots(1-x/n)} \\ &\qquad\times\frac{1}{n+1-x} \\ &\quad= \frac{1}{x(1-x^2)(1-x^2/4)\dots(1-x^2/n^2)}\cdot\frac{n}{n+1-x}. \end{aligned}$$

For $n\to+\infty$ the second factor, $n/(n+1-x)$, approaches 1. Hence, we have

$$\begin{aligned} \lim_{n\to+\infty} g_n(x)g_n(1-x) &= \lim_{n\to+\infty}\frac{1}{x(1-x^2)(1-x^2/4)\dots(1-x^2/n^2)} \\ &= \frac{1}{x\prod_{k=1}^{\infty}(1-x^2/k^2)} = \frac{\pi}{\pi x\prod_{k=1}^{\infty}(1-x^2/k^2)} \\ &= \frac{\pi}{\sin \pi x}. \end{aligned}$$

The last equality is a consequence of Theorem 9.1. Thus, we have

$$\begin{aligned}\Gamma(x)\Gamma(1-x) &= \lim_{n\to+\infty} g_n(x) \lim_{n\to+\infty} g_n(1-x)\\ &= \lim_{n\to+\infty} g_n(x)g_n(1-x) = \frac{\pi}{\sin \pi x}\end{aligned}$$

if x is not an integer. This completes the proof.

PROB. 10.3. Prove: If n is a positive integer, then

$$\Gamma\left(\tfrac{1}{2}-n\right) = \frac{(-2)^n\sqrt{\pi}}{1\cdot 3\cdot 5\cdots(2n-1)}.$$

We now prove an important asymptotic formula due to Stirling.

Theorem 10.2 (Stirling's Formula). *We have*

$$n! \sim \sqrt{2\pi}\, n^{n+1/2}e^{-n} \qquad \textit{as} \quad n\to+\infty. \tag{10.13}$$

This formula is useful in approximating to $n!$ for large values of n. Before proving this formula, we state and prove some preliminary lemmas.

Lemma 10.1.* *Let g be defined as*

$$g(x) = \left(x+\frac{1}{2}\right)\ln\left(1+\frac{1}{x}\right) - 1 \qquad \textit{for} \quad x>0. \tag{10.14}$$

g satisfies

$$0 < g(x) \leqslant \frac{1}{12x} - \frac{1}{12(x+1)}. \tag{10.15}$$

PROOF. Use Prob. IX.5.5 to obtain

$$\operatorname{Arctanh} y = \frac{1}{2}\ln\frac{1+y}{1-y} = y + \frac{y^3}{3} + \frac{y^5}{5} + \dots \qquad \text{for} \quad |y|<1.$$

Let $y = 1/(2x+1)$, where $x>0$. We obtain

$$\frac{1}{2}\ln\left(1+\frac{1}{x}\right) = \frac{1}{2x+1} + \frac{1}{3(2x+1)^3} + \frac{1}{5(2x+1)^5} + \cdots.$$

Multiply both sides of the above by $(2x+1)$ and then subtract 1 from both sides. This yields

$$\begin{aligned} g(x) = \left(x+\frac{1}{2}\right)\ln\left(1+\frac{1}{x}\right) - 1 &= \frac{1}{3(2x+1)^2} + \frac{1}{5(2x+1)^4} + \cdots\\ &= \sum_{k=1}^{\infty} \frac{1}{(2k+1)(2x+1)^{2k}}\\ &= \frac{1}{(2x+1)^2}\sum_{k=1}^{\infty}\frac{1}{(2k+1)(2x+1)^{2(k-1)}} > 0.\end{aligned}$$

* E. Artin, *loc. cit.*

Since $k \geqslant 1$ in the above, we have $2k + 1 \geqslant 3$. Therefore, for each term in the last series, we have

$$\frac{1}{(2k+1)(2x+1)^{2(k-1)}} \leqslant \frac{1}{3(2x+1)^{2(k-1)}}.$$

It follows that

$$0 < g(x) \leqslant \frac{1}{3(2x+1)^2} \sum_{k=1}^{\infty} \frac{1}{\left[(2x+1)^2\right]^{k-1}}. \tag{10.16}$$

The series on the right is a geometric series with ratio $r = 1/(2x+1)^2$. It, therefore, sums to

$$\frac{1}{1 - 1/(2x+1)^2} = \frac{(2x+1)^2}{(2x+1)^2 - 1} = \frac{(2x+1)^2}{4x(x+1)}$$
$$= \frac{(2x+1)^2}{4}\left(\frac{1}{x} - \frac{1}{x+1}\right).$$

Thus, (10.16) implies that

$$0 < g(x) \leqslant \frac{1}{12x} - \frac{1}{12(x+1)}. \tag{10.17}$$

This completes the proof.

Lemma 10.2.* *If* g *is defined as in* (10.14), *then the series*

$$\sum_{n=0}^{\infty} g(x+n), \qquad x > 0, \tag{10.18}$$

converges to a real number $\mu(x)$, *thereby defining a function* μ *on* $(0; +\infty)$. *This function* μ, *which is defined as*

$$\mu(x) = \sum_{n=0}^{\infty} g(x+n) \qquad \text{for} \quad x > 0, \tag{10.19}$$

is convex and satisfies

$$0 < \mu(x) \leqslant \frac{1}{12x} \qquad \text{for} \quad x > 0. \tag{10.20}$$

Proof. By the definition of g and the last lemma, the terms of the series (10.19) are all positive. We also know from that lemma that

$$0 < g(x+k) \leqslant \frac{1}{12}\left(\frac{1}{x+k} - \frac{1}{x+k+1}\right)$$

for each nonnegative integer k. (10.21)

*E. Artin, *loc. cit.*

Hence, the nth partial sum S_n of the series (10.18) satisfies

$$0 < S_n = \sum_{k=0}^{n} g(x+k) \leqslant \frac{1}{12} \sum_{k=0}^{n} \left(\frac{1}{x+k} - \frac{1}{x+k+1} \right)$$

$$= \frac{1}{12}\left(\frac{1}{x} - \frac{1}{x+n+1} \right) < \frac{1}{12x} \tag{10.22}$$

for $x > 0$. Thus, for each $x > 0$ the terms of (10.18) are positive, its partial sum sequence is bounded, so that the series converges. Writing $\mu(x)$ for the sum, we have

$$0 < \mu(x) \leqslant \frac{1}{12x} \qquad \text{for} \quad x > 0. \tag{10.23}$$

This proves that the function μ defined in (10.19) satisfies (10.20). We prove below that μ is a convex function.

We have

$$g(x+n) = \left(x+n+\frac{1}{2} \right) \ln\left(1 + \frac{1}{x+n} \right) - 1,$$

where $x > 0$ for each nonnegative integer n. Simple calculations show that

$$g''(x+n) = \frac{1}{2(x+n)^2(x+n+1)^2} > 0$$

$$\text{for} \quad x > 0 \quad \text{and each nonnegative integer } n.$$

This proves that each term $g(x+n)$ in the series (10.19) is convex (Prob. VIII.2.2). It follows that its partial sums are convex. Since the limit of a sequence of convex functions is convex (Prob. VIII.2.14), the sum $\mu(x)$ is convex. This completes the proof.

Lemma 10.3.* *Let g be the function defined in* (10.14) *and μ the series*

$$\mu(x) = \sum_{n=0}^{\infty} g(x+n) \qquad \textit{for} \quad x > 0 \tag{10.24}$$

dealt with in Lemma 10.2. *The function f defined as*

$$f(x) = x^{x-1/2} e^{-x} e^{\mu(x)} \qquad \textit{for} \quad x > 0 \tag{10.25}$$

is log-convex and satisfies the functional equation

$$f(x+1) = xf(x) \qquad \textit{for} \quad x > 0. \tag{10.26}$$

Moreover, the function h defined as $f(x)/f(1)$ is identical with the gamma function on $(0; +\infty)$ *and we have*

$$x^{x-1/2} e^{-x} e^{\mu(x)} = f(1)\Gamma(x) = a\Gamma(x) \qquad \textit{for} \quad x > 0, \tag{10.27}$$

where $f(1) = a$ is a positive *constant.*

* E. Artin, *loc. cit.*

Proof. It follows from the definition of f that $f(x) > 0$ for $x > 0$ and that

$$\frac{f(x+1)}{f(x)} = \frac{(x+1)^{x+1/2}e^{-x-1}e^{\mu(x+1)}}{x^{x-1/2}e^{-x}e^{\mu(x)}} = x\left(1+\frac{1}{x}\right)^{x+1/2}e^{-1}e^{\mu(x+1)-\mu(x)} \tag{10.28}$$

for $x > 0$. Clearly,

$$\mu(x) - \mu(x+1) = \sum_{n=0}^{\infty} g(x+n) - \sum_{n=0}^{\infty} g(x+n+1)$$
$$= \sum_{n=0}^{\infty} g(x+n) - \sum_{n=1}^{\infty} g(x+n) = g(x),$$

so that

$$\mu(x) - \mu(x+1) = \left(x+\frac{1}{2}\right)\ln\left(1+\frac{1}{x}\right) - 1.$$

This and (10.28) imply that

$$\frac{f(x+1)}{f(x)} = x\left(1+\frac{1}{x}\right)^{x+1/2}e^{-1}e^{1-(x+1/2)\ln(1+1/x)}$$
$$= x\left(1+\frac{1}{x}\right)^{x+1/2}\frac{1}{(1+1/x)^{x+1/2}}$$
$$= x,$$

yielding (10.26). Since μ is convex, e^{μ} is log-convex. Now define the function A as

$$A(x) = x^{x-1/2}e^{-x} \qquad \text{for} \quad x > 0.$$

It is easily checked that $(\ln A(x))'' > 0$ for $x > 0$ and, hence, that the function A is log-convex. It follows that the product Ae^{μ} is log-convex (cf. Prob. VIII.7.1). Thus, f is log-convex.

The function h, where

$$h(x) = \frac{f(x)}{f(1)} \qquad \text{for} \quad x > 0,$$

is a positive multiple of a log-convex function and is, therefore, also log-convex. As can be verified, h satisfies

$$h(x+1) = xh(x) \qquad \text{for} \quad x > 0$$

and has the property $h(1) = 1$. By Theorem IX.9.1, h is identical with the gamma function. Writing $f(1) = a$, we have $f(x) = ah(x)$, so that

$$x^{x-1/2}e^{-x}e^{\mu(x)} = a\Gamma(x) \qquad \text{for} \quad x > 0,$$

where a is a positive constant. The proof is now complete.

We are now in a position to prove Stirling's formula.

Proof of Theorem 10.2. Use the notation adopted in the last three lemmas and begin with (10.27). After replacing x there by a positive integer n, we obtain

$$n^{n-1/2}e^{-n}e^{\mu(n)} = a\Gamma(n).$$

Multiply both sides by n and arrive at

$$n^{n+1/2}e^{-n}e^{\mu(n)} = an\Gamma(n) = a\Gamma(n+1) = an!$$

so that

$$e^{\mu(n)} = \frac{an!\,e^n}{n^{n+1/2}}. \tag{10.29}$$

By Lemma 10.2, $0 < \mu(n) \leqslant 1/12n$. Hence,

$$1 < e^{\mu(n)} \leqslant e^{1/12n}.$$

This and (10.29) imply that

$$1 < \frac{an!\,e^n}{n^{n+1/2}} \leqslant e^{1/12n} \qquad \text{for each positive integer } n. \tag{10.30}$$

Since $\lim_{n\to+\infty} e^{1/12n} = 1$, this yields

$$\lim_{n\to+\infty} \frac{n!\,e^n}{n^{n+1/2}} = \frac{1}{a}. \tag{10.31}$$

Let x_n be the sequence defined as

$$x_n = \frac{n!\,e^n}{n^{n+1/2}} \qquad \text{for each } n. \tag{10.32}$$

We just proved (see (10.31)) that

$$\lim_{n\to+\infty} x_n = \frac{1}{a}, \tag{10.33}$$

where $f(1) = a > 0$. Since $\langle x_n \rangle$ converges to the positive limit $1/a$, so does each of its subsequences. Hence, $\lim_{n\to+\infty} x_{2n} = 1/a$. Therefore,

$$\lim_{n\to+\infty} \frac{x_n^2}{x_{2n}} = \frac{1}{a}. \tag{10.34}$$

But

$$\frac{x_n^2}{x_{2n}} = \frac{(n!)^2 2^{2n}}{(2n)!\,\sqrt{n}}\sqrt{2} \qquad \text{for each } n. \tag{10.35}$$

By (10.1) and (10.34) this implies that

$$\frac{1}{a} = \lim_{n\to+\infty} \frac{x_n^2}{x_{2n}} = \sqrt{2} \lim_{n\to+\infty} \frac{(n!)^2 2^{2n}}{(2n)!\,\sqrt{n}} = \sqrt{2\pi}.$$

This evaluates a. Using this result in (10.31) we find that

$$\lim_{n\to+\infty} \frac{n!\,e^n}{n^{n+1/2}} = \sqrt{2\pi} \tag{10.36}$$

or that

$$\lim_{n \to +\infty} \frac{n!}{\sqrt{2\pi}\, e^{-n} n^{n+1/2}} = 1.$$

This proves Theorem 10.2.

Remark 10.1. In the notation used in the last three lemmas we have by the proof of Stirling's formula

$$a = f(1) = \frac{1}{\sqrt{2\pi}} . \tag{10.37}$$

By the definition of f (see (10.25)), this implies that

$$\frac{1}{\sqrt{2\pi}} = e^{-1} e^{\mu(1)} \quad \text{and, therefore,} \quad \mu(1) = \ln \frac{e}{\sqrt{2\pi}} . \tag{10.38}$$

By (10.20) we have $\mu(1) > 0$. Hence, (10.38) yields

$$e > \sqrt{2\pi} . \tag{10.39}$$

Using (10.19) of Lemma 10.2, we obtain

$$\ln \frac{e}{\sqrt{2\pi}} = \mu(1) = \sum_{n=0}^{\infty} g(1+n) = \sum_{n=1}^{\infty} g(n). \tag{10.40}$$

This and the definition of g in (10.14) yield

$$\sum_{n=1}^{\infty} \left[\left(n + \frac{1}{2} \right) \ln\left(1 + \frac{1}{n} \right) - 1 \right] = \ln \frac{e}{\sqrt{2\pi}} . \tag{10.41}$$

PROB. 10.4. Prove: $\Gamma(x) \sim \sqrt{2\pi}\, e^{-x} x^{x-1/2}$ as $x \to +\infty$.

11. Evaluation of Certain Constants Associated with the Gamma Function

We can now evaluate the constants a_p defined in Prob. IX.9.3 and the limits associated with them there and in Probs. 9.4–9.6. In Prob. IX.9.3 we defined

$$a_p = p\Gamma\left(\frac{1}{p}\right)\Gamma\left(\frac{2}{p}\right) \cdots \Gamma\left(\frac{p}{p}\right), \tag{11.1}$$

where p is a positive integer. In Prob. IX.9.5 we found that

$$a_p = p \lim_{n \to +\infty} \frac{(n!)^p p^{np}}{(np)!\, n^{(p-1)/2}} . \tag{11.2}$$

Now consider the sequence $\langle x_n \rangle$, where

$$x_n = \frac{n!\, e^n}{n^{n+1/2}} \qquad \text{for each } n. \tag{11.3}$$

This sequence was defined in (10.32). In (10.36) we saw that

$$\lim x_n = \sqrt{2\pi}\,. \tag{11.4}$$

Since $\langle x_{np}\rangle = \langle x_p, x_{2p}, \ldots \rangle$ is a subsequence of $\langle x_n\rangle$, we have $x_{np} \to \sqrt{2\pi}$ as $n \to +\infty$. Therefore,

$$\frac{x_n^p}{x_{np}} \to \left(\sqrt{2\pi}\,\right)^{(p-1)} = (2\pi)^{(p-1)/2} \qquad \text{as} \quad n \to +\infty. \tag{11.5}$$

On the other hand, by the definition of x_n, we know that

$$\frac{x_n^p}{x_{np}} = \frac{(n!)^p e^{pn}}{n^{pn+p/2}} \cdot \frac{(np)^{np+1/2}}{(np)!\, e^{pn}} = \frac{(n!)^p p^{np} p^{1/2}}{(np)!\, n^{(p-1)/2}}\,.$$

From this it follows readily that

$$p^{-1/2}\frac{x_n^p}{x_{np}} = \frac{(n!)^p p^{np}}{(np)!\, n^{(p-1)/2}}\,. \tag{11.6}$$

Now take limits as $n \to +\infty$. Use (11.5) to obtain the limit of the left-hand side and (11.2) to obtain the limit of the right-hand side and conclude that

$$p^{-1/2}(2\pi)^{(p-1)/2} = \frac{a_p}{p}\,.$$

This implies that

$$a_p = p^{1/2}(2\pi)^{(p-1)/2}. \tag{11.7}$$

This combines with (11.1) to yield

$$p^{1/2}(2\pi)^{(p-1)/2} = p\Gamma\left(\frac{1}{p}\right)\Gamma\left(\frac{2}{p}\right)\cdots\Gamma\left(\frac{p}{p}\right). \tag{11.8}$$

For example, we have $a_1 = 1$. For $p = 2$, this yields

$$2^{1/2}(2\pi)^{1/2} = 2\Gamma(\tfrac{1}{2})$$

and, therefore, as already noted, that

$$\Gamma(\tfrac{1}{2}) = \sqrt{\pi}\,.$$

For $p = 3$, (11.8) yields

$$\Gamma(\tfrac{1}{3})\Gamma(\tfrac{2}{3}) = \frac{2\pi}{\sqrt{3}}\,. \tag{11.9}$$

Theorem 11.1 (Gauss' Multiplication Formula). *If $x > 0$ and p is a positive integer, then*

$$\Gamma\left(\frac{x}{p}\right)\Gamma\left(\frac{x+1}{p}\right)\cdots\Gamma\left(\frac{x+p-1}{p}\right) = \frac{(2\pi)^{(p-1)/2}}{p^{x-1/2}}\Gamma(x).$$

PROOF. Turn to Prob. IX.9.3 to obtain

$$p^x\Gamma\left(\frac{x}{p}\right)\Gamma\left(\frac{x+1}{p}\right)\dots\Gamma\left(\frac{x+p-1}{p}\right)=a_p\Gamma(x).$$

The conclusion follows from this and (11.7).

Corollary (Legendre's Relation). *If* $x>0$, *then*

$$\Gamma\left(\frac{x}{2}\right)\Gamma\left(\frac{x+1}{2}\right)=\frac{\sqrt{\pi}}{2^{x-1}}\Gamma(x).$$

PROOF. Use Gauss' Multiplication Formula with $p=2$ (cf. Prob. IX.9.6).

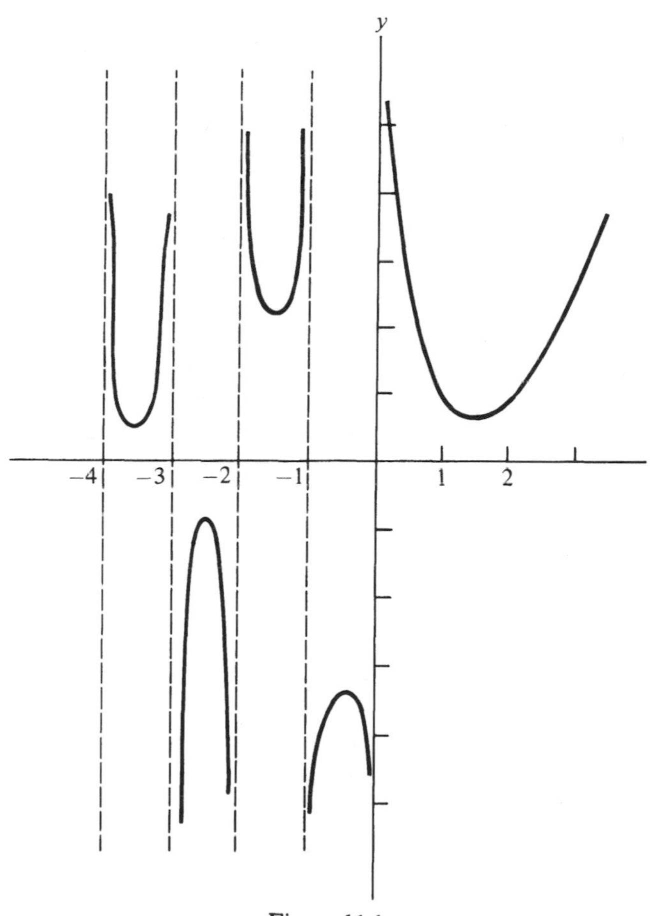

Figure 11.1

CHAPTER XI
More on Series: Sequences and Series of Functions

1. Introduction

We began studying infinite series in Chapter IV. They were used to define the function exp and the sine and cosine functions. Not enough properties of exp and the trigonometric functions were derived in Chapter IV to use them in examples and problems illustrating the theorems on infinite series gathered there. As for the natural logarithm, this function was first defined in a later chapter, so we did not consider its series in Chapter IV. In this chapter, among other things, we fill this gap in our development of introductory analysis. We ask the reader to review the material in Chapter IV.

EXAMPLE 1.1. Consider the infinite series

$$\sum_{n=2}^{\infty} \frac{1}{\ln n}. \tag{1.1}$$

(Here we sum from $n = 2$ on because $\ln 1 = 0$.) We first note that $0 < \ln n < n - 1 < n$ for $n \geqslant 2$ (explain) so that

$$\frac{1}{n} < \frac{1}{\ln n} \qquad \text{for} \quad n \geqslant 2. \tag{1.2}$$

Comparing (1.1) with the harmonic series, we gather from (1.2) that the series (1.1) diverges since the harmonic series does.

Prob. 1.1(a). Prove: If $p \in \mathbb{R}$, then

$$\sum_{n=2}^{\infty} \frac{1}{(\ln n)^p}$$

diverges.

Prob. 1.1(b). Prove: $\sum(1 - (\ln n)/n)$ diverges.

Example 1.2. We test

$$\sum_{n=1}^{\infty} \frac{\sin(\pi/n)}{n} \tag{1.3}$$

for convergence. We have

$$\frac{\sin(\pi/n)/n}{1/n^2} = \frac{1}{n}\pi\frac{\sin(\pi/n)}{\pi/n^2} = \pi\frac{\sin(\pi/n)}{\pi/n} \to \pi \quad \text{as} \quad n \to +\infty.$$

Since $\sum_{n=1}^{\infty}(1/n^2)$ converges, this implies that the series (1.3) converges (see the Corollary of Theorem IV.4.3).

Remark 1.1. The use of the ratio and root tests for convergence of series is fairly straightforward. The reader can review this material by turning to Section IV.5.

Prob. 1.2. Test $\sum a_n$ for convergence if for each n

(1) $a_n = e^{1/n}/n^2$,
(2) $a_n = \sqrt{n+1} - \sqrt{n}$,
(3) $a_n = (\sqrt{n+1} - \sqrt{n})/n$,
(4) $a_n = (\sqrt{n+1} - \sqrt{n})/n^2$,
(5) $a_n = (\ln^2 n)/n^2$,
(6) $a_n = \sqrt[n]{p} - 1, p > 0$,
(7) $a_n = (1 - (\ln n)/n)^n$.

Prob. 1.3. Test $\sum a_n$ for convergence if

(1) $a_n = 1/(\ln n)^n$ for $n \geqslant 2$,
(2) $a_n = 2^n n!/n^n$ for each n.

2. Cauchy's Condensation Test

The test which we give in this section for convergence of series is applicable to infinite series whose terms are nonnegative and monotonically decreasing. We first prove a lemma.

Lemma 2.1. *If $\langle a_n \rangle$ is a monotonically decreasing sequence of nonnegative numbers, then, for each positive integer n, the nth partial sum of $\sum_{n=1}^{\infty} a_n$ satisfies*

$$2S_{2^n} \geqslant \sum_{k=0}^{n} 2^k a_{2^k} \tag{2.1a}$$

and

$$S_{2^n-1} \leqslant \sum_{k=0}^{n-1} 2^k a_{2^k}. \tag{2.1b}$$

PROOF. Each part will be proved by induction. We prove (2.1a) first. First note that

$$S_2 = a_1 + a_2 \geqslant \frac{a_1}{2} + a_2$$

so that

$$2S_2 \geqslant a_1 + 2a_2 = \sum_{k=0}^{1} 2^k a_{2^k},$$

thus, (2.1a) holds for $n = 1$. Similarly, since $a_3 \geqslant a_4$, we have $a_3 + a_4 \geqslant 2a_4$, so that

$$S_{2^2} = S_4 = a_1 + a_2 + a_3 + a_4 \geqslant \frac{a_1}{2} + a_2 + 2a_4,$$

implying

$$2S_{2^2} \geqslant a_1 + 2a_2 + 2^2 a_4 = \sum_{k=0}^{2} 2^k a_{2^k}.$$

This proves (2.1a) for $n = 2$. Assume that (2.1a) holds for some positive integer n. Consider

$$S_{2^{n+1}} = S_{2^n} + \sum_{k=2^n+1}^{2^{n+1}} a_k.$$

On the right we have $k \leqslant 2^{n+1}$ in the second term and, therefore, $a_k \geqslant a_{2^{n+1}}$ for all such k. Therefore,

$$S_{2^{n+1}} = S_{2^n} + \sum_{k=2^n+1}^{2^{n+1}} a_k \geqslant S_{2^n} + a_{2^{n+1}} \sum_{k=2^n+1}^{2^{n+1}} 1. \tag{2.2}$$

We have

$$\sum_{k=2^n+1}^{2^{n+1}} 1 = 2^{n+1} - 2^n = 2^n.$$

This and (2.2) imply that

$$S_{2^{n+1}} \geqslant S_{2^n} + 2^n a_{2^{n+1}}.$$

Multiply through by 2 to obtain

$$2S_{2^{n+1}} \geqslant 2S_{2^n} + 2^{n+1} a_{2^{n+1}} \geqslant \sum_{k=0}^{n} 2^k a_{2^k} + 2^{n+1} a_{2^{n+1}}.$$

(The second inequality is a consequence of the induction hypothesis.) It follows that

$$2S_{2^{n+1}} \geqslant \sum_{k=0}^{n+1} 2^k a_{2^k}.$$

Thus, induction on n proves (2.1a).

We prove (2.1b). Note that

$$S_{2^1-1} = S_1 = a_1 = \sum_{k=0}^{0} 2^k a_{2^k}$$

and since $a_2 \geqslant a_3$, that $a_2 + a_3 \leqslant 2a_2$ and

$$S_{2^2-1} = S_3 = a_1 + a_2 + a_3 \leqslant a_1 + 2a_2 = \sum_{k=0}^{1} 2^k a_{2^k}.$$

This proves (2.1b) for $n = 1$ and $n = 2$. Assume that (2.1b) holds for some positive integer n and consider

$$S_{2^{n+1}-1} = S_{2^n-1} + \sum_{k=2^n}^{2^{n+1}-1} a_k. \tag{2.3}$$

The second term on the right consists of a sum such that the k satisfies $k \geqslant 2^n$. Since the sequence $\langle a_n \rangle$ is decreasing, this implies that $a_k \leqslant a_{2^n}$ for such k and, therefore, that

$$\sum_{k=2^n}^{2^{n+1}-1} a_k \leqslant a_{2^n} \sum_{k=2^n}^{2^{n+1}-1} 1 = a_{2^n}\left[(2^{n+1}-1) - (2^n-1)\right] = 2^n a_{2^n}.$$

Using (2.3) and the induction hypothesis we see that

$$S_{2^{n+1}-1} \leqslant S_{2^n-1} + 2^n a_{2^n} \leqslant \sum_{k=0}^{n-1} 2^k a_{2^k} + 2^n a_{2^n} = \sum_{k=0}^{n} 2^k a_{2^k}.$$

Invoking induction, we see that (2.1b) holds for all positive integers n.

Theorem 2.1 (Cauchy's Condensation Test). *If $\langle a_n \rangle$ is a monotonically decreasing sequence of nonnegative numbers, then the series*

$$\sum_{n=1}^{\infty} a_n \quad \text{and} \quad \sum_{n=1}^{\infty} 2^n a_{2^n}$$

converge together or diverge together.

PROOF. The theorem is an easy consequence of Lemma 2.1 and we ask the reader to carry out the details.

Remark 2.1. The reason for calling the last theorem a "condensation" test is that establishment of convergence or divergence of a series by means of it is accomplished by deleting infinitely many of its terms.

EXAMPLE 2.1. We use Cauchy's Condensation Test to test the convergence of the series

$$\sum_{n=2}^{\infty} \frac{1}{n \ln n}.$$

By Theorem 2.1 this series diverges or converges according as does the series

$$\sum_{n=1}^{\infty} \frac{2^k}{2^k \ln 2^k} = \sum_{k=1}^{\infty} \frac{1}{k \ln 2} = \frac{1}{\ln 2} \sum_{k=1}^{\infty} \frac{1}{k}.$$

Since the last series on the right is a positive multiple of the harmonic series and consequently diverges, the first series diverges.

PROB. 2.1. Use Cauchy's Condensation Test to prove that

$$\sum_{n=1}^{\infty} \frac{1}{n^p}, \qquad p > 0,$$

converges if $p > 1$ and diverges if $p \leqslant 1$. (Cf. Examples IV.1.3 and IV.3.2.)

PROB. 2.2. Prove that the series

$$\sum_{n=2}^{\infty} \frac{1}{n(\ln n)^p}$$

converges if $p > 1$ and diverges if $p \leqslant 1$ (cf. Example 2.1).

PROB. 2.3. Prove:

$$\sum_{n=3}^{\infty} \frac{1}{(n \ln n)(\ln \ln n)}$$

diverges.

PROB. 2.4. Prove that the series

$$\sum_{n=3}^{\infty} \frac{1}{(n \ln n)(\ln \ln n)^p}, \qquad p > 0,$$

diverges if $p > 1$ and converges if $p \leqslant 1$.

PROB. 2.5. Prove: If $n > 1$ where n is an integer, then

$$\frac{n}{2} < 1 + \frac{1}{2} + \frac{1}{3} + \cdots + \frac{1}{2^n - 1} < n.$$

3. Gauss' Test

Theorem 3.1. *If $\sum a_n$ has positive terms and we can express a_n / a_{n+1}, for each n, as*

$$\frac{a_n}{a_{n+1}} = 1 + \frac{r}{n} + \frac{A_n}{n^\lambda}, \tag{3.1}$$

where $\lambda > 1$ and $\langle A_n \rangle$ is a bounded sequence, then a_n converges if $r > 1$ and diverges if $r \leqslant 1$.

PROOF. First assume that $r \neq 1$ so that $r < 1$ or $r > 1$, and apply the modified Raabe test (corollary of Theorem IV.6.2). We can write (3.1) as

$$n\left(\frac{a_n}{a_{n+1}} - 1\right) = r + \frac{A_n}{n^{\lambda-1}} \qquad \text{for each } n.$$

Hence,

$$\lim_{n\to+\infty} n\left(\frac{a_n}{a_{n+1}} - 1\right) = \lim_{n\to+\infty}\left(r + \frac{A_n}{n^{\lambda-1}}\right) = r.$$

The modified Raabe test yields the conclusion in this case.

Assume that $r = 1$ so that (3.1) now has the form

$$\frac{a_n}{a_{n+1}} = 1 + \frac{1}{n} + \frac{A_n}{n^{\lambda}} \qquad \text{for each } n. \tag{3.2}$$

This can be written

$$n\frac{a_n}{a_{n+1}} - (n+1) = \frac{A_n}{n^{\lambda-1}} \qquad \text{for each } n.$$

We now multiply both sides by $\ln n$ and then we add to both sides $(n+1)\ln n - (n+1)\ln(n+1)$. The result is

$$(n\ln n)\frac{a_n}{a_{n+1}} - (n+1)\ln(n+1) = (n+1)\ln\frac{n}{n+1} + A_n\frac{\ln n}{n^{\lambda-1}}$$

$$= \ln\frac{1}{(1+1/n)^{n+1}} + A_n\frac{\ln n}{n^{\lambda-1}}. \tag{3.3}$$

As $n \to +\infty$, the first term in the last expression approaches $\ln e^{-1} = -1$, while the second term approaches 0. Hence,

$$\lim_{n\to+\infty}\left[(n\ln n)\frac{a_n}{a_{n+1}} - (n+1)\ln(n+1)\right] = -1. \tag{3.4}$$

Therefore, there exists a positive integer N such that

$$(n\ln n)\frac{a_n}{a_{n+1}} - (n+1)\ln(n+1) < 0 \qquad \text{for} \quad n > N. \tag{3.5}$$

The sequence $\langle b_n \rangle$, where $b_n = n\ln n$, has positive terms for $n > 1$. Moreover $\sum_{n=2}^{\infty} b_n^{-1}$ diverges (Example 2.1). By Kummer's test (Theorem IV.6.1), $\sum a_n$ diverges. This proves the conclusion for $r = 1$. The proof is complete.

EXAMPLE 3.1 (The Hypergeometric Series). The infinite series

$$1 + \frac{ab}{1!\,c}x + \frac{a(a+1)b(b+1)}{2!\,c(c+1)}x^2$$

$$+ \frac{a(a+1)(a+2)b(b+1)(b+2)}{3!\,c(c+1)(c+2)}x^3 + \cdots, \tag{3.6}$$

where a, b, and c are neither 0 nor negative integers, is called the *hypergeometric* series. (If c is equal to 0 or a negative integer the series is not defined. If a or b is equal to 0 or a negative integer, then the series terminates.) We first test for convergence using the ratio test. We can write the nth term a_n as

$$a_n = \frac{\binom{a+n-1}{n}\binom{b+n-1}{n}}{\binom{c+n-1}{n}} x^n. \tag{3.7}$$

Since

$$\frac{a_{n+1}}{a_n} = \frac{(a+n)(b+n)}{(n+1)(c+n)} x, \tag{3.8}$$

we have

$$\lim_{n\to+\infty} \frac{|a_{n+1}|}{|a_n|} = |x| \tag{3.9}$$

and conclude that the series converges absolutely for $|x| < 1$ and diverges for $|x| > 1$. Next we investigate the case $|x| = 1$.

Consider the hypergeometric series for $x = 1$. We have from (3.8)

$$\frac{a_n}{a_{n+1}} = \frac{(n+1)(n+c)}{(a+n)(b+n)} = \frac{n^2+(c+1)n+c}{n^2+(a+b)n+ab} \quad \text{for each } n. \tag{3.10}$$

We wish to write this in the form (3.1) so that we can apply Theorem 3.1. By (3.10),

$$n\left(\frac{a_n}{a_{n+1}} - 1\right) = n\left[\frac{n^2+(c+1)n+c}{n^2+(a+b)n+ab} - 1\right]$$
$$= \frac{(c+1-a-b)n^2+(c-ab)n}{n^2+(a+b)n+ab}. \tag{3.11}$$

"Dividing out" in the last expression we find that it can be written

$$\frac{(c+1-a-b)n^2+(c-ab)n}{n^2+(a+b)n+ab} = c+1-a-b+\frac{An+B}{n^2+(a+b)n+ab}, \tag{3.12}$$

where

$$A = c - ab - (a+b)(c+1-a-b) \quad \text{and} \quad B = ab(a+b-c-1). \tag{3.13}$$

Using (3.11) and (3.12) we can write

$$\frac{a_n}{a_{n+1}} = 1 + \frac{c+1-a-b}{n} + \frac{An+B}{n[n^2+(a+b)n+ab]}, \tag{3.14}$$

where A and B are defined as in (3.13) and are independent of n. We have

$$\frac{An+B}{n^3+(a+b)n^2+abn}=\frac{1}{n^2}\left(\frac{A+B/n}{1+(a+b)/n+ab/n^2}\right). \tag{3.15}$$

Putting

$$A_n=\frac{A+B/n}{1+(a+b)/n+ab/n^2} \qquad \text{for each } n \tag{3.16}$$

and noting that $A_n \to A$ as $n \to +\infty$, we conclude that the sequence $\langle A_n \rangle$ is bounded. Thus, (3.14) can be written

$$\frac{a_n}{a_{n+1}}=1+\frac{c+1-a-b}{n}+\frac{A_n}{n^2}, \tag{3.17}$$

where $\langle A_n \rangle$ is a bounded sequence of real numbers. Gauss' test (Theorem 3.1) tells us that the hypergeometric series (3.6) converges for $x=1$ if $c+1-(a+b)>1$ and diverges if $c+1-(a+b) \leqslant 1$. Accordingly, the series

$$1+\frac{ab}{1!\,c}+\frac{a(a+1)b(b+1)}{2!\,c(c+1)}+\frac{a(a+1)(a+2)b(b+1)(b+2)}{3!\,c(c+1)(c+2)}+\cdots \tag{3.18}$$

converges if $a+b<c$ and diverges if $a+b \geqslant c$.

We now investigate the convergence of (3.6) when $x=-1$. In this case, by (3.8)

$$\frac{a_n}{a_{n+1}}=-\frac{(n+1)(c+n)}{(a+n)(b+n)}=-\frac{n^2+(c+1)n+c}{n^2+(a+b)n+ab}. \tag{3.19}$$

This implies that $\lim(a_n/a_{n+1})=-1<0$. From this we conclude that a positive integer N_1 exists such that $a_n/a_{n+1}<0$ for $n>N_1$. Accordingly, for $n>N_1$, the terms have opposite signs. Gauss' test cannot be used here. However, the situation can be analyzed from another point of view.

We add 1 to both sides of (3.19) and obtain

$$\frac{a_n}{a_{n+1}}+1=\frac{(a+b-c-1)n+ab-c}{n^2+(a+b)n+ab}. \tag{3.20}$$

If $a+b>c+1$, then there exists a positive integer N_2 such that for $n>N_2$ the numerator and denominator on the right-hand side of (3.20) are both positive. Combining this information with that at the end of the last paragraph, we have for $n>\max\{N_1,N_2\}$ that

$$1>\frac{a_n}{a_{n+1}}+1>0$$

which implies that

$$\frac{|a_n|}{|a_{n+1}|}<1 \qquad \text{for} \quad n>\max\{N_1,N_2\}.$$

In turn, this implies that $\lim a_n \neq 0$ and, therefore, that the series diverges if $a + b - c > 1$.

Now consider the case $a + b < c + 1$, so that $a + b - c - 1 < 0$. This implies that for sufficiently large n the fraction on the right in (3.20) is negative (explain). Since also $a_n / a_{n+1} < 0$ for sufficiently large n, it follows that for sufficiently large n

$$\left| \frac{a_n}{a_{n+1}} \right| = -\frac{a_n}{a_{n+1}} > 1. \tag{3.21}$$

This implies that for sufficiently large n

$$|a_n| > |a_{n+1}|. \tag{3.22}$$

Thus, there exists an N such that for $n > N$ the a_n's alternate in sign and their absolute values decrease monotonically. We prove that $a_n \to 0$ as $n \to +\infty$ holds also (in the present case). We rewrite (3.19) as

$$\frac{a_{n+1}}{a_n} = -\frac{n^2 + (a+b)n + ab}{n^2 + (c+1)n + c}.$$

By (3.21),

$$0 < \frac{n^2 + (a+b)n + ab}{n^2 + (c+1)n + c} = -\frac{a_{n+1}}{a_n} = \left| \frac{a_{n+1}}{a_n} \right| < 1 \tag{3.23}$$

for sufficiently large n.

Calculations similar to the ones already performed prove that we can write

$$-\frac{a_{n+1}}{a_n} = \frac{n^2 + (a+b)n + ab}{n^2 + (c+1)n + c} = 1 - \frac{c+1-a-b}{n} + \frac{E_n}{n^2} \qquad \text{for each } n,$$

where $\langle E_n \rangle$ is a bounded sequence. Because of (3.23) this implies that for sufficiently large n

$$0 < 1 - \frac{c+1-a-b}{n} + \frac{E_n}{n^2} = \frac{|a_{n+1}|}{|a_n|} < 1. \tag{3.24}$$

We can also show that we have

$$\frac{E_n}{n^2} \leqslant \frac{|E_n|}{n^2} < \frac{c+1-a-b}{n} \qquad \text{for sufficiently large } n. \tag{3.25}$$

We conclude that there exists a positive integer N such that (3.24) and (3.25) both hold for $n > N$. We have

$$|a_n| = |a_N| \frac{|a_{N+1}|}{|a_N|} \frac{|a_{N+2}|}{|a_{N+1}|} \cdots \frac{|a_n|}{|a_{n-1}|} \qquad \text{for} \quad n > N$$

and, therefore, by (3.24),

$$|a_n| = |a_N| \prod_{k=N+1}^{n} \left[1 - \left(\frac{c+1-a-b}{k} - \frac{E_k}{k^2} \right) \right] \qquad \text{for} \quad n > N. \tag{3.26}$$

Now the infinite product

$$\prod_{k=N+1}^{\infty}\left[1-\left(\frac{c+1-a-b}{k}-\frac{E_k}{k^2}\right)\right] \tag{3.27}$$

converges if and only if

$$\sum_{k=N+1}^{\infty}\left(\frac{c+1-a-b}{k}-\frac{E_k}{k^2}\right) \tag{3.28}$$

converges (Theorem X.7.3). But

$$\sum_{k=N+1}^{\infty}\frac{c+1-a-b}{k}\quad\text{diverges and}\quad\sum_{k=N+1}^{\infty}\frac{E_k}{k^2}\quad\text{converges.}$$

(Explain.) Therefore, (3.28) diverges. This implies that the infinite product (3.27) diverges. Because of (3.24) and (3.25), we have

$$0<\frac{c+1-a-b}{k}-\frac{E_k}{k^2}<1\qquad\text{for}\quad k>N.$$

Because of the divergence of the infinite product (3.27) this tells us that it diverges to 0 (Remark X.7.2). It follows from (3.26) that $a_n \to 0$ as $n \to +\infty$. The information we have thus far about $\sum_{n=N+1}^{\infty} a_n$ is that it is an alternating series and, therefore, converges. Consequently $\sum_{n=0}^{\infty} a_n$ converges. We conclude that if $a+b<c+1$, then the hypergeometric series converges for $x=-1$.

We examine the hypergeometric series for $x=-1$ if $a+b=c+1$. In this case we have

$$\frac{a_{n+1}}{a_n}=-1-\frac{E_n}{n^2}\qquad\text{for each } n. \tag{3.29}$$

Since $\langle E_n\rangle$ is bounded, there exists an $M>0$ such that $-E_n \leqslant |E_n| \leqslant M$ for all n. Hence, $-M \leqslant E_n$ for all n. Let N be a positive integer such that $N^2>M$. We have

$$-\frac{M}{N^2}\leqslant\frac{E_N}{N^2},\qquad -\frac{M}{(N+1)^2}\leqslant\frac{E_{N+1}}{(N+1)^2},\ \ldots$$

and, therefore,

$$0<1-\frac{M}{N^2}\leqslant 1+\frac{E_N}{N^2}=\frac{|a_{N+1}|}{|a_N|},$$

$$0<1-\frac{M}{(N+1)^2}\leqslant 1+\frac{E_{N+1}}{(N+1)^2}=\frac{|a_{N+2}|}{|a_{N+1}|},$$

$$\vdots$$

$$0<1-\frac{M}{(n-1)^2}\leqslant 1+\frac{E_{n-1}}{(n-1)^2}=\frac{|a_n|}{|a_{n-1}|}$$

for $n > N$. Consequently,

$$|a_n| = |a_N| \frac{|a_{N+1}|}{|a_N|} \frac{|a_{N+2}|}{|a_{N+1}|} \cdots \frac{|a_n|}{|a_{n-1}|}$$

$$\geqslant |a_N|\left(1 - \frac{M}{N^2}\right)\left(1 - \frac{M}{(N+1)^2}\right) \cdots \left(1 - \frac{M}{(n-1)^2}\right) \tag{3.30}$$

for $n > N$. Now $|a_N| > 0$. The factor of $|a_N|$ in (3.29) is a product of nonzero factors and is a partial product of a convergent infinite product. Therefore, this factor approaches a positive limit P as $n \to +\infty$. From this and (3.29) it follows that $\lim a_n \neq 0$ and, hence, that the series diverges. We summarize the results of this example in:

Theorem 3.2. *The hypergeometric series* (3.6) *converges if* $|x| < 1$ *and diverges if* $|x| > 1$. *If* $x = 1$, *then* (3.6) *converges if* $a + b < c$ *and diverges if* $a + b \geqslant c$. *If* $x = -1$, *then* (3.6) *converges if* $a + b - c < 1$ *and diverges if* $a + b - c \geqslant 1$.

A function defined by the hypergeometric series is written as $F(a,b,c;x)$ and is called a *hypergeometric* function. Thus,

$$F(a,b,c;x) = \sum_{n=0}^{\infty} \frac{\binom{a+n-1}{n}\binom{b+n-1}{n}}{\binom{c+n-1}{n}} x^n \tag{3.31}$$

whenever the series on the right converges. Many of the functions we have already encountered are hypergeometric functions.

Prob. 3.1. Prove: If $|x| < 1$, then $F(a,b,b;x) = (1-x)^{-a}$.

Prob. 3.2. Prove: (1) If $0 < |x| < 1$, then

$$F(1,1,2;-x) = \frac{\ln(1+x)}{x};$$

(2) $F(1,1,2;-1)$ converges and $F(1,1,2;-1) = \ln 2$.

Prob. 3.3. Prove: If $|x| < |b|$, then

$$F\left(a,b,a;\frac{x}{b}\right) = \left(1 - \frac{x}{b}\right)^{-b}.$$

Note, that if we fix x and take b such that $b > |x|$, then

$$\lim_{b \to +\infty} F\left(a,b,a;\frac{x}{b}\right) = \lim_{b \to +\infty}\left(1 - \frac{x}{b}\right)^{-b} = \frac{1}{e^{-x}} = e^x.$$

Prob. 3.4. Prove: (1) $F(a,b,a;1)$ converges if $b < 0$ and in this case $F(a,b,a;1) = 0$. (2) If $b \geqslant 0$, then $F(a,b,a;1)$ diverges.

Prob. 3.5. Prove: (1) If $b < 1$, then $F(a,b,a;\,-1)$ converges and in this case $F(a,b,a;\,-1) = 2^{-b}$. (2) If $b \geqslant 1$, then $F(a,b,a;\,-1)$ diverges.

4. Pointwise and Uniform Convergence

Let $\langle f_n \rangle$ be a sequence of real-valued functions all having a common domain D and let f be a real-valued function defined on $A \subseteq D$, where $A \neq \emptyset$. If for each $x \in A$, we have

$$\lim_{n \to +\infty} f_n(x) = f(x), \tag{4.1}$$

then we say that $\langle f_n \rangle$ *converges pointwise* to f on A.

Actually, we have been dealing with this notion all along. For example, in one of our early results on limits we proved:

$$\lim_{n \to +\infty} x^n = 0 \qquad \text{for} \quad |x| < 1.$$

Here the sequence of functions was $\langle (\ )^n \rangle$, and the limit function was the constant function whose value is 0 for all x such that $|x| < 1$. Similarly, we defined the exponential function exp by means of

$$\exp x = \lim_{n \to +\infty} \sum_{k=1}^{n} \frac{x^k}{k!} \qquad \text{for} \quad x \in \mathbb{R}.$$

Here the sequence of functions was $\langle f_n \rangle$, where for each n

$$f_n(x) = \sum_{k=0}^{n} \frac{x^k}{k!} = 1 + x + \frac{x^2}{2!} + \cdots + \frac{x^n}{n!}.$$

For each fixed $x \in \mathbb{R}$, the sequence of real numbers $\langle f_n(x) \rangle$ converges and $\exp x$ is defined as its limit; the sequence $\langle f_n \rangle$ here converges pointwise to exp.

In terms of ϵ's and N's, if $\langle f_n \rangle$ converges pointwise to f on A, then for $x \in A$ and $\epsilon > 0$ there exists an N such that

$$|f_n(x) - f(x)| < \epsilon.$$

Usually, the N which exists for each $\epsilon > 0$ depends on x. If for each $\epsilon > 0$, the N depends on ϵ only we say the convergence of $\langle f_n \rangle$ to f on A is *uniform*. We give an "official" definition.

Def. 4.1. Let $\langle f_n \rangle$ be a sequence of real-valued functions all having a common domain D, and let f be defined on $A \subseteq D$ where $A \neq \emptyset$. We say that $\langle f_n \rangle$ converges *uniformly* to f on A if and only if, for each $\epsilon > 0$, there exists an N, depending only on $\epsilon > 0$ such that if $n > N$, then

$$|f_n(x) - f(x)| < \epsilon \qquad \text{holds for all} \quad x \in A. \tag{4.2}$$

(See Fig. 4.1.) When $\langle f_n \rangle$ converges uniformly to f on D, then we say simply that $\langle f_n \rangle$ converges uniformly to f, often writing $f_n \to f$ as $n \to +\infty$.

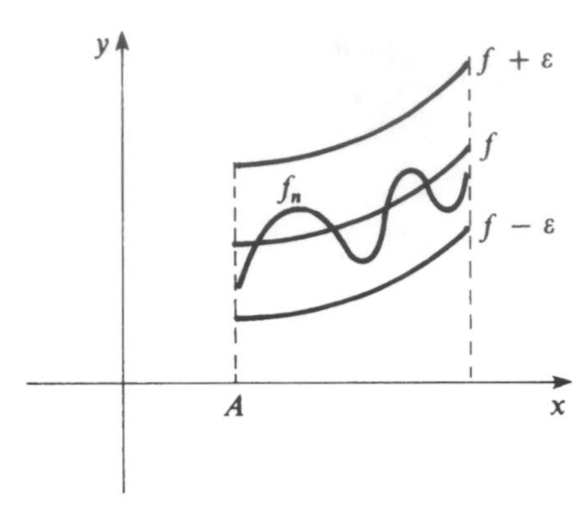

Figure 4.1

Remark 4.1. We often dispense with $\langle\ \rangle$ and write "f_n converges to f."

Remark 4.2. Clearly, if f_n converges uniformly to f then it converges to it pointwise.

Remark 4.3. From the definition of uniform convergence (Def. 4.1) it should be clear that we may assume, when necessary, that the N in that definition is a positive integer.

Remark 4.4. It also follows from Def. 4.1 that when $f_n \to f$ as $n \to +\infty$ uniformly on a set A, then an N exists such that if $n > N$, then the difference $f_n - f$ is bounded on A (see (4.2) and Theorem 4.1 below).

Theorem 4.1. *Let D be the common domain of a sequence of $\langle f_n \rangle$ and a function f. Suppose that there exists an N such that for $n > N$, $f_n - f$ is bounded on D. Define M_n by means of*

$$M_n = \sup_{x \in D} |f_n(x) - f(x)| \qquad \textit{for} \quad n > N. \tag{4.3}$$

If $\lim_{n\to+\infty} M_n = 0$ for the sequence $\langle M_n \rangle$ defined in (4.3), *then f_n converges to f uniformly. Conversely, if f_n converges to f uniformly, then there exists an N such that if $n > N$, then $f_n - f$ is bounded (on D) and the sequence $\langle M_n \rangle$ defined by means of* (4.3) *is such that $\lim_{n\to+\infty} M_n = 0$.*

Proof. Suppose that an N exists such that for $n > N$, $f_n - f$ is bounded and $\lim M_n = 0$ for the sequence $\langle M_n \rangle$ defined in (4.3). Let $\epsilon > 0$ be given. There exists an N_1 such that if $n > N_1$, then $0 \leqslant M_n < \epsilon$. N_1 depends on ϵ alone, since the M_n are constants. For $n > N_1$,

$$|f_n(x) - f(x)| \leqslant M_n < \epsilon \qquad \text{for all} \quad x \in D.$$

This implies that f_n converges uniformly (on D).

Conversely, assume that f_n converges to f uniformly (on D). Let $\epsilon > 0$ be given. There exists an N such that if $n > N$, then

$$|f_n(x) - f(x)| < \frac{\epsilon}{2} \qquad \text{for all} \quad x \in D.$$

Thus, for $n > N$, $f_n - f$ is bounded on D and

$$0 \leqslant M_n = \sup_{x \in D} |f_n(x) - f(x)| \leqslant \epsilon/2 < \epsilon.$$

This implies that $\lim M_n = 0$.

EXAMPLE 4.1. For each positive integer n define f_n as

$$f_n(x) = x^n \qquad \text{for} \quad |x| < 1.$$

This sequence converges pointwise to the constant function 0 on $(-1; 1)$. However, for each n we have

$$M_n = \sup_{-1<x<1} |x^n - 0| = \sup_{-1<x<1} |x|^n = 1.$$

Thus, $\lim M_n = 1$. By Theorem 4.1 we see that f_n does not converge to its limit function uniformly on D. (If it did we would have $\lim M_n = 0$.) We depict f_1, f_2, and f_3 on $[0, 1)$ in Fig. 4.2.

EXAMPLE 4.2. Sometimes the convergence of f_n to f is not uniform on one set but is uniform on another set. Let us consider the f_n's of the last example but examine their convergence on the set $A = [-x_0, x_0]$, where x_0 is some fixed number such that $0 < x_0 < 1$. Write

$$M_n^* = \sup_{-x_0<x<x_0} |x^n - 0| = \sup_{-x_0<x<x_0} |x|^n = |x_0|^n.$$

Here $M_n^* = |x_0|^n \to 0$ as $n \to +\infty$. Hence, by Theorem 4.1, $f_n \to 0$ uniformly on A.

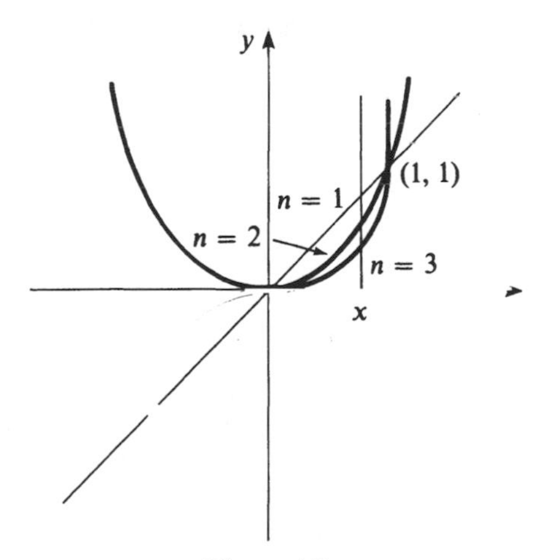

Figure 4.2

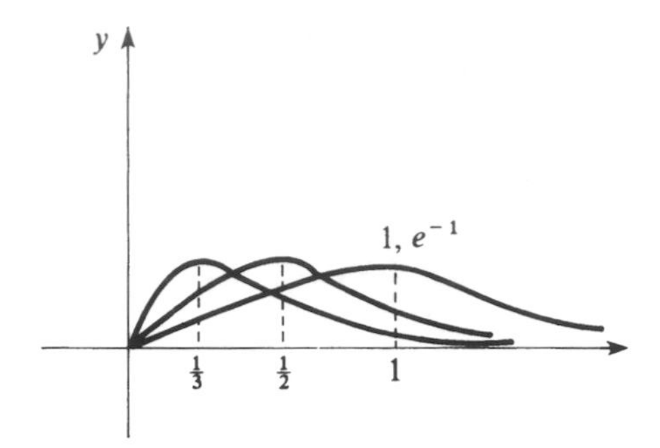

Figure 4.3

EXAMPLE 4.3. For each positive integer n, let the sequence $\langle f_n \rangle$ be defined by means of

$$f_n(x) = nxe^{-nx} \qquad \text{for} \quad x \geqslant 0.$$

Clearly, $f_n(0) = 0$ and $\lim_{n\to+\infty} f_n(x) = 0$, for $x > 0$. Hence,

$$\lim_{n\to+\infty} f_n(x) = \lim_{n\to+\infty} \frac{nx}{e^{nx}} = 0.$$

Thus, this sequence $\langle f_n \rangle$ tends to the constant function 0 pointwise on $A = [0, +\infty)$. The maximum of f_n on $[0, +\infty)$ occurs at $x = 1/n$ and is equal to $f_n(1/n) = e^{-1}$ for each n. Therefore,

$$M_n = \sup_{x \geqslant 0} |f_n(x) - 0| = \sup_{x \geqslant 0} |nxe^{-x}| = \max_{x \geqslant 0} (nxe^{-nx}) = e^{-1},$$

so that $\lim_{n\to+\infty} M_n = e^{-1}$. As in Example 4.1, we conclude that the convergence is not uniform (see Fig. 4.3).

PROB. 4.1. Let $f_n(x) = 1/(1 + x^{2n})$ for each positive integer n, where $x \in \mathbb{R}$. What is the pointwise limit of this sequence? State whether or not the convergence is uniform.

PROB. 4.2. Discuss the pointwise and uniform convergence of f_n on $[0, +\infty)$, where

(1) $f_n(x) = 1/(n + x)$;
(2) $f_n(x) = nx^2/(1 + nx)$;
(3) $f_n(x) = nx/(1 + n^2x)$;
(4) $f_n(x) = (\sin nx)/n$.

Before we show why the notion of uniform convergence is important we state Cauchy's criterion for uniform convergence.

Theorem 4.2. *Let $\langle f_n \rangle$ be a sequence of functions defined on some domain D. A necessary and sufficient condition for this sequence to converge uniformly to some function f on D is that for each $\epsilon > 0$, there exists an N depending on ϵ*

only such that if $m > N$ and $n > N$, then

$$|f_m(x) - f_n(x)| < \epsilon \qquad \textit{for all} \quad x \in D. \tag{4.4}$$

Proof. We prove the "if" part and ask the reader (see Prob. 4.3) to prove the "only if" part.

Assume that for each $\epsilon > 0$ there exists an N depending on ϵ only such that if $m > N$ and $n > N$ then (4.4) holds. Fix $x \in D$. Since (4.4) holds for the sequence of numbers $\langle f_n(x)\rangle$, this sequence is a Cauchy sequence of real numbers. Consequently, it has a unique limit. Write this limit as $f(x)$. The function f defined in this manner is the pointwise limit of f_n. We prove that the convergence is uniform. Let $\epsilon > 0$ be given. There exists an N depending on only ϵ such that if $m > N$ and $n > N$, then

$$|f_m(x) - f_n(x)| < \frac{\epsilon}{2} \qquad \text{for all} \quad x \in D.$$

Fix $n > N$ and take some $x \in D$. We know that $\lim_{m\to+\infty} f_m(x) = f(x)$. Therefore,

$$|f(x) - f_n(x)| = \lim_{m\to+\infty} |f_m(x) - f_n(x)| \leqslant \frac{\epsilon}{2} < \epsilon,$$

where $x \in D$. Here N depends only on ϵ and for $n > N$ we have

$$|f(x) - f_n(x)| < \epsilon \qquad \text{for} \quad x \in D.$$

Thus, $f_n \to f$ uniformly on D.

Prob. 4.3. Prove: If $\langle f_n\rangle$ converges uniformly to f on $\mathfrak{D}$, then for each $\epsilon > 0$, there exists an N such that if $m > N$ and $n > N$, then

$$|f_m(x) - f_n(x)| < \epsilon \qquad \text{for all} \quad x \in D.$$

(This completes the proof of Theorem 4.2.)

Remark 4.5. It is clear that we may assume, if necessary, that the N in Theorem 4.2 is a positive integer.

The reason uniform convergence is important is that when $\langle f_n\rangle$ converges uniformly to the function f, some of the properties common to all the functions f_n are transmitted to the limit function f. We shall see this in the next two theorems. We first define what we mean by the *uniform boundedness* of a sequence of functions.

Def. 4.2. A sequence $\langle f_n\rangle$ of real-valued functions is said to be *uniformly bounded* on a set D if and only if there exists an $M > 0$ such that for each n

$$|f_n(x)| \leqslant M \qquad \text{for all} \quad x \in D. \tag{4.5}$$

For example, let f_n be defined as

$$f_n(x) = \frac{1}{1 + x^{2n}} \qquad \text{for each} \quad x \in \mathbb{R} \quad \text{and for each } n.$$

For each n

$$|f_n(x)| = \frac{1}{1 + x^{2n}} \leqslant 1 \qquad \text{for all} \quad x \in \mathbb{R}.$$

Hence, this sequence of functions is uniformly bounded on $\mathbb{R}$.

Theorem 4.3. *If $\langle f_n \rangle$ converges uniformly to f on a set D and each f_n is bounded on D, then $\langle f_n \rangle$ is uniformly bounded on D and the limit function f is bounded on D.*

Proof. Since the convergence on D is uniform, there exists a positive integer N not depending on x such that if $m > N$ and $n > N$, then

$$|f_m(x) - f_n(x)| < 1 \qquad \text{for} \quad x \in D. \tag{4.6}$$

Fix $m > N$. By hypothesis, f_m is bounded on D. Hence, there exists an $M > 0$ such that $|f_m(x)| \leqslant M$ for all $x \in D$. Take $n > N$ and $x \in D$. We have

$$|f_n(x)| \leqslant |f_m(x) - f_n(x)| + |f_m(x)| < 1 + M.$$

Thus,

$$|f_n(x)| < 1 + M \qquad \text{for all} \quad x \in D \quad \text{and for} \quad n > N.$$

Hence, the sequence $\langle f_{N+1}, f_{N+2}, \ldots \rangle$ is uniformly bounded on D by $1 + M$. But $f_1, \ldots, f_N$ are each bounded on D, and there exist M_1, $M_2, \ldots, M_N$ such that for each $k \in \{1, \ldots, N\}$

$$|f_k(x)| \leqslant M_k \qquad \text{for all} \quad x \in D.$$

Let $M^* = \max\{M_1, \ldots, M_N, 1 + M\}$. Clearly, for each positive integer n, we have

$$|f_n(x)| \leqslant M^* \qquad \text{for} \quad x \in D.$$

It follows that $\langle f_n \rangle$ is uniformly bounded on D. Since f_n converges uniformly to f on D, it converges to f pointwise there and, by (4.6),

$$|f(x)| = \lim_{n \to +\infty} |f_n(x)| \leqslant M^* \qquad \text{for each} \quad x \in D.$$

This proves that f is bounded on D, as claimed.

Example 4.4. Here we exhibit a sequence of bounded functions converging pointwise but not uniformly to a function f on D whose limit function is not bounded on D. Define

$$f_n(x) = \frac{n^2}{1 + n^2x^2} \qquad \text{for each positive integer } n \text{ and for} \quad 0 < x \leqslant 1$$

Clearly,

$$|f_n(x)| \leqslant n^2 \qquad \text{for each positive integer } n \text{ and for} \quad x \in D = (0, 1].$$

Thus, each f_n is bounded on D. We have

$$\lim_{n\to+\infty} f_n(x) = \lim_{n\to+\infty} \frac{n^2}{1+n^2x^2}$$
$$= \lim_{n\to+\infty} \frac{1}{1/n^2+x^2} = \frac{1}{x^2} \qquad \text{for} \quad 0 < x \leqslant 1.$$

The limit function is f, where $f(x) = x^{-2}$ for $0 < x \leqslant 1$, and f is not bounded even though each f_n is. Note that

$$|f_n(x) - f(x)| = \left| \frac{n^2}{1+n^2x^2} - \frac{1}{x^2} \right|$$
$$= \frac{1}{x^2(1+nx^2)} \qquad \text{for each } n \text{ and for} \quad x \in (0,1].$$

Thus, none of the differences $f_n - f$ is bounded on D. By Theorem 4.1, this sequence does not converge uniformly to f.

Example 4.5. The sequence $\langle f_n \rangle$, where

$$f_n(x) = \frac{1}{x} + \frac{x}{n} \qquad \text{for each } n \text{ and for all } x \text{ in } (0,1],$$

converges uniformly to f, where $f(x) = x^{-1}$ for $x \in (0, 1]$. This is so because

$$M_n = \sup_{0<x\leqslant 1} |f_n(x) - f(x)| = \sup_{0<x\leqslant 1} \left| \frac{x}{n} \right| = \frac{1}{n}$$

and $\lim M_n = 0$. However, none of the f_n is bounded. Thus, uniform convergence alone is not sufficient to guarantee uniform boundedness.

Another property transmitted to the uniform limit of a sequence of functions is continuity.

Theorem 4.4. *If all the functions of a sequence $\langle f_n \rangle$ are continuous on a set D and f_n converges uniformly to the function f on D, then the limit function f is continuous on D.*

Proof. Let $x_0 \in D$ and let $\epsilon > 0$ be given. Since the convergence of f_n to f is uniform (by hypothesis), there exists an N depending only on ϵ such that if $n > N$, then

$$|f_n(x) - f(x)| < \frac{\epsilon}{3} \qquad \text{for all} \quad x \in D. \tag{4.7}$$

Fix $n > N$ and note that

$$|f_n(x_0) - f(x_0)| < \frac{\epsilon}{3}\,. \tag{4.8}$$

By hypothesis, f_n is continuous on D, and, thus, in particular, at x_0. There exists a $\delta > 0$ such that if $x \in D$ and $|x - x_0| < \delta$, then

$$|f_n(x) - f_n(x_0)| < \frac{\epsilon}{3}\,. \tag{4.9}$$

Now take $x \in D$ and $|x - x_0| < \delta$. Inequality (4.7), (4.8), and (4.9) hold for such x. Hence,

$$\begin{aligned} |f(x) - f(x_0)| &\leqslant |f(x) - f_n(x)| + |f_n(x) - f_n(x_0)| + |f_n(x_0) - f(x_0)| \\ &< \frac{\epsilon}{3} + \frac{\epsilon}{3} + \frac{\epsilon}{3} = \epsilon. \end{aligned}$$

Thus, for each $\epsilon > 0$, there exists a $\delta > 0$ such that if $x \in D$ and $|x - x_0| < \delta$, then $|f(x) - f(x_0)| < \epsilon$. This implies that f is continuous at x_0. We proved that f is continuous at each $x_0 \in D$ and, hence, that f is continuous on D.

Remark 4.6. Mere pointwise continuity of a sequence of continuous function to a limit function does not guarantee the continuity of the limit function. For each positive integer n define

$$g_n(x) = x^n \qquad \text{for} \quad 0 \leqslant x \leqslant 1.$$

Here $D = [0, 1]$. We have

$$g(x) = \begin{cases} \lim_{n \to +\infty} x^n = 0 & \text{for} \quad 0 \leqslant x < 1 \\ \lim_{n \to +\infty} x^n = 1 & \text{for} \quad x = 1. \end{cases}$$

The sequence $\langle g_n \rangle$ converges pointwise to g on $[0, 1] = D$. Each g_n is continuous but g is not continuous on D because it lacks continuity at $x = 1$. The sequence does not converge *uniformly* to g.

Note that the pointwise limit of a sequence of continuous functions may be continuous even though the convergence is not uniform. The example of $\langle f_n \rangle$ where $f_n(x) = x^n$ for $0 < x < 1$ for each n illustrations this. Here f_n converges pointwise to the constant function 0 on $(0; 1)$. The limit function and all the f_n's are continuous on $(0; 1)$.

Remark 4.7. Theorem 4.4 states that if each f_n is continuous on D and f_n converges uniformly to f on D, then f is continuous. Thus, if for each n,

$$\lim_{x \to x_0} f_n(x) = f_n(x_0) \qquad \text{for} \quad x_0 \in D \tag{4.10}$$

and

$$\lim_{n \to +\infty} f_n(x) = f(x) \qquad \text{for} \quad x \in D \tag{4.11}$$

and the convergence is uniform, then

$$\lim_{x \to x_0} f(x) = f(x_0). \tag{4.12}$$

Because of (4.11), this yields

$$\lim_{x \to x_0} \left(\lim_{n \to +\infty} f_n(x) \right) = f(x_0). \tag{4.13}$$

Taking limits as $n \to +\infty$ in (4.10), we have

$$\lim_{n \to +\infty} \left(\lim_{x \to x_0} f_n(x) \right) = \lim_{n \to +\infty} f_n(x_0) = f(x_0).$$

This and (4.13) imply that

$$\lim_{x\to x_0}\left(\lim_{n\to+\infty} f_n(x)\right) = \lim_{n\to+\infty}\left(\lim_{x\to x_0} f(x)\right) = f(x_0). \tag{4.14}$$

Thus, Theorem 4.4 states that the uniform convergence of f_n to f permits the interchangeability of the order of limits indicated in (4.14).

PROB. 4.4. Prove: If f_n converges uniformly to f on a set D, then f_n converges uniformly to f on any nonempty subset of D.

PROB. 4.5. Prove: If f_n converges uniformly to f on D and g_n also converges uniformly to g on D, then (1) the sum sequence $\langle f_n + g_n\rangle$ converges uniformly to $f+g$ on D and (2) the product sequence $\langle f_n g_n\rangle$ converges uniformly to fg on D, provided each f_n and g_n is bounded on D.

PROB. 4.6.* Prove: If f_n converges uniformly to f on a set D and $\langle x_n\rangle$ is a sequence of elements of D such that $x_n \to x$ as $n\to+\infty$, where also $x\in D$, then

$$f_n(x_n)\to f(x) \qquad \text{as} \quad n\to+\infty.$$

PROB. 4.7.† For each positive integer n define

$$f_n(x) = \frac{nx}{1+nx} \qquad \text{for} \quad x\in[0,+\infty).$$

Note that

$$\lim_{n\to+\infty} f_n(x) = f(x) = \begin{cases} 0 & \text{if} \quad x=0 \\ 1 & \text{if} \quad x>0. \end{cases}$$

Let $x_n = 1/n$ for each positive integer n. Note that $\lim_{n\to+\infty} f(1/n) = 1/2$ and we have $\lim_{n\to+\infty} f_n(x_n) \neq f(0)$, in spite of the fact that $\lim_{n\to+\infty} x_n = 0$. Reconcile this result with Prob. 4.6.

5. Applications to Power Series

The notions of pointwise and uniform convergence defined in the last section for sequences of functions are readily extended to infinite series. Let $\sum f_n$ be an infinite series of functions all defined on a set D, and let $\langle S_n\rangle$ be the sequence of partial sums of the series, that is, $S_n = f_1 + \cdots + f_n$ for each n. Let S also be defined on D. We say that $\sum_{n=1}^{\infty} f_n$ *converges pointwise* to S on D if and only if S_n converges pointwise to S on D, i.e., if and only if

$$S_n(x)\to S(x) \qquad \text{for each} \quad x\in D.$$

*J. Bass, *Exercises in Mathematics*, Academic Press, New York, 1966, pp. 54–55.

†*Ibid.*

We then write

$$S(x) = \sum_{k=1}^{\infty} f_k(x) \qquad \text{for each } x,$$

and call S the *sum* of the series. When S_n converges uniformly to S on D, we say that the series $\sum_{n=1}^{\infty} f_n$ *converges uniformly* to its sum S on D.

Prob. 5.1. Prove: If $\langle f_n \rangle$ is a sequence of functions all of which are defined on D and continuous there and the series $\sum_{n=1}^{\infty} f_n$ converges uniformly to its sum S on D, then S is continuous on D.

To test the uniform convergence of series we often use *Weierstrass' M-test*.

Theorem 5.1. *If $\langle f_n \rangle$ is a sequence of functions all of which are defined in a set D and $\langle M_n \rangle$ is a sequence of positive numbers such that for each n*

$$|f_n(x)| \leqslant M_n \qquad \text{for each } n \text{ and for all} \quad x \in D \tag{5.1}$$

and such that $\sum_{n=1}^{\infty} M_n$ converges, then the infinite series $\sum_{n=1}^{\infty} f_n$ converges uniformly to its sum on D.

Proof. Let $\epsilon > 0$ be given. Since $\sum_{n=1}^{\infty} M_n$ converges, there exists an N, depending only on ϵ, such that for $m > n > N$,

$$\sum_{k=n+1}^{m} M_k < \epsilon. \tag{5.2}$$

Let $\langle S_n \rangle$ be the sequence of partial sums of the series. Bearing in mind (5.1) and (5.2), we obtain for $m > n > N$,

$$|S_m(x) - S_n(x)| = \left| \sum_{k=n+1}^{m} f_k(x) \right| \leqslant \sum_{k=n+1}^{n} |f_k(x)|$$

$$\leqslant \sum_{k=n+1}^{m} M_k < \epsilon \qquad \text{for all} \quad x \in D.$$

This implies that S_n converges uniformly on D and hence that the series converges uniformly on D.

We now apply some of our theory to the study of *power series*.

A power series is a series of the form

$$\sum_{n=0}^{\infty} a_n(x - x_0)^n = a_0 + a_1(x - x_0) + a_2(x - x_0)^2 + \cdots \tag{5.3}$$

where $\langle a_n \rangle$ is a sequence of constants and x_0 is some fixed number. x_0 is called the *center* of the series. We refer to (5.3) as a power series in $x - x_0$.

When $x_0 = 0$, (5.3) has the form

$$\sum_{n=0}^{\infty} a_n x^n = a_0 + a_1 x + a_2 x^2 + \cdots. \tag{5.4}$$

This is a power series in x.

The development in this chapter will be confined to *real* series. Thus, the a_n's in (5.3), which we call the *coefficients* of the series will always be real. This applies to x and x_0 also.

As examples we have the series

$$e^x = 1 + x + \frac{x^2}{2!} + \frac{x^3}{3!} + \cdots, \qquad x \in \mathbb{R}, \tag{5.5}$$

$$\sin x = x - \frac{x^3}{3!} + \frac{x^5}{5!} - \frac{x^7}{7!} + \cdots, \qquad x \in \mathbb{R}, \tag{5.6}$$

$$\cos x = 1 - \frac{x^2}{2!} + \frac{x^4}{4!} - \frac{x^6}{6!} + \cdots, \qquad x \in \mathbb{R}, \tag{5.7}$$

$$\ln(1+x) = x - \frac{x^2}{2} + \frac{x^3}{3} - \frac{x^4}{4} + \cdots, \qquad \text{where } -1 < x \leqslant 1. \tag{5.8}$$

The series

$$\ln x = (x-1) - \frac{(x-1)^2}{2} + \frac{(x-1)^3}{3} - \frac{(x-1)^4}{4} + \cdots, \tag{5.9}$$

where $0 < x \leqslant 2$, is an example of one *not* centered at $x = 0$.

Note that when $x = x_0$, the power series (5.3) converges trivially to a_0.

Theorem 5.2. *Given the power series $\sum_{n=0}^{\infty} a_n x^n$ and $x_1 \neq 0$. If the series converges for $x = x_1$, then it converges absolutely for all x such that $|x| < |x_1|$. However, if the series diverges for $x = x_1$, then it diverges for all x such that $|x| > |x_1|$.*

PROOF. Assume $\sum_{n=0}^{\infty} a_n x_1^n$ converges. This implies $\lim_{n\to+\infty}(a_n x_1^n) = 0$ and, hence, that the sequence $\langle a_n x_1^n \rangle$ is bounded. There exists an $M > 0$ such that, for all n,

$$|a_n x_1^n| < M. \tag{5.10}$$

Take x such that $|x| < |x_1|$, so that

$$0 \leqslant r = \frac{|x|}{|x_1|} < 1,$$

and consider the series $\sum_{n=0}^{\infty} |a_n x^n|$. Since, by (5.10),

$$|a_n x^n| = |a_n x_1^n| \left| \frac{x}{x_1} \right|^n < M r^n \qquad \text{for all } n$$

and $\sum_{n=0}^{\infty} M r^n$ converges (why?), it follows that $\sum_{n=0}^{\infty} |a_n x^n|$ converges. Thus, $\sum_{n=0}^{\infty} a_n x^n$ converges absolutely.

Now assume $\sum_{n=0}^{\infty} a_n x_1^n$ diverges and take x such that $|x| > |x_1|$. If $\sum_{n=0}^{\infty} a_n x^n$ converges, it would follow by what was just proved that $\sum_{n=0}^{\infty} a_n x_1^n$ converges, contradicting the assumption. We, therefore, conclude that $\sum_{n=0}^{\infty} a_n x^n$ diverges. This completes the proof.

Remark 5.1. The power series (5.5), (5.6), and (5.7) converge for all x. On the other hand, (5.8) is a power series in x converging for $|x| < 1$ and diverging for $|x| > 1$. Thus, some power series in x converge for all x and some converge for certain $x \neq 0$ and diverge for other $x \neq 0$. All power series in x converge for $x = 0$. We exhibit a power series in x converging for $x = 0$ only. Consider

$$\sum_{n=0}^{\infty} n!\, x^n = 1 + x + 2!\, x + 3!\, x^3 + \cdots \tag{5.11}$$

for some $x \neq 0$. Write the nth term as a_n. We have $a_n = n!\, x^n$ for each n. Applying the ratio test to $\sum_{n=0}^{\infty} |a_n|$ we have

$$\frac{|a_{n+1}|}{|a_n|} = (n+1)|x| \to +\infty \qquad \text{as} \quad n \to +\infty.$$

There exists a positive integer N such that if $n > N$, then

$$\frac{|a_{n+1}|}{|a_n|} > 1.$$

We have $0 < |a_N| < |a_{N+1}| < \ldots$ so that $\lim_{n\to+\infty} a_n = \lim_{n\to+\infty} n!\, x^n \neq 0$. Hence, the series (5.11) diverges for $x \neq 0$. This means that $x = 0$ is the only point at which our series converges.

Theorem 5.3. *Given the power series $\sum_{n=0}^{\infty} a_n x^n$. Let*

$$L = \overline{\lim_{n\to+\infty}} \sqrt[n]{|a_n|}\ . \tag{5.12}$$

We have $0 \leqslant L \in \mathbb{R}^*$. If (1) $L = +\infty$, then the series converges for $x = 0$ only. If (2) $L = 0$, then the series converges absolutely for all $x \in \mathbb{R}$. If (3) $0 < L < +\infty$, then the series converges absolutely if $|x| < L^{-1}$ and diverges if $|x| > L^{-1}$.

PROOF. Apply the root test (Theorem IV.5.2) to the series $\sum_{n=0}^{\infty} |a_n x^n|$ to test for absolute convergence. Write

$$A_n = |a_n x^n| \qquad \text{for each } n.$$

We know that

$$\overline{\lim_{n\to+\infty}} \sqrt[n]{A_n} = \overline{\lim_{n\to+\infty}} \sqrt[n]{|a_n x^n|} = |x| \overline{\lim_{n\to+\infty}} \sqrt[n]{|a_n|} = |x| L. \tag{5.13}$$

By the root test and (5.13) we conclude: $\sum |a_n x^n|$ converges if $|x| L < 1$ and diverges if $|x| L > 1$. Examine the case $|x| L > 1$, where $\sum_{n=0}^{\infty} |a_n x^n|$ diverges. In this case we see that $\overline{\lim} \sqrt[n]{A_n} > 1$. It follows that $\sqrt[n]{A_n} > 1$ for infinitely many n. This implies that $|a_n x^n| = A_n > 1$ for infinitely many n and, therefore, that $\lim a_n x^n \neq 0$. It follows that our (original) series diverges. Accordingly we may state: $\sum a_n x^n$ converges absolutely if $|x|L < 1$ and diverges if $|x|L > 1$. If $L = +\infty$, then, for $x \neq 0$, $|x|L > 1$ and our series diverges. In this case we see that our series converges for $x = 0$ only. If $L = 0$, then $|x|L = 0 < 1$ for each $x \in \mathbb{R}$ and, therefore, our series converges for all $x \in \mathbb{R}$. Finally assume that $0 < L < +\infty$. If $|x| < L^{-1}$, we have $|x|L < 1$ and the series converges absolutely. If $|x| > L^{-1}$, then $|x|L > 1$ and the series diverges. This completes the proof.

Corollary. *Given the power series in $\sum_{n=0}^{\infty} a_n(x - x_0)^n$, define L as in the above theorem. If* (1) *$L = +\infty$, then the series converges for $x = x_0$ only. If* (2) *$L = 0$, then the series converges for all $x \in \mathbb{R}$. If* (3) *$0 < L < +\infty$, then the series converges absolutely if $|x - x_0| < L^{-1}$ and diverges if $|x - x_0| < L^{-1}$.*

PROOF. Write $x' = x - x_0$ so that $\sum_{n=0}^{\infty} a_n(x - x_0)^n = \sum_{n=0}^{\infty} a_n x'^n$. Apply the theorem to $\sum_{n=0}^{\infty} a_n x'^n$. When applied to $\sum_{n=0}^{\infty} a_n x'^n$, the three alternatives of the theorem give rise to the three alternatives of this corollary.

Remark 5.2. Consider the power series $\sum_{n=0}^{\infty} a_n(x - x_0)^n$. When $L = \overline{\lim}_{n\to+\infty} \sqrt[n]{|a_n|}$ is such that $0 < L < +\infty$, the last corollary states that the series converges for all x such that $|x - x_0| < L^{-1}$, i.e., in the interval $(x_0 - L^{-1}; x_0 + L^{-1})$ and diverges for x such that $|x - x_0| > L^{-1}$, i.e., for x in $(-\infty; L^{-1}) \cup (L^{-1}; +\infty)$. The interval $(x_0 - L^{-1}; x_0 + L^{-1})$ is called the *interval of convergence* and L^{-1} is called the *radius of convergence* of the series. We repeat that R, where

$$R = L^{-1} = \frac{1}{\overline{\lim}_{n\to+\infty} \sqrt[n]{(a_n)}}, \tag{5.14}$$

is the radius of convergence of the series and $(x_0 - R; x_0 + R)$ is its interval of convergence (see Fig. 5.1). The above-mentioned corollary yields no information about the endpoints $x_0 \pm R$ of the interval of convergence. We shall see below that some series exist which converge at neither of these endpoints; others at both endpoints. There are also series converging only at one of the endpoints.

The three cases delineated in the corollary of Theorem 5.2 can be subsumed under one if we define $1/0 = +\infty$. We do this just in this corollary. Accordingly, if $L = 0$, then (5.14) yields $R = +\infty$ and we say

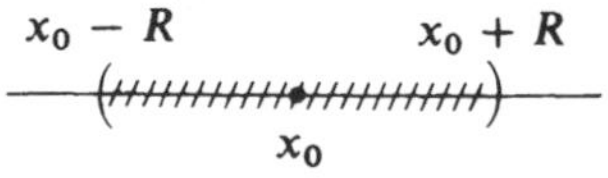

$x_0 - R$ $x_0 + R$

x_0

Figure 5.1

that the radius of convergence is equal to $+\infty$. We then have $x_0 - R = x_0 - \infty = -\infty$ and $x_0 + R = x_0 + \infty = +\infty$ so that, in this case, the interval of convergence becomes $(x_0 - R; x_0 + R) = (-\infty; +\infty) = \mathbb{R}$. This is consistent with case (2) of the corollary. When $L = +\infty$, (5.14) yields $R = 0$ and we say the radius of convergence is $R = 0$. In this case, the interval of convergence is $(x_0 - R; x_0 + R) = (x_0; x_0) = \emptyset$. Thus, in this case, there is no *interval* of convergence, and the series converges only for $x = x_0$. This is consistent with alternative (1) of the corollary. Formula (5.14) is called the *Cauchy–Hadamard* formula for the radius of convergence of a power series.

Theorem 5.4. *If the power series $\sum_{n=0}^{\infty} A_n(x - x_0)^n$ has nonzero coefficients and $\lim_{n\to+\infty}(|A_{n+1}|/|A_n|)$ exists in $\mathbb{R}^*$, then the radius of convergence R of the series is given by*

$$R = \lim_{n\to+\infty} \frac{|A_n|}{|A_{n+1}|} . \tag{5.15}$$

PROOF. We recall Theorem IV.5.3. It states that if the series $\sum a_n$ has positive terms, then

$$\underline{\lim}\, \frac{a_{n+1}}{a_n} \leqslant \underline{\lim}\, \sqrt[n]{a_n} \leqslant \overline{\lim}\, \sqrt[n]{a_n} \leqslant \overline{\lim}\, \frac{a_{n+1}}{a_n} . \tag{5.16}$$

The proof of Theorem IV.5.3 shows that these inequalities concern themselves with the *sequence* $\langle a_n \rangle$ of positive numbers only and do not depend on the assumption that the a_n's are the terms of the series $\sum a_n$. It follows from (5.16) that if $l = \lim(a_{n+1}/a_n)$ exists in $\mathbb{R}^*$, then $\lim \sqrt[n]{a_n}$ exists and is equal to l so that $\lim(a_{n+1}/a_n) = l = \lim \sqrt[n]{a_n}$. Apply this to the sequence $\langle |A_n| \rangle$. We obtain

$$\lim \frac{|A_{n+1}|}{|A_n|} = \lim \sqrt[n]{|A_n|} . \tag{5.17}$$

If $\lim(|A_{n+1}|/|A_n|) > 0$, then by (5.17) and Lemma III.8.2 we have

$$\lim \frac{|A_n|}{|A_{n+1}|} = \frac{1}{\lim \sqrt[n]{|A_n|}} = R.$$

If $\lim(|A_{n+1}|/|A_n|) = 0$, then we have

$$\lim \frac{|A_n|}{|A_{n+1}|} = +\infty = R$$

in this case also.

EXAMPLE 5.1. The geometric series

$$\sum_{n=0}^{\infty} x^n = 1 + x + x^2 + \cdots$$

has the interval of convergence $(-1; 1)$ since $\lim_{n\to+\infty} \sqrt[n]{|a_n|} = \lim_{n\to+\infty} 1 = 1$ and

$$R = \lim_{n\to+\infty} \frac{1}{\sqrt[n]{|a_n|}} = 1.$$

EXAMPLE 5.2. The logarithmic series

$$\sum_{n=1}^{\infty} (-1)^{n+1} \frac{x^n}{n} = x - \frac{x^2}{2} + \frac{x^3}{3} - \frac{x^4}{4} + \cdots$$

has radius of convergence $R = 1$. Here $|a_n| = 1/n$ for each positive integer n so

$$\sqrt[n]{|a_n|} = \frac{1}{\sqrt[n]{n}} \qquad \text{for each } n.$$

Since $\sqrt[n]{n} \to 1$, we see that

$$R = \lim \frac{1}{\sqrt[n]{|a_n|}} = \lim \sqrt[n]{n} = 1.$$

The interval of convergence is $(-1; 1)$. This series converges for $x = 1$ and diverges for $x = -1$.

EXAMPLE 5.3. One can readily calculate that the series

$$\sum_{n=1}^{\infty} \frac{x^n}{n^2}$$

has radius of convergence $R = 1$. Its interval of convergence is $(-1; 1)$ and it also converges for $x = 1$ and $x = -1$.

The set of all x at which a power series $\sum a_n(x - x_0)^n$ converges will be called its *domain of convergence*. The domain of convergence may include the endpoints of the interval of convergence I, and so may differ from it. For each x in the domain of convergence the power series converges to its sum $P(x)$. The function P defined as

$$P(x) = \sum_{n=0}^{\infty} a_n(x - x_0)^n \qquad \text{for } x \text{ in the domain of convergence of the power series} \tag{5.18}$$

is the sum of the power series. A power series converges pointwise to its sum P in the domain of convergence. In the next theorem we prove that P is continuous on the *interval* of convergence. Before stating this theorem we point out that if the power series converges for $x = x_0$ only, its sum is certainly continuous since its domain consists of x_0 only and the latter is, therefore, an isolated point (Remark VI.1.1) of this domain (see Example VI.1.1).

Theorem 5.5. *If the power series* (5.18) *has a radius of convergence* $R > 0$, *and* a, b *are real numbers such that* $x_0 - R < a < b < x_0 + R$, *then it converges uniformly on* $[a,b]$ *to its sum* P. *This sum is a continuous function on the interval of convergence* $(x_0 - R; x_0 + R)$ *of the power series.*

PROOF. The power series converges absolutely at a and at b, which are points in the interval of convergence. Thus,

$$\sum_{n=0}^{\infty} |a_n||a - x_0|^n \quad \text{and} \quad \sum_{n=0}^{\infty} |a_n||b - x_0|^n \tag{5.19}$$

converge. Let

$$M = \max\{|a - x_0|, |b - x_0|\}. \tag{5.20}$$

Clearly, $M > 0$. Assume that $a \leqslant x \leqslant b$ and obtain $-M \leqslant a - x_0 \leqslant x - x_0 \leqslant b - x_0 \leqslant M$, and, hence, $|x - x_0| \leqslant M$. This proves $[a,b] \subseteq [x_0 - M, x_0 + M]$. Since the two series in (5.19) converge, we know that

$$\sum_{n=0}^{\infty} |a_n| M^n \quad \text{converges.} \tag{5.21}$$

Take x such that $|x - x_0| \leqslant M$. For such x we have

$$|a_n(x - x_0)^n| = |a_n||x - x_0|^n \leqslant |a_n| M^n.$$

This and (5.21) imply that $\sum_{n=0}^{\infty} a_n(x - x_0)^n$ converges uniformly on the set of all x such that $|x - x_0| \leqslant M$, and, hence, on $[x_0 - M, x_0 + M]$. Since $[a,b] \subseteq [x_0 - M, x_0 + M]$, we conclude (Prob. 4.4) that the power series (5.15) converges uniformly on $[a,b]$. The partial sum sequence $\langle P_n \rangle$ of the power series consists of polynomials in x, and, consequently, of functions which are continuous on $[a,b]$ and by what was just proved, converge uniformly to the sum P of the power series on $[a,b]$. By Theorem 4.4, P is continuous on $[a,b]$.

Remark 5.3. The situation is described by saying that the sum P of the power series is continuous on every bounded closed subinterval $[a,b]$ contained in the interval $(x_0 - R; x_0 + R)$ of convergence. We prove that P is continuous on the interval of convergence. Take $x_1 \in (x_0 - R; x_0 + R)$ so that $x_0 - R < x_1 < x_0 + R$. There exist real numbers a and b such that $x_0 - R < a < x_1 < b < x_0 + R$. Since P is continuous on $[a,b]$ and $x_1 \in [a, b]$, it is continuous at x_1. Thus, P is continuous for each $x_1 \in (x_0 - R; x_0 + R)$ and, therefore, it is continuous on $(x_0 - R; x_0 + R)$.

Having shown that the sum of a power series is continuous on the interval of convergence, it is natural to ask whether or not it is continuous at an endpoint of this interval when it converges at such a point. We shall deal with this question later. We also postpone the discussion of uniform convergence and differentiability. Instead we continue discussing uniform convergence and continuity in the next two sections. In the next section we exploit Theorem 4.4 to prove the existence of a continuous function on $\mathbb{R}$ which is nowhere differentiable.

6. A Continuous But Nowhere Differentiable Function

Weierstrass was the first to give an example of a function which was continuous for all $x \in \mathbb{R}$ and differentiable nowhere. Here we present an example of such a function due to Van der Waerden.

The above-mentioned function is constructed with the aid of the function $\langle\ \rangle^*$, the distance to the nearest integer function defined in Example V.2.4. Recall: for each $x \in \mathbb{R}$, $\langle x\rangle^*$ is defined as

$$\langle x\rangle^* = \text{the distance from } x \text{ to the integer nearest } x. \tag{6.1}$$

(The graph of $\langle\ \rangle^*$ is drawn in Fig. V.2.1.)

In Example V.2.4, it is shown that

$$\langle x\rangle^* = \tfrac{1}{2} - \left|\tfrac{1}{2} - x + [x]\right| \qquad \text{for} \quad x \in \mathbb{R}. \tag{6.2}$$

By means of this formula one can easily prove that $\langle\ \rangle^*$ is a continuous function. To see this, we first recall that $[\ ]$, the greatest integer function, is continuous for each x which is not an integer and is continuous from the right for all x, including integers. It is left to prove that $\langle\ \rangle^*$ is continuous from the left at $x = n$, where n is an integer. First note that $[n] = n$ and, therefore, that

$$\langle n\rangle^* = \tfrac{1}{2} - \left|\tfrac{1}{2} - n + [n]\right| = \tfrac{1}{2} - \left|\tfrac{1}{2} - n + n\right| = 0. \tag{6.3}$$

Note that $[x] \to n - 1$ as $x \to n-$. Using this and (6.2) we have

$$\begin{aligned}\langle x\rangle^* = \tfrac{1}{2} - \left|\tfrac{1}{2} - x + [x]\right| &\to \tfrac{1}{2} - \left|\tfrac{1}{2} - n + (n-1)\right| \\ &= \tfrac{1}{2} - \left|-\tfrac{1}{2}\right| = 0 = \langle n\rangle^*,\end{aligned}$$

as $x \to n-$. Thus, we have

$$\lim_{x \to n-} \langle x\rangle^* = \langle n\rangle^* = \lim_{x \to n+} \langle x\rangle^*.$$

This means that $\langle\ \rangle^*$ is continuous at each integer n. Since it is also continuous for each x which is not an integer, it is continuous for all $x \in \mathbb{R}$.

Prob. 6.1. Let m be some integer. Define $g: \mathbb{R} \to \mathbb{R}$ by means of

$$g(x) = \langle mx\rangle^* \qquad \text{for all} \quad x \in \mathbb{R}.$$

Prove that g is continuous on $\mathbb{R}$.

To follow the reasoning it is useful to recall (Probs. V.2.15 and V.2.16) that

$$\langle x+1\rangle^* = \langle x\rangle^* \tag{6.4a}$$

and that

$$0 \leqslant \langle x \rangle^* \leqslant \tfrac{1}{2} \quad \text{for} \quad x \in \mathbb{R}. \tag{6.4b}$$

We now define Van der Waerden's function as

$$f(x) = \sum_{n=0}^{\infty} \frac{\langle 10^n x \rangle^*}{10^n} \tag{6.5}$$

By (6.4b), we know that

$$\left| \frac{\langle 10^n x \rangle^*}{10^n} \right| \leqslant \frac{1}{2 \cdot 10^n} \quad \text{for each nonnegative integer } n. \tag{6.6}$$

Since

$$\sum_{n=0}^{\infty} \frac{1}{2 \cdot 10^n}$$

obviously converges, it follows from (6.6) and the Weierstrass M-test for uniform convergence that the series on the right converges uniformly to its sum f on $\mathbb{R}$. The partial sum functions S_n are given by

$$S_n(x) = \sum_{k=0}^{n} \frac{\langle 10^k x \rangle^*}{10^k} \quad \text{for each } n \text{ and all} \quad x \in \mathbb{R}.$$

Each S_n is continuous on $\mathbb{R}$ since it is a finite sum of constant multiples of functions which are continuous on $\mathbb{R}$ (Prob. 6.1). Thus, the sum f of the series in (6.5) is a limit of a sequence of functions continuous on $\mathbb{R}$ which converges to it uniformly on $\mathbb{R}$. Therefore, f is continuous on $\mathbb{R}$. In the next paragraph we prove that f is differentiable for no $x \in \mathbb{R}$.

We take $x \in \mathbb{R}$ and write it in its nonterminating decimal form so that

$$x = N + .a_1 a_2 \ldots, \tag{6.7}$$

where N is an integer and $a_1 a_2, \ldots$ are digits. Construct the sequence $\langle h_m \rangle$, where for each m

$$h_m = \begin{cases} -\dfrac{1}{10^m} & \text{if} \quad a_m = 4 \text{ or } 9 \\[2ex] \dfrac{1}{10^m} & \text{if} \quad a_m \text{ is neither 4 nor 9.} \end{cases} \tag{6.8}$$

Form

$$\frac{f(x + h_m) - f(x)}{h_m} = \sum_{m=0}^{\infty} \frac{\langle 10^n(x + h_m) \rangle^* - \langle 10^n x \rangle^*}{10^n h_m}. \tag{6.9}$$

We examine the numerators of the terms of the last series for $n \geqslant m$ and then for $n < m$. If $n \geqslant m$, then

$$10^n(x + h_m) = 10^n x + 10^n h_m = 10^n x \pm 10^{n-m}. \tag{6.10}$$

Since the function $\langle\ \rangle^*$ is periodic with period 1 (see (6.4a)) and 10^{n-m} is an integer, (6.10) gives us

$$\langle 10^n(x + h_m) \rangle^* = \langle 10^n x \rangle^* \quad \text{for} \quad n \geqslant m.$$

Therefore, the numerators of the terms of the series on the right in (6.9) all vanish for $n \geqslant m$, and (6.9) becomes

$$\begin{aligned}\frac{f(x+h_m)-f(x)}{h_m} &= \sum_{n=0}^{m-1} \frac{\langle 10^n(x+h_m)\rangle^* - \langle 10^n x\rangle^*}{10^n h_m} \\ &= \sum_{n=0}^{m-1} \pm 10^{m-n}(\langle 10^n(x+h_m)\rangle^* - \langle 10^n x\rangle^*).\end{aligned} \tag{6.11}$$

In the sum on the right here, we have $n < m$. To evaluate this sum we first multiply both sides of (6.7) by 10^n and obtain

$$10^n x = 10^n N + a_1 a_2 \ldots a_n . a_{n+1} a_{n+2} \cdots . \tag{6.12}$$

Here $a_1 a_2 \ldots a_n$ is not the product of the digits $a_1, a_2, \ldots, a_n$. It is rather the decimal representation of the integer $M = a_1 10^{n-1} + a_2 10^{n-2} + \cdots + a_n$. To avoid confusion, we write this integer as $M = (a_1 a_2 \ldots a_n)$, and then write $I_1 = 10^n N + (a_1 a_2 \ldots a_n)$ and $I_2 = I_1 + 1$. Thus, (6.12) can be written

$$10^n x = I_1 + .a_{n+1} a_{n+2} \cdots , \tag{6.13}$$

where I_1 is an integer. It follows from this that the integer nearest $10^n x$ is I_1 or $I_2 = I_1 + 1$, and that the distance from $10^n x$ to the nearest integer is

$$\langle 10^n x\rangle^* = \begin{cases} .a_{n+1} a_{n+2} \cdots & \text{if} \quad .a_{n+1} a_{n+2} \cdots \leqslant \frac{1}{2} \\ 1 - .a_{n+1} a_{n+2} \cdots & \text{if} \quad .a_{n+1} a_{n+2} \cdots > \frac{1}{2}. \end{cases} \tag{6.14}$$

Now

$$10^n(x+h_m) = 10^n x + 10^n h_m = I_1 + .a_{n+1} a_{n+2} \cdots \pm 10^{-(m-n)}. \tag{6.15}$$

Since $n < m$ here, we have $a_m = a_{n+(m-n)}$, where $m - n > 0$. We see that (6.15) can be written

$$\begin{aligned} 10^n(x+h_m) &= I_1 + a_{n+1} 10^{-1} + a_{n+2} 10^{-2} + \cdots \\ &\quad + a_m 10^{-(m-n)} + \cdots \pm 10^{-(m-n)} \\ &= I_1 + a_{n+1} 10^{-1} + a_{n+2} 10^{-2} + \cdots \\ &\quad + (a_m \pm 1) 10^{-(m-n)} + \cdots . \end{aligned} \tag{6.16}$$

The decimal $.a_{n+1} a_{n+2} \ldots$ is nonterminating.

Assume that $.a_{n+1} a_{n+2} \cdots \leqslant 1/2 = .4999 \ldots$, so that $a_{n+1} \leqslant 4$. If $a_m = 4$ or 9, we have from (6.8) that $10^n h_m = -10^n h_m = -10^{-(m-n)}$. Therefore, the term $(a_m \pm 1)10^{-(m-n)}$ in (6.16) is equal to $(a_m - 1)10^{-(m-n)}$, and $a_m - 1 = 3$ or 8. This and (6.16) imply that

$$I_1 < 10^n(x+h_m) < I_1 + .a_{n+1} a_{n+2} \cdots \leqslant I_1 + \tfrac{1}{2}.$$

It follows from this that I_1 is the integer nearest $10^n(x+h_m)$, and (6.16) yields

$$\langle 10^n(x+h_m)\rangle^* = .a_{n+1} a_{n+2} \cdots \pm 10^{-(m-n)}. \tag{6.17}$$

If $a_m \neq 4$ and $a_m \neq 9$, then (6.8) implies that $10^n h_m = 10^{-(m-n)}$. Therefore, the term $(a_m \pm 1)10^{-(m-n)}$ in (6.16) is equal to $(a_m + 1)10^{-(m-n)}$ and $a_m + 1 \leqslant 9$. We then have

$$I_1 < 10^n(x + h_m) \leqslant I_1 + .4999\ldots = I_1 + \tfrac{1}{2}.$$

Again the integer nearest $10^n(x + h_m)$ is I_1. Hence, (6.17) holds here also.

Assume that $.a_{n+1}a_{n+2}\ldots > 1/2$. This implies $a_{n+1} \geqslant 5$. Let $a_m = 4$ or 9, we know that $10^n h_m = -10^{-(m-n)}$, so the term $(a_m \pm 1)10^{-(m-n)}$ in (6.16) is equal to $(a_m - 1)10^{-(m-n)}$. This implies that

$$10^n(x + h_m) < I_1 + .a_{n+1}a_{n+2}\ldots \leqslant I_1 + 1. \tag{6.18}$$

Suppose $m > n + 1$. Since $a_{n+1} \geqslant 5$, we also have

$$I_1 + \tfrac{1}{2} \leqslant 10^n(x + h_m). \tag{6.19}$$

This and (6.18) imply that

$$I_1 + \tfrac{1}{2} \leqslant 10^n(x + h_m) < I_1 + 1,$$

from which we conclude that

$$\langle 10^n(x + h_m)\rangle^* = 1 - .a_{n+1}a_{n+2}\ldots \mp 10^{-(m-n)}. \tag{6.20}$$

Suppose $m = n + 1$, then $a_m = a_{n+1} \geqslant 5$. Since $a_m = 4$ or 9, we have $a_{n+1} = a_m = 9$. Again,

$$I_1 + \tfrac{1}{2} < 10^n(x + h_m) < I_1 + 1,$$

and (6.20) holds in this case also. If $a_m \neq 4$ and $a_m \neq 9$, it is not difficult to see that (6.20) still remains valid.

We summarize and state: If $n < m$, then

$$\langle 10^n(x + h_m)\rangle^* = \begin{cases} .a_{n+1}a_{n+2}\ldots \pm 10^{-(m-n)} & \text{if } .a_{n+1}a_{n+2}\ldots \leqslant \tfrac{1}{2} \\ 1 - .a_{n+1}a_{n+2}\ldots \mp 10^{-(m-n)} & \text{if } .a_{n+1}a_{n+2}\ldots > \tfrac{1}{2}. \end{cases}$$

This, in conjunction with (6.14), leads to: If $n < m$, then

$$\langle 10^n(x + h_m)\rangle^* - \langle 10^n x\rangle^* = \begin{cases} \pm 10^{-(m-n)} & \text{if } .a_{n+1}a_{n+2}\ldots \leqslant \tfrac{1}{2} \\ \mp 10^{-(m-n)} & \text{if } .a_{n+1}a_{n+2}\ldots > \tfrac{1}{2}. \end{cases}$$

Therefore, if $n < m$, then

$$\pm 10^{n-m}(\langle 10^n(x + h_m)\rangle^* - \langle 10^n x\rangle^*) = \pm 1.$$

Substituting these in the sum on the right-hand side of (6.11), we have

$$\frac{f(x + h_m) - f(x)}{h_m} = \sum_{n=0}^{m-1} e_n, \tag{6.21}$$

where $e_n = \pm 1$, for $0 \leqslant n \leqslant m - 1$ for each positive integer m. Here $h_m \to 0$, as $m \to +\infty$ by the construction of $\langle h_m\rangle$. However, the limit in (6.21) does not exist as $m \to +\infty$ since on the right there the sum is an odd integer if $m - 1$ is even, and an even integer if $m - 1$ is odd. Therefore, the sequence of numbers on the left cannot be a Cauchy sequence and diverges. Thus, f is differentiable for no $x \in \mathbb{R}$, as claimed.

7. The Weierstrass Approximation Theorem

In 1885 Weierstrass proved that a function which is continuous on a bounded closed interval can be approximated uniformly on this interval by polynomials. This result has many applications. We present here a proof due to Bernstein. We first state some preliminary results and invite the reader to prove some of these himself.

PROB. 7.1. Prove: If n is a nonnegative integer, then

$$\sum_{k=0}^{n}\binom{n}{k}x^k(1-x)^{n-k}=1 \quad \text{for} \quad x\in\mathbb{R}. \tag{7.1}$$

Lemma 7.1. *If $x\in\mathbb{R}$ and n is a positive integer, then*

$$\sum_{k=1}^{n}k\binom{n}{k}x^k(1-x)^{n-k}=nx. \tag{7.2}$$

PROOF. We have

$$\begin{aligned}
&\sum_{k=1}^{n}k\binom{n}{k}x^k(1-x)^{n-k}\\
&=x\sum_{k=1}^{n}k\binom{n}{k}x^{k-1}(1-x)^{n-1-(k-1)}\\
&=x\sum_{k=1}^{n}k\frac{n(n-1)\cdots(n-k+1)}{k!}x^{k-1}(1-x)^{n-1-(k-1)}\\
&=nx\sum_{k=1}^{n}\frac{(n-1)(n-2)\cdots(n-1-(k-1)+1)}{(k-1)!}x^{k-1}(1-x)^{n-1-(k-1)}\\
&=nx\sum_{k=1}^{n}\binom{n-1}{k-1}x^{k-1}(1-x)^{n-1-(k-1)}.
\end{aligned}$$

We now reindex in the last sum by writing $j=k-1$, then use Prob. 7.1, and conclude that

$$\sum_{k=1}^{n}\binom{n-1}{k-1}x^{k-1}(1-x)^{n-1-(k-1)}=\sum_{j=0}^{n-1}\binom{n-1}{j}x^j(1-x)^{n-1-j}=1.$$

We see from this and the last set of equalities that

$$\begin{aligned}
\sum_{k=1}^{n}k\binom{n}{k}x^k(1-x)^{n-k}&=nx\sum_{k=1}^{n}\binom{n-1}{k-1}x^{k-1}(1-x)^{n-1-(k-1)}\\
&=nx(1)=nx.
\end{aligned} \tag{7.3}$$

Remark 7.1. The last result is also valid when $n = 0$ for then both sides of (7.2) vanish. Hence, Lemma 7.1 can be used also when $x \in \mathbb{R}$ and n is a nonnegative integer.

Lemma 7.2. *If $x \in \mathbb{R}$ and n is a positive integer, then*

$$\sum_{k=1}^{n} k^2 \binom{n}{k} x^k (1-x)^{n-k} = nx(1 - x + nx). \tag{7.4}$$

PROOF. As in the last lemma, we have

$$\sum_{k=1}^{n} k^2 \binom{n}{k} x^k (1-x)^{n-k} = nx \sum_{k=1}^{n} k \binom{n-1}{k-1} x^{k-1} (1-x)^{n-k}.$$

Now reindex in the sum on the right by writing $j = k - 1$ to obtain

$$\begin{aligned}\sum_{k=1}^{n} k^2 \binom{n}{k} x^k (1-x)^{n-k} &= nx \sum_{j=0}^{n-1} (j+1) \binom{n-1}{j} x^j (1-x)^{n-1-j} \\ &= nx \left(\sum_{j=0}^{n-1} j \binom{n-1}{j} x^j (1-x)^{n-1-j} \right. \\ &\qquad \left. + \sum_{j=0}^{n-1} \binom{n-1}{j} x^j (1-x)^{n-1-j} \right) \\ &= nx((n-1)x + 1) = nx(1 - x + nx).\end{aligned}$$

PROB. 7.2. Prove: If $x \in \mathbb{R}$ and n is a positive integer, then

$$\sum_{k=0}^{n} (k - nx)^2 \binom{n}{k} x^k (1-x)^{n-k} = nx(1-x). \tag{7.5}$$

(Hint: expand inside the summation sign to obtain $(k - nx)^2 = k^2 - 2nkx + n^2x^2$, then apply the last two results.)

PROB. 7.3. Prove: If $x \in \mathbb{R}$ and n is a positive integer, then

$$\sum_{k=0}^{n} (k - nx)^2 \binom{n}{k} x^k (1-x)^{n-k} \leqslant \frac{n}{4}.$$

Def. 7.1. Let f be defined on $[0, 1]$. The *Bernstein polynomial* of order n for f, written as $B_n(f; x)$, is defined as

$$B_n(f; x) = \sum_{k=0}^{n} f\left(\frac{k}{n}\right) \binom{n}{k} x^k (1-x)^{n-k}. \tag{7.6}$$

EXAMPLE 7.1. Let u_2 be the function defined as

$$u_2(x) = x^2 \qquad \text{for} \quad x \in [0, 1].$$

By (7.6), the above definition, and Lemma 7.2, we have

$$\begin{aligned} B_n(u_2; x) &= \sum_{k=0}^{n} u_2\left(\frac{k}{n}\right)\binom{n}{k} x^k(1-x)^{n-k} \\ &= \sum_{k=0}^{n}\left(\frac{k}{n}\right)^2\binom{n}{k} x^k(1-x)^{n-k} \\ &= \frac{1}{n^2}\sum_{k=1}^{n} k^2\binom{n}{k} x^k(1-x)^{n-k} \\ &= \frac{1}{n^2} nx[1-x+nx] \\ &= \frac{x}{n}[1-x+nx] \\ &= \frac{x(1-x)}{n} + x^2 \end{aligned}$$

for $x \in [0, 1]$.

Prob. 7.4. Let $u_0 = 1$ for $x \in [0, 1]$. Prove that

$$B_n(u_0, x) = 1 \quad \text{for} \quad x \in [0, 1].$$

Prob. 7.5. What is the Bernstein polynomial $B_n(u_1; x)$ for u_1, where $u_1(x) = x$ for all $x \in [0, 1]$?

Theorem 7.1. *If f is continuous on* $[0, 1]$, *then the sequence* $\langle B_n \rangle$ *of its Bernstein polynomials converges uniformly to f on* $[0, 1]$.

Proof. We wish to prove that for each $\epsilon > 0$ there exists an N, depending only on ϵ, such that if $n > N$, then

$$|B_n(f; x) - f(x)| < \epsilon \quad \text{for all} \quad x \in [0, 1]. \tag{7.7}$$

Using Prob. 7.1, we have for $x \in [0, 1]$,

$$\begin{aligned} |B_n(f; x) - f(x)| &= \left| B_n(f; x) - f(x)\sum_{k=0}^{n}\binom{n}{k} x^k(1-x)^{n-k} \right| \\ &= \left| \sum_{k=0}^{n} f\left(\frac{k}{n}\right)\binom{n}{k} x^k(1-x)^{n-k} \right. \\ &\qquad \left. - f(x)\sum_{k=0}^{n}\binom{n}{k} x^k(1-x)^{n-k} \right| \\ &= \left| \sum_{k=0}^{n}\left(f\left(\frac{k}{n}\right) - f(x)\right)\binom{n}{k} x^k(1-x)^{n-k} \right| \\ &\leqslant \sum_{k=0}^{n}\left| f\left(\frac{k}{n}\right) - f(x)\right|\binom{n}{k} x^k(1-x)^{n-k}. \end{aligned}$$

Thus, for each n,

$$|B_n(f;x) - f(x)| \leqslant \sum_{k=0}^{n} \left| f\left(\frac{k}{n}\right) - f(x) \right| \binom{n}{k} x^k (1-x)^{n-k} \qquad \text{for} \quad x \in [0,1]. \tag{7.8}$$

Since f is continuous on the bounded closed interval $[0, 1]$, it is bounded there. Therefore, there exists an $M > 0$ such that

$$|f(x)| \leqslant M \qquad \text{for all} \quad x \in [0,1]. \tag{7.9}$$

Now let $\epsilon > 0$ be given. Since f is continuous on the bounded closed interval $[0, 1]$, it is uniformly continuous there. Therefore, there exists a $\delta > 0$ depending only on ϵ such that

$$|f(x') - f(x'')| < \frac{\epsilon}{4} \quad \text{for } x' \text{ and } x'' \text{ in } [0,1] \text{ and } |x' - x''| < \delta. \tag{7.10}$$

We take $x \in [0, 1]$. Corresponding to the positive integer n, we consider the set $\omega_{n+1} = \{0, 1, \ldots, n\}$ of nonnegative integers. We partition ω_{n+1} into the sets A_n and B_n such that

$$A_n = \left\{ k \in \omega_{n+1} \,\middle|\, \left|\frac{k}{n} - x\right| < \delta \right\} \quad \text{and} \quad B_n = \left\{ k \in \omega_{n+1} \,\middle|\, \left|\frac{k}{n} - x\right| \geqslant \delta \right\}.$$

Clearly, $A_n \cap B_n = \emptyset$ and $A_n \cup B_n = \omega_{n+1}$. We can now express the sum on the right-hand side of inequality (7.8) as

$$\begin{aligned} &\sum_{k=0}^{n} \left| f\left(\frac{k}{n}\right) - f(x) \right| \binom{n}{k} x^k (1-x)^{n-k} \\ &\qquad = \sum_{k \in A_n} \left| f\left(\frac{k}{n}\right) - f(x) \right| \binom{n}{k} x^k (1-x)^{n-k} \\ &\qquad\quad + \sum_{k \in B_n} \left| f\left(\frac{k}{n}\right) - f(x) \right| \binom{n}{k} x^k (1-x)^{n-k}. \end{aligned} \tag{7.11}$$

Let $k \in A_n$. Because of (7.10), we have

$$\left| f\left(\frac{k}{n}\right) - f(x) \right| < \frac{\epsilon}{4}\,.$$

Hence, the first sum on the right-hand side can be estimated as follows:

$$\begin{aligned} \sum_{k \in A_n} \left| f\left(\frac{k}{n}\right) - f(x) \right| \binom{n}{k} x^k (1-x)^{n-k} &\leqslant \frac{\epsilon}{4} \sum_{k \in A_n} \binom{n}{k} x^k (1-x)^{n-k} \\ &\leqslant \frac{\epsilon}{4} \sum_{k=0}^{n} \binom{n}{k} x^k (1-x)^{n-k} \\ &= \frac{\epsilon}{4} 1 \\ &= \frac{\epsilon}{4}\,. \end{aligned} \tag{7.12}$$

We now turn to the second sum on the right-hand side of (7.11). If $k \in B_n$, then $k/n \in [0,1]$. Since x is also in $[0,1]$, we see that

$$\left|f\left(\frac{k}{n}\right) - f(x)\right| \leqslant \left|f\left(\frac{k}{n}\right)\right| + |f(x)| \leqslant 2M \qquad \text{for} \quad k \in B_n. \tag{7.13}$$

Thus, for $k \in B_n$, we have

$$\sum_{k \in B_n} \left|f\left(\frac{k}{n}\right) - f(x)\right| \binom{n}{k} x^k (1-x)^{n-k} \leqslant 2M \sum_{k \in B_n} \binom{n}{k} x^k (1-x)^{n-k} \tag{7.14}$$

and

$$\left|\frac{k}{n} - x\right| \geqslant \delta.$$

This yields

$$\frac{(k-nx)^2}{n^2\delta^2} \geqslant 1 \qquad \text{for} \quad k \in B_n. \tag{7.15}$$

Consider the sum on the right-hand side of (7.14). From the above, using Prob. 7.3, we conclude that

$$\begin{aligned}
\sum_{k \in B_n} \binom{n}{k} x^k (1-x)^{n-k} &\leqslant \sum_{k \in B_n} \frac{(k-nx)^2}{n^2\delta^2} \binom{n}{k} x^k (1-x)^{n-k} \\
&= \frac{1}{n^2\delta^2} \sum_{k \in B_n} (k-nx)^2 \binom{n}{k} x^k (1-x)^{n-k} \\
&\leqslant \frac{1}{n^2\delta^2} \sum_{k=0}^{n} (k-nx)^2 \binom{n}{k} x^k (1-x)^{n-k} \\
&\leqslant \frac{1}{n^2\delta^2} \frac{n}{4} = \frac{1}{4n\delta^2}.
\end{aligned}$$

This and (7.14) imply that

$$\sum_{k \in B_n} \left|f\left(\frac{k}{n}\right) - f(x)\right| \binom{n}{k} x^k (1-x)^{n-k} \leqslant \frac{2M}{4n\delta^2} = \frac{M}{2n\delta^2}. \tag{7.16}$$

Therefore, by (7.8), (7.11), (7.12), and (7.16),

$$|B_n(f;x) - f(x)| \leqslant \frac{\epsilon}{4} + \frac{M}{2n\delta^2} \qquad \text{for each } n \text{ and for} \quad x \in [0,1]. \tag{7.17}$$

Take n such that $n > M/\epsilon\delta^2$. For such n, because of (7.17), we have

$$|B_n(f;x) - f(x)| \leqslant \frac{\epsilon}{4} + \frac{\epsilon}{2} = \frac{3}{4}\epsilon < \epsilon \qquad \text{for all} \quad x \in [0,1]. \tag{7.18}$$

This completes the proof.

Theorem 7.2 (Weierstrass' Approximation Theorem). *If f is continuous on the bounded closed interval $[a,b]$, then for each $\epsilon > 0$, there exists a polyno-*

mial P such that

$$|f(x) - P(x)| < \epsilon \qquad \textit{for all} \quad x \in [a, b]. \tag{7.19}$$

PROOF. Let $\epsilon > 0$ be given. For $u \in [0, 1]$, we obviously have $a \leqslant a + u(b - a) \leqslant b$. Define $g : [0, 1] \to \mathbb{R}$ as

$$g(u) = f(a + u(b - a)), \qquad \text{for} \quad u \in [0, 1].$$

Clearly, g is continuous on $[0, 1]$. By Theorem 7.1, we know that an n exists such that the Bernstein polynomial $B_n(g; u)$ satisfies

$$|B_n(g; u) - g(u)| < \epsilon \qquad \text{for all} \quad u \in [0, 1]. \tag{7.20}$$

Now take $x \in [a, b]$ so that $a \leqslant x \leqslant b$ and

$$0 \leqslant \frac{x - a}{b - a} \leqslant 1.$$

Substitute $(x - a)/(b - a)$ for u in (7.20). Since

$$g\left(\frac{x - a}{b - a}\right) = f(x),$$

we see that

$$\left|B_n\left(g; \frac{x - a}{b - a}\right) - f(x)\right| < \epsilon \qquad \text{for} \quad x \in [a, b]. \tag{7.21}$$

Put

$$P(x) = B_n\left(g; \frac{x - a}{b - a}\right). \tag{7.22}$$

P is a polynomial in x and, in view of (7.21), it satisfies (7.19).

This theorem can be used to prove that a continuous periodic function of period 2π can be approximated uniformly by trigonometric polynomials (Def. X.6.1), a result also proved by Weierstrass. We state Weierstrass' result in Theorem 7.4 below. Before doing this we prove:

Theorem 7.3. *If f is continuous on the interval $[0, \pi]$, then for each $\epsilon > 0$, there exists an even trigonometric polynomial T such that*

$$|f(x) - T(x)| < \epsilon \qquad \textit{for all} \quad x \in [0, \pi]. \tag{7.23}$$

PROOF. The cosine function maps $[0, \pi]$ *onto* $[-1, 1]$ in a one-to-one way. Write $x = \operatorname{Arccos} y$ for $y \in [-1, 1]$. We have $0 \leqslant \operatorname{Arccos} y \leqslant \pi$. Define g as the composite $f \circ \operatorname{Arccos}$. Thus,

$$g(y) = f(\operatorname{Arccos} y) \qquad \text{for} \quad y \in [-1, 1]. \tag{7.24}$$

g is the composite of continuous functions and, therefore, continuous. Hence, if $\epsilon > 0$ is given, there exists a polynomial P such that

$$|g(y) - P(y)| < \epsilon \qquad \text{for all} \quad y \in [-1, 1].$$

P is of the form

$$P(y) = a_0 y^n + a_1 y^{n-1} + \cdots + a_n,$$

so that

$$|f(\operatorname{Arccos} y) - (a_0 y^n + a_1 y^{n-1} + \cdots + a_n)| < \epsilon \qquad \text{for all} \quad y \in [-1,1]. \tag{7.24'}$$

Since $y = \cos x$, we obtain from this

$$|f(x) - (a_0 \cos^n x + a_1 \cos^{n-1} x + \cdots + a_n)| < \epsilon \qquad \text{for all} \quad x \in [0,1]. \tag{7.25}$$

By Theorem X.6.1, each $\cos^k x$ for $k \in \{0, 1, 2, \ldots, n\}$ can be written as a finite sum of even trigonometric functions. It is clear from the definition of even trigonometric polynomials that any finite sum of constant multiples of such trigonometric polynomials is also an even trigonometric polynomial. Therefore, T, where

$$T(x) = a_0 \cos^n x + a_1 \cos^{n-1} x + \cdots + a_n,$$

is an even trigonometric polynomial. Substituting in (7.25), we obtain (7.23).

Corollary. *If f is an even function, continuous on $\mathbb{R}$ and periodic with period 2π, then for each $\epsilon > 0$ there exists an even trigonometric polynomial T such that*

$$|f(x) - T(x)| < \epsilon \qquad \textit{for all} \quad x \in \mathbb{R}. \tag{7.26}$$

Proof. If $\epsilon > 0$ is given, there exists, by the theorem, an even trigonometric polynomial T such that

$$|f(x) - T(x)| < \epsilon \qquad \text{for all} \quad x \in [0, \pi]. \tag{7.27}$$

It is clear that T is also an even function since it is a sum of constant multiples of cosines. Take $x \in [-\pi, 0]$ so that $-x \in [0, \pi]$. From (7.27) we have

$$|f(x) - T(x)| = |f(-x) - T(-x)| < \epsilon \qquad \text{for} \quad x \in [-\pi, 0]. \tag{7.28}$$

It follows that $|f(x) - T(x)| < \epsilon$ holds for all $x \in [-\pi, \pi]$.

Now take $x \in \mathbb{R}$. There exists an integer n such that

$$-\pi \leqslant x - 2n\pi < \pi.$$

Since f and T are periodic with period 2π and $x - 2n\pi \in [-\pi, \pi)$, we obtain

$$|f(x) - T(x)| = |f(x - 2n\pi) - T(x - 2n\pi)| < \epsilon \qquad \text{for} \quad x \in \mathbb{R}.$$

This completes the proof.

The corollary can be phrased as follows: An even function which is continuous on $\mathbb{R}$ and is periodic with period 2π can be approximated uniformly by (even) trigonometric polynomials.

PROB. 7.6. Prove: If T is a trigonometric polynomial, then g, defined as

$$g(x) = T(x)\sin x \qquad \text{for} \quad x \in \mathbb{R},$$

can also be written as a trigonometric polynomial (Hint: note the identities

$$\sin x \cos nx = \frac{\sin(n+1)x - \sin(n-1)x}{2}$$

and

$$\sin x \sin nx = \frac{\cos(n-1)x - \cos(n+1)x}{2}\Bigg).$$

PROB. 7.7. Prove: Let a be some constant. If T is a trigonometric polynomial, then so is the function h, where $h(x) = T(x+a)$ for all $x \in \mathbb{R}$.

Theorem 7.4.* (Weierstrass). *If f is continuous on $\mathbb{R}$ and periodic with period 2π, then for each $\epsilon > 0$ there exists a trigonometric polynomial T such that*

$$|f(x) - T(x)| < \epsilon \qquad \textit{for all} \quad x \in \mathbb{R}. \tag{7.29}$$

PROOF. Define the functions g and h by means of

$$g(x) = \frac{f(x)+f(-x)}{2}, \qquad h(x) = \frac{f(x)-f(-x)}{2}\sin x \qquad \text{for} \quad x \in \mathbb{R}. \tag{7.30}$$

It is easy to see that g and h are even functions and from the hypothesis on f that they are periodic with period 2π.

Let $\epsilon > 0$ be given. By the last corollary, there exist even trigonometric polynomials T_1 and T_2 such that

$$|g(x) - T_1(x)| < \frac{\epsilon}{8} \quad \text{and} \quad |h(x) - T_2(x)| < \frac{\epsilon}{8} \qquad \text{for} \quad x \in \mathbb{R}. \tag{7.31}$$

Multiply the first inequality by $\sin^2 x$, the second by $|\sin x|$, and obtain

$$|g(x)\sin^2 x - T_1(x)\sin^2 x| \leqslant \frac{\epsilon}{8}\sin^2 x \leqslant \frac{\epsilon}{8} \qquad \text{for} \quad x \in \mathbb{R} \tag{7.32}$$

and

$$|h(x)\sin x - T_2(x)\sin x| \leqslant \frac{\epsilon}{8}|\sin x| \leqslant \frac{\epsilon}{8} \qquad \text{for} \quad x \in \mathbb{R}. \tag{7.33}$$

By Prob. 7.6 applied twice, the product $T_1(x)\sin^2 x$ can be written as a trigonometric polynomial in x. The product $T_2(x)\sin x$ can also be written

* I. Nathanson, *Theory of Functions of a Real Variable*, Frederick Ungar Publishing Co., New York, 1955, Vol. I, p. 111.

that way. Let T_3 be defined as

$$T_3(x) = T_1(x)\sin^2 x + T_2(x)\sin x \qquad \text{for} \quad x \in \mathbb{R}. \tag{7.34}$$

T_3 can also be written as a trigonometric polynomial. Hence, by (7.33) and (7.34) we have

$$|g(x)\sin^2 x + h(x)\sin x - T_3(x)| \leqslant \frac{\epsilon}{4} \qquad \text{for} \quad x \in \mathbb{R}, \tag{7.35}$$

where T_3 is a trigonometric polynomial. Using (7.30) we see that

$$g(x)\sin^2 x + h(x)\sin x = f(x)\sin^2 x \qquad \text{for} \quad x \in \mathbb{R}.$$

Substituting this in (7.35) we have

$$|f(x)\sin^2 x - T_3(x)| \leqslant \frac{\epsilon}{4} \qquad \text{for} \quad x \in \mathbb{R}. \tag{7.36}$$

Define g_1 as

$$g_1(x) = f\left(x - \frac{\pi}{2}\right) \qquad \text{for all} \quad x \in \mathbb{R}.$$

g_1 is periodic with period 2π (show this). It is also continuous on $\mathbb{R}$. Hence, we can manipulate g_1 in the same way that we did g above and prove that there exists a trigonometric polynomial T_4 such that

$$\left|f\left(x - \frac{\pi}{2}\right)\sin^2 x - T_4(x)\right| = \left|g_1(x)\sin^2 x - T_4(x)\right| \leqslant \frac{\epsilon}{4}$$

for $x \in \mathbb{R}$.

We replace x here by $x + \pi/2$ and obtain

$$\left|f(x)\cos^2 x - T_4\left(x + \frac{\pi}{2}\right)\right| \leqslant \frac{\epsilon}{4} \qquad \text{for} \quad x \in \mathbb{R}. \tag{7.37}$$

But by Prob. 7.7, $T_4(x + \pi/2)$ can be written as a trigonometric polynomial in x. Define T_5 as $T_5(x) = T_4(x + \pi/2)$ for $x \in \mathbb{R}$ and substitute in (7.38) to conclude that

$$|f(x)\cos^2 x - T_5(x)| \leqslant \frac{\epsilon}{4} \qquad \text{for all} \quad x \in \mathbb{R}. \tag{7.37$'$}$$

Use this and (7.36) to arrive at

$$\begin{aligned} |f(x) - (T_3(x) + T_4(x))| &= |f(x)\sin^2 x + f(x)\cos^2 x - (T_3(x) + T_4(x))| \\ &\leqslant |f(x)\sin^2 x - T_3(x)| + |f(x)\cos^2 x - T_5(x)| \\ &\leqslant \frac{\epsilon}{4} + \frac{\epsilon}{4} \\ &= \frac{\epsilon}{2} \end{aligned}$$

for all $x \in \mathbb{R}$. $T_5 = T_3 + T_4$ is a trigonometric polynomial and by the last equality we have

$$|f(x) - T_5(x)| \leqslant \frac{\epsilon}{2} < \epsilon \qquad \text{for all} \quad x \in \mathbb{R}.$$

This completes the proof.

8. Uniform Convergence and Differentiability

Suppose $\langle f_n \rangle$ is a sequence of functions each of which is differentiable on some set D. For each n, let f_n' be the derivative of f_n. We refer to the sequence $\langle f_n' \rangle$ of derivatives as the *derived* sequence of the sequence $\langle f_n \rangle$. Does the convergence of $\langle f_n \rangle$ imply the convergence of its derived sequence or conversely? If both $\langle f_n \rangle$ and its derived sequence $\langle f_n' \rangle$ converge, how are their limit functions related? We deal with these questions in this section. First we consider some examples.

EXAMPLE 8.1. For each n, define f_n as

$$f_n(x) = x + n \qquad \text{for} \quad x \in \mathbb{R}.$$

It is clear that for no $x \in \mathbb{R}$ does $\langle f_n \rangle$ converge. Since

$$f_n'(x) = 1 \qquad \text{for} \quad x \in \mathbb{R},$$

we have $\lim_{n \to +\infty} f_n'(x) = 1$ for each $x \in \mathbb{R}$. The sequence $\langle f_n' \rangle$ consists of terms each of which is the constant function 1 on $\mathbb{R}$ and converges (even uniformly) to the constant function 1. Thus, we have an example of a sequence which does not converge even though its derived sequence converges uniformly.

EXAMPLE 8.2. For each n, define g_n, as

$$g_n(x) = \frac{\sin nx}{n} \qquad \text{for} \quad x \in \mathbb{R}.$$

Since, for each n,

$$|g_n(x)| = \left| \frac{\sin nx}{n} \right| \leqslant \frac{1}{n} \qquad \text{for} \quad x \in \mathbb{R},$$

we have $\lim_{n \to +\infty} g_n(x) = 0$ for $x \in \mathbb{R}$. Thus, $g_n \to 0$ on $\mathbb{R}$. Also, for each n,

$$g_n'(x) = \cos nx \qquad \text{for} \quad x \in \mathbb{R}.$$

The derived sequence $\langle g_n' \rangle$ does not converge on $\mathbb{R}$. It does not converge for $x = \pi$ and $x = \pi/2$, for example. Here we have an example of a sequence which converges uniformly on a set, but whose derived sequence does not converge on that set.

Theorem 8.1.* *If $\langle f_n \rangle$ consists of functions all of which are differentiable on the bounded closed interval $[a,b]$ and for some $c \in [a,b]$ the sequence $\langle f_n(c) \rangle$ of constants converges and if its derived sequence $\langle f_n' \rangle$ converges uniformly on $[a,b]$, then* (1) *$\langle f_n \rangle$ converges uniformly to a function f on $[a,b]$ which is differentiable on $[a,b]$ and* (2) *$f'(x) = \lim_{n \to +\infty} f_n'(x)$ for $x \in [a,b]$.*

PROOF. We first prove that $\langle f_n \rangle$ converges uniformly on $[a,b]$. Let $\epsilon > 0$ be given. Since $\langle f_n' \rangle$ converges uniformly on $[a,b]$, there exists an N_1, depend-

*Burril and Knudsen, *Real Variables*, Holt, Rinehart, Winston, New York, 1969, p. 240.

ing only on ϵ, such that if $m > N_1$ and $n > N_1$, then

$$|f_m'(x) - f_n'(x)| < \frac{\epsilon}{2(b-a)} \qquad \text{for all} \quad x \in [a,b]. \tag{8.1}$$

There exists an N_2, depending only on ϵ, such that if $m > N_2$ and $n > N_2$, then

$$|f_m(c) - f_n(c)| < \frac{\epsilon}{2}. \tag{8.2}$$

Let $N = \max\{N_1, N_2\}$. N depends only on ϵ because N_1 and N_2 have this property. Take $m > N$ and $n > N$ and apply the Mean-Value Theorem to $f_m - f_n$. For $x \in [a,b]$, there exists an x_0 between x and c such that

$$f_m(x) - f_n(x) - (f_m(c) - f_n(c)) = (f_m'(x_0) - f_n'(x_0))(x-c). \tag{8.3}$$

Because of (8.1), we have $|f_m'(x_0) - f_n'(x_0)| < \epsilon/2(b-a)$. This and (8.3) imply that

$$|f_m(x) - f_n(x) - (f_m(c) - f_n(c))| \leqslant \frac{\epsilon}{2(b-a)}|x-c|. \tag{8.4}$$

Since x and c are in $[a,b]$, we have $|x-c|/(b-a) \leqslant 1$. From this and (8.4) we have

$$|f_m(x) - f_n(x) - (f_m(c) - f_n(c))| \leqslant \frac{\epsilon}{2}. \tag{8.5}$$

Note next that

$$|f_m(x) - f_n(x)| \leqslant |f_m(x) - f_n(x) - (f_m(c) - f_n(c))| + |f_m(c) - f_n(c)|.$$

This, (8.5), and (8.2) imply that

$$|f_m(x) - f_n(x)| < \epsilon.$$

This holds for any $m > N$ and $n > N$ and for $x \in [a,b]$. From the Cauchy criterion for uniform convergence it follows that $\langle f_n \rangle$ converges uniformly on $[a,b]$. Define f by means of

$$f(x) = \lim_{n \to +\infty} f_n(x) \qquad \text{for} \quad x \in [a,b]. \tag{8.6}$$

Since the convergence of f_n to f is uniform on $[a,b]$ and each f_n is continuous there, we know that f is continuous on $[a,b]$. We prove that f is differentiable on $[a,b]$.

Take some $x_1 \in [a,b]$. For each n define F_n as

$$F_n(x) = \begin{cases} \dfrac{f_n(x) - f_n(x_1)}{x - x_1} & \text{if} \quad x \neq x_1, \quad x \in [a,b] \\ f_n'(x_1) & \text{if} \quad x = x_1. \end{cases} \tag{8.7}$$

Since f_n is differentiable, F_n is continuous on $[a,b]$ and $\langle F_n \rangle$ is a sequence of functions each of which is continuous on $[a,b]$. By hypothesis, the sequence $\langle f_n'(x_1) \rangle$ converges. Let

$$L = \lim_{n \to +\infty} f_n'(x_1). \tag{8.8}$$

Define F on $[a,b]$ as

$$F(x)=\begin{cases}\dfrac{f(x)-f(x_1)}{x-x_1} & \text{if} \quad x\neq x_1, \quad x\in[a,b], \\ L & \text{if} \quad x=x_1.\end{cases} \tag{8.9}$$

The sequence $\langle F_n\rangle$ converges pointwise to F on $[a,b]$ since for $x\neq x_1$,

$$\lim_{n\to+\infty}F_n(x)=\lim_{n\to+\infty}\frac{f_n(x)-f_n(x_1)}{x-x_1}=\frac{f(x)-f(x_1)}{x-x_1}$$

and

$$\lim_{n\to+\infty}F_n(x_1)=\lim_{n\to+\infty}f_n'(x_1)=L.$$

We prove that this convergence of F_n to F is uniform on $[a,b]$. Note that for each m and n and $x\neq x_1$, $x\in[a,b]$, we have

$$F_m(x)-F_n(x)=\frac{f_m(x)-f_n(x)-(f_m(x_1)-f_n(x_1))}{x-x_1}.$$

By the Mean-Value Theorem applied to f_m-f_n, there exists a τ between x and x_1 and, therefore, in $[a,b]$ such that

$$\frac{f_m(x)-f_n(x)-(f_m(x_1)-f_n(x_1))}{x-x_1}=f_m'(\tau)-f_n'(\tau).$$

This implies that if $x\neq x_1$, then

$$|F_m(x)-F_n(x)|=|f_m'(\tau)-f_n'(\tau)|. \tag{8.10}$$

For $x=x_1$ we have

$$|F_m(x_1)-F_n(x_1)|=|f_m'(x_1)-f_n'(x_1)|. \tag{8.11}$$

$\langle f_n'\rangle$ converges uniformly on $[a,b]$. Therefore, using the Cauchy criterion for uniform convergence, (8.10) and (8.11) imply that $\langle F_n\rangle$ converges uniformly on $[a,b]$. It follows that $F=\lim_{n\to+\infty}F_n$ is continuous on $[a,b]$. We have

$$\lim_{x\to x_1}F(x)=F(x_1)=L$$

and, hence,

$$\lim_{x\to x_1}\frac{f(x)-f(x_1)}{x-x_1}=L.$$

Thus, f is differentiable at x_1 and

$$f'(x_1)=L=\lim_{n\to+\infty}f_n'(x_1).$$

This holds for each x_1 in $[a,b]$. Therefore, f is differentiable in $[a,b]$ and

$$f'(x)=\lim_{n\to+\infty}f_n'(x) \qquad \text{for} \quad x\in[a,b],$$

where $f(x)=\lim_{n\to+\infty}f_n(x)$ for $x\in[a,b]$. This completes the proof.

Corollary. *If* $\lim_{n\to+\infty} f_n = f$ *pointwise on some interval* I, *where each* f_n *is differentiable on* I *and the derived sequence* $\langle f_n' \rangle$ *converges uniformly on every bounded closed subinterval of* I, *then* f *is differentiable on* I *and* $\lim_{n\to+\infty} f_n' = f'$ *pointwise on* I.

PROOF. Exercise.

If all the terms of the sequence $\langle f_n \rangle$ are differentiable on some set D, then it is said to be *termwise* differentiable on D if its derived sequence converges on D and

$$\lim f_n' = f',$$

where $f = \lim_{n\to+\infty} f_n$ on D. A termwise differentiable sequence of functions is also said to be differentiable *term by term*, or *element by element*. If $\langle f_n \rangle$ is termwise differentiable, then

$$\lim_{n\to+\infty} \frac{df_n(x)}{dx} = \frac{df(x)}{dx},$$

where $f(x) = \lim_{n\to+\infty} f_n(x)$ so that

$$\lim_{n\to+\infty} \frac{df_n(x)}{dx} = \frac{d}{dx} \lim_{n\to+\infty} f_n(x).$$

Accordingly, we say that $\langle f_n \rangle$ is termwise differentiable if we can interchange the order of the operations of differentiation and passing to the limit.

EXAMPLE 8.3. For each positive integer n define f_n as

$$f_n(x) = \frac{2x}{1 + n^2x^2} \qquad \text{for} \quad x \in [0, +\infty).$$

Inspection shows that $\langle f_n \rangle$ converges to 0 pointwise on $[0, +\infty)$. We can even show that the convergence to 0 is uniform. To see this note that $(1 - nx)^2 \geqslant 0$ for all x and n so that for each n

$$0 \leqslant \frac{2x}{1 + n^2x^2} \leqslant \frac{1}{n} \qquad \text{for all} \quad x \in [0, +\infty).$$

Thus, $0 \leqslant f_n(x) \leqslant 1/n$ for all $x \in [0, +\infty)$ and for each n. We see from this that $f_n \to 0$ as $n \to +\infty$ uniformly on $[0, +\infty)$. Each f_n is differentiable on $[0, +\infty)$, and

$$f_n'(x) = \frac{2(1 - n^2x^2)}{(1 + n^2x^2)^2} \qquad \text{for} \quad x \in [0, +\infty).$$

We have $f_n'(0) = 2$ and, for $x > 0$,

$$f_n'(x) = \frac{2}{n^2} \frac{(1/n^2 - x^2)}{(1/n^2 + x^2)^2}$$

so that $f_n'(x) \to 0$ as $n \to +\infty$ for $x > 0$. Thus, $\lim_{n\to+\infty} f_n'(x) = g(x)$, where

$$g(x) = \begin{cases} 2 & \text{if} \quad x = 0 \\ 0 & \text{if} \quad x > 0. \end{cases}$$

The convergence of f_n' to g as $n \to +\infty$ is not uniform (explain). We have

$$\frac{d}{dx}\Big(\lim_{n\to+\infty} f_n(x)\Big) = \frac{d}{dx}(0) = 0 \qquad \text{for all} \quad x \geqslant 0,$$

while

$$\Big(\lim_{n\to+\infty} \frac{d}{dx} f_n(x)\Big)_{x=0} = 2,$$

so on $[0, +\infty)$

$$\frac{d}{dx}\Big(\lim_{n\to+\infty} f_n(x)\Big) \neq \lim_{n\to+\infty} \frac{df_n(x)}{dx}$$

and $\langle f_n \rangle$ is not differentiable termwise on $[0, +\infty)$.

We state series formulations of Theorem 8.1 and its corollary.

Theorem 8.2. *If the terms of the infinite series $\sum_{n=1}^{\infty} f_n$ are all differentiable functions on the bounded closed interval $[a,b]$ and for some $x_0 \in [a,b]$, $\sum_{n=1}^{\infty} f_n(x_0)$ converges and the derived series $\sum_{n=1}^{\infty} f_n'$ converges uniformly on $[a,b]$, then* (1) *$\sum_{n=1}^{\infty} f_n$ converges uniformly to a sum function S on $[a,b]$ which is differentiable there and* (2)

$$S'(x) = \sum_{n=1}^{\infty} f_n'(x) \qquad \textit{for} \quad x \in [a,b].$$

Proof. Exercise.

Corollary. *If the terms of the series $\sum_{n=1}^{\infty} f_n$ are all differentiable on some interval and the series converges pointwise to some function S on I, and the derived series $\sum f_n'$ converges uniformly on every bounded subinterval of I, then S is differentiable on I and $S' = \sum f_n'$ pointwise on I.*

Proof. Exercise.

Example 8.4. Let

$$S(x) = \sum_{n=1}^{\infty} \frac{\sin nx}{n^3} \qquad \text{for} \quad x \in \mathbb{R}. \tag{8.12}$$

Since for each $n \geqslant 1$

$$\left| \frac{\sin x}{n^3} \right| \leqslant \frac{1}{n^3} \qquad \text{for all} \quad x \in \mathbb{R}$$

and $\sum_{n=1}^{\infty}(1/n^3)$ converges, we conclude that series in (8.12) converges uniformly to its sum function S on $\mathbb{R}$. The terms of the series

$$\sum_{n=1}^{\infty} \frac{\cos nx}{n^2} \quad \text{for} \quad x \in \mathbb{R} \tag{8.13}$$

are the derivatives of the terms of the series (8.12) and this series also converges uniformly on $\mathbb{R}$. Therefore, the series (8.13) converges uniformly on every bounded closed interval. By the corollary of Theorem 8.2, S is differentiable on $\mathbb{R}$ and

$$S'(x) = \sum_{n=1}^{\infty} \frac{\cos nx}{n^2} \quad \text{for} \quad x \in \mathbb{R}.$$

Application to the Gamma Function

We can now obtain information about the differentiability of the gamma function. We use the Weierstrass product representation of $\Gamma(x)$ confining ourselves to $x > 0$ at first. In Example X.7.4 we saw that

$$\Gamma(x) = e^{-\gamma x} \frac{1}{x} \prod_{k=1}^{\infty} \frac{e^{x/k}}{1 + x/k}, \tag{8.14}$$

where γ is the Euler–Mascheroni constant

$$\gamma = \lim_{n \to +\infty} \left(1 + \frac{1}{2} + \cdots + \frac{1}{n} - \ln n\right). \tag{8.15}$$

By Theorem IX.8.1 we know that $\Gamma(x) > 0$ if $x > 0$. Assume that $x > 0$ so that $\Gamma(x) > 0$. For each positive integer n we have

$$\prod_{k=1}^{n} \frac{e^{x/k}}{1 + x/k} > 0$$

and

$$\lim_{n \to +\infty} \ln \prod_{k=1}^{n} \frac{e^{x/k}}{1 + x/k} = \ln \prod_{k=1}^{\infty} \frac{e^{x/k}}{(1 + x/k)}. \tag{8.16}$$

This and (8.14) imply that

$$\begin{aligned} \ln \Gamma(x) &= -\gamma x - \ln x + \ln \prod_{k=1}^{\infty} \frac{e^{x/k}}{(1 + x/k)} \\ &= -\gamma x - \ln x + \lim_{n \to +\infty} \sum_{k=1}^{n} \left(\frac{x}{k} - \ln\left(1 + \frac{x}{k}\right)\right) \\ &= -\gamma x - \ln x + \sum_{k=1}^{\infty} \left(\frac{x}{k} - \ln\left(1 + \frac{x}{k}\right)\right). \end{aligned}$$

Thus,

$$\ln \Gamma(x) = -\gamma x - \ln x + \sum_{k=1}^{\infty} \left(\frac{x}{k} - \ln\left(1 + \frac{x}{k}\right)\right) \quad \text{for} \quad x > 0. \tag{8.17}$$

The series

$$\phi(x) = \sum_{k=1}^{\infty}\left(\frac{x}{k} - \ln\left(1 + \frac{x}{k}\right)\right) \tag{8.18}$$

converges. Its derived series is

$$\sum_{k=1}^{\infty}\left(\frac{1}{k} - \frac{1}{x+k}\right) \qquad \text{for} \quad x > 0. \tag{8.19}$$

We prove that it converges uniformly on every interval of the form $(0, b]$, where $b > 0$. Let $0 < x \leqslant b$. For each term of (8.19) we have

$$0 < \frac{1}{k} - \frac{1}{x+k} = \frac{x}{k(x+k)} = \frac{x}{kx + k^2} < \frac{x}{k^2} \leqslant \frac{b}{k^2}. \tag{8.20}$$

This holds for $k = 1, 2, 3, \ldots$. The series

$$\sum_{k=1}^{\infty} \frac{b}{k^2}$$

converges. It follows from (8.20), using the Weierstrass M-test, that the series (8.19) converges uniformly on the intervals $(0, b]$ for each $b > 0$. Therefore, if $0 < a < b$, this series converges uniformly on $[a, b]$. Hence, it converges uniformly on every bounded subinterval of the interval $(0; +\infty)$. Applying the corollary of Theorem 8.2, we conclude that the series (8.18) is termwise differentiable on $(0; +\infty)$ and

$$\phi'(x) = \sum_{k=1}^{\infty}\left(\frac{1}{k} - \frac{1}{x+k}\right) \qquad \text{for} \quad x > 0. \tag{8.20$'$}$$

Turning to (8.17), we obtain from this that $\ln \Gamma(x)$ is differentiable on $(0; +\infty)$. But then $\Gamma(x)$ is differentiable there. Moreover, we may differentiate both sides of (8.17) to obtain

$$\frac{\Gamma'(x)}{\Gamma(x)} = -\gamma - \frac{1}{x} + \sum_{k=1}^{\infty}\left(\frac{1}{k} - \frac{1}{x+k}\right) \qquad \text{for} \quad x > 0. \tag{8.21}$$

We wish to extend the validity of this equality to include negative values of x that are not negative integers. Towards this end, we write the series on the right-hand side of (8.21) as $H(x)$ so that

$$H(x) = \sum_{k=1}^{\infty}\left(\frac{1}{k} - \frac{1}{x+k}\right) \tag{8.22}$$

and also define

$$G(x) = \sum_{k=0}^{\infty}\left(\frac{1}{k+1} - \frac{1}{x+k}\right). \tag{8.23}$$

Suppose the last series converges for some $x \in \mathfrak{D}(\Gamma) = \mathbb{R} - \{0, -1, -2, \ldots\}$. We then have

$$G(x) = 1 - \frac{1}{x} + \sum_{k=1}^{\infty}\left(\frac{1}{k+1} - \frac{1}{x+k}\right)$$

so that

$$G(x) + \frac{1}{x} = 1 + \sum_{k=1}^{\infty}\left(\frac{1}{k+1} - \frac{1}{x+k}\right) \tag{8.24}$$

for such x. Note that

$$1 = \sum_{k=1}^{\infty}\left(\frac{1}{k} - \frac{1}{k+1}\right).$$

Substitute this series for 1 in (8.24) and obtain

$$\begin{aligned} G(x) + \frac{1}{x} &= \sum_{k=1}^{\infty}\left(\frac{1}{k} - \frac{1}{k+1}\right) + \sum_{k=1}^{\infty}\left(\frac{1}{k+1} - \frac{1}{x+k}\right) \\ &= \sum_{k=1}^{\infty}\left(\frac{1}{k} - \frac{1}{x+k}\right) \\ &= H(x) \end{aligned}$$

so that

$$G(x) + \frac{1}{x} = H(x). \tag{8.25}$$

Observe that we may reindex in the series $H(x)$ and write

$$\begin{aligned} H(x) &= \sum_{k=1}^{\infty}\left(\frac{1}{k} - \frac{1}{x+k}\right) = \sum_{k=0}^{\infty}\left(\frac{1}{k+1} - \frac{1}{x+k+1}\right) \\ &= \sum_{k=0}^{\infty}\left(\frac{1}{k+1} - \frac{1}{(x+1)+k}\right) = G(x+1). \end{aligned}$$

This and (8.25) yield

$$G(x) + \frac{1}{x} = H(x) = G(x+1) \tag{8.26}$$

so that

$$G(x+1) - G(x) = \frac{1}{x}. \tag{8.27}$$

We conclude from (8.26) that if $G(x)$ converges for some $x \in \mathfrak{D}(\Gamma)$, then $H(x)$ converges, and so does $G(x+1)$. Also if $G(y)$ converges for some $y \in \mathfrak{D}(\Gamma)$, then $G(y-1)$ and $H(y-1)$ both converge.

By (8.20′) and (8.22) we have $H(x) = \phi'(x)$. We already proved that the series $\phi'(x) = H(x)$ converges for $x > 0$. Thus, $G(x)$ converges for $x > 0$. Now assume that $-1 < x < 0$ so that $0 < x + 1$. $G(x+1)$ then converges. By the concluding remarks of the last paragraph we see that $G(x)$ and $H(x)$ converge. It follows that $G(x)$ and $H(x)$ converge for $-1 < x < 0$. Continue and assume $-2 < x < -1$ so that $-1 < x + 1 < 0$. This implies that $G(x+1)$ converges and, therefore, that $G(x)$ and $H(x)$ converge for $-2 < x < -1$. Obviously this procedure can be continued inductively, leading to the conclusion that $G(x)$ and $H(x) = G(x+1)$ converge for $-n < x < -n+1$ for every positive integer n. These, of course, also converge for $x > 0$. Thus, $G(x)$ and $H(x) = G(x+1)$ converge on $\mathfrak{D}(\Gamma) = \mathbb{R} - \{0, -1, -2, \dots\}$. All this extends the meaning of the right-hand side of (8.21) to include negative values of x which are not negative integers. It is still left to prove that Eq. (8.21) persists for such values of x. We begin the proof in the next paragraph.

Assume $-1 < x < 0$ so that $x + 1 > 0$. From $\Gamma(x+1) = x\Gamma(x)$, we have

$$\Gamma(x) = \frac{\Gamma(x+1)}{x}. \tag{8.28}$$

The right-hand side is differentiable. We know this because we proved that Γ is differentiable for $x > 0$. Differentiating both sides of (8.28) we obtain

$$\begin{aligned}\Gamma'(x) &= \frac{\Gamma'(x+1)}{x} - \frac{\Gamma(x+1)}{x^2} \\ &= \frac{\Gamma'(x+1)}{x} - \frac{\Gamma(x)}{x}.\end{aligned}$$

Dividing both sides by $\Gamma(x)$ and rearranging terms we get

$$\frac{\Gamma'(x+1)}{\Gamma(x+1)} - \frac{\Gamma'(x)}{\Gamma(x)} = \frac{1}{x} \qquad \text{for} \quad -1 < x < 0. \tag{8.29}$$

This also holds for $x > 0$, as the reader could show.

Because of (8.22) and (8.25), we can write (8.21) as

$$\frac{\Gamma'(x)}{\Gamma(x)} = -\gamma + G(x) \qquad \text{for} \quad x > 0. \tag{8.30}$$

Retaining the assumption $-1 < x < 0$, we can replace x by $x + 1$ in (8.30) to obtain

$$\frac{\Gamma'(x+1)}{\Gamma(x+1)} = -\gamma + G(x+1) \qquad \text{for} \quad -1 < x < 0. \tag{8.31}$$

Now use (8.27) and (8.29) and substitute in (8.30). We have

$$\frac{1}{x} + \frac{\Gamma'(x)}{\Gamma(x)} = -\gamma + \frac{1}{x} + G(x).$$

Therefore, (8.30) holds for $-1 < x < 0$. It now follows that Eq. (8.21) holds for $x \in (-1; 0) \cup (0; +\infty)$. This procedure can be used repeatedly to obtain that (8.21) holds for $x \in \mathfrak{D}(\Gamma) = \mathbb{R} - \{0, -1, -2, \dots\}$. In sum, we have the following result:

Theorem 8.3. *If* $x \in \mathfrak{D}(\Gamma) = \mathbb{R} - \{0, -1, -2, \dots\}$, *then* Γ *is differentiable and*

$$\frac{\Gamma'(x)}{\Gamma(x)} = -\gamma - \frac{1}{x} + \sum_{k=1}^{\infty}\left(\frac{1}{k} - \frac{1}{x+k}\right), \tag{8.32}$$

the series on the right-hand side being convergent for $x \in \mathfrak{D}(\Gamma)$.

Prob. 8.1. Prove that Γ is twice differentiable for $x \in \mathfrak{D}(\Gamma)$ and that

$$\frac{d}{dx}\frac{\Gamma'(x)}{\Gamma(x)} = \sum_{k=1}^{\infty} \frac{1}{(x+k)^2} \qquad \text{for} \quad x \in \mathfrak{D}(\Gamma).$$

Prob. 8.2. Prove that Γ has derivatives of all orders and that for each positive integer n

$$\frac{d^n}{dx^n}\frac{\Gamma'(x)}{\Gamma(x)} = \sum_{k=0}^{\infty} (-1)^{n+1} \frac{n!}{(x+k)^{n+1}} \qquad \text{for} \quad x \in \mathfrak{D}(\Gamma).$$

Prob. 8.3. Let G be the function defined in (8.23). Prove: $G(1) = 0$, $G(2) = 1$, and that if n is a positive integer, then

$$G(n+1) = 1 + \frac{1}{2} + \cdots + \frac{1}{n}.$$

Prob. 8.4. Define

$$\psi(x) = \frac{\Gamma'(x)}{\Gamma(x)} \qquad \text{for} \quad x \in \mathfrak{D}(\Gamma).$$

Prove that $\psi(1) = -\gamma$, where γ is the Euler–Mascheroni constant. Also prove that $\psi(1/2) = -\gamma - 2\ln 2$.

Prob. 8.5. Using the notation of Prob. 8.4, prove: If $0 < x < 1$, then

$$\psi(1-x) - \psi(x) = \pi \cot \pi x$$

(Hint: use Theorem X.10.1).

An Application to the Trigonometric Functions

Theorem X.9.1 states that

$$\sin x = x \prod_{n=1}^{\infty} \left(1 - \frac{x^2}{\pi^2 n^2}\right) \qquad \text{for} \quad x \in \mathbb{R}.$$

Squaring, we have

$$\sin^2 x = x^2 \prod_{n=1}^{\infty} \left(1 - \frac{x^2}{\pi^2 n^2}\right)^2 \tag{8.33}$$

and, therefore, that

$$\ln \sin^2 x = \ln x^2 + \sum_{n=1}^{\infty} \ln\left(1 - \frac{x^2}{\pi^2 n^2}\right)^2 \qquad \text{if} \quad x \notin \{0, \pm\pi, \pm 2\pi, \ldots\}. \tag{8.34}$$

Write the series on the right as

$$\psi(x) = \sum_{n=1}^{\infty} \ln\left(1 - \frac{x^2}{\pi^2 n^2}\right)^2 \tag{8.35}$$

and its derived series as

$$h(x) = \sum_{n=1}^{\infty} \frac{-4x}{n^2\pi^2 - x^2} \qquad \text{for} \quad x \notin \{\pm\pi, \pm 2\pi, \dots\}. \tag{8.36}$$

We seek information about the uniformity of the convergence of the series $h(x)$. Take some $a > 0$ and $|x| \leqslant a$. Let N be some positive integer such that $2a/\pi < N$. This implies that $|x| < \pi N/2$ and $x^2 < N^2\pi^2/4$. For each $n > N$, we have $x^2 < n^2\pi^2/4$. For such n's we have

$$\left|\frac{-4x}{n^2\pi^2 - x^2}\right| = \frac{4|x|}{n^2\pi^2 - x^2} \leqslant 4a \frac{1}{3n^2\pi^2/4 + n^2\pi^2/4 - x^2} < \frac{16a}{3\pi^2 n^2}. \tag{8.37}$$

Since the series

$$\sum_{n=N+1}^{\infty} \frac{16a}{3\pi^2 n^2}$$

converges, (8.37) implies that the series

$$\sum_{n=N+1}^{\infty} \frac{-4x}{n^2\pi^2 - x^2}$$

converges uniformly for $|x| \leqslant a$. It follows that the series

$$\psi_1(x) = \sum_{n=N+1}^{\infty} \ln\left(1 - \frac{x^2}{\pi^2 n^2}\right) \tag{8.38}$$

is differentiable term by term and its derivative is

$$\psi_1'(x) = \sum_{n=N+1}^{\infty} \frac{-4x}{n^2\pi^2 - x^2} \tag{8.39}$$

for $|x| \leqslant a$. But the sum

$$\psi_2(x) = \sum_{n=1}^{N} \ln\left(1 - \frac{x^2}{n^2\pi^2}\right)^2 \tag{8.40}$$

is a finite sum of differentiable terms if $x \notin \{\pm\pi, \pm 2\pi, \dots\}$ and its derivative is

$$\psi_2'(x) = \sum_{n=1}^{N} \frac{-4x}{n^2\pi^2 - x^2}. \tag{8.41}$$

Returning to (8.35), we now have

$$\psi(x) = \psi_2(x) + \psi_1(x) \tag{8.42}$$

and that ψ is differentiable with

$$\psi'(x) = \psi_2'(x) + \psi_1'(x) = \sum_{n=1}^{\infty} \frac{-4x}{n^2\pi^2 - x^2} = h(x) \tag{8.43}$$

for $|x| \leqslant a$, $x \notin \{\pm\pi, \pm 2\pi, \ldots\}$. But a is arbitrary. It follows that (8.43) holds for all $x \notin \{\pm\pi, \pm 2\pi, \ldots\}$. Therefore, we may differentiate (8.34) to obtain

$$\frac{d\ln\sin^2 x}{dx} = \frac{d\ln x^2}{dx} + \sum_{n=1}^{\infty} \frac{4x}{x^2 - n^2\pi^2}.$$

We obtain after dividing by 2 that

$$\cot x = \frac{1}{x} + \sum_{n=1}^{\infty} \frac{2x}{x^2 - n^2\pi^2}, \qquad x \notin \{0, \pm\pi, \pm 2\pi, \ldots\}. \tag{8.44}$$

It is also easy to see that replacing x by πx in the last equality gives us

$$\pi\cot\pi x = \frac{1}{x} + \sum_{n=1}^{\infty} \frac{2x}{x^2 - n^2}, \qquad x \notin \{0, \pm 1, \pm 2, \ldots\}. \tag{8.45}$$

PROB. 8.6. Obtain

$$\ln\cos^2 x = \sum_{n=1}^{\infty} \ln\left(1 - \frac{x^2}{\left(n - \frac{1}{2}\right)^2\pi^2}\right)^2.$$

Justify differentiating the above term by term and arrive at

$$\tan x = \sum_{n=1}^{\infty} \frac{2x}{\left(n - \frac{1}{2}\right)^2\pi^2 - x^2} \qquad \text{for} \quad x \notin \left\{\pm\frac{\pi}{2}, \pm\frac{3}{2}, \ldots\right\}.$$

PROB. 8.7. Note that if $x \notin \{0, \pm\pi, \pm 2\pi, \ldots\}$, then $\csc x = \tan(x/2) + \cot x$. Prove:

$$\csc x = \frac{1}{x} + \sum_{n=1}^{\infty} (-1)^{n+1} \frac{2x}{n^2\pi^2 - x^2}.$$

9. Application to Power Series

Theorem 5.5 tells us that a power series having radius of convergence $R > 0$ is continuous in its interval of convergence. We now investigate its differentiability there.

Theorem 9.1. *The two power series*

$$(1) \quad \sum_{n=0}^{\infty} a_n(x - x_0)^n \quad \text{and} \quad (2) \quad \sum_{n=1}^{\infty} n a_n (x - x_0)^{n-1}$$

have the same radius of convergence and interval of convergence.

PROOF. Multiply the second series by $x - x_0$ to obtain the power series

$$(3) \qquad \sum_{n=1}^{\infty} n a_n (x - x_0)^n.$$

The power series (2) converges for some x if and only if the power series (3) converges for that x. Hence, (2) and (3) have the same radius of convergence. We prove that (1) and (3) have the same radius of convergence.

For each positive integer n, we have $\sqrt[n]{|a_n|} \leqslant \sqrt[n]{|na_n|}$. This implies that

$$\overline{\lim}\sqrt[n]{|a_n|} \leqslant \overline{\lim}\sqrt[n]{|na_n|}. \tag{9.1}$$

Now recall that $\lim \sqrt[n]{n} = 1$ (Example III.6.6). By Theorem III.8.4 we have

$$\overline{\lim}\sqrt[n]{|na_n|} \leqslant \overline{\lim}\left(\sqrt[n]{n}\sqrt[n]{|a_n|}\right) \leqslant \overline{\lim}\sqrt[n]{n}\;\overline{\lim}\sqrt[n]{|a_n|} = \overline{\lim}\sqrt[n]{|a_n|}.$$

This and (9.1) imply that

$$\overline{\lim}\sqrt[n]{|a_n|} = \overline{\lim}\sqrt[n]{|na_n|}. \tag{9.2}$$

By the definition of the radius of convergence of a power series, this implies that series (1) and (3) have the same radius of convergence. Hence, series (1) and (2) have the same radius of convergence. It now follows that (1) and (2) have the same interval of convergence.

Remark 9.1. Note that series (2) in the last theorem is the derived series of series (1). We, therefore, may rephrase this theorem as

Theorem 9.1′. *A power series has the same radius of convergence and interval of convergence as its derived series.*

Theorem 9.2. *If* $\sum_{n=0}^{\infty} a_n(x-x_0)^n$ *has a nonzero radius of convergence* R *and*

$$S(x) = \sum_{n=0}^{\infty} a_n(x-x_0)^n \quad \text{for} \quad x \in (x_0 - R; x_0 + R), \tag{9.3}$$

then S *is differentiable on* $(x_0 - R; x_0 + R)$ *and*

$$S'(x) = \sum_{n=1}^{\infty} na_n(x-x_0)^{n-1}. \tag{9.4}$$

Proof. By Theorem 5.5, the power series $\sum_{n=0}^{\infty} a_n(x-x_0)^n$ converges to its sum $S(x)$ for $x \in (x_0 - R; x_0 + R)$. By Theorem 9.1′, the derived series of our series has the same interval of convergence, $(x_0 - R; x_0 + R)$, as our series. Therefore, by Theorem 5.5, the derived power series converges uniformly on every bounded closed subinterval of $(x_0 - R; x_0 + R)$. By the corollary of Theorem 8.2, the series $\sum_{n=0}^{\infty} a_n(x-x_0)^n$ is differentiable termwise on $(x_0 - R; x_0 + R)$, and equation (9.4) holds. This completes the proof.

Prob. 9.1. Prove:

$$\frac{d}{dx}\left(\frac{e^x - 1}{x}\right) = \sum_{n=1}^{\infty} \frac{n}{(n+1)!} x^{n-1}.$$

Corollary. *If the power series*

$$S(x)=\sum_{n=0}^{\infty} a_n(x-x_0)^n \tag{9.5}$$

has a positive radius of convergence R, then its sum S has derivatives of all orders on the interval of convergence $(x_0-R; x_0+R)$. Moreover, the kth derivative $S^{(k)}$ of the sum S is obtained by differentiating the series term by term k times. In other words, if $x\in(x_0-R; x_0+R)$, then

$$S^{(k)}(x)=\sum_{n=k}^{\infty} n(n-1)\cdots(n-k+1)a_n(x-x_0)^{n-k} \tag{9.6}$$

for each positive integer k.

Proof. Exercise. Use induction on k.

Remark 9.2. We extend the validity of the last corollary to include $k=0$ by interpreting $S^{(0)}(x)$ as $S(x)$ and then by writing the expression $n(n-1)\cdots(n-k+1)$ in the sum appearing in (9.6) as $(n)_k$ (see Section II.6 on factorials). We recall that $(n)_0=1$. With these extensions (9.6) becomes for $k=0$

$$S^{(0)}(x)=\sum_{n=0}^{\infty}(n)_0 a_n(x-x)^n=\sum_{n=0}^{\infty}a_n(x-x_0)^n=S(x),$$

which is true in view of the convention just adopted. Accordingly, under the hypothesis of the last corollary, we have for the sum $S(x)$:

$$S^{(k)}(x)=\sum_{n=k}^{\infty}(n)_k a_n(x-x_0)^{n-k} \tag{9.7}$$

for each nonnegative integer k. Here, of course,

$$\begin{aligned}(n)_0&=1,\\ (n)_k&=n(n-1)\cdots(n-k+1)\qquad \text{if}\quad k\geqslant 1.\end{aligned} \tag{9.8}$$

Remark 9.3. An alternate form for (9.6) is obtained from

$$\begin{aligned}S^{(k)}(x)&=\sum_{n=k}^{\infty}n(n-1)\cdots(n-k+1)a_n(x-x_0)^{n-k}\\ &=\sum_{n=k}^{\infty}k!\,\frac{n(n-1)\cdots(n-k+1)}{k!}a_n(x-x_0)^{n-k}\\ &=k!\sum_{n=k}^{\infty}\binom{n}{k}a_n(x-x_0)^{n-k}\\ &=k!\sum_{n=k}^{\infty}\binom{n}{n-k}a_n(x-x_0)^{n-k}.\end{aligned}$$

This is equivalent to

$$\frac{S^{(k)}(x)}{k!} = \sum_{n=k}^{\infty} \binom{n}{n-k} a_n (x - x_0)^{n-k}. \tag{9.9}$$

Now reindex, in the above, by writing $m = n - k$ and $n = m + k$. The above becomes

$$\frac{S^{(k)}(x)}{k!} = \sum_{m=0}^{\infty} \binom{m+k}{m} a_{m+k} (x - x_0)^{m} \tag{9.10}$$

and is valid for all nonnegative integers k (Remark 9.2). This may be written

$$\frac{S^{(k)}(x)}{k!} = a_k + \sum_{m=1}^{\infty} \binom{m+k}{m} a_{m+k} (x - x_0)^{m}.$$

Substitute x_0 for x and obtain

$$\frac{S^{(k)}(x_0)}{k!} = a_k, \tag{9.11}$$

where k is a nonnegative integer. This proves:

Theorem 9.3. *If the power series $S(x) = \sum_{n=0}^{\infty} a_n (x - x_0)^n$ has a radius of convergence $R > 0$, then*

$$a_n = \frac{S^{(n)}(x_0)}{n!} \quad \textit{for each nonnegative integer } n,$$

and

$$S(x) = \sum_{n=0}^{\infty} \frac{S^{(n)}(x_0)}{n!} (x - x_0)^n \quad \textit{for each } x \textit{ in the interval of convergence.} \tag{9.11'}$$

Each power series is, therefore, the Taylor series of its sum.

Remark 9.4. Clearly, Theorem 9.2 and its corollary retain their validity when the interval of convergence mentioned in them is replaced by any interval $(x_0 - r; x_0 + r)$, where $0 < r \leqslant R$. This is so because such an interval is a subset of the interval of convergence.

Def. 9.1. We say that a function f can be *expanded in a power series about the point* x_0 if and only if an $r > 0$ and a power series in $x - x_0$ exist such that for each x in the interval $(x_0 - r; x_0 + r)$, $f(x)$ is the sum of the power series evaluated at x. When f can be expanded in a power series about x_0, we call f *analytic* at x_0. In other words f is analytic at x_0 if and only if an

$r > 0$ and a power series $\sum a_n(x - x_0)^n$ exist such that

$$f(x) = \sum_{n=0}^{\infty} a_n(x - x_0)^n \qquad \text{for each} \quad x \in (x_0 - r; x_0 + r).$$

When f is analytic at $x_0 = 0$, we say that it is *analytic at the origin.*

Remark 9.5. When f is analytic at a point, it necessarily has derivatives of all orders at the point. However, Example IX.5.3 shows that the converse is false. There we saw that the function f defined as

$$f(x) = \begin{cases} e^{-(1/x^2)} & \text{for} \quad x \neq 0 \\ 0 & \text{for} \quad x = 0 \end{cases}$$

has derivatives of all orders at $x = 0$. As a matter of fact, we proved $f^{(n)}(0) = 0$ for all nonnegative integers n. f is not analytic at 0. Indeed, if f were analytic at $x = 0$, there would exist an $r > 0$ and a power series $\sum_{n=0}^{\infty} a_n x^n$ in x such that

$$f(x) = \sum_{n=0}^{\infty} a_n x^n \qquad \text{for } x \text{ in } (x_0 - r; x_0 + r).$$

By Theorem 9.3, we would have

$$a_n = \frac{f^{(n)}(0)}{n!} = 0 \qquad \text{for all nonnegative integers } n, \tag{9.12}$$

so we would have $f(x) = 0$ for all x in $(x_0 - r; x_0 + r)$. This is impossible in view of the fact that

$$f(x) = e^{-(1/x^2)} > 0 \qquad \text{for} \quad x \neq 0.$$

Thus, functions exist having derivatives of all orders at some point in their domain which are not analytic at this point.

EXAMPLE 9.1. A polynomial P on $\mathbb{R}$, where

$$P(x) = a_0 x^n + a_1 x^{n-1} + \cdots + a_{n-1} x + a_n \qquad \text{for each} \quad x \in \mathbb{R},$$

is certainly analytic at the origin. P is the sum of a power series in x for which the coefficients of the terms containing x^m are all equal to 0 for $m > n$. The power series collapses to a finite sum $P(x)$ for all $x \in \mathbb{R}$ and certainly converges for each $x \in \mathbb{R}$ to $P(x)$. By Theorem VII.4.1, we have for P: If $a \in \mathbb{R}$, then

$$P(x) = \sum_{k=0}^{n} \frac{P^{(k)}(a)}{k!}(x - a)^k \qquad \text{for all} \quad x \in \mathbb{R}.$$

This shows that P is analytic not only at the origin but it is analytic for each $a \in \mathbb{R}$.

PROB. 9.2. Prove: If in some interval $I = (x_0 - r; x_0 + r)$, where $r > 0$, f has derivatives of all orders, then f is analytic at x_0 if and only if the remainder $R_{n+1}(x_0, x)$ in Taylor's formula of order n (see Remarks IX.4.1–4.2) approaches zero as $n \to +\infty$ for each $x \in I$.

PROB. 9.3. Prove: If in some interval $I = (x_0 - r; x_0 + r)$, where $r > 0$, f has derivatives of all orders and there exists an $M > 0$ such that $|f^{(n)}(x)| \leqslant M$ for all $x \in I$ for each positive integer n, then f is analytic at x_0 (use Taylor's formula of order n with the Lagrange form of the remainder. This is found in Remark IX.4.4).

The results in the last two problems supply us with methods of proving the analyticity of a function f at a point x_0. To use Prob. 9.2 we must obtain Taylor's formula of order n for f at x_0 with the remainder $R_{n+1}(x_0, x)$ and attempt to show that $R_{n+1}(x_0, x) \to 0$ as $n \to +\infty$. This method was used in the examples in Sections IX.5 and IX.6. There we used somewhat different terminology, however. To relate the terminology in Chapter IX to the one used here, we observe that a power series in $x - x_0$ is the Taylor series of its sum $S(x)$ (Theorem 9.3). From this observation it follows that the definition of analyticity at a point may be phrased as follows: A function f having derivatives of all orders at a point x_0 is analytic at x_0 if and only if there exists an interval $(x_0 - r; x_0 + r)$, where $r > 0$, such that the Taylor series of f at x_0 represents $f(x)$ for each x in that interval. We recall that the Taylor series of f at x_0 represents $f(x)$ when $f(x)$ is its sum for x.

It is often not practical to use the method of Prob. 9.2 to prove analyticity of a function at a point. To use Taylor's formula of order n with any form of the remainder one needs to calculate all the derivates $f^{(n)}(x)$ of f when these exist. This often leads to rather cumbersome formulas. In the examples below we use other methods.

EXAMPLE 9.2. We proved in Chapter IX that

$$\frac{1}{1-x} = 1 + x + x^2 + \cdots = \sum_{n=0}^{\infty} x^n \qquad \text{for all} \quad x \in (-1; 1). \tag{9.13}$$

Accordingly, we say that the function f, defined as

$$f(x) = \frac{1}{1-x} \qquad \text{for} \quad x \neq 1, \tag{9.14}$$

is analytic at the origin. The proof of (9.13) did not use Taylor's formula with the remainder. Instead, we used the identity

$$\frac{1-x^n}{1-x} = 1 + x + \cdots + x^{n-1} \qquad \text{for} \quad x \neq 1$$

for each positive integer n. From this we obtained

$$\left| \frac{1}{1-x} - (1 + x + \cdots + x^{n-1}) \right| = \frac{|x|^n}{1-x}$$

and that for $|x| < 1$, the left-hand side, and, therefore, also the right-hand side, tend to zero as $n \to +\infty$. This constituted a proof of (9.13). f is not analytic at $x = 1$ since it is not defined there. Even if we assigned to f some value at 1, the resulting function would not be differentiable at 1 and, therefore, would not be analytic there.

Is f analytic at $x_0 \neq 1$? To investigate this, we note that for $x_0 \neq 1$,

$$f(x) = \frac{1}{1 - x_0 - (x - x_0)} = \frac{1}{1 - x_0} \frac{1}{1 - (x - x_0)/(1 - x_0)} . \tag{9.15}$$

Let x be such that $|x - x_0| < |1 - x_0|$ so that $|x - x_0|/|1 - x_0| < 1$ and

$$\frac{1}{1 - (x - x_0)/(1 - x_0)} = 1 + \frac{x - x_0}{1 - x_0} + \left(\frac{x - x_0}{1 - x_0} \right)^2 + \cdots .$$

Using (9.15), we conclude that

$$f(x) = \frac{1}{1 - x} = \frac{1}{1 - x_0} + \frac{1}{(1 - x_0)^2}(x - x_0) + \frac{1}{(1 - x_0)^3}(x - x_0)^2 + \cdots$$

$$= \sum_{n=0}^{\infty} \frac{1}{(1 - x_0)^{n+1}} (x - x_0)^n \tag{9.16}$$

for $|x - x_0| < |1 - x_0|$, where $x_0 \neq 1$. We conclude that our f is analytic at $x_0 \neq 1$. It is clear that the series on the right in (9.16) converges in the interval $(x_0 - r; x_0 + r)$, where $r = (1 - x_0)$. Taylor's theorem with the remainder was not used here. As a matter of fact, we can use (9.16) to evaluate $f^{(n)}(x_0)$. By Theorem 9.3, (9.16) implies that

$$f^{(n)}(x_0) = n!\, a_n = \frac{n!}{(1 - x_0)^{n+1}} .$$

We have shown that f is not only analytic at the origin, but it is analytic at each real number other than 1. We call a function *analytic on a set* if it is analytic at each point of the set. In Section 10 we shall prove that if a function is analytic at a point, then it is analytic in some ϵ-neighborhood of the point.

PROB. 9.4. Prove that f is analytic at the origin if

(1) $f(x) = (1 + x^2)/(1 - x)$, $x \neq 1$;
(2) $f(x) = 1/(1 + x + x^2)$;
(3) $f(x) = 1/(1 - x)^2$;
(4) $f(x) = \sqrt{1 - x^2}$, $|x| < 1$ (Hint: see Prob. IX.6.3).

Theorem 9.4. *If*

$$\sum_{n=0}^{\infty} a_n (x - x_0)^n = 0 \qquad \textit{for all} \quad x \in (x_0 - r; x_0 + r), \quad r > 0,$$

then $a_n = 0$ for all nonnegative integers n.

PROOF. Let

$$f(x) = \sum_{n=0}^{\infty} a_n(x - x_0)^n \qquad \text{for all} \quad x \in (x_0 - r; x_0 + r).$$

By the hypothesis, $f(x) = 0$ for all x in the interval $(x_0 - r; x_0 + r)$. Hence, $f^{(n)}(x) = 0$ for all $x \in (x_0 - r; x_0 + r)$ and for each nonnegative integer n. On the other hand, the hypothesis also implies that the series has a nonzero radius of convergence. f is the sum of the series on the given interval. Hence, by Theorem 9.3

$$0 = \frac{f^{(n)}(x_0)}{n!} = a_n \qquad \text{for all nonnegative integers } n,$$

as claimed.

Corollary. *If some $r > 0$ exists such that*

$$\sum_{n=0}^{\infty} a_n(x - x_0)^n = \sum_{n=0}^{\infty} b_n(x - x_0)^n \qquad \textit{for all} \quad x \in (x_0 - r; x_0 + r),$$

then $a_n = b_n$ for all nonnegative integers n.

PROOF. Exercise.

This corollary is known as the *identity* or the *uniqueness* theorem for power series. Its use in applications is often referred to as the method of *undetermined coefficients*. We illustrate it in the next example. In that example we present another means of proving the analyticity of a function at a point.

EXAMPLE 9.3. The method of testing for analyticity given here can be presented more briefly after integration is studied. We test f, where

$$f(x) = \operatorname{Arctan} x \qquad \text{for each} \quad x \in \mathbb{R},$$

for analyticity at the origin. Thus, we seek a power series $\sum a_n x^n$ such that

$$f(x) = \operatorname{Arctan} x = \sum_{n=0}^{\infty} a_n x^n \qquad \text{for some interval} \quad (-r; r) = I. \tag{9.17}$$

If such a series exists, then it must be differentiable term by term on I, so

$$f'(x) = \frac{1}{1 + x^2} = \sum_{n=1}^{\infty} n a_n x^{n-1} = \sum_{n=0}^{\infty} (n + 1) a_{n+1} x^n \qquad \text{for all} \quad x \in I. \tag{9.18}$$

But

$$\frac{1}{1 + x^2} = \frac{1}{1 - (-x^2)} = 1 - x^2 + x^4 - x^6 + \cdots = \sum_{n=0}^{\infty} (-1)^n x^{2n}$$

$$\text{for all} \quad x \in (-1; 1).$$

Equating the two right-hand sides, we have

$$\sum_{n=0}^{\infty}(n+1)a_{n+1}x^n=\sum_{n=0}^{\infty}(-1)^n x^{2n} \qquad \text{for} \quad x\in(-1;1)\cap(r;r). \tag{9.19}$$

The intersection $(-1;1)\cap(-r;r)$ is again an interval $(-r_1;r_1)$, where $r_1=\min\{1,r\}$. It will be convenient to replace the "dummy variable" n in the second series by k, so that (9.19) can be written

$$\sum_{n=0}^{\infty}(n+1)a_{n+1}x^n=\sum_{k=0}^{\infty}(-1)^k x^{2k}. \tag{9.20}$$

We apply the uniqueness theorem for the power series above and "equate coefficients." When $n=2k$, $k\in\{0,1,2,\ldots\}$, the corresponding coefficients on opposite sides of (9.20) are equal. Hence,

$$(2k+1)a_{2k+1}=(-1)^k \qquad \text{for all} \quad k\in\{0,1,2,\ldots\}. \tag{9.21}$$

When $n=2k-1$, $k\in\{1,2,\ldots\}$, the coefficient of x^{2k-1} on the right-hand side is equal to 0, and the corresponding coefficient on the left is $([2k-1]+1)a_{[2k-1]+1}=2ka_{2k}$. Equating these coefficients we have

$$2ka_{2k}=0 \qquad \text{for all} \quad k\subset\{1,2,3,\ldots\},$$

so that $a_{2k}=0$ for all positive integers k. By (9.21) we have

$$a_{2k+1}=\frac{(-1)^k}{2k+1} \qquad \text{for all nonnegative integers } k.$$

The series we are looking for is

$$\sum_{n=0}^{\infty}a_n x^n=\sum_{k=0}^{\infty}(-1)^k\frac{x^{2k+1}}{2k+1}=x-\frac{x^3}{3}+\frac{x^5}{5}-\frac{x^7}{7}+\cdots. \tag{9.22}$$

By the ratio test for absolute convergence, this series converges absolutely for $|x|<1$ and diverges for $|x|>1$. The interval of convergence of the series is $(-1;1)$. The question that needs to be answered now is: does it converge to our f?

Write the sum of the series as $\phi(x)$ so that

$$\phi(x)=\sum_{k=0}^{\infty}(-1)^k\frac{x^{2k+1}}{2k+1} \qquad \text{for all} \quad x\in(-1;1). \tag{9.23}$$

By termwise differentiation, we have

$$\phi'(x)=\sum_{k=0}^{\infty}(-1)^k x^{2k}=\frac{1}{1+x^2}=(\operatorname{Arctan}x)' \qquad \text{for all} \quad x\in(-1;1). \tag{9.24}$$

By Theorem VII.6.3, we obtain from (9.24) that a constant c exists such that

$$\phi(x)=\operatorname{Arctan}x+c \qquad \text{for all} \quad x\in(-1;1). \tag{9.25}$$

But $\phi(0)=0$ holds as is seen from (9.23). Substituting 0 for x in (9.25) we see that $0=\phi(0)=\operatorname{Arctan}0+c=c$ and that $c=0$. We can now write

$\phi(x) = \operatorname{Arctan} x$ for all $x \in (-1; 1)$. This leaves us with the result

$$\operatorname{Arctan} x = x - \frac{x^3}{3} + \frac{x^5}{5} - \frac{x^7}{7} + \cdots \qquad \text{for} \quad x \in (-1; 1). \tag{9.26}$$

This proves that the Arctan function is analytic at the origin. We shall see later that this method can be shortened considerably. Since the validity of what was done here is confined to the *open* interval $(-1; 1)$, we cannot, as yet, say anything about the endpoints except that the series on the right converges at both endpoints (why?).

Prob. 9.5. Prove:

$$\operatorname{Arctanh} x = x + \frac{x^3}{3} + \frac{x^5}{5} + \cdots \qquad \text{if} \quad -1 < x < 1.$$

10. Analyticity in a Neighborhood of x_0. Criteria for Real Analyticity

We first prove:

Theorem 10.1. *A function f is analytic at a point x_0 in its domain if and only if there exist a real $r > 0$ and a positive real number $A(r)$, not depending on x, such that f possesses derivatives of all orders on $(x_0 - r; x_0 + r)$ and*

$$|f^{(n)}(x)| \leqslant \frac{rA(r)n!}{(r - |x - x_0|)^{n+1}} \qquad \textit{for all} \quad x \in (x_0 - r; x_0 + r) \tag{10.1}$$

and for each nonnegative integer n.

Proof. Assume the existence of an r and an $A(r)$ and the validity of (10.1) for each nonnegative integer n. Take x such that $0 < |x - x_0| < r/2$ so that

$$\frac{|x - x_0|}{r - |x - x_0|} < 1. \tag{10.2}$$

By Taylor's theorem of order n with the Lagrange form of the remainder we have

$$f(x) = \sum_{k=0}^{n} \frac{f^{(k)}(x_0)}{k!}(x - x_0)^k + \frac{f^{(n+1)}(c)}{(n+1)!}(x - x_0)^{n+1}.$$

This and (10.1) imply that

$$\left| f(x) - \sum_{k=0}^{n} \frac{f^{(k)}(x_0)}{k!}(x - x_0)^k \right| \leqslant \frac{rA(r)(n+1)!}{(r - |x - x_0|)^{n+1}} \, \frac{|x - x_0|^{n+1}}{(n+1)!}$$

$$= rA(r)\left(\frac{|x - x_0|}{r - |x - x_0|} \right)^{n+1} \tag{10.3}$$

Since (10.2) holds, we have

$$\left(\frac{|x-x_0|}{r-|x-x_0|}\right)^{n+1} \to 0 \qquad \text{as} \quad n \to +\infty.$$

Hence, by (10.3), we conclude that

$$f(x) = \sum_{k=0}^{\infty} \frac{f^{(k)}(x_0)}{k!}(x-x_0)^k \tag{10.4}$$

for $|x-x_0| < r/2$. Therefore, f is analytic at x_0.

Conversely, assume that f is analytic at x_0. Therefore, f is expandable in a power series about x_0 which has a nonzero radius of convergence R and is infinitely differentiable in the interval of convergence $(x_0 - R; x_0 + R)$. Thus, a power series $\sum a_n(x-x_0)^n$ exists such that

$$f(x) = \sum_{n=0}^{\infty} a_n(x-x_0)^n \qquad \text{for all} \quad x \in (x_0 - R; x_0 + R). \tag{10.5}$$

Using Remark 9.3 we have for each nonnegative integer k,

$$f^{(k)}(x) = k! \sum_{n=0}^{\infty} \binom{n+k}{k} a_{n+k}(x-x_0)^n.$$

Taking absolute values, we obtain

$$|f^{(k)}(x)| \leqslant k! \sum_{n=0}^{\infty} \binom{n+k}{n} |a_{n+k}||x-x_0|^n. \tag{10.6}$$

Now take r such that $0 < r < R$ so that $x_0 + r$ is in the interval of convergence of the power series in (10.5). Then the series

$$\sum_{n=0}^{\infty} |a_n| r^n$$

converges and we conclude that $|a_n| r^n \to 0$ as $n \to +\infty$. This implies that there exists a positive real number $A(r)$, such that for each nonnegative integer n

$$|a_n| r^n < A(r)$$

holds. Using this in (10.6), we arrive at

$$\begin{aligned} |f^{(k)}(x)| &= k!\, r^{-k} \sum_{n=0}^{\infty} \binom{n+k}{n} |a_{n+k}| \frac{|x-x_0|^n}{r^n} r^{n+k} \\ &\leqslant k!\, r^{-k} A(r) \sum_{n=0}^{\infty} \binom{n+k}{n} \left|\frac{x-x_0}{r}\right|^n. \end{aligned} \tag{10.7}$$

By Prob. IX.6.4, we know that

$$\frac{1}{(1-|(x-x_0)/r|)^{k+1}} = \sum_{n=0}^{\infty} \binom{n+k}{n} \left|\frac{x-x_0}{r}\right|^n$$

and, hence, by (10.7) that

$$|f^{(k)}(x)| \leqslant k!\,r^{-k}\frac{A(r)}{(1-|x-x_0|/r)^{k+1}} = \frac{rA(r)k!}{(r-|x-x_0|^{k+1}}$$

for each nonnegative integer k. Replacing k by n yields (10.1).

We apply this result to obtain:

Theorem 10.2. *If f is analytic at a point x_0 in its domain, then there exists an $r > 0$ such that f is analytic on the open interval $(x_0 - r; x_0 + r)$.*

PROOF. By Theorem 10.1, we know that there exists an $r > 0$ and a positive $A(r)$ such that f possesses derivatives of all orders on $(x_0 - r; x_0 + r)$ and

$$|f^{(n)}(x)| \leqslant \frac{rA(r)n!}{(r-|x-x_0|)^{n+1}} \qquad \text{for all} \quad x \in (x_0 - r; x_0 + r) \tag{10.8}$$

for each nonnegative integer n. Fix $x_1 \in (x_0 - r; x_0 + r)$ so that $|x_1 - x_0| < r$ and $r - |x_1 - x_0| > 0$. Let

$$r_1 = \tfrac{1}{2}\min\{|x_1 - x_0|, r - |x_1 - x_0|\}. \tag{10.9}$$

We have

$$0 < r_1 \leqslant \tfrac{1}{2}|x_1 - x_0| \quad \text{and} \quad 0 < r_1 \leqslant \tfrac{1}{2}(r - |x_1 - x_0|). \tag{10.10}$$

We first prove that $(x_1 - r_1; x_1 + r_1) \subseteq (x_0 - r; x_0 + r)$. Assume that $x \in (x_1 - r_1; x_1 + r_1)$ so that $|x - x_1| < r_1$. For such x,

$$|x - x_0| \leqslant |x - x_1| + |x_1 - x_0| < r_1 + |x_1 - x_0|. \tag{10.11}$$

If (1) $|x_1 - x_0| \leqslant r/2$, then, by the first inequality in (10.10),

$$r_1 + |x_1 - x_0| \leqslant \tfrac{1}{2}|x_1 - x_0| + |x_1 - x_0| = \tfrac{3}{2}|x_1 - x_0| \leqslant \tfrac{3}{4}r < r. \tag{10.12}$$

If (2) $r/2 < |x_1 - x_0| < r$, then, by the second inequality in (10.10),

$$\begin{aligned} r_1 + |x_1 - x_0| &\leqslant \frac{1}{2}(r - |x_1 - x_0|) + |x_1 - x_0| \\ &= \frac{r}{2} + \frac{|x_1 - x_0|}{2} < \frac{r}{2} + \frac{r}{2} = r. \end{aligned} \tag{10.13}$$

In either case,

$$r_1 + |x_1 - x_0| < r. \tag{10.14}$$

This and (10.11) imply that $|x - x_0| < r$. We proved: If $|x - x_1| < r_1$, then $|x - x_0| < r$ and, hence, $(x_1 - r_1; x_1 + r_1) \subseteq (x_0 - r; x_0 + r)$.

Now assume that $|x - x_1| < r_1$. By what was just proved, we know that $x \in (x_0 - r; x_0 + r)$ and, hence, that (10.8) holds for such x. Since (10.14) holds, we know that $|x_1 - x_0| < r - r_1$. Since

$$|x - x_0| - |x - x_1| \leqslant |x_1 - x_0| < r - r_1,$$

we have for our x that

$$0 < r_1 - |x - x_1| < r - |x - x_0|.$$

Hence, for each nonnegative integer n,

$$\frac{1}{(r - |x - x_0|)^{n+1}} < \frac{1}{(r_1 - |x - x_1|)^{n+1}} \qquad \text{for} \quad x \in (x_1 - r_1 ; x_1 + r_1).$$

This and (10.8) imply that

$$|f^{(n)}(x)| \leqslant \frac{rA(r)n!}{(r - |x - x_0|)^{n+1}} < \frac{rA(r)n!}{(r_1 - |x - x_1|)^{n+1}} \tag{10.15}$$

for $x \in (x_1 - r_1 ; x_1 + r_1)$ and for each nonnegative integer n. Thus, for each nonnegative integer n,

$$|f^{(n)}(x)| < \frac{rA(r)n!}{(r_1 - |x - x_1|)^{n+1}} = \frac{r_1((r/r_1)A(r))n!}{(r_1 - |x - x_1|^{n+1}}$$

for $x \in (x_1 - r_1 ; x_1 + r_1)$.

Writing $B(r_1) = (r/r_1)A(r)$. We conclude that for each nonnegative integer n, we have an $r_1 > 0$ and a positive $B(r_1)$ such that

$$|f^{(n)}(x)| \leqslant \frac{r_1 B(r_1)n!}{(r_1 - |x - x_1|)^{n+1}} \qquad \text{for all} \quad x \in (x_1 - r_1 ; x_1 + r_1)$$

and for each nonnegative integer n. By Theorem 10.1, we conclude that f is analytic at x_1.

This proves that f is analytic for each $x_1 \in (x_0 - r; x_0 + r)$ and, therefore, that it is analytic on $(x_0 - r; x_0 + r)$.

CHAPTER XII

Sequences and Series of Functions II

1. Arithmetic Operations with Power Series

We consider power series $\sum a_n(x - x_0)^n$ and $\sum b_n(x - x_0)^n$ with respective radii of convergence R_a and R_b and write $R = \min\{R_a, R_b\}$. We also assume that $R > 0$ so that $R_a \geqslant R > 0$ and $R_b \geqslant R > 0$.

For each $x \in (x_0 - R; x_0 + R)$, each of the above power series converges and so does $\sum(a_n \pm b_n)(x - x_0)^n$. We have

$$\sum_{n=0}^{\infty} a_n(x - x_0)^n \pm \sum_{n=0}^{\infty} b_n(x - x_0)^n = \sum_{n=0}^{\infty} (a_n \pm b_n)(x - x_0)^n. \quad (1.1)$$

As for the products of the above series, each converges absolutely for $x \in (x_0 - R; x_0 + R)$. They can, therefore, be multiplied by using Cauchy products (Def. IV.7.1 and Theorem IV.7.1). We have

$$\sum_{n=0}^{\infty} a_n(x - x_0)^n \sum_{n=0}^{\infty} b_n(x - x_0)^n = \sum_{n=0}^{\infty} c_n, \quad (1.2)$$

where for each n

$$c_n = \sum_{k=0}^{n} a_k(x - x_0)^k b_{n-k}(x - x_0)^{n-k} = (x - x_0)^n \sum_{k=0}^{n} a_k b_{n-k}. \quad (1.3)$$

Thus,

$$\sum_{n=0}^{\infty} a_n(x - x_0)^n \sum_{n=0}^{\infty} b_n(x - x_0)^n = \sum_{n=0}^{\infty} \left(\sum_{k=0}^{n} a_k b_{n-k} \right)(x - x_0)^n$$

$$\text{for} \quad x \in (x_0 - R; x_0 + R). \quad (1.4)$$

(Here it is useful to note that

$$\sum_{k=0}^{n} a_k b_{n-k} = \sum_{k=0}^{n} a_{n-k} b_k \tag{1.5}$$

holds for each n.)

Some examples of the product of power series were given in Sections IV.7 and IV.8. We consider further examples.

EXAMPLE 1.1. Suppose the power series in x, $\sum a_n x^n$, has a radius of convergence $R > 0$. Let $\rho = \min\{R, 1\}$, so that $0 < \rho \leqslant R$ and $0 < \rho \leqslant 1$. Assume that $|x| < \rho$. This implies $|x| < 1$ and $|x| < r$. Hence, the geometric series

$$\frac{1}{1-x} = 1 + x + x^2 + \cdots = \sum_{n=0}^{\infty} x^n$$

and the series we began with converge absolutely, and

$$\frac{1}{1-x} \sum_{n=0}^{\infty} a_n x^n = \sum_{n=0}^{\infty} x^n \sum_{n=0}^{\infty} a_n x^n = \sum_{n=0}^{\infty} \left(\sum_{k=0}^{n} a_k \right) x^n \quad \text{for} \quad |x| < \rho. \tag{1.6}$$

The first few terms of the product are

$$\frac{1}{1-x} \sum_{n=0}^{\infty} a_n x^n = a_0 + (a_0 + a_1)x + (a_0 + a_1 + a_2)x^2 + \cdots.$$

For example, by (1.6), we have

$$\frac{1}{(1-x)^2} = \frac{1}{1-x} \frac{1}{1-x} = \frac{1}{1-x} \sum_{n=0}^{\infty} x^n = \sum_{n=0}^{\infty} \left(\sum_{k=0}^{n} 1 \right) x^n = \sum_{n=0}^{\infty} (n+1) x^n.$$

EXAMPLE 1.2. We expand the function f given by

$$f(x) = \frac{\ln(1+x)}{1+x} \quad \text{for} \quad x > -1 \tag{1.7}$$

in a power series about the origin. If $|x| < 1$, then

$$\frac{1}{1+x} = \sum_{n=0}^{\infty} (-1)^n x^n \quad \text{and} \quad \ln(1+x) = \sum_{n=1}^{\infty} (-1)^{n+1} \frac{x^n}{n}.$$

We divide both sides of the second equation by x to obtain

$$\frac{\ln(1+x)}{x} = \sum_{n=1}^{\infty} (-1)^{n+1} \frac{x^{n-1}}{n} = \sum_{n=0}^{\infty} (-1)^n \frac{x^n}{n+1}.$$

Using Cauchy products, we have

$$\frac{1}{1+x}\,\frac{\ln(1+x)}{x} = \sum_{n=0}^{\infty}(-1)^n x^n \sum_{n=0}^{\infty}(-1)^n \frac{x^n}{n+1}$$
$$= \sum_{n=0}^{\infty}\left(\sum_{k=0}^{n}(-1)^{n-k}\frac{(-1)^k}{k+1}\right)x^n$$
$$= \sum_{n=0}^{\infty}(-1)^n\left(\sum_{k=0}^{n}\frac{1}{k+1}\right)x^n$$

so that

$$\frac{\ln(1+x)}{1+x} = \sum_{n=0}^{+\infty}(-1)^n\left(\sum_{k=0}^{n}\frac{1}{k+1}\right)x^{n+1}$$
$$= x - \left(1+\frac{1}{2}\right)x^2 + \left(1+\frac{1}{2}+\frac{1}{3}\right)x^3 - \left(1+\frac{1}{2}+\frac{1}{3}+\frac{1}{4}\right)x^4 + \cdots \tag{1.8}$$

for $|x| < 1$.

Prob. 1.1. Expand in a power series about the origin:

(1) $1/(1-x)^3$,
(2) $1/(1+x)^2$,
(3) $1/(1+x)^3$,
(4) $(\ln(1-x))/(1-x)$.

Prob. 1.2. We write a repeating decimal

$$N + .d_1d_2 \ldots d_kd_1d_2 \ldots d_k \ldots,$$

where N is an integer and $d_1, d_2, \ldots, d_k$ are digits as

$$N + .\overline{d_1d_2 \ldots d_k}\,.$$

For example

$$\frac{7}{33} = .212121\ldots = .\overline{21}\,.$$

Prove:

(1) $1/9^2 = 1/81 = .\overline{012345679}$,
(2) $1/(99)^2 = .\overline{0001020304 \ldots 95969799}$.

We now consider division of power series. We state the results in terms of power series in x. Extensions to power series in $x - x_0$ can be carried out in the obvious manner.

Let $\sum a_n x^n$ be a power series in x with nonzero radius of convergence R and sum

$$g(x) = \sum_{n=0}^{\infty} a_n x^n \qquad \text{for all} \quad x \in (-R; R). \tag{1.9}$$

We assume that $a_0 \neq 0$. The reciprocal h of g is given by

$$\begin{aligned} h(x) &= \frac{1}{g(x)} = \frac{1}{a_0 + a_1 x + a_2 x^2 + \cdots} \\ &= \frac{1}{a_0} \frac{1}{1 + a_1/a_0 + (a_2/a_0)x^2 + \cdots}. \end{aligned} \tag{1.10}$$

To deal with h, it suffices to examine the factor of $1/a_0$. We, therefore, examine series of the form $1 + \sum_{n=1}^{\infty} c_n x_n$.

Theorem 1.1. *If*

$$g(x) = \sum_{n=0}^{\infty} c_n x^n, \tag{1.11}$$

where $c_0 = 1$ and the power series on the right has radius of convergence $R > 0$, then there exists an $r > 0$ such that $g(x) \neq 0$ for $|x| < r$ and a power series

$$h(x) = \sum_{n=0}^{\infty} b_n x^n, \tag{1.12}$$

which converges for $|x| < r$ such that

$$g(x)h(x) = \sum_{n=0}^{\infty} c_n x^n \sum_{n=0}^{\infty} b_n x^n = 1 \qquad \textit{for all} \quad x \in (-r; r).$$

Moreover, the coefficients b_n of the series in (1.12) satisfy the equations

$$\begin{cases} c_0 b_0 = 1 \\ \displaystyle\sum_{k=0}^{n} c_k b_{n-k} = 0 \qquad \textit{for each positive integer } n \end{cases} \tag{1.13}$$

and are uniquely determined by them.

Proof. The existence of a sequence $\langle b_n \rangle_{n \geqslant 0}$ satisfying (1.13) can be proved by induction on n. Since $c_0 = 1$, obviously $b_0 = 1$. For $n = 1$, the system (1.13) is

$$\begin{aligned} c_0 b_0 &= 1, \\ c_0 b_1 + c_1 b_0 &= 0. \end{aligned}$$

Clearly $b_0 = 1$, $b_1 = -c_1$ satisfy this system. If $b_0, b_1, \ldots, b_n$ satisfy the system

$$\begin{aligned}
&c_0b_0 = 1,\\
&c_0b_1 + c_1b_0 = 0,\\
&\vdots\\
&c_0b_n + c_1b_{n-1} + \cdots + c_nb_0 = 0,
\end{aligned}$$

where n is some positive integer, then $b_0, b_1, \ldots, b_n, b_{n+1}$, where

$$b_{n+1} = -c_1b_n - c_2b_{n-1} - \cdots - c_nb_1 - c_{n+1}$$

will satisfy the system

$$\begin{aligned}
&c_0b_0 = 1,\\
&c_0b_1 + c_1b_0 = 0,\\
&\vdots\\
&c_0b_{n+1} + c_1b_n + \cdots + c_nb_1 + c_{n+1}b_0 = 0.
\end{aligned}$$

Thus, there exists a sequence $\langle b_n \rangle$ satisfying the system of equations (1.13). We prove the uniqueness of this sequence later.

If a sequence $\langle b_n \rangle$ satisfies (1.13), then the power series $\sum b_n x^n$ has a nonzero radius of convergence. We prove this now. By hypothesis the series in (1.11) has a nonzero radius of convergence. Therefore,

$$\overline{\lim}\ \sqrt[n]{|c_n|} < +\infty$$

and the sequence $\langle \sqrt[n]{|c_n|} \rangle$ is bounded. An $M > 0$ exists such that

$$\sqrt[n]{|c_n|} \leqslant M \quad \text{and, therefore,} \quad |c_n| \leqslant M^n$$

for each positive integer n. Now, $b_1 = -c_1$ so that $|b_1| \leqslant |c_1| \leqslant M$. Also $b_2 = c_0b_2 = -(c_1b_1 + c_2b_0) = -(c_1b_1 + c_2)$ and, therefore,

$$|b_2| \leqslant |c_1||b_1| + |c_2| \leqslant M^2 + M^2 = 2M^2.$$

We prove that

$$|b_n| \leqslant 2^{n-1}M^n \qquad \text{for each positive integer } n \tag{1.14}$$

by using complete induction on n. Assume that (1.14) holds for $m \in \{1, 2, \ldots, n-1\}$ (we already saw that it holds for $n = 1$ and $n = 2$), where n is some positive integer. Since

$$\begin{aligned}
b_n = c_0b_n &= -(c_1b_{n-1} + c_2b_{n-2} + \cdots + c_{n-1}b_1 + c_nb_0)\\
&= -(c_1b_{n-1} + c_2b_{n-2} + \cdots + c_{n-1}b_1 + c_n),
\end{aligned}$$

we have

$$\begin{aligned}
|b_n| &\leqslant |c_1||b_{n-1}| + |c_2||b_{n-2}| + \cdots + |c_{n-1}||b_1| + |c_n| \\
&\leqslant M(2^{n-2}M^{n-1}) + M^2(2^{n-3}M^{n-2}) + \cdots + M^{n-1}M + M^n \\
&= M^n(2^{n-2} + 2^{n-3} + \cdots + 1 + 1) \\
&= M^n\left(\frac{2^{n-1}-1}{2-1} + 1\right) \\
&= 2^{n-1}M^n.
\end{aligned}$$

Invoking the principle of complete induction, we conclude that (1.14) holds for each positive integer n. The series

$$1 + \sum_{n=1}^{\infty} M^n 2^{n-1} x^n = 1 + \frac{1}{2}\sum_{n=1}^{\infty}(2Mx)^n$$

converges if $|2Mx| < 1$, i.e., if

$$|x| < \frac{1}{2M}. \tag{1.15}$$

Since

$$|b_n x^n| = |b_n||x|^n \leqslant 2^{n-1}M^n|x|^n$$

holds for each positive integer n, it follows that the series $\sum b_n x^n$ converges for x satisfying (1.15). We conclude that the radius of convergence of $\sum b_n x^n$ is not equal to 0. We now know that there exists an $r_1 > 0$ such that the series (1.12) converges on $(-r_1; r_1)$. Put $r = \min\{r_1, R\}$. Then $0 < r \leqslant r_1$, $0 < r \leqslant R$, and both of the series (1.11) and (1.12) converge on the interval $(-r; r)$. We multiply the two series and obtain, in view of (1.13),

$$\begin{aligned}
g(x)h(x) &= \sum_{n=0}^{\infty} c_n x^n \sum_{n=0}^{\infty} b_n x^n \\
&= \sum_{n=0}^{\infty}\left(\sum_{k=0}^{n} c_k b_{n-k}\right)x^n \\
&= c_0 b_0 + \sum_{n=1}^{\infty}\left(\sum_{n=0}^{n} c_k b_{n-k}\right)x^n \\
&= 1 + 0 \\
&= 1
\end{aligned}$$

for $x \in (-r; r)$. This proves that $g(x) \neq 0$ for all $x \in (-r; r)$, that the series with sum h converges for all $x \in (-r; r)$, and that $g(x)h(x) = 1$ for such x.

It remains to prove that the sequence $\langle b_n \rangle$ is uniquely determined by (1.13). Assume that $\langle b'_n \rangle$ also is a sequence satisfying (1.13). The series

$$h_1(x) = \sum_{n=0}^{\infty} b'_n x^n$$

converges for all $x \in (-r; r)$ and satisfies $g(x)h_1(x) = 1$ for all such x. We then conclude that $g(x)h_1(x) = g(x)h(x)$ for all $x \in (-r; r)$ and, therefore, that $h_1(x) = h(x)$ for such x. This implies that

$$\sum_{n=0}^{\infty} b_n x^n = \sum_{n=0}^{\infty} b'_n x^n \qquad \text{for all} \quad x \in (-r; r). \tag{1.16}$$

By the identity theorem for power series (Corollary of Theorem XI.9.4) we conclude from (1.16) that $b_n = b'_n$ for all n.

Corollary 1. *If f is analytic at the origin and $f(0) \neq 0$, then its reciprocal h, defined as*

$$h(x) = \frac{1}{f(x)} \qquad \textit{for all } x \textit{ such that} \quad g(x) \neq 0, \tag{1.17}$$

is also analytic at the origin.

PROOF. By hypothesis, there exists a power series $\sum a_n x^n$ and an $r_1 > 0$ such that

$$f(x) = \sum_{n=0}^{\infty} a_n x^n \qquad \text{for all} \quad x \in (-r_1; r_1).$$

Here $a_0 = f(0) \neq 0$ and

$$f(x) = a_0\left(1 + \sum_{n=1}^{\infty} \frac{a_n}{a_0} x^n\right) \qquad \text{for all} \quad x \in (-r_1; r_1).$$

The power series

$$g(x) = 1 + \sum_{n=1}^{\infty} \frac{a_n}{a_0} x^n$$

converges for all $x \in (-r_1; r_1)$ and we have $f(x) = a_0 g(x)$ for $x \in (-r_1; r_1)$. By the theorem, there exists an $r > 0$ and a power series $\sum b_n x^n$ such that

$$\frac{1}{g(x)} = \sum_{n=0}^{\infty} b_n x^n \qquad \text{for all} \quad x \in (-r; r).$$

Multiply through by $1/a_0$ and obtain

$$\frac{1}{f(x)} = \frac{1}{a_0 g(x)} = \sum_{n=0}^{\infty} \frac{b_n}{a_0} x^n \qquad \text{for all} \quad x \in (-r; r).$$

Thus, the function h in (1.17) is analytic at the origin. This completes the proof.

Prob. 1.3. Prove: If f is analytic at a point x_0 in its domain and $f(x_0) \neq 0$, then its reciprocal is analytic at x_0.

Corollary 2 (of Theorem 10.1). *If f and g are analytic at the origin and $g(0) \neq 0$, then f/g is also analytic at the origin.*

Proof. Exercise.

Prob. 1.4. Prove: If f and g are analytic at a point x_0 and $g(x_0) \neq 0$, then f/g is analytic at x_0.

Prob. 1.5. Prove: A rational function is analytic at each x_0 which is not a zero of the polynomial in its denominator.

2. Bernoulli Numbers

Consider the series

$$\frac{e^x - 1}{x} = 1 + \frac{x}{2!} + \frac{x^2}{3!} + \cdots = \sum_{n=0}^{\infty} \frac{x^n}{(n+1)!}. \tag{2.1}$$

The series on the right converges absolutely for all real x (why?). (The function on the left is not defined at the origin, but we extend its domain of definition to include the origin by assigning to the extended function the value 1 there. This is the value the series has at $x = 0$. The extension is continuous and infinitely differentiable at $x = 0$ since the latter point is in the interior of the interval of convergence of the series.) By Theorem 1.1, this (extended) function has a reciprocal B which is analytic at the origin. B is defined as

$$B(x) = \begin{cases} \dfrac{x}{e^x - 1} & \text{for } x \neq 0 \\ 1 & \text{for } x = 0. \end{cases} \tag{2.2}$$

There exists $r > 0$ such that at $x = 0$, i.e.,

$$B(x) = \frac{x}{e^x - 1} = \sum_{n=0}^{\infty} \frac{B^{(n)}(0)}{n!} x^n \quad \text{for all } x \in (-r; r). \tag{2.3}$$

We write

$$B_n = B^{(n)}(0) \qquad \text{for each } n \tag{2.4a}$$

and

$$B(x) = \frac{x}{e^x - 1} = \sum_{n=0}^{\infty} \frac{B_n}{n!} x^n \qquad \text{for} \quad -r < x < r. \tag{2.4b}$$

The B_n's are called the *Bernoulli numbers*, after their discoverer Jacob

Bernoulli. They can be evaluated with the aid of a certain recurrence relation that they satisfy.

Theorem 2.1. *If $\langle B_n \rangle$ is the sequence of Bernoulli numbers, then*

$$B_0 = 1 \quad \textit{and} \quad \sum_{k=0}^{n-1} \binom{n}{k} B_k = 0 \qquad \textit{for} \quad n \geqslant 2. \tag{2.5}$$

Proof. The numbers $B_n/n!$ are the coefficients of the power series in (2.4b). It is the reciprocal of the power series in (2.1). We write

$$c_n = \frac{1}{(n+1)!} \quad \text{and} \quad b_n = \frac{B_n}{n!} \qquad \text{for each } n.$$

Note the c_n's are the coefficients of the power series in (2.1). To determine the b_n's we use formulas (1.13) in Theorem 1.1. Before using these we observe that

$$\sum_{k=0}^{n} c_k b_{n-k} = \sum_{k=0}^{n} b_k c_{n-k}.$$

We, therefore, have

$$B_0 = \frac{1}{1!} \frac{B_0}{1!} = c_0 b_0 = 1 \tag{2.6a}$$

and

$$\sum_{k=0}^{n} \frac{B_k}{k!} \frac{1}{(n+1-k)!} = 0 \qquad \text{for each positive integer } n. \tag{2.6b}$$

Thus, $B_0 = 1$. We multiply both sides of (2.6b) by $(n+1)!$ to obtain

$$\sum_{k=0}^{n} \binom{n+1}{k} B_k = \sum_{k=0}^{n} \frac{(n+1)!}{k!(n+1-k)!} B_k = 0 \qquad \text{for each positive integer } n. \tag{2.7}$$

We then write $m = n + 1$ so that $m \geqslant 2$ and formulas (2.7) become

$$\sum_{k=0}^{m-1} \binom{m}{k} B_k = 0 \qquad \text{for} \quad m \geqslant 2.$$

These are really the second set of formulas in (2.5), written in terms of m instead of n. This completes the proof.

Remark 2.1. The formulas (2.5) can be written in terms of one set of formulas as follows:

$$\sum_{k=0}^{n-1} \binom{n}{k} B_k = \begin{cases} 1 & \text{if} \quad n = 1 \\ 0 & \text{if} \quad n \geqslant 2. \end{cases} \tag{2.8}$$

The Bernoulli numbers can now be evaluated one after the other from formulas (2.5) (or formulas (2.8)). For example, for $n = 2$, we have

$$\sum_{k=0}^{1} \binom{2}{k} B_k = 0 \quad \text{or} \quad B_0 + 2B_1 = 0.$$

Since $B_0 = 1$, we obtain

$$B_1 = -\tfrac{1}{2}. \tag{2.9}$$

Similarly, for $n = 3$ we have

$$\sum_{k=0}^{2} \binom{3}{k} B_k = 0, \quad \text{and, therefore,} \quad B_0 + 3B_1 + 3B_2 = 0.$$

Since $B_0 = 1$ and $B_1 = -\frac{1}{2}$, we see that

$$B_2 = \tfrac{1}{6}. \tag{2.10}$$

PROB. 2.1. Show that

$$B_3 = 0, \quad B_4 = -\frac{1}{30}, \quad B_5 = 0, \quad B_6 = \frac{1}{42},$$

$$B_7 = 0, \quad B_8 = -\frac{1}{30}, \quad B_9 = 0, \quad B_{10} = \frac{5}{66}.$$

PROB. 2.2. Prove: If $n = 0$ or $n \geqslant 2$, then

$$B_n = \sum_{k=0}^{n} \binom{n}{k} B_k.$$

Remark 2.2. A mnemonic device for remembering formula (2.5) for the Bernoulli numbers is to replace each B_n in it by B^n. We then obtain

$$\sum_{k=0}^{n-1} \binom{n}{k} B^k \equiv 0 \qquad \text{for} \quad n \geqslant 2 \tag{2.11a}$$

which can be written, in turn,

$$(B+1)^n - B^n \equiv 0 \qquad \text{if} \quad n \geqslant 2. \tag{2.11b}$$

This formula is not to be taken literally. Its sense is, that after expanding, each B^k should be replaced by B_k. The symbolic B in (2.11) is called an "umbra" and the symbol $\equiv$ is used to express symbolic equivalence, where we write $B^k \equiv B_k$ in (2.11).*

Theorem 2.2. *If n is an odd integer, $n > 2$, then $B_n = 0$.*

*Andrew Guinand, The umbral method: A survey of elementary mnemonic and manipulative uses, *The American Mathematical Monthly*, (1979), 181–184.

PROOF. Assume that $-r < x < r$, where r is the positive number in (2.4). Since $B_2 = -\frac{1}{2}$, (2.4b) implies that

$$\frac{x}{e^x - 1} + \frac{x}{2} = 1 + \sum_{n=2}^{\infty} \frac{B_n}{n!} x^n.$$

This can be written

$$\frac{x}{2} \frac{e^x + 1}{e^x - 1} = 1 + \sum_{n=2}^{\infty} \frac{B_n}{n!} x^n. \tag{2.12}$$

Since

$$\frac{e^x + 1}{e^x - 1} = \frac{e^{x/2} + e^{-(x/2)}}{e^{x/2} - e^{-(x/2)}} = \frac{\cosh(x/2)}{\sinh(x/2)} = \coth \frac{x}{2},$$

(2.12) becomes

$$\frac{x}{2} \coth \frac{x}{2} = 1 + \sum_{n=2}^{\infty} \frac{B_n}{n!} x^n \qquad \text{for} \quad -r < x < r. \tag{2.13}$$

The function defined by the left-hand side is even. Hence,

$$1 + \sum_{n=2}^{\infty} \frac{B_n}{n!} (-x)^n = 1 + \sum_{n=2}^{\infty} \frac{B_n}{n!} x^n$$

so that

$$1 + \sum_{n=2}^{\infty} (-1)^n \frac{B_n}{n!} x^n = 1 + \sum_{n=2}^{\infty} \frac{B_n}{n!} x^n \qquad \text{for} \quad -r < x < r.$$

Equating coefficients, we obtain

$$(-1)^n B_n = B_n \qquad \text{for} \quad n \geqslant 2. \tag{2.14}$$

We conclude that if n is odd and > 2, then $B_n = 0$, as claimed.

Remark 2.3. Write the odd integer $n > 2$ as $n = 2k + 1$, where $k \geqslant 1$. Then

$$B_{2k+1} = 0 \qquad \text{for each positive integer } k. \tag{2.15}$$

We can now rewrite (2.4b), (2.12), and (2.13) as

$$\frac{x}{e^x - 1} = 1 - \frac{x}{2} + \sum_{n=1}^{\infty} \frac{B_{2k}}{(2k)!} x^{2k}, \tag{2.16}$$

$$\frac{x}{2} \frac{e^x + 1}{e^x - 1} = \frac{x}{2} \coth \frac{x}{2} = 1 + \sum_{k=1}^{\infty} \frac{B_{2k}}{(2k)!} x^{2k} = \sum_{k=0}^{\infty} \frac{B_{2k}}{(2k)!} x^{2k}, \tag{2.17}$$

each formula being valid for $-r < x < r$.

Remark 2.4. We replace $x/2$ by x in (2.17) and obtain

$$x \coth x = \sum_{k=0}^{\infty} \frac{2^{2k} B_{2k}}{(2k)!} x^{2k} \qquad \text{for} \quad -\frac{r}{2} < x < \frac{r}{2}. \tag{2.18}$$

Prob. 2.3. Show

$$x \coth x = 1 + \frac{1}{3}x^2 - \frac{1}{45}x^4 + \frac{2}{945}x^6 - \frac{1}{4725}x^8 \dots .$$

Prob. 2.4. Prove:

$$\tanh x = 2\coth 2x - \coth x.$$

Then prove

$$\tanh x = \sum_{k=1}^{\infty} \frac{2^{2k}(2^{2k}-1)}{(2k)!} B_{2k} x^{2k-1}$$

$$= x - \frac{1}{3}x^3 + \frac{2}{15}x^5 - \frac{17}{315}x^7 + \cdots$$

for $|x| < r/4$. (The r here is defined in (2.4).)

Prob. 2.5. Prove:

$$\coth x - \tanh \frac{x}{2} = \frac{1}{\sinh x}.$$

Then prove that

$$\frac{x}{\sinh x} = -\sum_{k=0}^{\infty} \frac{(2^{2k}-2)B_{2k}}{(2k)!} x^{2k}$$

$$= 1 - \frac{x^2}{6} + \frac{7}{360}x^4 - \frac{31}{15120}x^6 + \cdots$$

for $|x| < r/2$.

Remark 2.5. Some writers call the numbers B_k', where

$$B_k' = (-1)^{k+1}B_{2k} \qquad \text{for} \quad k \geqslant 1, \tag{2.19}$$

the Bernoulli numbers. With this definition and in view of Prob. 2.1, we have, for example,

$$B_1' = \frac{1}{6}, \qquad B_2' = \frac{1}{30}, \qquad B_3' = \frac{1}{42}.$$

Theorem 2.3. *If k is a positive integer, then*

$$(-1)^{k+1}B_{2k} = B_k' > 0. \tag{2.20}$$

Proof.* Simple calculations show that

$$\frac{x}{e^x+1} = \frac{x}{e^x-1} - \frac{2x}{e^{2x}-1}. \tag{2.21}$$

*L. J. Mordell, Signs of the Bernoulli numbers, *The American Mathematical Monthly*, (1973).

Use (2.4b) for both terms on the right to arrive at

$$\frac{x}{e^x+1}=\sum_{n=0}^{\infty}\frac{(1-2^n)B_n}{n!}x^n \qquad \text{for} \quad -\frac{r}{2}<x<\frac{r}{2}. \tag{2.22}$$

Multiply both sides by $x/(e^x-1)$. If $-r/2<x<r/2$, then

$$\frac{x}{2}\frac{2x}{e^{2x}-1}=\frac{x^2}{(e^x+1)(e^x-1)}=\sum_{n=0}^{\infty}\frac{(1-2^n)B_n}{n!}x^n\sum_{n=0}^{\infty}\frac{B_n}{n!}x^n$$

$$=\sum_{n=0}^{\infty}\left(\sum_{k=0}^{n}\frac{(1-2^k)B_k}{k!}\frac{B_{n-k}}{(n-k)!}\right)x^n. \tag{2.23}$$

For the coefficient of x^n on the right-hand side of this formula and for all nonnegative n we have

$$\sum_{k=0}^{n}\frac{1-2^k}{k!(n-k)!}B_kB_{n-k}=\frac{1}{n!}\sum_{k=0}^{n}(1-2^k)\binom{n}{k}B_kB_{n-k}. \tag{2.24}$$

Note that if $n=0$, then both sides of the above vanish, so (2.23) can be written

$$\frac{x}{2}\frac{2x}{e^{2k}-1}=\sum_{n=1}^{\infty}\left(\sum_{k=0}^{n}(1-2^k)\binom{n}{k}B_kB_{n-k}\right)\frac{x^n}{n!} \qquad \text{if} \quad -\frac{r}{2}<x<\frac{r}{2}. \tag{2.25}$$

Now use (2.4b) for the factor $2x/(e^{2x}-1)$ on the left here to obtain

$$\frac{x}{2}\frac{2x}{e^{2x}-1}=\sum_{n=0}^{\infty}2^{n-1}\frac{B_n}{n!}x^{n+1}=\sum_{n=1}^{\infty}\frac{2^{n-2}}{(n-1)!}B_{n-1}x^n, \tag{2.26}$$

where $-r/2<x<r/2$. Comparing coefficients on the right in (2.25) and (2.26) we arrive at

$$\frac{2^{n-2}}{(n-1)!}B_{n-1}=\frac{1}{n!}\sum_{k=0}^{n}(1-2^k)\binom{n}{k}B_kB_{n-k}=\frac{1}{n!}\sum_{k=1}^{n}(1-2^k)\binom{n}{k}B_kB_{n-k} \tag{2.27}$$

for each positive integer n. (The 0th term in the middle expression vanishes. This is why the rightmost sum begins with $k=1$.) If n is even and >2, then $n>3$ and $n-1>2$ so that $n-1$ is odd and >2. For such n, Theorem 1.2 states that $B_{n-1}=0$. Putting $n=2m$, where m is a positive integer, we, therefore, conclude from (2.27) that

$$0=\sum_{k=1}^{2m}(1-2^k)\binom{2m}{k}B_kB_{2m-k} \qquad \text{if} \quad m>1.$$

But $B_{2m-1}=0$ when $m>1$. Hence, in the sum on the right we can omit the term corresponding to $k=1$, and the last equation is equivalent to

$$0=\sum_{k=2}^{2m}(1-2^k)\binom{2m}{k}B_kB_{2m-k}, \qquad m\geqslant 2.$$

Here the terms for which k is odd vanish. We, therefore, write $k = 2j$, $j \in \{1, 2, \ldots, m\}$, and obtain

$$\sum_{j=1}^{m} (1 - 2^{2j}) \binom{2m}{2j} B_{2j} B_{2(m-j)} = 0. \tag{2.28}$$

We isolate the last term and write

$$\sum_{j=1}^{m-1} (1 - 2^{2j}) \binom{2m}{2j} B_{2j} B_{2(m-j)} + (1 - 2^{2m}) B_{2m} = 0$$

so that

$$(2^{2m} - 1) B_{2m} = \sum_{j=1}^{m-1} (1 - 2^{2j}) \binom{2m}{2j} B_{2j} B_{2(m-j)} \cdot \tag{2.29}$$

Now $B_k' = (-1)^{k+1} B_{2k}$ for each positive integer. Substitute this in (2.29). This yields

$$(2^{2m} - 1)(-1)^{m+1} B_m' = \sum_{j=1}^{m-1} (1 - 2^{2j}) \binom{2m}{2j} (-1)^{j+1} (-1)^{m-j+1} B_j' B_{m-j}' \cdot \tag{2.30}$$

We divide both sides of the above by $(-1)^{m+1}$. The powers of -1 in each term on the right are now $(-1)^{j+1}(-1)^j = -1$. Then we note that $(1 - 2^{2j})(-1) = 2^{2j} - 1$ and finally obtain from (2.30) that

$$(2^{2m} - 1) B_m' = \sum_{j=1}^{m-1} (2^{2j} - 1) \binom{2m}{2j} B_j' B_{m-j}', \qquad m \geqslant 2. \tag{2.31}$$

This puts us in a position to use induction on m. By Remark 2.5, we know that $B_1' > 0$ and $B_2' > 0$. Assume that $B_1', B_2', \ldots, B_{m-1}'$ are all positive for some $m \geqslant 2$. Then (2.31) implies that B_m' is positive. By the principle of complete induction and the fact that $B_1' > 0$ we conclude that $B_m' > 0$ for all positive integers m. This completes the proof.

The next two lemmas will help us prove:

Theorem 2.4. *If $|x| < r/2$, where r is the positive number defined in* (2.4b), *then*

$$x \cot x = \sum_{n=0}^{\infty} (-1)^n \frac{2^{2n} B_{2n}}{(2n)!} x^{2n}. \tag{2.32}$$

(*See formula* (2.18).)

Lemma 2.1. *If*

$$\sum_{n=0}^{\infty} a_n \quad \textit{and} \quad \sum_{n=0}^{\infty} b_n \tag{2.34}$$

converge absolutely, then

$$\sum_{n=0}^{\infty}(-1)^n a_n \sum_{n=0}^{\infty}(-1)^n b_n = \sum_{n=0}^{\infty}(-1)^n\left(\sum_{k=0}^{n} a_k b_{n-k}\right). \tag{2.35}$$

PROOF. The last two series on the left in (2.35) converge absolutely because the series in (2.34) do. Forming their Cauchy product, we have

$$\begin{aligned}\sum_{n=0}^{\infty}(-1)^n a_n \sum_{n=0}^{\infty}(-1)^n b_n &= \sum_{n=0}^{\infty}\left(\sum_{k=0}^{n}(-1)^k a_k (-1)^{n-k} b_{n-k}\right)\\ &= \sum_{n=0}^{\infty}(-1)^n\left(\sum_{k=0}^{n} a_k b_{n-k}\right).\end{aligned}$$

Remark 2.6. Consider the particular case where the series in the lemma are power series

$$A(x) = \sum_{n=0}^{\infty} a_n x^n \quad \text{and} \quad B(x) = \sum_{n=0}^{\infty} b_n x^n$$

with radii of convergence $R_a > 0$ and $R_b > 0$. Then it is easy to see that the lemma implies that

$$\sum_{n=0}^{\infty}(-1)^n a_n x^n \sum_{n=0}^{\infty}(-1)^n b_n x^n = \sum_{n=0}^{\infty}(-1)^n\left(\sum_{k=0}^{n} a_k b_{n-k}\right)x^n \tag{2.36}$$

for $|x| < R = \min\{R_a, R_b\}$.

Lemma 2.2. *If the series*

$$g(x) = \sum_{n=0}^{\infty} c_n x^n,$$

where $c_0 = 1$, has the series

$$h(x) = \sum_{n=0}^{\infty} b_n x^n$$

as its reciprocal on the interval $(-r; r)$, $r > 0$, then the series

$$g^*(x) = \sum_{n=0}^{\infty}(-1)^n c_n x^n$$

has the series

$$h^*(x) = \sum_{n=0}^{\infty}(-1)^n b_n x^n$$

as its reciprocal on $(-r; r)$.

Proof. For x such that $|x| < r$, g^* and h^* converge absolutely and we have by Remark 2.6

$$\begin{aligned} g^*(x)h^*(x) &= \sum_{n=0}^{\infty} (-1)^n c_n x^n \sum_{n=0}^{\infty} (-1)^n b_n x^n \\ &= \sum_{n=0}^{\infty} (-1)^n \left(\sum_{k=0}^{\infty} c_k b_{n-k} \right) x^n \\ &= c_0 b_0 + \sum_{n=1}^{\infty} (-1)^n \left(\sum_{k=0}^{n} c_k b_{n-k} \right) x^n. \end{aligned}$$

Since h is the reciprocal of g on $(-r; r)$, we know by Theorem 1.1 that Eqs. (1.13) hold. We, therefore, obtain from the above that if $|x| < r$, then

$$g^*(x)h^*(x) = c_0 b_0 + 0 = c_0 b_0 = 1.$$

This completes the proof.

We now give a proof of Theorem 2.4. Consider the series for the hyperbolic sine

$$\sinh x = x + \frac{x^3}{3!} + \frac{x^5}{5!} + \cdots = \sum_{n=0}^{\infty} \frac{x^{2n+1}}{(2n+1)!} = x \sum_{n=0}^{\infty} \frac{x^{2n}}{(2n+1)!}.$$

This series converges on $\mathbb{R} = (-\infty; +\infty)$. Divide both sides by x to obtain

$$\frac{\sinh x}{x} = \sum_{n=0}^{\infty} \frac{x^{2n}}{(2n+1)!} = 1 + \frac{x^2}{3!} + \frac{x^4}{5!} + \cdots \tag{2.37}$$

for $x \neq 0$. We extend the domain of the function on the left to include 0 by assigning to it at 0 the value at 0 of the series on the right. This value is 1. The resulting function is analytic at 0 since the right-hand side is. The reciprocal of this function has a power series expansion at 0 which is found in Prob. 2.5. We have from there that

$$\frac{x}{\sinh x} = \sum_{n=0}^{\infty} \frac{(2 - 2^{2n}) B_{2n}}{(2n)!} x^{2n} \quad \text{for} \quad |x| < \frac{r}{2}, \tag{2.38}$$

where r is defined in (2.4b). We also have

$$\frac{\sin x}{x} = \sum_{n=0}^{\infty} (-1)^n \frac{x^{2n}}{(2n+1)!} = 1 - \frac{x^2}{3!} + \frac{x^4}{5!} - \frac{x^6}{7!} + \cdots. \tag{2.39}$$

This expansion is valid on $\mathbb{R}$ when we extend the domain of the function on the left to include 0 by assigning to it, at 0, the value that the series on the right has at 0. This value is 1. We now apply Lemma 2.2 to the function g and g^* defined as

$$g(x) = \frac{\sinh x}{x} \quad \text{and} \quad g^*(x) = \frac{\sin x}{x}. \tag{2.40}$$

The power series for these have coefficients which are related to each other as are the coefficients of g and g^* in that lemma. In view of (2.38) it follows that the reciprocal of g^* has the series expansion

$$\frac{x}{\sin x} = \sum_{n=0}^{\infty} (-1)^n \frac{(2-2^{2n})B_{2n}}{(2n)!} = \sum_{n=0}^{\infty} (-1)^{n+1} \frac{(2^{2n}-2)B_{2n}}{(2n)!} x^{2n} \tag{2.41}$$

for $|x| < r/2$. We recall that

$$\cosh x = \sum_{n=0}^{\infty} \frac{x^{2n}}{(2n)!} \tag{2.42a}$$

and

$$\cos x = \sum_{n=0}^{\infty} (-1)^n \frac{x^{2n}}{(2n)!} \tag{2.42b}$$

for all $x \in \mathbb{R}$. We know from (2.18) that

$$\cosh x\left(\frac{x}{\sinh x}\right) = x \coth x = \sum_{n=0}^{\infty} \frac{2^{2n}B_{2n}}{(2n)!} x^{2n} \qquad \text{for} \quad |x| < \frac{r}{2}. \tag{2.43}$$

Use Remark (2.6) with

$$A(x) = \cosh x \quad \text{and} \quad B(x) = \frac{x}{\sinh x}$$

and take note of (2.42a) and (2.42b) and also of (2.38) and (2.41) and obtain from the remark cited that

$$x \cot x = \cos x\left(\frac{x}{\sin x}\right) = \sum_{n=0}^{\infty} (-1)^n \frac{2^{2n}B_{2n}}{(2n)!} x^{2n} \qquad \text{for} \quad |x| < \frac{r}{2}.$$

Remark 2.7. In the process of proving the last theorem we proved (see (2.41)) that

$$\frac{x}{\sin x} = \sum_{n=0}^{\infty} (-1)^{n+1} \frac{(2^{2n}-2)B_{2n}}{(2n)!} x^{2n} \qquad \text{for} \quad |x| < \frac{r}{2}. \tag{2.44}$$

Remark 2.8. The radius of convergence ρ of the series (2.44) is positive. If $\rho > \pi$, the series would converge for $x = \pi$. This is impossible since it would then follow from (2.44) that

$$1 = \left(\sum_{n=0}^{\infty} (-1)^{n+1} \frac{(2^{2n}-2)B_{2n}}{(2n)!} x^{2n}\right) \frac{\sin x}{x}$$

at $x = \pi$. Hence, $0 < \rho \leqslant \pi$. It follows that $r/2 \leqslant \rho \leqslant \pi$ and, hence, that $0 < r \leqslant 2\pi$.

PROB. 2.6. Prove that

$$\tan x = \cot x - 2 \cot 2x$$

and show that

$$\tan x = \sum_{n=1}^{\infty} (-1)^{n+1} \frac{2^{2n}(2^{2n}-1)B_{2n}}{(2n)!} x^{2n-1}$$

for $|x| < r/4$ (cf. Prob. 2.4) and that the coefficients in the series are all positive.

3. An Application of Bernoulli Numbers

We shall evaluate the sums

$$S_r(n) = \sum_{k=1}^{n} k^r = 1^r + 2^r + \cdots + n^r, \tag{3.1}$$

where n is a positive integer and r is a nonnegative integer. We already know that

$$S_0(n) = n \tag{3.2a}$$

and that

$$S_1(n) = \frac{n(n+1)}{2}. \tag{3.2b}$$

One method of evaluating $S_1(n)$ is to note that

$$\sum_{k=1}^{n} \left((k+1)^2 - k^2\right) = (2^2 - 1^2) + (3^2 - 2^2) + \cdots + \left((n+1)^2 - n^2\right)$$
$$= (n+1)^2 - 1$$

and that

$$\sum_{k=1}^{n} \left((k+1)^2 - k^2\right) = \sum_{k=1}^{n} (2k+1) = 2\sum_{k=1}^{n} k + \sum_{k=1}^{n} 1 = 2S_1(n) + n$$

so that the two right-hand sides above are equal. Therefore,

$$2S_1(n) + n = (n+1)^2 - 1 \quad \text{or} \quad 2S_1(n) = (n+1)^2 - (n+1) = n(n+1).$$

From this (3.2b) follows after solving for $S_1(n)$.

This method can be used to obtain $S_r(n)$ for $r = 1, 2, 3, \ldots$. In fact,

$$\sum_{k=1}^{n} \left((k+1)^r - k^r\right) = (n+1)^r - 1,$$
$$\sum_{k=1}^{n} \left((k+1)^r - k^r\right) = \sum_{k=1}^{n} \left(\sum_{j=0}^{r} \binom{r}{j} k^j - k^r \right) = \sum_{k=1}^{n} \left(\sum_{j=0}^{r-1} \binom{r}{j} k^j \right) \tag{3.3}$$

so that

$$\sum_{k=1}^{n} \left(\sum_{j=0}^{r-1} \binom{r}{j} k^j \right) = (n+1)^r - 1. \tag{3.4}$$

Consider the left side of (3.4). Keeping j fixed and summing first with respect to k and then with respect to j, we have from (3.4)

$$\sum_{j=0}^{r-1}\binom{r}{j}\left(\sum_{k=1}^{n} k^j\right) = (n+1)^r - 1. \tag{3.5}$$

This gives us

$$\sum_{j=0}^{r-1}\binom{r}{j} S_j(n) = (n+1)^r - 1. \tag{3.6}$$

Equation (3.6) can be used to successively calculate $S_j(n)$. For example, for $r = 3$ we have

$$\sum_{j=0}^{2}\binom{3}{j} S_j(n) = (n+1)^3 - 1$$

or

$$S_0(n) + 3S_1(n) + 3S_2(n) = (n+1)^3 - 1.$$

Using formulas (3.2) we obtain

$$n + 3\,\frac{n(n+1)}{2} + 3S_2(n) = (n+1)^3 - 1.$$

Solving for $S_2(n)$ we get

$$\begin{aligned} S_2(n) &= \frac{1}{3}\left((n+1)^3 - (n+1) - \frac{3}{2}n(n+1)\right) \\ &= \frac{1}{3}(n+1)\left((n+1)^2 - \frac{3}{2}n - 1\right) \\ &= \frac{1}{6}(n+1)(2n^2+n) \\ &= \frac{n(n+1)(2n+1)}{6}, \end{aligned}$$

i.e., that

$$S_2(n) = \frac{n(n+1)(2n+1)}{6}. \tag{3.7}$$

Prob. 3.1. Use (3.7) and formulas (3.2) to obtain

$$S_3(n) = \left(\frac{n(n+1)}{2}\right)^2.$$

Prob. 3.2. Prove that

$$S_4(n) = \frac{n(n+1)(2n+1)}{30}(3n^2+3n-1).$$

Prob. 3.3. Prove: For each nonnegative integer r, there exists a polynomial of $P_{r+1}(n)$ in n such that $S_r(n) = P_{r+1}(n)$.

Prob. 3.4. Prove: If $r \geqslant 1$, then for each positive integer n, $n(n+1)$ divides $S_r(n)$.

We shall now obtain a variant of Jacob Bernoulli's formula for $S_r(n)$ which is expressed in terms of Bernoulli numbers. We begin with

$$1 + e^x + \cdots + e^{nx} = \frac{e^{(n+1)x} - 1}{e^x - 1} \qquad \text{for} \quad x \in \mathbb{R}. \tag{3.8}$$

(The validity of this formula for $x = 0$ is to be interpreted in the sense that the function on the right is assigned the value $n + 1$ at $x = 0$. This is its limit as $x \to 0$. On the left in (3.8) we have a *finite* sum of functions each of which is analytic at 0. We may, therefore, substitute the power series expansion in x for each e^{kx} there and then add the terms involving x^r to obtain

$$\begin{aligned} 1 + e^x + e^{2x} + \cdots + e^{nx} &= 1 + \sum_{k=1}^{n} e^{kx} \\ &= 1 + \sum_{k=1}^{n} \left(\sum_{r=0}^{\infty} k^r \frac{x^r}{r!} \right) \\ &= 1 + \sum_{r=0}^{\infty} \frac{x^r}{r!} \left(\sum_{k=1}^{n} k^r \right) \\ &= 1 + \sum_{r=0}^{\infty} S_r(n) \frac{x^r}{r!} \\ &= 1 + n + \sum_{r=1}^{\infty} S_r(n) \frac{x^r}{r!} \end{aligned} \tag{3.9}$$

for all $x \in \mathbb{R}$.

As for the function on the right-hand side of (3.8), we can write it as

$$\frac{e^{(n+1)x} - 1}{e^x - 1} = (n+1) \frac{e^{(n+1)x} - 1}{(n+1)x} \cdot \frac{x}{e^x - 1} \tag{3.10}$$

and then use the power series expansion in x for the last two factors on the right. These are

$$\frac{e^{(n+1)x} - 1}{(n+1)x} = \sum_{r=0}^{\infty} \frac{(n+1)^r x^r}{(r+1)!} \tag{3.11a}$$

and

$$\frac{x}{e^x - 1} = \sum_{r=0}^{\infty} \frac{B_r}{r!} x^r. \tag{3.11b}$$

(Here B_r is the rth Bernoulli number (see (2.4b).) Now substitute into (3.10). This becomes

$$\frac{e^{(n+1)x}-1}{e^x-1} = (n+1)\sum_{r=0}^{\infty}\frac{(n+1)^r}{(r+1)!}x^r\sum_{r=0}^{\infty}\frac{B_r}{r}x^r$$

$$= (n+1)\sum_{r=0}^{\infty}\left[\sum_{k=0}^{r}\left(\frac{(n+1)^k}{(k+1)!}\,\frac{B_{r-k}}{(r-k)!}\right)x^r\right]$$

$$= (n+1)\left[1+\sum_{r=1}^{\infty}\left(\sum_{k=0}^{r}\frac{(n+1)^k}{(k+1)!}\,\frac{B_{r-k}}{(r-k)!}\right)x^r\right].$$

From this (3.8) and (3.9), we have

$$\sum_{r=1}^{\infty}S_r(n)\frac{x^r}{r!} = (n+1)\sum_{r=1}^{\infty}\left(\sum_{k=0}^{r}\frac{(n+1)^k}{(k+1)!}\,\frac{B_{r-k}}{(r-k)!}\right)x^r.$$

Equating coefficients, we obtain

$$\frac{S_r(n)}{r!} = \sum_{k=0}^{r}\frac{(n+1)^{k+1}}{k+1}\,\frac{B_{r-k}}{k!\,(r-k)!} \qquad \text{for} \quad r \geqslant 1.$$

Multiply both sides by $r!$ and use the expression for the binomial coefficients to obtain from this

$$S_r(n) = \sum_{k=0}^{r}\frac{(n+1)^{k+1}}{k+1}\binom{r}{k}B_{r-k} \qquad \text{for} \quad r \geqslant 1 \tag{3.12}$$

for each positive integer n. This is the formula we were aiming at.

Checking, we find from (3.12): If $r = 1$, then

$$S_1(n) = \sum_{k=0}^{1}\frac{(n+1)^{k+1}}{(k+1)}\binom{1}{k}B_{1-k}$$

$$= (n+1)B_1 + \frac{(n+1)^2}{2}B_0 = \frac{(n+1)^2}{2} - \frac{n+1}{2} = \frac{n(n+1)}{2}$$

and

$$S_3(n) = \sum_{k=0}^{3}\frac{(n+1)^{k+1}}{k+1}\binom{3}{k}B_{3-k}$$

$$= (n+1)B_3 + \frac{(n+1)^2}{2}3B_2 + \frac{(n+1)^3}{3}3B_1 + \frac{(n+1)^4}{4}B_0$$

$$= \frac{(n+1)^2}{4} - \frac{(n+1)^3}{2} + \frac{(n+1)^4}{4} = \left(n\frac{(n+1)}{2}\right)^2.$$

PROB. 3.5. Prove:

(1) $S_5(n) = 1^5 + 2^5 + 3^5 + \cdots + n^5 = n^2(n+1)^2(2n^2+2n-1)/12$,
(2) $S_6(n) = 1^6 + 2^6 + 3^6 + \cdots + n^6 = n(n+1)(2n+1)(3n^4+6n^3-3n+1)/42$.

The Euler Numbers

The hyperbolic cosine series is

$$\cosh x = \sum_{n=0}^{\infty} \frac{x^{2n}}{(2n)!} = 1 + \frac{x^2}{2!} + \frac{x^4}{4!} + \cdots \qquad \text{for all} \quad x \in \mathbb{R}. \tag{3.13}$$

Hence, its reciprocal is analytic at the origin. Write

$$\frac{1}{\cosh x} = \sum_{n=0}^{\infty} E_n \frac{x^n}{n!} = E_0 + E_1 x + E_2 \frac{x^2}{2} + \cdots. \tag{3.14}$$

The numbers $E_0, E_1, E_2, \ldots$ are called the *Euler* numbers. We see readily that the function defined by the expression on the left in (3.14) is even so that

$$E_{2k-1} = 0 \qquad \text{for each positive integer } k. \tag{3.15}$$

Hence, we may write

$$\frac{1}{\cosh x} = \sum_{n=0}^{\infty} \frac{E_{2n}}{(2n)!} x^{2n}. \tag{3.16}$$

PROB. 3.6. Prove: If $\langle E_n \rangle$ is the sequence of Euler numbers, then

$$E_0 = 1 \quad \text{and} \quad \sum_{k=0}^{n} \binom{2n}{2k} E_{2n-2k} = 0 \qquad \text{for} \quad n \geqslant 1 \tag{3.17}$$

and show

$$E_2 = -1, \quad E_4 = 5, \quad E_6 = -61, \quad E_8 = 1385, \quad E_{10} = -50{,}521.$$

PROB. 3.7. Prove:

$$\frac{1}{\cos x} = \sum_{n=0}^{\infty} (-1)^n \frac{E_{2n}}{(2n)!} x^{2n}.$$

4. Infinite Series of Analytic Functions

The series

$$S(x) = \frac{1}{1-x} + \frac{1}{(1-x)^2} + \frac{1}{(1-x)^3} + \cdots \tag{4.1}$$

can be written

$$S(x) = \frac{1}{1-x}\left(1 + \frac{1}{1-x} + \frac{1}{(1-x)^2} + \cdots\right).$$

The second factor is a geometric series in $u = 1/(1-x)$. Hence, if

$$\frac{1}{|1-x|} < 1,$$

that is, if $x < 0$ or $x > 2$, then (4.1) converges to

$$S(x) = \frac{1}{1-x}\,\frac{1}{1 - 1/(1-x)} = -\frac{1}{x}.$$

Write each term as an infinite series.

$$\frac{1}{1-x} = 1 + x + x^2 + \cdots,$$

$$\frac{1}{(1-x)^2} = 1 + 2x + 3x^2 + \cdots,$$

$$\frac{1}{(1-x)^3} = 1 + 3x + 6x^2 + \cdots,$$

$$\vdots$$

It is clear that we could not add columnwise and then write the sum as a power series in x; for then we would have

$$\begin{aligned} &1 + 1 + \cdots \\ &\quad + (1 + 2 + 3 + \cdots)x \\ &\quad + (1 + 3 + 6 + \cdots)x^2 \\ &\quad + \cdots, \end{aligned}$$

which is impossible. To add columnwise and then sum again we need a condition guaranteeing that the columnwise addition should yield convergent columns.

Theorem 4.1. *Given a sequence $\langle f_i \rangle$ of power series such that for each nonnegative integer i*

$$f_i(x) = \sum_{n=0}^{\infty} a_{in} x^n \tag{4.2}$$

converges for all x such that $|x| < R$, $R > 0$. Assume that for each i the series

$$u_i(r) = \sum_{n=0}^{\infty} |a_{in}| r^n \tag{4.3a}$$

and

$$\sum_{i=0}^{\infty} u_i(r) \tag{4.3b}$$

converges for all r with $0 < r < R$. *Then* (1) *for each n*

$$A_n = \sum_{i=0}^{\infty} a_{in} \tag{4.4}$$

converges absolutely; (2) *the series*

$$\sum_{n=0}^{\infty} A_n x^n \tag{4.5a}$$

and

$$\sum_{i=0}^{\infty} f_i(x) \tag{4.5b}$$

converge absolutely and uniformly for $|x| \leqslant r$, *and*

$$\sum_{n=0}^{\infty} A_n x^n = \sum_{i=0}^{\infty} f_i(x), \tag{4.6}$$

i.e.,

$$\sum_{n=0}^{\infty}\left(\sum_{i=0}^{\infty} a_{in}\right)x^n = \sum_{i=0}^{\infty}\left(\sum_{n=0}^{\infty} a_{in}x^n\right). \tag{4.7}$$

Proof. The series in (4.3b) consists of nonnegative terms and converges. Hence, its sequence $\langle S_n(r)\rangle$ of partial sums is bounded. For each m, we have

$$\begin{aligned} S_m(r) = \sum_{k=0}^{m} u_k(r) &= \sum_{k=0}^{m}\left(\sum_{n=0}^{\infty}|a_{kn}|r^n\right) \\ &= \sum_{n=0}^{\infty}\left(\sum_{k=0}^{m}|a_{kn}|\right)r^n \end{aligned} \tag{4.8}$$

since $S_m(r)$ is a finite sum of nonnegative terms. An $M > 0$ exists such that $S_m(r) \leqslant M$ for all m. It follows that for each m and n,

$$r^n \sum_{k=1}^{m} |a_{kn}| \leqslant M$$

or

$$\sum_{k=1}^{m} |a_{kn}| \leqslant \frac{M}{r^n}.$$

This implies that $\sum_{i=0}^{\infty}|a_{in}|$ converges and, hence, that for each n the series (4.4) converges absolutely. Put

$$B_n = \sum_{i=0}^{\infty} |a_{in}|, \tag{4.9}$$

and consider the series

$$\sum_{n=0}^{\infty} B_n r^n.$$

We prove that this is a Cauchy sequence. We know that the series (4.3b) converges. Hence, its sequence $\langle S_n(r)\rangle$ of partial sums is a Cauchy sequence.

Assume that $\epsilon > 0$ is given. By what was just mentioned, there exists an N such that if $m > j > N$, then

$$\sum_{i=j+1}^{m} u_i(r) < \frac{\epsilon}{4}.$$

This implies that

$$\sum_{n=0}^{\infty}\left(\sum_{i=j+1}^{m}|a_{in}|\right)r^n = \sum_{i=j+1}^{m}\left(\sum_{n=0}^{\infty}|a_{in}|r^n\right)$$
$$= \sum_{i=j+1}^{m} u_i(r) < \frac{\epsilon}{4} \qquad \text{for} \quad m > j > N.$$

We conclude from this that if $\nu > k$, then

$$\sum_{n=k+1}^{\nu}\left(\sum_{i=j+1}^{m}|a_{in}|\right)r^n \leqslant \sum_{n=0}^{\infty}\left(\sum_{i=j+1}^{m}|a_{in}|\right)r^n < \frac{\epsilon}{4}.$$

We fix ν and k and let $m \to +\infty$ and obtain from this

$$\sum_{n=k+1}^{\nu}\left(\sum_{i=j+1}^{\infty}|a_{in}|\right)r^n \leqslant \frac{\epsilon}{4} < \frac{\epsilon}{2} \qquad \text{for} \quad \nu > k \quad \text{and} \quad j > N. \tag{4.10}$$

Now consider

$$\sum_{i=1}^{j} u_i(r) = \sum_{i=1}^{j}\left(\sum_{n=0}^{\infty}|a_{in}|r^n\right) = \sum_{n=0}^{\infty}\left(\sum_{n=0}^{j}|a_{in}|\right)r^n.$$

The series on the right converges. Hence, it is a Cauchy sequence. Therefore, there exists an N_1 such that if $\nu > k > N_1$, then

$$\sum_{n=k+1}^{\nu}\left(\sum_{i=1}^{j}|a_{in}|\right)r^n < \frac{\epsilon}{2}. \tag{4.11}$$

Adding this to (4.10) we find that if $\nu > k > N_1$, then

$$\sum_{n=k+1}^{\nu}\left(\sum_{i=0}^{\infty}|a_{in}|\right)r^n < \epsilon.$$

This states that

$$\sum_{n=k+1}^{\nu} B_n r^n < \epsilon \qquad \text{if} \quad \nu > k > N_1.$$

We conclude that $\sum B_n r^n$ is a convergent series. Since $|A_n| \leqslant B_n$ for each n, it now follows that if $|x| \leqslant r$, then $\sum_{m=0}^{\infty} A_n x^n$ converges absolutely. By the Weierstrass M-test for uniform convergence it follows that the convergence is also uniform on $[-r; r]$.

We now prove that (4.6) holds for $|x| \leqslant r$. Given $\epsilon > 0$ we use the known convergence of $\sum_{i=0}^{\infty} u_i(r)$ to obtain an N such that if $n > N$, then

$$\sum_{i=n+1}^{\infty} u_i(r) = \left| \sum_{i=n+1}^{\infty} u_i(r) \right| < \frac{\epsilon}{2}$$

or that

$$\sum_{i=n+1}^{\infty} \left(\sum_{\nu=0}^{\infty} |a_{i\nu}| r^{\nu} \right) < \frac{\epsilon}{2} \qquad \text{for} \quad n > N.$$

It follows from this that for each m,

$$\sum_{i=n+1}^{\infty} \left(\sum_{\nu=0}^{m} |a_{i\nu}| r^{\nu} \right) \leqslant \sum_{i=n+1}^{\infty} \left(\sum_{\nu=0}^{\infty} |a_{i\nu}| r^{\nu} \right) < \frac{\epsilon}{2} \qquad \text{for} \quad n < N. \tag{4.12}$$

But

$$\sum_{i=n+1}^{\infty} \left(\sum_{\nu=0}^{m} |a_{i\nu}| r^{\nu} \right) = \sum_{\nu=0}^{m} \left(\sum_{i=n+1}^{\infty} |a_{i\nu}| \right) r^{\nu}.$$

This and (4.12) imply that for each m,

$$\sum_{\nu=0}^{m} \left(\sum_{i=n+1}^{\infty} |a_{i\nu}| \right) r^{\nu} < \frac{\epsilon}{2} \qquad \text{for} \quad n > N.$$

Now let $m \to +\infty$. We conclude that

$$\sum_{\nu=0}^{\infty} \left(\sum_{i=n+1}^{\infty} |a_{i\nu}| \right) r^{\nu} \leqslant \frac{\epsilon}{2} < \epsilon \qquad \text{for} \quad n > N. \tag{4.13}$$

We conclude from this that for $n > N$ and $|x| \leqslant r$

$$\begin{aligned}
\left| \sum_{\nu=0}^{\infty} A_{\nu} x^{\nu} - \sum_{i=0}^{n} f_i(x) \right| &= \left| \sum_{\nu=0}^{\infty} A_{\nu} x^{\nu} - \sum_{i=0}^{n} \left(\sum_{\nu=0}^{\infty} a_{i\nu} x^{\nu} \right) \right| \\
&= \left| \sum_{\nu=0}^{\infty} A_{\nu} x^{\nu} - \sum_{r=0}^{\infty} \left(\sum_{i=0}^{n} a_{i\nu} \right) x^{\nu} \right| \\
&= \left| \sum_{\nu=0}^{\infty} \left(A_{\nu} - \sum_{i=0}^{n} a_{i\nu} \right) x^{\nu} \right| \leqslant \sum_{\nu=0}^{\infty} \left| A_{\nu} - \sum_{i=0}^{n} a_{i\nu} \right| |x|^{\nu} \\
&= \sum_{\nu=0}^{\infty} \left| \sum_{i=n+1}^{\infty} a_{i\nu} \right| |x|^{\nu} \leqslant \sum_{\nu=0}^{\infty} \left(\sum_{i=n+1}^{\infty} |a_{i\nu}| \right) r^{\nu} < \epsilon.
\end{aligned}$$

It now follows that

$$\sum_{\nu=0}^{\infty} A_{\nu} x^{\nu} = \lim_{n \to +\infty} \sum_{i=0}^{n} f_i(x) = \sum_{i=0}^{\infty} f_i(x),$$

the convergence being uniform on $[-r; r]$. This completes the proof.

Example 4.1. We knew that certain series converged but we could not sum them. Now we can use the above theorem to obtain such sums. By formula XI.8.45,

$$\pi \cot \pi x = \frac{1}{x} + \sum_{n=1}^{\infty} \frac{2x}{x^2 - n^2} \qquad \text{if } n \text{ is not an integer.} \tag{4.14}$$

We write this as

$$x\pi \cot \pi x = 1 - 2 \sum_{n=1}^{\infty} \frac{x^2}{n^2 - x^2}. \tag{4.15}$$

For each $n \geqslant 1$ we have

$$\frac{x^2}{n^2 - x^2} = \frac{x^2}{n^2}\left(\frac{1}{1 - x^2/n^2}\right) = \frac{x^2}{n^2}\left(1 + \frac{x^2}{n^2} + \frac{x^4}{n^4} + \cdots\right)$$

$$= \sum_{k=1}^{\infty} \frac{x^{2k}}{n^{2k}} \qquad \text{for} \quad |x| < n. \tag{4.16}$$

Hence, if $|x| < 1$, we have

$$\sum_{n=1}^{\infty} \frac{x^2}{n^2 - x^2} = \sum_{n=1}^{\infty}\left(\sum_{k=1}^{\infty} \frac{x^{2k}}{n^{2k}}\right). \tag{4.17}$$

We apply Theorem 4.1 with

$$f_n(x) = \frac{x^2}{n^2 - x^2} = \sum_{k=1}^{\infty} \frac{x^{2k}}{n^{2k}} \qquad |x| < 1,$$

for $n \in \{1, 2, 3, \ldots\}$. These power series have terms which are all positive and converge absolutely and their sum converges; this is seen from (4.17) and (4.15). Using (4.7) in Theorem 4.1, we have

$$\sum_{n=1}^{\infty} \frac{x^2}{n^2 - x^2} = \sum_{n=1}^{\infty}\left(\sum_{k=1}^{\infty} \frac{x^{2k}}{n^{2k}}\right) = \sum_{k=1}^{\infty}\left(\sum_{n=1}^{\infty} \frac{1}{n^{2k}}\right)x^{2k}.$$

Substituting for this into (4.15), we obtain

$$x\pi \cot \pi x = 1 - 2\sum_{k=1}^{\infty}\left(\sum_{n=1}^{\infty} \frac{1}{n^{2k}}\right)x^{2k} \qquad \text{for} \quad |x| < 1.$$

Replacing $x\pi$ here by x, we have

$$x \cot x = 1 - 2\sum_{k=1}^{\infty}\left(\sum_{n=1}^{\infty} \frac{1}{n^{2k}}\right)\frac{x^{2k}}{\pi^{2k}} \qquad \text{for} \quad |x| < \pi. \tag{4.18}$$

Turning to Theorem 2.4, we recall that

$$x \cot x = \sum_{k=0}^{\infty} (-1)^k \frac{2^{2k} B_{2k}}{(2k)!} x^{2k}$$

$$= 1 + \sum_{k=1}^{\infty} (-1)^k \frac{2^{2k} B_{2k}}{(2k)!} x^{2k} \qquad \text{for} \quad |x| < \frac{r}{2}. \tag{4.19}$$

Take $|x| < \min\{\pi, r/2\}$. Compare (4.18) and (4.19) and obtain

$$\sum_{k=1}^{\infty} (-1)^k \frac{2^{2k} B_{2k}}{(2k)!} x^{2k} = -2\sum_{k=1}^{\infty}\left(\sum_{n=1}^{\infty} \frac{1}{n^{2k}}\right)\frac{x^{2k}}{\pi^{2k}} \qquad \text{for} \quad |x| < \min\left\{\pi, \frac{r}{2}\right\}.$$

Equating coefficients yields

$$(-1)^k \frac{2^{2k}B_{2k}}{(2k)!} = -\frac{2}{\pi^{2k}} \sum_{n=1}^{\infty} \frac{1}{n^{2k}} \qquad \text{for} \quad k \geqslant 1.$$

This implies that

$$\zeta(2k) = \sum_{n=1}^{\infty} \frac{1}{n^{2k}} = (-1)^{k+1} \frac{(2\pi)^{2k}B_{2k}}{2(2k)!} \qquad \text{for} \quad k \geqslant 1. \tag{4.20}$$

Thus, for $k = 1$, we have

$$\zeta(2) = \sum_{n=1}^{\infty} \frac{1}{n^2} = \frac{(2\pi)^2 B_2}{2(2!)} = \frac{\pi^2}{6}. \tag{4.21}$$

Prob. 4.1. Prove:

(1) $\zeta(4) = \sum_{n=1}^{\infty} 1/n^4 = \pi^4/90$,
(2) $\zeta(6) = \sum_{n=1}^{\infty} 1/n^6 = \pi^6/945$.

Prob. 4.2(a). Use the equality

$$\tan x = \sum_{n=1}^{\infty} \frac{2x}{(n-\frac{1}{2})^2\pi^2 - x^2},$$

where $x \notin \{\pm\pi/2, \pm 3\pi/2, \dots\}$ (see Prob. XI.8.6), and the result in Prob. 2.6, where we learned

$$\tan x = \sum_{n=1}^{\infty} (-1)^{n+1} \frac{2^{2n}(2^{2n}-1)B_{2n}}{(2n)!} x^{2n-1} \qquad \text{for} \quad |x| < \frac{r}{4},$$

to obtain: If k is a positive integer, then

$$\sum_{n=1}^{\infty} \frac{1}{(2n-1)^{2k}} = 1 + \frac{1}{3^{2k}} + \frac{1}{5^{2k}} + \cdots = (-1)^{k+1} \frac{2^{2k}-1}{2(2k)!} B_{2k}\pi^{2k}.$$

(b) Obtain

$$1 + \frac{1}{3^2} + \frac{1}{5^2} + \cdots = \frac{\pi^2}{8} \quad \text{and} \quad 1 + \frac{1}{3^4} + \frac{1}{5^4} + \cdots = \frac{\pi^4}{96}.$$

Prob. 4.3(a). Prove: If k is a positive integer, then

$$\sum_{n=1}^{\infty} \frac{(-1)^{k+1}}{n^{2k}} = 1 - \frac{1}{2^{2k}} + \frac{1}{3^{2k}} - \frac{1}{4^{2k}} + \cdots$$

$$= (-1)^{k+1} \frac{2^{2k-1}-1}{(2k)!} \pi^{2k} B_{2k}.$$

(b) Obtain:

$$1-\frac{1}{2^2}+\frac{1}{3^2}-\frac{1}{4^2}+\cdots=\frac{\pi^2}{12},$$

$$1-\frac{1}{2^4}+\frac{1}{3^4}-\frac{1}{4^4}+\cdots=\frac{7}{720}\pi^4.$$

Remark 4.1. This remark deals with the radius of convergence of some of the series involving Bernoulli numbers. Again we recall formula (2.32):

$$x\cot x=\sum_{n=0}^{\infty}(-1)^n\frac{2^{2n}B_{2n}}{(2n)!}x^{2n}\qquad\text{for}\quad |x|<\frac{r}{2},$$

where r is defined in (2.4b). The coefficient a_n of x^n in this series is

$$a_n=\begin{cases}0 & \text{if}\quad n\text{ is odd}\\ a_{2k}=(-1)^k\dfrac{B_{2k}2^{2k}}{(2k)!} & \text{if}\quad n=2k,\quad k\in\{0,1,2,\dots\}.\end{cases}\tag{4.22}$$

By Theorem 2.3, we know that $(-1)^{k+1}B_{2k}>0$, so

$$|a_{2k}|=(-1)^{k+1}\frac{B_{2k}2^{2k}}{(2k)!}\qquad\text{for}\quad k\in\{0,1,2,\dots\}.$$

By (4.20), we have

$$\frac{2}{\pi^{2k}}\sum_{n=1}^{\infty}\frac{1}{n^{2k}}=(-1)^{k+1}\frac{B_{2k}2^{2k}}{(2k)!}=|a_{2k}|,\tag{4.23}$$

where $k\in\{1,2,3,\dots\}$. We know that $2<\pi<4$ and

$$\frac{1}{2}=\sin\frac{\pi}{6}<\frac{\pi}{6}$$

so that $3<\pi<4$ and $9<\pi^2<16$. This implies that

$$\frac{3}{2}<\frac{\pi^2}{6}<\frac{8}{3}.$$

Hence,

$$1<\sum_{n=1}^{\infty}\frac{1}{n^{2k}}\leqslant\sum_{n=1}^{\infty}\frac{1}{n^2}=\frac{\pi^2}{6}<\frac{8}{3}\qquad\text{for}\quad k\in\{1,2,3,\dots\}.$$

Thus,

$$1<\zeta(2k)<\frac{8}{3}\quad\text{and}\quad 1<\sqrt[2k]{\zeta(2k)}<\left(\frac{8}{3}\right)^{1/2k}.$$

Since $\lim_{k\to+\infty}(8/3)^{1/2k}=1$, this implies that

$$\lim_{k\to+\infty}\sqrt[2k]{\zeta(2k)}=1.\tag{4.24}$$

It follows from this by (4.23) that

$$\varlimsup_{n\to+\infty}\sqrt[n]{|a_n|}=\lim_{k\to+\infty}\sqrt[2k]{|a_{2k}|}=\lim_{k\to+\infty}\left[\sqrt[2k]{\frac{2}{\pi^{2k}}}\ \sqrt[2k]{\zeta(2k)}\right]=\frac{1}{\pi}\,. \tag{4.25}$$

We see from this that the radius of convergence for the series in (2.32) is π.
Since (4.23) can be written

$$\frac{2}{\pi^{2k}}\zeta(2k)=(-1)^{k+1}\frac{B_{2k}2^{2k}}{(2k)!}\,,$$

we have

$$\lim_{k\to+\infty}\left[2\sqrt[2k]{(-1)^{k+1}\frac{B_{2k}}{(2k)!}}\right]=\lim_{k\to+\infty}\sqrt[2k]{(-1)^{k+1}\frac{B_{2k}2^{2k}}{(2k)!}}$$

$$=\lim_{k\to+\infty}\sqrt[2k]{\frac{2}{\pi^{2k}}\zeta(2k)}=\frac{1}{\pi}$$

and, therefore, that

$$\lim_{k\to+\infty}\sqrt[2k]{(-1)^{k+1}\frac{B_{2k}}{(2k)!}}=\frac{1}{2\pi}\,. \tag{4.26}$$

From this it follows that the radius of convergence of the series in (2.16) is 2π.

PROB. 4.4. Prove that the radius of convergence of the series in Prob. 2.4 is $\pi/2$.

PROB. 4.5. Prove: $2(2k)!\,(2\pi)^{-2k}<|B_{2k}|$ for $k\geqslant 1$ and that $\lim_{k\to+\infty}|B_{2k}| = +\infty$

5. Abel's Summation Formula and Some of Its Consequences

Let $\langle x_n\rangle=\langle x_0,x_1,x_2,\ldots\rangle$ be a sequence of complex numbers. Define Δx_n as

$$\Delta x_n=x_{n+1}-x_n \qquad \text{for each nonnegative integer } n. \tag{5.1}$$

We call Δx_n the nth *successive difference* of the sequence $\langle x_n\rangle$. The operation Δ assigns to the sequence $\langle x_n\rangle$ the sequence $\langle\Delta x_n\rangle=\langle\Delta x_0,\Delta x_1,$

$\Delta x_2, \ldots \rangle = \langle x_1 - x_0, x_2 - x_1, x_3 - x_2, \ldots \rangle$. The operation Δ on sequences is analogous to the differentiation operation on functions. To see this let us apply Δ to the product sequence $\langle x_n y_n \rangle$. We observe that for each nonnegative integer k,

$$x_{k+1}y_{k+1} - x_k y_k = x_{k+1}(y_{k+1} - y_k) + y_k(x_{k+1} - x_k). \tag{5.2}$$

In terms of the Δ operation this can be expressed as

$$\Delta(x_k y_k) = x_{k+1}\Delta y_k + y_k \Delta x_k, \tag{5.3}$$

reminding us of

$$\frac{dfg}{dx} = f\frac{dg}{dx} + g\frac{df}{dx}. \tag{5.4}$$

It is also clear that

$$\Delta(x_k y_k) = y_{k+1}\Delta x_k + x_k \Delta y_k. \tag{5.5}$$

Next we observe that if $m \geqslant n+1$, then

$$\begin{aligned} \sum_{k=n+1}^{m} \Delta x_k &= \sum_{k=n+1}^{m} (x_{k+1} - x_k) \\ &= (x_{n+2} - x_{n+1}) + (x_{n+3} - x_{n+2}) + \cdots + (x_{m+1} - x_m) \end{aligned}$$

so that

$$\sum_{k=n+1}^{m} \Delta x_k = x_{m+1} - x_{n+1}. \tag{5.6}$$

If we define

$$\sum_{k=n+1}^{n} x_k = 0, \tag{5.7}$$

formula (5.6) retains its validity for $m = n$. Now apply $\sum$ to both sides of (5.3) to obtain

$$x_{m+1}y_{m+1} - x_{n+1}y_{n+1} = \sum_{k=n+1}^{m} \Delta(x_k y_k) = \sum_{k=n+1}^{m} (x_{k+1}\Delta y_k + y_k \Delta x_k)$$

or

$$x_{m+1}y_{m+1} - x_{n+1}y_{n+1} = \sum_{k=n+1}^{m} (x_{k+1}\Delta y_k) + \sum_{k=n+1}^{m} (y_k \Delta x_k). \tag{5.8}$$

This implies that

$$\sum_{k=n+1}^{m} (x_{k+1}\Delta y_k) = x_{m+1}y_{m+1} - x_{n+1}y_{n+1} - \sum_{k=n+1}^{m} (y_k \Delta x_k). \tag{5.9}$$

We now consider sequences $\langle a_n \rangle$ and $\langle b_n \rangle$ of complex numbers. With $\sum a_n$ we associate its partial sum sequence $\langle S_n \rangle$. We have

$$S_n = \sum_{k=1}^{n} a_k \qquad \text{for each nonnegative integer } n. \tag{5.10}$$

Here we define

$$S_0 = \sum_{k=1}^{0} a_k = 0. \tag{5.11}$$

Note that if $k \geqslant 1$, then

$$\Delta S_{k-1} = S_k - S_{k-1} = a_k. \tag{5.12}$$

In view of (5.11), we have

$$\Delta S_0 = S_1 - S_0 = S_1 = a_1. \tag{5.13}$$

From (5.12) we obtain by reindexing

$$\sum_{k=n+2}^{m+1} a_k b_k = \sum_{k=n+2}^{m+1} b_k a_k = \sum_{k=n+2}^{m+1} b_k \Delta S_{k-1} = \sum_{k=n+1}^{m} b_{k+1} \Delta S_k. \tag{5.14}$$

This implies that

$$a_{m+1}b_{m+1} - a_{n+1}b_{n+1} + \sum_{k=n+1}^{m} a_k b_k = \sum_{k=n+2}^{m+1} a_k b_k = \sum_{k=n+1}^{m} b_{k+1} \Delta S_k. \tag{5.15}$$

Apply (5.9) to the right-hand side and obtain

$$\sum_{k=n+1}^{m} b_{k+1} \Delta S_k = b_{m+1} S_{m+1} - b_{n+1} S_{n+1} - \sum_{k=n+1}^{m} S_k \Delta b_k.$$

This and (5.15) imply that

$$a_{m+1}b_{m+1} - a_{n+1}b_{n+1} + \sum_{k=n+1}^{m} a_k b_k = b_{m+1} S_{m+1} - b_{n+1} S_{n+1} - \sum_{k=n+1}^{m} S_k \Delta b_k.$$

Solving for $\sum a_k b_k$, we have

$$\sum_{k=n+1}^{m} a_k b_k = b_{m+1}(S_{m+1} - a_{m+1}) - b_{n+1}(S_{n+1} - a_{n+1}) - \sum_{k=n+1}^{m} S_k \Delta b_k$$

and finally that

$$\sum_{k=n+1}^{m} a_k b_k = b_{m+1} S_m - b_{n+1} S_n - \sum_{k=n+1}^{m} S_k \Delta b_k. \tag{5.16}$$

This formula is called *Abel's partial sum formula*. In particular, for $n = 0$, (5.16) becomes

$$\sum_{k=1}^{m} a_k b_k = b_{m+1} S_m - \sum_{k=1}^{m} S_k \Delta b_k \qquad \text{for each nonnegative integer } m. \tag{5.17}$$

We apply Abel's formulas (5.16) and (5.17) to obtain criteria for the convergence of certain series. Throughout this section we adhere to the notation just adopted.

Theorem 5.1. *$\sum a_n b_n$ converges if the sequence $\langle b_{n+1}S_n\rangle$ and the series*

$$\sum_{k=1}^{\infty} S_k(b_{k+1} - b_k)$$

converge.

Proof. The theorem is a consequence of (5.17) when we replace the m there by n and let $n \to +\infty$.

Def. 5.1. If $\langle b_n\rangle$ is a sequence of real numbers, we say it is of *bounded variation* if the series $\sum(b_{n+1} - b_n)$ converges absolutely.

Remark 5.1. Any real sequence which is monotonic and bounded is necessarily of bounded variation for

$$|b_{k+1} - b_k| = \begin{cases} b_{k+1} - b_k & \text{if } \langle b_n\rangle \text{ is increasing} \\ b_k - b_{k+1} & \text{if } \langle b_n\rangle \text{ is decreasing} \end{cases} \tag{5.18}$$

for each positive integer k. Hence,

$$\sum_{k=1}^{n} |b_{k+1} - b_k| = \begin{cases} b_{n+1} - b_1 & \text{if } \langle b_n\rangle \text{ is increasing} \\ b_1 - b_{n+1} & \text{if } \langle b_n\rangle \text{ is decreasing.} \end{cases} \tag{5.19}$$

In either case, if $\langle b_n\rangle$ is monotonic,

$$\sum_{k=1}^{n} |b_{k+1} - b_k| = |b_{n+1} - b_1| \qquad \text{for each positive integer } n. \tag{5.20}$$

If, besides being monotonic, $\langle b_n\rangle$ is also bounded, then there exists an $M > 0$ such that $|b_n| \leqslant M$ for all n. Therefore, by (5.20),

$$\sum_{k=1}^{n} |b_{k+1} - b_k| = |b_{n+1} - b_1| \leqslant |b_n| + |b_1| \leqslant 2M$$

for each positive integer n. It follows that $\sum|b_{n+1} - b_n|$ converges (explain) and, therefore, that $\langle b_n\rangle$ is of bounded variation. We proved:

Theorem 5.2. *If $\langle b_n\rangle$ is a bounded and monotonic sequence, then it is of bounded variation.*

Theorem 5.3 (DuBois–Reymond's Test). *$\sum a_n b_n$ converges if* (1) *$\sum a_n$ converges and* (2) *$\langle b_n\rangle$ is of bounded variation.*

Proof. This theorem is actually a corollary of Theorem 5.1. Since $\sum a_n$ converges by hypothesis, its partial sums are bounded. Hence, an $M > 0$ exists such that $|S_n| \leqslant M$. (Here, of course, $S_n = \sum_{k=1}^{n} a_k$ for each n.) Since $\langle b_n\rangle$ is of bounded variation, the series $\sum(b_{n+1} - b_n)$ converges absolutely. Let $\epsilon > 0$ be given. There exists an N such that if $m > n > N$, then

$$\sum_{k=m+1}^{m} |b_k - b_{k+1}| < \frac{\epsilon}{M}.$$

It follows that if $m > n > N$, then

$$\left|\sum_{k=n+1}^{m} S_k(b_k - b_{k+1})\right| \leqslant M \sum_{k=1}^{n} |b_k - b_{k+1}| < M \frac{\epsilon}{M} = \epsilon.$$

By the Cauchy convergence criterion for series, we conclude that the series

$$\sum_{n=1}^{\infty} S_n(b_{n+1} - b_n) = -\sum_{n=1}^{\infty} S_n(b_n - b_{n+1}) \quad \text{converges.} \tag{5.21}$$

Now

$$\sum_{k=1}^{n} (b_k - b_{k+1}) = b_1 - b_{n+1}$$

and the fact that $\sum(b_k - b_{k+1})$ converges imply that the sequence $\langle b_{n+1}\rangle$ converges. This and the convergence of $\langle S_n\rangle$ imply that the sequence $\langle b_{n+1}S_n\rangle$ converges. Using Theorem 5.1, we conclude from this and (5.21) that $\sum a_n b_n$ converges.

Corollary (Abel's Test). *If $\sum a_n$ converges and $\langle b_n\rangle$ is monotonic and bounded, then $\sum a_n b_n$ converges.*

Proof. Exercise.

Prob. 5.1. A series of the form

$$\sum_{n=1}^{\infty} \frac{a_n}{n^x} \tag{5.22}$$

is called a *Dirichlet* series. Note that if $x_0 \in \mathbb{R}$ and $x > x_0$, then the sequence $\langle n^{-(x-x_0)}\rangle$ decreases monotonically to 0. Prove: If for some $x_0 \in \mathbb{R}$ the series

$$\sum_{n=1}^{\infty} \frac{a_n}{n^{x_0}}$$

converges, then the series in (5.22) converges for each $x > x_0$.

Prob. 5.2. Prove: If $\sum a_n$ converges, then so do the series

(1) $\sum_{n=1}^{\infty} a_n/n$,
(2) $\sum_{n=2}^{\infty} a_n/\ln n$,
(3) $\sum_{n=1}^{\infty} (1 + 1/n)^n a_n$.

Prob. 5.3. Prove: If the Dirichlet series (5.22) diverges for some $x = x_1$, then it diverges for all x such that $x < x_1$.

Theorem 5.4 (Dedekind's Test). *$\sum a_n b_n$ converges if the sequence $\langle S_n\rangle$ of partial sums of $\sum a_n$ is bounded, $\langle b_n\rangle$ is of bounded variation and* $\lim b_n = 0$.

PROOF. As in the proof of Theorem 5.3, one can prove that the hypothesis implies that the series

$$\sum_{n=1}^{\infty} S_n(b_{n+1} - b_n)$$

converges. The boundedness of $\langle S_n \rangle$ and $\lim b_n = 0$ imply that

$$\lim_{n\to+\infty} S_n b_{n+1} = 0.$$

Theorem 5.1 can now be used to complete the proof.

Corollary (Dirichlet's Test). *$\sum a_n b_n$ converges if the sequence of partial sums of $\sum a_n$ is bounded and $\langle b_n \rangle$ converges monotonically to* 0.

PROOF. Exercise.

EXAMPLE 5.1. We apply Dedekind's test to the series

$$\sum_{m=1}^{\infty} b_n \sin nx, \tag{5.22'}$$

where $\langle b_n \rangle$ is of bounded variation and $\lim b_n = 0$ to prove that it converges for all x. If x is an integral multiple of π then there is nothing to prove, for then all the terms of the series vanish and the series converges trivially to 0. Hence, if we fix some x which is not an integral multiple of π, then, by Prob. X.6.1, we have

$$\sum_{k=1}^{n} \sin kx = \frac{\sin(nx/2)\sin((n+1)x/2)}{\sin(x/2)}. \tag{5.23}$$

It follows from this that

$$\left| \sum_{k=1}^{n} \sin kx \right| \leqslant \frac{1}{\sin(x/2)}.$$

Hence, the partial sums of the series

$$\sum_{k=1}^{\infty} \sin kx$$

are bounded. Since $\langle b_n \rangle$ is of bounded variation and $\lim b_n = 0$, it follows from Theorem 5.4 that the series (5.22′) also converges for values of x that are not integral multiples of π. Thus, the series (5.22′) converges for all $x \in \mathbb{R}$.

PROB. 5.4. Prove: If $\langle b_n \rangle$ is of bounded variation and $\lim b_n = 0$, then the series

$$\sum_{n=0}^{\infty} b_n \cos nx$$

converges for each x which is not an integral multiple of 2π.

6. More Tests for Uniform Convergence

Weierstrass's M-test for uniform convergence, though quite practical and easy to use, is limited in scope. It requires, among other things, that the series it is testing is absolutely convergent. The tests of the last section can be easily modified to yield more delicate tests for uniform convergence.

Throughout this section we write the nth partial sum corresponding to x of the series of functions

$$S(x) = \sum_{n=1}^{\infty} a_n(x)$$

as $S_n(x)$. By S_n we shall mean the nth partial sum function of this series.

Theorem 6.1. *$\sum a_n(x)b_n(x)$ converges uniformly on $D \subseteq \mathbb{R}$ provided that the series*

$$\sum_{n=1}^{\infty} S_n(x)(b_{n+1}(x) - b_n(x)) = \sum_{k=1}^{\infty} S_k(x)\Delta b_n(x) \tag{6.1}$$

converges uniformly on D, and the sequence of functions $\langle b_{n+1}S_n \rangle$ converges uniformly on D.

PROOF. Applying (5.17), we have for each $x \in D$,

$$\sum_{k=1}^{n} a_k(x)b_k(x) = b_{n+1}(x)S_n(x) - \sum_{k=1}^{n} S_k(x)\Delta b_k(x). \tag{6.2}$$

The right side of (6.2) converges uniformly on D (why?). Hence, so does its left side.

Theorem 6.2 (Dubois–Reymond's Test for Uniform Convergence). *The series $\sum a_n(x)b_n(x)$ converges uniformly on the set $D \subseteq \mathbb{R}$ if $\sum a_n(x)$ and $\sum |b_{n+1}(x) - b_n(x)|$ converge uniformly and the sequence of functions $\langle b_n \rangle$ is uniformly bounded on D.*

PROOF. Let

$$S(x) = \sum_{k=1}^{\infty} a_k(x) \qquad \text{for} \quad x \in D.$$

If $m > n$, we have for $x \in D$.

$$S(x)(b_{m+1}(x) - b_{n+1}(x)) = \sum_{k=n+1}^{m} S(x)(b_{k+1}(x) - b_k(x))$$

$$= \sum_{k=n+1}^{m} S(x)\Delta b_k(x). \tag{6.2'}$$

By Abel's summation formula (5.16), we have

$$\sum_{k=n+1}^{m} a_k(x)b_k(x) = b_{m+1}(x)S_m(x) - b_{n+1}(x)S_n(x) - \sum_{k=n+1}^{m} S_k(x)\Delta b_k(x).$$

Because of (6.2) this can be written

$$\sum_{k=n+1}^{m} a_k(x)b_k(x) = b_{m+1}(x)(S_m(x) - S(x)) - b_{n+1}(x)(S_n(x) - S(x))$$
$$- \sum_{k=n+1}^{m} (S_k(x) - S(x))\Delta b_k(x). \tag{6.3}$$

By the hypothesis, there exists an $M > 0$ such that $|b_j(x)| \leqslant M$ holds for all j and $x \in D$. Because of (6.3), this implies that if $m > n$ and $x \in D$, then

$$\left|\sum_{k=n+1}^{m} a_k(x)b_k(x)\right| \leqslant M(|S_m(x) - S(x)| + |S_n(x) - S(x)|)$$
$$+ \sum_{k=n+1}^{n} |S_k(x) - S(x)||\Delta b_k(x)|. \tag{6.4}$$

Since $\sum a_n(x)$ converges uniformly on D, there exists an N_1, not depending on x, such that if $k > N_1$, then

$$|S_k(x) - S(x)| < 1 \qquad \text{for all} \quad x \in D.$$

It follows from this and (6.4) that if $m > n > N_1$, then

$$\left|\sum_{k=n+1}^{m} a_k(x)b_k(x)\right| \leqslant M(|S_m(x) - S(x)| + |S_n(x) - S(x)|)$$
$$+ \sum_{k=n+1}^{m} |\Delta b_k(x)| \tag{6.5}$$

for all $x \in D$. Given $\epsilon > 0$, we have from the uniform convergence of $\sum a_n(x)$ and $\sum |b_{n+1}(x) - b_n(x)| = \sum |\Delta b_k(x)|$ that there exist N_2 and N_3, depending on ϵ only, such that for all $x \in D$

$$|S_m(x) - S(x)| < \frac{\epsilon}{2M}, \qquad |S_n(x) - S(x)| < \frac{\epsilon}{2M}$$

if $m > n > N_2$ and

$$\sum_{k=n+1}^{m} |\Delta b_k(x)| < \frac{\epsilon}{2}$$

if $m > n > N_3$. Take $m > n > N = \max\{N_1, N_2, N_3\}$. The above and (6.5) imply that for all $x \in D$

$$\left|\sum_{k=n+1}^{m} a_k(x)b_k(x)\right| < \epsilon \qquad \text{for} \quad n > N.$$

By Cauchy's criterion for uniform convergence this implies that $\sum a_n(x)$ $b_n(x)$ converges uniformly on D.

Theorem 6.3 (Modified Abel* Test for Uniform Convergence). *The series* $\sum a_n(x)b_n(x)$ *converges uniformly on a set* $D \subseteq \mathbb{R}$ *provided that* $\sum a_n(x)$ *converges uniformly on D and there exists an* $M > 0$ *such that*

$$\sum_{n=1}^{\infty} |b_{n+1}(x) - b_n(x)| \leqslant M \quad \textit{and} \quad |b_1(x)| \leqslant M \qquad \textit{for all} \quad x \in D.$$

PROOF. By hypothesis,

$$S(x) = \sum_{n=1}^{\infty} a_n(x) \qquad \text{converges uniformly on } D.$$

We now turn to (6.3) in the proof of Theorem 6.2. This also holds here for all $x \in D$. It follows from (6.3) that if $m > n$, then

$$\left|\sum_{k=n+1}^{m} a_k(x)b_k(x)\right| \leqslant |b_{n+1}(x)|\,|S_m(x) - S(x)| + |b_{n+1}(x)|\,|S_n(x) - S(x)| + \sum_{k=n+1}^{m} |S_k(x) - S(x)|\,|\Delta b_k(x)| \tag{6.6}$$

for all $x \in D$.

From the hypothesis on the sequence $\langle b_n \rangle$ and the fact that

$$b_{n+1}(x) = b_1(x) + \sum_{k=1}^{n} (b_{k+1}(x) - b_k(x))$$

holds for each $x \in D$, we obtain

$$|b_{n+1}(x)| \leqslant |b_1(x)| + \sum_{k=1}^{n} |b_{k+1}(x) - b_k(x)| \leqslant M + M = 2M.$$

This and the hypothesis on b_1 imply that

$$|b_n(x)| \leqslant 2M \qquad \text{for all} \quad x \in D \quad \text{and all } n.$$

In turn, this and (6.6) imply: If $m \geqslant n$, then

$$\left|\sum_{k=n+1}^{m} a_k(x)b_k(x)\right| \leqslant 2M(|S_m(x) - S(x)| + |S_n(x) - S(x)|) + \sum_{k=n+1}^{m} |S_k(x) - S(x)|\,|\Delta b_k(x)| \tag{6.7}$$

for all $x \in D$.

Let $\epsilon > 0$ be given. Since $\sum a_n(x)$ converges uniformly on D, there exists an N depending on ϵ only such that if $n > N$, then

$$|S_n(x) - S(x)| < \frac{\epsilon}{4M} \qquad \text{for all} \quad x \in D.$$

* W. Fulks (*Advanced Calculus*, 2nd Edition, John Wiley and Sons, New York, 1978 (Theorem (14.4b)) calls this the modified Dirichlet Test for uniform convergence.

It follows from this and (6.7) that if $m > n > N$, then

$$\left|\sum_{k=n+1}^{m} a_k(x)b_k(x)\right| \leqslant 2M\left(\frac{\epsilon}{4M}+\frac{\epsilon}{4M}\right)+\frac{\epsilon}{4M}\sum_{k=n+1}^{m}|b_{k+1}(x)-b_k(x)|$$

$$\leqslant \frac{\epsilon}{2}+\frac{\epsilon}{4M}(M)=\frac{3}{4}\epsilon<\epsilon$$

for all $x \in D$. This implies that $\sum a_n(x)b_n(x)$ converges uniformly on D.

Corollary 1 (Abel's Test for Uniform Convergence). *$\sum a_n(x)b_n(x)$ converges uniformly on D provided that $\sum a_n(x)$ converges uniformly on D, the sequence of functions $\langle b_n\rangle$ is uniformly bounded on D, and for each $x \in D$, the sequence $\langle b_n(x)\rangle$ is monotonic.*

PROOF. By the uniform boundedness condition on $\langle b_n\rangle$, there exists an $M > 0$ such that

$$|b_n(x)| \leqslant M \qquad \text{for all } n \text{ and all} \quad x \in D. \tag{6.8}$$

Since $\langle b_n(x)\rangle$ is monotonic for each $x \in D$, we have for each n

$$\sum_{k=1}^{n}|b_{k+1}(x)-b_k(x)| = |b_1(x)-b_n(x)| \qquad \text{for each} \quad x \in D.$$

Because of (6.8), this implies that

$$\sum_{k=1}^{n}|b_{k+1}(x)-b_k(x)| \leqslant 2M \qquad \text{for each } n \text{ and all} \quad x \in D.$$

The terms on the left here are all nonnegative, so the limit as $n \to +\infty$ exists and we have

$$\sum_{k=1}^{\infty}|b_{k+1}(x)-b_k(x)| \leqslant 2M = k \qquad \text{for all} \quad x \in D.$$

Also

$$|b_1(x)| \leqslant M < k \qquad \text{for all} \quad x \in D.$$

The conditions of the hypothesis of the theorem are now seen to be fulfilled and we conclude that $\sum a_n(x)b_n(x)$ converges uniformly on D.

An important and interesting consequence of this result is:

Corollary 2 (Abel's Limit Theorem). *If $\langle a_n\rangle$ is a sequence of real constants and $\sum a_n$ converges, then*

$$\lim_{x\to 1-}\sum_{n=0}^{\infty} a_n x^n = \sum_{n=0}^{\infty} a_n. \tag{6.9}$$

Similarly, if $\sum(-1)^n a_n$ converges, then

$$\lim_{x\to(-1)+}\sum_{n=0}^{\infty} a_n x^n = \sum_{n=0}^{\infty}(-1)^n a_n. \tag{6.10}$$

PROOF. We first write

$$\sum_{n=0}^{\infty} a_n x^n = \sum_{n=1}^{\infty} a_{n-1} x^{n-1}$$

so as to be in a position to apply Corollary 1. Putting

$$A_n = a_{n-1} \quad \text{and} \quad B_n(x) = x^{n-1}$$

for each positive integer n, we have

$$\sum_{n=0}^{\infty} a_n x^n = \sum_{n=1}^{\infty} A_n B_n(x).$$

The series $\sum_{n=0}^{\infty} a_n = \sum_{n=1}^{\infty} A_n$ is a converging series of constant functions. As such, it converges uniformly on $\mathbb{R}$. For each $x \in [0, 1]$, the sequence $\langle B_n(x) \rangle$ is monotonically decreasing. Since $|B_n(x)| = |x^{n-1}| \leqslant 1$ for all $x \in [0, 1]$ for each positive integer n, $\langle B_n \rangle$ is also uniformly bounded on $[0, 1]$. By Corollary 1, the series

$$\sum_{n=0}^{\infty} a_n x^n = \sum_{n=1}^{\infty} A_n B_n(x)$$

converges uniformly on $[0, 1]$. The partial sum sequence of $\sum a_n x^n$ is a sequence of polynomials on $\mathbb{R}$. As such, it is a sequence of functions which is continuous on $[0, 1]$. Hence, the sum $S(x)$ of the series is continuous on $[0, 1]$ and

$$\lim_{x \to 1+} S(x) = S(1).$$

This is equivalent to (6.9).

As for the second part of the corollary, we consider

$$T(y) = \sum_{n=0}^{\infty} (-1)^n a_n y^n.$$

Since now $\sum (-1)^n a_n$ converges, we have by the first part

$$\lim_{y \to 1-} T(y) = T(1) = \sum_{n=0}^{\infty} (-1)^n a_n .$$

Putting $y = -x$, we have $y \to 1-$ if and only if $x \to (-1)+$. Hence,

$$\lim_{x \to (-1)+} \sum_{n=0}^{\infty} a_n x^n = \lim_{x \to (-1)+} \sum_{n=0}^{\infty} (-1)^n a_n (-x)^n$$

$$= \lim_{y \to 1-} \sum_{n=0}^{\infty} (-1)^n a_n y^n = \sum_{n=0}^{\infty} (-1)^n a_n .$$

EXAMPLE 6.1. In Example XI.9.3 we proved (see formula XI.9.26) that

$$\operatorname{Arctan} x = x - \frac{x^3}{3} + \frac{x^5}{5} - \frac{x^7}{7} + \cdots \qquad \text{for all} \quad x \in (-1; 1). \quad (6.11)$$

The series obtained by replacing x by 1 on the right is

$$1-\frac{1}{3}+\frac{1}{5}-\frac{1}{7}+\cdots.$$

This series converges. By Abel's limit theorem and (6.11) we may conclude that

$$\begin{aligned}\frac{\pi}{4}&=\lim_{x\to 1-}\operatorname{Arctan} x=\lim_{x\to 1-}\left(x-\frac{x^3}{3}+\frac{x^5}{5}-\frac{x^7}{7}+\cdots\right)\\&=1-\frac{1}{3}+\frac{1}{5}-\frac{1}{7}+\cdots\end{aligned}$$

so that we have

$$\frac{\pi}{4}=1-\frac{1}{3}+\frac{1}{5}-\frac{1}{7}+\cdots. \tag{6.12}$$

The series on the right is called the *Gregory Series*.

Remark 6.1. If the power series $\sum a_n x^n$ has radius of convergence $0<R<+\infty$ and $\sum a_n R^n$ converges, then

$$\lim_{x\to R-}\sum_{n=0}^{\infty} a_n x^n=\sum_{n=0}^{\infty} a_n R^n.$$

Indeed, by Abel's limit theorem,

$$\lim_{y\to 1-}\sum_{n=0}^{\infty} a_n R^n y^n=\sum_{n=0}^{\infty} a_n R^n.$$

Put $x=Ry$. Since $x\to R-$, if and only if $y\to 1-$, we have

$$\lim_{x\to R-}\sum_{n=0}^{\infty} a_n x^n=\lim_{x\to R-}\sum_{n=0}^{\infty} a_n R^n\left(\frac{x}{R}\right)^n=\lim_{y\to 1-}\sum_{n=0}^{\infty} a_n R^n y^n=\sum_{n=0}^{\infty} a_n R^n.$$

Similarly, if $\sum(-1)^n a_n R^n$ converges, we can conclude that

$$\lim_{x\to(-R)+}\sum_{n=0}^{\infty}(-1)^n a_n x^n=\sum_{n=0}^{\infty}(-1)^n a_n R^n.$$

Theorem 6.4 (Dedekind's Test for Uniform Convergence). *$\sum a_n(x)b_n(x)$ converges uniformly on a set $D\subseteq\mathbb{R}$ provided that the sequence $\langle S_n(x)\rangle$ of partial sums of $\sum a_n(x)$ is uniformly bounded on D, $\sum|b_{k+1}(x)-b_k(x)|$ converges uniformly on D, and $b_n(x)\to 0$ as $n\to+\infty$ uniformly on D.*

Proof. Since the partial sum sequence $\langle S_n(x)\rangle$ of $\sum a_n(x)$ is bounded uniformly on D, there exists an $M>0$ such that

$$|S_n(x)|\leqslant M \qquad \text{for all } n \text{ and all } \quad x\in D.$$

If $m>n$, this implies that

$$\left|\sum_{k=n+1}^{n} S_k(x)(b_{k+1}(x)-b_k(x))\right|\leqslant M\sum_{k=n+1}^{m}|b_{k+1}(x)-b_k(x)|$$

for all $x \in D$. Since $\sum|b_{n+1}(x) - b_n(x)|$ is uniformly convergent on D, the last inequality implies that $\sum S_n(x)(b_{n+1}(x) - b_n(x))$ is uniformly convergent on D. Also,

$$|b_{n+1}(x)S_n(x)| \leqslant M|b_{n+1}(x)| \qquad \text{for all} \quad x \in D.$$

Since $\langle b_n(x)\rangle$ converges uniformly to 0 on D, this implies that the sequence of functions $\langle b_{n+1}S_n\rangle$ converges uniformly to 0 on D. By Theorem 6.1 it follows that $\sum a_n(x)b_n(x)$ converges uniformly on D.

Corollary (Dirichlet's Test for Uniform Convergence). *$\sum a_n(x)b_n(x)$ converges uniformly on D provided that the partial sum sequence $\langle S_n(x)\rangle$ of $\sum a_n(x)$ is uniformly bounded on D, $\langle b_n(x)\rangle$ is monotonic for each $x \in D$, and the sequence $\langle b_n(x)\rangle$ converges to 0 uniformly on D.*

PROOF. Since $\langle b_n(x)\rangle$ is monotonic for each $x \in D$, it follows that if $m \geqslant n$, then

$$\sum_{k=n+1}^{m} |b_{k+1}(x) - b_k(x)| = |b_{m+1}(x) - b_{n+1}(x)| \qquad \text{for all} \quad x \in D.$$

But, by hypothesis $\langle b_n(x)\rangle$ converges uniformly on D. We, therefore, obtain from the last equality that $\sum|b_{n+1}(x) - b_n(x)|$ is uniformly convergent on D. The partial sum sequence $\langle S_n(x)\rangle$ of $\sum a_n(x)$ is uniformly bounded on D. The hypothesis of Theorem 6.4 is now satisfied and $\sum a_n(x)b_n(x)$ converges uniformly on D.

EXAMPLE 6.2. We examine the series

$$\sum_{n=1}^{\infty} b_n \cos nx,$$

where $\langle b_n\rangle$ converges monotonically to 0. We write

$$a_n(x) = \cos nx \qquad \text{for each } n$$

and examine the partial sums

$$\sum_{k=1}^{n} a_k(x) = \sum_{k=1}^{n} \cos kx.$$

By Prob. X.6.2, we have: If n is a positive integer, then

$$\sum_{k=1}^{n} \cos kx = \frac{\sin(x/2)\cos((n+1)/2)x}{\sin(x/2)}$$

if x is not an integral multiple of 2π,

so that for each positive integer n

$$\left|\sum_{k=1}^{n} \cos kx\right| \leqslant \frac{1}{|\sin(x/2)|} \qquad \text{if } x \text{ is not an integral multiple of } 2\pi.$$

We take δ such that $0 < \delta < \pi$ and $0 < \delta \leqslant x \leqslant 2\pi - \delta$. We have

$$0 < \sin\frac{\delta}{2} \leqslant \sin\frac{x}{2} \qquad \text{if} \quad \delta \leqslant x \leqslant \pi$$

and

$$0 < \sin\frac{\delta}{2} = \sin\left(\pi - \frac{\delta}{2}\right) \leqslant \sin\frac{x}{2} \qquad \text{if} \quad \pi \leqslant x \leqslant 2\pi - \delta.$$

In either case,

$$\sin\frac{x}{2} \geqslant \sin\frac{\delta}{2} > 0.$$

Equivalently,

$$0 < \frac{1}{\sin(x/2)} \leqslant \frac{1}{\sin(\delta/2)} \qquad \text{if} \quad x \in [\delta, 2\pi - \delta].$$

Thus, we have: for each positive integer n,

$$\left|\sum_{k=1}^{n} \cos kx\right| \leqslant \frac{1}{\sin(\delta/2)} \qquad \text{for} \quad x \in [\delta, 2\pi - \delta].$$

It follows that the partial sum sequence of the series

$$\sum_{n=1}^{\infty} \cos nx$$

is uniformly bounded on $[\delta; 2\pi - \delta]$.

On the other hand, by our assumption, the sequence $\langle b_n \rangle$ of constant functions b_n is monotonic for each x and converges uniformly to 0 on $\mathbb{R}$. By Dirichlet's test, we conclude that the series

$$\sum_{n=1}^{\infty} b_n \cos nx$$

converges uniformly on $[\delta, 2\pi - \delta]$. Since the partial sums of this series are all continuous, the sum of the above series is continuous on $[\delta, 2\pi - \delta]$, where $0 < \delta < 2\pi - \delta$.

Prob. 6.1. Prove: If $\langle b_n \rangle$ is a real sequence which converges monotonically to 0, then the series

$$\sum_{n=1}^{\infty} b_n \sin nx$$

converges uniformly on $[\delta, 2\pi - \delta]$, where $0 < \delta < \pi$.

CHAPTER XIII
The Riemann Integral I

1. Darboux Integrals

Until now we dealt with one type of limit process. The different aspects of it that we treated were variations on the same theme. In integration we encounter a different type of limit.

We begin with a bounded closed interval $I = [a,b]$ and assign to it a number called its *length*. The *length* $L([a,b])$ of $[a,b]$, is defined as

$$L([a,b]) = b - a. \tag{1.1}$$

By a partition P of $[a,b]$ we mean a finite sequence $\langle x_0, x_1, \ldots, x_n \rangle$ of points of $[a,b]$ such that

$$a = x_0 < x_1 < \cdots < x_{n-1} < x_n = b \tag{1.2}$$

(see Fig. 1.1). The partition P effects a partitioning of $[a,b]$ into subintervals $I_1 = [x_0, x_1]$, $I_2 = [x_1, x_2], \ldots, I_n = [x_{n-1}, x_n]$ such that

$$\bigcup_{i=1}^{n} [x_{i-1}, x_i] = [a,b]. \tag{1.3}$$

Since $I_i = [x_{i-1}, x_i]$ for each $i \in \{1, 2, \ldots, n\}$, we have $L(I_i) = x_i - x_{i-1}$. We also write

$$\Delta x_i = x_i - x_{i-1} = L(I_i) \qquad \text{for each} \quad i \in \{1, 2, \ldots, n\}. \tag{1.4}$$

Note that

$$\sum_{i=1}^{n} L(I_i) = \sum_{i=1}^{n} \Delta x_i = \sum_{i=1}^{n} (x_i - x_{i-1}) = x_n - x_0 = b - a$$
$$= L([a,b]) = L\left(\bigcup_{i=1}^{n} I_i\right). \tag{1.5}$$

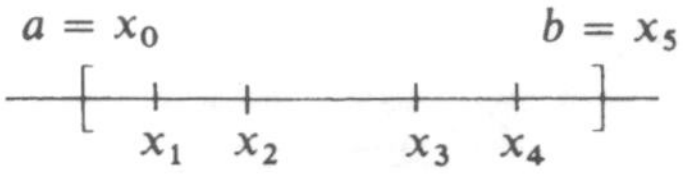

$a = x_0$ $b = x_5$

x_1 x_2 x_3 x_4

Figure 1.1

By abuse of language, we sometimes refer to the finite sequence of intervals $\langle I_1, I_2, \ldots, I_n \rangle$ induced by the partition P as a partition of $[a,b]$, and to the intervals I_i as the subintervals of the partition.

EXAMPLE 1.1. Given $[a,b]$, where $a \in \mathbb{R}$, $b \in \mathbb{R}$, and $a < b$, define

$$x_i = a + i\frac{b-a}{n}.$$

Here $i \in \{1, 2, \ldots, n\}$, and n is some positive integer.

$$x_0 = a, \qquad x_1 = a + \frac{b-a}{n}, \ldots, \qquad x_n = a + n\frac{b-a}{n} = b.$$

For each $i \in \{1, \ldots, n\}$,

$$\Delta x_i = x_i - x_{i-1} = a + i\frac{b-a}{n} - \left(a + (i-1)\frac{b-a}{n}\right) = \frac{b-a}{n}.$$

All subintervals of this partition have the same length.

Def. 1.1. By the *norm* $\|P\|$ of a partition $P = \langle x_0, x_1, \ldots, x_n \rangle$ of $[a,b]$ we mean

$$\|P\| = \max\{\Delta x_1, \ldots, \Delta x_n\} = \max_{1 \leqslant i \leqslant n} \Delta x_i. \tag{1.6}$$

For example, in the partition of Example 1.1 we have

$$\|P\| = \max_{1 \leqslant i \leqslant n} \Delta x_i = \max_{1 \leqslant i \leqslant n} \frac{b-a}{n} = \frac{b-a}{n}.$$

EXAMPLE 1.2. The subintervals of a partition need not all have the same length. Let a and b be real numbers such that $0 < a < b$ and n be some positive integer and let

$$q = \left(\frac{b}{a}\right)^{1/n}$$

so that $q > 1$. Let

$$x_i = a\left(\frac{b}{a}\right)^{i/n} \qquad \text{for} \quad i \in \{0, 1, \ldots, n\}.$$

Clearly, $x_0 = a$ and $x_n = b$ and

$$\Delta x_i = x_i - x_{i-1} = a\left(\frac{b}{a}\right)^{i/n} - a\left(\frac{b}{a}\right)^{(i-1)/n} = a\left(\frac{b}{a}\right)^{(i-1)/n}\left(\left(\frac{b}{a}\right)^{1/n} - 1\right) > 0$$

for each $i \in \{1, 2, \ldots, n\}$. We have

$$a = x_0 < x_1 < \cdots < x_{n-1} < x_n = b,$$

but the lengths of different subintervals are not equal.

Prob. 1.1. Show that for the partition P of Example 1.2, we have

$$\|P\| = b^{1-1/n}(b^{1/n} - a^{1/n}).$$

What follows will apply to a real-value function f defined on $[a,b]$ and bounded there. We write

$$m = \inf_{[a,b]} f \quad \text{and} \quad M = \sup_{[a,b]} f. \tag{1.7}$$

Clearly,

$$m \leqslant f(x) \leqslant M \qquad \text{if} \quad x \in [a,b]. \tag{1.8}$$

On the subintervals $I_i = [x_{i-1}, x_i]$ of the partition $[a,b]$, f inherits the boundedness property. We write

$$m_i = \inf_{I_i} f \quad \text{and} \quad M_i = \sup_{I_i} f \qquad \text{for} \quad i \in \{1, \ldots, n\}. \tag{1.9}$$

Since $[x_{i-1}, x_i] \subseteq [a,b]$, we have $f([x_{i-1}, x_i]) \subseteq f([a,b])$. Hence, $x \in [x_{i-1}, x_i]$ implies that

$$m \leqslant m_i \leqslant f(x) \leqslant M_i \leqslant M, \tag{1.10}$$

where $i \in \{1, 2, \ldots, n\}$. It will sometimes be necessary to emphasize that m, m_i, M_i, and M are associated with a particular function. We then write these as $m(f)$, $m_i(f)$, $M_i(f)$, and $M(f)$.

When $f(x) \geqslant 0$ for $x \in [a,b]$, then the products $m_i \Delta x_i$ and $M_i \Delta x_i$ are interpreted as the areas of rectangles based on $I_i = [x_{i-1}, x_i]$ with respective "heights" m_i and M_i (see Fig. 1.2) for a given partition P of $[a,b]$.

Def. 1.2. If $P = \langle x_0, x_1, \ldots, x_n \rangle$ is a partition of $[a,b]$ and f is bounded on $[a,b]$, then the sums

$$\sum_{i=1}^{n} m_i \Delta x_i \quad \text{and} \quad \sum_{i=1}^{n} M_i \Delta x_i$$

will be called respectively the *lower* and *upper Darboux sums* of f on $[a,b]$

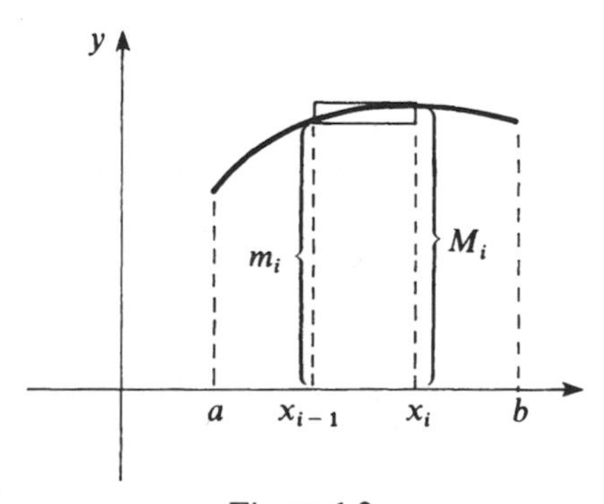

Figure 1.2

for the partition P and will be written as $\underline{S}(f,P)$ and $\overline{S}(f,P)$. Thus,

$$\underline{S}(f,P)=\sum_{i=1}^{n} m_i\Delta x_i \quad\text{and}\quad \overline{S}(f,P)=\sum_{i=1}^{n} M_i\Delta x_i\,. \tag{1.11}$$

Prob. 1.2. Prove: If f is bounded on $[a,b]$, then for each partition P of $[a,b]$, we have

$$m(b-a)\leqslant \underline{S}(f,P)\leqslant \overline{S}(f,P)\leqslant M(b-a). \tag{1.12}$$

Def. 1.3. By a *refinement* of a partition P of $[a,b]$ we mean a partition P' of $[a,b]$ such that $P\subseteq P'$.

Lemma 1.1. *If f is bounded on $[a,b]$ and P' is a refinement of the partition P of $[a,b]$, then*

$$\underline{S}(f,P)\leqslant \underline{S}(f,P')\leqslant \overline{S}(f,P')\leqslant \overline{S}(f,P). \tag{1.13}$$

(*In short, a refinement does not decrease the lower Darboux sum and does not increase the upper one.*)

Proof. Let $P=\langle x_0,x_1,\ldots,x_n\rangle$ and $P'=\langle x_0',x_1',\ldots,x_m'\rangle$ be partitions of $[a,b]$, where P' is a refinement of P. Then

$$a=x_0<x_1<\cdots<x_{n-1}<x_n=b \tag{1.14}$$

and

$$a=x_0'<x_1'<\cdots<x_{m-1}'<x_m'=b$$

and $P\subseteq P'$.

Assume that $I_i=[x_{i-1},x_i]$ is a subinterval of $P\subseteq P'$. Then $x_{i-1}\in P'$ and $x_i\in P'$ so that j and k exist such that $x_{i-1}=x_j'$ and $x_i=x_k'$. Clearly, j and k are nonnegative integers such that $j<k$ and, therefore, $j\leqslant k-1$. Hence,

$$x_{i-1}=x_j'\leqslant x_{k-1}'<x_k'=x_i\,.$$

Thus, either (1) $j=k-1$ or (2) $j<k-1$. In the first case,

$$[x_{i-1},x_i]=[x_{k-1}',x_k']. \tag{1.15}$$

In the second case, $x_{i-1}=x_j'<x_{k-1}'<x_k'=x_i$ and there is at least one partition point of P' between x_{i-1} and x_i. Take all the partition points, say $x_{j+1}',x_{j+2}',\ldots,x_{k-1}'$, of P' which are between x_{i-1} and x_i. We have

$$x_{i-1}=x_j'<x_{j+1}'<\cdots<x_{k-1}'<x_k'=x_i\,. \tag{1.16}$$

The points x_j', $x_{j+1}',\ldots,x_{k-1}'$, x_k' constitute a partition of the interval $I_i=[x_{i-1},x_i]$, in this case. Write m_ν' and M_ν' for the infimum and supremum of f on the subinterval $[x_{\nu-1}',x_\nu']$ of P'. When $j=k-1$, (1.15) holds. In that case,

$$m_i\Delta x_i=m_k'\Delta x_k' \quad\text{and}\quad M_i\Delta x_i=M_k'\Delta x_k'\,. \tag{1.17}$$

If $j < k - 1$, then (1.16) holds. Write $\underline{S}'_{I_i}$ and $\overline{S}'_{I_i}$ for the lower and upper Darboux sums of f restricted to the subinterval $I_i = [x_{i-1}, x_i]$ and due to the partitioning of I_i by $x'_j, x'_{j+1}, \ldots, x'_{k-1}, x'_k$. We have

$$m_i \Delta x_i = m_i(x_i - x_{i-1}) \leqslant \underline{S}'_{I_i} \leqslant \overline{S}'_{I_i} \leqslant M_i(x_i - x_{i-1}) = M_i \Delta x_i . \quad (1.18)$$

Here $\underline{S}'_{I_i}$ appears as a sum in $\underline{S}(f, P')$ of f corresponding to the partition P' and, similarly, $\overline{S}'_{I_i}$ appears as a sum in $\overline{S}(f, P')$. What was done in I_i can be repeated for all subintervals of P. Forming the Darboux sums for P and P' by using (1.17) and (1.18) we obtain

$$\underline{S}(f, P) \leqslant \underline{S}(f, P') \leqslant \overline{S}(f, P') \leqslant \overline{S}(f, P),$$

as claimed.

Lemma 1.2. *Let f be bounded over $[a, b]$. For any two partitions P and P'' of $[a, b]$ we have*

$$\underline{S}(f, P) \leqslant \overline{S}(f, P''). \quad (1.19)$$

PROOF. Let $P' = P \cup P''$. P' is now a partition of $[a, b]$. Since $P \subseteq P'$ and $P'' \subseteq P'$, P' is a refinement of P and P''. By Lemma 1.1, we have

$$\underline{S}(f, P) \leqslant \underline{S}(f, P') \leqslant \overline{S}(f, P') \leqslant \overline{S}(f, P''),$$

and the conclusion follows.

PROB. 1.3. Prove: If f is bounded on $[a, b]$ and B is some upper bound of $|f|$ on $[a, b]$, then for any partitions P and P' of $[a, b]$, we have

$$-B(b - a) \leqslant \underline{S}(f, P) \leqslant \overline{S}(f, P') \leqslant B(b - a)$$

and, therefore,

$$0 \leqslant \overline{S}(f, P') - \underline{S}(f, P) \leqslant 2B(b - a).$$

Note that when f is bounded on $[a, b]$, each lower Darboux sum of f for a partition P is never greater than any upper Darboux sum for f on $[a, b]$. Hence, if G is the set of all lower Darboux sums of f over $[a, b]$, then G is bounded from above by each $\overline{S}(f, P)$ (Lemma 1.2). Since G is certainly not empty, it has a real supremum. Similarly, the set H of all upper Darboux sums of f over $[a, b]$ is bounded from below by any lower Darboux sum of f over $[a, b]$ and, therefore, has an infimum.

Def. 1.4. If f is bounded on $[a, b]$, then the infimum of H, the set of all upper Darboux sums of f over $[a, b]$, is called the *upper Darboux integral* of f over $[a, b]$ and is written

$$\overline{\int_a^b} f(x)\, dx.$$

The supremum of the set G of all lower Darboux sums of f over $[a,b]$ is called the *lower Darboux integral* of f over $[a,b]$ and is written

$$\underline{\int_a^b} f(x)\,dx.$$

Remark 1.1. To guarantee the existence of each of

$$\underline{\int_a^b} f(x)\,dx \quad \text{and} \quad \overline{\int_a^b} f(x)\,dx,$$

it suffices to assume that f is bounded on $[a,b]$.

Lemma 1.3. *If f is bounded on $[a,b]$ and P and P'' are partitions of $[a,b]$, then*

$$\underline{S}(f,P) \leqslant \underline{\int_a^b} f(x)\,dx \leqslant \overline{\int_a^b} f(x)\,dx \leqslant \overline{S}(f,P'').$$

PROOF. The first inequality is an immediate consequence of the definition of the lower Darboux integral of f over $[a,b]$. Similarly the last inequality follows directly from the definition of the upper Darboux integral of f over $[a,b]$. We prove the inequality

$$\underline{\int_a^b} f(x)\,dx \leqslant \overline{\int_a^b} f(x)\,dx. \tag{1.20}$$

Since $\underline{S}(f,P) \leqslant \overline{S}(f,P')$ holds for any partitions P and P' of $[a,b]$, any upper Darboux $\overline{S}(f,P')$ is an upper bound for the set G of all lower Darboux sums of f over $[a,b]$. This implies

$$\underline{\int_a^b} f(x)\,dx \leqslant \overline{S}(f,P') \tag{1.21}$$

since on the left we have the supremum of G. But (1.21) holds for all upper Darboux sums of f over $[a,b]$. Accordingly, the left side of (1.21) is a lower bound for the set H of all upper Darboux sums of f over $[a,b]$ and, therefore, cannot exceed the infimum of H. This implies (1.20). The proof is complete.

Def. 1.5. If f is bounded over $[a,b]$, then it is called *Darboux integrable* over $[a,b]$ if and only if

$$\underline{\int_a^b} f(x)\,dx = \overline{\int_a^b} f(x)\,dx. \tag{1.22}$$

When f is Darboux integrable over $[a,b]$, we call the common value in (1.22) the *Darboux integral* of f over $[a,b]$ and write it as

$$\int_a^b f(x)\,dx.$$

Thus,

$$\int_a^b f(x)\,dx = \underline{\int_a^b} f(x)\,dx = \overline{\int_a^b} f(x)\,dx. \tag{1.23}$$

We also define

$$\int_a^a f(x) = 0 \tag{1.24a}$$

and

$$\int_b^a f(x)\,dx = -\int_a^b f(x)\,dx \qquad \text{if} \quad a < b. \tag{1.24b}$$

If f is Darboux integrable over $[a,b]$, then we say briefly that it is *D-integrable* over $[a,b]$.

PROB. 1.4. Prove: If f is bounded over and B is some upper bound of $|f|$, then

$$0 \leqslant \overline{\int_a^b} f(x)\,dx - \underline{\int_a^b} f(x)\,dx \leqslant 2B(b-a)$$

(see Prob. 1.3).

Do D-integrable functions exist? We exhibit such a function.

Theorem 1.1. *If k is some real number and $f(x) = k$ for all $x \in [a,b]$, that is, if f is the constant function having value k on $[a,b]$, then f is D-integrable over $[a,b]$ and*

$$\int_a^b k\,dx = k(b-a).$$

PROOF. Let $P = \langle x_0, x_1, \ldots, x_n \rangle$ be some partition of $[a,b]$. Then

$$m_i = \inf_{x_{i-1} \leqslant x \leqslant x_i} f(x) = \inf_{x_{i-1} \leqslant x \leqslant x_i} k = k$$

and

$$M_i = \sup_{x_{i-1} \leqslant x \leqslant x_i} f(x) = \sup_{x_{i-1} \leqslant x \leqslant x_i} k = k.$$

Hence,

$$\underline{S}(f,P) = \sum_{i=1}^{n} m_i \Delta x_i = \sum_{i=1}^{n} k \Delta x_i = k(b-a)$$

and

$$\overline{S}(f,P) = \sum_{i=1}^{n} M_i \Delta x_i = \sum_{i=1}^{n} k \Delta x_i = k(b-a).$$

It follows that

$$\underline{\int_a^b} k\,dx = k(b-a) = \overline{\int_a^b} k\,dx,$$

that is, our function is D-integrable over $[a,b]$ and

$$\int_a^b k\,dx = k(b-a).$$

Are there functions which are not D-integrable?

EXAMPLE 1.3. Take the Dirichlet function D and restrict it to the interval $[0,1]$. Thus,

$$D(x) = \begin{cases} 1 & \text{if } x \text{ is rational and } \quad 0 \leqslant x \leqslant 1 \\ 0 & \text{if } x \text{ is irrational and } \quad 0 \leqslant x \leqslant 1. \end{cases}$$

Let $P = \langle x_0, x_1, \ldots, x_n \rangle$ be any partition of $[0,1]$. We have

$$m_i = \inf_{x_{i-1} \leqslant x \leqslant x_i} D(x) = 0 \quad \text{and} \quad M_i = \sup_{x_{i-1} \leqslant x \leqslant x_i} D(x) = 1$$

(explain) for each subinterval $[x_{i-1}, x_i]$ of P. Therefore,

$$\underline{S}(D,P) = \sum_{i=1}^n m_i \Delta x_i = \sum_{i=1}^n 0 \Delta x_i = 0$$

and

$$\overline{S}(D,P) = \sum_{i=1}^n M_i \Delta x_i = \sum_{i=1}^n 1 \Delta x = 1 - 0 = 1.$$

These hold for any partition P of $[0,1]$. It follows that

$$\underline{\int_0^1} D(x)\,dx = 0 \quad \text{and} \quad \overline{\int_0^1} D(x) = 1.$$

Therefore, D is not D-integrable over $[0,1]$.

To produce more D-integrable functions we prove the following criterion for D-integrability.

Theorem 1.2. *If f is bounded over $[a,b]$, then f is D-integrable over $[a,b]$ if and only if for each $\epsilon > 0$ there exists a partition P of $[a,b]$ such that*

$$\overline{S}(f,P) - \underline{S}(f,P) < \epsilon. \tag{1.25}$$

PROOF. Assume that for each $\epsilon > 0$, there exists a partition P of $[a,b]$ for which (1.25) holds. For any partition P of $[a,b]$ we have

$$\underline{S}(f,P) \leqslant \underline{\int_a^b} f(x)\,dx \leqslant \overline{\int_a^b} f(x)\,dx \leqslant \overline{S}(f,P)$$

and, hence,

$$0 \leqslant \overline{\int_a^b} f(x)\,dx - \underline{\int_a^b} f(x)\,dx \leqslant \bar{S}(f,P) - \underline{S}(f,P). \tag{1.26}$$

Let $\epsilon > 0$ be given so that for some partition P' of $[a,b]$, we have $\bar{S}(f,P') - \underline{S}(f,P') < \epsilon$. Using (1.26) we obtain

$$0 \leqslant \overline{\int_a^b} f(x)\,dx - \underline{\int_a^b} f(x)\,dx \leqslant \bar{S}(f,P') - \underline{S}(f,P') < \epsilon.$$

Thus,

$$0 \leqslant \overline{\int_a^b} f(x)\,dx - \underline{\int_a^b} f(x)\,dx < \epsilon. \tag{1.27}$$

This holds for any given $\epsilon > 0$. It follows that

$$\underline{\int_a^b} f(x) = \overline{\int_a^b} f(x)\,dx$$

(explain) and, therefore, that f is D-integrable over $[a,b]$.

Conversely, assume that f is D-integrable over $[a,b]$ so that

$$\underline{\int_a^b} f(x)\,dx = \overline{\int_a^b} f(x)\,dx = \int_a^b f(x)\,dx.$$

Let $\epsilon > 0$ be given. We know that

$$A = \underline{\int_a^b} f(x)\,dx - \frac{\epsilon}{2} < \underline{\int_a^b} f(x)\,dx \quad \text{and} \quad \overline{\int_a^b} f(x)\,dx < \overline{\int_a^b} f(x)\,dx + \frac{\epsilon}{2} = B.$$

Here A is not an upper bound for the set G of lower Darboux sums of f, so a partition P' of $[a,b]$ exists such that the lower Darboux sum $\underline{S}(f,P')$ corresponding to P' satisfies

$$\underline{\int_a^b} f(x)\,dx - \frac{\epsilon}{2} < \underline{S}(f,P'). \tag{1.28}$$

Similarly, B is not a lower bound for the set H of upper Darboux sums of f, and, therefore, a partition P'' of $[a,b]$ exists such that

$$\bar{S}(f,P'') < \overline{\int_a^b} f(x) + \frac{\epsilon}{2}. \tag{1.29}$$

Let $P = P' \cup P''$ so that P is a refinement of P' and of P'' and, therefore,

$$\underline{S}(f,P') \leqslant \underline{S}(f,P) \leqslant \bar{S}(f,P) \leqslant \bar{S}(f,P'').$$

This, (1.28) and (1.29) imply

$$\underline{\int_a^b} f(x)\,dx - \frac{\epsilon}{2} < \underline{S}(f,P) \leqslant \bar{S}(f,P) < \overline{\int_a^b} f(x)\,dx + \frac{\epsilon}{2}.$$

Now use this and the equality of the upper and lower integrals to obtain for the partition P

$$\bar{S}(f,P)-\underline{S}(f,P)<\int_a^b f(x)\,dx+\frac{\epsilon}{2}-\left(\int_a^b f(x)\,dx-\frac{\epsilon}{2}\right)=\epsilon.$$

This completes the proof.

Theorem 1.3. *If f is monotonic on $[a,b]$, then f is D-integrable over $[a,b]$.*

Proof. The monotonicity of f on the bounded closed interval implies its boundedness (why?). Assume, for definiteness, that f is monotonically increasing on $[a,b]$. Let $P=\langle x_0,x_1,\ldots,x_n\rangle$ be a partition of $[a,b]$. For the subinterval $[x_{i-1},x_i]$ of P we have

$$f(x_{i-1})\leqslant f(x)\leqslant f(x_i)\qquad \text{for all}\quad x\in I_i=[x_{i-1},x_i],$$

$$m_i=\inf_{I_i} f=\min_{I_i} f=f(x_{i-1}),$$

and

$$M_i=\sup_{I_i} f=\max_{I_i} f=f(x_i).$$

Therefore,

$$\sum_{i=1}^n f(x_{i-1})\Delta x_i=\underline{S}(f,P)\leqslant\bar{S}(f,P)=\sum_{i=1}^n f(x_i)\Delta x_i$$

and

$$\bar{S}(f,P)-\underline{S}(f,P)=\sum_{i=1}^n (f(x_i)-f(x_{i-1}))\Delta x_i\,. \tag{1.30}$$

If f is constant on $[a,b]$, then we know it is D-integrable there. Assume f is not constant on $[a,b]$. Hence, $f(a)<f(b)$. Take a partition P such that

$$\|P\|<\frac{\epsilon}{f(b)-f(a)}$$

and obtain

$$\Delta x_i\leqslant\|P\|<\frac{\epsilon}{f(b)-f(a)}$$

for all $i\in\{1,\ldots,n\}$. Using (1.30), this implies that

$$\begin{aligned}\bar{S}(f,P)-\underline{S}(f,P)&=\sum_{i=1}^n (f(x_i)-f(x_{i-1}))\Delta x_i\\&<\sum_{i=1}^n (f(x_i)-f(x_{i-1}))\frac{\epsilon}{f(b)-f(a)}\\&=(f(b)-f(a))\frac{\epsilon}{f(b)-f(a)}\\&=\epsilon.\end{aligned}$$

Thus, the integrability criterion in Theorem 1.2 is satisfied and we conclude that f is D-integrable over $[a,b]$.

We leave the case where f is monotonic decreasing to the reader (Prob. 1.5).

PROB. 1.5. Complete the proof of Theorem 1.3 by proving: If f is monotonically decreasing on $[a,b]$, then f is D-integrable there.

EXAMPLE 1.4. Let a and b be real numbers such that $0 < a < b$ and let f be defined as

$$f(x) = \frac{1}{x^2} \quad \text{for} \quad x \in [a,b].$$

Let $P = \langle x_0, x_1, \ldots, x_n \rangle$ be any partition of $[a,b]$. Since $0 < a \leqslant x_{i-1} \leqslant x \leqslant x_i$ for $x \in I_i = [x_{i-1}, x_i]$, we have

$$0 < x_{i-1}^2 \leqslant x^2 \leqslant x_i^2$$

and, therefore, that

$$\frac{1}{x_i^2} \leqslant \frac{1}{x^2} \leqslant \frac{1}{x_{i-1}^2} \quad \text{for} \quad x \in I_i.$$

Clearly,

$$m_i = \inf_{I_i} f = \frac{1}{x_i^2} \quad \text{and} \quad M_i = \sup_{I_i} f = \frac{1}{x_{i-1}^2}$$

and

$$\underline{S}(f,P) = \sum_{i=1}^{n} \frac{1}{x_i^2} \Delta x_i = \sum_{i=1}^{n} \frac{1}{x_i^2}(x_i - x_{i-1}),$$

$$\overline{S}(f,P) = \sum_{i=1}^{n} \frac{1}{x_{i-1}^2} \Delta x_i = \sum_{i=1}^{n} \frac{1}{x_{i-1}^2}(x_i - x_{i-1}).$$

Note that, for each $i \in \{1,2,\ldots,n\}$, we have

$$\frac{1}{x_i^2} < \frac{1}{x_i x_{i-1}} < \frac{1}{x_{i-1}^2}$$

so that

$$\underline{S}(f,P) = \sum_{i=1}^{n} \frac{1}{x_i^2}(x_i - x_{i-1}) < \sum_{i=1}^{n} \frac{x_i - x_{i-1}}{x_i x_{i-1}}$$

$$< \sum_{i=1}^{n} \frac{1}{x_{i-1}^2}(x_i - x_{i-1}) = \overline{S}(f,P).$$

This implies that

$$\underline{S}(f,P) < \sum_{i=1}^{n} \left(\frac{1}{x_{i-1}} - \frac{1}{x_i} \right) < \overline{S}(f,P)$$

for each partition P of $[a,b]$. This is equivalent to

$$\underline{S}(f,P) < \frac{1}{a} - \frac{1}{b} < \overline{S}(f,P)$$

for each partition P of $[a,b]$. This yields

$$\underline{\int_a^b} \frac{1}{x^2}\,dx \leqslant \frac{1}{a} - \frac{1}{b} \leqslant \overline{\int_a^b} \frac{1}{x^2}\,dx. \tag{1.31}$$

But f is monotonic decreasing and, therefore, D-integrable on $[a,b]$. It follows that the upper and lower integrals in (1.31) are equal, their common value being the Darboux integral of f over $[a,b]$. This and (1.31) imply that

$$\int_a^b \frac{1}{x^2}\,dx = \frac{1}{a} - \frac{1}{b}. \tag{1.32}$$

PROB. 1.6. Suppose that $0 < a < b$ and that f is defined as

$$f(x) = \frac{1}{x^3} \qquad \text{for} \quad x \in [a,b].$$

Let $P = \langle x_0, x_1, \ldots, x_n \rangle$ be a partition of $[a,b]$. Show that

$$\frac{2}{x_i^3} < \frac{1}{x_i^2 x_{i-1}} + \frac{1}{x_i x_{i-1}^2} < \frac{2}{x_{i-1}^3}$$

holds for the points of P. Prove that

$$\int_a^b \frac{1}{x^3}\,dx = \frac{1}{2}\left(\frac{1}{a^2} - \frac{1}{b^2}\right).$$

EXAMPLE 1.5. Once it is known that the function is D-integrable and we merely wish to evaluate its integral we need not deal with all of the partitions of the interval. It suffices to choose certain special sequences of partitions. Thus, consider the function f, where

$$f(x) = x^2$$

for $x \in [0,b]$, $b > 0$. On this interval f is monotonic increasing and, therefore, D-integrable. We wish to evaluate

$$\int_0^b x^2\,dx.$$

Take a sequence $\langle P_n \rangle$ of partitions of $[0,b]$ such that

$$x_i = i\frac{b}{n} \qquad \text{for} \quad i \in \{0, 1, \ldots, n\}.$$

We have

$$\langle x_0, x_1, \ldots, x_n \rangle = \langle 0, \frac{b}{n}, \frac{2b}{n}, \ldots, \frac{n-1}{n}b, b \rangle$$

and

$$[x_{i-1}, x_i] = \left[\frac{i-1}{n} b, \frac{i}{n} b \right] \quad \text{and} \quad \Delta x_i = \frac{b}{n}.$$

For the interval $I_i = [x_{i-1}, x_i]$ we have

$$m_i = \inf_{x \in I_i} x^2 = \left(\frac{i-1}{n} b \right)^2 \quad \text{and} \quad M_i = \sup_{x \in I_i} x^2 = \left(\frac{i}{n} b \right)^2.$$

Therefore,

$$\sum_{i=1}^{n} m_i \Delta x_i = \sum_{i=1}^{n} \left(\frac{i-1}{n} b \right)^2 \frac{b}{n} \quad \text{and} \quad \sum_{i=1}^{n} M_i \Delta x_i = \sum_{n=1}^{n} \left(\frac{i}{n} b \right)^2 \frac{b}{n}.$$

This implies that

$$\underline{S}(f, P_n) = \frac{b^3}{n^3} \sum_{i=1}^{n} (i-1)^2 \quad \text{and} \quad \overline{S}(f, P_n) = \frac{b^3}{n^3} \sum_{i=1}^{n} i^2.$$

Now, as is seen from Prob. II.4.6(b),

$$\sum_{i=1}^{n} (i-1)^2 = \frac{(n-1)n(2n-1)}{6} \quad \text{and} \quad \sum_{i=1}^{n} i^2 = \frac{n(n+1)(2n+1)}{6}.$$

Thus,

$$\underline{S}(f, P_n) = \frac{b^3}{n^3} \frac{(n-1)n(2n-1)}{6} = \frac{b^3}{6} \left(1 - \frac{1}{n} \right) \left(2 - \frac{1}{n} \right)$$

and

$$\overline{S}(f, P_n) = \frac{b^3}{n^3} \frac{n(n+1)(2n+1)}{6} = \frac{b^3}{6} \left(1 + \frac{1}{n} \right) \left(2 + \frac{1}{n} \right).$$

Therefore,

$$\frac{b^3}{6} \left(1 - \frac{1}{n} \right) \left(2 - \frac{1}{n} \right) \leqslant \int_0^b \frac{1}{x^2} dx \leqslant \frac{b^3}{6} \left(1 + \frac{1}{n} \right) \left(2 + \frac{1}{n} \right)$$

for each positive integer n. Taking limits as $n \to +\infty$, we obtain

$$\frac{b^3}{3} = \int_0^b x^2 \, dx.$$

Prob. 1.7. Assume $b > 0$. Evaluate, using Darboux sums:

(1) $\int_0^b x^3 \, dx$,

(2) $\int_0^b x^4 \, dx$,

(3) $\int_0^b e^x \, dx$.

Prob. 1.8. Prove: The functions f and g defined as

$$f(x) = \begin{cases} 0 & \text{if } a \leqslant x < b \\ f(b) \neq 0 & \text{if } x = b \end{cases}$$

and

$$g(x) = \begin{cases} 0 & \text{if } a < x \leqslant b \\ g(a) \neq 0 & \text{if } x = a, \end{cases}$$

are D-integrable and

$$\int_a^b f(x)\,dx = 0 = \int_a^b g(x)\,dx.$$

Theorem 1.4. *If f is continuous on $[a,b]$, then f is D-integrable over $[a,b]$.*

PROOF. The continuity of f on the bounded closed interval $[a,b]$ implies its boundedness and uniform continuity there. Let $\epsilon > 0$ be given. There exists a $\delta > 0$ such that if ξ and η are points of $[a,b]$ with $|\xi - \eta| < \delta$, then

$$|f(\xi) - f(\eta)| < \frac{\epsilon}{b-a}.$$

Take a partition $P = \langle x_0, x_1, \ldots, x_n \rangle$ such that $\|P\| < \delta$. For each subinterval $[x_{i-1}, x_i]$ of P (since f is continuous on that bounded closed interval) f has a maximum and a minimum there. Let $f(\xi_i)$ and $f(\eta_i)$ be the respective maximum and minimum of f on $[x_{i-1}, x_i]$. Thus, ξ_i and η_i are in $[x_{i-1}, x_i]$ and

$$f(\xi_i) = m_i, \qquad f(\eta_i) = M_i.$$

We have $|\xi_i - \eta_i| \leqslant \Delta x_i \leqslant \|P\| < \delta$ and

$$M_i - m_i = |M_i - m_i| = |f(\xi_i) - f(\eta_i)| < \frac{\epsilon}{b-a}$$

for each $i \in \{1, 2, \ldots, n\}$. Hence,

$$\bar{S}(f,P) - \underline{S}(f,P) = \sum_{i=1}^{n} (M_i - m_i)\Delta x_i < \frac{\epsilon}{b-a} \sum_{i=1}^{n} \Delta x_i = \frac{\epsilon}{b-a}(b-a) = \epsilon.$$

By Theorem 1.2, f is D-integrable over $[a,b]$.

Lemma 1.4. *If f is bounded over $[a,b]$ and $a < c < b$, then*

$$\underline{\int_a^c} f(x)\,dx + \underline{\int_c^b} f(x)\,dx = \underline{\int_a^b} f(x)\,dx \tag{1.33}$$

and

$$\overline{\int_a^c} f(x)\,dx + \overline{\int_c^b} f(x)\,dx = \overline{\int_a^b} f(x)\,dx. \tag{1.34}$$

PROOF. We prove (1.33) and leave the proof of (1.34) to the reader (Prob. 1.9). Let $P = \langle x_0, x_1, \ldots, x_n \rangle$ be a partition of $[a,b]$. Let $[x_{k-1}, x_k]$ be the subinterval of P containing c. If c is not one of the endpoints x_{k-1} or x_k, define $P' = P \cup \{c\}$. If c is one of x_{k-1} or x_k, define $P' = P$. P' is now a partition of $[a,b]$ containing the point c and since $P \subseteq P'$, P' is a refinement of P. Let $P'_{[a,c]}$ and $P'_{[c,b]}$ be the subsets of P' forming respectively partitions of $[a,c]$ and $[c,b]$. We have

$$P \subseteq P' = P'_{[a,c]} \cup P'_{[c,b]}$$

and

$$\begin{aligned}\underline{S}(f,P) \leqslant \underline{S}(f,P') &= \underline{S}(f,P'_{[a,c]}) + \underline{S}(f,P'_{[c,b]})\\ &\leqslant \underline{\int_a^c} f(x)\,dx + \underline{\int_c^b} f(x)\,dx.\end{aligned}$$

Thus,

$$\underline{S}(f,P) \leqslant \underline{\int_a^c} f(x)\,dx + \underline{\int_c^b} f(x)\,dx$$

for any partition P of $[a,b]$. The number on the right is an upper bound for the set G of all lower Darboux sums of f over $[a,b]$. Since $\underline{\int_a^b} f(x)\,dx$ is a supremum of G, this implies that

$$\underline{\int_a^b} f(x)\,dx \leqslant \underline{\int_a^c} f(x)\,dx + \underline{\int_c^b} f(x)\,dx. \tag{1.35}$$

We now prove that this inequality can be reversed. Let $\epsilon > 0$ be given. We have

$$\underline{\int_a^c} f(x)\,dx - \frac{\epsilon}{2} < \underline{\int_a^c} f(x)\,dx \quad \text{and} \quad \underline{\int_c^b} f(x)\,dx - \frac{\epsilon}{2} < \underline{\int_c^b} f(x)\,dx.$$

Therefore, there exist partitions $P_{[a,c]}$ of $[a,c]$ and $P_{[c,b]}$ of $[b,c]$ such that

$$\underline{\int_a^c} f(x)\,dx - \frac{\epsilon}{2} < \underline{S}(f,P_{[a,c]}) \quad \text{and} \quad \underline{\int_c^b} f(x)\,dx - \frac{\epsilon}{2} < \underline{S}(f,P_{[c,b]}).$$

Adding, we obtain

$$\underline{\int_a^c} f(x)\,dx + \underline{\int_c^b} f(x)\,dx - \epsilon < \underline{S}(f,P_{[a,c]}) + \underline{S}(f,P_{[c,b]}). \tag{1.36}$$

Put $P' = P_{[a,c]} \cup P_{[c,b]}$. P' is a partition of $[a,b]$. Moreover,

$$\underline{S}(f,P_{[a,b]}) + \underline{S}(f,P_{[c,b]}) = \underline{S}(f,P').$$

This and (1.36) imply that

$$\underline{\int_a^c} f(x)\,dx + \underline{\int_c^b} f(x)\,dx - \epsilon < \underline{S}(f,P') \leqslant \underline{\int_a^b} f(x)\,dx.$$

Therefore,

$$\underline{\int_a^c} f(x)\,dx + \underline{\int_c^b} f(x)\,dx - \epsilon < \underline{\int_a^b} f(x)\,dx$$

for all $\epsilon > 0$. We conclude that

$$\underline{\int_a^c} f(x)\,dx + \underline{\int_c^b} f(x)\,dx \leqslant \underline{\int_a^b} f(x)\,dx. \tag{1.37}$$

This and (1.35) yield (1.33).

Prob. 1.9. Complete the proof of the last lemma by showing that (1.34) holds if f is bounded over $[a,b]$.

Theorem 1.5. *If $a < c < b$ for real numbers a, b, and c and f is D-integrable over $[a,c]$ and over $[c,b]$, then f is D-integrable over $[a,b]$ and*

$$\int_a^c f(x)\,dx + \int_c^b f(x)\,dx = \int_a^b f(x)\,dx.$$

Proof. Exercise.

Prob. 1.10. Prove: If $a \leqslant c \leqslant b$ is defined as

$$h(x) = \begin{cases} 0 & \text{if} \quad a \leqslant x \leqslant b, \quad x \neq c \\ h(c) \neq 0 & \text{if} \quad x = c, \end{cases}$$

then h is D-integrable over $[a,b]$ and $\int_a^b h(x)\,dx = 0$ (see Prob. 1.8).

Prob. 1.11. Prove: If $a < c < b$ and f is continuous on $[a,c]$ and on $[c,b]$, then f is D-integrable over $[a,b]$ and

$$\int_a^c f(x)\,dx + \int_c^b f(x)\,dx = \int_a^b f(x)\,dx.$$

Prob. 1.12. Prove: If f is D-integrable over $[a,b]$ and $a < c < b$, then f is D-integrable over $[a,c]$ and over $[c,b]$.

Prob. 1.13. Prove: If f is D-integrable over $[a,b]$, then it is D-integrable over every bounded, closed subinterval of $[a,b]$.

Remark 1.2. If f is integrable over all the intervals involved in

$$\int_a^c f(x)\,dx + \int_c^b f(x)\,dx = \int_a^b f(x)\,dx,$$

then this equality holds regardless of the order of a, b, and c. For example, suppose $c < b < a$, then

$$\int_c^b f(x)\,dx + \int_b^a f(x)\,dx = \int_c^a f(x)\,dx = -\int_a^c f(x)\,dx$$

and, therefore,

$$\int_a^c f(x)\,dx + \int_c^b f(x)\,dx + \int_b^a f(x)\,dx = 0.$$

This implies

$$\int_a^c f(x)\,dx + \int_c^b f(x)\,dx = -\int_b^a f(x)\,dx = \int_a^b f(x)\,dx.$$

Theorem 1.6. *If f is bounded on $[a,b]$ and is D-integrable over every closed interval $[c,d]$ such that $a < c < d < b$, then f is D-integrable over $[a,b]$.*

PROOF. Since f is bounded on $[a,b]$, there exists a $B > 0$ such that

$$|f(x)| \leqslant B \qquad \text{for all} \quad x \in [a,b].$$

Given $\epsilon > 0$, take δ such that

$$0 < \delta \leqslant \min\left\{ \frac{\epsilon}{4b}, \frac{b-a}{2} \right\}.$$

This implies that $0 < \delta \leqslant (b-a)/2$ and, hence, that $a < a + \delta \leqslant b - \delta < b$. We have

$$\overline{\int_a^b} f(x)\,dx = \overline{\int_a^{a+\delta}} f(x)\,dx + \int_{a+\delta}^{b-\delta} f(x)\,dx + \overline{\int_{b-\delta}^{b}} f(x)\,dx \tag{1.38}$$

and

$$\underline{\int_a^b} f(x)\,dx = \underline{\int_a^{a+\delta}} f(x)\,dx + \int_{a+\delta}^{b-\delta} f(x)\,dx + \underline{\int_{b-\delta}^{b}} f(x)\,dx. \tag{1.39}$$

Subtract (1.39) from (1.38) and obtain

$$0 \leqslant \overline{\int_a^b} f(x)\,dx - \underline{\int_a^b} f(x)\,dx = \left(\overline{\int_a^{a+\delta}} f(x)\,dx - \underline{\int_a^{a+\delta}} f(x)\,dx \right) + \left(\overline{\int_{b-\delta}^{b}} f(x)\,dx - \underline{\int_{b-\delta}^{b}} f(x)\,dx \right). \tag{1.40}$$

Apply the result in Prob. 1.4 to the two differences in parenthesis on the right and obtain

$$\left(\overline{\int_a^{a+\delta}} f(x)\,dx - \underline{\int_a^{a+\delta}} f(x)\,dx \right) + \left(\overline{\int_{b-\delta}^{b}} f(x)\,dx - \underline{\int_{b-\delta}^{b}} f(x)\,dx \right) < 2B\delta + 2B\delta = 4B\delta < \epsilon.$$

This and (1.40) imply that

$$0 \leqslant \overline{\int_a^b} f(x)\,dx - \underline{\int_a^b} f(x)\,dx < \epsilon$$

for each $\epsilon > 0$. It follows from this that

$$\overline{\int_a^b} f(x)\,dx = \underline{\int_a^b} f(x)\,dx.$$

This proves the theorem.

PROB. 1.14. Prove: If f is defined on $[-1,1]$ as

$$f(x) = \begin{cases} \sin\dfrac{1}{x} & \text{for} \quad x \neq 0 \quad \text{and} \quad x \in [-1,1] \\ 2 & \text{for} \quad x = 0, \end{cases}$$

then f is D-integrable over $[-1,1]$.

PROB. 1.15. Prove: If f is monotonic on $[a,b]$, then

$$\left| \int_a^b f(x)\,dx - \frac{b-a}{n} \sum_{k=1}^{n} f\left(a + k\frac{b-a}{n}\right) \right| \leqslant \frac{|f(b) - f(a)|}{n}(b-a)$$

for each positive integer n.

PROB. 1.16. Prove: If f is monotonic over $[a,b]$, then

$$\int_a^b f(x)\,dx$$

lies between $f(a)(b-a)$ and $f(b)(b-a)$.

2. Order Properties of the Darboux Integral

Lemma 2.1. *If f and g are bounded on $[a,b]$ and $f(x) \leqslant g(x)$ for all $x \in [a,b]$, then*

$$\underline{\int_a^b} f(x)\,dx \leqslant \underline{\int_a^b} g(x)\,dx \tag{2.1a}$$

and

$$\overline{\int_a^b} f(x)\,dx \leqslant \overline{\int_a^b} g(x)\,dx. \tag{2.1b}$$

PROOF. Let $m(f)$ be the infimum of f on $[a,b]$ and $M(f)$ its supremum there, and let $m(g)$ and $M(g)$ have similar meanings for g. We have

$$m(f) \leqslant f(x) \leqslant g(x) \leqslant M(g) \qquad \text{for all} \quad x \in [a,b].$$

It follows that

$$m(f) \leqslant m(g) \quad \text{and} \quad M(f) \leqslant M(g)$$

(explain).

Let $P = \langle x_0, x_1, \ldots, x_n \rangle$ be a partition of $[a,b]$. Apply the result just proved for the interval $[a,b]$ to each subinterval $[x_{i-1}, x_i]$ of P. We obtain

$$m_i(f) \leqslant m_i(g) \quad \text{and} \quad M_i(f) \leqslant M_i(g)$$

and, therefore,

$$\underline{S}(f,P) \leqslant \underline{S}(g,P) \quad \text{and} \quad \overline{S}(f,P) \leqslant \overline{S}(g,P).$$

It follows that

$$\underline{S}(f,P) \leqslant \underline{\int_a^b} g(x)\,dx \quad \text{and} \quad \overline{\int_a^b} f(x)\,dx \leqslant \overline{S}(g,P).$$

These hold for all partitions P of $[a,b]$. In turn, this implies that

$$\underline{\int_a^b} f(x)\,dx \leqslant \underline{\int_a^b} g(x)\,dx \quad \text{and} \quad \overline{\int_a^b} f(x)\,dx \leqslant \overline{\int_a^b} g(x)\,dx$$

(why?).

Theorem 2.1. *If f and g are D-integrable over $[a,b]$ and $f(x) \leqslant g(x)$ for all $x \in [a,b]$, then*

$$\int_a^b f(x)\,dx \leqslant \int_a^b g(x)\,dx.$$

PROOF. Exercise.

Corollary. *If f is D-integrable over $[a,b]$ and $f(x) \geqslant 0$ for all $x \in [a,b]$, then*

$$\int_a^b f(x)\,dx \geqslant 0.$$

PROOF. Exercise.

Theorem 2.2. *If f is continuous on $[a,b]$ and $f(x) \geqslant 0$ for all $x \in [a,b]$, but $f(x) \neq 0$ for some $x \in [a,b]$, then the strict inequality*

$$\int_a^b f(x)\,dx > 0$$

holds.

PROOF. Assume first that a c exists such that $a < c \leqslant b$ and $f(c) \neq 0$. By the hypothesis on f, we must have $f(c) > 0$. Since f is continuous at c, there exists a $\delta > 0$ such that if $x \in [a,b]$ and $c - \delta < x \leqslant c$, then $f(x) > 0$. Let $\lambda_1 = \max\{a, c-\delta\}$ so that $a \leqslant \lambda_1 < c$ and $c - \delta \leqslant \lambda_1 < c$. Take λ such that $\lambda_1 < \lambda < c$. It follows that $f(x) > 0$ for all $x \in [\lambda, c]$. Since $[\lambda, c] \subseteq [a, b]$, f is continuous on the bounded closed interval $[\lambda, c]$ and, therefore, has a minimum $f(\xi)$ there. Clearly, $f(\xi) > 0$. Now f is continuous on $[a,b]$ and, therefore, D-integrable over $[a,b]$, and

$$\int_a^b f(x)\,dx = \int_a^\lambda f(x)\,dx + \int_\lambda^c f(x)\,dx + \int_c^b f(x)\,dx. \tag{2.2}$$

But from $f(x) \geqslant 0$, $x \in [a,b]$, we conclude that

$$\int_a^\lambda f(x)\,dx \geqslant 0 \quad \text{and} \quad \int_c^b f(x)\,dx \geqslant 0.$$

This and (2.2) imply that

$$\int_a^b f(x)\,dx \geqslant \int_\lambda^c f(x)\,dx \geqslant f(\xi)(c-\lambda) > 0.$$

Here, the middle inequality is a consequence of Lemma 1.3 and the result cited in Prob. 1.2.

When a c exists such that $a \leqslant c < b$ and $f(c) \neq 0$, the conclusion follows by analogous reasoning. This completes the proof.

Remark 2.1. If f is continuous on $[a,b]$ and nonnegative there, but $f(x) \neq 0$ for some $x \in [a,b]$, then

$$\int_b^a f(x)\,dx = -\int_a^b f(x)\,dx < 0.$$

Prob. 2.1. Prove: If f is continuous on $[a,b]$ and nonnegative there and

$$\int_a^b f(x)\,dx = 0,$$

then $f(x) = 0$ for all $x \in [a,b]$.

Prob. 2.2. Prove: If f is continuous on $[a,b]$ and *nonpositive* there and

$$\int_a^b f(x)\,dx = 0,$$

then $f(x) = 0$ for all $x \in [a,b]$.

Prob. 2.3. Prove: If f is continuous on $[a,b]$ and

$$\int_a^b f^2(x)\,dx = 0,$$

then $f(x) = 0$ for all $x \in [a,b]$.

Lemma 2.2. *If f is bounded over $[a,b]$, then*: (1) $m(cf) = cm(f)$, *and* $M(cf) = cM(f)$ *for* $c > 0$; (2) $m(cf) = cM(f)$ *and* $M(cf) = cm(f)$ *for* $c < 0$.

Proof. Exercise.

Theorem 2.3. *If f is bounded and D-integrable over $[a,b]$ and c is some real constant, then so is cf and*

$$\int_a^b cf(x)\,dx = c\int_a^b f(x)\,dx.$$

PROOF. The theorem certainly holds if $c = 0$ (explain). Assume that $c > 0$.
Let $P = \langle x_0, x_1, \ldots, x_n\rangle$ be a partition of $[a,b]$. Apply Lemma 2.2(1) to each subinterval $[x_{i-1}, x_i]$ of P and conclude that

$$m_i(cf) = cm_i(f) \quad \text{and} \quad M_i(cf) = cM_i(f).$$

We obtain

$$\begin{aligned} c\,\underline{S}(f,P) &= c\sum_{i=1}^{n} m_i\Delta x_i = \sum_{i=1}^{n} cm_i\Delta x_i = \sum_{i=1}^{n} m_i(cf)\Delta x_i \\ &= \underline{S}(cf,P) \leqslant \underline{\int_a^b} cf(x)\,dx \end{aligned} \tag{2.3}$$

and

$$\begin{aligned} c\overline{S}(f,P) &= c\sum_{i=1}^{n} M_i\Delta x_i = \sum_{i=1}^{n} cM_i\Delta x_i = \sum_{i=1}^{n} M_i(cf)\Delta x_i \\ &= \overline{S}(cf,P) \geqslant \overline{\int_a^b} cf(x)\,dx. \end{aligned} \tag{2.4}$$

From (2.3) we conclude that

$$\underline{S}(f,P) \leqslant \frac{1}{c}\underline{\int_a^b} cf(x)\,dx.$$

This holds for any partition P of $[a,b]$. Hence

$$\int_a^b f(x)\,dx = \underline{\int_a^b} f(x)\,dx \leqslant \frac{1}{c}\underline{\int_a^b} cf(x)\,dx$$

and, therefore,

$$c\int_a^b f(x)\,dx \leqslant \underline{\int_a^b} cf(x)\,dx. \tag{2.5}$$

Similarly, we conclude from (2.4) that

$$c\int_a^b f(x)\,dx \geqslant \overline{\int_a^b} cf(x)\,dx. \tag{2.6}$$

Inequalities (2.5) and (2.6) imply that

$$\overline{\int_a^b} cf(x)\,dx \leqslant c\int_a^b f(x)\,dx \leqslant \underline{\int_a^b} cf(x)\,dx.$$

In turn, this implies (explain) that

$$\overline{\int_a^b} cf(x)\,dx = c\int_a^b f(x)\,dx = \underline{\int_a^b} cf(x)\,dx. \tag{2.7}$$

We conclude from this that cf is integrable and, moreover, that

$$\int_a^b cf(x) = c\int_a^b f(x)\,dx.$$

Now assume that $c < 0$. Again take a partition $P = \langle x_0, x_1, \ldots, x_n \rangle$ of $[a,b]$. Apply Lemma 2.2 to each subinterval $[x_{i-1}, x_i]$ of P and conclude that

$$m_i(cf) = cM_i(f) \quad \text{and} \quad M_i(cf) = cm_i(f).$$

Obtain

$$c\,\underline{S}(f,P) = \overline{S}(cf,P) \geqslant \overline{\int_a^b} cf(x)\,dx \tag{2.8}$$

and

$$c\overline{S}(f,P) = \underline{S}(cf,P) \leqslant \underline{\int_a^b} cf(x)\,dx, \tag{2.9}$$

and conclude from (2.8) and (2.9) that

$$\underline{S}(f,P) \leqslant \frac{1}{c}\overline{\int_a^b} cf(x)\,dx \quad \text{and} \quad \overline{S}(f,P) \geqslant \frac{1}{c}\underline{\int_a^b} cf(x)\,dx.$$

These hold for any partition P of $[a,b]$, so they imply respectively

$$\int_a^b f(x)\,dx = \underline{\int_a^b} f(x)\,dx \leqslant \frac{1}{c}\overline{\int_a^b} cf(x)\,dx \tag{2.10a}$$

and

$$\int_a^b f(x)\,dx = \overline{\int_a^b} f(x)\,dx \geqslant \frac{1}{c}\underline{\int_a^b} cf(x)\,dx \tag{2.10b}$$

and, hence,

$$\overline{\int_a^b} cf(x)\,dx \leqslant c\int_a^b f(x)\,dx \leqslant \underline{\int_a^b} cf(x)\,dx.$$

Reasoning as before, we obtain the conclusion in this case ($c < 0$) also. This completes the proof.

Corollary. *If f is bounded and D-integrable on $[a,b]$, then so is $-f$ and*

$$\int_a^b (-f(x))\,dx = -\int_a^b f(x)\,dx.$$

PROOF. Exercise.

We are now in a position to prove an important inequality concerning D-integrals.

Remark 2.2. Consider the function f defined on $[-1,1]$ as

$$f(x) = \begin{cases} 1 & \text{if } x \text{ is rational and} \quad -1 \leqslant x \leqslant 1 \\ -1 & \text{if } x \text{ is irrational and} \quad -1 \leqslant x \leqslant 1. \end{cases}$$

One sees readily that f is bounded but not D-integrable over $[a,b]$. However, $|f|$ is the constant function 1 on $[-1,1]$ and, therefore, D-integrable. Thus, $|f|$ may be D-integrable without f itself being so. In the next theorem (Theorem 2.4) we prove that if f is bounded and D-integrable, then so is $|f|$. We precede the proof with a definition and some lemmas.

Def. 2.1. If f is defined and real-valued on a set $A \neq \varnothing$, then its *fluctuation* $\Omega_A(f)$ on A, is defined as

$$\Omega_A(f) = \sup_A f - \inf_A f. \tag{2.11}$$

For example, let f be defined as

$$f(x) = \frac{1}{1+x^2} \qquad \text{for all} \quad x \in \mathbb{R}.$$

It is easily checked that

$$\sup_{\mathbb{R}} f = \max_{\mathbb{R}} f = 1 \quad \text{and} \quad \inf_{\mathbb{R}} f = 0.$$

Hence,

$$\Omega_{\mathbb{R}}(f) = 1 - 0 = 1.$$

Lemma 2.3. *If f is a real-valued function bounded on some set $A \neq \varnothing$, then*

$$\sup_{x_1, x_2 \in A} |f(x_1) - f(x_2)| = \Omega_A(f). \tag{2.12}$$

Proof. Write

$$D_A(f) = \sup_{x_1, x_2 \in A} |f(x_1) - f(x_2)|. \tag{2.13}$$

Note that if $x_1 \in A$ and $x_2 \in A$, then

$$\inf_A f \leqslant f(x_1) \leqslant \sup_A f \quad \text{and} \quad \inf_A f \leqslant f(x_2) \leqslant \sup_A f.$$

These imply that

$$|f(x_1) - f(x_2)| \leqslant \sup_A f - \inf_A f = \Omega_A(f).$$

This holds for any x_1 and x_2 in A and we obtain

$$D_A(f) \leqslant \Omega_A(f). \tag{2.14}$$

On the other hand, if $x_1 \in A$ and $x_2 \in A$, then

$$f(x_1) - f(x_2) \leqslant |f(x_1) - f(x_2)| \leqslant D_A(f)$$

so that

$$f(x_1) \leqslant f(x_2) + D_A(f) \qquad \text{for all} \quad x_1 \in A \quad \text{and each} \quad x_2 \in A.$$

This implies that

$$\sup_A f \leqslant f(x_2) + D_A(f) \qquad \text{for each} \quad x_2 \in A.$$

It follows from this that

$$\sup_A f - D_A(f) \leqslant f(x_2) \qquad \text{for each} \quad x_2 \in A.$$

Hence,

$$\sup_A f - D_A(f) \leqslant \inf_A f \quad \text{and} \quad \Omega_A \leqslant D_A(f).$$

In view of (2.13), this yields (2.12).

Lemma 2.4. *If f is bounded on A, then*

$$\Omega_A(|f|) \leqslant \Omega_A(f). \tag{2.15}$$

Proof. By the properties of absolute value and the last lemma we have: If $x_1 \in A$ and $x_2 \in A$, then

$$||f(x_1)| - |f(x_2)|| \leqslant |f(x_1) - f(x_2)| \leqslant \sup_{x_1, x_2 \in A} |f(x_1) - f(x_2)| = \Omega_A(f).$$

This and the last lemma imply that

$$\Omega_A(|f|) = \sup_{x_1, x_2 \in A} ||f(x_1)| - |f(x_2)|| \leqslant \Omega_A(f).$$

Theorem 2.4. *If f is bounded and D-integrable on $[a,b]$, then so is $|f|$ and*

$$\left|\int_a^b f(x)\,dx\right| \leqslant \int_a^b |f(x)|\,dx. \tag{2.16}$$

Proof. Since f is bounded on $[a,b]$, so is $|f|$. Let $P = \langle x_0, x_1, \ldots, x_n \rangle$ be any partition of $[a,b]$. Apply the last lemma to each subinterval $I_i = [x_{i-1}, x_i]$ of P. We have

$$\Omega_{I_i}(|f|) \leqslant \Omega_{I_i}(f)$$

which states that

$$\sup_{I_i} |f| - \inf_{I_i} |f| \leqslant \sup_{I_i} f - \inf_{I_i} f.$$

This can be written

$$M_i(|f|) - m_i(|f|) \leqslant M_i(f) - m_i(f).$$

Hence,

$$\overline{S}(|f|, P) - \underline{S}(|f|, P) \leqslant \overline{S}(f, P) - \underline{S}(f, P). \tag{2.17}$$

By hypothesis, f is D-integrable on $[a,b]$. By Theorem 1.2, (2.17) implies that $|f|$ is D-integrable on $[a,b]$ (prove this).

We complete the proof by proving (2.16). We have

$$-|f(x)| \leqslant f(x) \leqslant |f(x)| \qquad \text{for all} \quad x \in [a,b].$$

The functions involved in this inequality are all D-integrable. Hence,

$$-\int_a^b |f(x)|\,dx = \int_a^b -|f(x)|\,dx \leqslant \int_a^b f(x)\,dx \leqslant \int_a^b |f(x)|\,dx,$$

i.e.,

$$-\int_a^b |f(x)|\,dx \leqslant \int_a^b f(x)\,dx \leqslant \int_a^b |f(x)|\,dx.$$

This implies (2.16).

Prob. 2.4. Prove: If f is Darboux integrable on a bounded interval with endpoints t_1 and t_2 and B is an upper bound for the function $|f|$, then

$$\left|\int_{t_1}^{t_2} f(x)\,dx\right| \leqslant B|t_2 - t_1|.$$

3. Algebraic Properties of the Darboux Integral

Theorem 3.1. *If f and g are bounded and Darboux integrable on $[a,b]$, then so is $f+g$ and*

$$\int_a^b (f(x) + g(x))\,dx = \int_a^b f(x)\,dx + \int_a^b g(x)\,dx. \tag{3.1}$$

Proof. By hypothesis, f and g are both bounded on $[a,b]$. Hence, $f+g$ is. Let $P = \langle x_0, x_1, \ldots; x_n \rangle$ be a partition of $[a,b]$. For each subinterval $I_i = [x_{i-1}, x_i]$ of P, we have, in our usual notation,

$$m_i(f) \leqslant f(x) \leqslant M_i(f) \quad \text{and} \quad m_i(g) \leqslant g(x) \leqslant M_i(g) \qquad \text{for all} \quad x \in I_i.$$

These imply that

$$m_i(f) + m_i(g) \leqslant f(x) + g(x) \leqslant M_i(f) + M_i(g) \qquad \text{for all} \quad x \in I_i.$$

It follows that

$$m_i(f) + m_i(g) \leqslant m_i(f+g) \leqslant M_i(f+g) \leqslant M_i(f) + M_i(g)$$

for each $i \in \{1, 2, \ldots, n\}$. Forming Darboux sums, we have

$$\underline{S}(f,P) + \underline{S}(g,P) \leqslant \underline{S}(f+g,P) \leqslant \underline{\int_a^b} (f(x) + g(x))\,dx$$

and

$$\overline{\int_a^b} (f(x) + g(x))\,dx \leqslant \overline{S}(f+g,P) \leqslant \overline{S}(f,P) + \overline{S}(g,P)$$

so that

$$\underline{S}(f,P)+\underline{S}(f,P)\leqslant \underline{\int_a^b}(f(x)+g(x))\,dx \tag{3.2}$$

and

$$\overline{\int_a^b}(f(x)+g(x))\,dx\leqslant \overline{S}(f,P)+\overline{S}(g,P) \tag{3.3}$$

for each partition P of $[a,b]$.

Let $\epsilon>0$ be given. There exist partitions P' and P'' of $[a,b]$ such that

$$\int_a^b f(x)\,dx-\frac{\epsilon}{2}<\underline{S}(f,P') \tag{3.4a}$$

and

$$\int_a^b g(x)\,dx-\frac{\epsilon}{2}<\underline{S}(g,P''). \tag{3.4b}$$

Note, f and g are assumed to be D-integrable over $[a,b]$ so the integrals appearing on the left in (3.4) needn't be underlined with an underbar. Let $P=P'\cup P''$ so that P is a refinement of P' and also of P'' and, hence,

$$\underline{S}(f,P')\leqslant\underline{S}(f,P)\quad\text{and}\quad\underline{S}(g,P'')\leqslant\underline{S}(g,P).$$

These, together with (3.4), yield

$$\int_a^b f(x)\,dx-\frac{\epsilon}{2}<\underline{S}(f,P)\quad\text{and}\quad\int_a^b g(x)\,dx-\frac{\epsilon}{2}<\underline{S}(g,P).$$

After adding and using (3.2), these imply that

$$\int_a^b f(x)\,dx+\int_a^b g(x)\,dx-\epsilon<\underline{S}(f,P)+\underline{S}(g,P)\leqslant\underline{\int_a^b}(f(x)+g(x))\,dx.$$

Thus,

$$\int_a^b f(x)\,dx+\int_a^b g(x)\,dx-\epsilon<\underline{\int_a^b}(f(x)+g(x))\,dx.$$

This holds for each $\epsilon>0$. Therefore,

$$\int_a^b f(x)\,dx+\int_a^b g(x)\,dx\leqslant\underline{\int_a^b}(f(x)+g(x))\,dx. \tag{3.5}$$

Again, let $\epsilon>0$ be given. There exist partitions P' and P'' of $[a,b]$ such that

$$\overline{S}(f,P')<\int_a^b f(x)\,dx+\frac{\epsilon}{2}\quad\text{and}\quad\overline{S}(g,P'')<\int_a^b g(x)\,dx+\frac{\epsilon}{2}.$$

Proceeding as we did before and using (3.3) we obtain

$$\overline{\int_a^b}(f(x)+g(x))\,dx\leqslant\int_a^b f(x)\,dx+\int_a^b g(x)\,dx. \tag{3.6}$$

This and (3.5) yield

$$\overline{\int_a^b}(f(x)+g(x))\,dx \leqslant \underline{\int_a^b}(f(x)+g(x))\,dx.$$

Since the reverse inequality always holds, this implies that

$$\overline{\int_a^b}(f(x)+g(x))\,dx = \underline{\int_a^b}(f(x)+g(x))\,dx. \tag{3.7}$$

Thus, $f+g$ is D-integrable over $[a,b]$. Equation (3.1) now follows from (3.5), (3.6), and (3.7). This completes the proof.

In view of this theorem and Theorem 2.3, we have:

Theorem 3.2. *If c and d are real constants and both f and g are bounded and D-integrable over $[a,b]$, then so is $cf+dg$ and*

$$\int_a^b (cf(x)+dg(x))\,dx = c\int_a^b f(x)\,dx + d\int_a^b g(x)\,dx.$$

PROOF. Exercise.

PROB. 3.1. Prove: Let f and g be continuous on $[a,b]$. If $f(x) \leqslant g(x)$ for all $x \in [a,b]$, but $f(x) < g(x)$ holds for some $x \in [a,b]$, then the strict inequality

$$\int_a^b f(x)\,dx < \int_a^b g(x)\,dx$$

holds.

PROB. 3.2. Prove: If n is a nonnegative integer, then

$$\int_0^{\pi/2} \sin^{n+1}x\,dx < \int_0^{\pi/2} \sin^n x\,dx.$$

Theorem 3.3.* *If f is bounded on $[a,b]$ and its values are changed at finitely many points of $[a,b]$, then this changes neither its D-integrability nor its D-integral.*

PROOF. Assume that f is D-integrable on $[a,b]$ and its value is changed at some $c \in [a,b]$. Let g be the new function. Then

$$g(x) = \begin{cases} f(x) & \text{for } x \neq c, \quad x \in [a,b] \\ g(c) \neq f(x) & \text{for } x = c. \end{cases}$$

Define h as $h(x) = g(x) - f(x)$ for each $x \in [a,b]$. We have

$$h(x) = \begin{cases} 0 & \text{if } x \neq c, \quad x \in [a,b] \\ g(c) - f(c) \neq 0 & \text{if } x = c. \end{cases}$$

*See Section XIV.5 below.

h is certainly bounded on $[c,b]$. By Prob. 1.10, it is Darboux integrable and

$$\int_a^b h(x)\,dx = 0.$$

Now $g = f + h$ on $[a,b]$, where f and h are bounded and D-integrable, so

$$\int_a^b g(x)\,dx = \int_a^b (f(x) + h(x))\,dx = \int_a^b f(x)\,dx + \int_a^b h(x)\,dx = \int_a^b f(x)\,dx.$$

Thus, g has the same D-integral as f (over $[a,b]$). If f is not Darboux integrable, then the function g defined above is not Darboux integrable either (since $f = g - h$, the D-integrability of g would imply the D-integrability of f). To prove an analogous result for n points $c_1, \ldots, c_n$ rather than one point, we use induction on n. The reader should complete the proof by carrying out this induction.

Prob. 3.3. Prove: If f and g are bounded and Darboux integrable on $[a,b]$, then the functions m and M, defined as

$$m(x) = \min\{f(x), g(x)\} \quad \text{and} \quad M(x) = \max\{f(x), g(x)\}$$

for each $x \in [a,b]$, are also bounded and Darboux integrable on $[a,b]$ (Hint: see Prob. I.13.17).

Prob. 3.4. Prove: If f is bounded and Darboux integrable on $[a,b]$, then so are f^+ and f^-, where these are defined as

$$f^+(x) = \max\{0, f(x)\} \quad \text{and} \quad f^-(x) = \min\{0, f(x)\}$$

for each $x \in [a,b]$.

We now consider products of Darboux integrable functions. First we prove:

Theorem 3.4. *If f is bounded and Darboux integrable on $[a,b]$, then so is f^2.*

Proof. Suppose f is real-valued and bounded on a set $A \neq \emptyset$. Then

$$m_A(|f|) \leqslant |f(x)| \leqslant M_A(|f|) \qquad \text{for all} \quad x \in A. \tag{3.8}$$

Here $m_A(|f|)$ and $M_A(|f|)$ are defined as

$$m_A(|f|) = \inf_A |f| \quad \text{and} \quad M_A(|f|) = \sup_A |f|.$$

Since $|f(x)| \geqslant 0$ for all $x \in A$, it follows that $0 \leqslant m_A(|f|)$. Hence, upon squaring in (3.8) we obtain

$$m_A^2(|f|) \leqslant |f(x)|^2 \leqslant m_A^2(|f|) \qquad \text{for all} \quad x \in A. \tag{3.9}$$

Since $|f(x)|^2 = f^2(x)$, (3.9) can be written

$$m_A^2(|f|) \leqslant f^2(x) \leqslant M_A^2(|f|). \tag{3.10}$$

Now let $P = \langle x_0, x_1, \ldots, x_n \rangle$ be a partition of $[a,b]$. By (3.10) we obtain for each subinterval $[x_{i-1}, x_i]$ of P

$$m_i^2(|f|) \leqslant m_i(f^2) \leqslant M_i(f^2) \leqslant M_i^2(|f|).$$

Hence, for each $i \in \{1, 2, \ldots, n\}$, we have

$$\begin{aligned} M_i(f^2) - m_i(f^2) &\leqslant M_i^2(|f|) - m_i^2(|f|) \\ &= (M_i(|f|) + m_i(|f|))(M_i(|f|) - m_i(|f|)). \end{aligned} \tag{3.11}$$

Let B be some upper bound of $|f|$ on $[a,b]$. We have

$$M_i(|f|) + m_i(|f|) \leqslant 2M_i(|f|) \leqslant 2B$$

for each $i \in \{1, 2, \ldots, n\}$. We use (3.11) and form Darboux sums for f^2 and obtain

$$\bar{S}(f^2, P) - \underline{S}(f^2, P) \leqslant 2B\big(\bar{S}(|f|, P) - \underline{S}(|f|, P)\big). \tag{3.12}$$

By hypothesis, f is bounded and Darboux integrable over $[a,b]$. By Theorem 2.4, this implies that $|f|$ is also bounded and Darboux integrable over $[a,b]$. Using (3.12) and Theorem 1.2 it is easy to see that the Darboux integrability of $|f|$ implies that of f^2. Since $|f|$ is Darboux integrable over $[a,b]$, it follows that so is f^2.

Corollary. *If f and g are bounded and Darboux integrable over $[a,b]$, then so is their product fg.*

Proof. As the reader can prove, this is a consequence of

$$fg = \tfrac{1}{2}\big[(f+g)^2 - (f-g)^2\big].$$

Prob. 3.5. Prove: If there exist real numbers b and B such that

$$0 < b \leqslant f(x) \leqslant B \qquad \text{for all} \quad x \in [a,b]$$

and f is Darboux integrable over $[a,b]$, then $1/f$ is bounded and Darboux integrable of $[a,b]$.

4. The Riemann Integral

We can also approach the elementary theory of the integral via *Riemann sums*.

Def. 4.1. If f is defined on $[a,b]$, $P = \langle x_0, x_1, \ldots, x_n \rangle$ is a partition of $[a,b]$, and $\xi_1, \xi_2, \ldots, \xi_n$ are real numbers such that for each $i \in \{1, \ldots, n\}$

$$\xi_i \in [x_{i-1}, x_i], \tag{4.1}$$

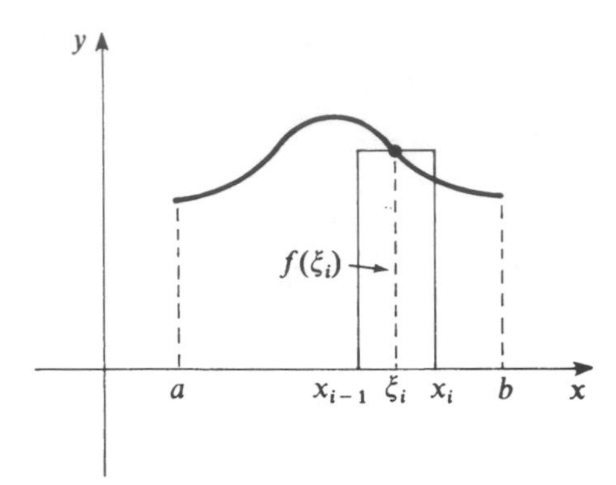

Figure 4.1

then we write

$$R(f, \xi, P) = \sum_{i=1}^{n} f(\xi_i)\Delta x_i \tag{4.2}$$

and call this sum a *Riemann sum* of f over $[a, b]$.

The motivation for treating Riemann sums arises from considering the special case where $f(x) \geqslant 0$ for all $x \in [a, b]$. In this case each $f(\xi_i)\Delta x_i$ in the Riemann sum (4.2) can be interpreted as the area of a rectangle based on the subinterval $I_i = [x_{i-1}, x_i]$ whose "height" is $f(\xi_i)$ (see Fig. 4.1). The Riemann sum (4.2) is then the sum of n such rectangles. This sum is an approximation to the area of the set of points in the x, y plane which are below the graph of f, above the x-axis and between the vertical lines $x = a$ and $x = b$ (Fig. 4.2).

Remark 4.1. The ξ in $R(f, \xi, P)$ is actually an ordered n-tuple $\xi = \langle \xi_1, \xi_2, \ldots, \xi_n \rangle$ whose coordinates $\xi_1, \xi_2, \ldots, \xi_n$ are all subject to conditions (4.1).

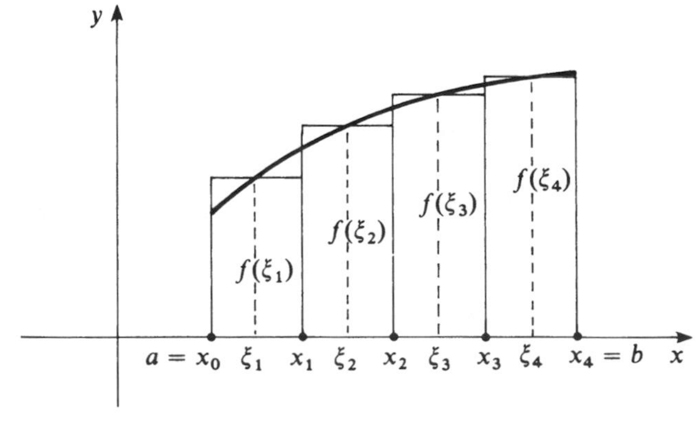

Figure 4.2

Remark 4.2. Given an f defined on $[a,b]$ and a partition $P = \langle x_0, x_1, \ldots, x_n \rangle$, there are two Darboux sums associated with P, the upper one $\overline{S}(f,P)$ and the lower sum $\underline{S}(f,P)$. However, for the same partition P, there could be infinitely many Riemann sums associated with P. In general, each choice of $\xi_1, \xi_2, \ldots, \xi_n$ subject to $\xi_i \in [x_{i-1}, x_i]$ for $i \in \{1, 2, \ldots, n\}$, gives rise to a different Riemann sum.

Def. 4.2. A function f defined on a bounded closed interval $[a,b]$ is called *Riemann Integrable* over $[a,b]$ if and only if some real number J exists such that for each $\epsilon > 0$, there exists a $\delta > 0$ such that for all partitions $P = \langle x_0, x_1, \ldots, x_n \rangle$ of $[a,b]$ with $\|P\| < \delta$ and points $\xi_1, \xi_2, \ldots, \xi_n$ such that $\xi_i \in [x_{i-1}, x_i]$ for all $i \in \{1, 2, \ldots, n\}$ we have

$$|R(f,\xi,P) - J| < \epsilon. \tag{4.3}$$

If f is Riemann integrable over $[a,b]$, then we write

$$\lim_{\|P\| \to 0} R(f,\xi,P) = J \quad \text{or} \quad \lim_{\|P\| \to 0} \sum_{i=1}^{n} f(\xi_i)\Delta x_i = J, \tag{4.4}$$

where J is the number satisfying (4.3). The number J is then called the *Riemann Integral* of f over $[a,b]$.

Theorem 4.1. *If f is Riemann integrable over $[a,b]$, then there is exactly one J such that* (4.4) *holds.*

PROOF. We know by Def. 4.2 that there exists at least one J such that (4.4) holds. Suppose (4.4) holds also for J', that is, suppose we have

$$\lim_{\|P\| \to 0} R(f,\xi,P) = J \quad \text{and} \quad \lim_{\|P\| \to 0} R(f,\xi,P) = J'.$$

Assume that $\epsilon > 0$ is given. By Def. 4.2, there exist: (1) a $\delta_1 > 0$ such that if $P = \langle x_0, x_1, \ldots, x_n \rangle$ is a partition with $\|P\| < \delta_1$ and $\xi_1, \ldots, \xi_n$ are such that $\xi_i \in [x_{i-1}, x_i]$ for all $i \in \{1, 2, \ldots, n\}$, then

$$|R(f,\xi,P) - J| < \frac{\epsilon}{2}, \tag{4.5}$$

and (2) a $\delta_2 > 0$ such that if $P' = \langle x'_0, x'_1, \ldots, x'_m \rangle$ is a partition of $[a,b]$ with $\|P'\| < \delta_2$ and $\xi'_1, \xi'_2, \ldots, \xi'_m$ are such that $\xi'_i \in [x'_{i-1}, x'_i]$ for all $i \in \{1, 2, \ldots, m\}$, then

$$|R(f,\xi',P') - J'| < \frac{\epsilon}{2}. \tag{4.6}$$

Let $\delta = \min\{\delta_1, \delta_2\}$ and $P'' = \langle x''_0, x''_1, \ldots, x''_q \rangle$ be a partition of $[a,b]$ such that $\|P''\| < \delta$. Choose $\xi''_i \in [x''_{i-1}, x''_i]$ for each $i \in \{1, \ldots, q\}$. Since, necessarily, $\|P''\| < \delta_1$ and $\|P''\| < \delta_2$, it follows that

$$|R(f,\xi'',P'') - J| < \frac{\epsilon}{2} \quad \text{and} \quad R(f,\xi'',P'') - J'| < \epsilon.$$

By properties of absolute value, these imply that

$$|J - J'| \leqslant |J - R(f,\xi'',P'')| + |R(f,\xi'',P'') - J'| < \frac{\epsilon}{2} + \frac{\epsilon}{2} = \epsilon$$

Thus, $|J - J'| < \epsilon$ holds for all $\epsilon > 0$, implying that $|J - J'| \leqslant 0$. This can only occur if $J - J' = 0$, i.e., $J = J'$. This completes the proof.

Before forming a Darboux sum of f over $[a,b]$, we must assume that f is bounded over $[a,b]$. On the other hand, a Riemann sum can be formed even if f is not bounded over $[a,b]$. However, we can prove that if f is Riemann integrable over $[a,b]$, then it is bounded there. This is the content of the next theorem.

Theorem 4.2. *If f is Riemann integrable over $[a,b]$, then it is necessarily bounded on $[a,b]$.*

PROOF. Suppose that f is not bounded over $[a,b]$ but is Riemann integrable. There exists a real J and a $\delta > 0$ such that if $P = \langle x_0, x_1, \ldots, x_n \rangle$ is a partition of $[a,b]$ with $\|P\| < \delta$ and $\xi_1, \xi_2, \ldots, \xi_n$ are points such that $\xi_1 \in [x_{i-1}, x_i]$ for each $i \in \{1, \ldots, n\}$, then

$$\left| \sum_{i=1}^{n} f(\xi_i) \Delta x_i - J \right| < 1. \tag{4.7}$$

Fix such a partition P. Since f is not bounded on $[a,b]$, then a subinterval $[x_{k-1}, x_k] = I_k$ of P exists on which f is not bounded (why?). We have

$$|f(\xi_k)\Delta x_k| - \left| \sum_{\substack{i=1 \\ i \neq k}}^{n} f(\xi_i)\Delta x_i - J \right| \leqslant \left| \sum_{i=1}^{n} f(\xi_i)\Delta x_i - J \right| < 1$$

so that

$$|f(\xi_k)| < \frac{1}{\Delta x_k} \left[1 + \left| \sum_{\substack{i=1 \\ i \neq k}}^{n} f(\xi_i)\Delta x_i - J \right| \right]. \tag{4.8}$$

Fix the ξ_i's for $i \neq k$, $i \in \{1, \ldots, n\}$. Then the number on the right is fixed. On the left in (4.8) we can take any ξ_k such that $x_{k-1} \leqslant \xi_k \leqslant x_k$. This implies that f is bounded on I_k, giving us a contradiction. We must, therefore, conclude that if f is Riemann integrable over $[a,b]$, then it is bounded there.

We will now prove that if a function is bounded, then it is Riemann integrable over $[a,b]$ if and only if it is D-integrable and that the two integrals are equal when they exist. We first prove:

Theorem 4.3. *If f is Riemann integrable over $[a,b]$, then it is D-integrable over $[a,b]$ and*

$$\lim_{\|P\| \to 0} R(f, \xi, P) = \int_a^b f(x)\, dx. \tag{4.9}$$

(Here, the integral on the right is the Darboux integral.)

Proof. By hypothesis, f is Riemann integrable on $[a,b]$. By Theorem 4.2, f is bounded on $[a,b]$. This implies that the upper and lower Darboux integrals

$$\overline{\int_a^b} f(x)\,dx \quad \text{and} \quad \underline{\int_a^b} f(x)\,dx$$

are real numbers. Let $\epsilon > 0$ be given. Since f is Riemann integrable over $[a,b]$, there exists a $\delta > 0$ such that if $P = \langle x_0, x_1, \ldots, x_n \rangle$ is a partition of $[a,b]$ with $\|P\| < \delta$ and $\xi_1, \ldots, \xi_n$ are points such that $\xi_1 \in I_i = [x_{i-1}, x_i]$ for each $i \in \{1, \ldots, n\}$, then

$$\left| \sum_{i=1}^{n} f(\xi_i)\Delta x_i - J \right| < \frac{\epsilon}{2}.$$

Hence

$$J - \frac{\epsilon}{2} < \sum_{i=1}^{n} f(\xi_i)\Delta x_i < J + \frac{\epsilon}{2}. \tag{4.10}$$

For each i, write the supremum of f on I_i as M_i and the infimum of f on I_i as m_i. There exists a $\xi_i' \in I_i$ such that

$$M_i - \frac{\epsilon}{2(b-a)} < f(\xi_i').$$

Multiply by Δx_i and sum to obtain

$$\sum_{i=1}^{n} M_i \Delta x_i - \frac{\epsilon \sum_{i=1}^{n} \Delta x_i}{2(b-a)} < \sum_{i=1}^{n} f(\xi_i')\Delta x_i.$$

This implies, in view of (4.10), that

$$\overline{S}(f,P) - \frac{\epsilon}{2} < \sum_{i=1}^{n} f(\xi_i')\Delta x_i < J + \frac{\epsilon}{2}.$$

Thus,

$$\overline{\int_a^b} f(x)\,dx \leqslant \overline{S}(f,P) < J + \epsilon$$

and, hence,

$$\overline{\int_a^b} f(x)\,dx < J + \epsilon. \tag{4.11}$$

This holds for each $\epsilon > 0$. We conclude that

$$\underline{\int_a^b} f(x)\,dx \leqslant \overline{\int_a^b} f(x)\,dx \leqslant J. \tag{4.12}$$

Similarly, for each i there is an $\eta_i \in I_i$ such that

$$f(\eta_i) < m_i + \frac{\epsilon}{2(b-a)}.$$

We multiply each of these by Δx_i and sum to obtain

$$\sum_{i=1}^{n} f(\eta_i)\Delta x_i < \underline{S}(f,P) + \frac{\epsilon}{2}.$$

Using (4.10), this implies that

$$J - \frac{\epsilon}{2} < \underline{S}(f,P) + \frac{\epsilon}{2} \leq \underline{\int_a^b} f(x)\,dx + \frac{\epsilon}{2}$$

so that

$$J < \underline{\int_a^b} f(x)\,dx + \epsilon.$$

This holds for each $\epsilon > 0$. We conclude from this that

$$J \leq \underline{\int_a^b} f(x)\,dx \leq \overline{\int_a^b} f(x)\,dx.$$

This and (4.12) imply that

$$\underline{\int_a^b} f(x)\,dx = J = \overline{\int_a^b} f(x)\,dx.$$

We conclude that f is D-integrable over $[a,b]$ and (4.9) holds.

We prepare for a proof of a converse of this by proving:

Lemma 4.1. *If f is bounded on $[a,b]$, then for each $\epsilon > 0$ there exists a $\delta > 0$ such that if P is a partition of $[a,b]$ with $\|P\| < \delta$, then*

$$\underline{\int_a^b} f(x)\,dx - \epsilon < \underline{S}(f,P) \tag{4.13a}$$

and

$$\overline{S}(f,P) < \overline{\int_a^b} f(x)\,dx + \epsilon. \tag{4.13b}$$

Proof. We prove (4.13a). Let $\epsilon > 0$ be given. There exists a partition $P' = \langle x_0', x_1', \ldots, x_m' \rangle$ of $[a,b]$ such that

$$\underline{\int_a^b} f(x)\,dx - \frac{\epsilon}{2} < \underline{S}(f,P'). \tag{4.14}$$

We fix the partition P'. P' consists of m subintervals. (This m will be a fixed integer in what follows.) Since f is bounded on $[a,b]$ by hypothesis, there exists a $B > 0$ such that $|f(x)| \leq B$ for all $x \in [a,b]$. Take a δ_1 such that

$$0 < \delta_1 \leq \frac{\epsilon}{4mB}, \tag{4.15}$$

and any partition $P = \langle x_0, x_1, \ldots, x_n \rangle$ of $[a,b]$ such that $\|P\| < \delta_1$. Let $P'' = P \cup P'$. P'' is a refinement of P and of P', so

$$\underline{\int_a^b} f(x)\,dx - \frac{\epsilon}{2} < \underline{S}(f,P') \leq \underline{S}(f,P''). \tag{4.16}$$

Write the points of P'' as $z_0, z_1, z_2, \ldots, z_r$. We wish to estimate the difference $\underline{S}(f,P'') - \underline{S}(f,P)$.

A term of $\underline{S}(f,P)$ is of the form $m_i\Delta x_i$, where m_i is the infimum of f on $[x_{i-1},x_i]$, and a term of $\underline{S}(f,P'')$ is of the form $m_\nu''\Delta z_\nu$, where m_ν'' is the infimum of f on $[z_{\nu-1},z_\nu]$. Since $x_{i-1}\in P''$ and $x_i\in P''$, for each i there exists a $z_{j-1}\in P''$ and a $z_k\in P''$ such that $x_{i-1}=z_{j-1}$ and $x_i=z_k$, this implies that $j-1<k$ and, therefore, that $j\leqslant k$. If $j=k$, then

$$[x_{i-1},x_i]=[z_{j-1},z_k]=[z_{j-1},z_j]$$

and

$$m_i\Delta x_i = m_j''\Delta z_j\,. \tag{4.17}$$

If $j<k$, then $x_{i-1}=z_{j-1}<z_j<z_k=x_i$. Hence, there exist points z_j, $z_{j+1},\ldots,z_{k-1}$ such that

$$x_{i-1}=z_{j-1}<z_j<\cdots<z_{k-1}<z_k=x_i\,. \tag{4.18}$$

All the points here constitute a partition of the subinterval $[x_{i-1},x_i]$ of the partition P, and

$$-B\Delta x_i\leqslant m_i\Delta x_i\leqslant\sum_{\nu=j}^{k}m_\nu''\Delta z_\nu\leqslant B\Delta x_i\,.$$

This implies that

$$\sum_{\nu=j}^{k}m_\nu''\Delta z_\nu-m_i\Delta x_i\leqslant 2B\Delta x_i\,. \tag{4.19}$$

But $\Delta x_i\leqslant\|P\|<\delta_1$ so that $2B\Delta x_i<2B\delta_1$. This, (4.19), and (4.15) imply that

$$\sum_{\nu=j}^{k}m_\nu''\Delta z_\nu-m_i\Delta x_i<\frac{\epsilon}{2m}\,. \tag{4.20}$$

Consider $\underline{S}(f,P'')-\underline{S}(f,P)$. We see that if $j=k$, then the contribution to it from $x_{i-1}=z_{j-1}$, $x_i=z_k$ vanishes because (4.17) holds in this case. If $j<k$, the contribution to $\underline{S}(f,P'')-\underline{S}(f,P)$ from the points in the inequalities (4.18) appears on the left of (4.20) and is estimated there. Note that when $j<k$, the points $z_j,\ldots,z_{k-1}$ between $x_{i-1}=z_{j-1}$ and $x_i=z_k$ are all points of the partition P'. But $j<k$ occurs at most m times. In fact, if this were not the case, then (4.18) would occur more than m times and P' would consist of more than m subintervals, which is impossible. Adding up all the contributions to $\underline{S}(f,P'')-\underline{S}(f,P)$ from all the subintervals of P, we obtain in view of (4.20),

$$\underline{S}(f,P'')-\underline{S}(f,P)<m\frac{\epsilon}{2m}=\frac{\epsilon}{2}\,. \tag{4.21}$$

This implies $\underline{S}(f,P'')<\underline{S}(f,P)+\epsilon/2$. Using (4.16), we conclude from this that

$$\underline{\int_a^b} f(x)\,dx-\frac{\epsilon}{2}<\underline{S}(f,P)+\frac{\epsilon}{2}\,. \tag{4.22}$$

Thus, for each partition of P of $[a,b]$, with $\|P\| < \delta_1$, we have

$$\underline{\int_a^b} f(x)\,dx - \epsilon < \underline{S}(f,P). \tag{4.22'}$$

We now ask the reader to prove that there exists a $\delta_2 > 0$ such that if P is a partition of $[a,b]$ with $\|P\| < \delta_2$, then

$$\overline{S}(f,P) < \overline{\int_a^b} f(x)\,dx + \epsilon. \tag{4.23}$$

(The required proof is similar to the previous one.) Finally, taking $\delta = \min\{\delta_1, \delta_2\}$ and partitions P such that $\|P\| < \delta$, we conclude that both (4.13a) and (4.13b) hold for all such P.

Theorem 4.4. *If f is bounded on $[a,b]$ and D-integrable then it is also Riemann integrable and*

$$\lim_{\|P\|\to 0} R(f,\xi,P) = \int_a^b f(x)\,dx.$$

Proof. Since f is D-integrable on $[a,b]$, we have

$$\int_a^b f(x)\,dx = \underline{\int_a^b} f(x)\,dx = \overline{\int_a^b} f(x)\,dx.$$

Let $\epsilon > 0$ be given. By Lemma 4.1, there exists a $\delta > 0$ such that if $\|P\| < \delta$, then

$$\int_a^b f(x) - \epsilon < \underline{S}(f,P) \leqslant \overline{S}(f,P) < \int_a^b f(x)\,dx + \epsilon. \tag{4.24}$$

Suppose that $P = \langle x_0, x_1, \ldots, x_n \rangle$, where $\|P\| < \delta$. For each $i \in \{1, \ldots, n\}$ we have, in the usual notation,

$$m_i \leqslant f(\xi_i) \leqslant M_i \qquad \text{for} \quad \xi_i \in [x_{i-1}, x_i]$$

and, hence,

$$\underline{S}(f,P) \leqslant \sum_{i=1}^{n} f(\xi_i)\Delta x_i \leqslant \overline{S}(f,P).$$

Because of (4.24), this implies that

$$\int_a^b f(x)\,dx - \epsilon < \sum_{i=1}^{n} f(\xi_i)\Delta x_i < \int_a^b f(x)\,dx + \epsilon.$$

Therefore,

$$\left| R(f,\xi,P) - \int_a^b f(x)\,dx \right| < \epsilon \qquad \text{for} \quad \|P\| < \delta.$$

Hence, the conclusion holds.

We can combine Theorems 4.3 and 4.4 to obtain:

Theorem 4.5. *If f is bounded on $[a,b]$, then f is D-integrable on $[a,b]$ if and only if it is Riemann integrable there and the two integrals are equal.*

Remark 4.3. Because of Theorem 4.5, we no longer distinguish between the Darboux and Riemann integrals and refer to a function as being either Riemann integrable or not. When a function is Riemann integrable, we often abbreviate and call it R-integrable. The Riemann integral

$$\int_a^b f(x)\,dx$$

is also termed the *definite* integral of f over $[a,b]$, and the function f is called the *integrand.*

It is often desirable to consider a sequence $\langle P_m \rangle = \langle P_1, P_2, \ldots \rangle$ of partitions of an interval $[a,b]$. In this case we need a notation indicating which partition a point belongs to and which point of the partition it is. This suggests the use of double subscripts. Thus, x_{im} stands for the ith point of the mth partition of the sequence. Different partitions may consist of a different number of subintervals. The number of subintervals in P_m will be written as n_m. Thus, for example,

$$P_1 = \langle x_{01}, x_{11}, \ldots, x_{n_1 1} \rangle,$$

$$P_2 = \langle x_{02}, x_{12}, \ldots, x_{n_2 2} \rangle,$$

$$\vdots$$

$$P_m = \langle x_{0m}, x_{1m}, \ldots, x_{n_m m} \rangle.$$

The ith subinterval of P_m is $I_{im} = [x_{(i-1)m}, x_{im}]$ and its length is $\Delta x_{im} = x_{(i-1)m} - x_{im}$. When Riemann sums are used, we select some $\xi_{im} \in I_{im}$. Corresponding to these ξ's in the mth partition, we have the Riemann sum

$$R(f, \xi_m, P_m) = \sum_{i=1}^{n_m} f(\xi_{im}) \Delta x_{im}.$$

Here $\langle \xi_m \rangle$ is a sequence of ordered n_m-tuples, where

$$\xi_m = \langle \xi_{1m}, \xi_{2m}, \ldots, \xi_{n_m n} \rangle,$$

and $\xi_{im} \in I_{im}$.

As an example, we take the interval $[a,b]$ and

$$P_m = \left\langle a, a + \frac{b-a}{m}, a + 2\frac{b-a}{m}, \ldots, a + m\frac{b-a}{m} \right\rangle.$$

Here

$$x_{im} = a + i\frac{b-a}{m}$$

where $i \in \{0, 1, \ldots, m\}$ and $n_m = m$,

$$\Delta x_{im} = \frac{b-a}{m}, \qquad i \in \{0, 1, \ldots, m\},$$

with

$$\xi_{im} = a + \frac{i-1}{m} b - a \quad \text{or} \quad \xi_{im} = a + i\frac{b-a}{m}.$$

PROB. 4.1. Let f be Riemann integrable over $[a,b]$. Let $\langle P_m \rangle$ be a sequence of partitions of $[a,b]$, and $\langle \xi_m \rangle$ a sequence of ordered n_m-tuples with $\xi_{im} \in I_{im} = [x_{(i-1)m}, x_{im}]$. Prove: If $\lim_{m\to+\infty}\|P_m\| = 0$, then

$$\lim_{m\to+\infty} R(f, \xi_m, P_m) = \int_a^b f(x)\,dx.$$

PROB. 4.2. If f is Riemann integrable over $[a,b]$, then

$$\lim_{n\to+\infty} \frac{b-a}{n} \sum_{i=1}^{n} f\left(a + i\frac{b-a}{n}\right)$$

$$= \int_a^b f(x)\,dx = \lim_{n\to+\infty}\left(\frac{b-a}{n} \sum_{i=1}^{n} f\left(a + (i-1)\frac{b-a}{n}\right)\right).$$

PROB. 4.3. Use Prob. 4.2 to prove: If $b > 0$, then

$$\int_0^b \cos x\,dx = \sin b \quad \text{and} \quad \int_0^b \sin x\,dx = 1 - \cos b.$$

5. Primitives

Def. 5.1. If f and G are defined on an interval I and

$$\frac{dG(x)}{dx} = f(x) \qquad \text{for all} \quad x \in I, \tag{5.1}$$

then G is called a *primitive* of f on I. We shall also use this terminology when I is replaced by some set S which is open in $\mathbb{R}$. Primitives are also called *antiderivatives*.

EXAMPLE 5.1. By Example VII.I.3, we have, if $\beta \neq -1$, then

$$\frac{d}{dx}\left(\frac{x^{\beta+1}}{\beta+1}\right) = x^{\beta} \qquad \text{for all} \quad x \in (0; +\infty). \tag{5.2}$$

Therefore, if $\beta \neq -1$, then the function G defined as $G(x) = x^{\beta+1}/(\beta+1)$ for $x \in (0; +\infty)$ is a primitive on $(0; +\infty)$ of f defined as $f(x) = x^{\beta}$ for $x \in (0; +\infty)$. If $\beta > 0$, so that $\beta + 1 > 1$, then the domains of G and f, when extended to include 0, by defining $G(0) = 0 = f(0)$, are still related in the same way; that is, G is a primitive of f on the closed interval $[0, +\infty)$.

If G is a primitive of f on an open set S and c is some constant, then $(G(x)+c)' = G'(x) = f(x)$ for $x \in S$ and therefore $G + c$ is also a primitive of f on S. Accordingly, once a function has a primitive on an open set, then it has infinitely many primitives there. We also observe that if F and G are primitives of a function f on an interval I, then $F'(x) = f(x) = G'(x)$ for $x \in I$ and therefore by Theorem VII.6.3 a constant c exists such that $F(x) = G(x) + c$ for $x \in I$. Thus two primitives of a function on an interval differ by a constant there.

Notation and Terminology

The traditional notation for a primitive of f is

$$\int f(x)\,dx.$$

Thus, if G is known to be a primitive of f on some interval I, then by the last paragraph,

$$\int f(x)\,dx = G(x) + c \qquad \text{for} \quad x \in I,$$

where c is some constant.

A primitive of f is also called an *integral* of f, but this should not be confused with

$$\int_a^b f(x)\,dx,$$

which is the Riemann integral of f over $[a,b]$. The relation between primitives and Riemann integrals is revealed by the *Fundamental Theorem of the Calculus*.

To "integrate" f is to find a primitive of f. For example, if $\beta \neq -1$, then

$$\int x^{\beta}\,dx = \frac{x^{\beta+1}}{\beta+1} + c \qquad \text{for} \quad x > 0. \tag{5.3}$$

If $\beta = -1$, then

$$\int \frac{1}{x}\,dx = \ln|x| + c \qquad \text{for} \quad x \neq 0. \tag{5.4}$$

By Example VII.3.2, this holds for $x > 0$. To see that it also holds for $x < 0$, assume that $x < 0$ and note that

$$\frac{d\ln|x|}{dx} = \frac{d\ln(-x)}{dx} = \frac{1}{-x}\,\frac{d(-x)}{dx} = \frac{1}{x}\,.$$

Since

$$\frac{d\ln|x|}{dx} = \frac{1}{x} \qquad \text{for} \quad x \neq 0,$$

(5.4) holds.

We list some more primitives.

$$\int e^x dx = e^x + c \qquad x \in \mathbb{R} \qquad \text{(Example VII.1.1)} \tag{5.5}$$

$$\int \cos x\, dx = \sin x + c, \qquad x \in \mathbb{R} \qquad \text{(Example VII.1.2)} \tag{5.6}$$

$$\int \sin x\, dx = -\cos x + c, \qquad x \in \mathbb{R} \qquad \text{(Prob. VII.1.1)} \tag{5.7}$$

$$\int \cosh x\, dx = \sinh x + c, \qquad x \in \mathbb{R} \qquad \text{(Prob. VII.3.4)} \tag{5.8}$$

$$\int \sinh x\, dx = \cosh x + c \tag{5.9}$$

$$\int \sec^2 x\, dx = \tan x + c, \qquad x \neq (4n \pm 1)\frac{\pi}{2}, \quad \text{where } n \text{ is an integer} \qquad \text{(Prob. VII.3.7)} \tag{5.10}$$

$$\int \sec x \tan x\, dx = \sec x + c, \qquad x \neq (4n \pm 1)\frac{\pi}{2}, \quad \text{where } n \text{ is an integer} \qquad \text{(Prob. VII.3.7)} \tag{5.11}$$

$$\int \csc^2 x\, dx = -\cot x + c, \qquad x \neq n, \quad \text{where } n \text{ is an integer} \qquad \text{(Prob. VII.3.8)} \tag{5.12}$$

$$\int \csc x \cot x\, dx = -\csc x + c, \qquad x \neq n, \quad \text{where } n \text{ is an integer} \qquad \text{(Prob. VII.3.8)} \tag{5.13}$$

$$\int \operatorname{sech}^2 x\, dx = \tanh x + c, \qquad x \in \mathbb{R} \qquad \text{(Prob. VII.3.9)} \tag{5.14}$$

$$\int \operatorname{sech} x \tanh x\, dx = -\operatorname{sech} x + c, \qquad x \in \mathbb{R} \qquad \text{(Prob. VII.3.9)} \tag{5.15}$$

$$\int \operatorname{csch}^2 x\, dx = -\coth x + c, \qquad x \neq 0 \qquad \text{(Prob. VII.3.10)} \tag{5.16}$$

$$\int \operatorname{csch} x \coth x\, dx = -\operatorname{csch} x + c, \qquad x \neq 0 \qquad \text{(Prob. VII.3.10)} \tag{5.17}$$

$$\int \frac{1}{\sqrt{1-x^2}}\, dx = \operatorname{Arcsin} x + c, \qquad |x| < 1 \qquad \text{(Remark VIII.6.1)} \tag{5.18}$$

$$\int \frac{1}{\sqrt{x^2-1}}\, dx = \cosh^{-1} x + c = \ln\left(x + \sqrt{x^2-1}\right) + c, \qquad |x| > 1 \qquad \text{(Prob. VII.6.4)} \tag{5.19}$$

$$\int \frac{1}{\sqrt{1+x^2}}\, dx = \sinh^{-1} x + c = \ln\left(x + \sqrt{1+x^2}\right) + c, \qquad x \in \mathbb{R} \qquad \text{(Example VII.6.1)} \tag{5.20}$$

$$\int \frac{1}{1+x^2}\,dx = \operatorname{Arctan} x + c, \qquad x \in \mathbb{R} \qquad \text{(Prob. VIII.6.10)} \quad (5.21)$$

$$\int \frac{1}{1-x^2}\,dx = \tanh^{-1}x + c = \frac{1}{2}\ln\frac{1+x}{1-x} + c \qquad \text{if} \quad |x|<1$$

$$= \coth^{-1}x + c = \frac{1}{2}\ln\frac{x+1}{x-1} \qquad \text{if} \quad |x|>1$$

$$\text{(Prob. VII.6.4)} \qquad (5.22a)$$

$$\int \frac{1}{1-x^2}\,dx = \frac{1}{2}\ln\left|\frac{1+x}{1-x}\right| + c \qquad (5.22b)$$

$$\int \frac{1}{x\sqrt{x^2-1}}\,dx = \operatorname{Arcsec}|x| + c, \qquad |x|>1 \qquad \text{(Prob. VIII.6.17)} \quad (5.23)$$

$$\int \frac{1}{x\sqrt{1-x^2}}\,dx = -\operatorname{sech}^{-1}|x| + c = -\ln\frac{1+\sqrt{1-x^2}}{|x|}, \qquad 0<|x|<1$$

$$\text{(Prob. VII.6.4)} \qquad (5.24)$$

$$\int \frac{1}{x\sqrt{1+x^2}}\,dx = -\operatorname{csch}^{-1}|x| + c = -\ln\frac{1+\sqrt{1+x^2}}{|x|} + c, \qquad x \neq 0$$

$$\text{(Prob. VII.6.4).} \qquad (5.25)$$

Prob. 5.1. Prove: If f_1 and f_2 have primitives on some set S, where S is either an open set or an interval, then so does $f_1 + f_2$ and

$$\int (f_1(x) + f_2(x))\,dx = \int f_1(x)\,dx + \int f_2(x)\,dx.$$

Prob. 5.2. Prove: If f has a primitive on a set S, where S is either an open set or an interval, and c is some constant, then

$$\int cf(x)\,dx = c\int f(x)\,dx.$$

Prob. 5.3. Prove: If the functions $f_1, f_2, \ldots, f_n$ have primitives on an interval I and $c_1, c_2, \ldots, c_n$ are constants, then the function $c_1 f_1 + c_2 f_2 + \cdots + c_n f_n$ has a primitive on I and

$$\int (c_1 f_1(x) + c_2 f_2(x) + \cdots + c_n f_n(x))\,dx$$
$$= c_1\int f_1(x) + c_2\int f_2(x)\,dx + \cdots + c_n\int f_n(x)\,dx.$$

Example 5.2. If n is a nonnegative integer, then

$$\int (a_0 x^n + a_1 x^{n-1} + \cdots + a_n)\,dx = a_0\frac{x^{n+1}}{n+1} + a_1\frac{x^n}{n} + \cdots + a_n x + c. \qquad (5.26)$$

It follows that each polynomial on $\mathbb{R}$ has a primitive on $\mathbb{R}$ and that its primitive is obtained as in (5.30).

EXAMPLE 5.3. Using $\tan^2 x = \sec^2 x - 1$, we have

$$\int \tan^2 x\, dx = \int (\sec^2 x - 1)\, dx = \int \sec^2 x\, dx - \int dx = \tan x - x + c.$$

Substitution Formula for Primitives

Theorem 5.1. *If G is a primitive of f on an interval J and v is a function which is differentiable on some interval I whose range is in J, then the composite $G \circ v$ is a primitive of $(f \circ v) dv/dx$ on I. In symbols,*

$$\int f(v(x)) \frac{dv(x)}{dx}\, dx = G(v(x)) + c \qquad \text{for} \quad x \in I. \tag{5.27}$$

PROOF. By hypothesis,

$$G'(u) = f(u) \qquad \text{for} \quad u \in J.$$

Also, according to the hypothesis, $x \in I$ implies $v(x) \in J$. By the chain rule, the composite $G \circ v$ is differentiable on I and

$$(G \circ v)'(x) = (G(v(x))' = G'(v(x))v'(x) = f(v(x)) \frac{dv(x)}{dx}$$

for $x \in I$. In the notation for primitives, this yields (5.27).

Notation. It will be convenient to introduce the following notation. If G is a primitive of f and v is a function such that $\mathfrak{R}(v) \subseteq \mathfrak{D}(G)$, then we define

$$\int^{v(x)} f(u)\, du = G(v(x)) \qquad \text{for} \quad x \in \mathfrak{D}(v). \tag{5.28}$$

In this notation, (5.27) becomes

$$\int f(v(x)) \frac{dv(x)}{dx}\, dx = \int^{v(x)} f(u)\, du + c \qquad \text{for} \quad x \in i. \tag{5.29}$$

This is easier to remember if it is written

$$\int f(v(x)) \frac{dv(x)}{dx}\, dx = \int^{v(x)} f(v)\, dv + c \qquad \text{for} \quad x \in I. \tag{5.30}$$

If f is the constant function 1, this becomes

$$\int \frac{dv(x)}{dx}\, dx = \int^{v(x)} dv + c = v(x) + c \qquad \text{for} \quad x \in I. \tag{5.31}$$

This is consistent with the fact that v is a primitive of $dv(x)/dx$ on the interval I.

Def. 5.2. As a mnemonic device for remembering (5.30), if v is a differentiable function on the interval I, we call

$$\frac{dv(x)}{dx}dx$$

the *differential* of v on I and write it as dv. Thus,

$$\frac{dv(x)}{dx}dx = dv \qquad \text{for} \quad x \in I. \tag{5.32}$$

The meaning this has for us is that it summarizes the behavior of the symbol dx as described by (5.30) and (5.31).

For example, the differential of $v(x) = x^2 + 2x$ for $x \in \mathbb{R}$ is $d(x^2 + 2x) = (2x + 2)dx$ for $x \in \mathbb{R}$.

Remark 5.1. The form of dv in (5.32) remains the same under a differentiable change of variables. Thus, if $x = \phi(u)$, where ϕ is a differentiable function, and $v^*(u) = v(\phi(u))$, then by the chain rule

$$\begin{aligned} dv^* &= \frac{dv^*(u)}{du}du = \frac{dv(\phi(u))}{du}du - v'(\phi(u))\phi'(u)du \\ &= v'(x)\frac{dx(u)}{du}du = v'(x)\,dx = \frac{dv(x)}{dx}dx = dv. \end{aligned}$$

More briefly,

$$\frac{dv^*(u)}{du}du = \frac{dv(x)}{dx}dx. \tag{5.33}$$

Use of the Substitution Formula for Obtaining Primitives

EXAMPLE 5.4. We evaluate

$$\int \frac{x}{\sqrt{1-x^2}}dx.$$

We use the substitution $v(x) = 1 - x^2$ for $-1 \leqslant x \leqslant 1$ so that

$$-2xdx = \frac{d(1-x^2)}{dx}dx = dv.$$

From (5.30), we obtain

$$\begin{aligned} \int \frac{x}{\sqrt{1-x^2}}dx &= \int -\frac{1}{2}\,\frac{-2x}{\sqrt{1-x^2}}dx = -\frac{1}{2}\int (1-x^2)^{-1/2}\frac{d(1-x^2)}{dx}dx \\ &= -\frac{1}{2}\int^{1-x^2} v^{-1/2}\,dv. \end{aligned} \tag{5.34}$$

Noting that

$$\int v^{-1/2}\,dv = \frac{v^{-1/2+1}}{-\frac{1}{2}+1} + c = 2v^{1/2} + c,$$

we obtain from this and (5.34)

$$\int \frac{x}{\sqrt{1-x^2}}\,dx = -\frac{1}{2}\int^{1-x^2} v^{-1/2}\,dv$$

$$= -\frac{1}{2}\left(2(1-x^2)^{1/2} + c\right) = -(1-x^2)^{1/2} + c_1,$$

$c_1 = -\frac{1}{2}c$.

EXAMPLE 5.5. If $a > 0$, then

$$\int a^x\,dx = \int e^{x\ln a}\,dx = \frac{1}{\ln a}\int e^{x\ln a}\frac{d(x\ln a)}{dx}\,dx = \frac{1}{\ln a}\int^{x\ln a} e^v\,dv$$

$$= \frac{1}{\ln a}(e^{x\ln a} + c) = \frac{1}{\ln a}(a^x + c) = \frac{a^x}{\ln a} + c_1.$$

PROB. 5.4. Show: If $a \neq 0$, then

(1) $\int(1/(a^2+x^2))\,dx = (1/a)\operatorname{Arctan}(x/a) + c$,

(2) $\int(1/\sqrt{a^2-x^2}\,)\,dx = \operatorname{Arcsin}(x/a) + c$.

EXAMPLE 5.6. If v is a differentiable function on an interval I and $v(x) \neq 0$ for $x \in I$, then

$$\int \frac{1}{v(x)}\frac{dv(x)}{dx}\,dx = \int^{v(x)} \frac{1}{v}\,dv + c = \ln|v(x)| + c. \tag{5.35}$$

EXAMPLE 5.7. We illustrate the result in the last example by finding

$$\int \sec x\,dx.$$

Multiply and divide the integrand $\sec x$ by $\sec x + \tan x$ to obtain

$$\int \sec x\,dx = \int \frac{\sec^2 x + \sec x\tan x}{\sec x + \tan x}\,dx = \int \frac{1}{\sec x + \tan x}\frac{d}{dx}(\sec x + \tan x)\,dx$$

$$= \ln|\sec x + \tan x| + c = \ln\left|\frac{1+\sin x}{\cos x}\right| + c.$$

PROB. 5.5. Show that

$$\int \csc x\,dx = \ln\left|\frac{\sin x}{1+\cos x}\right| + c.$$

Prob. 5.6. Integrate

(1) $$\int \frac{e^x}{1+e^x}\,dx,$$

(2) $$\int \frac{1}{1+e^x}\,dx,$$

(3) $$\int \frac{\sin x}{1+\cos x}\,dx,$$

(4) $$\int \frac{1}{x}(\ln x)^n\,dx, \qquad n \neq 1,$$

(5) $$\int \frac{1}{x\ln x}\,dx,$$

(6) $$\int \cot x\,dx,$$

(7) $$\int \tan x\,dx,$$

(8) $$\int \frac{e^{\sqrt{x}}}{\sqrt{x}}\,dx.$$

Prob. 5.7. Show that

(1) $\int f(a+x)\,dx = \int^{a+x} f(v)\,dv + c$,
(2) $\int f(ax)\,dx = 1/a \int^{ax} f(v)\,dv + c$, $a \neq 0$.

Prob. 5.8. Integrate

(1) $$\int \frac{1}{1+\cos x}\,dx,$$

(2) $$\int \frac{1}{1+\sin x}\,dx.$$

Example 5.8. When f is of the form $f(x) = g(\sin x)\cos x$, then

$$\int f(x)\,dx = \int g(\sin x)\cos x\,dx = \int g(\sin x)\frac{d\sin x}{dx}\,dx$$
$$= \int^{\sin x} g(v)\,dv + c. \tag{5.36}$$

Also

$$\int h(\cos x)\sin x\,dx = \int h(\cos x)\frac{d\cos x}{dx}\,dx = \int^{\cos x} h(v)\,dv + c. \tag{5.37}$$

Example 5.9. As a special case of the last example, we have

$$\int \cos^4 x\sin x\,dx = -\int \cos^4 x\frac{d\cos x}{dx}\,dx = -\frac{\cos^5 x}{5} + c.$$

EXAMPLE 5.10. When the integrand is of the form $\sin^n x \cos^m x$, where one of m or n is an odd positive integer, then the integration can be performed with the aid of the idea in Example 5.8. For example, we may write

$$\sin^2 x \cos^3 x = \sin^2 x \cos^2 x \cos x = \sin^2 x(1 - \sin^2 x)\cos x,$$

then

$$\begin{aligned}\int \sin^2 x \cos^3 x \, dx &= \int \sin^2 x(1 - \sin^2 x)\cos x \, dx \\ &= \int \sin^2 x \cos x \, dx - \int \sin^4 x \cos x \, dx \\ &= \frac{\sin^3 x}{3} - \frac{\sin^5 x}{5} + c.\end{aligned}$$

PROB. 5.9. Integrate

(1) $\int \cos^3 x \, dx$,
(2) $\int \sin^3 x \, dx$,
(3) $\int \cos^5 x \, dx$,
(4) $\int \sqrt[3]{\cos x} \sin^3 x \, dx$.

EXAMPLE 5.10′. The trigonometric identities

$$\sin^2 x = \frac{1 - \cos 2x}{2} \quad \text{and} \quad \cos^2 x = \frac{1 + \cos 2x}{2}$$

are useful in dealing with even powers of sine or cosine. Thus,

$$\begin{aligned}\int \sin^2 x \, dx &= \int \frac{1 - \cos 2x}{2} dx = \frac{1}{2} \int (1 - \cos 2x) \, dx = \frac{1}{2}\left(x - \frac{\sin 2x}{2}\right) + c \\ &= \frac{x}{2} - \frac{\sin 2x}{4} + c\end{aligned}$$

and

$$\int \cos^2 x \, dx = \int \frac{1 + \cos 2x}{2} dx = \frac{1}{2} \int (1 + \cos 2x) \, dx = \frac{x}{2} + \frac{\sin 2x}{4} + c.$$

If the integrand is an even power of sine or cosine as in

$$\int \cos^4 x \, dx,$$

then

$$\begin{aligned}\int \cos^4 x \, dx &= \int (\cos^2 x)^2 \, dx = \int \left(\frac{1 + \cos 2x}{2}\right)^2 dx \\ &= \frac{1}{4} \int (1 + 2\cos 2x + \cos^2 2x) \, dx = \frac{1}{4} \int \left(1 + 2\cos 2x + \frac{1 + \cos 4x}{2}\right) dx \\ &= \frac{1}{8} \int (3 + 4\cos 2x + \cos 4x) \, dx = \frac{1}{8}\left(3x + 2\sin 2x + \frac{\sin 4x}{4}\right) + c.\end{aligned}$$

Prob. 5.10. Integrate

(1) $\int \sin^4 x\, dx$,
(2) $\int \cos^6 x\, dx$,
(3) $\int \sin^6 x\, dx$,
(4) $\int \sin^2 x \cos^4 x\, dx$.

Prob. 5.11. Show that

$$\cosh^2 x = \frac{1+\cosh 2x}{2} \quad \text{and} \quad \sinh^2 x = \frac{\cosh 2x - 1}{2}$$

and that

(1) $$\int \cosh^2 x\, dx = \frac{1}{2}\left(x + \frac{\sinh 2x}{2}\right) + c,$$

(2) $$\int \sinh^2 x\, dx = \frac{1}{2}\left(\frac{\sinh 2x}{2} - x\right) + c.$$

Example 5.11 (A Reduction Formula). If $n \neq 1$, then

$$\int \tan^n x\, dx = \int \tan^{n-2} x \tan^2 x\, dx = \int \tan^{n-2} x(\sec^2 x - 1)\, dx$$

$$= \int \tan^{n-2} x \sec^2 x\, dx - \int \tan^{n-2} x\, dx = \frac{\tan^{n-1} x}{n-1} - \int \tan^{n-2} x\, dx$$

or

$$\int \tan^n x\, dx = \frac{\tan^{n-1} x}{n-1} - \int \tan^{n-2} x\, dx, \tag{5.38}$$

where $n \neq -1$. This is an example of a reduction formula. If n is a positive integer such that $n > 1$, then it leads to the integration of a power of tan which is two degrees lower.

Prob. 5.12. Integrate

(1) $\int \tan^5 x\, dx$,
(2) $\int \tan^6 x\, dx$,
(3) $\int \sec^4 x\, dx$,
(4) $\int \sec^6 x\, dx$.

Example 5.12. The substitution $v = (b/a)\tan x$ can be used to integrate

$$\int \frac{1}{a^2\cos^2 x + b^2 \sin^2 x}\, dx,$$

where $a \neq 0$ and $b \neq 0$. We have

$$\int \frac{1}{a^2\cos^2 x + b^2\sin^2 x}\, dx = \frac{1}{ab}\int \frac{1}{1+((b/a)\tan x)^2}\,\frac{b}{a}\sec^2 x\, dx$$

$$= \frac{1}{ab}\operatorname{Arctan}\left(\frac{b}{a}\tan x\right) + c.$$

EXAMPLE 5.13. The primitives

(1) $\int \cos mx \cos nx\, dx$,
(2) $\int \sin mx \sin nx\, dx$,
(3) $\int \sin mx \cos nx\, dx$

play an important role in the theory of Fourier series. If $m^2 \neq n^2$, then we use the trigonometric identities

(1) $\cos mx \cos nx = \frac{1}{2}(\cos(m+n)x + \cos(m-n)x)$,
(2) $\sin mx \sin nx = \frac{1}{2}(\cos(m-n)x - \cos(m+n)x)$,
(3) $\sin mx \cos nx = \frac{1}{2}(\sin(m+n)x + \sin(m-n)x)$.

Hence,

$$\int \cos mx \cos nx\, dx = \frac{1}{2}\int (\cos(m+n)x + \cos(m-n)x)\, dx$$

$$= \frac{1}{2}\left(\frac{\sin(m+n)x}{m+n} + \frac{\sin(m-n)x}{m-n}\right) + c, \tag{5.39}$$

$$\int \sin mx \sin nx\, dx = \frac{1}{2}\int (\cos(m-n)x - \cos(m+n)x)\, dx$$

$$= \frac{1}{2}\left(\frac{\sin(m-n)x}{m-n} - \frac{\sin(m+n)x}{m+n}\right) + c \tag{5.40}$$

$$\int \sin mx \cos nx\, dx = \frac{1}{2}\int (\sin(m+n)x + \sin(m-n)x)\, dx$$

$$= -\frac{1}{2}\left(\frac{\cos(m+n)x}{m+n} + \frac{\cos(m-n)x}{m-n}\right) + c. \tag{5.41}$$

The case where $m^2 = n^2 \neq 0$ is left to the reader (Prob. 5.13).

PROB. 5.13. Show: If $m^2 = n^2 \neq 0$ so that (1), (2), and (3) of the last example become respectively

(1′) $\int \cos^2 nx\, dx$,
(2′) $\int \sin^2 nx\, dx$,
(3′) $\int \sin nx \cos nx\, dx$,

then we have

$$\int \cos^2 nx\, dx = \frac{1}{2}\left(x + \frac{\sin 2nx}{2n}\right) + c,$$

$$\int \sin^2 nx\, dx = \frac{1}{2}\left(x - \frac{\sin 2nx}{2n}\right) + c,$$

$$\int \sin nx \cos nx\, dx = -\frac{\cos 2nx}{4n} + c.$$

Further techniques for obtaining primitives will be given later. We proceed to the next section where it is shown how to use them to evaluate Riemann integrals.

6. Fundamental Theorem of the Calculus

Theorem 6.1 (Fundamental Theorem of the Calculus). *If f is Riemann integrable over an interval $[a,b]$ and has a primitive G on $[a,b]$, then*

$$\int_a^b f(x)\,dx = G(b) - G(a). \tag{6.1}$$

PROOF. By the definition of primitive, we have from the hypothesis

$$G'(x) = f(x) \qquad \text{for all} \quad x \in [a,b]. \tag{6.2}$$

Let $P = \langle x_0, x_1, \ldots, x_n \rangle$ be a partition of $[a,b]$. Then

$$\sum_{i=1}^{n} (G(x_i) - G(x_{i-1})) = G(x_n) - G(x_0) = G(b) - G(a). \tag{6.3}$$

By the Mean-Value Theorem for derivatives there exists in each subinterval $[x_{i-1}, x_i]$ of P a ξ_i such that $G(x_i) - G(x_{i-1}) = G'(\xi_i)\Delta x_i = f(\xi_i)\Delta x_i$. Because of (6.3), this implies that

$$\sum_{i=1}^{n} f(\xi_i)\Delta x_i = G(b) - G(a). \tag{6.4}$$

Writing m_i and M_i for the infimum and supremum of f on $[x_{i-1}, x_i]$ for each i, we have

$$\underline{S}(f,P) = \sum_{i=1}^{n} m_i \Delta x_i \leqslant \sum_{i=1}^{n} f(\xi_i)\Delta x_i \leqslant \sum_{i=1}^{n} M_i \Delta x_i = \overline{S}(f,P)$$

so that (6.4) implies

$$\underline{S}(f,P) \leqslant G(b) - G(a) \leqslant \overline{S}(f,P). \tag{6.5}$$

Here P is any partition of $[a,b]$. This implies that

$$\underline{\int_a^b} f(x)\,dx \leqslant G(b) - G(a) \leqslant \overline{\int_a^b} f(x)\,dx. \tag{6.6}$$

But according to the hypothesis, f is Riemann integrable over $[a,b]$. This guarantees that f is bounded and Darboux integrable over $[a,b]$ and that

$$\int_a^b f(x)\,dx = \underline{\int_a^b} f(x)\,dx = \overline{\int_a^b} f(x)\,dx.$$

This and (6.6) yield (6.1).

Notation. When G is defined on a bounded closed interval whose endpoints are a and b, we often write

$$G(b)-G(a)=G(x)\big|_a^b. \tag{6.7}$$

Hence, (6.1) can be written

$$\int_a^b f(x)\,dx=G(x)\big|_a^b, \tag{6.8}$$

where $G'(x)=f(x)$ for $x\in[a,b]$.

EXAMPLE 6.1. If $\beta \geqslant 0$, then

$$\int_0^1 x^\beta dx=\frac{x^{\beta+1}}{\beta+1}\bigg|_0^1=\frac{1^{\beta+1}-0^{\beta+1}}{\beta+1}=\frac{1}{\beta+1}. \tag{6.9}$$

Remark 6.1. Theorem 6.1 does not state that a Riemann integrable function necessarily has a primitive on $[a,b]$. It merely states that *if* it has a primitive on $[a,b]$, then this primitive may be used to evaluate its Riemann integral as indicated in (6.1). A function may be Riemann integrable on an interval $[a,b]$ without having a primitive there. For example, let f be defined as

$$f(x)=\begin{cases}-1 & \text{if } 0\leqslant x\leqslant 1\\ 1 & \text{if } 1<x\leqslant 2.\end{cases}$$

f is defined and monotonic increasing on $[0,2]$, so it is Riemann integrable. On the other hand, this f has no primitive on $[0,2]$. If there were a primitive G of f on $[a,b]$, then f would be a derivative of G on $[0,2]$. By Theorem VII.7.2, f would necessarily have the intermediate-value property on $[0,2]$. But f clearly does not have the intermediate-value property on $[0,2]$ (explain) and we have a contradiction. Hence, f has no primitive on $[0,2]$.

We also note that a function may have a primitive and not be Riemann integrable. Define G as

$$G(x)=\begin{cases}0 & \text{if } x=0,\\ x^2\sin\dfrac{1}{x^2} & \text{if } x\neq 0 \text{ and } x\in[-1,1],\end{cases}$$

and f as

$$f(x)=G'(x)=\begin{cases}0 & \text{if } x=0\\ -\dfrac{2}{x}\cos\dfrac{1}{x^2}+2x\sin\dfrac{1}{x^2} & \text{if } x\neq 0,\ x\in[-1,1].\end{cases}$$

f has G as its primitive on $[-1,1]$. But f is not Riemann integrable on $[a,b]$ since it is not bounded there. (We recall that Theorem 4.2 states that a function which is Riemann integrable on a bounded closed interval is necessarily bounded there.)

Remark 6.2. The fundamental theorem of the calculus implies:

Corollary (of Theorem 6.1). *If G is differentiable on $[a,b]$ and its derivative G' is R-integrable there, then*

$$\int_a^b \frac{dG(x)}{dx}\,dx = G(b) - G(a). \tag{6.10}$$

PROOF. This follows from the fact that G is a primitive of G' on $[a,b]$, the assumption that G' is R-integrable on $[a,b]$, and Theorem 6.1.

Next we present a sufficient condition for a function to have a primitive on an interval.

Indefinite Integrals

Def. 6.1. Let f be defined on an interval I and let $a \in I$. If f is R-integrable on every bounded closed subinterval of I, then the function F defined as

$$F(x) = \int_a^x f(t)\,dt \qquad \text{for each} \quad x \in I \tag{6.11}$$

is called an *indefinite integral* of f on I.

We give an example. We have

$$\frac{d\ln t}{dt} = \frac{1}{t} \qquad \text{for} \quad t \in (0; +\infty).$$

For each $x > 0$ and $x \neq 1$, the f defined as $f(t) = 1/t$, for all t in the interval whose endpoints are x and 1, is continuous. Hence,

$$\int_1^x \frac{1}{t}\,dt = \ln t\big|_1^x = \ln x - \ln 1 = \ln x \tag{6.11'}$$

for all $x \in (0; +\infty)$. (Because of (1.24) this relation also holds for $x = 1$.) Accordingly, the natural logarithm function is an indefinite integral of the f just defined. In some developments this is used as the definition of the natural logarithm.

If f has a primitive G on an interval I and $a \in I$, then the indefinite integral F of Def. 6.1 can be evaluated by the fundamental theorem of the calculus and (1.24) to yield

$$F(x) = \int_a^x f(t)\,dt = G(x) - G(a) \qquad \text{for all} \quad x \in I \tag{6.12}$$

and we have

$$F'(x) = \frac{d}{dx}(G(x) - G(a)) = G'(x) \qquad \text{for all} \quad x \in I.$$

In this case, the indefinite integral F is also a primitive of f in I and we have

$$\int f(x)\,dx = c + \int_a^x f(t)\,dt \qquad \text{for all} \quad x \in I, \tag{6.13}$$

where c is some constant.

Formula (6.13) makes sense only if f has a primitive on I and is R-integrable there. However, the indefinite integral of f may exist on an interval I (it suffices to require that f is R-integrable on every bounded closed subinterval of I) and still not possess a primitive on an I. To see this, define f on $I = \mathbb{R}$ as

$$f(x) = \begin{cases} -1 & \text{if} \quad x < 0 \\ 1 & \text{if} \quad x \geqslant 0. \end{cases}$$

This f is R-integrable on every bounded closed subinterval in $\mathbb{R}$. Hence,

$$F(x) = \int_0^x f(t)\,dt$$

exists for each $x \in \mathbb{R} = I$, but f has no primitive on $\mathbb{R}$ since it does not have the intermediate value property on $\mathbb{R}$ (see Remark 6.1).

Remark 6.3. It is tempting to conclude that all the primitives of f, if f has any, can be obtained from (6.12) by using different a's in I and that the c in (6.13) is not necessary. This is not the case, however. Consider

$$\int_a^x \cos t\,dt = \sin x\big|_a^x = \sin x - \sin a \qquad \text{for} \quad x \in I. \tag{6.14}$$

Clearly, the primitive y, given by $y(x) = 100 + \sin x$ for $x \in \mathbb{R}$, cannot occur among those in (6.14) since $|-\sin a| \leqslant 1$ holds for any $a \in \mathbb{R}$.

Theorem 6.2. *If f is R-integrable on $[a, b]$, then the indefinite integral F given by*

$$F(x) = \int_a^x f(t)\,dt \qquad \text{for all} \quad x \in [a, b]$$

is (1) *differentiable from the left at each x_0 such that $a < x_0 \leqslant b$ for which f is continuous from the left, and* (2) *differentiable from the right at each x_0 such that $a \leqslant x_0 < b$ for which f is continuous from the right. In the respective cases we have*

$$f_L'(x_0) = f(x_0) \quad \text{and} \quad f_R'(x_0) = f(x_0). \tag{6.15}$$

Proof. Assume that $a \leqslant x_0 < b$ and that f is continuous from the right at x_0. First take h such that $0 < h < b - x_0$ so that $x_0 < x_0 + h \leqslant b$. We have

$$F(x_0 + h) = \int_a^{x_0 + h} f(t)\,dt = \int_a^{x_0} f(t)\,dt + \int_{x_0}^{x_0 + h} f(t)\,dt$$
$$= F(x_0) + \int_{x_0}^{x_0 + h} f(t)\,dt.$$

Hence,

$$\frac{F(x_0+h)-F(x_0)}{h}=\frac{1}{h}\int_{x_0}^{x_0+h} f(t)\,dt$$

and

$$\begin{aligned}\left|\frac{F(x_0+h)-F(x_0)}{h}-f(x_0)\right| &= \frac{1}{h}\left|\int_{x_0}^{x_0+h} f(t)\,dt-\int_{x_0}^{x_0+h} f(x_0)\,dt\right| \\ &= \frac{1}{h}\left|\int_{x_0}^{x_0+h}(f(t)-f(x_0))\,dt\right| \\ &\leqslant \frac{1}{h}\int_{x_0}^{x_0+h}|f(t)-f(x_0)|\,dt. \end{aligned} \tag{6.16}$$

Let $\epsilon>0$ be given. There exists a $\delta>0$ such that

$$|f(t)-f(x_0)|<\epsilon \qquad \text{for all } t \text{ such that} \quad x_0\leqslant t<x_0+\delta, \quad t\in[a,b]. \tag{6.17}$$

Now let $\delta_1=\min\{b-x_0,\delta\}$ and h such that $0<h<\delta_1$. This implies that $x_0<x_0+h\leqslant b$ and $x_0<x_0+h<x_0+\delta$. For all t such that $x_0\leqslant t\leqslant x_0+h$ we now have $x_0\leqslant t<x_0+\delta$, $t\in[a,b]$. It therefore follows from (6.17) that $|f(t)-f(x_0)|<\epsilon$ and that

$$\int_{x_0}^{x_0+h}|f(t)-f(x_0)|\,dt<\int_{x_0}^{x_0+h}\epsilon\,dt=\epsilon h$$

for $0<h<\delta_1$. This and (6.16) imply that

$$\left|\frac{F(x_0+h)-F(x_0)}{h}-f(x_0)\right|<\epsilon \qquad \text{for} \quad 0<h<\delta_1.$$

We conclude that

$$F_R'(x_0)=\lim_{h\to 0+}\frac{F(x_0+h)-F(x_0)}{h}=f(x_0).$$

This proves that F is differentiable from the right at x_0 and that the second equality in (6.15) holds if f is continuous from the right at x_0.

We leave the proof of (2) to the reader (Prob. 6.1).

Prob. 6.1. Complete the proof of the last theorem by proving: If f is R-integrable on $[a,b]$ and f is continuous from the left at x_0 such that $a<x_0\leqslant b$, then the indefinite integral F of f on $[a,b]$ (see above) is differentiable from the left at x_0 and $F_L'(x_0)=f(x_0)$.

Corollary 1 (of Theorem 6.2). *If f is R-integrable on $[a,b]$ and f is continuous at x_0, where $a<x_0<b$, then the indefinite integral F in Theorem 6.2 is differentiable at x_0 and $F'(x_0)=f(x_0)$.*

Proof. Exercise.

Corollary 2 (of Theorem 6.2). *If f is continuous on $[a,b]$, then the indefinite integral F given by*

$$F(x)=\int_a^x f(t)\,dt$$

is a primitive of f on $[a,b]$, i.e., we have

$$\frac{d}{dt}\int_a^x f(t)\,dt=f(x) \qquad \text{for} \quad x\in[a,b].$$

PROOF. The continuity of f on $[a,b]$ implies its R-integrability. Corollary 1 can now be applied to give the conclusion.

Remark 6.4. The last corollary provides a sufficient condition for a function f to have a primitive on an interval I, namely, if f is continuous on an interval I, then it is R-integrable on every bounded subinterval of I. Therefore, an indefinite integral of f is differentiable on I and is a primitive of f on I.

When f is known to be R-integrable on $[a,b]$ we cannot conclude that an indefinite integral of f is differentiable. (See the discussion preceding Remark 6.3.) However, the next theorem states that, in this case, an indefinite integral of f is continuous.

Theorem 6.3. *If f is R-integrable on $[a,b]$, then the indefinite integral F of f given by*

$$F(x)=\int_a^x f(t)\,dt \qquad \text{for} \quad x\in[a,b]$$

is uniformly continuous on $[a,b]$.

PROOF. For x_1 and x_2 in $[a,b]$, we have

$$|F(x_2)-F(x_1)|=\left|\int_{x_1}^{x_2} f(t)\,dt\right|. \tag{6.18}$$

Because f is R-integrable on $[a,b]$, it is bounded there. Hence, a $B>0$ exists such that $|f(x)|\leqslant B$ for all $x\in[a,b]$. Let $\epsilon>0$ be given. Take x_1 and x_2 in $[a,b]$ such that

$$|x_2-x_1|<\frac{\epsilon}{B}.$$

This and (6.18) imply that

$$|F(x_2)-F(x_1)|\leqslant\begin{cases}\int_{x_1}^{x_2}|f(t)|\,dt & \text{if} \quad x_1\leqslant x_2\\ \int_{x_2}^{x_1}|f(t)|\,dt & \text{if} \quad x_2\leqslant x_1.\end{cases}$$

Therefore,

$$|F(x_2) - F(x_1)| \leqslant B|x_2 - x_1| < \epsilon \qquad \text{for } x_1 \text{ and } x_2 \text{ in } [a,b]$$

(explain). This completes the proof.

Prob. 6.2. Prove: If f is continuous on $[a,b]$ and u and v are differentiable functions on an interval $[c,d]$ which map this interval into the interval $[a,b]$, then

$$\frac{d}{dx}\int_{u(x)}^{v(x)} f(t)\,dt = f(v(x))v'(x) - f(u(x))u'(x)$$

for $x \in [c,d]$.

Prob. 6.3. Prove that

$$\lim_{x\to 0}\left(\frac{1}{x}\int_0^x e^{t^2}\,dt\right) = 1.$$

Prob. 6.4. Prove that

$$\lim_{x\to+\infty}\left(e^{-x^2}\int_0^x e^{t^2}\,dt\right) = 0.$$

Prob. 6.5. Evaluate

(a) $\lim_{n\to+\infty}\sum_{k=1}^{n} 1/(n+k)$,
(b) $\lim_{n\to+\infty}\sum_{k=1}^{n} n/(n^2+k^2)$,
(c) $\lim_{n\to+\infty}\sum_{k=1}^{n} k/(n^2+k^2)$.

Prob. 6.6. Prove: If f is positive and increasing on $[0,a]$, then the function G, defined as

$$G(x) = \frac{1}{x}\int_0^x f(t)\,dt$$

for $x \in (0;a)$, is increasing.

Prob. 6.7. Prove: If f is R-integrable over $[a,b]$ and has a primitive there, then

$$\frac{d}{dx}\int_x^a f(t)\,dt = -f(x).$$

Prob. 6.8. Evaluate

(1) $\int_a^b (x-a)(b-x)\,dx$,
(2) $\int_a^b (x-a)^2(x-b)^2\,dx$.

Prob. 6.9. Evaluate

(1) $\int_0^1 x^{.1}\,dx$,

(2) $\int_a^{2a}(\sqrt{a/x}+\sqrt{x/a})^2\,dx$, $a>0$.

Prob. 6.10. Evaluate

(1) $$\int_{-\pi/2}^{\pi/2}\cos x\,dx,$$

(2) $$\int_0^{\pi}\sin x\,dx,$$

(3) $$\int_0^{\sqrt{3}}\frac{1}{1+x^2}\,dx,$$

(4) $$\int_0^{1/\sqrt{2}}\frac{1}{\sqrt{1-x^2}}\,dx,$$

(5) $$\int_0^{1}\frac{1}{\sqrt{1+x^2}}\,dx,$$

(6) $$\int_2^{3}\frac{1}{\sqrt{x^2-1}}\,dx.$$

7. The Substitution Formula for Definite Integrals

As mentioned earlier, the Riemann integral

$$\int_a^b f(x)\,dx$$

is also called a *definite integral*.

In Theorem 6.1, we assumed that the integrand there has a primitive. The first theorem in this section gives conditions which guarantee the existence of primitives of certain integrands.

Theorem 7.1. *Let f be continuous on an interval J and v differentiable on an interval I, where $\mathfrak{R}(v)\subseteq J$ and $b\in J$. If G is the indefinite integral of f defined as*

$$G(u)=\int_b^u f(t)\,dt \qquad \textit{for} \quad u\in J, \tag{7.1}$$

then the composite $G\circ v$ is a primitive of $(f\circ v)v'$ on I and

$$\int f(v(x))\frac{dv(x)}{dx}\,dx=c+\int_b^{v(x)}f(t)\,dt, \tag{7.2}$$

where $x\in I$ and c is a constant.

PROOF. By the hypothesis on f and Remark 6.4 the function G defined in (7.1) is a primitive of f on J. Since $G'(v) = f(v)$ for $v \in J$, $\mathfrak{R}(v) \subseteq J$, and $J \subseteq \mathfrak{D}(G') \subseteq \mathfrak{D}(G)$, we know that $\mathfrak{R}(v) \subseteq \mathfrak{D}(G)$. By Theorem 5.1 we conclude that $G \circ v$ is a primitive of $(f \circ v)v'$ on I, and

$$\int f(v(x)) \frac{dv(x)}{dx} dx = C + G(v(x)) = c + \int_b^{v(x)} f(t)\, dt.$$

Remark 7.1. This theorem states conditions for the existence of a primitive of $(f \circ v)v'$. It yields no information about its R-integrability. The hypothesis only guarantees the R-integrability of f on the bounded closed subintervals of J. The hypothesis in the following theorem guarantees the R-integrability of $(f \circ v)v'$ on a certain interval.

Theorem 7.2. *Let f be continuous on an interval J with endpoints c and d. Let v be a function whose derivative is R-integrable on $[a,b]$ which maps $[a,b]$ onto the interval J where $v(a) = c$ and $v(b) = d$, then $(f \circ v)v'$ is R-integrable on $[a,b]$, and*

$$\int_a^b f(v(x)) \frac{dv(x)}{dx} dx = \int_{v(a)}^{v(b)} f(v)\, dv. \tag{7.3}$$

PROOF. The differentiability of v implies its continuity. Since f is continuous on J, v is continuous on $[a,b]$, and $\mathfrak{R}(v) \subseteq J$, the composite $f \circ v$ is continuous on $[a,b]$. This implies that $f \circ v$ is R-integrable on $[a,b]$. By the hypothesis on v' and the fact that the product of functions which are R-integrable on a bounded closed interval also are R-integrable, it follows that $(f \circ v)v'$ is R-integrable on $[a,b]$. It is left to evaluate the R-integral on the left of (7.3). By Theorem 7.1, we know that $G \circ v$, where G is the indefinite integral of f defined by

$$G(u) = \int_{v(a)}^{u} f(t)\, dt \qquad \text{for} \quad u \in J,$$

is a primitive of $(f \circ v)v'$ on $I = [a,b]$. By the Fundamental Theorem of the Calculus, we have

$$\begin{aligned}\int_a^b f(v(x)) \frac{dv(x)}{dx} dx &= G(v(x))\Big|_a^b \\ &= G(v(b)) - G(v(a)) \\ &= \int_{v(a)}^{v(b)} f(t)\, dt \\ &= \int_{v(a)}^{v(b)} f(v)\, dv.\end{aligned}$$

This proves the theorem.

EXAMPLE 7.1. To evaluate

$$\int_e^{e^2} \frac{1}{x \ln x}\, dx,$$

we use the substitution $v(x) = \ln x$, $x \in [e, e^2]$. v maps $[e, e^2]$ onto $[\ln e, \ln e^2] = [1, 2]$. Theorem 7.2 applies (explain) and

$$\int_e^{e^2} \frac{1}{x \ln x}\, dx = \int_e^{e^2} \frac{1}{\ln x} \frac{d \ln x}{dx}\, dx = \int_{\ln e}^{\ln e^2} \frac{1}{v}\, dv = \int_1^2 \frac{1}{v}\, dv$$
$$= \ln 2 - \ln 1 = \ln 2.$$

PROB. 7.1. Evaluate

(1) $\displaystyle\int_0^{1/2} \frac{x}{\sqrt{1-x^2}}\, dx,$

(2) $\displaystyle\int_0^{1/2} \frac{1}{\sqrt{1-x^2}}\, dx,$

(3) $\displaystyle\int_0^{1} \frac{1}{\sqrt{1-x^2}} \operatorname{Arcsin} x\, dx,$

(4) $\displaystyle\int_0^{1/2} \frac{x^2}{\sqrt{1-x^6}}\, dx,$

(5) $\displaystyle\int_{1/2}^{3} \frac{1}{\sqrt{2x+3}}\, dx,$

(6) $\displaystyle\int_0^{1} \frac{x}{(1+x^4)}\, dx.$

PROB. 7.2. Show: If $a \neq 0$, $b \neq 0$, then

$$\int_0^{\pi/4} \frac{1}{a^2 \cos^2 x + b^2 \sin^2 x}\, dx = \frac{1}{ab} \operatorname{Arctan} \frac{b}{a}.$$

One technique for obtaining the primitive

$$\int f(x)\, dx \tag{7.4}$$

is to make a substitution $x = \phi(u)$ and define

$$h(u) = f(\phi(u)) \frac{d\phi(u)}{du}. \tag{7.5}$$

One has in mind that by doing this it may be easier to find a primitive of h.

Suppose f is defined on some interval J. Introduce the substitution $x = \phi(u)$, where ϕ is continuous on an interval I, maps I into J, and is differentiable in the interior of I with $\phi'(u) \neq 0$ for u in the interior of I. It follows that ϕ has an universe ϕ^{-1} which maps the range of ϕ back into I

and

$$\frac{d\phi^{-1}(x)}{dx} = \frac{1}{\phi'(\phi^{-1}(x))} \qquad \text{for } x \text{ in the interior of the range of } \phi \quad (7.6)$$

(Theorem VII.6.5). Suppose also that the function h defined in (7.5) has a primitive G on I. We have

$$G'(u) = h(u) = f(\phi(u))\frac{d\phi(u)}{du} \qquad \text{for} \quad u \in I. \qquad (7.7)$$

This implies

$$\begin{aligned}\frac{d}{dx}G(\phi^{-1}(x)) &= G'(\phi^{-1}(x))\frac{d\phi^{-1}(x)}{dx}\\ &= G'(\phi^{-1}(x))\frac{1}{\phi'(\phi^{-1}(x))}\\ &= h(\phi^{-1}(x))\frac{1}{\phi'(\phi^{-1}(x))}\\ &= f(x)\phi'(\phi^{-1}(x))\frac{1}{\phi'(\phi^{-1}(x))}\\ &= f(x)\end{aligned}$$

for $x \in J$. In our notation for primitives this gives rise to

$$\begin{aligned}\int f(x)\,dx &= G(\phi^{-1}(x)) + c\\ &= \int^{\phi^{-1}(x)} h(u)\,du + c\\ &= \int^{\phi^{-1}(x)} f(\phi(u))\frac{d\phi(u)}{du}\,du + c,\end{aligned}$$

i.e., that

$$\int f(x)\,dx = \int^{\phi^{-1}(x)} f(\phi(u))\frac{d\phi(u)}{du}\,du + c \qquad \text{for} \quad x \in J. \qquad (7.8)$$

Example 7.2. We use (7.8) to obtain the primitive

$$\int \sqrt{a^2 - x^2}\,dx,$$

where $a > 0$. Here, of course, $-a \leqslant x \leqslant a$. Substitute $x = a\sin u$, $-\pi/2 \leqslant u \leqslant \pi/2$ so that

$$\sqrt{a^2 - x^2} = \sqrt{a^2 - a^2\sin^2 u} = a|\cos u| = a\cos u, \qquad dx = a\cos u\,du$$

and

$$u = \operatorname{Arcsin}\frac{x}{a}.$$

Note that $dx/du \neq 0$ for $-\pi/2 < u < \pi/2$. From (7.8) and Example 5.10′,

$$\begin{aligned}\int \sqrt{a^2 - x^2}\, dx &= \int^{\operatorname{Arcsin} x/a} (a\cos u)(a\cos u)\, du + c \\ &= a^2 \int^{\operatorname{Arcsin} x/a} \cos^2 u\, du + c \\ &= \frac{a^2}{2}(u + \sin u \cos u)\Big|^{u = \operatorname{Arcsin} x/a} + c \\ &= \frac{a^2}{2}\left(\operatorname{Arcsin}\frac{x}{a} + \frac{x}{a^2}\sqrt{a^2 - x^2}\right) + c.\end{aligned}$$

Prob. 7.3. Show: If $a > 0$, then

(1) $\int \sqrt{a^2 + x^2}\, dx = (a^2/2)(\sinh^{-1}(x/a) + (x/a^2)\sqrt{a^2 + x^2}) + c$
$\quad = (a^2/2)(\ln(x + \sqrt{a^2 + x^2}) + (x/a^2)\sqrt{a^2 + x^2}) + c_1,$

(2) $\int \sqrt{x^2 - a^2}\, dx = (a^2/2)(-\cosh^{-1}(x/a) + (x/a^2)\sqrt{x^2 - a^2}) + c$
$\quad = (a^2/2)(-\ln(x + \sqrt{x^2 - a^2}) + (x/a^2)\sqrt{x^2 - a^2}) + c_1.$

(Hint: in (1) use the substitution $x = a\sinh u$, $u \geqslant 0$, and in (2) use $x = a\cosh u$, $u \geqslant 0$. Also note that c_1 and c may differ).

Example 7.3. To obtain the primitive

$$\int \frac{1}{1+\sqrt{x}}\, dx,$$

we substitute $u = \sqrt{x}$, $x \geqslant 0$. Hence,

$$u^2 = x, \qquad \frac{1}{1+\sqrt{x}} = \frac{1}{1+u}, \qquad dx = 2u\, du$$

and

$$\begin{aligned}\int \frac{1}{1+\sqrt{x}}\, dx &= \int^{\sqrt{x}} \frac{1}{1+u} 2u\, du + c \\ &= 2\int^{\sqrt{x}} \frac{u}{1+u}\, du \\ &= 2\int^{\sqrt{x}} \left(1 - \frac{1}{1+u}\right) du \\ &= 2\left(\sqrt{x} - \ln(1+\sqrt{x})\right) + c.\end{aligned}$$

Prob. 7.4. Find

(1) $\int x\sqrt{x-1}\,dx,$

(2) $\int \frac{1}{x\sqrt{1+x^2}}\,dx.$

Corresponding to the technique for finding primitives discussed after Prob. 7.2, we have a theorem concerning definite integrals. This theorem is actually a corollary of Theorem 7.2.

Theorem 7.3. *Let f be continuous on the interval $J=[c,d]$, and ϕ continuously differentiable on a bounded closed interval I whose endpoints are a and b. If ϕ maps I onto J with $\phi(a)=c$ and $\phi(b)=d$ and $\phi'(u)\neq 0$ for u in the interior of I, then ϕ has an inverse ϕ^{-1} on J and*

$$\int_c^d f(x)\,dx=\int_{\psi^{-1}(c)}^{\phi^{-1}(d)} f(\phi(u))\,\frac{d\phi(u)}{du}\,du. \tag{7.9}$$

Proof. ϕ' is continuous on I and so it is R-integrable there. Hence, ϕ' is R-integrable on I. ϕ maps I onto $J=[c,d]$. Hence, Theorem 7.2 applies, and we have

$$\int_a^b f(\phi(u))\,\frac{d\phi(u)}{du}\,du=\int_{\phi(b)}^{\phi(a)} f(x)\,dx=\int_c^d f(x)\,dx. \tag{7.10}$$

By the hypothesis on ϕ, it has an inverse ϕ^{-1} on $J=[c,d]$. This inverse maps $[c,d]$ onto I and $\phi^{-1}(c)=a$, $\phi^{-1}(d)=b$. Hence, (7.10) can be written as (7.9) and the proof is complete.

Example 7.4. One way of evaluating

$$\int_0^a \sqrt{a^2+x^2}\,dx,$$

where $a>0$, is to introduce the substitution $x=a\sinh u$, $u\geqslant 0$. Since $\sinh^{-1}1=\ln(1+\sqrt{1^2+1})=\ln(1+\sqrt{2})$, we know that $1=\sinh\ln(1+\sqrt{2})$. Since $u\geqslant 0$, we have

$$\sqrt{a^2+x^2}=\sqrt{a^2+a^2\sinh^2 u}=a\cosh u$$

and $dx=a\cosh u\,du$. Also,

$$u=\sinh^{-1}\frac{x}{a},\quad u(0)=0,\quad\text{and}\quad u(a)=\sinh^{-1}\frac{a}{a}=\sinh^{-1}1=\ln(1+\sqrt{2}).$$

Therefore,

$$\int_0^a \sqrt{a^2 + x^2}\, dx = \int_0^{\sinh^{-1} 1} (a \cosh u)(a \cosh u)\, du$$

$$= a^2 \int_0^{\sinh^{-1} 1} \cosh^2 u\, du$$

$$= a^2 \int_0^{\ln(1+\sqrt{2})} \cosh^2 u\, du.$$

By Prob. 5.11, this implies that

$$\int_0^a \sqrt{a^2 + x^2}\, dx = a^2 \int_0^{\ln(1+\sqrt{2})} \cosh^2 u\, du$$

$$= \frac{a^2}{2}\left(u + \frac{\sinh 2u}{2}\right)\Bigg|_0^{\ln(1+\sqrt{2})}$$

$$= \frac{a^2}{2}(u + \sinh u \cosh u)\Big|_0^{\ln(1+\sqrt{2})}$$

$$= \frac{a^2}{2}\left(\ln(1+\sqrt{2}) + \sinh \ln(1+\sqrt{2}) \cosh \ln(1+\sqrt{2})\right)$$

$$= \frac{a^2}{2}\left(\ln(1+\sqrt{2}) + \sqrt{2}\right).$$

EXAMPLE 7.5 (The Substitution $u = \tan x/2$). We find the primitive

$$\int \frac{1}{3 + 4\cos x}\, dx$$

by using the substitution $x = 2 \operatorname{Arctan} u$. We obtain

$$u = \tan\frac{x}{2}, \qquad 1 + u^2 = \sec^2\frac{x}{2}, \qquad \cos^2\frac{x}{2} = \frac{1}{1+u^2}. \tag{7.11}$$

These imply that

$$\cos x = 2\cos^2\frac{x}{2} - 1 = \frac{2}{1+u^2} - 1 = \frac{1-u^2}{1+u^2} \tag{7.12}$$

and

$$\sin x = 2\sin\frac{x}{2}\cos\frac{x}{2} = 2\tan\frac{x}{2}\cos^2\frac{x}{2} = \frac{2u}{1+u^2}. \tag{7.13}$$

Also

$$dx = \frac{2}{1+u^2}\, du. \tag{7.14}$$

Substituting, we obtain

$$\int \frac{1}{3+4\cos x}\,dx = 2\int^{\tan(x/2)} \frac{1}{3+4((1-u^2)/(1+u^2))}\,\frac{1}{1+u^2}\,du$$

$$= 2\int^{\tan(x/2)} \frac{1}{7-u^2}\,du$$

$$= \frac{2}{7}\sqrt{7}\int^{\tan(x/2)} \frac{1}{1-\left(u/\sqrt{7}\right)^2}\,d\left(\frac{u}{\sqrt{7}}\right).$$

In short, using the constant of integration, we have

$$\int \frac{1}{3+4\cos x}\,dx = \frac{2\sqrt{7}}{7}\int^{\tan(x/2)} \frac{1}{1-\left(u/\sqrt{7}\right)^2}\,d\left(\frac{u}{\sqrt{7}}\right) + c. \tag{7.15}$$

By (5.22b), we know that

$$\int \frac{1}{1-\left(u/\sqrt{7}\right)^2}\,d\left(\frac{u}{\sqrt{7}}\right) = \frac{1}{2}\ln\left|\frac{1+u/\sqrt{7}}{1-u/\sqrt{7}}\right| = \frac{1}{2}\ln\left|\frac{\sqrt{7}+u}{\sqrt{7}-u}\right|. \tag{7.16}$$

This and (7.15) imply that

$$\int \frac{1}{3+4\cos x}\,dx = \frac{\sqrt{7}}{7}\ln\left|\frac{\sqrt{7}+\tan(x/2)}{\sqrt{7}-\tan(x/2)}\right| + c. \tag{7.17}$$

Since

$$\tan\frac{x}{2} = \frac{\sin x}{1+\cos x}, \tag{7.18}$$

we can obtain an alternate expression for this.

Remark 7.2. Let $R(u, v)$ be a rational expression in u and v, i.e., let

$$R(u, v) = \frac{P_0(u) + P_1(u)v + \cdots + P_n(u)v^n}{Q_0(u) + Q_1(u)v + \cdots + Q_m(u)v^m},$$

where m and n are nonnegative integers and $P_0, P_1, \ldots, P_n$; Q_0, Q_1, $\ldots$, Q_m are polynomials. Consider the integral

$$\int R(\cos x, \sin x)\,dx. \tag{7.19}$$

We say that a substitution $x = \phi(u)$ *rationalizes* this integral if it transforms it into

$$\int^{\phi^{-1}(x)} R^*(u)\,du, \tag{7.20}$$

where R^* is a rational function.

The substitution $x = 2\,\mathrm{Arctan}\,u$ rationalizes the integral (7.19). To see this, perform the manipulations carried out in Example 7.5 to obtain

$u = \tan(x/2)$ and

$$\cos x = \frac{1-u^2}{1+u^2}, \qquad \sin x = \frac{2u}{1+u^2}, \qquad dx = \frac{2}{1+u^2}\,du.$$

Then obtain

$$\int R(\cos x, \sin x)\,dx = \int^{\tan(x/2)} R\left(\frac{1-u^2}{1+u^2}, \frac{2u}{1+u^2}\right)\frac{2}{1+u^2}\,du + c.$$

It is easy to see that R^*, where

$$R^*(u) = R\left(\frac{1-u^2}{1+u^2}, \frac{2u}{1+u^2}\right)\frac{2}{1+u^2}$$

can be written

$$R^*(u) = \frac{P^*(u)}{Q^*(u)},$$

where P^* and Q^* are polynomials. It follows that the substitution $x = 2\operatorname{Arctan} u$ rationalizes (7.19).

Remark 7.3. In certain instances, the presence of a quadratic expression of the form $ax^2 + b + c$, $a \neq 0$, in the integrand is handled by "completing the square." Thus, we note that

$$ax^2 + bx + c = a\left[\left(x + \frac{b}{2a}\right)^2 + \frac{4ac - b^2}{4a^2}\right]. \tag{7.21}$$

How one proceeds further depends on the sign of $4ac - b^2$.

Consider, for example

$$\int \frac{x}{ax^2 + bx + c}\,dx, \qquad a \neq 0. \tag{7.22}$$

We obtain from (7.21),

$$\int \frac{x}{ax^2 + bx + c}\,dx = \frac{1}{a}\int \frac{x}{(x + b/2a)^2 + (4ac - b^2)/4a^2}\,dx. \tag{7.23}$$

We consider cases (1) $4ac - b^2 = 0$, (2) $4ac - b^2 > 0$, (3) $4ac - b^2 < 0$.

Case (1): Since $4ac - b^2 = 0$, (5.42) becomes

$$\int \frac{x}{ax^2 + bx + c}\,dx = \frac{1}{a}\int \frac{x}{(x + b/2a)^2}\,dx.$$

Substituting $u = x + (b/2a)$, we have $x = u - (b/2a)$ and $dx = du$. We obtain

$$\begin{aligned}\int \frac{x}{ax^2 + bx + c}\,dx &= \frac{1}{a}\int^{x+(b/2a)} \frac{u - (b/2a)}{u^2}\,du \\ &= \frac{1}{a}\int^{x+(b/2a)} \frac{1}{u}\,du - \frac{b}{2a^2}\int^{x+(b/2a)} \frac{1}{u^2}\,du \\ &= \frac{1}{a}\ln\left|x + \frac{b}{2a}\right| + \frac{b}{2a^2}\,\frac{1}{x + (b/2a)} + c.\end{aligned}$$

Case (2): We have $4ac - b^2 > 0$. Write

$$A = \frac{\sqrt{4ac - b^2}}{|a|}.$$

Proceeding from (7.23) we obtain

$$\int \frac{x}{ax^2 + bx + c}\,dx = \frac{1}{a}\int \frac{x}{(x + (b/2a))^2 + A^2}$$

$$= \frac{1}{a}\int^{x+(b/2a)} \frac{u - (b/2a)}{u^2 + A^2}\,du$$

$$= \frac{1}{a}\int^{x+(b/2a)} \frac{u}{u^2 + A^2}\,du - \frac{b}{2a^2}\int^{x+(b/2a)} \frac{1}{u^2 + A^2}\,du$$

$$= \frac{1}{2a}\ln\left[\left(x + \frac{b}{2a}\right)^2 + A^2\right]$$

$$- \frac{b}{2a^2A}\operatorname{Arctan}\frac{x + (b/2a)}{A} + c.$$

Case (3): We have $4ac - b^2 < 0$, so that $b^2 - 4ac > 0$. Write

$$B = \frac{\sqrt{b^2 - 4ac}}{|a|}.$$

We then obtain

$$\int \frac{x}{ax^2 + bx + c}\,dx = \frac{1}{a}\int \frac{x}{(x + (b/2a))^2 - B^2}\,dx$$

$$= \frac{1}{a}\int^{x+(b/2a)} \frac{u - (b/2a)}{u^2 - B^2}\,du$$

$$= \frac{1}{a}\int^{x+(b/2a)} \frac{u}{u^2 - B^2}\,du - \frac{b}{2a^2}\int^{x+(b/2a)} \frac{1}{u^2 - B^2}\,du$$

so that

$$\int \frac{x}{ax^2 + bx + c}\,dx = \frac{1}{2a}\ln\left[\left(x + \frac{b}{2a}\right)^2 - B^2\right]$$

$$+ \frac{b}{2a^2}\int^{x+(b/2a)} \frac{1}{B^2 - u^2}\,du$$

$$= \frac{1}{2a}\ln\left[\left(x + \frac{b}{2a}\right)^2 - B^2\right]$$

$$+ \frac{b}{4a^2B}\ln\left|\frac{B + (x + (b/2a))}{B - (x - (b/2a))}\right| + c.$$

PROB. 7.5. Integrate

(1) $$\int \frac{1}{3+4\sin x}\,dx,$$

(2) $$\int \frac{1}{4+3\cos x}\,dx,$$

(3) $$\int \frac{1}{1+x+x^2}\,dx,$$

(4) $$\int \sqrt{2x-x^2}\,dx,$$ complete the square under the square root,

(5) $$\int \frac{1}{\sqrt{x^2+6x+8}}\,dx,$$

(6) $$\int \sqrt{x^2+4x+5}\,dx,$$

(7) $$\int \frac{3x-2}{x^2-4x+5}\,dx.$$

8. Integration by Parts

Theorem 8.1. *If u and v are differentiable functions on an interval I and the product vu' has a primitive on I, then the product uv' has a primitive on I and*

$$\int u\frac{dv}{dx}\,dx = uv - \int v\frac{du}{dx}\,dx \qquad \textit{for} \quad x \in I. \tag{8.1}$$

PROOF. By hypothesis, the product uv is differentiable on I and

$$\frac{du(x)v(x)}{dx} = u(x)\frac{dv(x)}{dx} + v(x)\frac{du(x)}{dx} \qquad \text{for} \quad x \in I. \tag{8.2}$$

The hypothesis also states that uv' has a primitive on I. Let G_1 be this primitive. Then

$$\frac{dG_1(x)}{dx} = u(x)\frac{dv(x)}{dx} \qquad \text{for} \quad x \in I. \tag{8.3}$$

This and (8.2) imply that

$$\frac{d}{dx}(u(x)v(x) - G_1(x)) = v(x)\frac{du(x)}{dx} \qquad \text{for} \quad x \in I.$$

Thus, $uv - G_1$ is a primitive of vu' on I. Therefore,

$$\int v(x)\frac{du(x)}{dx}\,dx = u(x)v(x) - G_1(x) + c_1, \tag{8.4}$$

where c_1 is some constant. But $-G_1 + c_1$ is a primitive of $-uv'$ (see (8.3)). This and (8.4) imply that

$$\int v(x)\frac{du(x)}{dx}\,dx = v(x)v(x) - \int u(x)\frac{du(x)}{dx}\,dx \qquad \text{for} \quad x \in I. \tag{8.5}$$

We usually write (8.1) as

$$\int u\,dv = uv - \int v\,du. \tag{8.6}$$

When this formula or (8.1) is used to integrate a function we say that we are using *Integration by parts*.

Example 8.1. We perform the integration

$$\int x\cos x\,dx$$

using integration by parts. Let $u(x) = x$ and $dv(x) = \cos x\,dx$, so that $v(x) = \sin x$. Using (8.6),

$$\int x\cos x\,dx = \int x\,d\sin x = x\sin x - \int \sin x\,d(x) = x\sin x + \cos x + c.$$

Example 8.2. Sometimes we have to integrate by parts more than once. Thus, let us try to integrate

$$\int x^2\cos x\,dx.$$

Let $u(x) = x^2$, $dv(x) = \cos x\,dx$, and $v(x) = \sin x$. By (8.6),

$$\int x^2\cos x\,dx = \int x^2\,d(\sin x) = x^2\sin x - 2\int x\sin x\,dx. \tag{8.7}$$

In the second integral, put $u(x) = x$, $dv(x) = \sin x\,dx$ so that $v(x) = -\cos x\,dx$. From this and (8.7), we have

$$\begin{aligned}\int x^2\cos x\,dx &= x^2\sin x - 2\int x\,d(-\cos x)\\ &= x^2\sin x + 2\int x\,d\cos x\\ &= x^2\sin x + 2\Big(x\cos x - \int \cos x\,dx\Big)\\ &= x^2\sin x + 2x\cos x - 2\sin x + c.\end{aligned}$$

The reader is invited to check the right-hand side and show that it is actually a primitive of $x^2\cos x$.

EXAMPLE 8.3. We integrate

$$\int \sqrt{a^2 - x^2}\, dx$$

(see Example 7.2). We have

$$\int \sqrt{a^2-x^2}\,dx = \int \frac{a^2-x^2}{\sqrt{a^2-x^2}}\,dx = a^2\int \frac{1}{\sqrt{a^2-x^2}}\,dx - \int \frac{x^2}{\sqrt{a^2-x^2}}\,dx$$

$$= a^2 \operatorname{Arcsin}\frac{x}{a} - \int \frac{x^2}{\sqrt{a^2-x^2}}\,dx. \tag{8.8}$$

In the second integral on the right we use integration by parts. We have

$$\int \frac{x^2}{\sqrt{a^2-x^2}}\,dx = \int x \frac{x}{\sqrt{a^2-x^2}}\,dx. \tag{8.9}$$

Here, we put $u(x) = x$, $dv(x) = x(a^2 - x^2)^{-(1/2)}dx = -(1/2)(a^2 - x^2)^{-(1/2)}(-2x)\,dx$ so that $v(x) = -(1/2)(a^2 - x^2)^{1/2}(2) = -(a^2 - x^2)^{1/2}$. We have

$$\int \frac{x^2}{\sqrt{a^2-x^2}}\,dx = \int x\, d\left(-(a^2-x^2)^{1/2}\right) = -x(a^2-x^2)^{1/2} + \int \sqrt{a^2-x^2}\,dx.$$

This and (8.8) imply that

$$\int \sqrt{a^2-x^2}\,dx = a^2 \operatorname{Arcsin}\frac{x}{a} - \int \frac{x^2}{\sqrt{a^2-x^2}}\,dx$$

$$= a^2 \operatorname{Arcsin}\frac{x}{a} + x(a^2-x^2)^{1/2} - \int \sqrt{a^2-x^2}\,dx + c.$$

This implies that

$$2\int \sqrt{a^2-x^2}\,dx = a^2\left(\operatorname{Arcsin}\frac{x}{a} + \frac{x}{a^2}(a^2-x^2)^{1/2}\right) + c$$

so that

$$\int \sqrt{a^2-x^2}\,dx = \frac{a^2}{2}\left(\operatorname{Arcsin}\frac{x}{a} + \frac{x}{a^2}\sqrt{a^2-x^2}\right) + c_1.$$

EXAMPLE 8.4. We perform the integration

$$\int \ln x\, dx.$$

Let $u(x) = \ln x$, $dv(x) = dx$ so that $v(x) = x$. We have

$$\int \ln x\,dx = x\ln x - \int x\, d\ln x = x \ln x - \int x \frac{1}{x}\,dx = x\ln x - x + c.$$

Prob. 8.1. Integrate

(1) $\int x^n \ln x\, dx,$

(2) $\int \sec^3 x\, dx,$

(3) $\int \operatorname{Arcsin} x\, dx,$

(4) $\int \operatorname{Arctan} x\, dx,$

(5) $\int e^{ax} \cos bx\, dx,$

(6) $\int e^{ax} \sin bx\, dx,$

(7) $\int \frac{x}{e^x}\, dx,$

(8) $\int x^2 e^x\, dx,$

(9) $\int \ln\left(x+\sqrt{1+x^2}\right) dx,$

(10) $\int e^{\sqrt{x}}\, dx.$

We now give a Riemann integral counterpart of Theorem 8.1.

Theorem 8.2. *If u and v are differentiable on $[a,b]$, and u', v' are R-integrable, then*

$$\int_a^b u(x)\frac{dv(x)}{dx}\,dx = u(x)v(x)\Big|_a^b - \int_a^b v(x)\frac{du(x)}{dx}\,dx. \tag{8.10}$$

Proof. Note, $x \in [a,b]$ implies that

$$\frac{du(x)v(x)}{dx} = u(x)\frac{dv(x)}{dx} + v(x)\frac{du(x)}{dx}. \tag{8.11}$$

The right side here is a sum of R-integrable functions on $[a,b]$. As such, it is R-integrable on $[a,b]$ and its primitive is uv. By the Fundamental Theorem and properties of integrals,

$$u(x)v(x)\Big|_a^b = \int_a^b\left(u(x)\frac{dv(x)}{dx} + v(x)\frac{du(x)}{dx}\right)dx$$

$$= \int_a^b u(x)\frac{dv(x)}{dx}\,dx + \int_a^b v(x)\frac{du(x)}{dx}\,dx.$$

This yields (8.10).

EXAMPLE 8.5. We evaluate

$$\int_0^{\pi/2} \cos^2 x \, dx$$

using integration by parts. We have

$$\begin{aligned}
\int_0^{\pi/2} \cos^2 x \, dx &= \int_0^{\pi/2} \cos x \cos x \, dx = \int_0^{\pi/2} \cos x \frac{d \sin x}{dx} \, dx \\
&= \cos x \sin x \Big|_0^{\pi/2} - \int_0^{\pi/2} \sin x \frac{d \cos x}{dx} \, dx \\
&= \int_0^{\pi/2} \sin^2 x \, dx \\
&= \int_0^{\pi/2} (1 - \cos^2 x) \, dx \\
&= \int_0^{\pi/2} 1 \, dx - \int_0^{\pi/2} \cos^2 x \, dx \\
&= \frac{\pi}{2} - \int_0^{\pi/2} \cos^2 x \, dx.
\end{aligned}$$

This implies that

$$2 \int_0^{\pi/2} \cos^2 x \, dx = \frac{\pi}{2}$$

so that

$$\int_0^{\pi/2} \cos^2 x \, dx = \frac{\pi}{4} . \tag{8.11'}$$

EXAMPLE 8.6. The device used here can be used to obtain a reduction formula for integrals of the form

$$\int \cos^n x \, dx, \qquad \int \sin^n x \, dx.$$

We have

$$\begin{aligned}
\int \cos^n x \, dx &= \int \cos^{n-1} x \cos x \, dx = \int \cos^{n-1} x \frac{d \sin x}{dx} \, dx \\
&= \sin x \cos^{n-1} x - \int \sin x \big((n-1) \cos^{n-2} x (-\sin x) \big) dx \\
&= \sin x \cos^{n-1} x + (n-1) \int \sin^2 x \cos^{n-2} x \, dx \\
&= \sin x \cos^{n-1} x + (n-1) \int (1 - \cos^2 x) \cos^{n-2} x \, dx \\
&= \sin x \cos^{n-1} x + (n-1) \int \cos^{n-2} x \, dx - (n-1) \int \cos^n x \, dx
\end{aligned}$$

so that

$$n\int \cos^n x\,dx = \sin x \cos^{n-1}x + (n-1)\int \cos^{n-2}x\,dx$$

and finally that, if $n \neq 0$, then

$$\int \cos^n x\,dx = \frac{\sin x \cos^{n-1}x}{n} + \frac{n-1}{n}\int \cos^{n-2}x\,dx. \tag{8.12}$$

This implies that

$$\int_0^{\pi/2} \cos^n x\,dx = \frac{n-1}{n}\int_0^{\pi/2} \cos^{n-2}x\,dx. \tag{8.12$'$}$$

If $n = 2$, we obtain the result (8.11′). If $n = 3$, then

$$\int_0^{\pi/2} \cos^3 x\,dx = \frac{2}{3}\int_0^{\pi/2} \cos x\,dx = \frac{2}{3}. \tag{8.13}$$

By induction on m, we can prove: If m is a positive integer, then

$$\int_0^{\pi/2} \cos^{2m+1}x\,dx = \frac{2}{3}\cdot\frac{4}{5}\cdots\frac{2m-2}{2m-1}\,\frac{2m}{2m+1} \tag{8.14}$$

and

$$\int_0^{\pi/2} \cos^{2m}x\,dx = \frac{1}{2}\cdot\frac{3}{4}\cdots\frac{2m-1}{2m}\,\frac{\pi}{2}. \tag{8.15}$$

Prob. 8.2. Prove: If $n \neq 0$, then (1)

$$\int \sin^n x\,dx = -\frac{\cos x \sin^{n-1}x}{n} + \frac{n-1}{n}\int \sin^{n-2}x\,dx$$

and that (2)

$$\int_0^{\pi/2} \sin^n x\,dx = \frac{n-1}{n}\int_0^{\pi/2} \sin^{n-2}x\,dx.$$

Prob. 8.3. Prove: If n is an integer such that $n \geqslant 0$, then

$$\int_0^{\pi/2} \cos^n x\,dx = \int_0^{\pi/2} \sin^n x\,dx.$$

Prob. 8.4. Prove: If $n > 0$, then

$$\int \frac{1}{(1+x^2)^{n+1}}\,dx = \frac{x}{2n(1+x^2)^n} + \frac{2n-1}{2n}\int \frac{1}{(1+x^2)^n}\,dx$$

(Hint: begin with the integral on the right and use integration by parts with $u(x) = (1+x^2)^{-n}$, and $v(x) = x$).

9. Integration by the Method of Partial Fractions

We wish to discuss

$$\int \frac{P(x)}{Q(x)}\,dx, \tag{9.1}$$

where P and Q are polynomials.

$$\begin{aligned} P(x) &= a_0x^m + a_1x^{m-1} + \cdots + a_m, \qquad a_0 \neq 0, \\ Q(x) &= b_0x^n + b_1x^{n-1} + \cdots + b_n, \qquad b_0 \neq 0, \end{aligned} \tag{9.2}$$

m and n being nonnegative integers.

If $m \geqslant n$, we can perform a "long division" to express the rational function $R = P/Q$ as

$$R(x) = \frac{P(x)}{Q(x)} = h(x) + \frac{r(x)}{Q(x)},$$

where h is a polynomial and r is a polynomial of degree at most $n-1$. The h here is the quotient and r is the remainder upon dividing P by Q.

Since polynomials are readily integrated, (9.1) reduces the integration to that of a rational T, where

$$T(x) = \frac{r(x)}{Q(x)} \tag{9.3}$$

and r and Q are polynomials such that the degree of r is less than that of Q. We assume the last in the ensuing discussion.

Also, we assume that the leading coefficient a_0 of Q is equal to 1, and that the degree of Q is $n \geqslant 1$.

Case 1. Q has n distinct real zeros $x_1, x_2, \ldots, x_n$. It follows that $Q(x) = (x-x_1)\ldots(x-x_n)$. We prove that in this case there exist constants $A_1, \ldots, A_n$ such that

$$\frac{r(x)}{Q(x)} = \frac{A_1}{x-x_1} + \frac{A_2}{x-x_2} + \cdots + \frac{A_n}{x-x_n}. \tag{9.4}$$

Write the values r has at $x_1, \ldots, x_n$ as $y_1 = r(x_1), \ldots, y_n = r(x_n)$.

By Theorem VII.4.2, there exists exactly one polynomial P_1, of degree $n-1$ at most, which assumes these values at $x_1, \ldots, x_n$. Since r is of degree less than n, this implies that $r(x) = P(x)$ for all $x \in \mathbb{R}$. The polynomial P is set down in formula (VII.4.30). However, there is another form for this formula given in (VII.4.35) of Remark VII.4.4. Since the latter is easier to write out we use it here. The function g of Remark VII.4.4 is $g(x) = (x-x_1)\ldots(x-x_n)$, so it is identical with our Q.

According to (VII.4.35), we have (using our Q instead of the g there)

$$r(x) = P(x)$$

$$= \frac{y_1 Q(x)}{(x - x_1)Q'(x_1)} + \frac{y_2 Q(x)}{(x - x_2)Q'(x_2)} + \cdots + \frac{y_n Q(x)}{(x - x_n)Q'(x_n)}.$$

This yields

$$T(x) = \frac{r(x)}{Q(x)}$$

$$= \frac{r(x_1)}{(x - x_1)Q'(x_1)} + \frac{r(x_2)}{(x - x_2)Q'(x_2)} + \cdots + \frac{r(x_n)}{(x - x_n)Q'(x_n)}. \tag{9.5}$$

Putting

$$A_i = \frac{r(x_i)}{Q'(x_i)} \quad \text{for} \quad i \in \{1, \ldots, n\}, \tag{9.6}$$

we obtain (9.4) from (9.5).

Case 2. At least one of the zeros of Q is a real zero multiple of multiplicity $k > 1$. Clearly, $1 < k \leqslant n$. Let a be such a zero. We have

$$Q(x) = (x - a)^k h(x), \tag{9.7}$$

where $h(a) \neq 0$ and where h is of degree $n - k$. Thus,

$$\frac{r(x)}{Q(x)} = \frac{r(x)}{(x - a)^k h(x)}. \tag{9.8}$$

Let $A_1 = r(a)/h(a)$. We have

$$\frac{r(x)}{(x - a)^k h(x)} - \frac{A_1}{(x - a)^k} = \frac{r(x)h(a) - r(a)h(x)}{(x - a)^k h(x)h(a)}. \tag{9.9}$$

The numerator on the right is a polynomial of degree less than n and evidently has a as one of its zeros. By the Factor Theorem a polynomial S exists such that

$$r(x)h(a) - r(a)h(x) = (x - a)S(x).$$

The degree of S is less than $n - 1$. We substitute this into (9.9) and obtain

$$\frac{r(x)}{(x - a)^k h(x)} = \frac{A_1}{(x - a)^k} + \frac{(x - a)S(x)}{(x - a)^k h(x)h(a)}$$

$$= \frac{A_1}{(x - a)^k} + \frac{S^*(x)}{(x - a)^{k-1} h(x)},$$

where $S^*(x) = S(x)/h(a)$. In short, we have

$$\frac{r(x)}{(x-a)^k h(x)} = \frac{A_1}{(x-a)^k} + \frac{S^*(x)}{(x-a)^{k-1}h(x)}, \tag{9.10}$$

where $A_1 = r(a)/h(a)$, a constant, and S^* is a polynomial of degree less than $n-1$ ($n-1$ is the degree of $(x-a)^{k-1}h(x)$). Since $k-1>0$, we can apply the above procedure to

$$\frac{S^*(x)}{(x-a)^{k-1}h(x)}$$

to obtain

$$\frac{S^*(x)}{(x-a)^{k-1}h(x)} = \frac{A_2}{(x-a)^{k-1}} + \frac{S^{**}(x)}{(x-a)^{k-2}h(x)} \tag{9.11}$$

where $A_2 = S^*(a)/h(a)$, a constant, and S^{**} is a polynomial of degree less than $n-2$. From (9.10) we conclude (show this) that

$$A_2 = \frac{S^*(a)}{h(a)} = \frac{d}{dx}\frac{r(x)}{h(x)}\bigg|_{x=a}. \tag{9.12}$$

Equations (9.10) and (9.11) combine to yield

$$\frac{r(x)}{(x-a)^k h(x)} = \frac{A_1}{(x-a)^k} + \frac{A_2}{(x-a)^{k-1}} + \frac{S^{**}(x)}{(x-a)^{k-2}h(x)}. \tag{9.13}$$

We continue this procedure until we finally arrive at

$$\begin{aligned}\frac{r(x)}{Q(x)} &= \frac{r(x)}{(x-a)^k h(x)} \\ &= \frac{A_1}{(x-a)^k} + \frac{A_2}{(x-a)^{k-1}} + \cdots + \frac{A_k}{x-a} + \frac{g(x)}{h(x)},\end{aligned} \tag{9.14}$$

where $A_1, \ldots, A_k$ are constants, h and g are polynomials such that the degree of h is $n-k$, the degree of g is less than that of h, and where $h(a) \neq 0$. Multiply both sides of (9.14) by $(x-a)^k$ to obtain

$$\frac{r(x)}{h(x)} = A_1 + A_2(x-a) + \cdots + A_k(x-a)^{k-1} + \frac{g(x)}{h(x)}(x-a)^k. \tag{9.15}$$

Substitute a for x to obtain

$$\frac{r(a)}{h(a)} = A_1. \tag{9.16}$$

Thus, A_1 is uniquely determined by r/h. We note that in

$$f(x) = \frac{g(x)(x-a)^k}{h(x)},$$

a is a zero of multiplicity k of the numerator. Hence, $f, f', f'', \ldots, f^{k-1}$ all vanish at $x = a$. It follows from (9.15) that

$$\left.\frac{d}{dx}\frac{r(x)}{h(x)}\right|_{x=a} = A_2, \qquad \left.\frac{1}{2!}\frac{d^2}{dx^2}\frac{r(x)}{h(x)}\right|_{x=a} = A_3, \ldots, \tag{9.17}$$

$$\left.\frac{1}{(k-1)!}\frac{d^{k-1}}{dx^{k-1}}\frac{r(x)}{h(x)}\right|_{x=a} = A_k,$$

and that not only is A_1 uniquely determined by r/h but so are $A_2, \ldots, A_k$. Using (9.14) we also see that the g there is uniquely determined by r/h.

If Q has k real zeros $x_1, \ldots, x_k$ and is of the form

$$Q(x) = (x - x_1)^{n_1}(x - x_2)^{n_2} \ldots (x - x_k)^{n_k}, \tag{9.18}$$

where $n = n_1 + n_2 + \cdots + n_k$ and $n_1, n_2, \ldots, n_k$ are positive integers, then we write

$$h(x) = (x - x_2)^{n_2} \ldots (x - x_k)^{n_k}. \tag{9.19}$$

Assuming that r is a polynomial of degree less than n, we use (9.14), with x_1 replacing a and n_1 replacing k, to obtain

$$\begin{aligned}\frac{r(x)}{Q(x)} &= \frac{r(x)}{(x - x_1)^{n_1}h(x)}\\ &= \frac{A_{11}}{(x - x_1)^{n_1}} + \frac{A_{12}}{(x - x_1)^{n_1 - 1}} + \cdots + \frac{A_{1n_k}}{x - x_1} + \frac{g(x)}{h(x)}.\end{aligned} \tag{9.20}$$

Here $A_{11}, A_{12}, \ldots, A_{1n_k}$ are uniquely determined constants, g is a polynomial of degree less than that of h, and $h(x_1) \neq 0$. We can then apply the procedure to g/h and continue until all the factors in (9.19) are exhausted and finally obtain

$$\begin{aligned}\frac{r(x)}{Q(x)} &= \frac{r(x)}{(x - x_1)^{n_1}(x - x_2)^{n_2} \ldots (x - x_k)^{n_k}}\\ &= \frac{A_{11}}{(x - x_1)^{n_1}} + \frac{A_{12}}{(x - x_1)^{n_1 - 1}} + \cdots + \frac{A_{1n_1}}{x - x_1}\\ &\quad + \frac{A_{21}}{(x - x_2)^{n_2}} + \frac{A_{22}}{(x - x_2)^{n_2 - 1}} + \cdots + \frac{A_{2n_2}}{x - x_2}\\ &\quad + \cdots + \frac{A_{k1}}{(x - x_k)^{n_k}} + \frac{A_{k2}}{(x - x_k)^{n_k - 1}} + \cdots + \frac{A_{kn_k}}{x - x_k},\end{aligned} \tag{9.21}$$

where all the A's here are uniquely determined constants.

Case 3. The Q in (9.3) has an imaginary zero α. Since Q is a real polynomial, $\bar{\alpha}$, the conjugate of α, is also a zero of q. This implies that $x - \alpha$ and $x - \bar{\alpha}$ are factors of Q and, hence, the product $(x - \alpha)(x - \bar{\alpha})$

$= x^2 - (\alpha + \bar{\alpha})x + \alpha\bar{\alpha}$ is a factor of Q. We write this quadratic factor as $x^2 + bx + c$, where $b = -(\alpha + \bar{\alpha})$ and $c = \alpha\bar{\alpha}$. We have $(\alpha + \bar{\alpha})^2 - 4\alpha\bar{\alpha} = b^2 - 4ac < 0$ and hence $(\alpha - \bar{\alpha})^2 < 0$, so that $\alpha - \bar{\alpha} \neq 0$. We assume that

$$Q(x) = (x^2 + bx + c)^p u(x), \tag{9.22}$$

where u is a real polynomial of degree $n - 2$, p is a positive integer, and $u(\alpha) \neq 0$. This, of course, implies that $u(\bar{\alpha}) \neq 0$. We prove* that unique real constants A_1 and B_1 and a real polynomial S of degree less than $n - 2$ exists such that

$$\begin{aligned} T(x) = \frac{r(x)}{Q(x)} &= \frac{r(x)}{(x^2 + bx + c)^p u(x)} \\ &= \frac{A_1 x + B_1}{(x^2 + bx + c)^p} + \frac{S(x)}{(x^2 + bx + c)^{p-1} u(x)} . \end{aligned} \tag{9.23}$$

We first prove that there exist unique real constants A_1 and B_1 such that the polynomial equation

$$r(x) - u(x)(A_1 x + B_1) = 0 \tag{9.24}$$

has α and $\bar{\alpha}$ as roots.

First assume that α is a root of (9.24). It follows that (9.24) also has $\bar{\alpha}$ as a root and that

$$A_1\alpha + B_1 = \frac{r(\alpha)}{u(\alpha)} = v(\alpha), \qquad A_1\bar{\alpha} + B_1 = \frac{r(\bar{\alpha})}{u(\bar{\alpha})} = v(\bar{\alpha}). \tag{9.25}$$

Subtracting the second of these equations from the first and dividing by $\alpha - \bar{\alpha} \neq 0$ gives us A_1 in terms of α. We then obtain B_1 from $B_1 = v(\alpha) - A_1\alpha$. Thus,

$$\begin{aligned} A_1 &= \frac{v(\alpha) - v(\bar{\alpha})}{\alpha - \bar{\alpha}} = \frac{v(\alpha) - \overline{v(\alpha)}}{\alpha - \bar{\alpha}} , \\ B_1 &= \frac{\alpha v(\bar{\alpha}) - \bar{\alpha} v(\alpha)}{\alpha - \bar{\alpha}} = \frac{\alpha v(\bar{\alpha}) - \overline{\alpha v(\bar{\alpha})}}{\alpha - \bar{\alpha}} . \end{aligned} \tag{9.26}$$

We see that A_1 and B_1 are real. Using these values of A_1 and B_1, it is easy to see from (9.26) that (9.25) and, hence, (9.24), has α and $\bar{\alpha}$ as roots. This implies that the product $(x - \alpha)(x - \bar{\alpha}) = x^2 + bx + c$ is a factor of $r(x) - u(x)(Ax + B)$, i.e.,

$$r(x) - u(x)(A_1 x + B_1) = (x^2 + bx + c)S(x), \tag{9.27}$$

where S is a polynomial of degree less than $n - 2$. Hence,

$$\frac{r(x) - u(x)(A_1 x + B_1)}{(x^2 + bx + c)^p} = \frac{(x^2 + bx + c)S(x)}{(x^2 + bx + c)^p} = \frac{S(x)}{(x^2 + bx + c)^{p-1}} .$$

*Hugh J. Hamilton, The partial fraction decomposition of a rational function, *Mathematics Magazine*, May, 1972, 117–119.

This yields (9.23). Thus, real constants A_1 and B_1 and a real polynomial S exist such that (9.23) holds. The uniqueness of A_1 and B_1 follows from the fact that if (9.23) holds, so must (9.24) and, therefore, (9.26). The latter determine A_1 and B_1 uniquely.

If $p = 1$, we have from (9.23)

$$\frac{r(x)}{(x^2+bx+c)u(x)} = \frac{A_1x+B_1}{x^2+bx+c} + \frac{S(x)}{u(x)},$$

where $u(\alpha) \neq 0$. If $p > 1$, we return to (9.23). Repeated applications of this procedure lead to the existence of unique real constants $A_2, B_2, \ldots, A_p, B_p$, and a real polynomial S^* such that, together with A_1 and B_1, we have

$$\frac{r(x)}{(x^2+bx+c)^p u(x)} = \frac{A_1x+B_1}{(x^2+bx+c)^p} + \frac{A_2x+B_2}{(x^2+bx+c)^{p-1}} + \cdots + \frac{A_px+B_p}{x^2+bx+c} + \frac{S^*(x)}{u(x)} \tag{9.28}$$

for all $x \in \mathbb{R}$ for which $u(x) \neq 0$.

We now combine all these cases. Q may be of the form

$$Q(x) = (x-x_1)^{n_1} \cdots (x-x_k)^{n_k}(x^2+b_1x+c_1)^{p_1}(x^2+b_2x+c_2)^{p_2} \cdots (x^2+b_jx+c_j)^{p_j}, \tag{9.29}$$

where $x_1, \ldots, x_k$ are real, the quadratic polynomials $x^2+b_1x+c_1$, $x^2+b_2x+c_2, \ldots, x^2+b_jx+c_j$ have respective distinct pairs of imaginary zeros $\alpha_1, \bar{\alpha}_1, \alpha_2, \bar{\alpha}_2, \ldots, \alpha_j, \bar{\alpha}_j$, and where $n = n_1 + \cdots + n_k + 2p_1 + \cdots 2p_j$, $n_1, \ldots, n_k, p_1, \ldots, p_j$ being positive integers. If r is a real polynomial of degree $< n$, then

$$\begin{aligned}\frac{r(x)}{Q(x)} &= \left[\frac{A_{11}}{(x-x_1)^{n_1}} + \cdots + \frac{A_{in_1}}{x-x_1}\right] + \cdots \\ &+ \left[\frac{A_{k1}}{(x-x_k)^{n_k}} + \cdots + \frac{A_{kn_k}}{x-x_k}\right] \\ &+ \left[\frac{C_{11}x+D_{11}}{(x^2+b_1x+c_1)^{p_1}} + \cdots + \frac{C_{1p_1}x+D_{1p_1}}{x^2+b_1x+c_1}\right] + \cdots \\ &+ \left[\frac{C_{j1}x+D_{j1}}{(x^2+b_jx+c_j)^{p_j}} + \cdots + \frac{C_{jp_j}x+D_{jp_j}}{x^2+b_jx+c_j}\right],\end{aligned} \tag{9.30}$$

where the A's, C's, and D's are uniquely determined real constants.

EXAMPLE 9.1. To evaluate

$$\int \frac{2x^2+3x+1}{x^4+6x^3+11x^2+6x}\,dx,$$

we write $Q(x)=x^4+6x^3+11x^2+6x=x(x^3+6x^2+11x+6)$. By inspection we see that $x=-1$ is a zero of $x^3+6x^2+11x+6$. Hence,

$$\begin{aligned}Q(x)&=x(x^3+6x^2+11x+6)=x(x+1)(x^2+5x+6)\\&=x(x+1)(x+2)(x+3).\end{aligned}$$

There exist unique constants A, B, C, and D such that

$$\begin{aligned}\frac{2x^2+3x+1}{x^4+6x^3+11x^2+6x}&=\frac{2x^2+3x+1}{x(x+1)(x+2)(x+3)}\\&=\frac{A}{x}+\frac{B}{x+1}+\frac{C}{x+2}+\frac{D}{x+3}\end{aligned}\tag{9.31}$$

for all $x\notin\{0,-1,-2,-3\}$. Multiply both sides by $Q(x)=x(x+1)(x+2)(x+3)$ to obtain

$$\begin{aligned}2x^2+3x+1&=A(x+1)(x+2)(x+3)+Bx(x+2)(x+3)\\&\quad+Cx(x+1)(x+3)+Dx(x+1)(x+2).\end{aligned}$$

Next let $x\to 0$, $x\to -1$, $x\to -2$, $x\to -3$ successively to obtain

$$A=\tfrac{1}{6},\qquad B=0,\qquad C=\tfrac{3}{2},\qquad D=-\tfrac{5}{3}.$$

These imply

$$\begin{aligned}\int\frac{2x^2+3x+1}{x^4+6x^3+11x^2+6x}\,dx&=\frac{1}{6}\int\frac{1}{x}\,dx+\frac{3}{2}\int\frac{1}{x+2}\,dx-\frac{5}{3}\int\frac{1}{x+3}\,dx\\&=\frac{1}{6}\ln|x|+\frac{3}{2}\ln|x+2|-\frac{5}{3}\ln|x+3|+C\\&=\ln(|x|^{1/6}|x+2|^{3/2}/|x+3|^{5/3})+C\\&=\ln\left[\sqrt[6]{\left|\frac{x(x+2)^9}{(x+3)^{10}}\right|}\right]+C.\end{aligned}$$

The answer could have also be deduced by using (9.6) with $x_1=0$, $x_2=-1$, $x_3=-2$, and $x_4=-3$.

EXAMPLE 9.2. Integrate

$$\int\frac{x^4+3}{(x+1)^2(x^2+1)}\,dx.\tag{9.32}$$

In the integrand, the degree of the numerator and denominator are equal.

We divide out to obtain

$$\frac{x^4+3}{(x+1)^2(x^2+1)} = \frac{x^4+3}{x^4+2x^3+2x^2+2x+1}$$

$$= 1 - \frac{2x^3+2x^2+2x-2}{x^4+2x^3+2x^2+2x+1}. \tag{9.33}$$

Hence,

$$\int \frac{x^4+3}{(x+1)^2(x^2+1)}\,dx = x - \int \frac{2x^3+2x^2+2x-2}{(x+1)^2(x^2+1)}\,dx. \tag{9.34}$$

We evaluate the second integral I_2 on the right using the partial fraction expansion

$$\frac{2x^3+2x^2+2x-2}{(x+1)^2(x^2+1)} = \frac{A}{x+1} + \frac{B}{(x+1)^2} + \frac{Cx+D}{x^2+1}.$$

Multiplying both sides by $(x+1)^2(x^2+1)$, we have

$$\begin{aligned} 2x^3+2x^2+2x-2 &= A(x+1)(x^2+1) + B(x^2+1) + (x+1)^2(Cx+D) \\ &= (A+C)x^3 + (A+B+2C+D)x^2 \\ &\quad + (A+C+2D)x + A+B+D. \end{aligned}$$

Equating coefficients, we have

$$\begin{aligned} A \quad + C \quad\quad &= 2, \\ A + B + 2C + D &= 2, \\ A \quad + C + 2D &= 2, \\ A + B \quad\quad + D &= -2. \end{aligned}$$

Solving for A, B, C, and D we obtain

$$A = 0, \qquad B = -2, \qquad C = 2, \qquad D = 0.$$

Hence,

$$I_2 = \int \frac{2x^3+2x^2+2x-2}{(x+1)^2(x^2+1)}\,dx = \int \left(\frac{2x}{x^2+1} - \frac{2}{(x+1)^2} \right) dx$$

$$= \ln(1+x^2) + \frac{2}{x+1} + C.$$

This and (9.34) imply that

$$\int \frac{x^4+3}{(x+1)^2(x+1)^2}\,dx = x - \ln(1+x^2) - \frac{2}{x+1} + C'$$

$$= x + \ln\frac{1}{1+x^2} - \frac{2}{x+1} + C'.$$

Prob. 9.1. Find

(1) $$\int \frac{x}{(x+1)(x^2-1)}\,dx,$$

(2) $$\int \frac{1}{(x+a)(x+b)}\,dx,$$

(3) $$\int \frac{5x^3+2}{x^3-5x^2+4x}\,dx,$$

(4) $$\int \frac{1}{(x^2-4x+3)(x^2+4x+5)}\,dx,$$

(5) $$\int \frac{1}{1+x^3}\,dx,$$

(6) $$\int \frac{1}{(1+x^2)^2}\,dx,$$

(7) $$\int \frac{1}{(x+1)(x^2+x+1)^2}\,dx,$$

(8) $$\int \frac{x^4}{x^4-1}\,dx,$$

(9) $$\int \frac{1}{(x^4-1)^2}\,dx,$$

(10) $$\int \frac{1}{x^4+1}\,dx \left(\text{Hint: note } (1+\sqrt{2}\,x+x^2)(1-\sqrt{2}\,x+x^2)=1+x^4\right).$$

CHAPTER XIV

The Riemann Integral II

1. Uniform Convergence and R-Integrals

We now consider the legitimacy of passing to the limit under the integral sign. If the sequence $\langle f_n \rangle$ of R-integrable functions converges to a limit f on an interval $[a,b]$ does it necessarily follow that

$$\lim_{n\to+\infty} \int_a^b f_n(x)\,dx = \int_a^b f(x)\,dx? \tag{1.1}$$

The question has an affirmative answer provided the convergence is uniform.

Theorem 1.1. *If $\langle f_n \rangle$ is a sequence of R-integrable functions on $[a,b]$ which converges uniformly to some function f, then f is R-integrable on $[a,b]$ and* (1.1) *holds.*

PROOF. Each f_n is bounded on $[a,b]$. Since the convergence of $\langle f_n \rangle$ to f is uniform on $[a,b]$, the sequence is uniformly bounded and the limit function f is bounded on $[a,b]$ (Theorem XI.4.3). We prove next that f is R-integrable on $[a,b]$.

Let $P = \langle x_0, x_1, \ldots, x_m \rangle$ be some partition of $[a,b]$. Write the suprema and infima of f and each f_n on each subinterval $I_i = [x_{i-1}, x_i]$ of P as

$$M_i(f),\quad M_i(f_n),\quad m_i(f),\quad m_i(f_n), \tag{1.2}$$

respectively. We have

$$\begin{aligned}\bar{S}(f,P)-\underline{S}(f,P)&=\sum_{i=1}^{m}(M_i(f)-m_i(f))\Delta x_i\\&=\sum_{i=1}^{m}(M_i(f)-M_i(f_n))\Delta x_i\\&\quad+\sum_{i=1}^{m}(M_i(f_n)-m_i(f_n))\Delta x_i\\&\quad+\sum_{i=1}^{m}(m_i(f_n)-m_i(f))\Delta x_i\\&\leqslant\sum_{i=1}^{n}|M_i(f)-M_i(f_n)|\Delta x_i\\&\quad+\bar{S}(f_n,P)-\underline{S}(f_n,P)+\sum_{i=1}^{n}|m_i(f)-m_i(f_n)|\Delta x_i\end{aligned}\tag{1.3}$$

for each n.

Let $\epsilon > 0$ be given. There exists an N such that if $n > N$, then

$$|f_n(x)-f(x)|<\frac{\epsilon}{3(b-a)}\qquad\text{for all}\quad x\in[a,b].\tag{1.4}$$

This implies that

$$\sup_{x\in I_i}|f_n(x)-f(x)|\leqslant\frac{\epsilon}{3(b-a)}\qquad\text{for each } i.$$

But for each i

$$|M_i(f)-M_i(f_n)|\leqslant\sup_{x\in I_i}|f(x)-f_n(x)|$$

and

$$|m_i(f)-m_i(f_n)|\leqslant\sup_{x\in I_i}|f(x)-f_n(x)|$$

(prove these). Therefore, for $n > N$, we have

$$\sum_{i=1}^{m}|M_i(f)-M_i(f_n)|\Delta x_i\leqslant\frac{\epsilon}{3(b-a)}(b-a)=\frac{\epsilon}{3}\tag{1.5}$$

and

$$\sum_{i=1}^{m}|m_i(f)-m_i(f_n)|\Delta x_i\leqslant\frac{\epsilon}{3(b-a)}(b-a)=\frac{\epsilon}{3}.\tag{1.6}$$

By hypothesis, each f_n is R-integrable on $[a,b]$. Fix an f_n with $n > N$. There exists a partition P' of $[a,b]$ such that

$$\bar{S}(f_n,P')-\underline{S}(f_n,P')<\frac{\epsilon}{3}.\tag{1.7}$$

Since (1.5) and (1.6) hold for any partition P of $[a,b]$, they hold for the partition P' in particular. We use (1.3), with P' replacing P, (1.5), (1.6), and (1.7) to conclude that

$$\bar{S}(f,P') - \underline{S}(f,P') < \frac{\epsilon}{3} + \frac{\epsilon}{3} + \frac{\epsilon}{3} = \epsilon. \tag{1.8}$$

This proves that for each $\epsilon > 0$, there exists a partition P' of $[a,b]$ for which $\bar{S}(f,P') - \underline{S}(f,P') < \epsilon$ holds, and that f is R-integrable on $[a,b]$.

We now prove that (1.1) holds. Given $\epsilon > 0$, there exists an N such that if $n > N$, then

$$|f_n(x) - f(x)| < \frac{\epsilon}{b-a} \qquad \text{for all} \quad x \in [a,b].$$

Therefore, if $n > N$, then

$$\left|\int_a^b f_n(x)\,dx - \int_a^b f(x)\,dx\right| = \left|\int_a^b (f_n(x) - f(x)\,dx\right|$$
$$\leqslant \int_a^b |f_n(x) - f(x)|\,dx < \frac{\epsilon}{b-a}(b-a) = \epsilon.$$

This implies that (1.1) holds.

Corollary 1. *If $\langle u_n \rangle$ is a sequence of R-integrable functions on $[a,b]$ and the series $\sum u_n(x)$ converges uniformly to a sum $S(x)$ on $[a,b]$, then S is R-integrable on $[a,b]$ and*

$$\int_a^b \left(\sum_{n=1}^{\infty} u_n(x)\right) dx = \sum_{n=1}^{\infty} \int_a^b f_n(x)\,dx. \tag{1.9}$$

Proof. Exercise.

Corollary 2. *If $\langle f_n \rangle$ is a sequence of continuous functions on $[a,b]$ which converges uniformly to a function f on $[a,b]$, then f is R-integrable in $[a,b]$ and*

$$\lim_{n\to+\infty} \int_a^b f_n(x)\,dx = \int_a^b f(x)\,dx. \tag{1.10}$$

Proof. Exercise.

Corollary 3. *If each u_n in Corollary 1 is continuous on $[a,b]$, then (1.9) holds.*

Remark 1.1. We give an example of a sequence $\langle f_n \rangle$ of functions which converges pointwise to a function f on $[0,1]$ but for which (1.1) is false there. For each positive integer n we define

$$f_n(x) = 2n^2 x e^{-nx} \qquad \text{for} \quad x \in [0,1]$$

for each positive integer n. Clearly, $f_n(0) = 0$ for all n. Take x such that

$0 < x \leqslant 1$. For such x

$$f_n(x) = 2(nx)^2 e^{-nx} \frac{1}{x} = \frac{2}{x} \frac{(nx)^2}{e^{nx}}$$

for each n. Hence also

$$\lim_{n \to +\infty} f_n(x) = 0 \qquad \text{for} \quad x \in (0, 1].$$

Thus, $\langle f_n \rangle$ converges pointwise to the constant function $f = 0$ on $[0, 1]$. On the other hand,

$$\int_0^1 f_n(x)\,dx = 2n^2 \int_0^1 x e^{-nx} dx = 2 - \frac{2n+2}{e^n}$$

so that

$$\lim_{n \to +\infty} \int_0^1 f_n(x)\,dx = 2 \neq 0 = \int_0^1 f(x)\,dx.$$

Thus, (1.1) is false for this sequence. It follows from Theorem 1.1 that this sequence does not converge uniformly on $[0, 1]$.

Example 1.1. In Example XI.9.3 we proved that

$$\operatorname{Arctan} x = x - \frac{x^3}{3} + \frac{x^5}{5} - \frac{x^7}{7} + \cdots \qquad \text{for} \quad |x| < 1. \tag{1.11}$$

This result can be established much more easily with the aid of the corollaries of Theorem 1.1.

We observe that

$$\frac{1}{1+t^2} = 1 - t^2 + t^4 - t^6 + \cdots \qquad \text{for} \quad |t| < 1. \tag{1.12}$$

The series on the right is a power series expansion about the origin of the function on the left. Hence, it converges uniformly on every bounded closed interval $[-r, r]$, where $0 < r < 1$. By Corollary 3 we can integrate the series term by term to obtain for each $x \in [-r; r]$

$$\operatorname{Arctan} x = \int_0^x \frac{1}{1+t^2}\,dt = x - \frac{x^3}{3} + \frac{x^5}{5} - \frac{x^7}{7} + \cdots.$$

It follows that this expansion is valid for $|x| < 1$. We can then use Abel's limit theorem (Corollary 2 of Theorem XII.6.3) to extend its validity to the closed interval $|x| \leqslant 1$ and to obtain

$$\frac{\pi}{4} = 1 - \frac{1}{3} + \frac{1}{5} - \frac{1}{7} + \cdots \tag{1.13}$$

(See Example XII.6.1). The series on the right sums to $\pi/4$. Hence, it can be used to evaluate π. Although this result is aesthetically pleasing, it is not very practical. The nth term of the series on the right side of (1.13) is $(-1)^{n+1}a_n$, where $a_n = 1/(2n-1)$. The series is clearly alternating. By

Theorem IV.2.1, we see that

$$\left|\frac{\pi}{4}-\left(1-\frac{1}{3}+\frac{1}{5}-\frac{1}{7}+\cdots+(-1)^{n+1}\frac{1}{2n-1}\right)\right|<\frac{1}{2n+1}$$

and, hence,

$$\left|\pi-\left(4-\frac{4}{3}+\frac{4}{5}-\frac{4}{7}+\cdots+(-1)^{n+1}\frac{4}{2n-1}\right)\right|<\frac{4}{2n+1}.$$

For $n = 500$, this states that the error in approximating π by summing the first 500 terms of the series

$$4-\frac{4}{3}+\frac{4}{5}-\frac{4}{7}+\cdots \tag{1.14}$$

is less than $4/1001 < 4/1000 = 0.004 < 0.005$. To guarantee an approximation of π to the nearest one-hundredth we must sum at least the first 500 terms of the series. We obtain next a more rapidly converging series expansion for π.

Let u and v satisfy

$$-\frac{\pi}{2}<u<\frac{\pi}{2},\quad -\frac{\pi}{2}<v<\frac{\pi}{2},\quad \frac{\pi}{2}<u+v<\frac{\pi}{2}.$$

Put $\tan u = s$ and $\tan v = t$. Then

$$\tan(u+v)=\frac{\tan u+\tan v}{1-\tan u\tan v}=\frac{s+t}{1-st}$$

so that

$$u+v=\operatorname{Arctan}(\tan(u+v))=\operatorname{Arctan}\frac{s+t}{1-st}$$

or

$$\operatorname{Arctan}s+\operatorname{Arctan}t=\operatorname{Arctan}\frac{s+t}{1-st}. \tag{1.15}$$

By using $s=\frac{1}{2}$, $t=\frac{1}{3}$, we obtain from this

$$\frac{\pi}{4}=\operatorname{Arctan}1=\operatorname{Arctan}\tfrac{1}{2}+\operatorname{Arctan}\tfrac{1}{3}. \tag{1.16}$$

By (1.11),

$$\operatorname{Arctan}\tfrac{1}{2}=\tfrac{1}{2}-\tfrac{1}{3}\left(\tfrac{1}{2}\right)^3+\tfrac{1}{5}\left(\tfrac{1}{2}\right)^5-\tfrac{1}{7}\left(\tfrac{1}{2}\right)^7+\cdots$$

and

$$\operatorname{Arctan}\tfrac{1}{3}=\tfrac{1}{3}-\tfrac{1}{3}\left(\tfrac{1}{3}\right)^2+\tfrac{1}{5}\left(\tfrac{1}{3}\right)^5-\tfrac{1}{7}\left(\tfrac{1}{3}\right)^7+\cdots.$$

These and (1.16) imply that

$$\begin{aligned}\pi&=4\left(\frac{1}{2}+\frac{1}{3}\right)-\frac{4}{3}\left(\frac{1}{2^3}+\frac{1}{3^3}\right)+\frac{4}{5}\left(\frac{1}{2^5}+\frac{1}{3^5}\right)-\frac{4}{7}\left(\frac{1}{2^7}+\frac{1}{3^7}\right)+\cdots\\&=\sum_{n=1}^{\infty}(-1)^{n+1}\frac{4}{2n-1}\left(\frac{1}{2^{2n-1}}+\frac{1}{3^{2n-1}}\right).\end{aligned} \tag{1.17}$$

The series on the right is alternating (check this). Its nth term is $(-1)^{n+1}a_n$, where

$$a_n = \frac{4}{2n-1}\left(\frac{1}{2^{2n-1}} + \frac{1}{3^{2n-1}}\right).$$

Writing S_n for the nth partial sum of the series for each n, we obtain (Theorem IV.2.1),

$$\begin{aligned} |\pi - S_n| &< \frac{4}{2n+1}\left(\frac{1}{2^{2n+1}} + \frac{1}{3^{2n+1}}\right) \\ &< \frac{4}{2n+1}\left(\frac{1}{2^{2n+1}} + \frac{1}{2^{2n+1}}\right) \\ &= \frac{1}{2n+1}\,\frac{1}{4^{n-1}}\,. \end{aligned}$$

For $n = 4$, we have

$$|\pi - S_4| < \frac{1}{9}\cdot\frac{1}{64} = \frac{1}{576} < \frac{1}{500} = 0.002.$$

To evaluate π with the aid of (1.17) to the nearest hundredth it suffices to sum its first four terms. (Compare this with the result obtained using series (1.14).)

PROB. 1.1. Prove:

$$\operatorname{Arctan}\tfrac{1}{2} = \operatorname{Arctan}\tfrac{1}{3} + \operatorname{Arctan}\tfrac{1}{7}$$

and

$$\frac{\pi}{4} = 2\operatorname{Arctan}\tfrac{1}{3} + \operatorname{Arctan}\tfrac{1}{7}\,.$$

PROB. 1.2. Prove: If $|x| \leqslant 1$, then

$$\operatorname{Arcsin} x = x + \frac{1}{2}\cdot\frac{x^3}{3} + \frac{1\cdot 3}{2\cdot 4}\cdot\frac{x^5}{5} + \frac{1\cdot 3\cdot 5}{2\cdot 4\cdot 6}\cdot\frac{x^7}{7} + \cdots$$

and, therefore, that

$$\frac{\pi}{2} = 1 + \frac{1}{2}\cdot\frac{1}{3} + \frac{1\cdot 3}{2\cdot 4}\cdot\frac{1}{5} + \frac{1\cdot 3\cdot 5}{2\cdot 4\cdot 6}\cdot\frac{1}{7} + \cdots.$$

PROB. 1.3. Prove:

$$\frac{1}{2!} + \frac{2}{3!} + \frac{3}{4!} + \frac{4}{5!} + \cdots = 1.$$

EXAMPLE 1.2. We prove that

$$\lim_{n\to+\infty}\int_0^{\pi/2}\frac{\sin 2nx}{\sin x}\,dx = \frac{\pi}{2}\,. \tag{1.18}$$

When x is not a multiple of π and n is a positive integer, then (Prob. X.6.3)

$$\frac{\sin 2nx}{\sin x} = 2 \sum_{k=1}^{n} \cos(2k-1)x. \tag{1.19}$$

The integrand is undefined at 0. To deal with this we proceed as follows. By L'Hopital's rule, we have

$$\lim_{x \to 0} \frac{\sin 2nx}{\sin x} = 2n.$$

The function f_n defined on $[0, \pi/2]$ as

$$f_n(x) = \begin{cases} 2n & \text{if } x = 0 \\ \dfrac{\sin 2nx}{\sin x} & \text{if } 0 < x \leqslant \dfrac{\pi}{2}, \end{cases}$$

for each positive integer n is continuous on $[0, \pi/2]$ and, therefore, R-integrable there. With this interpretation, we are to evaluate

$$\int_0^{\pi/2} \frac{\sin 2nx}{\sin x}\, dx.$$

(All integrals of the form

$$\int_0^a f(x) \frac{\sin nx}{\sin x}\, dx \quad \text{or} \quad \int_0^a f(x) \frac{\sin nx}{x}\, dx$$

are called *Dirichlet's Integrals*.) By (1.19), we have

$$\begin{aligned}
\int_0^{\pi/2} \frac{\sin 2nx}{\sin x}\, dx &= 2\int_0^{\pi/2} \left(\sum_{k=1}^{n} \cos(2k-1)x \right) dx \\
&= 2 \sum_{k=1}^{n} \int_0^{\pi/2} \cos(2k-1)x\, dx \\
&= 2 \sum_{k=1}^{n} \left. \frac{\sin(2k-1)x}{(2k-1)} \right|_0^{\pi/2} \\
&= 2 \sum_{k=1}^{n} - \frac{\cos k\pi}{2k-1} \\
&= 2 \sum_{k=1}^{n} (-1)^{k+1} \frac{1}{2k-1} \\
&= 2\left(1 - \tfrac{1}{3} + \tfrac{1}{5} - \tfrac{1}{7} + \cdots (-1)^{n+1} \frac{1}{2n-1}\right).
\end{aligned}$$

By (1.13), we obtain from this

$$\begin{aligned}
&\lim_{n \to +\infty} \int_0^{\pi/2} \frac{\sin 2nx}{\sin x}\, dx \\
&= \lim_{n \to +\infty} 2\left(1 - \tfrac{1}{3} + \tfrac{1}{5} - \tfrac{1}{7} + \cdots + (-1)^{n+1} \frac{1}{2n-1}\right) \\
&= 2\frac{\pi}{4} = \frac{\pi}{2}.
\end{aligned}$$

Prob. 1.4. Use (1.19) to obtain: If x is not a multiple of π, then for each positive integer n

$$\frac{\sin 2nx \cos x}{\sin x} = 2\sum_{k=1}^{n} \cos((2k-1)x)\cos x$$
$$= \sum_{k=1}^{n} (\cos 2kx + \cos(2k-2)x).$$

Prove: If n is a positive integer, then

$$\int_0^{\pi/2} \frac{\sin 2nx}{\sin x} \cos x \, dx = \frac{\pi}{2}.$$

Prob. 1.5. Prove: If n is a positive integer, then

(a) $$\int_0^1 \frac{1-(1-x)^n}{x}\, dx = \int_0^1 \frac{1-x^n}{1-x}\, dx = 1 + \frac{1}{2} + \cdots + \frac{1}{n}$$

(b) $$\binom{n}{1} - \binom{n}{2}\frac{1}{2} + \cdots + (-1)^{n+1}\binom{n}{n} = \sum_{k=1}^{n} (-1)^{k+1}\frac{1}{k}\binom{n}{k}$$
$$= 1 + \frac{1}{2} + \cdots + \frac{1}{n}.$$

Prob. 1.6. Prove: Let $\langle f_n \rangle$ be a sequence of functions that are R-integrable on $[a,b]$. If $\langle f_n \rangle$ converges uniformly to a function f on $[a,b]$, then, for each g which is R-integrable on $[a,b]$, we have

$$\lim_{n\to+\infty} \int_a^b f_n(x)g(x)\, dx = \int_a^b f(x)g(x)\, dx$$

(see Prob. XI.4.6).

Prob. 1.7. Prove: If f is continuous on $[a,b]$ and for each positive integer n, we have

$$\int_a^b f(x)x^n dx = 0,$$

then $f(x) = 0$ for all $x \in [a,b]$ (Hint: use Weierstrass' approximation theorem XI.7.2 and attempt to prove that the hypothesis implies

$$\int_a^b f^2(x)\, dx = 0).$$

Prob. 1.8. Prove: if $x \in \mathbb{R}$, then

(a) $\int_0^x \cos t^2\, dt = \sum_{n=0}^{\infty} (-1)^n x^{4n+1}/(4n+1)(2n)!$,
(b) $\int_0^x e^{-t^2}\, dt = \sum_{n=0}^{\infty} (-1)^n x^{2n+1}/(2n+1)n!$.

2. Mean-Value Theorems for Integrals

Theorem 2.1. *If f and g are R-integrable functions on $[a,b]$ such that g does not change sign on $[a,b]$ (either $g(x) \geqslant 0$ for all $x \in [a,b]$ or $g(x) \leqslant 0$ for all $x \in [a,b]$) and if m and M are respectively the infimum and supremum of f on $[a,b]$, then there exists a μ such that $m \leqslant \mu \leqslant M$ and*

$$\int_a^b f(x)g(x)\,dx = \mu \int_a^b g(x)\,dx. \tag{2.1}$$

Proof. Consider the case $g(x) \geqslant 0$ for all $x \in [a,b]$. We have

$$m \leqslant f(x) \leqslant M \quad \text{and, therefore,} \quad mg(x) \leqslant f(x)g(x) \leqslant Mg(x) \tag{2.2}$$

for all $x \in [a,b]$. It follows from the second set of inequalities that

$$m\int_a^b g(x)\,dx \leqslant \int_a^b f(x)g(x)\,dx \leqslant M\int_a^b g(x)\,dx. \tag{2.3}$$

If $\int_a^b g(x)\,dx = 0$, then the middle integral is also equal to 0, and, therefore, (2.1) holds trivially for any μ. If $\int_a^b g(x)\,dx > 0$, then (2.3) implies that

$$m \leqslant \frac{\int_a^b f(x)g(x)\,dx}{\int_a^b g(x)\,dx} \leqslant M.$$

Writing

$$\mu = \frac{\int_a^b f(x)g(x)\,dx}{\int_a^b g(x)\,dx},$$

we obtain (2.1) with $m \leqslant \mu \leqslant M$.

Corollary 1. *If f is continuous on $[a,b]$ and g is an R-integrable function which does not change sign on $[a,b]$, then there exists a c such that $a \leqslant c \leqslant b$ and*

$$\int_a^b f(x)g(x)\,dx = f(c)\int_a^b g(x)\,dx. \tag{2.4}$$

Proof. Since f is continuous on $[a,b]$, it is R-integrable there. Hence, the theorem guarantees the existence of a μ such that $m \leqslant \mu \leqslant M$ and such that (2.1) holds. (Here, of course, m and M are the infimum and supremum of f on $[a,b]$.) Since f is a function which is continuous on the bounded closed interval $[a,b]$, it has a minimum $f(x_0)$ and a maximum $f(x_1)$ there. Thus, $m = f(x_0)$ and $M = f(x_1)$, where $x_0 \in [a,b]$ and $x_1 \in [a,b]$. We have $f(x_0) = m \leqslant \mu \leqslant M = f(x_1)$. The continuity of f on $[a,b]$ implies that it has the intermediate-value property there. Hence, for some c in $[a,b]$, $f(c) = \mu$. This yields (2.4) with $c \in [a,b]$.

PROB. 2.1. Prove: If f is continuous on a bounded closed interval with endpoints a and b, then a c exists between a and b such that

$$\int_a^b f(x)\,dx = f(c)(b-a). \tag{2.5}$$

Remark 2.1. Corollary 1 of Theorem 2.1 is often called the *first mean-value theorem of the integral calculus.*

Remark 2.2. The result cited in Prob. 2.1 can be used to prove that if f is continuous on $[a,b]$, then the indefinite integral F defined as

$$F(x) = \int_a^x f(t)\,dt, \qquad a \leqslant x \leqslant b,$$

is a primitive of f on $[a,b]$; that is, F is differentiable on $[a,b]$ and

$$F'(x) = f(x) \qquad \text{for} \quad x \in [a,b].$$

(See Corollary 2 of Theorem XIII.6.2.) For proof take $x \in [a,b]$ and h such that $h \neq 0$, $x + h \in [a,b]$. By the result in Prob. 2.1, there exists a c between x and $x + h$ such that

$$F(x+h) - F(x) = \int_x^{x+h} f(t)\,dt = f(c)h$$

so that

$$\frac{F(x+h) - F(x)}{h} = f(c).$$

But $\lim_{h\to 0} f(c) = f(x)$ (why?). Hence,

$$F'(x) = \lim_{h\to 0} \frac{F(x+h) - F(x)}{h} = f(x) \qquad \text{for} \quad x \in [a,b].$$

The Second Mean-Value Theorem for Integrals

Theorem 2.2. *If f is monotonically decreasing and nonnegative on $[a,b]$ and g is R-integrable on $[a,b]$, then there exists a $c \in [a,b]$ such that*

$$\int_a^b f(x)g(x)\,dx = f(a)\int_a^c g(x)\,dx. \tag{2.6}$$

PROOF.* Since it is monotonic on $[a,b]$, f is R-integrable on $[a,b]$ (Theorem XIII.1.3). By hypothesis, g is R-integrable on $[a,b]$. It follows (Corollary of Theorem XIII.3.4) that the product fg is R-integrable on $[a,b]$. Suppose

* H. S. Carslaw, *Introduction to the Theory of Fourier Series and Integrals*, Dover, New York, 1930, pp. 110–112.

$f(a) = 0$. Since f is monotonically decreasing and nonnegative on $[a,b]$, it follows that $f(x) = 0$ for all $x \in [a,b]$. In this case the integral on the left-hand side in (2.1) is equal to 0 and the right-hand side there is equal to 0 for any $c \in [a,b]$. Thus, the conclusion holds in this case.

Suppose $f(a) > 0$. Given $\epsilon > 0$, there exists a partition $P = \langle x_0, \ldots, x_n \rangle$ of $[a,b]$ such that

$$\int_a^b f(x)g(x)\,dx - \frac{\epsilon}{3} < \underline{S}(fg,P) \leqslant \overline{S}(fg,P) < \int_a^b f(x)g(x)\,dx + \frac{\epsilon}{3} \tag{2.7}$$

and

$$\int_a^b g(x)\,dx - \frac{\epsilon}{3f(a)} < \underline{S}(g,P) \leqslant \overline{S}(g,P) \leqslant \int_a^b g(x)\,dx + \frac{\epsilon}{3f(a)} \tag{2.8}$$

(using refinements if necessary). Put

(a) $\sigma = \sum_{i=1}^{n} f(x_{i-1})g(x_{i-1})\,\Delta x_i$,
(b) $c_i = g(x_{i-1})\,\Delta x_i$ and $d_i = \sum_{k=1}^{i} c_k$ for each $i \in \{1, \ldots, n\}$,

so that $c_i = d_i - d_{i-1}$ for each $i \in \{1, \ldots, n\}$. We have $\underline{S}(fg,P) \leqslant \sigma \leqslant \underline{S}(fg,P)$. Therefore, by (2.7) it follows that

$$\left| \sigma - \int_a^b f(x)g(x)\,dx \right| < \frac{\epsilon}{3}.$$

In turn, this implies that

$$\sigma - \frac{\epsilon}{3} < \int_a^b f(x)g(x)\,dx < \sigma + \frac{\epsilon}{3}. \tag{2.9}$$

Continuing, we have

$$\begin{aligned}
\sigma &= \sum_{i=1}^{n} f(x_{i-1})c_i \\
&= f(x_0)c_1 + \sum_{i=2}^{n} f(x_{i-1})c_i \\
&= f(x_0)d_1 + \sum_{i=2}^{n} f(x_{i-1})(d_i - d_{i-1}) \\
&= f(x_0)d_1 + f(x_1)(d_2 - d_1) + f(x_2)(d_3 - d_2) + \cdots \\
&\quad + f(x_{n-1})(d_n - d_{n-1}) \\
&= (f(x_0) - f(x_1))d_1 + (f(x_1) - f(x_2))d_2 + \cdots \\
&\quad + (f(x_{n-2}) - f(x_{n-1}))d_{n-1} + f(x_{n-1})d_n \\
&= \sum_{i=1}^{n-1} (f(x_{i-1}) - f(x_i))d_i + f(x_{n-1})d_n.
\end{aligned} \tag{2.10}$$

Let $d_q = \max_{1 \leqslant i \leqslant n} d_i$ and $d_p = \min_{1 \leqslant i \leqslant n} d_i$. Since f is monotonically decreasing and positive, we know that $f(x_{i-1}) - f(x_i) \geqslant 0$ for all $i \in$

$\{1, \ldots, n-1\}$, but $f(x_0) > 0$. Hence, we conclude from (2.10) that

$$d_p\left(\sum_{i=1}^{n-1}(f(x_{i-1}) - f(x_i)) + f(x_{n-1})\right) \leqslant \sigma$$
$$\leqslant d_q\left(\sum_{i=1}^{n-1}(f(x_{i-1}) - f(x_i)) + f(x_{n-1})\right)$$

or that

$$d_p f(a) \leqslant \sigma \leqslant d_q f(a). \tag{2.11}$$

Consider $d_q = \sum_{k=1}^{q} c_k = \sum_{i=1}^{q} g(x_{i-1})\Delta x_i$. The points $x_0, x_1, \ldots, x_{q+1}$ constitute a partition of the subinterval $[a, x_{q+1}]$ of $[a, b]$. Write this partition as $P^* = \langle x_0, x_1, \ldots, x_{q+1}\rangle$. Clearly,

$$\underline{S}(g, P^*) \leqslant d_q \leqslant \bar{S}(g, P^*) \quad \text{and} \quad \underline{S}(g, P^*) \leqslant \int_a^{x_{q+1}} g(x)\,dx \leqslant \bar{S}(g, P^*). \tag{2.12}$$

Note also that $\bar{S}(g, P^*) - \underline{S}(g, P^*) \leqslant \bar{S}(g, P) - \underline{S}(g, P)$ (explain). It follows from this and (2.8) that

$$0 \leqslant \bar{S}(g, P^*) - \underline{S}(g, P^*) < \frac{2\epsilon}{3f(a)}.$$

This implies after taking note of the second inequality in (2.12) that

$$d_q = \sum_{i=1}^{q} g(x_{i-1})\,\Delta x_i \leqslant \bar{S}(g, P^*) < \underline{S}(g, P^*) + \frac{2\epsilon}{3f(a)}$$
$$< \int_a^{x_{q+1}} g(x)\,dx + \frac{2\epsilon}{3f(a)}.$$

From this and (2.11) we conclude that

$$\sigma \leqslant f(a)d_q < f(a)\int_a^{x_{q+1}} g(x)\,dx + \frac{2\epsilon}{3}.$$

A similar analysis of d_p yields

$$f(a)\int_a^{x_{p+1}} g(x)\,dx - \frac{2\epsilon}{3} < f(a)d_p \leqslant \sigma.$$

Hence,

$$f(a)\int_a^{x_{p+1}} g(x)\,dx - \frac{2\epsilon}{3} < \sigma < f(a)\int_a^{x_{q+1}} g(x)\,dx + \frac{2\epsilon}{3}. \tag{2.13}$$

Combining this with (2.9) we obtain

$$f(a)\int_a^{x_{p+1}} g(x)\,dx - \epsilon < \int_a^b f(x)g(x)\,dx < f(a)\int_a^{x_{q+1}} g(x)\,dx + \epsilon. \tag{2.14}$$

Define F by means of

$$F(x) = \int_a^x g(t)\,dt \qquad \text{for} \quad x \in [a, b].$$

F is continuous on $[a,b]$ (Theorem XIII.6.3). Hence, it has a maximum $M = F(x_1)$ and a minimum $m = F(x_0)$ on $[a,b]$. Accordingly,

$$\int_a^{x_0} g(t)\,dt = F(x_0) \leqslant \int_a^{x_{p+1}} g(t)\,dt \quad \text{and} \quad \int_a^{x_{q+1}} g(t)\,dt \leqslant F(x_1) = \int_a^{x_1} g(t)\,dt.$$

By (2.14), we obtain from these

$$f(a)F(x_0) - \epsilon < \int_a^b f(x)g(x)\,dx < f(a)F(x_1) + \epsilon.$$

This holds for any $\epsilon > 0$. Hence,

$$f(a)F(x_0) \leqslant \int_a^b f(x)g(x)\,dx \leqslant f(a)F(x_1).$$

Thus, we have

$$F(x_0) \leqslant \frac{\int_a^b f(x)g(x)\,dx}{f(a)} \leqslant F(x_1).$$

But F is continuous on $[a,b]$ and, therefore, has the intermediate-value property there. Hence, there exists a c between x_0 and x_1 and, hence, in $[a,b]$ such that

$$\int_a^c g(x)\,dx = F(c) = \frac{\int_a^b f(x)g(x)\,dx}{f(a)}.$$

Thus, the conclusion follows in the case $f(a) > 0$.

The following theorem has a proof similar to the last one:

Theorem 2.3. *If f is monotonic increasing and nonnegative on $[a,b]$ and g is R-integrable on $[a,b]$, then there exists a $c \in [a,b]$ such that*

$$\int_a^b f(x)g(x)\,dx = f(b)\int_c^b g(x)\,dx.$$

PROB. 2.2. Prove: Theorem 2.3.

A more symmetric version of the last two theorems is obtained by retaining the monotonicity of f but dropping the requirement that it be of fixed sign on $[a,b]$.

Theorem 2.4. *If f is monotonic and g is R-integrable on $[a,b]$, then there exists a $c \in [a,b]$ such that*

$$\int_a^b f(x)g(x)\,dx = f(a)\int_a^c g(x)\,dx + f(b)\int_c^b g(x)\,dx. \tag{2.15}$$

PROOF. Let f be monotonic increasing. Consider h, where $h(x) = f(b) - f(x)$ for $x \in [a,b]$. The function h is nonnegative and monotonic decreasing on $[a,b]$. By Theorem 2.2, there exists a $c \in [a,b]$ such that

$$\int_a^b h(x)g(x)\,dx = h(a)\int_a^c g(x)\,dx.$$

Hence,

$$\int_a^b (f(b)-f(x))g(x)\,dx = (f(b)-f(a))\int_a^c g(x)\,dx$$

and

$$\int_a^b f(x)g(x)\,dx = f(b)\left[\int_a^b g(x)\,dx - \int_a^c g(x)\,dx\right] + f(a)\int_a^c g(x)\,dx$$

$$= f(a)\int_a^c g(x)\,dx + f(b)\int_c^b g(x)\,dx.$$

If f is monotonic decreasing, define h as $h(x) = f(x) - f(b)$ for $x \in [a,b]$ and argue as before, using Theorem 2.3.

Remark 2.3. For an application of the last theorem see the proof of Theorems XV.3.3 and XV.3.4 in Chapter XV.

3. Young's Inequality and Some of Its Applications

Theorem 3.1. *If f has a continuous and nonzero derivative f' on $[a,b]$, then*

$$bf(b) - af(a) = \int_a^b f(x)\,dx + \int_{f(a)}^{f(b)} f^{-1}(x)\,dx. \tag{3.1}$$

PROOF. By the hypothesis on f, f is strictly monotonic on $[a,b]$ and has a strictly monotonic and differentiable inverse f^{-1} on the interval with endpoints $f(a)$ and $f(b)$. Integration by parts yields

$$\int_a^b xf'(x)\,dx = xf(x)\Big|_a^b - \int_a^b f(x)\,dx = bf(b) - af(a) - \int_a^b f(x)\,dx. \tag{3.2}$$

On the other hand by the substitution formula for definite integrals (Theorem XIII.7.2),

$$\int_a^b xf'(x)\,dx = \int_a^b f^{-1}(f(x))\frac{df(x)}{dx}\,dx = \int_{f(a)}^{f(b)} f^{-1}(v)\,dv.$$

This and (3.2) yield (3.1).

Theorem 3.2 (Young's Inequality). *If f has a continuous and positive derivative on $[0,c]$ $(c > 0)$ and $f(0) = 0$, then for $a \in [0,c]$, $b \in [0, f(c)]$, we have*

$$ab \leqslant \int_0^a f(x)\,dx + \int_0^b f^{-1}(x)\,dx. \tag{3.3}$$

The equality holds if and only if $b = f(a)$.

PROOF. f is necessarily strictly increasing on $[0, c]$ so that $f(x) > f(0) = 0$ for $0 < x \leqslant c$. f has a continuous and strictly increasing inverse defined on $[0, f(c)]$ and $f^{-1}(0) = 0 < f^{-1}(x)$ for $0 < x \leqslant f(c)$. By Theorem 3.1, we have: If $z \in [0, c]$, then

$$zf(z) = \int_0^z f(x)\,dx + \int_0^{f(z)} f^{-1}(x)\,dx. \tag{3.4}$$

Assume that $a \in [0, c]$ and $b \in [0, f(c)]$. There are three cases to consider: (1) $0 \leqslant f(a) < b$, (2) $f(a) = b$, (3) $0 \leqslant b < f(a)$. In case (1) we note that $x \geqslant f(a)$ implies $f^{-1}(x) \geqslant f^{-1}(f(a)) = a$ so that

$$\int_{f(a)}^b f^{-1}(x)\,dx > a(b - f(a)). \tag{3.5}$$

Using (3.4) with $z = a$, we see that

$$\begin{aligned}\int_0^a f(x)\,dx + \int_0^b f^{-1}(x)\,dx &= \int_0^a f(x)\,dx + \int_0^{f(a)} f^{-1}(x)\,dx + \int_{f(a)}^b f^{-1}(x)\,dx \\ &= af(a) + \int_{f(a)}^b f^{-1}(x)\,dx \\ &> af(a) + a(b - f(a)) = ab.\end{aligned}$$

Thus, (3.3) holds in case (1). In case (2), the equality in (3.3) holds in view of (3.4). Finally, in case (3) we have $f(a) > b \geqslant 0$ so that $a > f^{-1}(b) \geqslant 0$. Here $x \geqslant f^{-1}(b)$ implies that $f(x) \geqslant f(f^{-1}(b)) = b$ and

$$\begin{aligned}\int_{f^{-1}(b)}^a f(x)\,dx > f(f^{-1}(b))\big(a - f^{-1}(b)\big) &= b\big(a - f^{-1}(b)\big) \\ &= ab - bf^{-1}(b). \end{aligned}\tag{3.6}$$

This implies that

$$\begin{aligned}\int_0^a f(x)\,dx + \int_0^b f^{-1}(x)\,dx &= \int_0^{f^{-1}(b)} f(x)\,dx + \int_{f^{-1}(b)}^a f(x)\,dx + \int_0^b f^{-1}(x)\,dx \\ &> \int_0^{f^{-1}(b)} f(x)\,dx + \int_0^b f^{-1}(x)\,dx + ab - bf^{-1}(b).\end{aligned}$$

In view of (3.4) with $z = f^{-1}(b)$ and $b = f(z)$, the last expression reduces to ab. Therefore, (3.3) follows in this case also. Here cases (1) and (3) yield the strict inequality in (3.3). Hence, we conclude that the equality in (3.3) holds in case (2) only. This completes the proof.

Remark 3.1. Young's inequality furnishes another proof of part (a) of Theorem II.12.4 and even extends its validity to the case where the r mentioned there is real and not necessarily rational. Thus, we can prove: Suppose r is *real*, $r^2 \neq r$, and $s = r/(r - 1)$ so that

$$\frac{1}{r} + \frac{1}{s} = 1 \tag{3.7}$$

and A and B are positive real numbers. Then $r > 1$ implies

$$AB \leqslant \frac{A^r}{r} + \frac{B^s}{s}. \tag{3.8}$$

Equality holds in (3.8) if and only if $B = A^{r-1}$.

To prove the last statement we define f as $f(x) = x^{r-1}$ for $x \geqslant 0$. The inverse of f is f^{-1}, where

$$f^{-1}(x) = x^{1/(r-1)} \qquad \text{for} \quad x \geqslant 0.$$

By (3.3) we have

$$\begin{aligned} AB &\leqslant \int_0^A x^{r-1} dx + \int_0^B x^{1/(r-1)} dx \\ &= \frac{A^r}{r} + \frac{B^{1/(r-1)+1}}{1/(r-1)+1} \\ &= \frac{A^r}{r} + \frac{B^{r/(r-1)}}{r/(r-1)} \\ &= \frac{A^r}{r} + \frac{B^s}{s}. \end{aligned}$$

The equality holds here if and only if $B = f(A) = A^{r-1}$.

Remark 3.2. The inequality of the last remark has as a consequence part (a) of Hölder's inequality (Theorem II.12.5). Hence, we have: Suppose r is *real*, $r > 1$, and $s = r/(r-1)$ so that

$$\frac{1}{r} + \frac{1}{s} = 1,$$

then for any nonnegative real numbers $a_1, \ldots, a_n; b_1, \ldots, b_n$,

$$\sum_{i=1}^{n} a_i b_i \leqslant \left(\sum_{i=1}^{n} a_i^r \right)^{1/r} \left(\sum_{i=1}^{n} b_i^s \right)^{1/s}. \tag{3.9}$$

A consequence of this result is part (a) of Minkowski's inequality (Theorem II.12.7). Hence, we have the following: Suppose $r > 1$. If $a_1, \ldots, a_n; b_1, \ldots, b_n$ are nonnegative real numbers, then

$$\left(\sum_{i=1}^{n} (a_i + b_i)^r \right)^{1/r} \leqslant \left(\sum_{i=1}^{n} a_i^r \right)^{1/r} + \left(\sum_{i=1}^{n} b_i^r \right)^{1/r}. \tag{3.10}$$

The Holder and Minkowski inequalities have integral analogues. For example, if $r = 2$, the following result holds:

If f and g are R-integrable over $[a, b]$, then so are f^2, g^2, and fg (Theorem XIII.3.4 and its corollary) and we have

$$\left| \int_a^b f(x) g(x)\, dx \right| \leqslant \sqrt{\int_a^b f^2(x)\, dx}\ \sqrt{\int_a^b g^2(x)\, dx}\ . \tag{3.11}$$

Prob. 3.1. Prove (3.11) (Hint: first consider

$$\int_a^b \left(tf^2(x) - g(x)\right)^2 dx = At^2 - 2Bt + C,$$

where

$$A = \int_a^b f^2(x)\,dx, \quad B = \int_a^b f(x)g(x)\,dx, \quad C = \int_a^b g^2(x)\,dx,$$

and observe that in the present case $At^2 - 2Bt + C \geqslant 0$ holds for all real t). State conditions for equality to hold in (3.11).

Prob. 3.2. Prove: If f and g are R-integrable over $[a, b]$, then

$$\left(\int_a^b (f(x) + g(x))^2 dx\right)^{1/2} \leqslant \left(\int_a^b f^2(x)\,dx\right)^{1/2} + \left(\int_a^b g^2(x)\,dx\right)^{1/2}.$$

State conditions for the equality to hold.

Prob. 3.3. Prove: If f, g, and h are R-integrable over $[a, b]$, then

$$\left(\int_a^b (f(x) - h(x))^2 dx\right)^{1/2} \leqslant \left(\int_a^b (f(x) - g(x))^2 dx\right)^{1/2} + \left(\int_a^b (g(x) - h(x))^2 dx\right)^{1/2}.$$

4. Integral Form of the Remainder in Taylor's Theorem

Taylor's Theorem with Schlömlich's form of the remainder (Theorem IX.4.2) states that if f and its derivatives up to and including order n are continuous on an interval I and $f^{(n+1)}(x)$ exists at least for x in the interior of I, then for distinct a and x in I, there exists a c between a and x such that

$$f(x) = \sum_{k=0}^{n} \frac{f^{(k)}(a)}{k!}(x-a)^k + \frac{f^{(n+1)}(c)}{(p+1)n!}(x-c)^{n-p}(x-a)^{p+1}. \tag{4.1}$$

Here n is a nonnegative integer and $p > -1$. The last term is Schlömlich's form of the remainder in Taylor's formula of order n. We write it as

$$R_{n+1}(a, x) = \frac{f^{(n+1)}(c)}{(p+1)n!}(x-c)^{n-p}(x-a)^{p+1}. \tag{4.2}$$

For $p = n$ the remainder becomes

$$R_{n+1}(a,x) = \frac{f^{(n+1)}(c)}{(n+1)!}(x-a)^{n+1}, \tag{4.3}$$

which is Lagrange's form of the remainder. For $p = 0$, (4.2) becomes

$$R_{n+1}(a,x) = \frac{f^{(n+1)}(c)}{n!}(x-c)^n(x-a), \tag{4.4}$$

the Cauchy form of the remainder.

If we strengthen the conditions on f by requiring the continuity of the $(n+1)$th derivative, then we can obtain an integral form of the remainder.

Theorem 4.1. *If f and its derivatives up to and including those of order $n+1$ are continuous on an interval I and $a \in I$, then for each x in I, we have*

$$f(x) = \sum_{k=0}^{n} \frac{f^{(k)}(a)}{k!}(x-a)^k + \frac{1}{n!}\int_a^x (x-t)^n f^{(n+1)}(t)\,dt. \tag{4.5}$$

Here n is some nonnegative integer.

Proof. We use induction on n. Let $n = 0$. Then f' is continuous on I and, for x and a in I,

$$f(x) = f(a) + \int_a^x f'(t)\,dt. \tag{4.6}$$

This means that the theorem holds for $n = 0$. Assume that the theorem holds for some nonnegative integer n and that f and all its derivatives up to and including those of order $n+2$ are continuous on I. Thus, (4.5) holds. Using integration by parts, we have, for x and a in I,

$$\begin{aligned}\int_a^x \frac{(x-t)^n}{n!} f^{(n+1)}(t)\,dt &= \int_a^x f^{(n+1)}(t)\frac{d}{dt}\left(-\frac{(x-t)^{n+1}}{(n+1)!}\right)dt \\ &= -f^{(n+1)}(t)\frac{(x-t)^{n+1}}{(n+1)!}\Bigg|_a^x \\ &\quad + \frac{1}{(n+1)!}\int_a^x (x-t)^{n+1} f^{(n+2)}(t)\,dt \\ &= f^{(n+1)}(a)\frac{(x-a)^{n+1}}{(n+1)!} \\ &\quad + \frac{1}{(n+1)!}\int_a^x (x-t)^{n+1} f^{(n+2)}(t)\,dt.\end{aligned}$$

We can now substitute in (4.5) to obtain

$$f(x) = \sum_{k=0}^{n} \frac{f^{(k)}(a)}{k!}(x-a)^k + \frac{f^{(n+1)}(a)}{(n+1)!}(x-a)^{n+1}$$
$$+ \frac{1}{(n+1)!}\int_a^x (x-t)^{n+1} f^{(n+2)}(t)\,dt$$
$$= \sum_{k=0}^{n+1} \frac{f^{(k)}(a)}{k!} + \frac{1}{(n+1)!}\int_a^x (x-t)^{n+1} f^{(n+2)}(t)\,dt.$$

This means that if the theorem, therefore, holds for n, then it holds for $n+1$. Since it also holds for $n = 0$ because of (4.6), we obtain by the principle of mathematical induction that it holds for all nonnegative integers n. This completes the proof.

PROB. 4.1. Prove: If f is continuous on some interval I and x and a are in I, then

(a) $$\int_a^x \left(\int_a^t f(u)\,du\right) dt = \int_a^x (x-t) f(t)\,dt,$$

(b) $$\int_a^x \left(\int_a^{t_n}\left(\cdots\left(\int_a^{t_1} f(u)\,du\right) dt_1 \ldots\right) dt_n = \frac{1}{n!}\int_a^x (x-t)^n f(t)\,dt.\right.$$

5. Sets of Measure Zero. The Cantor Set

We saw (Theorem XIII.3.3) that a function defined on a bounded closed interval $[a,b]$ can have its value changed at finitely many points of $[a,b]$ without this affecting its R-integrability or its R-integral. It follows from this that if f is defined on $[a,b]$ except on a finite nonempty set $S \subseteq [a,b]$ and there exists an extension g of f to all of $[a,b]$ which is R-integrable on $[a,b]$, then all extensions of f to all of $[a,b]$ are R-integrable and have the same R-integral on $[a,b]$. Hence, we call f R-integrable on $[a,b]$ if (1) it is defined there except possibly on a finite set $S \subseteq [a,b]$ and (2) it has an extension g to all of $[a,b]$ which is R-integrable on $[a,b]$ in the sense of Chapter XIII. If f is R-integrable in the sense just defined, then its R-integral is defined as that of some extension of f to $[a,b]$. For example, the integrand in the integral

$$\int_0^1 \frac{\sin x}{x}\,dx$$

is defined on $[0,1]$ except at $x = 0$. However, this integrand has an extension to $[0,1]$ which is continuous and, therefore, R-integrable in the sense of Chapter XIII. Hence, we say that this integrand is R-integrable (in the sense of the last paragraph) on $[0, 1]$.

Under what conditions does a function defined on the bounded open interval $(a; b)$ have an extension to $[a, b]$ which is R-integrable? Theorem XIII.1.6 points to the answer. Let f be bounded on the bounded open interval $(a; b)$ and R-integrable on every closed subinterval $[c, d]$ of $(a; b)$. Define $f(a)$ and $f(b)$ in some way to obtain an extension of f which is bounded on $[a, b]$ and R-integrable on every closed subinterval $[c, d]$ of $[a, b]$ such that $a < c < d < b$. Then, by Theorem XIII.1.6, this extension of f to all of $[a, b]$ is R-integrable.

An f which is continuous on the bounded open interval $I = (a; b)$ and has finite one-sided limits $f(a+)$, $f(b-)$ at the endpoints of I has a continuous extension g to all of $[a, b]$. In fact, it suffices to define $g(a) = f(a+)$ and $g(b) = f(b-)$ and $g(x) = f(x)$ for $a < x < b$. This extension g is continuous on the bounded closed interval $[a, b]$ and, therefore, bounded and R-integrable on $[a, b]$.

These ideas lead us to the notion of a *piecewise continuous* function on $[a, b]$. We call f *piecewise continuous* on $[a, b]$ if a partition $P = \langle x_0, x_1, \ldots, x_n \rangle$ of $[a, b]$ exists such that f is continuous on each open interval $I_i = (x_{i-1}, x_i)$ of P and has finite one-sided limits $f(x_{i-1}+)$, $f(x_i-)$ at the two endpoints of I_i for each $i \in \{1, \ldots, n\}$. It follows from what was said at the beginning of this paragraph that:

Theorem 5.1. *If f is piecewise continuous on $[a, b]$, then it is R-integrable on $[a, b]$.*

To investigate these matters further it is necessary to introduce the notion of a set of *measure zero*.

Def. 5.1. A set $S \subseteq \mathbb{R}$ is said to have *measure zero* if and only if for each $\epsilon > 0$, there exists a sequence $\langle I_n \rangle$ of open intervals such that $S \subseteq \bigcup_{n=1} I_n$ and

$$\sum_{n=1}^{\infty} L(I_n) < \epsilon,$$

where $L(I_i)$ is the length of I_i.

The sequence $\langle I_n \rangle$ in this definition is said to *cover* S; if A is a set and $\mathbb{S}$ is a set or family or sequence of sets such that $A \subseteq \cup \mathbb{S}$, then we say that $\mathbb{S}$ *covers* A or that it constitutes a *cover* of A. We also refer to $\sum L(I_n)$ as the *total length* of the sequence $\langle I_n \rangle$. Using this terminology, Def. 5.1 can be phrased as: A set $S \subseteq \mathbb{R}$ is said to have *measure zero* if and only if for each $\epsilon > 0$, there exists a sequence $\langle I_n \rangle$ of open intervals covering S whose total length is less than ϵ.

EXAMPLE 5.1. We show that a denumerable set (Def. II.10.7) $S = \{x_1, x_2, \ldots\}$ is of measure zero. Let $\epsilon > 0$ be given. Corresponding to each $x_i \in S$ take the open interval

$$I_i = \left(x_i - \frac{\epsilon}{2^{i+2}}; x_i + \frac{\epsilon}{2^{i+2}}\right).$$

Note each $x_i \in I_i \subseteq \bigcup_{k=1}^{\infty} I_k$, so $S \subseteq \bigcup_{k=1}^{\infty} I_k$. Also,

$$L(I_i) = \frac{\epsilon}{2^{i+2}} + \frac{\epsilon}{2^{i+2}} = \frac{\epsilon}{2^{i+1}}$$

so that

$$\sum_{i=1}^{\infty} L(I_i) = \epsilon \sum_{i=1}^{\infty} \frac{1}{2^{i+1}} = \frac{\epsilon}{2^2} \frac{1}{1-\frac{1}{2}} = \frac{\epsilon}{2} < \epsilon.$$

By Def. 5.1, S has measure zero.

Theorem 5.2. *If a set $S \subseteq \mathbb{R}$ is such that for each $\epsilon > 0$, there exists a finite sequence $\langle I_1, I_2, \ldots, I_n \rangle$ of open intervals covering S such that*

$$\sum_{i=1}^{n} L(I_i) < \epsilon,$$

then S has measure zero.

PROOF. Given $\epsilon > 0$, we must prove that there exists an infinite sequence $\langle J_k \rangle$ of intervals covering S whose total length is less than ϵ. But the hypothesis guarantees the existence of a finite sequence of intervals $\langle I_1, \ldots, I_n \rangle$ such that

$$S \subseteq \bigcup_{k=1}^{n} I_k \quad \text{and} \quad \sum_{i=1}^{n} L(I_i) < \frac{\epsilon}{2}.$$

Using this n, let $I_n = (a_n; b_n)$. Construct the sequence $\langle J_k^* \rangle$ as follows: For each positive integer k, let $J_k^* = (b_n - \epsilon/2^{k+2}; b_n + \epsilon/2^{k+2})$. We have

$$L(J_k^*) = \frac{\epsilon}{2^{k+1}} \quad \text{and} \quad \sum_{k=1}^{\infty} L(J_k^*) = \epsilon\left(\frac{1}{2^2} + \frac{1}{2^3} + \cdots\right) = \frac{\epsilon}{2}.$$

Now consider the new sequence $\langle J_k \rangle$ defined as $J_1 = I_1, \ldots, J_n = I_n$, $J_{n+1} = J_1^*, J_{n+2} = J_2^*, \ldots$. Clearly,

$$S \subseteq \bigcup_{k=1}^{\infty} J_k \quad \text{and} \quad \sum_{k=1}^{\infty} L(J_k) = \sum_{k=1}^{n} L(I_k) + \sum_{k=n+1}^{\infty} L(J_k^*) < \frac{\epsilon}{2} + \frac{\epsilon}{2} = \epsilon.$$

Thus, for each $\epsilon > 0$, there exists an infinite sequence $\langle J_k \rangle$ of open intervals whose total length is less than ϵ so that S has measure zero, as contended.

PROB. 5.1. Prove: Each finite set of real numbers has measure zero.

A Noncountable Set of Measure Zero

A noncountable set is one which is neither finite nor denumerable. Hence, it is an infinite nondenumerable set. Being infinite, such a set contains a denumerable subset (Theorem II.10.3). The following lemma proves the existence of noncountable sets.

Lemma 5.1. *The interval* $(0, 1]$ *in* $\mathbb{R}$ *is a noncountable set.*

PROOF. For each positive integer n, we have $1/n \in (0, 1]$. Therefore, $\{1, 1/2, 1/3, \ldots\} \subseteq (0, 1]$. $(0, 1]$ contains an infinite set and is, therefore, infinite.

We present a proof, due essentially to Cantor, that $(0, 1]$ is nondenumerable. The proof is indirect. Assume $I = (0, 1]$ is denumerable. Let $f: \mathbb{Z}_+ \to I$ be a one-to-one correspondence between $\mathbb{Z}_+$ and I. Write $x_i = f(i)$ for each $i \in \mathbb{Z}_+$. Let

$$x_i = .d_{1i}d_{2i} \ldots \qquad \text{for each} \quad i \in \mathbb{Z}_+$$

be the nonterminating decimal representation of x_i (see Example IV.3.1 and Theorem III.10.4). Construct the real number $x = .d_1d_2 \ldots$ as follows. Examine the ith digit d_{ii} of x_i and define

$$d_i = \begin{cases} 4 & \text{if} \quad d_{ii} = 5 \\ 5 & \text{if} \quad d_{ii} \neq 5. \end{cases}$$

Thus, $d_i \neq d_{ii}$ for all i. The decimal $x = .d_1d_2 \ldots$ is nonterminating and $x \in (0, 1]$. By the assumption on f, there exists a positive integer j such that $x = x_j$. This implies that the nonterminating decimal representations of x and x_j are the same and $d_j = d_{jj}$—an impossibility. Since I is not finite, and, as just proved, nondenumerable, it is noncountable.

Lemma 5.2. *The interval* $(0, 1]$ *and the open interval* $(0; 1)$ *are equipotent sets.*

PROOF. We prove the existence of a one-to-one correspondence $f: (0; 1) \to (0, 1]$. Consider the sets

$$A = \left\{ \frac{n}{n+1} \mid n \in \mathbb{Z}_+ \right\} = \left\{ \frac{1}{2}, \frac{2}{3}, \frac{3}{4}, \ldots \right\},$$

$$B = A \cup \{1\} = \left\{ 1, \frac{1}{2}, \frac{2}{3}, \frac{3}{4}, \ldots \right\}.$$

These sets are both denumerable. Hence, they are equipotent and there exists a one-to-one correspondence $g: A \to B$ between A and B. We also know that

$$A \subset (0; 1) \quad \text{and} \quad B \subset (0, 1]$$

and that

$$(0;1) - A = (0,1] - B.$$

Let $C = (0,1) - A = (0,1] - B$. We have

$$A \cup C = (0,1) \quad \text{and} \quad B \cup C = (0,1],$$

with $A \cap C = \emptyset = B \cap C$. Define $f: (0;1) \to (0,1]$ as follows:

$$f(x) = \begin{cases} g(x) & \text{for} \quad x \in A \\ x & \text{for} \quad x \in C, \end{cases}$$

where g is the one-to-one correspondence between A and B mentioned above. f maps A *onto* B and C onto itself and is one-to-one and is, therefore, a one-to-one correspondence between $(0,1) = A \cup C$ and $(0,1] = B \cup C$. Thus, $(0;1)$ and $(0,1]$ are equipotent as claimed.

Theorem 5.3. *The half-open interval* $(0;1]$ *and the set* $\mathbb{R}$ *of real numbers are equipotent.*

PROOF. By the last lemma, we have $(0,1] \simeq (0;1)$. The function g defined as $g(x) = \pi x - \pi/2$ is a one-to-one correspondence between $(0;1)$ and $(-\pi/2;\pi/2)$. Therefore, $(0;1) \simeq (-\pi/2;\pi/2)$. The tangent function $f(x) = \tan x$ for $-\pi/2 < x < \pi/2$ (see Fig. 5.1) maps $(-\pi/2;\pi/2)$ onto $\mathbb{R}$ and is one-to-one. Thus, we see that $(-\pi/2;\pi/2) \simeq \mathbb{R}$. We have

$$(0,1] \simeq (0;1), \qquad (0;1) \simeq (-\pi/2;\pi/2), \qquad (-\pi/2;\pi/2) \simeq \mathbb{R}.$$

These imply that $(0,1]$ and $\mathbb{R}$ are equipotent and that we have $(0,1] \simeq \mathbb{R}$.

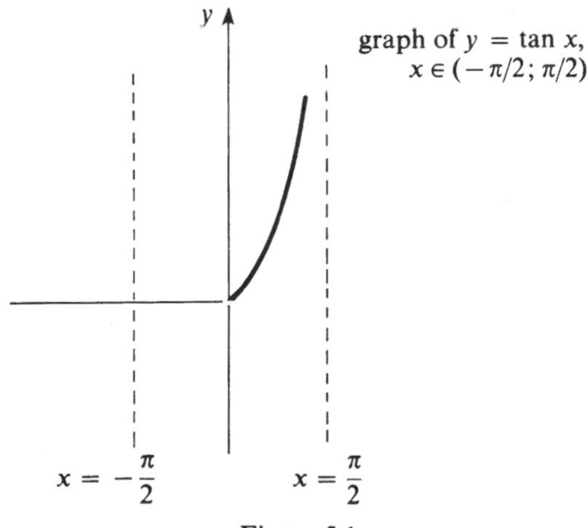

Figure 5.1

Def. 5.2. Any set S such that $S \simeq \mathbb{R}$ is said to have *cardinal* $\aleph$. (Recall Def. II.10.7.) There, a denumerable set was said to have cardinal $\aleph_0$. $\aleph$ is the first letter of the Hebrew alphabet.

PROB. 5.2. Prove: $[0, 1] \simeq [0; 1)$, $(0, 1] \simeq [0, 1)$, $[0, 1] \simeq (0; 1)$.

PROB. 5.3. Prove: If a and b are real numbers such that $a < b$, then $(a, b] \simeq \mathbb{R}$, $(-\infty, a] \simeq \mathbb{R}$, $[a, +\infty) \simeq \mathbb{R}$.

PROB. 5.4. Prove: A set containing a noncountable subset is noncountable (Hint: see Prob. II.10.6).

Remark 5.1. The proof that the interval $(0, 1]$ is not countable relied on the decimal representation of real numbers. This representation uses 10 as a *base*. Any positive integer $b > 1$ can be used as a base to obtain a system of numerals representing the real numbers. Put $\omega_b = \{0, 1, \ldots, b-1\}$. For example, if $b = 2$, then $\omega_2 = \{0, 1\}$; if $b = 3$, then $\omega_3 = \{0, 1\}$. If b is greater than 10, say, $b = 12$, use t for 10 and e for 11 and obtain $\omega_{12} = \{0, 1, 2, 3, 4, 5, 6, 7, 8, 9, t, e\}$. If N is some positive integer, then there exist integers $a_1, a_2, \ldots, a_n$, where n is some positive integer such that $a_1 \in \omega_b$, $a_2 \in \omega_b, \ldots, a_n \in \omega_b$ and

$$N = a_n b^{n-1} + a_{n-1} b^{n-1} + \cdots + a_2 b + a_1 .$$

We then write

$$N = (a_m a_{n-1} \ldots a_2 a_1)_b .$$

For example, if $b = 2$ so that $\omega_2 = \{0, 1\}$, then

$$2 = 1 \cdot 2 + 0, \qquad 2^2 = 1 \cdot 2^2 + 0 \cdot 2 + 0, \qquad 2^3 = 1 \cdot 2^3 + 0 \cdot 2^2 + 0 \cdot 2 + 0,$$

etc., and

$$2 = 10_2, \qquad 2^2 = 100_2, \qquad 2^3 = 100_2 .$$

If $N = 29$, then

$$29 = 16 + 8 + 4 + 1 = 1 \cdot 2^4 + 1 \cdot 2^3 + 1 \cdot 2^2 + 0 \cdot 2 + 0 + 1 = 111001_2 .$$

The system of numerals with 2 as a base is called the *binary* system. The *ternary* system uses 3 as a base. We have $\omega_3 = \{0, 1, 2\}$ and

$$3 = 1 \cdot 3 + 0 = 10_3, \qquad 3^2 = 1 \cdot 3^2 + 0 \cdot 3 + 0 = 100_3,$$

$$3^3 = 1 \cdot 3^3 + 0 \cdot 3^2 + 0 \cdot 3 + 0 = 1000_3 .$$

In this system we have

$$3 = 10_3, \qquad 4 = 1 \cdot 3 + 1 = 11_3, \qquad 5 = 1 \cdot 3 + 2 = 12_3,$$

$$6 = 2 \cdot 3 + 0 = 20_3, \qquad 7 = 2 \cdot 3 + 1 = 21_3, \qquad 8 = 2 \cdot 3 + 2 = 22_3,$$

$$9 = 3^2 = 100_3, \qquad 10 = 101_3, \qquad 11 = 102_3,$$

etc. We then have

$$29 = 27 + 2 = 1 \cdot 3^2 + 0 \cdot 3^2 + 0 \cdot 3 + 2 = 1002_3$$

and

$$64 = 54 + 9 + 1 = 2 \cdot 3^3 + 1 \cdot 3^2 + 0 \cdot 3 + 1 = 2101_3.$$

In connection with the representation of real numbers in base $b > 1$, we have an analogue of Theorem III.10.4:

Theorem. *Let b be an integer, $b > 1$. If $x \in \mathbb{R}$, there exists a unique integer N, and a unique sequence $\langle a_n \rangle$ of elements of $\omega_b = \{0, 1, \ldots, b-1\}$ (digits in base b) such that (1) $a_n \neq 0$ for infinitely many n and (2) if $\langle r_n \rangle$ is the sequence defined as*

$$r_n = N + \sum_{k=1}^{n} a_k b^{-k} \qquad \text{for each } n, \tag{5.1}$$

Then

$$r_n < x \leqslant r_n + b^{-n} \tag{5.2}$$

so that

$$|x - r_n| \leqslant b^{-n}$$

for each n.

Remark 5.2. An analogue of Example IV.3.1: If $\langle a_n \rangle$ is a sequence of elements of $\omega_b = \{0, 1, \ldots, b-1\}$, where b is an integer > 1, then the series

$$S = N + \sum_{n=1}^{\infty} a_n b^{-n} = N + \frac{a_1}{b} + \frac{a_2}{b^2} + \cdots \tag{5.3}$$

has the sequence $\langle r_n \rangle$, where

$$r_n = N + \sum_{k=1}^{n} a_k b^{-k} = N + \frac{a_1}{b} + \frac{a_2}{b^2} + \cdots + \frac{a_n}{b^n} \qquad \text{for each } n \tag{5.4}$$

as its sequence of partial sums. We write

$$(.a_1 a_2 \ldots a_n)_b = \sum_{k=1}^{n} a_k b^{-k}, \tag{5.5}$$

so that $r_n = N + (.a_1 a_2 \ldots a_n)_b$ for each n. We also write the series (5.3) as

$$S = N + (.a_1 a_2 \ldots)_b. \tag{5.6}$$

Each real number x has a unique representation in base b,

$$x = N + (.a_1 a_2 \ldots)_b. \tag{5.7}$$

Here N is an integer and infinitely many of the a_n's are nonzero. If $b = 2$ we call this the nonterminating *binary* representation of x and when $b = 3$

the nonterminating *ternary* representation of x. In general, we call (5.7) the nonterminating *b-adic* representation of x. As examples consider

$$\frac{1}{3}=\frac{1}{2^2}+\frac{1}{2^4}+\frac{1}{2^6}+\cdots \quad \text{and} \quad \frac{1}{2}=\frac{1}{3}+\frac{1}{3^2}+\frac{1}{3^3}+\cdots. \tag{5.8}$$

These hold because

$$\frac{1}{2^2}+\frac{1}{2^4}+\frac{1}{2^6}+\cdots=\frac{1}{2^2}\left(1+\frac{1}{2^2}+\frac{1}{2^4}+\cdots\right)=\frac{1}{2^2}\,\frac{1}{1-1/2^2}=\frac{1}{3}$$

and

$$\frac{1}{3}+\frac{1}{3^2}+\cdots=\frac{1}{3}\left(1+\frac{1}{3}+\frac{1}{3^2}+\cdots\right)=\frac{1}{3}\,\frac{1}{1-\frac{1}{3}}=\frac{1}{2}.$$

We, therefore, write

$$\frac{1}{3}=.010101\ldots_2 \quad \text{and} \quad \frac{1}{2}=.111\ldots_3.$$

If in $(.a_1a_2\ldots)_b$ there is a solidly repeating block of digits, then we indicate this by placing a bar over the repeating block. For instance, we write

$$\tfrac{1}{3}=.010101\ldots_2=.\overline{01}_2 \quad \text{and} \quad \tfrac{1}{2}=.111\ldots_3=.\overline{1}_3.$$

We now present an example of a noncountable set of measure zero.

Example 5.2 (The Cantor Set). Begin with the bounded closed interval $S_0=[0,1]$. Remove its open middle third $(\frac{1}{3};\frac{2}{3})$ and denote the remaining set as S_1 so that $S_1=[0,\frac{1}{3}]\cup[\frac{2}{3},1]$. Next remove the open middle thirds of each of the subintervals of S_1 to obtain $S_2=[0,\frac{1}{9}]\cup[\frac{2}{9},\frac{1}{3}]\cup[\frac{2}{3},\frac{7}{9}]\cup[\frac{8}{9},1]$. Continue this inductively (see Fig. 5.2) to obtain the sequence $\langle S_n\rangle$ of sets where

$$\begin{aligned} S_0&=[0,1],\\ S_1&=[0,\tfrac{1}{3}]\cup[\tfrac{2}{3},1],\\ S_2&=[0,\tfrac{1}{9}]\cup[\tfrac{2}{9},\tfrac{1}{3}]\cup[\tfrac{2}{3},\tfrac{7}{9}]\cup[\tfrac{8}{9},1].\\ &\vdots \end{aligned}$$

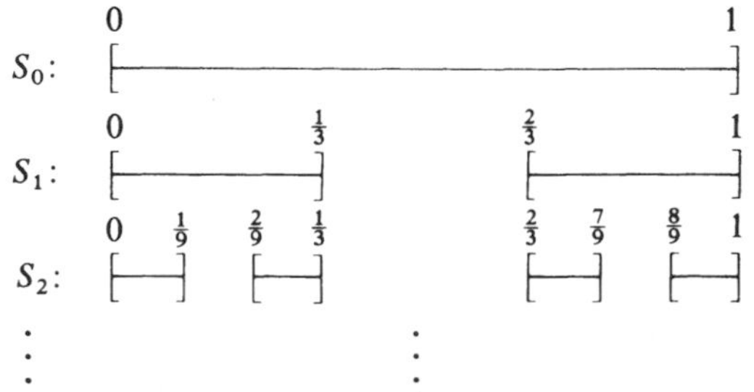

Figure 5.2

The Cantor set is defined as

$$C = \bigcap_{n=1}^{\infty} S_n . \tag{5.9}$$

Each S_n is a finite union of bounded closed intervals and is, therefore, a closed set (Prob. VI.7.9). C is the intersection of a family of closed sets in $\mathbb{R}$ and is, therefore, a closed (Theorem VI.7.3) set. This implies that any converging sequence $\langle x_n \rangle$ of elements of C converges to an element of C. C is not empty since the endpoints of the subintervals of each S_n are in C. For example, 0, $\frac{1}{9}$, $\frac{2}{9}$, $\frac{1}{3}$, $\frac{2}{3}$, $\frac{7}{9}$, $\frac{8}{9}$ are in C. In Theorem 5.4 below we prove that C is noncountable and in Theorem 5.5 we prove that C has measure zero.

Theorem 5.4 will be proved by using the ternary representation of the real numbers in $[0, 1]$. We agree to write zero as 0. When a number has both a terminating and a nonterminating ternary representation we agree to use the nonterminating form except when the terminating representation ends in 2. Thus, we write $\frac{1}{3}$ as $.0\bar{2}_3$ rather than as $.1_3$ and we write $\frac{2}{3}$ as $.2_3$ rather than as $.1\bar{2}_3$.

Prob. 5.5. Prove: If $x = (.a_1a_2a_3 \ldots)_3$ is the ternary representation of $x \in [0, 1]$, where $.0\bar{2}_3 < x < .2_3$, then $a_1 = 1$, and from among $a_2, a_3, \ldots$ at least one $a_i \neq 0$ and at least one $a_j \neq 2$.

Prob. 5.6. Prove the converse of Prob. 5.5 above; that is, prove: if $x = (.1a_2a_3 \ldots)_3$ is in ternary form, where at least one of $a_2, a_3, \ldots$ is $\neq 0$ and at least one is $\neq 2$., then $.0\bar{2}_3 < x < .2_3$.

Remark 5.3. Since $x \in S_1 = [0, \frac{1}{3}] \cup [\frac{2}{3}, 1]$ if and only if $0 \leqslant x \leqslant .0\bar{2}_3$ or $.2_3 \leqslant x \leqslant .\bar{2}_3$, it follows from the last two problems that $x \in S_1$ if and only if it has a ternary representation $(.a_1a_2a_3 \ldots)_3$ such that $a_1 \in \{0, 2\}$. In forming S_2, we remove the open middle thirds $(.00\bar{2}_3; .02_3)$ of $[0, .0\bar{2}_3]$ and $(.20\bar{2}_3; .22_3)$ of $[.2_3, .\bar{2}_3]$ so that

$$S_2 = \left[0, .00\bar{2}_3\right] \cup \left[.02_3, .0\bar{2}_3\right] \cup \left[.2_3, .20\bar{2}_3\right] \cup \left[.22_3, .\bar{2}_3\right].$$

We see from this that $x \in S_2$ if and only if x has one of the ternary representations

$$(.00a_3a_4 \ldots)_3, \quad (.02a_3a_4 \ldots), \quad (.20a_3a_4 \ldots)_3, \quad (.22a_3a_4)_3 .$$

This amounts to saying that $x \in S_2$ if and only if x has a ternary representation $(.a_1a_2a_3 \ldots)_3$ such that $a_1 \in \{0, 2\}$ and $a_2 \in \{0, 2\}$.

Continuing inductively, we find that $x \in S_k$ if and only if x has a ternary representation $(.a_1a_2 \ldots)_3$ such that $a_1 \in \{0, 2\}$, $a_2 \in \{0, 2\}, \ldots, a_k \in \{0, 1\}$ etc. We conclude that $x \in C$, the Cantor set, if and only if it has a ternary representation $(.a_1a_2a_3 \ldots)_3$ such that $a_i \in \{0, 2\}$ for all i. This is equivalent to saying that $x \in C$ if and only if it has a ternary representation $(.a_1a_2a_3 \ldots)_3$ such that $a_i = 2b_i$, where $b_i \in \{0, 1\}$ for all i.

Theorem 5.4. *The Cantor set is noncountable.*

PROOF. Let B be the subset of C of elements $x \in C$ having a *nonterminating* ternary representation $(.a_1a_2a_3 \ldots)_3$ such that $a_i \in \{0, 2\}$ for all positive integers i and let $A = \{0\} \cup B$. This omits all those elements of C having a *terminating* ternary representation that ends with 2. Now define f: $A \to [0, 1]$ as follows: (1) $f(0) = 0$; (2) if $x \in A$ so that $x = (.a_1a_2 \ldots)_3$, where this ternary representation of x is nonterminating and each $a_i \in \{0, 2\}$, then define $f(x)$ as the y in $[0, 1]$ whose binary (base 2) representation is $(.b_1b_2 \ldots)_2$, where $b_i = a_i/2$ for each 0. Thus, $f(x) = (.b_1b_2 \ldots)_2$ where on the right we have a nonterminating binary representation of y. For example,

$$f(\tfrac{1}{3}) = f(.0\bar{2}_3) = .0\bar{1}_2 = \tfrac{1}{2},$$

$$f(\tfrac{1}{4}) = f(.\overline{02}_3) = .\overline{01}_2 = .010101 \ldots_2 = \frac{1}{2^2} + \frac{1}{2^4} + \frac{1}{2^6} + \cdots = \frac{1}{3},$$

$$f(\tfrac{1}{9}) = f(.00\bar{2}_3) = .00\bar{1}_2 = \frac{1}{2^3} + \frac{1}{2^4} + \frac{1}{2^5} + \cdots = \frac{1}{4}.$$

The f defined here is one-to-one on A (why?). We prove that it maps A *onto* $[0, 1]$. Assume that $y \in [0, 1]$. If $y = 0$, then $f(0) = 0$. If $0 < y \leqslant 1$, let $(.b_1b_2 \ldots)_2$ be the nonterminating *binary* representation of y. We have $b_i \in \{0, 1\}$ for all i with infinitely many of the b_i's $\neq 0$. Now take $(.a_1a_2 \ldots)_3$ with $a_i = 2b_i$ for each i. The x whose ternary representation is $(.a_1a_2 \ldots)_3$ is in A and $f(x) = y$. Thus, each y in $[0, 1]$ is the image of an x in A. We conclude $A \simeq [0, 1]$ and, hence, that A is noncountable. But $A \subseteq C$. By Prob. 5.4, C is noncountable.

This result is surprising. After removing so much of $[0, 1]$ in order to form the Cantor set C, one does not expect to have noncountably many elements left there. In the next theorem we prove that C has measure zero.

Theorem 5.5. *The Cantor set has measure zero.*

PROOF. The Cantor set C was defined in Example 5.2 in formula (5.9). Each set S_n of Example 5.2 is the union of 2^n bounded closed subintervals each of length 3^{-n}. Given $\epsilon > 0$, each of the subintervals of S_n can be enclosed in an interval of length $3^{-n} + 2^{-n-1}\epsilon$. [For example, in $S_1 = [0, \frac{1}{3}] \cup [\frac{2}{3}, 1]$, we enclose $[0, \frac{1}{3}]$ in $G_1 = (-2^{-3}\epsilon; \frac{1}{3} + 2^{-3}\epsilon)$ and $[\frac{2}{3}, 1]$ in $G_2 = (\frac{2}{3} - 2^{-3}\epsilon; 1 + 2^{-3}\epsilon)$. We have $S_1 \subseteq G_1 \cup G_2$ and $L(G_1) = L(G_2) = 3^{-1} + 2^{-2}\epsilon$.] Thus S_n—and hence C—can be covered by a finite sequence of open intervals whose total length is $2^n(3^{-n} + 2^{-n-1}\epsilon) = (\frac{2}{3})^n + \epsilon/2$. For sufficiently large n we have $(\frac{2}{3})^n < \epsilon/2$. Hence, for each $\epsilon > 0$, there exists a finite sequence of open intervals of total length $(\frac{2}{3})^n + \epsilon/2 < \epsilon/2 + \epsilon/2 = \epsilon$ covering C. By Theorem 5.2, C has measure zero.

Remark 5.4. Sets of measure zero are of significance for the theory of Riemann integration because of the following theorem (which we shall not prove here):

Theorem. *A function which is bounded on a bounded closed interval* $[a, b]$ *is R-integrable if and only if the set of points in* $[a, b]$ *at which it is discontinuous has measure zero.*

CHAPTER XV

Improper Integrals. Elliptic Integrals and Functions

1. Introduction. Definitions

When f is R-integrable over $[a, b]$, then its indefinite integral F, defined as

$$F(x) = \int_a^x f(t)\,dt \qquad \text{for} \quad x \in [a, b], \tag{1.1}$$

is continuous on $[a, b]$ (Theorem XIII.6.3). Hence,

$$\lim_{x \to b-} \int_a^x f(t)\,dt = \int_a^b f(t)\,dt. \tag{1.2}$$

Similarly, under the above condition on f,

$$\lim_{x \to a+} \int_x^b f(t)\,dt = \int_a^b f(t)\,dt. \tag{1.3}$$

However, it is possible for the limits in (1.2) and (1.3) to exist and be finite without f being R-integrable on the bounded closed interval $[a, b]$. For example, let $f(x) = 1/\sqrt{1 - x^2}$ for $0 \leqslant x < 1$. Any extension of this function to the closed interval $[0, 1]$ is unbounded and, therefore, not R-integrable. Nevertheless,

$$\lim_{c \to 1-} \int_0^c \frac{1}{\sqrt{1 - x^2}}\,dx = \lim_{c \to 1-} (\operatorname{Arcsin} c) = \frac{\pi}{2}\,. \tag{1.4}$$

Thus, it is possible for the limit

$$\lim_{c \to b-} \int_a^c f(x)\,dx$$

to exist and be finite even though f is not R-integrable on $[a, b]$.

We are confronted with the same situation when we have an unbounded interval $[a, +\infty)$ or $(-\infty, b]$. The Riemann integral is not defined on such intervals, but there are functions f for which the limit

$$\lim_{B \to +\infty} \int_a^B f(x)\, dx$$

exists and is finite; e.g., let f be given by $f(x) = x^{-2}$ for $x \in [1, +\infty)$. The R-integral of this f is not defined on $[1, +\infty)$, but

$$\lim_{B \to +\infty} \int_1^B \frac{1}{x^2}\, dx = \lim_{B \to +\infty} \left(-\frac{1}{x} \Big|_1^B \right) = \lim_{B \to +\infty} \left(1 - \frac{1}{B} \right) = 1. \tag{1.5}$$

We note that if f is not R-integrable on the bounded closed interval $[a,b]$ but is R-integrable on every interval $[a,c]$ such that $a \leqslant c < b$, then f is necessarily unbounded on $[a,b)$. For, if f were bounded on $[a,b)$, then we could extend its definition by defining it at b in some manner. The resulting extension would then be bounded on $[a,b]$. By the previous assumptions on f and Theorem XIII.1.6 this extension would be R-integrable on $[a,b]$. Hence, f would be R-integrable in the extended sense of Section XIV.5, contradicting the original assumption on f.

We are thus led to the notion of an improper integral.

Def. 1.1(a). Assume that $a < b$, where $a \in \mathbb{R}$ and $b \in \mathbb{R}^*$. If f is R-integrable on every interval $[a,c]$ such that $a < c < b$, but not R-integrable on $[a,b]$ (either because $b \in \mathbb{R}$ and f is not bounded on $[a,b)$ or because $b = +\infty$), then we call

$$\int_a^b f(x)\, dx \tag{1.6}$$

an *improper integral*. Specifically we say that it is *improper at b*. We refer to b as a *singularity* of f. We then define (1.6) as

$$\int_a^b f(x)\, dx = \lim_{c \to b-} \int_a^c f(x)\, dx, \tag{1.7}$$

If the limit on the right in (1.7) is finite, then we call (1.6) a *convergent* integral. If the limit on the right in (1.7) is infinite or fails to exist, then we call (1.6) a *divergent* integral. Convergent integrals are said to *converge* and divergent ones to *diverge*. When (1.6) converges we say that f is *improperly* integrable on $[a,b]$ at b.

We give a similar definition of (1.6) if the singularity of f is at a.

Def. 1.1(b). Assume that $a < b$, where $a \in \mathbb{R}^*$ and $b \in \mathbb{R}$. If f is R-integrable on every interval $[c,b]$ such that $a < c < b$ but not R-integrable on $[a,b]$ (either because $a \in \mathbb{R}$ and f is unbounded on $(a,b]$ or because

$a = -\infty$), then we call (1.6) an *improper* integral on $[a,b]$ at a. In this case (1.6) is defined as

$$\int_a^b f(x)\,dx = \lim_{c \to a+} \int_c^b f(x)\,dx \tag{1.8}$$

when the limit on the right exists. If the limit on the right in (1.8) is finite, we call (1.6) a *convergent* integral and if it is infinite or fails to exist, then we call (1.8) a *divergent* integral. When (1.6) converges and the f there has a singularity at a, then we say that f is *improperly* integrable on $[a,b]$ at a.

For example, in view of (1.4) and (1.5),

$$\int_0^1 \frac{1}{\sqrt{1-x^2}}\,dx = \frac{\pi}{2} \tag{1.9a}$$

and

$$\int_1^{+\infty} \frac{1}{x^2}\,dx = 1, \tag{1.9b}$$

both integrals being improper and convergent.

PROB. 1.1. Prove:

$$\int_{-1}^0 \frac{1}{\sqrt{1-x^2}}\,dx = \frac{\pi}{2}.$$

PROB. 1.2. Prove: If f is improperly integrable on $[a,b]$ at a or at b and $a < c < b$, then

$$\int_a^b f(x)\,dx = \int_a^c f(x)\,dx + \int_c^b f(x)\,dx.$$

EXAMPLE 1.1. We examine

$$\int_0^1 \frac{1}{x^p}\,dx \qquad \text{for} \quad p \in \mathbb{R}. \tag{1.10}$$

The integrand is continuous on $[c, 1]$ for all c such that $0 < c \leqslant 1$ but is not defined at $x = 0$. We consider cases.

Case 1. $p \leqslant 0$. In this case the integrand is bounded on $(0, 1]$ and, when extended by defining it at $x = 0$ as 0 when $p < 0$ and as 1 when $p = 0$, the extension is continuous and, therefore R-integrable on $[0, 1]$.

Case 2. $0 < p < 1$. The integrand is unbounded on $(0, 1]$ and $1 - p > 0$.

We have

$$\int_0^1 \frac{1}{x^p}\,dx = \lim_{c\to 0+} \int_c^1 x^{-p}dx = \lim_{c\to 0+} \frac{x^{1-p}}{1-p}\bigg|_c^1$$

$$= \lim_{c\to 0+}\left(\frac{1}{1-p} - \frac{c^{1-p}}{1-p}\right) = \frac{1}{1-p}$$

so that, for $p < 1$,

$$\int_0^1 \frac{1}{x^p}\,dx = \frac{1}{1-p}\,. \tag{1.11}$$

Case 3. $p > 1$. In this case $p - 1 > 0$ and, therefore,

$$\int_0^1 \frac{1}{x^p}\,dx = \lim_{c\to 0+}\left(\frac{1}{1-p} - \frac{c^{1-p}}{1-p}\right) = \lim_{c\to 0+}\left(\frac{1}{(p-1)c^{p-1}} - \frac{1}{1-p}\right) = +\infty.$$

The integral diverges.

Case 4. $p = 1$. Here

$$\int_0^1 \frac{1}{x^p}\,dx = \int_0^1 \frac{1}{x}\,dx = \lim_{c\to 0+}\int_c^1 \frac{1}{x}\,dx = \lim_{c\;0+}(-\ln c) = +\infty.$$

Accordingly,

$$\int_0^1 \frac{1}{x^p}\,dx = \begin{cases} \dfrac{1}{1-p} & \text{if} \quad p < 1 \\ +\infty & \text{if} \quad p \geqslant 1. \end{cases} \tag{1.12}$$

Example 1.2. We show that

$$\int_0^{+\infty} e^{-x}\,dx = 1. \tag{1.13}$$

We have

$$\int_0^{+\infty} e^{-x}\,dx = \lim_{c\;+}\int_0^c e^{-x}\,dx = \lim_{c\to+\infty}(1 - e^{-c}) = 1.$$

Prob. 1.3. Prove:

$$\int_1^{\infty} \frac{1}{x^p}\,dx = \begin{cases} +\infty & \text{if} \quad p \leqslant 1 \\ \dfrac{1}{p-1} & \text{if} \quad p > 1. \end{cases}$$

Prob. 1.4. Prove:

$$\int_0^{+\infty} \frac{1}{1+x^2}\,dx = \frac{\pi}{2}\,.$$

Prob. 1.5. Show that the integrals

$$\int_0^{+\infty} \cos x\,dx \quad \text{and} \quad \int_0^{+\infty} \sin x\,dx$$

diverge.

Remark 1.1. If f is R-integrable over $[a, b]$, then so is f^2. This property fails to carry over to improper integrals. Thus (see Example 1.1),

$$\int_0^1 \frac{1}{\sqrt{x}}\,dx \quad \text{converges, but} \quad \int_0^1 \left(\frac{1}{\sqrt{x}}\right)^2 dx = \int_0^1 \frac{1}{x}\,dx \quad \text{diverges.}$$

Def. 1.2. Let $a < c < b$, where $a \in \mathbb{R}^*$ and $b \in \mathbb{R}^*$. If the integrals

$$\int_a^c f(x)\,dx, \qquad \int_c^b f(x)\,dx \tag{1.14}$$

are improper, where either the first is improper at a and the second at b or both are improper at c and each is convergent, then we define $\int_a^b f(x)\,dx$ as

$$\int_a^b f(x)\,dx = \int_a^c f(x)\,dx + \int_c^b f(x)\,dx \tag{1.15}$$

and say that it is convergent. If at least one of the integrals on the right diverge, we call $\int_a^b f(x)\,dx$ *divergent*. It is also convenient to use (1.15) when the integrals on the right are in $\mathbb{R}^*$ and their sum is defined in $\mathbb{R}^*$.

As in the Riemann integrable case, if $b < a$, then define

$$\int_a^b f(x)\,dx = -\int_b^a f(x)\,dx \tag{1.16}$$

when the integral on the right is improper and is in $\mathbb{R}^*$.

EXAMPLE 1.3. For any $p \in \mathbb{R}$, we have

$$\int_0^{+\infty} \frac{1}{x^p} = \int_0^1 \frac{1}{x^p}\,dx + \int_1^{+\infty} \frac{1}{x^p}\,dx.$$

This integral diverges because one of the integrals on the right is divergent.

EXAMPLE 1.4. Consider

$$\int_{-1}^1 \frac{1}{\sqrt{1-x^2}}\,dx.$$

By (1.9a) and Prob. 1.1, we have

$$\int_{-1}^1 \frac{1}{\sqrt{1-x^2}}\,dx = \int_{-1}^0 \frac{1}{\sqrt{1-x^2}}\,dx + \int_0^1 \frac{1}{\sqrt{1-x^2}}\,dx = \frac{\pi}{2} + \frac{\pi}{2} = \pi. \tag{1.17}$$

Remark 1.2. If the singularity of the integrand is in the interior of the interval of integration, say, at c where $a < c < b$, and we wish to examine

$$\int_a^b f(x)\,dx = \int_a^c f(x)\,dx + \int_c^b f(x)\,dx$$

for convergence or divergence, then we must evaluate the limits on the right-hand side, needed for the evaluation of the improper integral, separately. This means that

$$\int_a^b f(x)\,dx = \lim_{h\to 0+} \int_a^{c-h} f(x)\,dx + \lim_{k\to 0+} \int_{c+k}^b f(x)\,dx, \tag{1.18}$$

rather than

$$\int_a^b f(x)\,dx = \lim_{h\to 0} \left(\int_a^{c-h} f(x)\,dx + \int_{c+h}^b f(x)\,dx \right). \tag{*}$$

Using the latter may result in calling a divergent integral convergent. In fact, the limit on the right-hand side of (*) may exist even though none of the limits on the right-hand side of (1.18) exists. For example, the singularity in

$$\int_0^3 \frac{1}{(x-1)^3}\,dx \tag{1.19}$$

occurs at $x = 1$. We check for convergence by writing

$$\begin{aligned} \int_0^3 \frac{1}{(x-1)^3}\,dx &= \int_0^1 \frac{1}{(x-1)^3}\,dx + \int_1^3 \frac{1}{(x-1)^3}\,dx \\ &= \lim_{h\to 0+} \int_0^{1-h} \frac{1}{(x-1)^3} + \lim_{k\to 0+} \int_{1+k}^3 \frac{1}{(x-1)^3}\,dx. \end{aligned} \tag{1.20}$$

Evaluation of the limits on the right shows that each of them diverges so that our integral diverges. On the other hand, if we try to evaluate (1.19) by writing the extreme right-hand side as in (*), we obtain

$$\begin{aligned} &\lim_{h\to 0} \left(\int_0^{1-h} \frac{1}{(x-1)^3}\,dx + \int_{1+h}^3 \frac{1}{(x-1)^3}\,dx \right) \\ &= \lim_{h\to 0} \left[\left(-\frac{1}{2(x-1)^2} \Big|_0^{1-h} \right) + \left(-\frac{1}{2(x-1)^2} \Big|_{1+h}^3 \right) \right] \\ &= \lim_{h\to 0} \left[-\frac{1}{2h^2} + \frac{1}{2} - \left(-\frac{1}{2\cdot 2^2} + \frac{1}{2h^2} \right) \right] = \frac{5}{8}. \end{aligned}$$

This answer makes no sense in view of the divergence of (1.19).

If $\int_a^b f(x)\,dx$ diverges but the limit in (*) exists, then we call that limit the *Cauchy principal* value of $\int_a^b f(x)\,dx$. We shall not deal with this concept.

2. Comparison Tests for Convergence of Improper Integrals

We now obtain some criteria for the convergence and divergence of improper integrals of nonnegative functions. We will state these for improper integrals of the form

$$\int_a^{+\infty} f(x)\,dx. \tag{2.1}$$

The reader can then adapt them to the case where the interval of integration is bounded and there is a singularity of the integrand at one of the endpoints of the interval.

Theorem 2.1. *If f is nonnegative and R-integrable over $[a, B]$ for all $B > a$, then the improper integral of f over $[a, +\infty)$ converges if and only if its indefinite integral F, defined as*

$$F(x) = \int_a^x f(t)\,dt \qquad \text{for} \quad x \in [a, +\infty), \tag{2.2}$$

is bounded on $[a, +\infty)$. In general, for an f satisfying the hypothesis, we have

$$\sup_{x \geqslant a} \int_a^x f(t)\,dt = \int_a^{+\infty} f(t)\,dt. \tag{2.3}$$

The values in (2.3) *are in* $\mathbb{R}^*$.

PROOF. Since f is nonnegative for $x \geqslant a$, F is a nonnegative monotonically increasing function. This implies that $\lim_{x\to+\infty} F(x)$ exists in $\mathbb{R}^*$ and that

$$\sup_{x \geqslant a} F(x) = \lim_{x\to+\infty} F(x) = \int_a^{+\infty} f(x)\,dx. \tag{2.4}$$

This is equivalent to (2.3).

If F is bounded on $[a, +\infty)$, then there exists a real M such that $F(x) \leqslant M$ for all $x \geqslant a$. Then (2.4) implies that $\lim_{x\to+\infty} F(x)$ exists, is finite, and

$$\sup_{x \geqslant a} f(x) = \int_a^{+\infty} f(x)\,dx \leqslant M < +\infty.$$

On the other hand, if F is not bounded on $[a, +\infty)$ then, using properties of monotonically increasing functions, we see that $\lim_{x\to+\infty} F(x) = +\infty$ and, hence, that

$$\int_a^{+\infty} f(x)\,dx = +\infty.$$

In this case the integral diverges. It follows that if the integral converges, then F must be bounded on $[a, +\infty)$. This completes the proof.

Theorem 2.2. *If $0 \leqslant f(x) \leqslant g(x)$ for $x \geqslant a$ and f and g are both R-integrable over $[a, B]$ for all real B such that $B > a$, then convergence of the improper integral of g over $[a, +\infty)$ implies the convergence of the improper integral of f over $[a, +\infty)$ and the divergence of the improper integral of f over $[a, +\infty)$ implies the divergence of the integral of g over $[a, +\infty)$.*

PROOF. The conclusion follows from Theorem 2.1 and the fact that the hypothesis implies that

$$0 \leqslant \int_a^B f(x)\,dx \leqslant \int_a^B g(x)\,dx \leqslant \int_a^{+\infty} g(x)\,dx$$

for all B such that $a < B < +\infty$.

PROB. 2.1. Assume that (i) f and g are nonnegative on $[a, +\infty)$ and R-integrable on $[a, B]$ for all B such that $a < B < +\infty$ and (ii) $f = O(g)$ as $x \to +\infty$ (see Section IX.3 for the meaning of the O symbol). Prove that the convergence of the integral of g over $[a, +\infty)$ implies the convergence of the integral of f and the divergence of the integral of f implies the divergence of the integral of g.

There is also a Cauchy criterion for the convergence of improper integrals. (See Theorem 3.1 below).

EXAMPLE 2.1 (Euler's Second Integral). Consider G, where for each α we have

$$G(\alpha) = \int_0^{+\infty} t^{\alpha-1}e^{-t}dt. \tag{2.5}$$

The integral on the right is known as *Euler's Second Integral*. We test for convergence. If $\alpha - 1 < 0$, there is a singularity of the integrand at the lower limit of integration. Hence, we consider

$$G(\alpha) = \int_0^1 t^{\alpha-1}e^{-t}dt + \int_1^{+\infty} t^{\alpha-1}e^{-t}dt. \tag{2.6}$$

We denote the first integral on the right by I_1 and the second by I_2. We prove that I_2 converges by comparing its integrand with t^{-2}. Note that for $\alpha \in \mathbb{R}$,

$$\lim_{t\to+\infty} \frac{t^{\alpha-1}e^{-t}}{t^{-2}} = \lim_{t\to+\infty} \frac{t^{\alpha+1}}{e^t} = 0.$$

Using the "big O" notation we have from this that

$$t^{\alpha-1}e^{-t} = O(t^{-2}) \qquad \text{as} \quad t \to +\infty \tag{2.7}$$

(explain). Since the improper integral

$$\int_1^{\infty} t^{-2}dt$$

converges on $[1, +\infty)$ (Prob. 1.3), we obtain from Prob. 2.1 that I_2 converges for all $\alpha \in \mathbb{R}$. We now examine

$$I_1 = \int_0^1 t^{\alpha-1}e^{-t}dt. \tag{2.8}$$

If $\alpha - 1 < 0$, then the singularity is at the origin. We compare the integrand with g, where $g(t) = t^{\alpha-1}$ and note that

$$\lim_{t\to 0+} \frac{t^{\alpha-1}e^{-t}}{t^{\alpha-1}} = 1.$$

By Prob. IX.3.6,

$$t^{\alpha-1}e^{-t} = O(t^{\alpha-1}) \quad \text{and} \quad t^{\alpha-1} = O(t^{\alpha-1}e^{-t})$$

as $t\to 0$. This and Prob. 2.1 imply that the integrals (a) I_1 and (b) $\int_0^1 t^{\alpha-1}\,dt = \int_0^1 1/t^{1-\alpha}\,dt$ converge together or diverge together. Since the integral in (b) converges for $1-\alpha < 1$ and diverges for $1-\alpha \geqslant 1$, we see that I_1 converges for $\alpha > 0$ and diverges for $\alpha \leqslant 0$. Combining the results obtained for I_1 and I_2, we have:

Theorem 2.3. *The integral* (2.5) *converges if and only if* $\alpha > 0$.

PROB. 2.2. Decide whether or not the following integrals converge and evaluate those that do.

(a) $\displaystyle\int_0^1 \frac{1}{\sqrt{x}}\,dx,$

(b) $\displaystyle\int_0^2 \sqrt[3]{2-x}\,dx,$

(c) $\displaystyle\int_{-\infty}^{\infty} \frac{1}{x^2+4x+5}\,dx,$

(d) $\displaystyle\int_1^{\infty} \frac{1}{x^3}\,dx,$

(e) $\displaystyle\int_0^1 \frac{1}{1-x^3}\,dx,$

(f) $\displaystyle\int_0^{1/2} \frac{1}{x\ln^2 x}\,dx,$

(g) $\displaystyle\int_0^{\infty} \frac{\operatorname{Arctan} x}{1+x^2}\,dx,$

(h) $\displaystyle\int_0^{\infty} xe^{-x}\,dx,$

(i) $\displaystyle\int_0^{\infty} xe^{-x^2}\,dx,$

(j) $\displaystyle\int_0^{\infty} e^{-x}\cos^2 x\,dx.$

PROB. 2.3. Test the following integrals for convergence:

(a) $\displaystyle\int_0^1 \frac{1}{x+\sqrt[3]{x}}\,dx,$

(b) $\displaystyle\int_0^\infty \frac{1}{(1+x)\sqrt{x}}\,dx,$

(c) $\displaystyle\int_0^\infty \frac{1}{1+e^x}\,dx,$

(d) $\displaystyle\int_0^\infty e^{-x^2}dx,$

(e) $\displaystyle\int_2^\infty \frac{1}{x\sqrt{x^3-1}}\,dx.$

PROB. 2.4. Prove: If $a > 0$, then
(a) $\int_0^\infty e^{-ax}\cos bx\,dx = a/(a^2+b^2)$ and
(b) $\int_0^\infty e^{-ax}\sin bx = b/(a^2+b^2)$.

3. Absolute and Conditional Convergence of Improper Integrals

Theorem 3.1 (Cauchy Criterion for Convergence of Improper Integrals). *If f is R-integrable over $[a, B]$ for all B such that $a < B < +\infty$, then*

$$\int_a^{+\infty} f(x)\,dx$$

converges if and only if for each $\epsilon > 0$ there exists an $X \geqslant a$ such that $x'' > x' > X$ imply

$$\left|\int_{x'}^{x'} f(x)\,dx\right| < \epsilon. \tag{3.1}$$

PROOF. Apply the Cauchy criterion for the limits of a function (Theorem V.9.2) to the limit

$$\lim_{B\to+\infty}\int_0^B f(x)\,dx = \int_a^{+\infty} f(x)\,dx.$$

Remark 3.1. By making appropriate modifications the reader should extend the validity of Theorem 3.1 to include

(a) $\int_a^b f(x)dx$ and
(b) $\int_{-\infty}^b f(x)dx,$

where $a \in \mathbb{R}$ and $b \in \mathbb{R}$. Here, in the first case, we assume that f has a singularity at a only or at b only, and in the second that b is not a singularity of f.

EXAMPLE 3.1. Consider the integral

$$\int_0^\infty \frac{\sin x}{x}\,dx \tag{3.2}$$

—an integral treated by Dirichlet. Write

$$\int_0^\infty \frac{\sin x}{x}\,dx = \int_0^1 \frac{\sin x}{x}\,dx + \int_0^\infty \frac{\sin x}{x}\,dx. \tag{3.3}$$

The integrand in the first integral on the right is bounded on $(0, 1]$ and continuous for all $[c, 1]$ such that $0 < c \leqslant 1$. Hence, it can be extended to a function continuous on $[c, 1]$. Thus, the first integral is convergent. We examine the second integral on the right in (3.3) for convergence.

Note that

$$\sin x = 2\sin\frac{x}{2}\cos\frac{x}{2} = 2\frac{d}{dx}\left(\sin^2\frac{x}{2}\right). \tag{3.4}$$

Take $B > 1$. Using (3.4) and then integration by parts we obtain

$$\begin{aligned}\int_1^B \frac{\sin x}{x}\,dx &= 2\int_1^B \frac{1}{x}\,\frac{d\sin^2(x/2)}{dx}\,dx\\ &= 2\left[\frac{1}{x}\sin^2\frac{x}{2}\Big|_1^B + 2\int_1^B \frac{1}{x^2}\sin^2\frac{x}{2}\,dx\right]\\ &= \frac{2}{B}\sin^2\left(\frac{B}{2}\right) - 2\sin^2\frac{1}{2} + 2\int_1^B \frac{1}{x^2}\sin^2\frac{x}{2}\,dx.\end{aligned} \tag{3.5}$$

We examine the last integral on the right and note that if $x > 0$, then

$$0 \leqslant \frac{1}{x^2}\sin^2\frac{x}{2} \leqslant \frac{1}{x^2} \quad \text{and that} \quad \int_1^\infty \frac{1}{x^2}\,dx \quad \text{converges.}$$

This implies that

$$\int_1^\infty \frac{1}{x^2}\sin^2\frac{x}{2}\,dx$$

converges. This and

$$\lim_{B\to+\infty} \frac{2}{B}\sin^2\frac{B}{2} = 0$$

yield, after taking limits in (3.5) as $B \to +\infty$,

$$\int_1^{+\infty} \frac{\sin x}{x}\,dx = -2\sin^2\frac{1}{2} + 2\int_1^{+\infty} \frac{1}{x^2}\sin^2\frac{x}{2}\,dx, \tag{3.6}$$

where both integrals are convergent. We conclude that the integral (3.2) converges.

However, if absolute values are inserted in the integrand of (3.2) so that it becomes

$$\int_0^{+\infty} \frac{|\sin x|}{x} dx, \tag{3.7}$$

then we obtain a divergent integral. To see this let n be some positive integer. We have

$$\int_0^{n\pi} \frac{|\sin x|}{x} dx = \int_0^{\pi} \frac{|\sin x|}{x} dx + \int_{\pi}^{2\pi} \frac{|\sin x|}{x} dx + \cdots$$
$$+ \int_{(n-1)\pi}^{n\pi} \frac{|\sin x|}{x} dx. \tag{3.8}$$

If k is an integer such that $k > 1$ and $(k-1)\pi \leqslant x \leqslant k\pi$ so that

$$\frac{1}{k\pi} \leqslant \frac{1}{x} \leqslant \frac{1}{(k-1)\pi},$$

then we obtain

$$\int_{(k-1)\pi}^{k\pi} \frac{|\sin x|}{x} dx > \frac{1}{k\pi} \int_{(k-1)\pi}^{k\pi} |\sin x| dx = \frac{2}{k\pi}.$$

This and (3.5) yield

$$\int_0^{n\pi} \frac{|\sin x|}{x} dx > \frac{2}{\pi}\left(1 + \frac{1}{2} + \cdots + \frac{1}{n}\right) \tag{3.9}$$

for each positive integer n. From this it follows that the integral (3.7) diverges.

Prob. 3.1. Prove:

$$\int_0^{\infty} \frac{\sin^2 x}{x^2} dx = \int_0^{\infty} \frac{\sin x}{x} dx.$$

We now introduce the notion of absolute convergence for improper integrals.

Def. 3.1. Assume that $a \in \mathbb{R}^*$, $b \in \mathbb{R}^*$, and $a < b$. We call the improper integral $\int_a^b f(x)\,dx$ *absolutely convergent* over $[a, b]$ if and only if $\int_a^b |f(x)|\,dx$ converges. If the first of these integrals converges and the second does not, then we call the first one *conditionally convergent*.

For example, the integral (3.2) in Example 3.1 is conditionally convergent.

EXAMPLE 3.2. The integral

$$\int_1^\infty \frac{\cos x}{x^{3/2}}\,dx \tag{3.10}$$

is absolutely convergent. Since

$$\left|\frac{\cos x}{x^{3/2}}\right| \leqslant \frac{1}{x^{3/2}} \qquad \text{for} \quad x \geqslant 1$$

and $\int_1^\infty x^{-3/2}\,dx$ converges, the comparison test of Theorem 2.2 implies that (3.10) converges.

Theorem 3.2. *Absolutely convergent improper integrals converge.*

PROOF. Assume that $\int_a^{+\infty} f(x)\,dx$ converges absolutely so that $\int_a^{+\infty} |f(x)|\,dx$ converges. Let $\epsilon > 0$ be given. By Theorem 3.1, there exists an $X > a$ such that if x' and x'' are real numbers such that $x'' > x' > X$, then

$$\left|\int_{x'}^{x''} |f(x)|\,dx\right| < \epsilon.$$

It follows that

$$\left|\int_{x'}^{x''} f(x)\,dx\right| \leqslant \int_{x'}^{x''} |f(x)|\,dx = \left|\int_{x'}^{x''} |f(x)|\,dx\right| < \epsilon$$

for $x'' > x' > X$. By Theorem 3.1 it follows that $\int_a^{+\infty} f(x)\,dx$ converges.

The theorems on convergence of improper integrals resemble the theorems on convergence of real infinite series. We cite an example for which the resemblance fails.

EXAMPLE 3.3. We first prove that

$$\int_0^\infty \sin u^2\,du \tag{3.11}$$

converges. We have

$$\int_0^\infty \sin u^2\,du = \int_0^1 \sin u^2\,du + \int_1^\infty \sin u^2\,du. \tag{3.12}$$

Clearly, it suffices to consider only the convergence of the second integral on the right. We take $1 < B < +\infty$ and use the substitution $x = u^2$ for $u \geqslant 1$ to obtain

$$\int_1^B \sin u^2\,du = \frac{1}{2}\int_1^{B^2} \frac{\sin x}{\sqrt{x}}\,dx. \tag{3.13}$$

The integral on the right becomes, after using integration by parts,

$$\begin{aligned}\frac{1}{2}\int_1^{B^2}\frac{\sin x}{\sqrt{x}}\,dx &= \frac{1}{2}\int_1^{B^2} x^{-1/2}d(-\cos x)\\ &= -\frac{x^{-1/2}}{2}\cos x\Big|_1^{B^2} - \frac{1}{4}\int_1^{B^2} x^{-3/2}\cos x\,dx\\ &= \frac{\cos 1}{2} - \frac{\cos B^2}{2B} - \frac{1}{4}\int_1^{B^2} x^{-3/2}\cos x\,dx. \qquad (3.14)\end{aligned}$$

However, on the right we have, by Example 3.2, that the limit exists as $B \to +\infty$ and is finite. It follows that

$$\frac{1}{2}\int_1^{\infty}\frac{\sin x}{\sqrt{x}}\,dx = \frac{\cos 1}{2} - \frac{1}{4}\int_1^{\infty}\frac{\cos x}{x^{3/2}}\,dx, \qquad (3.15)$$

where both integrals converge. This implies that the integral on the left in (3.13) tends to a finite limit as $B \to +\infty$ and, hence, by (3.12), that our original integral (3.11) converges.

We remark that although the integral (3.11) converges, $\lim_{u\to+\infty}\sin u^2$ does not exist and, therefore, $\lim_{u\to+\infty}\sin u^2 \neq 0$. This contrasts with the behavior of infinite series. There the convergence of $\sum a_n$ implies that $a_n \to 0$ as $n \to +\infty$.

Prob. 3.2. Prove that

$$\int_0^{\infty}\cos x^2\,dx$$

converges.

Prob. 3.3. Let P and Q be polynomials of degrees m and n, respectively, where $n \geqslant m+2$. Let a be a number greater than the greatest zero of Q. Prove:

$$\int_a^{+\infty}\frac{P(x)}{Q(x)}\,dx \quad \text{converges absolutely.}$$

Prob. 3.4. Prove: If $p > 1$, then

(a) $\int_1^{+\infty}\sin x/x^p$ and
(b) $\int_1^{+\infty}\cos x/x^p\,dx$

converge absolutely.

Prob. 3.5. Prove: If $\int_a^{+\infty}f(x)\,dx$ converges, then, for each positive $c \in \mathbb{R}$, we have

$$\lim_{y\to+\infty}\int_y^{y+c} f(x)\,dx = 0.$$

PROB. 3.6. Prove: If $\int_a^{+\infty} f(x)dx$ converges absolutely, then

$$\left|\int_a^{+\infty} f(x)\,dx\right| \leqslant \int_a^{+\infty} |f(x)|\,dx.$$

PROB. 3.7. Assume that f is bounded on $[a,+\infty)$ ($a \in \mathbb{R}$) and R-integrable on every interval $[a,B]$ such that $a < B < +\infty$. Prove that if $\int_a^{+\infty} g(x)dx$ converges absolutely, then $\int_a^{+\infty} f(x)g(x)dx$ converges absolutely [Hint: use the Cauchy criterion for the convergence of improper integrals (Theorem 3.1)].

Theorem 3.3. *If f is monotonic and bounded on $[a,+\infty)$ ($a \in \mathbb{R}$),*

$$\lim_{x\to+\infty} f(x) = 0,$$

and g is bounded on $[a,+\infty)$ and R-integrable on every interval $[a,B]$ such that $a < B < +\infty$ and the G such that

$$G(x) = \int_a^x g(t)\,dt \qquad \textit{for} \quad x \in [a,+\infty)$$

is bounded on $[a,+\infty)$, then $\int_a^{+\infty} f(x)g(x)\,dx$ converges.

PROOF. First note that by Theorem XIV.2.4, for any x' and x'' such that $a \leqslant x' < x'' < +\infty$, there exists a $B \in [x',x'']$ such that

$$\int_{x'}^{x''} f(x)g(x)\,dx = f(x')\int_{x'}^{B} g(x)\,dx + f(x'')\int_{B}^{x''} g(x)\,dx. \tag{3.16}$$

Since G is bounded on $[a,+\infty)$, there exists an $M > 0$ such that

$$\left|\int_a^x g(t)\,dt\right| \leqslant M \qquad \text{for all} \quad x \in [a,+\infty).$$

This implies that, for B such that $a \leqslant B < +\infty$, we have

$$\begin{aligned}\left|\int_{x'}^{B} g(x)\,dx\right| &= \left|\int_a^B g(x)\,dx - \int_a^{x'} g(x)\,dx\right| \\ &\leqslant \left|\int_a^B g(x)\,dx\right| + \left|\int_a^{x'} g(x)\,dx\right| \leqslant 2M.\end{aligned} \tag{3.17}$$

Here x' is some value in $[a,+\infty)$. Similarly, for $x'' \in [a,+\infty)$, we have

$$\left|\int_B^{x''} g(x)\,dx\right| \leqslant 2M \qquad \text{if} \quad a \leqslant B < +\infty. \tag{3.18}$$

Let $\epsilon > 0$ be given. Since $f(x)\to 0$ as $x\to+\infty$, there exists an X such that

$$|f(x)| \leqslant \frac{\epsilon}{4M} \qquad \text{for} \quad x \geqslant X \geqslant a. \tag{3.19}$$

Now take x' and x'' such that $X \leqslant x' < x'' < +\infty$ so that a B exists in $[x', x'']$ for which (3.16) holds. This and (3.16)–(3.19) yield

$$\left| \int_{x'}^{x''} f(x)g(x)\,dx \right| < \epsilon \qquad \text{for} \quad X \leqslant x' < x'' < +\infty.$$

By Cauchy's criterion for the convergence of improper integrals, we conclude from this that $\int_a^{+\infty} f(x)g(x)\,dx$ converges.

Prob. 3.8. Prove: If $p > 0$, then

$$\int_1^{+\infty} \frac{\sin x}{x^p}\,dx \quad \text{and} \quad \int_1^{+\infty} \frac{\cos x}{x^p}\,dx$$

converge. (By Prob. 3.3, if $p > 1$, these integrals converge absolutely.)

Prob. 3.9. Prove the convergence of

$$\int_0^{+\infty} \frac{\cos ax - \cos bx}{x}\,dx.$$

Theorem 3.4. *If f is monotonic and bounded on $[a, +\infty)$ and g is R-integrable on every $[a, B]$ such that $a \leqslant B < +\infty$ and $\int_a^{+\infty} g(x)dx$ converges, then $\int_a^{+\infty} f(x)g(x)dx$ converges.*

Proof. As in the proof of the last theorem, we use Theorem XIV.2.4 to note that for any x' and x'' such that $a \leqslant x' < x'' < +\infty$ there exists a $B \in [x', x'']$ with

$$\int_{x'}^{x''} f(x)g(x)\,dx = f(x')\int_{x'}^{B} g(x)\,dx + f(x'')\int_{B}^{x''} g(x)\,dx. \tag{3.20}$$

The boundedness of f implies that some $M > 0$ exists such that $|f(x)| \leqslant M$ holds for all $x \in [a, +\infty)$. Hence,

$$|f(x')| \leqslant M \quad \text{and} \quad |f(x'')| \leqslant M \qquad \text{for } x' \text{ and } x'' \text{ in } [a, +\infty). \tag{3.21}$$

Let $\epsilon > 0$ be given. Since $\int_a^{+\infty} g(x)dx$ converges by hypothesis, there exists an $X \geqslant a$ such that for any c and d such that $X \leqslant c < d < +\infty$ we have

$$\left| \int_c^d g(x)\,dx \right| \leqslant \frac{\epsilon}{2M}. \tag{3.22}$$

It follows that for x' and x'' such that $X \leqslant x' < x''$ and $x' \leqslant B \leqslant x''$ we have

$$\left| \int_{x'}^{B} g(x)\,dx \right| < \frac{\epsilon}{2M} \quad \text{and} \quad \left| \int_{B}^{x''} g(x)\,dx \right| \leqslant \frac{\epsilon}{2M}.$$

This, (3.20), and (3.21) imply that for x' and x'' such that $X \leqslant x' < x''$

$< +\infty$ we have

$$\left|\int_{x'}^{x''} f(x)g(x)\,dx\right| < \epsilon.$$

Using Cauchy's criterion for the convergence of improper integrals, we conclude that $\int_a^{+\infty} f(x)g(x)\,dx$ converges.

PROB. 3.10. Prove:

(a) $$\int_1^\infty \frac{x}{1+x^2}\sin x dx,$$

(b) $$\int_0^\infty \frac{e^{-x}\sin x}{x}\,dx,$$

(c) $$\int_1^\infty \frac{(1-e^{-x})\cos x}{x}\,dx$$

converge.

4. Integral Representation of the Gamma Function

We learned in Example 2.1 that

$$G(\alpha) = \int_0^{+\infty} t^{\alpha-1}e^{-t}dt \quad \text{converges if and only if} \quad \alpha > 0. \tag{4.1}$$

We shall show that the G defined in (4.1) satisfies

$$G(\alpha+1) = \alpha G(\alpha) \qquad \text{for} \quad \alpha > 0. \tag{4.2}$$

Take B and $\epsilon > 0$ such that $0 < \epsilon < B < +\infty$. Integrating by parts we find that for $\alpha > 0$, we have

$$\int_\epsilon^B t^\alpha e^{-t}\,dt = \int_\epsilon^B t^\alpha d(-e^{-t}) = -t^\alpha e^{-t}\Big|_\epsilon^B + \alpha\int_\epsilon^B t^{\alpha-1}e^{-t}\,dt$$

$$= \frac{\epsilon^\alpha}{e^\epsilon} - \frac{B^\alpha}{e^B} + \alpha\int_\epsilon^B t^{\alpha-1}e^{-t}\,dt. \tag{4.2'}$$

Letting $\epsilon \to 0$ and $B \to +\infty$, we obtain

$$G(\alpha+1) = \int_0^\infty t^\alpha e^{-t}\,dt = \alpha\int_0^\infty t^{\alpha-1}e^{-t}\,dt = \alpha G(\alpha).$$

Thus, (4.1) is proved.

PROB. 4.1. Let G be defined by means of formula (4.1). Prove: $G(1) = 1 = G(2)$.

We establish a lemma which will help us prove that the function G defined in (4.1) is log-convex.

Lemma 4.1.* *If f is a real-valued function of two real variables t and x such that $a \leqslant t \leqslant b$ and x is in some interval I and if for each $x \in I$, f is continuous as a function of t, and for each $t \in [a,b]$, f is log-convex and differentiable as a function of x, then H, given by*

$$H(x) = \int_a^b f(t,x)\,dt \qquad \textit{for} \quad x \in I, \tag{4.3}$$

is log-convex.

Proof. For each positive integer n, let $P_n = \langle t_0, t_1, \ldots, t_n \rangle$ be the partition of $[a,b]$ defined as

$$t_0 = a, \qquad t_1 = a + \frac{b-a}{n}, \qquad t_2 = a + 2\frac{b-a}{n}, \ldots, t_n = b,$$

so that

$$\Delta t_n = \frac{b-a}{n} \quad \text{and} \quad t_i = a + i\Delta t_n \qquad \text{for} \quad i \in \{0, \ldots, n\}.$$

Corresponding to each n, construct the Riemann sum R_n, where

$$R_n(x) = \Delta t_n(f(a,x) + f(a + \Delta t_n, x) \ldots f(a + (n-1)\Delta t_n, x))$$

for $x \in I$. Each such R_n is a sum of log-convex functions and is, therefore, also log-convex. For each $x \in I$, f viewed as a function of t is continuous for $t \in [a,b]$ and is, therefore, R-integrable there. Now

$$\Delta t_n = \|P_n\| = \frac{b-a}{n} \to 0 \qquad \text{as} \quad n \to +\infty.$$

Hence, for $x \in I$,

$$R_n(x) \to \int_a^b f(t,x)\,dt = H(x) \qquad \text{as} \quad n \to +\infty.$$

This shows that H is a limit of log-convex functions on I. As such H itself is log-convex on I.

Theorem 4.1. *The function G defined in* (4.1), *called* Euler's second integral, *is identical with the gamma function. Thus,*

$$\Gamma(\alpha) = \int_0^{+\infty} t^{\alpha-1} e^{-t}\,dt \qquad \textit{for} \quad \alpha > 0. \tag{4.4}$$

Proof. We examine the integrand in the integral on the right. Define h as

$$h(t,\alpha) = t^{\alpha-1}e^{-t} \qquad \text{for} \quad t > 0, \quad \alpha > 0.$$

For each $t > 0$, h is a log-convex function of α. To see this, note that

$$\frac{d}{d\alpha}\ln h(t,\alpha) = \frac{d}{d\alpha}((\alpha - 1)\ln t - t) = \ln t$$

* E. Artin, The Gamma Function, *Loc. cit.*

and

$$\frac{d^2h}{d\alpha^2} = 0 \qquad \text{for} \quad \alpha > 0$$

and each t. Thus, let $\alpha > 0$, $0 < \epsilon \leqslant 1$, and $B \geqslant 1$. By Lemma 4.1 the functions f_ϵ and g_B, defined as

$$f_\epsilon(\alpha) = \int_\epsilon^1 t^{\alpha-1}e^{-t}\,dt \quad \text{and} \quad g_B(\alpha) = \int_1^B t^{\alpha-1}e^{-t}\,dt$$

are log-convex. But the limit of log-convex functions is log-convex. Hence, f and g, defined as

$$f(\alpha) = \lim_{\epsilon\to 0} f_\epsilon(\alpha) = \int_0^1 t^{\alpha-1}e^{-t}\,dt$$

and

$$g(\alpha) = \lim_{B\to+\infty} g_B(\epsilon) = \int_1^{+\infty} t^{\alpha-1}e^{-t}\,dt$$

for $\alpha > 0$, are log-convex. Finally, the sum $f + g$ of f and g is log-convex. This implies that, for $\alpha > 0$,

$$G(\alpha) = \int_0^1 t^{\alpha-1}e^{-t}\,dt + \int_1^{+\infty} t^{\alpha-1}e^{-t}\,dt$$

is log-convex.

We now recall that $G(1) = 1$ (Prob. 4.1) and that for $\alpha > 0$, $G(\alpha + 1) = \alpha G(\alpha)$ (see 4.2). Also, Theorem IX.9.1 states that the only function f with $f(1) = 1$ which is log-convex and satisfies the functional equation $f(\alpha + 1) = \alpha f(\alpha)$ for $\alpha > 0$ is the gamma function. Hence, $G(\alpha) = \Gamma(\alpha)$ for $\alpha > 0$. This proves (4.4).

Example 4.1. As an application of the last theorem, we prove that

$$\int_0^\infty e^{-t^2}\,dt = \frac{\sqrt{\pi}}{2}\,. \tag{4.5}$$

Proof. By formula (X.10.9) we have

$$\Gamma(\tfrac{1}{2}) = \sqrt{\pi}\,. \tag{4.6}$$

By the last theorem, this implies that

$$\int_0^\infty t^{-1/2}e^{-t}\,dt = \Gamma(\tfrac{1}{2}) = \sqrt{\pi}\,. \tag{4.7}$$

We now take B and ϵ such that $0 < \epsilon < B < +\infty$ and consider the integral

$$\int_\epsilon^B \frac{e^{-t}}{\sqrt{t}}\,dt. \tag{4.8}$$

We substitute $t = u^2$ for $u > 0$ so that $dt/du = 2u$. This yields

$$\int_{\epsilon}^{B} \frac{e^{-t}}{\sqrt{t}}\, dt = \int_{\epsilon^2}^{B^2} \frac{e^{-u^2}}{u} 2u\, du = 2\int_{\epsilon^2}^{B^2} e^{-u^2}\, du.$$

We let $B \to +\infty$ and obtain

$$\int_{\epsilon^2}^{+\infty} \frac{e^{-t}}{\sqrt{t}}\, dt = 2\int_{\epsilon^2}^{+\infty} e^{-u^2}\, du.$$

Next we let $\epsilon \to 0+$ and obtain, taking note of (4.7),

$$\sqrt{\pi} = \int_0^{+\infty} t^{-1/2} e^{-t}\, dt = 2\int_0^{+\infty} e^{-u^2}\, du.$$

This yields

$$\int_0^{+\infty} e^{-t^2}\, dt = \int_0^{+\infty} e^{-u^2}\, du = \frac{\sqrt{\pi}}{2},$$

as claimed.

Prob. 4.2. Prove:

$$\int_{-\infty}^{+\infty} e^{-t^2}\, dt = \sqrt{\pi}.$$

Prob. 4.3. Prove:

$$\int_0^1 \left(\ln \frac{1}{t}\right)^{\alpha - 1} dt = \Gamma(\alpha).$$

Prob. 4.4. Prove: If $a > 0$, then

$$\int_0^{\infty} e^{-at} t^{\alpha - 1}\, dt = \frac{\Gamma(\alpha)}{a^{\alpha}}.$$

5. The Beta Function

We begin with a problem for the reader:

Prob. 5.1. Prove: If $x > 0$ and $y > 0$, then

$$\int_0^1 t^{x-1}(1-t)^{y-1}\, dt$$

converges.

We now define a function B by means of

$$B(x, y) = \int_0^1 t^{x-1}(1-t)^{y-1}\,dt \qquad \text{for} \quad x > 0, \quad y > 0. \tag{5.1}$$

By the result cited in Prob. 5.1, this function is a real-valued function of x and y for $x > 0$, $y > 0$. It is called the *Beta function*. The integral on the right side of (5.1) is also called *Euler's first integral*.

PROB. 5.2. Prove:

$$B(x, y) = B(y, x) \qquad \text{for} \quad x > 0, \quad y > 0.$$

PROB. 5.3. Prove:

$$B(x, 1) = \frac{1}{x} \qquad \text{for} \quad x > 0 \quad \text{and} \quad B(1, y) = \frac{1}{y} \qquad \text{for} \quad y > 0.$$

PROB. 5.4. Prove: If $y > 0$ is fixed, then B in (5.1) is a log-convex function of x (Hint: see the proof of Theorem 4.1 where the log-convexity of Euler's second integral is demonstrated).

Lemma 5.1. *If* $x \in \mathbb{R}$ *and* $y \in \mathbb{R}$, *then*

$$B(x+1, y) = \frac{x}{x+y} B(x, y) \qquad \textit{for} \quad x > 0 \quad \textit{and} \quad y > 0. \tag{5.2}$$

PROOF. First take u and v such that $0 < u < v < 1$. The integrand appearing in $B(x+1, y)$ is R-integrable on the interval $[u, v]$. Using integration by parts, we obtain

$$\begin{aligned}
\int_u^v t^x(1-t)^{y-1}\,dt &= \int_u^v \left(\frac{t}{1-t}\right)^x (1-t)^{x+y-1}\,dt \\
&= \int_u^v \left(\frac{t}{1-t}\right)^x \frac{d}{dt}\left[-\frac{(1-t)^{x+y}}{x+y}\right] dt \\
&= -\left(\frac{t}{1-t}\right)^x \frac{(1-t)^{x+y}}{x+y}\Bigg|_u^v \\
&\quad + \frac{x}{x+y}\int_u^v \left(\frac{t}{1-t}\right)^{x-1} \frac{1}{(1-t)^2}(1-t)^{x+y}\,dt \\
&= -t^x \frac{(1-t)^y}{x+y}\Bigg|_u^v + \frac{x}{x+y}\int_u^v t^{x-1}(1-t)^{y-1}\,dt.
\end{aligned}$$

Let $v \to 1-$ and then $u \to 0+$ and obtain

$$B(x+1, y) = \frac{x}{x+y} \int_0^1 t^{x-1}(1-t)^{y-1}\, dt = \frac{x}{x+y} B(x, y),$$

as claimed.

Theorem 5.1. *The beta and gamma functions are related by*

$$B(x, y) = \frac{\Gamma(x)\Gamma(y)}{\Gamma(x+y)} \qquad \textit{for} \quad x > 0 \quad \textit{and} \quad y > 0. \tag{5.3}$$

Proof. We first construct the function G, given by

$$G(x, y) = B(x, y)\frac{\Gamma(x+y)}{\Gamma(y)} \qquad \text{for} \quad x > 0 \quad \text{and} \quad y > 0. \tag{5.4}$$

By the lemma and the properties of the gamma function, we have

$$G(x+1, y) = B(x+1, y)\frac{\Gamma(x+1+y)}{\Gamma(y)} = \frac{x}{x+y} B(x, y)\frac{(x+y)\Gamma(x+y)}{\Gamma(y)}$$

$$= xB(x, y)\frac{\Gamma(x+y)}{\Gamma(y)} = xG(x, y).$$

Thus, for fixed $y > 0$, we have

$$G(x+1, y) = xG(x, y) \qquad \text{if} \quad x > 0. \tag{5.5}$$

Also,

$$G(1, y) = B(1, y)\frac{\Gamma(1+y)}{\Gamma(y)} = B(1, y)\frac{y\Gamma(y)}{\Gamma(y)} = B(1, y)y. \tag{5.6}$$

Clearly, $B(1, y)y = 1$ holds (for $y > 0$) by Prob. 5.3. This and (5.6) imply that

$$G(1, y) = 1 \qquad \text{for} \quad y > 0. \tag{5.7}$$

Our aim is to prove

$$G(x, y) = \Gamma(x) \qquad \text{for all} \quad x > 0 \quad \text{and for each} \quad y > 0. \tag{5.8}$$

This will be done by first proving that for each fixed $y > 0$, the function G is a log-convex function of x. We know (by Prob. 5.4) that for each fixed $y > 0$, $B(x, y)$ is log-convex as a function of x. We also know that for each fixed $x > 0$, $\Gamma(x+y)$ is log-convex as a function of x (explain). Since the product of a log-convex function and a positive constant is a log-convex function (why?), it follows that as a function of x, $\Gamma(x+y)/\Gamma(y)$ is convex for each $y > 0$. Finally, since products of log-convex functions are log-convex, it follows that as a function of x, $G(x, y)$ is log-convex on $\mathbb{R}_+$ for each fixed $y > 0$. By what was proved (Eqs. (5.5) and (5.7)) and by

Theorem IX.9.1, it follows that

$$\Gamma(x) = B(x, y)\frac{\Gamma(x+y)}{\Gamma(y)} \qquad \text{for} \quad x>0 \quad \text{and} \quad y>0.$$

This proves (5.3).

PROB. 5.5. Prove:

$$B(x, y) = 2\int_0^{\pi/2} \sin^{2x-1}\theta \cos^{2y-1}\theta \, d\theta \qquad \text{if} \quad x>0, \quad y>0$$

(Hint: use the substitution $t = \sin^2\theta$ in (5.1)).

The result in Prob. 5.5 implies: If $\alpha > -\frac{1}{2}$, then

$$B\left(\frac{\alpha+1}{2}, \frac{1}{2}\right) = 2\int_0^{\pi/2} \sin^\alpha\theta \, d\theta, \tag{5.9a}$$

$$B\left(\frac{1}{2}, \frac{\alpha+1}{2}\right) = 2\int_0^{\pi/2} \cos^\alpha\theta \, d\theta. \tag{5.9b}$$

For $\alpha > -1$, we have

$$B\left(\frac{\alpha+1}{2}, \frac{1}{2}\right) = \frac{\Gamma((\alpha+1)/2)\Gamma(\frac{1}{2})}{\Gamma(\alpha/2+1)} = B\left(\frac{1}{2}, \frac{\alpha+1}{2}\right). \tag{5.10}$$

It follows (see (5.9)) that

$$\int_0^{\pi/2} \sin^\alpha\theta \, d\theta = \int_0^{\pi/2} \cos^\alpha\theta \, d\theta = \frac{\sqrt{\pi}}{\alpha} \frac{\Gamma((\alpha+1)/2)}{\Gamma(\alpha/2)}. \tag{5.11}$$

In (5.1) put $y = 1 - x$ and obtain from (5.3): If $0 < x < 1$, then

$$\int_0^1 t^{x-1}(1-t)^{-x} = B(x, 1-x) = \frac{\Gamma(x)\Gamma(1-x)}{\Gamma(1)} = \Gamma(x)\Gamma(1-x). \tag{5.12}$$

This, because of Theorem X.10.1, implies that

$$\int_0^1 t^{x-1}(1-t)^{-x} \, dt = \frac{\pi}{\sin x\pi} \qquad \text{if} \quad 0 < x < 1. \tag{5.13}$$

This result can also be expressed as

$$B(x, 1-x) = \frac{\pi}{\sin x\pi} \qquad \text{for} \quad 0 < x < 1. \tag{5.14}$$

In (5.13) substitute

$$u = \frac{t}{1-t}, \qquad 0 < t < 1$$

so that

$$u + 1 = \frac{1}{1-t}, \qquad t = \frac{u}{u+1}.$$

We have

$$0 < u < +\infty, \quad 0 < t < 1, \quad \text{and} \quad \frac{dt}{du} = \frac{1}{(1+u)^2}\,.$$

Therefore, for ϵ and T such that $0 < \epsilon < T < 1$, we obtain

$$\int_{\epsilon}^{T} t^{x-1}(1-t)^{-x}\,dt = \int_{\epsilon/(1-\epsilon)}^{T/(1-T)} \left(\frac{u}{1+u}\right)^{x-1}(1+u)^{x}\frac{1}{(1+u)^2}\,du$$

$$= \int_{\epsilon/(1-\epsilon)}^{T(1-T)} \frac{u^{x-1}}{1+u}\,du.$$

We now let $T \to 1-$ and then let $\epsilon \to 0+$ and obtain

$$B(x, 1-x) = \int_0^1 t^{x-1}(1-t)^{-x}\,dx = \int_0^{+\infty}\frac{u^{x-1}}{1+u}\,du \tag{5.15}$$

or

$$\int_0^{+\infty}\frac{u^{x-1}}{1+u}\,du = \frac{\pi}{\sin \pi x}\,. \tag{5.16}$$

PROB. 5.6. Prove: If $x > 0$, $y > 0$, and $r > 0$, then

$$\int_0^1 t^{x-1}(1-t^r)^{y-1}\,dt = \frac{1}{r}B\left(\frac{x}{r}, y\right).$$

PROB. 5.7. Prove:

$$\int_0^1 \frac{x^n}{\sqrt{1-x^2}}\,dx = \frac{\sqrt{\pi}}{2}\,\frac{\Gamma((n+1)/2)}{\Gamma((n+2)/2)}\,, \qquad n > 0$$

and

$$\int_0^1 \frac{1}{\sqrt{1-x^n}}\,dx = \frac{\sqrt{\pi}}{n}\,\frac{\Gamma\left(\frac{1}{n}\right)}{\Gamma\left(\frac{1}{n}+\frac{1}{2}\right)}\,.$$

EXAMPLE 5.1. Sometimes having information about just the convergence of an improper integral puts us in a position to evaluate the integral. Consider

$$\int_0^{\pi}\ln(\sin x)\,dx = \int_0^{\pi/2}\ln(\sin x)\,dx + \int_{\pi/2}^{\pi}\ln(\sin x)\,dx. \tag{5.17}$$

We have

$$\ln(\sin x) \to -\infty \qquad \text{as} \quad x \to 0+$$

and

$$\ln(\sin x) \to -\infty \qquad \text{as} \quad x \to \pi-.$$

Hence, the first of the integrals on the right is improper at $x = 0$, while the second is improper at $x = \pi$. We compare the integrand with $x^{-1/2}$. By L'Hôpital's rule we have

$$\lim_{x\to 0+} \frac{\ln(\sin x)}{x^{-1/2}} = -2\lim \frac{x^{3/2}\cos x}{\sin x} = -2\lim_{x\to 0+}\left(\frac{x}{\sin x}(\sqrt{x}\cos x)\right) = 0.$$

This implies that

$$\ln(\sin x) = O(x^{-1/2}) \quad \text{as} \quad x\to 0+. \tag{5.18}$$

From this we see that the first integral on the right in (5.17) converges.

As for the second integral on the right side of (5.17), we compare the integrand to $(x-\pi)^{-1/2}$. L'Hôpital's rule yields

$$\begin{aligned}\lim_{x\to\pi-} \frac{\ln(\sin x)}{(x-\pi)^{-1/2}} &= -2\lim_{x\to\pi-} \frac{(x-\pi)^{3/2}\cos x}{\sin x}\\ &= -2\lim_{x\to\pi-} \frac{x-\pi}{\sin x}\left((x-\pi)^{1/2}\cos x\right)\\ &= -2\lim_{x\to\pi-} \frac{x-\pi}{\sin x}\lim_{x\to\pi-}\left[(x-\pi)^{1/2}\cos x\right]\\ &= -2\left(\lim_{x\to\pi-}\frac{1}{\cos x}\right)\cdot 0\\ &= (-2)(-1)\cdot 0\\ &= 0.\end{aligned}$$

Thus,

$$\ln(\sin x) = O(x-\pi)^{-1/2} \quad \text{as} \quad x\to\pi-.$$

This implies that the second integral on the right in (5.17) also converges. It is easy to conclude that

$$\int_{\pi/2}^{\pi}\ln(\sin x)\,dx = \int_{\pi/2}^{\pi}\ln(\sin(\pi-x)\,dx = \int_0^{\pi/2}\ln(\sin u\,du) = \int_0^{\pi/2}\ln(\sin x)\,dx.$$

The second equality is obtained by using the substitution $u = \pi - x$. Thus, (5.17) gives rise to

$$\int_0^{\pi}\ln(\sin x)\,dx = 2\int_0^{\pi/2}\ln(\sin x)\,dx. \tag{5.19}$$

We begin anew with the left-hand side of (5.17) and note that

$$\int_0^{\pi}\ln(\sin x)\,dx = \int_0^{\pi}\ln\left(2\sin\frac{x}{2}\cos\frac{x}{2}\right)dx. \tag{5.20}$$

We take u and v such that $0 < u < v < \pi/2$ and use the substitution $t = x/2$, so that $dx = 2\,dt$. We have

$$\int_u^v\ln\left(2\sin\frac{x}{2}\cos\frac{x}{2}\right)dx = 2\int_{u/2}^{v/2}\ln(2\sin t\cos t)\,dt.$$

We let $u \to 0+$ and then $v \to \pi-$ to obtain

$$\int_0^{\pi} \ln\left(2 \sin \frac{x}{2} \cos \frac{x}{2}\right) dx = 2\int_0^{\pi/2} \ln(2 \sin t \cos t)\, dt$$
$$= 2\int_0^{\pi/2} \ln(2 \sin x \cos x)\, dx.$$

This and (5.20) yield

$$\int_0^{\pi} \ln(\sin x)\, dx = 2\int_0^{\pi/2} \ln(2 \sin x \cos x)\, dx$$
$$= 2\int_0^{\pi/2} (\ln 2 + \ln \sin x + \ln \cos x)\, dx$$
$$= \pi \ln 2 + 2\int_0^{\pi/2} \ln(\sin x)\, dx + 2\int_0^{\pi/2} \ln(\cos x)\, dx.$$

The two integrals on the lower right converge (why?). This and (5.19) imply that

$$2\int_0^{\pi/2} \ln(\sin x)\, dx = \pi \ln 2 + 2\int_0^{\pi/2} \ln(\sin x)\, dx + 2\int_0^{\pi/2} \ln(\cos x)\, dx. \quad (5.21)$$

We conclude from this that

$$\int_0^{\pi/2} \ln(\cos x)\, dx = -\frac{\pi}{2} \ln 2. \quad (5.22)$$

Note that $\cos x = \sin((\pi/2) - x)$. The substitution $t = (\pi/2) - x$ leads to

$$\int_0^{\pi/2} \ln(\cos x)\, dx = \int_0^{\pi/2} \ln(\sin x)\, dx. \quad (5.23)$$

Also,

$$\int_0^{\pi/2} \ln(\sin x)\, dx = -\frac{\pi}{2} \ln 2. \quad (5.24)$$

This, (5.23), and (5.24) yield

$$\int_0^{\pi} \ln(\sin x)\, dx = -\pi \ln 2. \quad (5.25)$$

PROB. 5.8. Obtain

$$\int_0^{\pi} \ln(1 - \cos x)\, dx = \int_0^{\pi} \ln(1 + \cos x)\, dx = -\pi \ln 2.$$

6. Evaluation of $\int_0^{+\infty}(\sin x)/x\,dx$

In Example 3.1 we proved that this integral converges conditionally. We now evaluate it. We first prove a lemma.

Lemma 6.1.* *If* $0 < x < \pi/2$, *then*

$$\sin^2 x > x^2 \cos x. \tag{6.1}$$

Proof. Define f as

$$f(x) = x - \sin x(\cos x)^{-1/2} \qquad \text{for} \quad |x| < \frac{\pi}{2}.$$

We take derivatives and obtain

$$f'(x) = 1 - \frac{\sin^2 x + 2\cos^2 x}{2(\cos x)^{3/2}} = 1 - \frac{1 + \cos^2 x}{2(\cos x)^{3/2}} \tag{6.2}$$

and

$$f''(x) = \frac{1}{4}(\cos x)^{-5/2}\sin x(\cos^2 x - 3) < 0 \qquad \text{if} \quad 0 < x < \frac{\pi}{2}. \tag{6.3}$$

This implies that f' is strictly decreasing on $[0, \pi/2)$. We conclude that $f'(x) < f'(0) = 0$ if $0 < x < \pi/2$. In turn, this implies that f is strictly decreasing on $[0, \pi/2)$ and we have: $f(x) < f(0) = 0$ for $0 < x < \pi/2$. This tells us that

$$x - \sin x(\cos x)^{-1/2} < 0 \qquad \text{if} \quad 0 < x < \frac{\pi}{2}.$$

Thus,

$$0 < x < \frac{\sin x}{(\cos x)^{1/2}} \qquad \text{if} \quad 0 < x < \frac{\pi}{2}.$$

This yields (6.1) for $0 < x < \pi/2$.

We now prove:

$$\int_0^\infty \frac{\sin x}{x}\,dx = \frac{\pi}{2}. \tag{6.4}$$

We begin by considering the sequence $\langle I_n \rangle$ of integrals

$$I_n = \int_0^{\pi/2} \sin(2nx)\left(\frac{1}{\sin x} - \frac{1}{x}\right)dx \qquad \text{for each positive integer } n. \tag{6.5}$$

These integrands are singular at $x = 0$. In fact, L'Hôpital's rule gives

$$\lim_{x\to 0+}\left(\frac{1}{\sin x} - \frac{1}{x}\right) = \lim_{x\to 0+}\frac{x - \sin x}{x\sin x} = \lim_{x\to 0+}\frac{1 - \cos x}{x\cos x + \sin x}$$

$$= \lim_{x\to 0+}\frac{\sin x}{2\cos x - x\sin x} = 0. \tag{6.6}$$

Hence, the integrands in the I_n's satisfy

$$\lim_{x\to 0+}\left[\sin 2nx\left(\frac{1}{\sin x} - \frac{1}{x}\right)\right] = 0.$$

For each positive integer n, the integrand in I_n has an extension f_n which is continuous on $[0, \pi/2]$ and is, therefore, R-integrable there.

*D. S. Mitronovic, *Analytic Inequalities*, p. 238, *Loc. cit.*

We take $\epsilon \in (0, \pi/2]$. By the discussion above, we have

$$\lim_{\epsilon\to 0+}\int_\epsilon^{\pi/2}\sin 2nx\left(\frac{1}{\sin x}-\frac{1}{x}\right)dx=\int_0^{\pi/2}\sin 2nx\left(\frac{1}{\sin x}-\frac{1}{x}\right)dx,$$

the limit here being finite (explain).

We use integration by parts and obtain for $\epsilon \in (0, /2]$

$$\begin{aligned}\int_\epsilon^{\pi/2}&\sin 2nx\left(\frac{1}{\sin x}-\frac{1}{x}\right)dx\\
&=\int_\epsilon^{\pi/2}\left(\frac{1}{\sin x}-\frac{1}{x}\right)\frac{d}{dx}\left(-\frac{\cos 2nx}{2n}\right)dx\\
&=-\frac{(-1)^n}{2n}\left(1-\frac{2}{\pi}\right)+\frac{\cos 2n\epsilon}{2n}\left(\frac{1}{\sin\epsilon}-\frac{1}{\epsilon}\right)\\
&\quad+\frac{1}{2n}\int_\epsilon^{\pi/2}\cos 2nx\left(\frac{1}{x^2}-\frac{\cos x}{\sin^2 x}\right)dx. \qquad (6.7)\end{aligned}$$

For $\epsilon\to 0+$, we have, as (6.6) shows, that

$$\frac{1}{\sin\epsilon}-\frac{1}{\epsilon}\to 0$$

and

$$\begin{aligned}\int_0^{\pi/2}\sin 2nx\left(\frac{1}{\sin x}-\frac{1}{x}\right)dx&=\frac{(-1)^{n+1}}{2n}\left(1-\frac{2}{\pi}\right)\\
&\quad+\frac{1}{2n}\int_0^{\pi/2}\cos 2nx\left(\frac{1}{x^2}-\frac{\cos x}{\sin^2 x}\right)dx. \qquad (6.8)\end{aligned}$$

This holds for each positive integer n. We examine the last integral. Using Lemma 6.1, we have for the integrand

$$\left|\cos 2nx\left(\frac{1}{x^2}-\frac{\cos x}{\sin^2 x}\right)\right|\leqslant\left|\frac{1}{x^2}-\frac{\cos x}{\sin^2 x}\right|=\frac{\sin^2 x-x^2\cos x}{x^2\sin^2 x}.$$

Applying L'Hôpital's rule several times we obtain

$$\lim_{x\to 0+}\left(\frac{\sin^2 x-x^2\cos x}{x^2\sin^2 x}\right)=0.$$

Accordingly, the integral on the right in (6.8) converges. Taking absolute values in (6.8) we obtain

$$\left|\int_0^{\pi/2}\sin 2nx\left(\frac{1}{\sin x}-\frac{1}{x}\right)dx\right|\leqslant\frac{1}{2n}\left(1-\frac{2}{\pi}\right)+\frac{1}{2n}\int_0^{\pi/2}\frac{\sin^2 x-x^2\cos x}{x^2\sin^2 x}$$

for each positive integer n. Now let $n\to+\infty$ and obtain

$$\lim_{n\to+\infty}\int_0^{\pi/2}\sin 2nx\left(\frac{1}{\sin x}-\frac{1}{x}\right)dx=0. \qquad (6.9)$$

By Example XIV.1.2, we know that

$$\lim_{n\to+\infty}\int_0^{\pi/2}\frac{\sin 2nx}{\sin x}dx=\frac{\pi}{2}.$$

Because of (6.9), this implies that

$$\lim_{n\to+\infty}\int_0^{\pi/2}\frac{\sin 2nx}{x}\,dx=\frac{\pi}{2}. \tag{6.10}$$

Using the substitution $u = 2nx$ we obtain, for each n,

$$\int_0^{\pi/2}\frac{\sin 2nx}{2nx}\,2n\,dx=\int_0^{n\pi}\frac{\sin u}{u}\,du.$$

This and (6.12) imply that

$$\frac{\pi}{2}=\lim_{n\to+\infty}\int_0^{\pi/2}\frac{\sin 2nx}{x}\,dx=\lim_{n\to+\infty}\int_0^{n\pi}\frac{\sin u}{u}\,du=\int_0^{+\infty}\frac{\sin u}{u}\,du,$$

proving (6.4)

PROB. 6.1. Prove: If n is a positive integer, then

(a) $$\int_0^{\pi/2}(\cos 2nx)\ln(\sin x)\,dx$$

converges and is equal to $-\pi/4n$. Prove:

(b) $$\int_0^{\pi/2}\cos 2nx\ln(\cos x)\,dx=(-1)^{n+1}\frac{\pi}{4n}.$$

7. Integral Tests for Convergence of Series

Sometimes improper integrals can be used to test for convergence or divergence of certain infinite series. This topic is usually presented together with the results on infinite series. Since we studied infinite series before the Riemann integral was even defined, we are treating this topic now.

Theorem 7.1. *If $\sum a_n$ is a series with positive decreasing terms and f is a monotonically decreasing function such that*

$$f(n)=a_n \qquad \textit{for all positive integers } n,$$

then the improper integral

$$\int_1^{+\infty}f(x)\,dx \tag{7.1}$$

and the infinite series $\sum a_n$ converge together or diverge together.

PROOF. Assume that $x\in[k,k+1]$, where k is some positive integer. The assumptions on f and on $\langle a_k\rangle$ imply that

$$a_k=f(k)\geqslant f(x)\geqslant f(k+1)=a_{k+1},$$

so that

$$a_k\geqslant f(x)\geqslant a_{k+1}.$$

Taking integrals on $[k, k+1]$, we see that

$$a_k = \int_k^{k+1} a_k\,dx \geqslant \int_k^{k+1} f(x)\,dx \geqslant \int_k^{k+1} a_{k+1}dx = a_{k+1}.$$

It follows that if k is a positive integer, then

$$a_k \geqslant \int_k^{k+1} f(x)\,dx \geqslant a_{k+1}. \tag{7.2}$$

Summing from $k = 1$ to $k = n$, we obtain for the nth partial sum S_n,

$$S_n = \sum_{k=1}^{n} a_k \geqslant \sum_{k=1}^{n} \int_k^{k+1} f(x)\,dx \geqslant \sum_{k=1}^{n} a_{k+1} = S_{n+1} - a_1.$$

This implies that

$$S_n \geqslant \int_1^{n+1} f(x)\,dx \geqslant S_{n+1} - a_1 \tag{7.3}$$

for each positive integer n.

Assume that $\sum a_n$ converges so that its partial sum sequence converges. From the inequality on the left in (7.3) it follows that the sequence $\langle y_n \rangle$, where

$$y_n = \int_1^n f(x)\,dx$$

for each n, is bounded. Since $f(x) \geqslant a_1 \geqslant 0$ for all $x \geqslant 1$ (explain), the sequence $\langle y_n \rangle$ is also monotonic increasing. We conclude that $\langle y_n \rangle$ converges. Since

$$\int_1^{\infty} f(x)\,dx = \lim_{n\to+\infty} \int_1^n f(x)\,dx = \lim_{n\to+\infty} y_n,$$

the integral (7.1) converges.

Conversely, if (7.1) converges, then $\langle y_n \rangle$ is a converging sequence and, therefore, bounded. The sequence $\langle y_n \rangle$ is also monotonically increasing. This and the right-hand inequality in (7.3) imply that $\langle S_n \rangle$ is bounded. The latter sequence is also monotonically increasing. We conclude that $\sum a_n$ converges.

We proved that $\sum a_n$ converges if and only if the integral (7.1) does. This is equivalent to saying the series $\sum a_n$ diverges if and only if the integral (7.1) does.

Theorem 7.2. *Let $\sum a_n$ be a series with positive decreasing terms and f a monotonically decreasing function such that $f(n) = a_n$ for all positive integers n. If $\langle S_n \rangle$ is the sequence of partial sums $\sum a_n$, then*

$$\lim_{n\to+\infty}\left(\int_1^n f(x)\,dx - S_n\right)$$

exists and is finite.

PROOF. Form the sequence $\langle T_n \rangle$, where for each positive integer n

$$T_n = \int_1^n f(x)\,dx - S_n .$$

For each positive integer n we have

$$T_{n+1} - T_n = \int_n^{n+1} f(x)\,dx - a_{n+1} . \tag{7.4}$$

Since (7.2) holds for each positive integer k, it follows that

$$T_{n+1} - T_n = \int_n^{n+1} f(x)\,dx - a_{n+1} \geqslant 0$$

for all n. Hence, the sequence $\langle T_n \rangle$ is monotonically increasing. Since (7.3) holds for each n, we have

$$0 \geqslant \int_1^{n+1} f(x)\,dx - S_n \geqslant a_{n+1} - a_1 .$$

Hence,

$$0 \geqslant T_{n+1} + a_{n+1} \geqslant a_{n+1} - a_1 .$$

Equivalently,

$$-a_{n+1} \geqslant T_{n+1} \geqslant -a_1$$

for each n. This and $T_1 = -a_1$ proves that the terms of $\langle T_n \rangle$ are negative. Hence, $\langle T_n \rangle$ is bounded from above by zero. Since it is also monotonically increasing, we conclude that $\langle T_n \rangle$ converges. Hence, the conclusion holds.

According to Theorem 7.1, the harmonic series

$$1 + \frac{1}{2} + \cdots + \frac{1}{n} + \cdots$$

diverges. Here the function f is $f(x) = 1/x$ for $x \geqslant 1$. Since

$$\int_1^{\infty} f(x)\,dx = \int_1^{\infty} \frac{1}{x}\,dx = \lim_{B \to +\infty} \int_1^{B} \frac{1}{x}\,dx = \lim_{B \to +\infty} \ln B = +\infty,$$

the harmonic series diverges.

On the other hand, if $p > 1$, then

$$\sum_{n=1}^{\infty} \frac{1}{n^p}$$

converges. This is so because

$$\int_1^{+\infty} \frac{1}{x^p}\,dx$$

converges for $p > 1$.

Theorem 7.2 also furnishes us with another proof of the fact that the sequence $\langle \gamma_n \rangle$, where

$$\gamma_n = 1 + \frac{1}{2} + \cdots + \frac{1}{n} - \ln n$$

for each positive integer n, converges. This is so because

$$\gamma_n = 1 + \frac{1}{2} + \cdots + \frac{1}{n} - \int_1^n \frac{1}{x}\,dx.$$

This converges according to Theorem 7.2.

PROB. 7.1. For what values of p do we have convergence and for which p do we have divergence in

(a) $\sum_{n=1} 1/n^p$,
(b) $\sum_{n=2} 1/n \ln^p n$,
(c) $\sum_{n=3} 1/n \ln n (\ln\ln n)^p$,
(d) $\sum_{n=3} 1/n \ln n (\ln\ln n)(\ln\ln\ln n)^p$?

8. Jacobian Elliptic Functions

We saw how infinite series can be used to define functions. We defined the sine and cosine functions by means of certain power series. It will now be shown that also indefinite integrals can be used to define functions.

We recall that

$$\operatorname{Arcsin} x = \int_0^x \frac{1}{\sqrt{1-t^2}}\,dt, \qquad -1 \leqslant x \leqslant 1. \tag{8.1}$$

This integral is improper at ± 1, and

$$\frac{\pi}{2} = \int_0^1 \frac{1}{\sqrt{1-t^2}}\,dt \quad \text{and} \quad -\frac{\pi}{2} = \int_0^{-1} \frac{1}{\sqrt{1-t^2}}\,dt. \tag{8.2}$$

A generalization of (6.1) is the function u given by

$$u(x,k) = \int_0^x \frac{1}{\sqrt{1-t^2}\sqrt{1-k^2t^2}}\,dt, \qquad -1 \leqslant x \leqslant 1, \tag{8.3}$$

where k is some fixed constant with $0 \leqslant k \leqslant 1$. If $k = 0$, we have

$$u(x,0) = \operatorname{Arcsin} x, \qquad -1 \leqslant x \leqslant 1. \tag{8.4}$$

When $k = 1$, we obtain

$$u(x,1) = \int_0^x \frac{1}{1-t^2}\,dt = \operatorname{Arctanh} x, \qquad -1 < x < 1. \tag{8.5}$$

If $x = \pm 1$, we obtain improper divergent integrals. This follows from

$$\lim_{x\to 1-} \int_0^x \frac{1}{1-t^2}\,dt = \lim_{x\to 1-} \operatorname{Arctanh} x = \lim_{x\to 1-} \left(\frac{1}{2} \ln \frac{1+x}{1-x} \right) = +\infty$$

and

$$\lim_{x \to (-1)+} \int_0^x \frac{1}{1-t^2}\, dt = \lim_{x \to (-1)+} \left(\frac{1}{2} \ln \frac{1+x}{1-x} \right) = -\infty.$$

For (8.3) to yield functions differing from Arctanh and Arcsin, we take $0 < k < 1$. For such values of k, (8.3) is called an *elliptic integral of the first kind*. This terminology is due to Legendre. He called the integrals

$$\int \frac{1}{\sqrt{1-x^2}\,\sqrt{1-k^2x^2}}\, dx, \tag{8.6}$$

$$\int \frac{1-k^2x^2}{\sqrt{1-x^2}\,\sqrt{1-k^2x^2}}\, dx, \tag{8.7}$$

$$\int \frac{1}{(1+nx^2)\sqrt{1-x^2}\,\sqrt{1-k^2x^2}}\, dx, \tag{8.8}$$

elliptic integrals of the *first, second, and third kind*, respectively. Since

$$\frac{1}{\sqrt{1-t^2}\,\sqrt{1-k^2t^2}} \leqslant \frac{1}{\sqrt{1-t^2}\,\sqrt{1-k^2}} \qquad \text{if} \quad |t| < 1 \quad \text{and} \quad 0 < k < 1, \tag{8.9}$$

and the integrals

$$\int_0^1 \frac{1}{\sqrt{1-t^2}}\, dt, \qquad \int_{-1}^0 \frac{1}{\sqrt{1-t^2}}\, dt$$

converge, it follows (in the notation of (8.3)) that the integrals

$$\begin{aligned} u(1,k) &= \int_0^1 \frac{1}{\sqrt{1-t^2}\,\sqrt{1-k^2t^2}}\, dt, \\ u(-1,k) &= \int_0^{-1} \frac{1}{\sqrt{1-t^2}\,\sqrt{1-k^2t^2}}\, dt, \qquad 0 < k < 1, \end{aligned} \tag{8.10}$$

converge. These integrals are called *complete* elliptic integrals of the first kind. k is called the *modulus* and $k' = \sqrt{1-k^2}$ is called the *complementary* modulus. We also put

$$K(k) = u(1,k) \qquad \text{if} \quad 0 < k < 1. \tag{8.11}$$

Here $K(k)$ for fixed k is analogous to $\pi/2$ which arises when $k = 0$ in (8.10).

We now derive some properties of u in (8.3). Using the substitution

$t=-s$, $dt=-ds$ in the integral in (8.3) we obtain

$$u(-x,k)=\int_0^{-x}\frac{1}{\sqrt{1-t^2}\sqrt{1-k^2t^2}}\,dt$$
$$=\int_x^0\frac{1}{\sqrt{1-s^2}\sqrt{1-k^2s^2}}\,ds=-u(x,k) \tag{8.12}$$

for $-1\leqslant x\leqslant 1$. Thus, u is an odd function of x. We have:

$$\begin{aligned} u(1,k)&=K(k),\\ u(0,k)&=0,\\ u(-1,k)&=-K(k), \end{aligned} \tag{8.13}$$

and

$$u(-x,k)=-u(x,k). \tag{8.14}$$

Since the integrand in (8.3) is positive, it is clear that

$$0<u(x,k)<K(k)\qquad\text{for}\quad 0<x<1 \tag{8.15}$$

and

$$-K(k)<u(x,k)<0\qquad\text{for}\quad -1<x<0 \tag{8.16}$$

(explain). Thus,

$$-K(k)\leqslant u\leqslant K(k)\qquad\text{for}\quad x\in[-1,1]. \tag{8.17}$$

By the definition of improper integrals, we have

$$\begin{aligned} \lim_{x\to 1-}u(x,k)&=u(1,k)=K(k),\\ \lim_{x\to(-1)+}u(x,k)&=u(-1,k)=-K(k). \end{aligned} \tag{8.18}$$

Thus, u is continuous on the closed interval $[-1,1]$ and its range is the interval $[-K,K]$. Also,

$$\frac{du}{dx}=\frac{1}{\sqrt{1-x^2}\sqrt{1-k^2x^2}}>0\qquad\text{for}\quad x\in(-1;1). \tag{8.19}$$

We conclude that u is strictly increasing on $[-1,1]$ and that it has a strictly increasing inverse defined on the range $[-K,K]$ of u, whose range is $[-1,1]$, the domain of u. We call this inverse the *modular sine* and write it as sn. For each $u\in[-K,K]$, $\operatorname{sn}(u,k)$ is the unique $x\in[-1,1]$ such that $u=u(x,k)$. This may be stated as:

$$x=\operatorname{sn}u\qquad\text{if and only if}\quad u=u(x,k) \tag{8.20}$$

for $u \in [-K, K]$. Since the function u itself is the inverse of its inverse, we may write

$$\operatorname{sn}^{-1}(x,k) = u(x,K) \quad \text{for} \quad x \in [-1,1]. \tag{8.21}$$

Thus,

$$\operatorname{sn}^{-1}(x,k) = \int_0^x \frac{1}{\sqrt{1-t^2}\sqrt{1-k^2t^2}}\,dt, \qquad -1 \leqslant x \leqslant 1. \tag{8.22}$$

In the new notation, we have

$$-K \leqslant \operatorname{sn}^{-1}(x,k) \leqslant K \quad \text{for} \quad x \in [-1,1], \tag{8.23}$$

$$\operatorname{sn}^{-1}(-1,k) = -K, \qquad \operatorname{sn}^{-1}(0,k) = 0, \qquad \operatorname{sn}^{-1}(1,k) = K, \tag{8.24}$$

and

$$\frac{d\operatorname{sn}^{-1}(x,k)}{dx} = \frac{1}{\sqrt{1-x^2}\sqrt{1-k^2x^2}} \quad \text{for} \quad x \in (-1;1). \tag{8.25}$$

Also,

$$\operatorname{sn}^{-1}(-x,k) = -\operatorname{sn}^{-1}(x,k), \tag{8.26}$$

$$\operatorname{sn}(-K,k) = -1, \qquad \operatorname{sn}(0,k) = 0, \qquad \operatorname{sn}(K,k) = 1, \tag{8.27}$$

and

$$-1 \leqslant \operatorname{sn}(u,k) \leqslant 1. \tag{8.28}$$

PROB. 8.1. Prove: $\operatorname{sn}(-u,k) = -\operatorname{sn}(u,k)$.

Other elliptic functions are:

$$\operatorname{cn}(u,k) = \sqrt{1-\operatorname{sn}^2(u,k)} \quad \text{and} \quad \operatorname{dn}(u,k) = \sqrt{1-k^2\operatorname{sn}^2(u,k)} \tag{8.29}$$

for $-K(k) \leqslant u \leqslant K(k)$. The function cn is called the *modular cosine*, while $\operatorname{dn} u$ is called *delta amplitude u*. The functions sn, cn, and dn are called *Jacobian elliptic functions*. It is often convenient to suppress their dependence on k. This we usually do from now on. In the theory, we define the so-called *complementary modulus* k' as

$$k' = \sqrt{1-k^2}, \qquad 0 < k < 1. \tag{8.30}$$

We have from the definitions:

$$\operatorname{cn}^2 u + \operatorname{sn}^2 u = 1, \qquad \operatorname{dn}^2 u + k^2\operatorname{sn}^2 u = 1, \tag{8.31}$$

$$\operatorname{cn}(-K) = 0 = \operatorname{cn} K, \qquad \operatorname{dn}(-K) = k' = \operatorname{dn} K, \tag{8.32}$$

and

$$\operatorname{cn} 0 = 1, \qquad \operatorname{dn} 0 = 1. \tag{8.33}$$

One also notes readily that

$$\operatorname{cn}(-u) = \operatorname{cn} u \quad \text{and} \quad \operatorname{dn}(-u) = \operatorname{dn} u \tag{8.34}$$

and that

$$0 < \operatorname{cn} u < 1, \qquad 0 < k' < \operatorname{dn} u < 1 \qquad \text{if} \quad 0 < |u| < K. \tag{8.35}$$

PROB. 8.2. Prove: If $-K \leqslant u \leqslant K$, then

(a) $k^2\operatorname{cn}^2 u + k'^2 = \operatorname{dn}^2 u$,
(b) $\operatorname{cn}^2 u + k'^2 \operatorname{sn}^2 u = \operatorname{dn}^2 u$.

To obtain differentiation formulas for the Jacobian elliptic functions we begin with sn. By theorems on derivatives of inverses of differentiable functions we have from (8.21), (8.25), and (8.29),

$$\frac{d \operatorname{sn} u}{du} = \frac{1}{(d \operatorname{sn}^{-1} x / dx)|_{x = \operatorname{sn} u}} = \left(\frac{1}{\sqrt{1 - \operatorname{sn}^2 u}\,\sqrt{1 - k^2 \operatorname{sn}^2 u}} \right)^{-1} = \operatorname{cn} u \operatorname{dn} u \tag{8.36}$$

for $-K < u < K$. Note that sn is continuous on $[-K, K]$. Hence,

$$\lim_{u \to K-} \frac{d \operatorname{sn} u}{du} = \lim_{u \to K-} \sqrt{1 - \operatorname{sn}^2 u}\,\sqrt{1 - k^2 \operatorname{sn}^2 u} = 0 = \operatorname{cn} K \operatorname{dn} K.$$

Similarly,

$$\lim_{u \to (-K)+} \frac{d \operatorname{sn} u}{du} = 0 = \operatorname{cn}(-K) \operatorname{dn}(-K).$$

In view of Prob. VII.7.6 we see that sn is differentiable from the left at $u = K$ and from the right at $u = -K$ and that (8.36) holds for $-K \leqslant u \leqslant K$. In sum,

$$\frac{d \operatorname{sn} u}{du} = \operatorname{cn} u \operatorname{dn} u \qquad \text{for} \quad -K \leqslant u \leqslant K. \tag{8.37}$$

PROB. 8.3. Prove: If $-K \leqslant u \leqslant K$, then

$$\frac{d(\operatorname{cn} u)}{du} = -\operatorname{sn} u \operatorname{dn} u, \tag{8.38a}$$

$$\frac{d(\operatorname{dn} u)}{du} = -k^2 \operatorname{sn} u \operatorname{cn} u. \tag{8.38b}$$

Theorem 8.1. *If s, c, and d are functions, defined on some interval I centered at 0, possibly $\mathbb{R}$ itself, such that*

$$\begin{aligned} s'(u) &= c(u)d(u), \\ c'(u) &= -s(u)d(u), \qquad u \in I \\ d'(u) &= -k^2 s(u)c(u), \qquad 0 \leqslant k \leqslant 1 \end{aligned} \tag{8.39}$$

and

$$s(0) = 0, \qquad c(0) = 1 = d(0), \tag{8.40}$$

then, for $u \in I$, we have

$$\begin{aligned} &(1) \qquad s^2(u) + c^2(u) = 1, \\ &(2) \qquad k^2 s^2(u) + d^2(u) = 1, \\ &(3) \qquad k^2 c^2(u) + k'^2 = d^2(u), \qquad k' = \sqrt{1-k^2}, \\ &(4) \qquad c^2(u) + k'^2 s^2(u) = d^2(u). \end{aligned} \tag{8.41}$$

PROOF. We prove (8.41)(1) and leave the proofs of the other formulas (8.41)(2), (3) and (4) to the reader (see Prob. 8.4 below).

We multiply the first equation in (8.39) by $s(u)$, the second by $c(u)$, and obtain

$$s(u)s'(u) + c(u)c'(u) = 0 \qquad \text{for all} \quad u \in I.$$

This implies that

$$\frac{d}{du}\left(\frac{s^2(u) + c^2(u)}{2}\right) = 0 \qquad \text{for all} \quad u \in I.$$

In turn, this implies that a constant a exists such that

$$s^2(u) + c^2(u) = 2a \qquad \text{for} \quad u \in I. \tag{8.42}$$

Substitute $u = 0$ here and use (8.40) to obtain

$$1 = 0 + 1 = s^2(0) + c^2(0) = 2a.$$

In view of (8.42) this implies that

$$s^2(u) + c^2(u) = 1 \qquad \text{for} \quad u \in I.$$

This proves (8.41)(1).

PROB. 8.4. Complete the proof of Theorem 8.1 by proving the relations (8.41)(2), (3), (4).

Prob. 8.5. Let s, d, and c be functions satisfying the hypothesis of Theorem 8.1 above. Prove that the relations

(1) $s'' = -s(d^2 + k^2c^2)$,
(2) $c'' = -c(d^2 - k^2s^2)$,
(3) $d'' = -k^2d(c^2 - s^2)$,

hold for $u \in I$, where I is the interval referred to in Theorem 8.1. Also prove that the relations

(4) $s'' = -(1 + k^2)s + 2k^2s^3$,
(5) $c'' = -(1 - 2k^2)c - 2k^2c^3$,
(6) $d'' = (2 - k^2)d - 2d^3$

hold for all $u \in I$.

Prob. 8.6.* Prove: If $0 < u < K$, then

$$\operatorname{cn} u < \operatorname{dn} u < 1 < \frac{u}{\operatorname{sn} u} < \frac{1}{\operatorname{cn} u}.$$

9. Addition Formulas

Formulas for $\operatorname{sn}(u + v)$, $\operatorname{cn}(u + v)$, and $\operatorname{dn}(u + v)$ are called *addition formulas*. We state these for functions s, c, and d satisfying the hypothesis of Theorem 8.1.

Theorem 9.1. *Let s, c, and d be functions, defined on an interval I centered at 0—possibly $\mathbb{R}$ itself—for which* (8.39) *and* (8.40) *hold for $u \in I$. Then*

(a) $s(u + v) = (s(u)c(v)d(v) + s(v)c(u)d(u))/(1 - k^2s^2(u)s^2(v))$,
(b) $c(u + v) = (c(u)c(v) - s(u)d(u)s(v)d(v))/(1 - k^2s^2(u)s^2(v))$,
(c) $d(u + v) = (d(u)d(v) - k^2s(u)c(u)s(v)c(v))/(1 - k^2s^2(u)s^2(v))$.

Proof. We begin by proving (a). Fix $\alpha \in I$, $u \in I$ and let $v = \alpha - u \in I$. Define

$$s_1(u) = s(u), \qquad s_2(u) = s(\alpha - u) = s(v). \tag{9.1}$$

*F. Bowman, *Introduction to Elliptic Functions*, Dover, New York, p. 11, Example 6.

Because of the hypothesis, we have from the above

$$\begin{aligned} \frac{ds_1(u)}{du} &= c(u)d(u), \\ \frac{ds_2(u)}{du} &= \frac{ds(v)}{du} = \frac{ds(v)}{dv}\frac{dv}{du} = -\frac{ds(v)}{dv} = -c(v)d(v). \end{aligned} \tag{9.2}$$

Therefore,

$$s_2\frac{ds_1}{du} - s_1\frac{ds_2}{du} = s(v)c(u)d(u) + s(u)c(v)d(v). \tag{9.3}$$

Also,

$$\begin{aligned} \frac{d}{du}\left(1 - k^2 s_1^2 s_2^2\right) &= \frac{d}{du}\left(1 - k^2(s_1 s_2)^2\right) = -2k^2(s_1 s_2)\left(s_2\frac{ds_1}{du} + s_1\frac{ds_2}{du}\right) \\ &= -2k^2 s(u)s(v)\big(s(v)c(u)d(u) - s(u)c(v)d(v)\big). \end{aligned} \tag{9.4}$$

Multiplying corresponding sides of (9.3) and (9.4) we obtain

$$\begin{aligned} &\left(s_2\frac{ds_1}{du} - s_1\frac{ds_2}{du}\right)\frac{d}{du}\left(1 - k^2 s_1 s_2^2\right) \\ &\quad = -2k^2 s_1 s_2\left(s^2(v)c^2(u)d^2(u) - s^2(u)c^2(v)d^2(v)\right) \\ &\quad = -2k^2 s_1 s_2\left(s_2^2\left(\frac{ds_1}{du}\right)^2 - s_1^2\left(\frac{ds_2}{du}\right)^2\right). \end{aligned} \tag{9.5}$$

Using the hypothesis and Theorem 8.1 we obtain

$$\left(\frac{ds_1}{du}\right)^2 = c^2(u)d^2(u) = \left(1 - s_1^2\right)\left(1 - k^2 s_1^2\right) = 1 - (1 + k^2)s_1^2 + k^2 s_1^4$$

and

$$\left(\frac{ds_2}{du}\right)^2 = c^2(v)d^2(v) = \left(1 - s_2^2\right)\left(1 - k^2 s_2^2\right) = 1 - (1 + k^2)s_2^2 + k^2 s_2^4.$$

These imply that

$$s_2^2\left(\frac{ds_1}{du}\right)^2 - s_1^2\left(\frac{ds_2}{du}\right)^2 = \left(s_2^2 - s_1^2\right)\left(1 - k^2 s_1^2 s_2^2\right). \tag{9.6}$$

Substituting this into the right-hand side of (9.5) we obtain

$$\left(s_2\frac{ds_1}{du} - s_1\frac{ds_2}{du}\right)\left(\frac{d}{du}\left(1 - k^2 s_1^2 s_2^2\right)\right) = -2k^2 s_1 s_2\left(s_2^2 - s_1^2\right)\left(1 - k^2 s_1^2 s_2^2\right). \tag{9.7}$$

The second equation in (9.2) implies that

$$\frac{d^2 s_2(u)}{du^2} = \frac{d}{du}\left(-\frac{ds(v)}{dv}\right) = \frac{d}{dv}\left(-\frac{ds}{dv}\right)\frac{dv}{du} = -\frac{d}{dv}\left(-\frac{ds(v)}{dv}\right)$$

$$= \frac{d^2 s(v)}{dv^2}. \tag{9.8}$$

By Prob. 8.5(4), we have

$$\frac{d^2 s_1}{du^2} = -(1+k^2)s_1 + 2k^2 s_1^3,$$

and, by (9.8),

$$\frac{d^2 s_2(u)}{du^2} = \frac{d^2 s(v)}{dv^2} = -(1+k^2)s(v) + 2k^2 s^3(v) = -(1+k^2)s_2 + 2k^2 s_2^3.$$

These imply that

$$s_2 \frac{d^2 s_1}{du^2} - s_1 \frac{d^2 s_2}{du^2} = -2k^2 s_1 s_2 (s_2^2 - s_1^2),$$

or that

$$\frac{d}{du}\left(s_2 \frac{ds_1}{du} - s_1 \frac{ds_2}{du}\right) = s_2 \frac{d^2 s_1}{du^2} - s_1 \frac{d^2 s_2}{du^2} = -2k^2 s_1 s_2 (s_2^2 - s_1^2). \tag{9.9}$$

Using this in the right-hand side of (9.7) gives

$$\left(s_2 \frac{ds_1}{du} - s_1 \frac{ds_2}{du}\right)\frac{d}{du}(1 - k^2 s_1^2 s_2^2) = \frac{d}{du}\left(s_2 \frac{ds_1}{du} - s_1 \frac{ds_2}{du}\right)(1 - k^2 s_1^2 s_2^2). \tag{9.10}$$

This is of the form

$$F\frac{dG}{du} - \frac{dF}{du}G = 0, \tag{9.11}$$

where

$$F = s_2 \frac{ds_1}{du} - s_1 \frac{ds_2}{du}, \qquad G = 1 - k^2 s_1^2 s_2^2.$$

This implies that

$$\frac{d}{du}\left(\frac{F}{G}\right) = 0 \qquad \text{for} \quad u \in I.$$

and, hence, that there is a constant c such that

$$\frac{F}{G} = c \qquad \text{for} \quad u \in I. \tag{9.12}$$

Therefore,

$$\frac{s_2(u)(ds_1(u)/du) - s_1(u)(ds_2(u)/du)}{1 - k^2 s_1^2(u) s_2^2(u)} = c, \qquad u \in I$$

which implies that for all $u \in I$,

$$\frac{s(\alpha - u)c(u)d(u) + s(u)c(\alpha - u)d(\alpha - u)}{1 - k^2 s^2(u) s^2(\alpha - u)} = c. \tag{9.13}$$

The constant c is determined from (8.40) by using $u = \alpha$, $s(0) = 0$, and $c(0) = 1 = d(0)$. We obtain

$$s(\alpha) = c.$$

By (9.13)

$$\frac{s(\alpha - u)c(u)d(u) + s(u)c(\alpha - u)d(\alpha - u)}{1 - k^2 s^2(u) s^2(\alpha - u)} = s(\alpha)$$

for $u \in I$. Putting $v = \alpha - u$ and $u + v = \alpha$, we obtain (a).

Next we prove (b). To simplify the formulas we put:

$$\begin{gathered} s_1(u) = s(u), \qquad d_1(u) = d(u), \qquad s_2(u) = s(\alpha - u) = s(v), \\ c_2(u) = c(\alpha - u) = c(v), \qquad d_2(u) = d(\alpha - u) = d(v) \end{gathered} \tag{9.14}$$

and

$$\Delta(u) = 1 - k^2 s^2(u) s^2(\alpha - u) = 1 - k^2 s_1^2(u) s_2^2(u). \tag{9.15}$$

From (1) of (8.41) in Theorem 8.1 and the already proved (a), we have

$$\begin{aligned} c^2(u + v) &= 1 - s^2(u + v) = 1 - \left(\frac{s(u)c(v)d(v) + s(v)c(u)d(u)}{\Delta} \right)^2 \\ &= \frac{\Delta^2 - (s(u)c(v)d(v) + s(v)c(u)d(u))^2}{\Delta^2}. \end{aligned} \tag{9.16}$$

Now,

$$1 - k^2 s_1^2 s_2^2 = c_1^2 + s_1^2 - k^2 s_1^2 s_2^2 = c_1^2 + s_1^2(1 - k^2 s_2^2) = c_1^2 + s_1^2 d_2^2. \tag{9.17}$$

Similarly,

$$1 - k^2 s_1^2 s_2^2 = c_2^2 + s_2^2 - k^2 s_1^2 s_2^2 = c_2^2 + s_2^2 d_1^2. \tag{9.18}$$

Hence,

$$\Delta^2 = \left(1 - k^2 s_1^2 s_2^2\right)^2 = \left(c_1^2 + s_1^2 d_2^2\right)\left(c_2^2 + s_2^2 d_1^2\right). \tag{9.19}$$

For u, v and $u + v$ in I we have, using (9.16) and (9.19),

$$\begin{aligned} c^2(u + v) &= \frac{\left(c_1^2 + s_1^2 d_2^2\right)\left(c_2^2 + s_2^2 d_1^2\right) - \left(s_1 c_2 d_2 + s_2 c_1 d_1\right)^2}{\Delta^2} \\ &= \frac{\left(c_1 c_2 - s_1 s_2 d_1 d_2\right)^2}{\Delta^2}. \end{aligned}$$

This implies that

$$c(u+v) = \pm \frac{c_1 c_2 - s_1 s_2 d_1 d_2}{\Delta} = \pm \frac{c(u)c(v) - s(u)s(v)d(u)d(v)}{1 - k^2 s^2(u) s^2(v)}.$$

Fixing $\alpha \in I$, we have for $u \in I$, $v = \alpha - u \in I$,

$$c(\alpha) = \pm \frac{c(u)c(\alpha - u) - s(u)s(\alpha - u)d(u)d(\alpha - u)}{1 - k^2 s^2(u) s^2(\alpha - u)}. \tag{9.20}$$

Using $u = \alpha$, we have

$$c(\alpha) = \pm \frac{c(\alpha)c(0) - s(\alpha)s(0)d(\alpha)d(0)}{1 - k^2 s^2(\alpha) s^2(0)} = \pm c(\alpha).$$

For $c(\alpha) \neq 0$, this implies that the minus sign cannot be used and (9.20) becomes

$$c(\alpha) = \frac{c(u)c(\alpha - u) - s(u)s(\alpha - u)d(u)d(\alpha - u)}{1 - k^2 s^2(u) s^2(\alpha - u)}$$

for u, α, and $\alpha - u = v \in I$. Using $\alpha = u + v$ yields (b).

We prove (c). This time, we notice from (9.17), (9.18), and Theorem 8.1, that

$$\begin{aligned}
\Delta = 1 - k^2 s_1^2 s_2^2 &= c_1^2 + s_1^2 d_2^2 = c_1^2 + s_1^2\left(k^2 c_2^2 + k'^2\right) \\
&= c_1^2 + s_1^2\left(k^2 c_2^2 + 1 - k^2\right) \\
&= 1 - k^2 s_1^2 + k^2 s_1^2 c_2^2 \\
&= d_1^2 + k^2 s_1^2 c_2^2 .
\end{aligned} \tag{9.21}$$

Similarly,

$$\Delta = d_2^2 + k^2 s_2^2 c_1^2 . \tag{9.22}$$

Therefore,

$$\Delta^2 = \left(d_1^2 + k^2 s_1^2 c_2^2\right)\left(d_2^2 + k^2 s_2^2 c_1^2\right). \tag{9.23}$$

By Theorem 8.1 and the already proved (a), we have

$$\begin{aligned}
d^2(u+v) = 1 - k^2 s^2(u+v) &= 1 - k^2\left(\frac{s_1 c_2 d_2 + s_2 c_1 d_1}{\Delta}\right)^2 \\
&= \frac{\Delta^2 - k^2(s_1 c_2 d_2 + s_2 c_1 d_1)^2}{\Delta^2}.
\end{aligned} \tag{9.24}$$

Examining the numerator on the right, we have from (9.22) and (9.23),

$$\begin{aligned}
\Delta^2 - k^2(s_1 c_2 d_2 + s_2 d_1 d_1)^2 &= \left(d_1^2 + k^2 s_1^2 c_2^2\right)\left(d_2^2 + k^2 s_2 c_1^2\right) \\
&\quad - k^2(s_1 c_2 d_2 + s_2 c_1 d_1)^2 \\
&= \left(d_1 d_2 - k^2 s_1 s_2 c_1 c_2\right)^2.
\end{aligned}$$

This and (9.24) imply that

$$d(u+v) = \pm \frac{d_1 d_2 - k^2 s_1 s_2 c_1 c_2}{\Delta} = \pm \frac{d(u)d(v) - k^2 s(u)s(v)c(u)c(v)}{1 - k^2 s^2(u) s^2(v)}$$

Here, we eliminate the minus sign by reasoning as we did in the proof of part (b) to obtain (c).

10. The Uniqueness of the s, c, and d in Theorem 8.1

Lemma 10.1. *If $a, b, c; a_1, b_1, c_1$ are real numbers, then*

$$\left[(ab - a_1 b_1)^2 + (bc - b_1 c_1)^2 + (ca - c_1 a_1)^2\right]^{1/2}$$
$$\leqslant \left[a^2 + b^2 + c^2 + a_1^2 + b_1^2 + c_1^2\right]^{1/2} \left[(a - a_1)^2 + (b - b_1)^2 + (c - c_1)^2\right]^{1/2}. \tag{10.1}$$

PROOF. By the properties of absolute value and Cauchy's inequality (Theorem II.12.6), we have

$$|ab - a_1 b_1| = |a(b - b_1) + b_1(a - a_1)| \leqslant \sqrt{a^2 + b_1^2}\,\sqrt{(a - a_1)^2 + (b - b_1)^2}\,,$$

$$|bc - b_1 c_1| \leqslant \sqrt{b^2 + c_1^2}\,\sqrt{(b - b_1)^2 + (c - c_1)^2}\,,$$

$$|ca - c_1 a_1| \leqslant \sqrt{c^2 + a_1^2}\,\sqrt{(a - a_1)^2 + (b - b_1)^2}\,.$$

Hence,

$$|ab - a_1 b_1| \leqslant \sqrt{a^2 + b_1^2}\,\sqrt{(a - a_1)^2 + (b - b_1)^2}$$
$$\leqslant \sqrt{a^2 + b_1^2}\,\sqrt{(a - a_1)^2 + (b - b_1)^2 + (c - c_1)^2}\,,$$
$$|bc - b_1 c_1| \leqslant \sqrt{b^2 + c_1^2}\,\sqrt{(a - a_1)^2 + (b - b_1)^2 + (c - c_1)^2}\,,$$
$$|ca - c_1 a_1| \leqslant \sqrt{c^2 + a_1^2}\,\sqrt{(a - a_1)^2 + (b - b_1)^2 + (c - c_1)^2}\,.$$

Squaring, adding, and then taking square roots, we arrive at (10.1).

Lemma 10.2. *If $\langle s, c, d\rangle$ and $\langle s_1, c_1, d_1\rangle$ are ordered triples of functions each of which satisfies the hypothesis of Theorem* 8.1 *on an interval I centered at* 0 *then, for each $u \in I$, we have*

$$\left[(cd - c_1 d_1)^2 + (sd - s_1 d_1)^2 + (sc - s_1 c_1)^2\right]^{1/2}\Big|_u$$
$$\leqslant 2\left[(s - s_1)^2 + (c - c_1)^2 + (d - d_1)^2\right]^{1/2}\Big|_u. \tag{10.2}$$

PROOF. By Lemma 10.2, we know that if $u \in I$, then

$$\left[(cd - c_1d_1)^2 + (sd - s_1d_1)^2 + (sc - s_1c_1)^2\right]^{1/2}\Big|_u$$
$$\leqslant \left[s^2 + c^2 + d^2 + s_1^2 + c_1^2 + d_1^2\right]^{1/2}\Big|_u$$
$$\cdot\left[(s - s_1)^2 + (c - c_1)^2 + (d - d_1)^2\right]^{1/2}\Big|_u. \tag{10.3}$$

By (8.41) of Theorem 8.1, we know that for $u \in I$,

$$s^2(u) + c^2(u) = 1 = s_1^2(u) + c_1^2(u). \tag{10.4}$$

We also know from there that

$$d^2(u) \leqslant 1, \qquad d_1^2(u) \leqslant 1 \qquad \text{for} \quad u \in I. \tag{10.5}$$

It follows that for $u \in I$ we have

$$s^2(u) + c^2(u) + d^2(u) \leqslant 1 + d^2(u) \leqslant 2$$

and

$$s_1^2(u) + c_1^2(u) + d_1^2(u) \leqslant 1 + d_1^2(u) \leqslant 2.$$

These and (10.3) imply (10.2).

The following lemma is used in the theory of ordinary differential equations to prove uniqueness of solutions of certain differential equations.

Lemma 10.3.* *Let $\epsilon > 0$. If f is continuous on $[a, a + \epsilon)$, differentiable on $(a; a + \epsilon)$, and if a constant L exists such that*

$$f'(x) \leqslant Lf(x) \tag{10.6}$$

for $x \in (a; a + \epsilon)$, then

$$f(x) \leqslant f(a)e^{L(x-a)} \qquad \textit{for} \quad x \in [a, a + \epsilon). \tag{10.7}$$

PROOF. Multiply both sides of (10.6) by e^{-Lx} and transpose to obtain

$$e^{-Lx}f'(x) - Le^{-Lx}f(x) \leqslant 0 \qquad \text{for} \quad x \in (a; a + \epsilon).$$

Hence,

$$\frac{d}{dx}\left(e^{-Lx}f(x)\right) \leqslant 0 \qquad \text{for} \quad x \in (a; a + \epsilon).$$

It follows that the function g, defined as

$$g(x) = e^{-Lx}f(x) \qquad \text{for} \quad x \in (a; a + \epsilon),$$

* Birkhoff and Rota, *Ordinary Differential Equations*, 2nd Ed., Blaisdell Publishing Co., Chap. 1, Lemma 2.

is monotonically decreasing on $[a, a+\epsilon)$. This implies that

$$f(x)e^{-Lx} \leqslant f(a)e^{-La} \qquad \text{for} \quad a \leqslant x < a + \epsilon.$$

Therefore, (10.7) holds.

Prob. 10.1. Prove: If f is continuous on $(a-\epsilon; a]$ $(\epsilon > 0)$, differentiable on $(a-\epsilon; a)$, and if a constant L exists such that

$$f'(x) \geqslant Lf(x) \qquad \text{for} \quad x \in (a-\epsilon; a),$$

then

$$f(x) \leqslant f(a)e^{L(x-a)} \qquad \text{for} \quad x \in (a-\epsilon; a].$$

Theorem 10.1. *If I is an interval centered at* 0, *then there exists at most one ordered triple $\langle s, c, d\rangle$ of functions s, c, d defined on I which satisfy the hypothesis of Theorem* 8.1.

Proof. Let $\langle s, c, d\rangle$ and $\langle s_1, c_1, d_1\rangle$ be ordered triples of functions satisfying the hypothesis of Theorem 8.1 on I. Define f as

$$f(u) = \left[(s-s_1)^2 + (c-c_1)^2 + (d-d_1)^2\right]\Big|_u \qquad \text{for} \quad u \in I. \tag{10.8}$$

Note that f is differentiable on I and that, in view of (8.40),

$$f(0) = 0. \tag{10.9}$$

Differentiating f, we have

$$f'(u) = 2\left[(s-s_1)(s'-s_1') + (c-c_1)(c'-c_1') + (d-d_1)(d'-d_1')\right]\Big|_u. \tag{10.10}$$

By the Cauchy–Schwartz inequality this implies that

$$f'(u) \leqslant 2\left[(s-s_1)^2 + (c-c_1)^2 + (d-d_1)^2\right]^{1/2}\Big|_u \left[(s'-s_1')^2 + (c'-c_1')^2 + (d'-d_1')^2\right]^{1/2}\Big|_u. \tag{10.11}$$

By (8.39) and the fact that $0 \leqslant k \leqslant 1$ we know that

$$\begin{aligned}
&\left[(s'-s_1')^2 + (c'-c_1')^2 + (d'-d_1')^2\right]^{1/2}\Big|_u \\
&\quad = \left[(cd-c_1d_1)^2 + (-sd+s_1d_1)^2 + k^4(sc-s_1c_1)^2\right]^{1/2}\Big|_u \\
&\quad \leqslant \left[(cd-c_1d_1)^2 + (sd-s_1d_1)^2 + (sc-s_1c_1)^2\right]^{1/2}\Big|_u.
\end{aligned}$$

Applying Lemma 10.2 to the right-hand side here we obtain

$$\left[(s'-s_1')^2+(c'-c_1')^2+(d'-d_1')^2\right]^{1/2}\Big|_u$$
$$\leqslant 2\left[(s-s_1)+(c-c_1)^2+(d-d_1)\right]^{1/2}\Big|_u,$$

and, hence, by (10.11),

$$|f'(u)| \leqslant 4\left[(s-s_1)^2+(c-c_1)^2+(d-d_1)^2\right]\Big|_u = 4f(u)$$

for $u \in I$. Thus,

$$-4f(u) \leqslant f'(u) \leqslant 4f(u) \qquad \text{for} \quad u \in I. \tag{10.12}$$

Now apply Lemma 10.3 to the inequality on the right. Since $f(0)=0$, we see that if $u \in I$ and $u \geqslant 0$, then

$$f(u) \leqslant e^{4u}f(0)=0. \tag{10.12$'$}$$

By definition, $f(u) \geqslant 0$ for $u \in I$. We conclude that

$$f(u)=0 \qquad \text{for all} \quad u \in I, \quad u \geqslant 0. \tag{10.13}$$

Using the inequality on the left-hand side of (10.12) and Prob. 10.1, we see that if $u \in I$ and $u \leqslant 0$, then

$$f(u) \leqslant e^{4u}f(0)=0. \tag{10.14}$$

Hence,

$$f(u)=0 \qquad \text{for} \quad u \in I, \quad u \leqslant 0. \tag{10.15}$$

Now (10.13) and (10.15) combine to yield $f(u)=0$ for all $u \in I$, and we see from (10.8) that

$$s(u)-s_1(u), \qquad c(u)=c_1(u), \qquad d(u)=d_1(u)$$

for $u \in I$. This completes the proof.

11. Extending the Definition of the Jacobi Elliptic Functions

Thus far, the Jacobi elliptic functions sn, dn, and cn are defined on $[-K,K]$. They satisfy the hypothesis of Theorem 8.1 (see (8.27), (8.33), (8.37), and Prob. 8.3). Hence, they satisfy the conclusion of Theorem 9.1 on $I=[-K,K]$. Because of this we can apply Theorem 9.1 to them with $s=\text{sn}$, $c=\text{cn}$, and $d=\text{dn}$ and we have: If u, v and $u+v$ are in $[-K,K]$, then

$$\operatorname{sn}(u+v)=\frac{\operatorname{sn}u\operatorname{cn}v\operatorname{dn}v+\operatorname{sn}v\operatorname{cn}u\operatorname{dn}u}{1-k^2\operatorname{sn}^2u\operatorname{sn}^2v}, \tag{11.1}$$

$$\operatorname{cn}(u+v)=\frac{\operatorname{cn}u\operatorname{cn}v-\operatorname{sn}u\operatorname{dn}u\operatorname{sn}v\operatorname{dn}v}{1-k^2\operatorname{sn}^2u\operatorname{sn}^2v}, \tag{11.2}$$

$$\operatorname{dn}(u+v)=\frac{\operatorname{dn}u\operatorname{dn}v-k^2\operatorname{sn}u\operatorname{cn}u\operatorname{sn}v\operatorname{cn}v}{1-k^2\operatorname{sn}^2u\operatorname{sn}^2v}, \tag{11.3}$$

We obtain "double angle" formulas for sn, cn, and dn by letting $v=u$ in the last equations. We have

$$\operatorname{sn}2u=\frac{2\operatorname{sn}u\operatorname{cn}u\operatorname{dn}u}{1-k^2\operatorname{sn}^4u}, \tag{11.4}$$

$$\operatorname{cn}2u=\frac{\operatorname{cn}^2u-\operatorname{sn}^2u\operatorname{dn}^2u}{1-k^2\operatorname{sn}^4u}, \tag{11.5}$$

$$\operatorname{dn}2u=\frac{\operatorname{dn}^2u-k^2\operatorname{sn}^2u\operatorname{cn}^2u}{1-k^2\operatorname{sn}^4u} \tag{11.6}$$

for $u\in[-K/2,K/2]$.

We extend the definitions of sn, cn, and dn from $[-K,K]$ to $[-2K,2K]$ by defining functions s, c, and d as

$$s(u)=\frac{2\operatorname{sn}(u/2)\operatorname{cn}(u/2)\operatorname{dn}(u/2)}{1-k^2\operatorname{sn}^4(u/2)}, \tag{11.7}$$

$$c(u)=\frac{\operatorname{cn}^2(u/2)-\operatorname{sn}^2(u/2)\operatorname{dn}^2(u/2)}{1-k^2\operatorname{sn}^4(u/2)}, \tag{11.8}$$

$$d(u)=\frac{\operatorname{dn}^2(u/2)-k^2\operatorname{sn}^2(u/2)\operatorname{cn}^2(u/2)}{1-k^2\operatorname{sn}^4(u/2)} \tag{11.9}$$

for $u\in[-2K,2K]$. Here we used (11.4), (11.5), and (11.6) with u replacing $2u$.

PROB. 11.1. Prove that the functions s, c, and d defined in (11.7), (11.8), and (11.9) satisfy the relations

(a) $s'(u)=c(u)d(u)$,
(b) $c'(u)=s(u)d(u)$,
(c) $d'(u)=-k^2s(u)c(u)$

for $u\in[-2K,2K]$, and that

(d) $s(0)=0$, $c(0)=1=d(0)$.

In view of the result cited in the last problem, we know by Theorem 10.1 that the s, c, and d defined in (11.7), (11.8), and (11.9) are the only functions defined on $[-2K,2K]$ satisfying the hypothesis of Theorem 8.1. Since the Jacobian elliptic functions sn, cn, and dn satisfy the hypothesis of

Theorem 8.1 on $[-K, K]$, the s, c, and d defined in (11.7), (11.8), and (11.9) agree with sn, cn, and dn on $[-K, K]$ and furnish us with respective extensions of the Jacobian elliptic functions on $[-2K, 2K]$. We, therefore, define sn, cn, and dn on $[-2K, 2K]$ by means of (11.7), (11.8), and (11.9). By Prob. 11.1, the extended Jacobian elliptic functions defined on $[-2K, 2K]$ satisfy the hypothesis of Theorem 8.1. By Theorem 9.1, these extended functions satisfy the addition formulas (a), (b), and (c) of that theorem. Because of this we can again use the "half-angle" formulas (11.7), (11.8), and (11.9) to obtain extensions of sn, cn, and dn to the interval $[-4K, 4K]$ in the same way in which we obtained extensions from $[-K, K]$ to $[-2K, 2K]$. This procedure can be applied inductively to any interval $[-2^nK, 2^nK]$ where n is a positive integer. The union of these intervals is $\mathbb{R}$ (explain). We now have sn, cn, and dn defined on all of $\mathbb{R}$. Moreover, they satisfy the addition formulas (a), (b), (c) of Theorem 9.1 on $\mathbb{R}$. Also, the extensions of the Jacobian elliptic functions were obtained in a manner preserving the property of satisfying the hypothesis of Theorem 8.1. We now have:

If u and v are in $\mathbb{R}$, then sn, cn, and dn satisfy (11.1), (11.2), and (11.3).

Prob. 11.2. Prove: If $u \in \mathbb{R}$, then

$$\operatorname{sn}(-u) = -\operatorname{sn} u, \qquad \operatorname{cn}(-u) = \operatorname{cn} u, \quad \text{and} \quad \operatorname{dn}(-u) = \operatorname{dn} u.$$

(Previously, these were known to hold only on $[-K, K]$.)

Prob. 11.3. Prove: If $u \in \mathbb{R}$ and $v \in \mathbb{R}$, then

(a) $$\operatorname{sn}(u - v) = \frac{\operatorname{sn} u \operatorname{cn} v \operatorname{dn} v - \operatorname{sn} v \operatorname{cn} u \operatorname{dn} u}{1 - k^2 \operatorname{sn}^2 u \operatorname{sn}^2 v},$$

(b) $$\operatorname{cn}(u - v) = \frac{\operatorname{cn} u \operatorname{cn} v + \operatorname{sn} u \operatorname{dn} u \operatorname{sn} v \operatorname{dn} v}{1 - k^2 \operatorname{sn}^2 u \operatorname{sn}^2 v},$$

(c) $$\operatorname{dn}(u - v) = \frac{\operatorname{dn} u \operatorname{dn} v + k^2 \operatorname{sn} u \operatorname{cn} u \operatorname{sn} v \operatorname{cn} v}{1 - k^2 \operatorname{sn}^2 u \operatorname{sn}^2 v}.$$

Prob. 11.4. Prove: If $u \in \mathbb{R}$, then

$$\operatorname{sn}(u \pm K) = \pm \frac{\operatorname{cn} u}{\operatorname{dn} u}, \qquad \operatorname{cn}(u \pm K) = \mp k' \frac{\operatorname{sn} u}{\operatorname{dn} u}, \qquad \operatorname{dn}(u \pm K) = \frac{k'}{\operatorname{dn} u},$$

and

$$\operatorname{sn}(2K) = 0, \qquad \operatorname{cn}(2K) = -1, \qquad \operatorname{dn}(2K) = 1.$$

Prob. 11.5. Prove: If $u \in \mathbb{R}$, then

$$\operatorname{sn}(u \pm 2K) = -\operatorname{sn} u, \qquad \operatorname{cn}(u \pm 2K) = -\operatorname{cn} u, \qquad \operatorname{dn}(u \pm 2K) = \operatorname{dn} u.$$

We see from the last relation that dn is periodic with period $2K$.

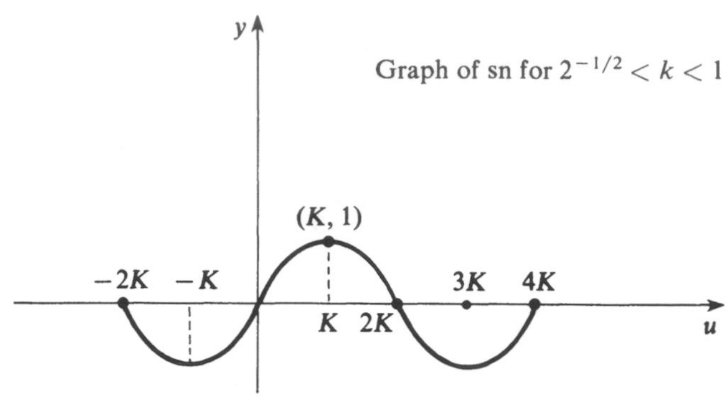

Figure 11.1

Proof. 11.6. Prove: If $u \in \mathbb{R}$, then

$$\operatorname{sn}(u + 4K) = \operatorname{sn} u \quad \text{and} \quad \operatorname{cn}(u + 4K) = \operatorname{cn} u.$$

Prob. 11.7. Prove: $\operatorname{dn} u > 0$ for all $u \in \mathbb{R}$ and, moreover, $0 < k' \leqslant \operatorname{dn} u \leqslant 1$.

Prob. 11.8. Prove: (a) If $K < u < 3K$, then $\operatorname{cn} u < 0$; (b) if $0 < u < 2K$, then $\operatorname{sn} u > 0$; (c) $\operatorname{sn} u < 0$ for $2K < u < 0$.

Prob. 11.9. Prove: (a) sn is strictly increasing on $[-K, K]$ and strictly decreasing on $[K, 3K]$; (b) sn is strictly concave on $[0, K]$ and strictly convex on $[-2K, 0]$ (see Fig. 11.1).

Prob. 11.10. Prove: (a) cn is strictly decreasing on $[0, 2K]$. (b) If $0 < k \leqslant 2^{-1/2}$, then cn is concave on $[-K, K]$ and convex on $[K, 3K]$. (c) If $2^{-1/2} < k < 1$, then cn is concave for u such that $-u_0 < u < u_0$, and convex for u such that $u_0 < u < K$, where $\operatorname{sn} u_0 = 1/k\sqrt{2}$ (see Fig. 11.2).

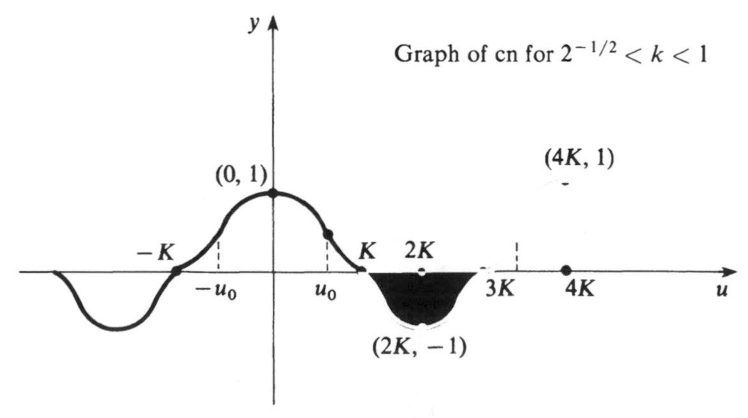

Figure 11.2

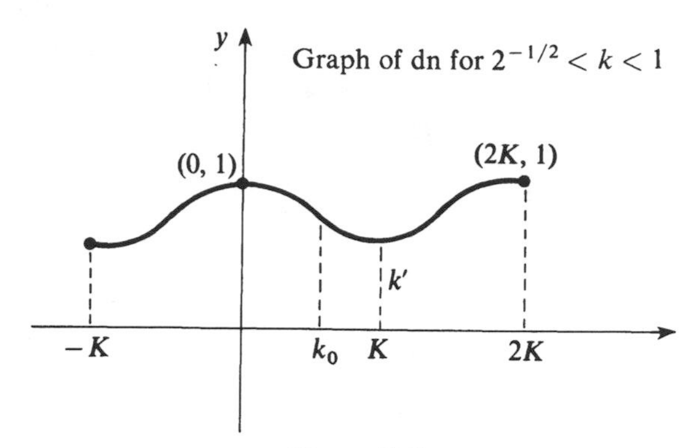

Figure 11.3

Prob. 11.11. Prove: (a) dn is strictly decreasing on $[0, K]$ and strictly increasing on $[K, 2K]$. (b) If u_0 is such that $\operatorname{sn} u_0 = 1/\sqrt{2} = \operatorname{cn} u_0$, where $0 < u_0 < K$, then dn is concave on $[-u_0, u_0]$ and convex on $[u_0, K]$ (see Fig. 11.3).

12. Other Elliptic Functions and Integrals

We go back to

$$\operatorname{sn}^{-1}(x,k) = \int_0^x \frac{1}{\sqrt{1-t^2}\sqrt{1-k^2t^2}}\,dt, \qquad -1 \leqslant x \leqslant 1. \tag{12.1}$$

Using the substitution $t = \sin\theta$, $\theta \in [-\pi/2, \pi/2]$ so that $dt = \cos\theta\, d\theta$, we have

$$\operatorname{sn}^{-1}(x,k) = \int_0^{\operatorname{Arcsin} x} \frac{1}{\sqrt{1-k^2\sin^2\theta}}\,d\theta. \tag{12.2}$$

One often writes $\phi = \operatorname{Arcsin} x$ and

$$F(k,\phi) = \int_0^{\phi} \frac{1}{\sqrt{1-k^2\sin^2\theta}}\,d\theta. \tag{12.3}$$

We, therefore, have

$$F(k,\phi) = F(k, \operatorname{Arcsin} x) = \operatorname{sn}^{-1}(x,k). \tag{12.4}$$

Other elliptic integrals are

$$E(k,\phi) = \int_0^{\phi} \sqrt{1-k^2\sin^2\theta}\,d\theta \tag{12.5}$$

and

$$\Pi(k,n,\phi) = \int_0^{\phi} \frac{1}{\sqrt{1+n\sin^2\theta}\sqrt{1-k^2\sin^2\theta}}\,d\theta. \tag{12.6}$$

These are different forms for the elliptic integrals of the second and third kind, respectively (see Section 8). Using the substitution $\theta = \operatorname{Arcsin} t$, $d\theta = (1-t^2)^{-1/2}\,dt$, we obtain

$$E(k,\phi) = \int_0^{\phi} \sqrt{1-k^2\sin^2\theta}\,d\theta = \int_0^{\sin\phi} \frac{\sqrt{1-k^2t^2}}{\sqrt{1-t^2}}\,dt$$

$$= \int_0^{\sin\phi} \frac{1-k^2t^2}{\sqrt{1-t^2}\sqrt{1-k^2t^2}}\,dt \tag{12.7}$$

and

$$\Pi(n,k,\phi) = \int_0^{\phi} \frac{1}{(1+n\sin^2\theta)\sqrt{1-k^2\sin^2\theta}}\,d\theta$$

$$= \int_0^{\sin\phi} \frac{1}{(1+nt^2)\sqrt{1-t^2}\sqrt{1-k^2t^2}}\,dt. \tag{12.8}$$

Writing $x = \sin\phi$, these become

$$E(k,\operatorname{Arcsin} x) = \int_0^{x} \frac{\sqrt{1-k^2t^2}}{\sqrt{1-t^2}}\,dt \tag{12.9}$$

and

$$\Pi(n,k,\operatorname{Arcsin} x) = \int_0^{x} \frac{1}{(1+nt^2)\sqrt{1-t^2}\sqrt{1-k^2t^2}}\,dt. \tag{12.10}$$

In (12.9) we use the substitution $t = \operatorname{sn} u$, $dt = \operatorname{cn} u \operatorname{dn} u\,du$ to obtain

$$E(k,\operatorname{Arcsin} x) = \int_0^{x} \frac{\sqrt{1-k^2t^2}}{\sqrt{1-t^2}}\,dt = \int_0^{\operatorname{sn}^{-1}x} \frac{\operatorname{dn} u}{\operatorname{cn} u} \operatorname{cn} u \operatorname{dn} u\,du$$

$$= \int_0^{\operatorname{sn}^{-1}x} \operatorname{dn}^2 u\,du = \int_0^{\operatorname{sn}^{-1}x} \operatorname{dn}^2 t\,dt. \tag{12.11}$$

Put $u = \operatorname{sn}^{-1}x$. Then $x = \operatorname{sn} u$, and we have

$$E(k,\operatorname{Arcsin}(\operatorname{sn} u)) = \int_0^{u} \operatorname{dn}^2 t\,dt. \tag{12.12}$$

Suppressing k, we define E by means of

$$E(u) = \int_0^{u} \operatorname{dn}^2 t\,dt. \tag{12.13}$$

Note that $E(0) = 0$.

PROB. 12.1. Prove:

$$\int_0^{u} \operatorname{sn}^2 t\,dt = \frac{u - E(u)}{k^2}.$$

The definite integral

$$E(K) = E(k, \text{Arcsin}(\text{sn}\, K)) = \int_0^K \text{dn}^2 t\, dt \tag{12.14}$$

is known as the *complete integral of the second kind*. Note that since $\text{sn}\, K = 1$, we have $\text{Arcsin}\,\text{sn}\, K = \pi/2$ and from (12.12) that

$$E(K) = E(k, \pi/2) = \int_0^K \text{dn}^2 t\, dt. \tag{12.15}$$

Because of this, we obtain from (12.9)

$$E(k, \pi/2) = E(K) = E(k, \text{Arcsin}\, 1) = \int_0^1 \frac{\sqrt{1 - k^2 t^2}}{\sqrt{1 - t^2}}\, dt.$$

PROB. 12.2. Prove:

$$F(k, \pi/2) = \int_0^{\pi/2} \frac{1}{\sqrt{1 - k^2 \sin^2\theta}}\, d\theta = u(1, k) = K.$$

We obtain next an addition formula for $E(u)$. First we ask the reader to solve the following two problems:

PROB. 12.3. Prove:

$$\text{dn}^2(c + u) - \text{dn}^2(c - u) = -4k^2 \frac{\text{sn}\, u\, \text{cn}\, u\, \text{dn}\, u\, \text{sn}\, c\, \text{cn}\, c\, \text{dn}\, c}{(1 - k^2 \text{sn}^2 u\, \text{sn}^2 c)^2}.$$

PROB. 12.4. Prove:

$$\text{sn}(c + u)\text{sn}(c - u) = \frac{\text{sn}^2 c - \text{sn}^2 u}{1 - k^2 \text{sn}^2 c\, \text{sn}^2 u}.$$

By (12.13) we obtain, after substituting and differentiating,

$$\frac{d}{du} E(c + u) = \text{dn}^2(c + u) \tag{12.16a}$$

and

$$\frac{d}{du} E(c - u) = -\text{dn}^2(c - u). \tag{12.16b}$$

This implies that

$$\int \left(\text{dn}^2(c + u) - \text{dn}^2(c - u)\right) du = C' + E(c + u) - E(c - u), \tag{12.17}$$

where C' is some constant.

In view of Prob. 12.3, it follows from this that

$$-\int 4k^2 \frac{\text{sn}\, u\, \text{cn}\, u\, \text{dn}\, u\, \text{sn}\, c\, \text{cn}\, c\, \text{dn}\, c}{(1 - k^2 \text{sn}^2 u\, \text{sn}^2 c)^2}\, du = C' + E(c - u) + E(c + u).$$

This implies that

$$\frac{2\operatorname{sn} c \operatorname{cn} c \operatorname{dn} c}{\operatorname{sn}^2 c} \int -2k^2 \frac{\operatorname{sn}^2 c \operatorname{sn} u \operatorname{cn} u \operatorname{dn} u}{(1 - k^2 \operatorname{sn}^2 u \operatorname{sn}^2 c)^2}\, du = C' + E(c-u) + E(c+u). \tag{12.18}$$

Since

$$\frac{d}{du}(1 - k^2 \operatorname{sn}^2 u \operatorname{sn}^2 c) = -2k^2 \operatorname{sn} u \operatorname{cn} u \operatorname{dn} u \operatorname{sn}^2 c,$$

(12.18) implies that

$$-\frac{2\operatorname{sn} c \operatorname{cn} c \operatorname{dn} c}{\operatorname{sn}^2 c} \cdot \frac{1}{1 - k^2 \operatorname{sn}^2 u \operatorname{sn}^2 c} = C'' + E(c-u) + E(c+u), \tag{12.19}$$

where C'' is some constant. Replacing u by c here yields

$$-\frac{2\operatorname{sn} c \operatorname{cn} c \operatorname{dn} c}{\operatorname{sn}^2 c} \cdot \frac{1}{1 - k^2 \operatorname{sn}^4 c} = C'' + E(2c). \tag{12.20}$$

Now subtract (12.20) from (12.19) and obtain

$$\begin{aligned} &E(c+u) + E(c-u) - E(2c) \\ &\quad = \frac{2\operatorname{sn} c \operatorname{cn} c \operatorname{dn} c}{\operatorname{sn}^2 c}\left(\frac{1}{1 - k^2 \operatorname{sn}^4 c} - \frac{1}{1 - k^2 \operatorname{sn}^2 u \operatorname{sn}^2 c}\right) \\ &\quad = \frac{2k^2 \operatorname{sn} c \operatorname{cn} c \operatorname{dn} c}{1 - k^2 \operatorname{sn}^4 c} \cdot \frac{\operatorname{sn}^2 c - \operatorname{sn}^2 u}{1 - k^2 \operatorname{sn}^2 u \operatorname{sn}^2 c}. \end{aligned}$$

Using Prob. 12.4, we obtain from this

$$E(c+u) + E(c-u) - E(2c) = \frac{2k^2 \operatorname{sn} c \operatorname{cn} c \operatorname{dn} c}{1 - k^2 \operatorname{sn}^4 c} \operatorname{sn}(c+u)\operatorname{sn}(c-u) \tag{12.21}$$

for u and c in $\mathbb{R}$. In view of (11.4), of which we now know that it holds for all $u \in \mathbb{R}$, (12.21) can be written

$$E(c+u) + E(c-u) - E(2c) = k^2 \operatorname{sn} 2c \operatorname{sn}(c+u) \operatorname{sn}(c-u). \tag{12.22}$$

We now put $x = c + u$, $y = c - u$, so that $x + y = 2c$ and obtain

$$E(x) + E(y) - E(x+y) = k^2 \operatorname{sn}(x+y) \operatorname{sn} x \operatorname{sn} y,$$

or that

$$E(x+y) = E(x) + E(y) - k^2 \operatorname{sn}(x+y) \operatorname{sn} x \operatorname{sn} y. \tag{12.23}$$

This is an addition formula for E.

We mention one more elliptic function. We put $E = E(K)$ (see (12.14)) and define Z by means of

$$Z(u) = \int_0^u \left(dn^2 t - \frac{E}{K}\right) dt = E(u) - \frac{E}{K} u. \tag{12.24}$$

This is known as *Jacobi's zeta function.*

From the definition of Z we obtain, in view of Prob. 8.3,

$$\frac{dZ}{du} = dn^2 u - \frac{E}{K}, \qquad \frac{d^2Z}{du^2} = -2k^2 \operatorname{sn} u \operatorname{cn} u \operatorname{dn} u. \tag{12.25}$$

PROB. 12.5. Prove: If $u \in \mathbb{R}$, then writing $E = E(K)$, we have

(a) $E(u + K) = E(u) + E - k^2(\operatorname{sn} u \operatorname{cn} u/\operatorname{dn} u)$,
(b) $E(2K) = 2E$,
(c) $E(u + 2K) = E(u) + 2E$,
(d) $E(-u) = -E(u)$.

PROB. 12.6. Prove: For the Jacobi zeta function Z, we have

(a) $Z(u + v) = Z(u) + Z(v) - k^2 \operatorname{sn} u \operatorname{sn} v \operatorname{sn}(u + v)$,
(b) $Z(0) = 0$, $z(K) = 0$, $Z(-u) = -Z(u)$,
(c) $Z(u + K) = Z(u) - k^2(\operatorname{sn} u \operatorname{cn} u/\operatorname{dn} u)$,
(d) $Z(u + 2K) = Z(u)$ for all $u \in \mathbb{R}$.

Bibliography

1. E. Artin, *The Gamma Function*, Holt, Rinehart and Winston, New York, 1964 (translated by M. Butler from the German original, *Einfübrung in Die Theorie Der Gamma Function*, B. G. Teubner, 1931)
2. F. Bowman, *Introduction to Elliptic Functions with Applications*, Dover, New York, 1961
3. N. Bourbaki, *Function D'une Variable Réelle* (Theorie Élémentaire), Hermann, Editeur Des Sciences et Des Arts, Paris, 1949
4. T. J. I. Bromwich, *Infinite Series* (Second Edition Revised), Macmillan & Co. Limited, London, 1942
5. Burrill and Knudsen, *Real Variables*, Holt, Rinehart and Winston, New York, 1969
6. G. Chrystal, *Textbook of Algebra*, *Vol. II*, Dover Edition
7. Courant and John, *Introduction to Calculus and Analysis*, Interscience Division, John Wiley and Sons, New York, 1965
8. Phillip Franklin, *Treatise on Advanced Calculus*, John Wiley and Sons, New York, 1940
9. Avmer Friedman, *Advanced Calculus*, Holt, Rinehart and Winston, New York, 1971
10. Watson Fulks, *Advanced Calculus* (Third Edition), John Wiley and Sons, New York, 1978
11. Eimar Hille, *Analysis*, *Vol. I*, Blaisdell Publishing Co., 1964

11a. E. W. Hobson, *The Theory of Functions of a Real Variable*, *Vol. II*, Dover Edition, New York, 1957

12. Konrad Knopf, *Theory and Application of Infinite Series*, Blackie and Sons, London, 1951
13. Konrad Knopf, *Infinite Sequences and Series*, Dover Edition, 1956
14. D. Mitronovic, *Analytic Inequalities*, Springer-Verlag, New York
15. Poly and Azëgo, *Aufgabe Der Analysis* (2 vols.) Dover
16. John Olmsted, *Advanced Calculus*, Appleton, Century, Crofts, New York, 1956

16a. James Pierpont, *Theory of Functions of Real Variables*, *Vol. II*, Ginn and Company, Boston, New York, Chicago, London, 1912

17. John F. Randolph, *Basic Real and Abstract Analysis*, Academic, NY, London, 1968
18. Walter Rudin, *Principles of Mathematical Analysis* (Second Edition), McGraw-Hill, New York, 1964

Index

9781461394839